Global Changes and Sustainable Development in Asian Emerging Market Economies Vol. 2

An Thinh Nguyen • Luc Hens
Editors

Global Changes and Sustainable Development in Asian Emerging Market Economies Vol. 2

Proceedings of EDESUS 2019

Springer

Editors
An Thinh Nguyen
University of Economics and Business
Vietnam National University
Hanoi City, Vietnam

Luc Hens
Vlaamse Instelling voor Technologisch
Onderzoek (VITO)
Keerbergen, Belgium

ISBN 978-3-030-81442-7 ISBN 978-3-030-81443-4 (eBook)
https://doi.org/10.1007/978-3-030-81443-4

This Springer imprint is published by the registered company Springer Nature Switzerland AG
The registered company address is: Gewerbestrasse 11, 6330 Cham, Switzerland

Contents

Chapter 1
The Sustainable Development of Green Space in the Tourism Zone of Moc Chau Mountains (Son La, Vietnam)

Phạm Anh Tuan and Duong Thi Loi

Abstract Green space is an inseparable part of ecotourism; it is also an important index of sustainable development. The aim of this research is to assess the change of green space quality through values of NDVI in the period of 2005–2018 in the Moc Chau national tourism zone, Son La. Four satellite images were preprocessed first to calculate the vegetation index. The NDVI was derived from post-processing Landsat data based on the difference between near-infrared and red bands, then classified to produce green maps with four classes: non-green, low green, moderate green, and dense green. This classification was applied according to the NDVI threshold values such as 0.2, 0.4, and 0.6. The results showed the remarkable reduction of the dense and moderate green while a significant increase in low green and non-green within 13 years (2005–2018). The green space tends to narrow continuously due to the negative impact from human beings. This result is considered as an important base in the management and planning of ecotourism sustainable development in the study area.

Keywords Green space · Sustainable development · Tourism zone · Moc Chau

1 Introduction

The term "green space" has been mentioned in many researches and projects as a strategy of sustainable development plans which may be started in the UK (Swanwick, 2003). Green areas are defined as coverage of natural or planted vegetation, comprised of vegetation and associated with natural elements (Fam et al., 2008). In general, green space is considered as public access areas covered by vegetated land or environment lands and brings a comfort to people. Water systems are also considered

P. A. Tuan
Tay Bac University, Son La, Vietnam

D. T. Loi (✉)
Hanoi National University of Education, Hanoi, Vietnam

A. T. Nguyen, L. Hens (eds.), *Global Changes and Sustainable Development in Asian Emerging Market Economies Vol. 2*, https://doi.org/10.1007/978-3-030-81443-4_1

as a part of green space. Green space plays an important role in physical, social, and mental development. It is a part of the environment, so the quality of green space will significantly affect the quality of life, cleans the air, adjusts urban climate, and eliminates noise (Duong et al., 2015), and contributes to encourage biodiversity of flora and fauna. Therefore, green space needs to be carefully researched and planned to maximize its contribution to the environment of urban areas while minimizing its negative aspects. The sustainable development of Green Space is a crucial strategy in urban development to give great and friendly living conditions for citizens.

Most recent studies tend to assess the green space based on density of land cover or on values of Normalized Difference Vegetation Index (NDVI) (Wei Li et al., 2015; Donovan, 2010). It can be found that NDVI is the most popular index in remote sensing to evaluate biomass. The difference in NDVI values is a reliable basis to assess the quality of the green and applied in many studies (Ahl, 2006; Jeevalakshmi, 2016). This index can be applied to determine the crop yields, green density of land use/land cover, or amount of biomass. Up to 12,633 results are found from ScienceDirect with keyword "NDVI," and more than 5800 results were found from this website with keyword "NDVI-green space" up to 2019. Vegetation index can be used as an indicator to quantify the greenness of plants within satellite data. There are several vegetation indices, but the most frequently used index is the Normalized Difference Vegetation Index (NDVI) (Jiaguo, 2005).

Moc Chau national tourism zone is located on Moc Chau plateau and has potentials for ecotourism development. However, population growth and *ineffective* land *management have caused much* negative consequence to green space. Land-use conversion processing is one of the reasons for reducing significantly forest area and ecological diversity in this study area. Understanding how to achieve development about economy while maintaining the importance of green space has become a major concern of managers at the Moc Chau national tourism zone. It was a dynamic and complex problem in sustainable urban development. To do so, understanding about standards of green space and conversion and change of green space in the study area is very important, thereby proposing the appropriate solutions and strategies for sustainable development. The goal of this research is to estimate the change of green space in the study area using Remote sensing (RS) and Geographic Information System (GIS) technology from 2005 to 2018.

2 Study Area

Moc Chau national tourism zone includes whole of Moc Chau and Van Ho districts, south of Son La province with the area of 206,150 hectares. Moc Chau tourism zone plays an important role in the development of Son La province, in particular, and mountainous northwestern region, in general. Moc Chau tourism zone shares its border with Hoa Binh province to the east and south-east, and Yen Chau district to the west and north-west, and Phu Yen district to the north. It shares its border with Thanh Hoa province to the south. The zone includes major tourism centers: the Moc

Fig. 1.1 Location of study area

Chau relaxation center, the Moc Chau ecotourism center, and the Moc Chau recreational center (Fig. 1.1).

3 Methodology

3.1 Data Use

The satellite images taken for the time periods of 2005, 2010, 2015, and 2018 were applied to assess the change of green space in the study area. Landsat 5 and Landsat 8 data were freely downloaded from website https://earthexplorer.usgs.gov/. Landsat 5 was launched on March 1, 1984, and observed the Earth until 15 January 2013 and carried the Multispectral Scanner (MSS) and the new Thematic Mapper (TM) sensors. Landsat 8 was launched on February 11, 2013, and carried Operational Land Imager and the Thermal Infrared Sensor (Survey, 2019). Both the Landsat 5 TM and Landsat 8 OLI-TIRS sensors have a spatial resolution of 30 m. The information of Landsat data is described in Table 1.1. To avoid cloud and unwanted shade-free imagery, we selected imageries of the October and November for this study. This is the transitional period of summer to winter in Moc Chau, Son La, so the sky is quite clean and less cloudy.

The overall methodology of this study is briefly shown below:

Table 1.1 Detailed information about satellite data

Satellite	Sensor	Path/Row	Acquisition date	Spatial resolution (m)
Landsat 8	OLI/TIRS	127/46	30/11/2018 21/10/2015	30
Landsat 5	TM	127/46	08/11/2010 09/10/2005	30

3.2 Preprocessing

Before the satellite image is used to calculate NDVI, they were processed to ensure the accuracy of the result. It includes the following steps.

- Geometric correction:
- All images were corrected to a common Universal Transverse Mercator (UTM) WGS- 84 Datum, Zone 48. The data used in this study need to be registered first through proper Ground Control Point (GCP) , in which, Landsat-8 scene of 2018 (path 127, row 46) was used as the reference image, and the other images of 2005, 2010, and 2015 were registered based on the image of 2018.
- Atmospheric correction:
- This process is done to eliminate the negative effects from the atmosphere, causing interpretation errors. It includes two steps: firstly, converting digital numbers (DNs) into radiance value by using standard calibration values to remove temporal differences in sensor calibration and in environmental factors between image acquisitions (López-Serrano et al., 2016). Secondly, the atmospheric conditions at the time of taking images were different, so we need to standardize the effect of the atmosphere (USGS, 2013). This process is applied on two bands: red and near-infrared at all imageries.
- The subset steps were also performed to reduce the size of the scenes that include only the study area and speed up processing. All these steps were processed with the support of ERDAS and ArcGIS 10.2 software.

3.3 NDVI Calculation

This index is calculated based on the difference between the near-infrared band and red band and expressed as Eq. (1.1):

$$(\text{NDVI}) = (\text{NIR} - \text{RED}) / (\text{NIR} + \text{RED}) \tag{1.1}$$

where: NIR = near-infrared band
RED = red band

Table 1.2 Values of NDVI

Values of NDVI	Description
<0.2	Non-green
0.2–0.4	Low green
0.4–0.6	Moderate green
>0.6	Dense green

3.4 *Defining the Indices*

The value range of NDVI is between −1 and +1. Higher values of NDVI present healthy vegetation in the area while lower values refer to sparse vegetation (Ravi Prakash Singh, 2016). The water, cloud, and snow reflect more in the visible band than in the near-infrared band, hence they are given at negative NDVI values, whereas, bare soil and rock often hold NDVI value of around zero. Based on the (USGS, n.d.) NDVI, value was classified into four classes: non-green, low green, moderate green, and dense green (Table 1.2). As the result, low green corresponds with NDVI values between 0.2 and 0.4, and it refers to shrub and grassland; moderate green values range between 0.4 and 0.6 representing agriculture land and orchard, while dense green corresponds with NDVI values ranging greater than 0.6 and referring to forest areas. Similarly, NDVI values less than 0.2 express water body, built-up and bare-land. It is assigned for non-green class.

The annual rate of change for each green class is calculated to recognize the change between a final year to initial year. It represents the magnitude of change between corresponding years was divided by the number of years. It was calculated using Eq. (1.2).

$$\text{Annual Rate of Change} = \frac{(\text{Final Year} - \text{Initial Year})}{\text{No of years}} \tag{1.2}$$

4 Results

NDVI in the period of 2005–2018 is shown in Fig. 1.2, in which, the value range of NDVI changes significantly and the upper value tends to be smaller. Accordingly, the value ranges of 2005, 2010, 2015, and 2018 are corresponded (from −0.46 to 0.76), (from −0.15 to 0.68), (from −0.19 to 0.65), and (from −0.13 to 0.61), respectively (Fig. 1.3). In order to avoid errors caused by the database, the data of the years are collected between October and November, the weather conditions and seasons are relatively homogeneous for the development of vegetation. Therefore, the decline in green space quality is determined by the impact of socioeconomic activities. This process is continuous and tends to be faster in recent years.

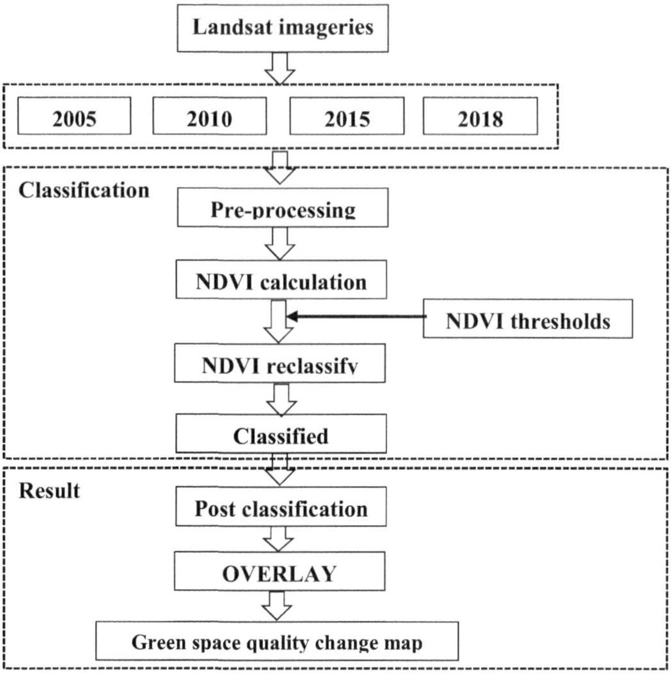

Fig. 1.2 Methodological flow chart

NDVI was categorized into four classes: non-green, low green, moderate green, and dense green as in Fig. 1.4. Table 1.3 indicates the area of classes derived from Landsat 2005, 2010, 2015, and 2018. It is found that from 2005 to 2018, there is an continuous increase in non-green and low green areas while the area at moderate and dense green tends to decrease dramatically. Even the dense green area was only 23.1 ha in 2015 and almost disappeared in 2018 with only 0.5 ha while this area was more than 77,000 ha in 2005. In addition, the area of non-green increased 4 times and the area of low green increased 10 times in 13 years (Fig. 1.5). By 2018, the low green keeps the largest area with approximately 70%, the second position is moderate green with 22.5%, and non-green also keeps to 7.8% while dense green areas almost disappeared. It can be seen that the rate of green space degradation is happening at high speed.

The matrix Table 1.4 shows the conversion process among the green space classes. Accordingly, many moderate greens and dense green areas were converted into non-green and low green. Specifically, up to 88420.32 ha (accounting for 88.4%) of the area was converted to low green. About 1776.8 ha (2.3%) dense green area was converted to non-green. More than 42,000 ha (54.6%) of this class changed to low green and more than 33,000 ha (approximately 43.1%) into the moderate green.

Table 1.5 indicates the severe degradation in green space quality in the study area from 2005 to 2018 through Annual Rate of Change. The data showed a negative

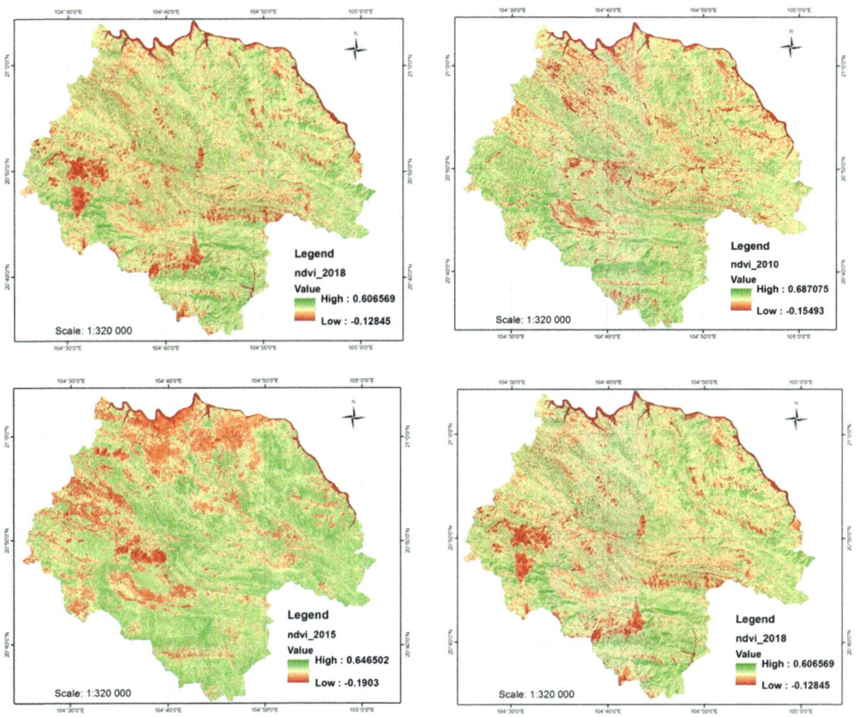

Fig. 1.3 NDVI in 2005, 2010, 2015, and 2018

change of moderate green (−84.4%) and dense green (−99.8%) while it also showed a positive change of non-green (25%) and low green (7.4%). The result showed that there was a considerable expansion of non-green and low green. That is the consequence of the spread of the built-up area and deforestation from 2005 to 2018. Especially, shifting cultivation has been still quite common in agricultural production, which causes an increase in soil erosion rapidly.

Figure 1.6 shows the degradation degree of green among administrative units. This change has taken place in all communes in the study area. Accordingly, the areas with the most fluctuations include Xuan Nha, Long Sap, Chiang Khua, and Muong Sang. Specifically, the area from green to non-green corresponds to the above communes: 441.6 ha, 140.2 ha, 121.9 ha, and 103.8 ha, respectively (Fig. 1.7a). In addition, there is a large area of moderate green being turned into non-green and low green in the period of 2005–2018. The communes with the most degradation levels are Xuan Nha, Muong Sang, and Tan Lap. The fastest change happened in Xuan Nha commune with the changed area from moderate green to non-green and low green are 948.1 ha and 11,896 ha, respectively. Similarly, the degradation from moderate green to non-green and low green also takes place quickly at some typical communes such as Muong Sang (915.5 ha and 6825.9 ha), Tan Lap (813.1 ha and 4637.8 ha), and Tan Hop (473.7 ha and 4637.2 ha). At Moc

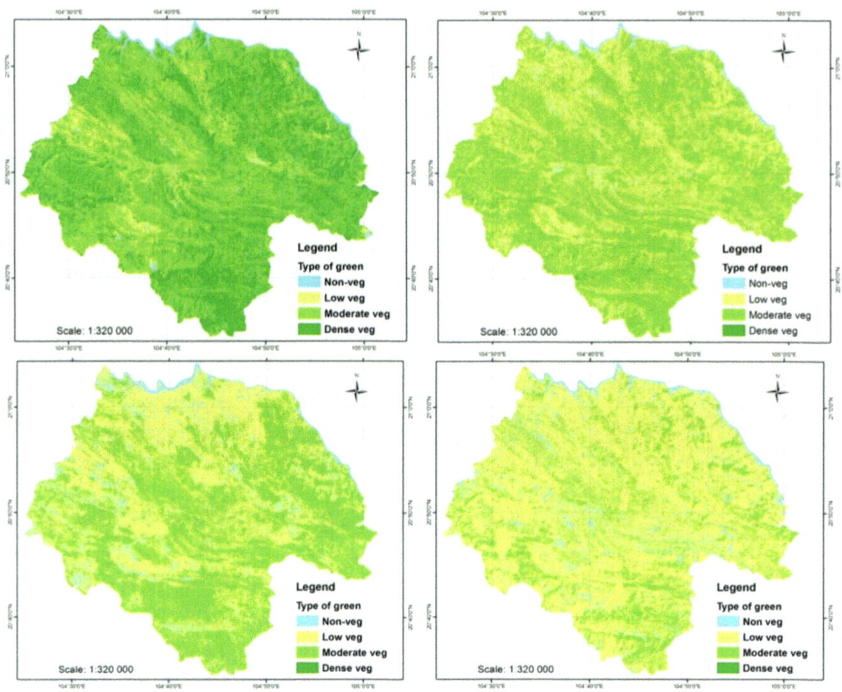

Fig. 1.4 Classified NDVI

Table 1.3 Area of classes derived from Landsat data

	2005		2010		2015		2018	
Type of green	Area (ha)	(%)	Area (ha)	(%)	Area (ha)	(%)	Area (ha)	(%)
Non veg	3661.6	1.8	4154.0	2.0	13499.9	6.6	15985.0	7.8
Low veg	13646.1	6.7	67652.5	33.2	87608.7	42.9	142115.4	69.7
Moderate veg	109312.0	53.6	130829.4	64.1	102846.7	50.4	45867.1	22.5
Dense veg	77344.2	37.9	1331.9	0.7	23.1	0.0	0.5	0.0

Chau town and Phieng Luong, the degradation is slower than other communes (Fig. 1.7b). The result is found that the process of change is concentrated significantly in communes located south and south-west of the study area.

1	Cho Long	7	Long Luong	13	Phien Luong	19	Tan Hop
2	Chieng Khoa	8	Long Sap	14	Quang Minh	20	Tan Lap
3	Chieng Hac	9	Muong Men	15	Quy Huong	21	Moc Chau town
4	Chieng Khua	10	Muong Sang	16	Song Khua	22	Van Ho
5	Chieng Yen	11	Muong Te	17	Suoi Bang	23	Xuan Nha
6	Hua Pang	12	Na Muong	18	To Mua		

Ordinal number and corresponding administrative unit

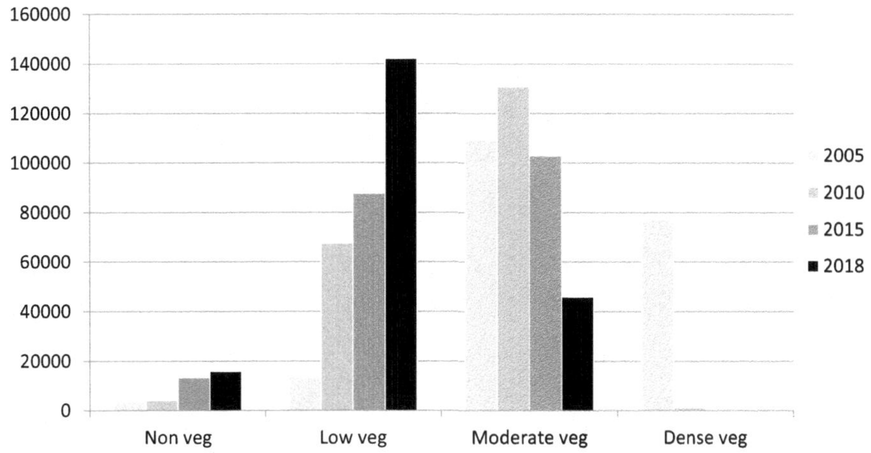

Fig. 1.5 Change of each vegetation class from 2005 to 2018

Table 1.4 Conversion matrix of green area in the period of 2005–2018

Year	2018				
2005	Non green	Low green	Moderated green	Dense green	Grand Total
Non green	2746.22	750.54	165.06		3661.82
Low green	0.26	10656.8	845.21		11502.27
Moderate green	0.17	88420.32	11561.74	0.09	99982.32
Dense green	1776.81	42262.95	33305.3	0.4	77345.46
Grand Total	4523.46	142090.61	45877.31	0.49	192491.87

Table 1.5 Green space quality change assessment based on time frame data (2005–2018)

	Years 2005–2018		
Types of green	Magnitude area (hectare)	% change	The annual rate of change
Non-green	(+) 915.6	(+) 25.0	(+) 70.4
Low green	(+) 845.8	(+) 7.4	(+) 65.1
Moderate green	(−) 88420.5	(−) 84.4	(−) 6801.5
Dense green	(−)77344.0	(−) 99.8	(−) 5949.5

(+) Sign denotes increase and (−) sign denotes decrease of magnitude of change of green space classes

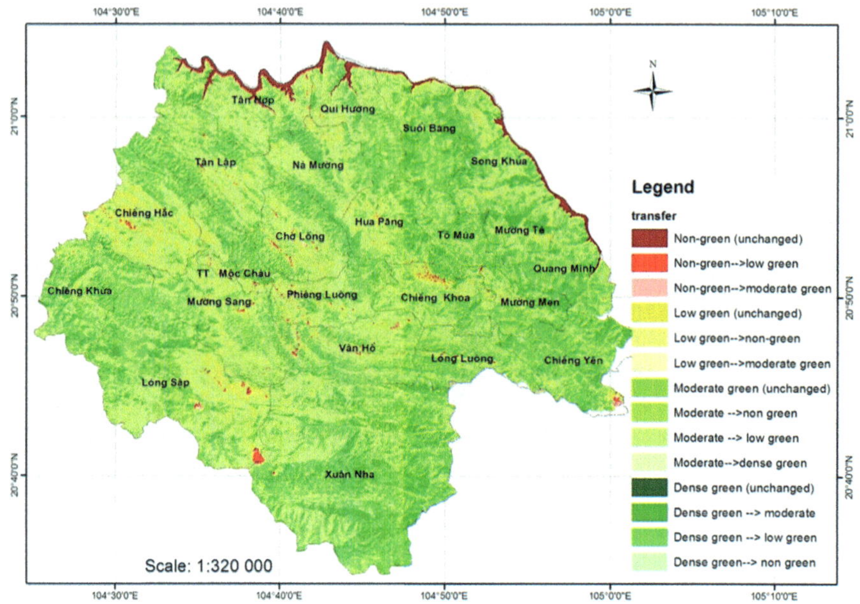

Fig. 1.6 Green space change map from 2005 to 2018. (**a**) From dense-level green to other levels. (**b**) From moderate-level green to other levels

5 Conclusion and Discussion

The use of multi-temporal satellite images brings positive results on the study of green space change. Accordingly, the dense green areas which represent forest have been under pressure from the surrounding population and economic activities. They are in serious decline and this type of land in many places is being converted into other land uses under the impact of industrialization and urbanization. Due to people's awareness and loose management, the vacant land area increased due to indiscriminate deforestation that causes negative consequences on the environment.

The trend of green space changes was indicated in this study, especially percentage increase in low green and non-green and reduce in dense green will be useful for decision-making to revert the situation to conserve the natural habitat of wildlife to ensure ecotourism development. Socioeconomic development has closely related to the changes in green space quality in the study area to ensure a sustainable development in the future. Therefore, the close combination between socioeconomic development and environmental protection is an important strategy in the spatial development orientation of the region. Maintaining and improving the quality of green space are considered as one of the priority solutions for the Moc Chau tourist area.

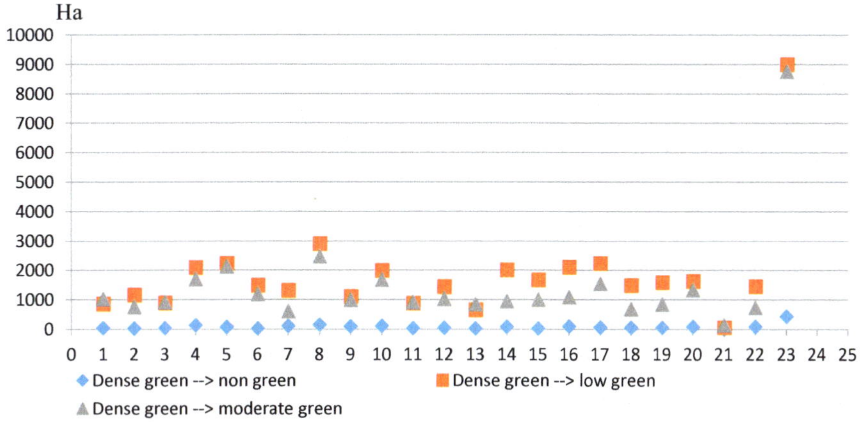

(a) From Dense level green to other levels

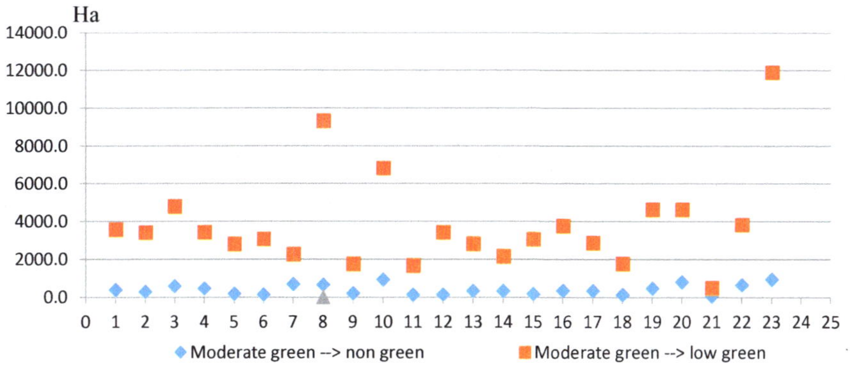

(b) From Moderate-level green to other levels

Fig. 1.7 Green space change by administrative units in period of 2005–2018

Acknowledgments This research was supported by the Ministry of Education and Training under Project No CT.2019.06.06.

References

Ahl, D. G. (2006). Monitoring spring canopy phenology of a deciduous broadleaf forest using MODIS. *Remote Sensing of Environment, 104*, 88–95.

Donovan, G. H. (2010). Trees in the city: Valuing street trees in Portland, Oregon. *Landscape and Urban Planning, 94*(2), 77–83.

Duong, T. L., Pham, A. T., & Gupta, K. (2015). Development of an index for assessment of urban green spaces at city level. *International Journal of Remote Sensing Application, 5*, 78–88.

Fam, D., E. Mosley, A. Lopes, L. Mathieson, J. Morison, G. Connellan (2008). *Irrigation of Urban Green Spaces: A review of the environmental, social and economic benefits*. Cooperative Research Centre for Irrigation Futures. Technical Report No. 04/08, Technologies Pty Ltd.

Jeevalakshmi, D.S.N. (2016). *Land cover classification based on NDVI using LANDSAT 8 Time series: A case study Tirupati Region*. International conference on Communication and Signal Processing, India, 6–8.

Jiaguo, W. C. (2005). Assessment of tropical forest degradation with canopy fractional cover from Landsat ETM+ and IKONOS imagery. *Earth Interactions, 9*(22), 1–17.

Li, W., Saphores, J.-D. M., & Gillespie, T. W. (2015). A comparison of the economic benefits of urban green spaces estimated with NDVI and with high-resolution land cover data. *Landscape and Urban Planning, 133*, 105–117.

López-Serrano, P. M., Corral-Rivas, J. J., Díaz-Varela, R. A., Álvarez-González, J. G., & López-Sánchez, C. A. (2016). Evaluation of radiometric and atmospheric correction algorithms for above ground forest biomass estimation using Landsat 5 TM data. *Remote Sensing, 8*, 1–19.

Ravi Prakash Singh, N. S. (2016). Normalized difference vegetation index (NDVI) based classification to assess the change in land use/land cover (LULC) in lower Assam, India. *International Journal of Advanced Remote Sensing and GIS, 5*(10), 1963–1970.

Survey, U.G. (2019). Retrieved from https://landsat.gsfc.nasa.gov/a-landsat-timeline/

Swanwick, C. D. (2003). Nature, role and value of green space in towns and cities: An overview. *Built Environment, 29*(2), 94–106.

USGS (2013). *Landsat missions: Frequently asked questions about the landsat missions*. USGS. Retrieved from http://www.gisagmaps.com/landsat-8-atco-guide/.

USGS (n.d.). NDVI, the Foundation for Remote Sensing Phenology. Retrieved from https://www.usgs.gov/land-resources/eros/phenology/science/ndvi-foundation-remote-sensing-phenology?qt-science_center_objects=0#qt-science_center_objects

Chapter 2
Wooden Funeral Sculptures of the Jrai and Bahnar in the Vietnam Central Highlands: Conservation or Destruction?

Nhan T. T. Ho

Abstract The wooden funeral sculptures of the Jrai and Bahnar in the Vietnam Central Highlands have been introduced to the world in many ethnographic studies since the early twentieth century by French researchers. For many reasons, there was a dearth of Vietnamese studies in this region not only during the French colonial period (1887–1954) but also in the Vietnam War (1954–1975) and throughout a decade after that. Except for a rare but major study conducted by the new government from 1986 to 1995 on various fields for exploiting local potentials and left a lot of valuable historical documents, the research gap on this topic continues to last until 2017. At the end of a 2-year research project since 2015, a report published by the local cultural agency showed that the art of wooden sculptures here was being recovered and well-developed. Although the number of new sculptures has increased recently as mentioned in the report, most of them seem very different from traditional sculptures in many respects. Through a comparative analysis of historical documents and fieldwork, this paper aims to point out the differences and identify the main characteristics that should not be confused between traditional wooden funeral sculptures and wooden artworks nowadays.

Keywords Jrai · Bahnar · Funeral sculpture · Vietnam Central Highlands

1 Introduction

The Central Highlands of Vietnam has an area of 54.470 km^2 (ISSCH, 2017; VCHSC, 2017), including four flat plateaus with various heights, located on the eastern slope of the Truong Son Mountains (while the western slopes are in the territory of Laos and Cambodia). Because of the height difference of the living area,

N. T. T. Ho (✉)
Van Lang University, HCM City, Vietnam
e-mail: nhan.htt@vlu.edu.vn

A. T. Nguyen, L. Hens (eds.), *Global Changes and Sustainable Development in Asian Emerging Market Economies Vol. 2*, https://doi.org/10.1007/978-3-030-81443-4_2

ethnic groups of people here were separated into relatively isolated subcultures, especially some large population groups such as the Jrai and the Bahnar are divided into even smaller groups. There are at least five subgroups of Jrai people that have been recognized through ethnographic studies, including Jrai Chor, Jrai H'drung, Jrai Arap, Jrai T'buăn, and Jrai Mthur (Nguyen, 2007, 2013), or the Bahnar with subgroups named Bahnar Kon Tum, Rơ Ngao, Jơlơng (Hơ lăng), Tơlô (Gơlar) (Dang, Cam, Tran, Le, & Ngo, 1981; Nguyen, 2013), Bahnar Roh, Bơnâm, Kon Kơđeh, and even more (Nguyen, 2007).

Despite different origins and divided into many local groups, the Jrai and Bahnar still share a lot of similar customs and practices, such as the grave-leaving ceremony (named "*pothi*" in the local language) in which the most beautiful folk artworks would be created as the last gifts to decedents. Among them, wooden sculptures used to be considered essential gifts in decorating the space around the tombs. These religious artworks have been mentioned in numerous studies on ethnography, society, and folklore, mainly focusing on the art of the two most populous ethnic groups due to their variety of styles and topics.

In the early twentieth century, when the French colonists began to exploit this area, they found that the indigenous people retained primitive characteristics with a lot of exciting traditional customs. Many rituals accompanied with folk visual arts were described in great detail in ethnographic studies of Jacques Dournes (Dam Bo, 2003; Dounes, 2006) and many other researchers who were priests (Dourisboure, 1873; Guerlach, 1884, 1887a, 1887b) or colonial government officials (Guilleminet, 1942, 1951, 1952, 1960; Maitre, 1909, 1912) but also attracted to culture here, in addition to ethnographic researchers (Condominas, 1957). Nearly all of those rituals are related to the crops and the human life cycle, including the grave-leaving ceremony (*pothi*).

Due to the limited circumstances of the war, most Vietnamese researchers have begun to study ethnic minorities in the Central Highlands only after 1975, except for a few studies about the Bahnar by Nguyen Dong Chi and Nguyen Kinh Chi under the control of the French government in Vietnam in the 1930s (Nguyen & Nguyen, 1937). At the time, after more than two decades from the Vietnam Independence 1975, the first series of fieldwork carried out in about 10 years in the North of the Central Highlands (1986–1996) organized by the local cultural agencies have left many valuable documents on indigenous culture and folk arts here, mainly on the two most populated ethnic groups of Jrai and Bahnar. Some studies by Tu Chi (Chi, 1986, 1995; Tran, 1986) - a nephew of Nguyen Dong Chi, one of the first Vietnamese researchers studied the minority people in the Central Highlands of Vietnam as mentioned above, Tran Phong (Tran, 1995, 2008), Ngo Van Doanh (Ngo, 1986, 1990, 1991a, 1991b), and some others (Nguyen, 1979; Phan & Nguyen, 1995) were remarkable.

Reviewing the research history, it was surprising to find that for about a decade from 2007 to 2017, there was no more in-depth study on neither wooden funerary sculptures nor grave-leaving rituals (pothi), except for a few presentations at some folklore preservation conferences, most of which are personal opinions based on data from some studies in the past. There were several books published after 2007, such as "*Wooden Sculptures in the Vietnam Central Highlands*" (Dao, 2007), "Festival in the Central Highlands of Vietnam" (Tran, 2008), but they were also based on photos collected from 1986 to 1995.

Although it seems that in research history there have been many Vietnamese and international studies on funeral sculpture of the Jrai and Bahnar so far, but due to many different views from ethnography or cultural studies rather than art study, the variation in style of sculptures which is an important aspect of art criticism was overlooked in research history. In order to fill this gap, a fairly long fieldwork from 2009 to 2014 on this topic was reported in a master thesis at HCMC University of Fine Arts and was highly appreciated for updating data as well as new findings for the funeral sculptures of the Jrai and Bahnar in the first half of the last decade (Ho, 2014). To continue this research, a paper published in the journal of Vietnamese Cultural Heritage in 2016 (Ho, 2016) tried to indicate the identity of funeral sculptures in the Central Highlands of Vietnam in general. Two years after that, another paper published in ACAHS 2018 in Japan (Ho, 2018) not only categorized the contents of those sculptures into five main groups of title (including: M*ourning emotions; sexual matter and lineage maintenance; daily/ceremonial activities; portraits/ monsters; animals and gifts*) but also classified different styles among local subgroups of the Jrai in particular.[1] Nevertheless, in another respect, the previous studies mentioned above have provided us with valuable historical images and descriptions for comparison because those sculptures were so unstable that most would decompose in the natural environment a few years after the ceremony.

In the two years from 2015 to 2017, the Department of Science and Technology of Gia Lai province has conducted a project on preserving and promoting the art of the Bahnar and Jrai folk wooden sculpture in the locality. According to the report (Hoang, 2017), the number of folk wooden sculptures of the Jrai (including statues in communal houses and tombs) was estimated to be about 1000, and the number of folk artisans was 297 with an average age of 40.[2] The corresponding statistics of the Bahnar people were 500, 261, and 50,[3] respectively. The leader of this project did believe that these sculptures are sufficient to prove the potential for the restoration and development of folk wooden sculpture in the Central Highlands in the near future. This statement was somewhat true; however, through fieldwork studies in the respective locations, we found a few quite serious issues in this project report that need to be reconsidered.

The main objectives of this paper are not only to recognize the success of a project after a decade of disruption in this topic, but also to analyze the issues to be considered, as well as propose some better solutions in preserving and promoting the art of wooden sculpture of the local ethnic minorities.

---

[1] Look up in the section of "3. Results" for more details.

[2] A survey on 12 districts, towns, and cities where the Jrai people live: Pleiku city, Ia Grai, Chu Pah, Chu Se, Chu Puh, Chu Prong, Duc Co, Phu Thien, Ia Pa, Ayun Pa, Krong Pa, Dak Doa (Hoang, 2017).

[3] A survey on 6 districts where the Bahnar people live: Dak Doa, Mang Yang, Dak Po, An Khe, Kbang, and Kong Chro, and a few communes in the districts of Chu Pah, Ia Pa, and Pleiku city, adjacent to the six mentioned districts (Hoang, 2017).

2 Methodology

The Jrai and Bahnarwere selected as representative models for the study of funeral sculpture in the Central Highlands of Vietnam for a number of reasons as follows: These are the two groups with the largest population among the ethnic groups in the Central Highlands of Vietnam that are separated into a lot of local subgroups; Because of the large area of residence covering the two provinces of Gia Lai and Kon Tum with different terrains, the subgroups have both similarities and variations due to different living conditions, which leads to a greater abundance of sculpture styles than other ethnic groups; For the same reason, in the research history, there have been a lot of Vietnamese and international studies on these two ethnic groups of people, leaving many ethnographic photos and descriptions that are sufficient to compare with the current situation of wooden sculptures recently in the locality.

Applying aesthetic theories proposed by internationally renowned experts in artistic criticism to analyze historical data from previous studies, this paper helps to generalize the basic characteristics of the funeral traditional sculptures.

According to previous studies, the content and style of sculptures created for the ritual (*pothi*) always have their reasons, so some spiritual characteristics of funeral sculptures applied to decorative sculptures may not be suitable. On the other hand, the opportunity of artistic exchange more easily and more frequently among subgroups as well as among different ethnic groups has made the sculpting styles no longer distinguish between ethnic groups nowadays. Therefore, based on the comparative method, this paper aims to show that wooden sculptures which are made recently for cultural display or tourism purposes have similarities and differences from the old ones, then discusses for some better solutions in preserving folk wooden sculptures in the Central Highlands of Vietnam nowadays.

3 Results

3.1 Some Basic Characteristics of Traditional Wooden Funeral Sculptures in the Vietnam Central Highlands (a Case Study on the Jrai and Bahnar as Representatives)

About the content of the traditional funeral sculptures, in some previous studies (Dao, 2007; Ho, 2014, 2016, 2018; Ngo, 1993; Tran, 1995), most researchers divided them into 5 groups based on the topics:

Mourning emotions (1): People sitting with their elbows touching their knees, hands embracing the sad face, or group of people comforting each other; ***Sexual matter and lineage maintenance (2):*** Men and women with magnified genitals, sometimes they are in mating status, or there are some pregnant women*;* ***Daily/ ceremonial activities (3):*** Daily activities such as pounding rice, carrying farm products, babysitting, folk games…, and activities in ceremonies such as drum beating, dancing...; ***Portraits/monsters (4):*** Portraits with long ears, jagged teeth, distorted faces; ***Animals and gifts (5):*** They preferred animals that could be easily seen in life such as mountain turtles, lizards, rabbits, peacocks, owls, dogs... to mythical creatures. Besides, there were some sculptures in the shape of buffalo horns, elephant ivories, and pots (Ho, 2016).

In a paper published in 2018 (Ho, 2018), another characteristic was proved that those five groups of topic were unevenly distributed throughout the region: While groups (1), (3), and (5) could be found in any cemetery, group (4) was less common and their distribution seemed not to follow any rule. In particular, group (2) was a special topic which only appeared in the residence of Jrai Arap subgroup in the last century. A hypothesis of the effect of natural conditions might help to explain is that the Jrai Arap subgroup was most inhabited in the northwest of Pleiku City, where Mount Chu Pah impeded the main flow of the Se San River, creating the Ia Ly waterfall. Before the1990s, it was the living area with the harshest natural environment in traditional conditions of the Central Highlands of Vietnam, where a lot of newborn babies and little kids died of diseases caused by weather or wild animals and other unpredictable causes, so the survival seems to become the most concerns in sculptures of the Jrai Arap (Fig. 2.1).

About the style of the traditional funeral sculptures, the differences between local groups within the same ethnic group were also demonstrated in a previous paper (Ho, 2016). According to this study, the residence of the Jrai mainly in the Gia Lai province can be divided into three smaller areas: Pleiku plateau in the center is the residence of Jrai H'drung people; the western and north-western to the border of

Fig. 2.1 A tomb of the Jrai Arap with all topic groups, of which most are sculptures belonging to the topic group (1) and (2) - Location: Chu Pah, Gia Lai province; Time: 1997; Source: Dao, 2007

Laos is the residence of Jrai T'buăn and Jrai Arap; the southern and south-eastern is the residence of Jrai Chor and Jrai Mthur. Before the end of the twentieth century, living conditions in these areas were very different.

The Pleiku plateau in the center is quite flat and has become the earliest urban area in the Central Highlands with the convenient transportation, so the Jrai H'drung here has had cultural exchanges with foreigners earlier than other groups, soon ignored traditional beliefs to accept Christianity and Protestantism. From the beginning of the 1990s, funeral sculptures have been seldom seen here except for a few ones about birds made of cement instead of wood.

The residence of the Jrai Arap was not a place of great interest to Westerners as well as the Cham people because of many reasons. There were some ruins of towers and stone inscriptions showed that the Cham may have once placed their governance in this area, or at least there was a close relationship between them in some other respects. However, those ruins were hardly seen in the upper land of the Jrai Arap, but mainly found in flat and lowland areas such as Ayun Pa (Yang Mum and Drang Lai towers, Quai King ancient sites), An Khê, Đak Đoa, Krong Pa, and some in Đak Lak province (Yang Prong tower). It leads to a hypothesis *that the group of Jrai Arap may have been relatively isolated by the natural conditions which, luckily, also helped them to keep the most pristine style compared with the others until the end of the twentieth century. By a sharp and powerful visual language created by axes, they made the sculptures of crying people with the structure of solid and hollow triangles in sequence, reminiscent of a similar layout on traditional brocade. If someone has never seen a sculpture of the Jrai Arap in a cemetery, he/she would have mistaken it for the artwork of a modern artist. The simplicity would make those sculptures not to be out of date even if they are exhibited in urban spaces nowadays.*

In comparison, the visual language in sculptures of the Jrai T'buăn was not as strong and sharp as the language in the artworks of the Jrai Arap. They had lived in the lower section of Po Ko river in aged forests with rather complicated terrain but still easier for living than the area of the Jrai Arap. Going further south, the sculptures of the Jrai T'buăn refused the visual language of linear shapes, became closer to the narrative description. They loved to decorate their artworks with more details of the eyes, mouth, and even details of costumes and jewelry. Most sculptures were painted with black charcoal or with colorful oil paint recently.

Leaving this upland for going downstream of the Ayun River to the foot of Chư A Thai mount until meeting the flow of the Pa River, there was a flat and rich land where the Jrai Chor had lived. From here, the Ayun river was renamed Ayun Pa, continued to flow southeast, received more water from the Bo river and then renamed once more Krong Pa, bending at the end of Tona Pass to the area where narrow plains and low gravelly hills interleaved each other, this was the residence of Gia Rai Mthur. The flat terrain with many rivers made it easier for transportation, so the sculptural styles of both Jrai Chor and Jrai Mthur with elaborate rounded details might be the result of influence from Cham styles.

About the structure, there were also some differences among subgroups. Most funeral sculptures in the upland were independent of the architecture, placed in a fence around the grave, while in the lowland, the relationship between sculptures and the tomb was more evident, in some cases, they became the pillars supporting the roof. With a simple form in the upland, most sculptures of the Jrai Arap separated into two parts: a pedestal (less than 1 m) and a statue with only one character, the height in total was around 1,5 m; The structure of sculptures in the lowland was much more complex (that the Jrai Mthur could be a representative case), with total height around 1,5–2 m or even more, separated into three parts: a basic pedestal (similar to the type of the Jrai Arap), an extra pedestal above was sculptured with geometric shapes, the main statue on the top. If the total height was under 2 m, the main statue might contain a group of characters (a dog beside a man carrying a baby, for example), but in case the sculpture became a pillar, it could be over 2 m and the extra pedestal became the main part, then the statue became very little or even disappeared (Table 2.1).

A lot of photos of funeral sculptures as described above can be found in many of the previous studies listed in the bibliography. In addition, the following table can help to easily visualize the style characteristics of each subgroup of the Jrai, and the distribution of topics in those subgroups:

3.2 Some Information May Lead to Misunderstandings in the Local Cultural agency's Report on the Current State of Wooden Sculpture of the Jrai and Bahnar Recently

Through many field trips at the same locations, the results showed that the number of funeral sculptures in the cemetery here has become extremely rare recently compared with historical studies around almost 2 decades (see the comparative photos and notes in Table 2.2), and it seems also less than the number in the report. The problem is that aside from the statistics, the report published by the Department of Science and Technology of Gia Lai province (Hoang, 2017) spent most of the remaining content on citing old studies for affirming the characteristics, classifications, and values of funerary sculptures in the past. Therefore, it might lead to a misunderstanding that most of the sculptures as mentioned are from cemeteries, and all the artisans mentioned above are local people who often make sculptures only for using in grave-leaving ceremonies.

Actually, these sculptures might have included the artifacts not only in cemeteries but also in all other places (such as communal houses, public spaces including museums and parks, and tourist service areas including cafes, restaurants, or hotels/resorts). Moreover, the statistics at places other than the cemeteries might have been the majority but have not been clarified here. Besides that, as far as we know, some of the artisans have become Catholics or Protestants, so they no longer participate in traditional rituals at all, but sometimes they also create sculptures to meet the request of cultural institutions, tourist festivals, or even of any customer.

Table 2.1 An overview of topic distribution and styles of funeral sculptures of the Jrai subgroups

<table>
<tr><th>Subgroups</th><th>Jrai H'drung</th><th>Jrai Arap</th><th>Jrai T'buăn</th><th>Jrai Chor</th><th>Jrai Mthur</th></tr>
<tr><td></td><td></td><td></td><td></td><td></td><td></td></tr>
<tr><td>Time</td><td>1995</td><td>1997</td><td>1995</td><td>1995</td><td>1995</td></tr>
<tr><td>Source</td><td>Dao, 2007</td><td>Dao, 2007</td><td>Tran, 1995</td><td>Tran, 1995</td><td>Tran, 1995</td></tr>
<tr><td rowspan="2">Descriptions</td><td rowspan="2">Rarely seen from 1990s; Most were made by cement; visual language: Realism but often enlarge the size of the objects (birds)</td><td rowspan="2">Includes all topics; Topic (2) was special concerned; Visual language: Sharp and strong with linear cubes created by large cuts of axes.</td><td rowspan="2">Includes nearly all topics except for topic (2); visual language: Some details are presented in charcoal black and engraved lines</td><td colspan="2">Includes all topics, but in group (2), there was often only a male or a female character (not both like Jrai Arap); visual language: Realism with elaborate rounded details.</td></tr>
<tr><td>The ratio of 1: 1, with a simple pedestal of about 1 meter high.</td><td>Miniature ratio, placed on top of a high pedestal (or a pillar), with a simple lower part and a linear engraved upper part.</td></tr>
<tr><td>Topic (1)</td><td>Not found</td><td>x</td><td>X</td><td>x</td><td>x</td></tr>
<tr><td>Topic (2)</td><td>Not found</td><td>x (essential, popular)</td><td>Not found</td><td>Rare (only male or female)</td><td>Rare (only male or female)</td></tr>
<tr><td>Topic (3)</td><td>Not found</td><td>x</td><td>X</td><td>x</td><td>x</td></tr>
<tr><td>Topic (4)</td><td>Not found</td><td>x</td><td>X</td><td>x</td><td>x</td></tr>
<tr><td>Topic (5)</td><td>x (most are birds)</td><td>x</td><td>X</td><td>x</td><td>x</td></tr>
</table>

3.3 *Is there Something Wrong in Dealing with Two Types of Sculptures?*

According to several experts of art criticism (Cagan, 2004; Freeland, 2001; Ocvirk, Wigg, & Le, 2006), the content and the form (style) of the artworks will be completely different according to their different purposes. It is unreasonable to combine both sculptures including funeral and decorative purposes into a preservation project. Therefore, to make this urgent, essential, and exciting project more realistic and accurate, it is necessary to separate the definition of "folk wood sculpture" here into

Table 2.2 Comparison of the number and quality of the sculptures in cemetery between the 1990s and 2010s

<table>
<tr><td></td><td></td></tr>
<tr><td>Location & time: Chu Pah, 1997
Group: Jrai Arap - Source: Dao, 2007</td><td>Location & time: Chu Pah, 2012
Group: Jrai Arap - source: Ho (field trip in 2012)</td></tr>
<tr><td colspan="2">Note: The number of sculptures in each tomb in 1997 was much more than in 2012; the topics and the postures of the characters were more diverse; it was also easier to see the new sculptures because pothi was an annual ritual.</td></tr>
<tr><td></td><td></td></tr>
<tr><td>Location & time: Pleiku, 1995
Group: Jrai H'drung - Source: Dao, 2007</td><td>Location & time: Pleiku, 2013
Group: Jrai H'drung - Ho (field trip in 2013)</td></tr>
<tr><td colspan="2">Note: The area of Pleiku city since the 1990s has only a few sculptures of birds made of cement. In 2013, almost all cemeteries only have graves of Christians shared the same design with a few rare remaining traditional tomb.</td></tr>
</table>

(continued)

Table 2.2 (continued)

Location & time: Kong Chro, 1993 Group: Bahnar Tolo - Source: Dao, 2007	Location & time: Kong Chro, 2018 Group: Bahnar Tolo - source: Ho (field trip in 2018)
Note: The tombs in Kong Chro district have not changed much in architecture form over the years, but the roof completely changed from using natural to industrial materials. But, the biggest change was that the funeral sculptures had completely disappeared, leaving only some small objects in group (5) carved on top of the pillars.	

Location & time: Krong Pa, 1993 Group: Jrai Mthur - Source: Dao, 2007	Location & time: Krong Pa, 2019 Group: Jrai Mthur - source: Ho (field trip in 2019)
Note: Although the form and materials for tombs have changed over the years, the topics and styles of the Jrai Mthur sculptures in Krong pa district have retained some basic characteristics of traditional sculptures, except for using industrial paints and a reduction in the number of artifacts on each tomb.	

at least two categories including funeral and decorative wood sculpture before giving solutions for them.

Because of not distinguishing the two types of those sculptures according to their purpose in the context of folklore (ritual or decorative purpose?) to meet the symbolic meaning of each type, the local cultural agencies have tried to organize annual sculptural training courses and competitions for restoring both at the same time. During these activities, all the topics of funeral statues in history, especially the three groups (1), (2), and (3), were always encouraged in new works of art, even if they were no longer relevant in the social context of the modern life. Though, in fact, among the three groups of topic, there is only group (3) based on daily activities/rituals could easily be developed in accordance with contemporary life, while the remaining two groups were closely linked to funerals with the meaning of condolence (group 1) and promoting the rebirth of decedents to the next step of the life circle (group 2), so it is difficult for them to be developed outside the context of the cemeteries.

All statues after those cultural activities were collected for display not only in some museums and local cultural agencies but also in many public spaces like parks. Unfortunately, it contributes to form a popular style of outdoor decoration that we can see in any space in the Central Highlands, not just in cultural spaces. From then on, several tourist companies also ordered artisans to carve similar statues to decorate their gardens/resorts. Let imagine that on the walkways in the relaxing gardens, visitors may be surprised to meet the statues suddenly appearing in the grove, sitting on the top of logs, holding their faces and crying; or sometimes there are statues of men and women showing their magnified genitals or being in mating status that could make the visitors confused (especially when traveling with children) because they cannot perceive and interpret the meaning of such "outlandish" sculptures in a completely unrelated context. These cases sometimes happen in new design spaces for tourism in the Central Highlands of Vietnam nowadays. In other words, it might result in improper ways of preserving and promoting folk arts.

When looking for data on wood sculpture conservation activities from provincial cultural agencies, we found that these activities have been regularly organized and has rather high budgeted, for example:

"International Festival of Gong Culture 2009" organized in Gia Lai province: A wood folk sculpture competition was one of 15 main activities in this event. According to the report of Gia Lai Provincial People's Committee (GLPPC, 2009), there were 26 artisans including the Bahnar (10) and the Jrai (16) participated in, created 21 statues in all topics such as human activities, animals, and gifts (in shapes of horns, ivories, pots, etc.). The organizers awarded all the artworks to encourage these folk artisans. There is no further information on how the statues were used after the event.

In another event named *"The week of Culture, Sports and Tourism"* organized the first time in Kon Plong District, Kon Tum province in 2013, a wood folk sculpture competition on a fairly large scale was held at a location nearby Mang Den Pass. After two weeks, 33 artisans from 8 districts, towns, and the city had finished 100 statues to celebrate the 100th birthday of Kon Tum province. As usual, all of the artisans were encouraged by different awards (artisan A Êk - the best prize for his beautiful artwork; artisan A Khăk - an award for overcoming many difficulties to participate in the contest; artisan A Lao who was born in 1930 - an award for being the oldest in the contest, …). All of these statues of which most of all are on topics of funerary sculpture, were displayed in an eco-resort nearby the Pa Sy waterfall (Kon Plong district) to serve tourists (come back to the problem "the artworks do not match the context" as mentioned above) (DCISKT, 2013).

We also found some reports of classes in which the teacher was a Jrai artisan named Ksor Nao, a very enthusiastic person in passing the skill of folk wood sculpture to the next generation. In August 2010, a class with 13 Jrai and Bahnar members who had learned and practiced for over a month under the instruction of artisan Ksor Nao and his assistant Ksor Krôh. About teaching methods, first of all, the instructor made a statue of the topic "mourning emotion," while the learners observed and then made a copy. During the time, all members were focusing on their own artworks, trying to imitate the instructor's sample, the instructor gave comments and corrected on each artwork. The following topics have been taught in this course: a monkey, a drummer, a peacock, a woman carrying a basket, a mother holding a baby, a couple of man and woman, and a water gourd. They had spent about 2 days to finish each lesson. At the end of the course, each learner chose one of their favorite topics and do their own artwork without any sample from the instructors.

After similar courses, local cultural institutions often hold contests to give more and more opportunities for the new artisans to practice repeating regularly, such as a contest in Chu Pah district (Gia Lai province) in 2011. All the costs for these activities were deducted from the budget of National Targets Program.

In addition to local activities, the Ministry of Culture, Sports and Tourism of Vietnam also organized annual events such as the week of *"Great National Unity - Vietnam's Cultural Heritage."* In 2013, *"Tay Nguyen folk wood sculpture"* was one of the main activities at the event in which participated nearly 30 artisans from many ethnic groups including Jrai, Bahnar, Gie Trieng, Ro Mam, Rhade, and M'nong. There were 76 wooden statues collected after the event, then displayed in the landscape of Dong Mo tourism destination (Vietnam National Village for Ethnic Culture and Tourism - Location: Son Tay town, Ba Vi district, Hanoi). Most of these artworks, as usual, closely followed the topics of funeral sculpture.

Certainly, it is easy to see that the activities mentioned above were extremely governmental efforts from the central to local levels and their success in many aspects: Firstly, these activities had encouraged the pride of the local people, which is really important because culture can be restored only when its owners really want to; Secondly, these activities had brought more opportunities to develop the basic skills of practicing folk arts for the local people, ensuring the career transmission from generation to generation. Finally, although the sculptures created in those courses and events have never been in cemeteries, they could be collected as "remake artifacts" in museums. For the reason of "matching the context" mentioned above, one important thing we would recommend here is: These sculptures need to be displayed in the right place (museums, folk art exhibitions, cultural agency offices...) with enough descriptions for visitors to understand the context before enjoying the beauty of the artworks.

A serious problem is that conservation without a relatively distinguish among ethnic groups has led to a mix of styles, so it is too difficult to recognize the sculptures were created by which groups or subgroups of ethnic people if no information sheets attached (see Table 2.3).

3.4 Challenges and Opportunities in Preserving each Type of Folk Wooden Sculpture

With the current policy of religious freedom, traditional beliefs in the Central Highlands will soon disappear as we have seen in many other places in the world. It is an inevitable consequence of social transformation. This prediction is based on the following data: According to the Census Steering Committee in 2009, there were 114.822 Catholics and 110.114 Protestants in total of 312.272 people in Gia Lai province. The distribution between urban and rural: 28.627 Catholics (32.5%) and 10.361 Protestants (11.8%) in a total of 87.880 urban residents; 86.195 Catholics (39%) and 99.753 Protestants (44.5%) in a total of 224.392 rural residents (PHCNSC, 2010a, 2010b). In addition, the survey found that the majority of indigenous people live in rural areas rather than in cities. In case of the Jrai, there were 368,926 rural residents (89.7%) and 42,349 urban residents (10.3%), in a total of 411,275 Jrai people living in this province. This distribution is similar to the Bahnar (IBID 2010). In other words, most of the indigenous people in the Central Highlands of Vietnam, especially in the rural area, became Catholics and Protestants (after a decade, the number might be much higher so far). In other words, the folk arts related to traditional beliefs, including sculptures, do not have much chance of survival unless they change to adapt to new conditions. During this time, they need to be supported by both government and nongovernmental organizations.

However, the predictable disappearance of funeral wood sculptures might still be at a future point. Coming back to the present time, during the fieldwork, we have seen quite a lot of families still holding grave-leaving ceremony (*pothi*) but they did not make or just made some small and simple wooden statues (Fig. 2.2) due to many reasons, in which the most serious are the lack of wood, the lack of skills, and the lack of historical knowledge about ancestral symbols.

The solution of providing wood for the local people to make statues in grave-leaving ceremonies is probably simple and easy based on the government's budget. Some solutions to support skills for the next generation have proven to be very effective through activities such as the courses and sculpture competitions mentioned above. The solution for the locals to improve their knowledge of the symbols their ancestors once valued, in our opinion, is the most difficult.

A serious problem in the conservation of local cultural agencies recently is in this point: They focus on restoring wood folk sculpture as a kind of funeral sculptures

Table 2.3 A few examples of funeral sculptures of some ethnic groups/subgroups that used to have typical characteristics to show that it is difficult to distinguish the artworks of different groups nowadays

Old funeral sculptures (before 2000)	New funeral sculptures (after 2000)	Wooden sculptures recently (for cultural festivals, tourism,…)

Jrai Arap in 1997 (Source: Dao, 2007*)*	*Jrai Arap in 2011 (Source: Ho 2011)*	*Jrai Arap in 2019 (Source:* Ho, 2019*)*
Note: The sculptures of the Jrai Arap were once sculpted in linear language and with no detail engravings; however, the new ones are in rounded language, with detailed engravings and colorized costumes like the Jrai T'buăn or some groups of Bahnar A la Kong before 2000		

Jrai T'buăn in 1993 (Source: Tran, 1995*)*	(almost disappeared)	*Jrai T'buăn in 2019 (Source:* Ho, 2019*)*
Note: The sculptures of the Jrai Arap keep all basic characteristics such as rounded language, colorized costumes, but more impressed by the industrial garish paints. However, the art of funeral sculpture of the Jrai T'buan has lost in reality.		

Table 2.3 (continued)

Old funeral sculptures (before 2000)	New funeral sculptures (after 2000)	Wooden sculptures recently (for cultural festivals, tourism,…)

Jrai Mthur in 1993 (Source: Tran, 1995*)*	*Jrai Mthur in 2019 (Source:* Ho, 2019*)*	(there is no sculpture of the Jrai Mthur in the display areas nowadays, probably because of their sculptures used to be placed on high pillars, which do not fit the visitor's vision)
Note: The Jrai Mthur have been keeping their style of funeral sculptures rather well so far, though the number of sculptures in the cemeteries is decreasing day by day		

Table 2.3 (continued)

Old funeral sculptures (before 2000)	New funeral sculptures (after 2000)	Wooden sculptures recently (for cultural festivals, tourism,…)

Jrai H'drung in 1995 (Source: Dao, 2007*)*	----- (almost disappeared)	*Jrai H'drung in 2019 (Source:* Ho, 2019*)*

Note: The Jrai H'drung have stopped making funeral sculptures since the late twentieth century, but now their artworks can be seen in new display areas. This means the style of their new statues is passed down from other groups.

Table 2.3 (continued)

Old funeral sculptures (before 2000)	New funeral sculptures (after 2000)	Wooden sculptures recently (for cultural festivals, tourism,…)

Bahnar A la Kong (1993) (Source: Tran, 1995*)*	*Bahnar A la Kong (2018) (Source:* Ho, 2018*)*	*Bahnar A la Kong (2019) (Source:* Ho, 2019*)*

Note: Funeral statues of Bahnar groups residing in the lowlands (A la Kong) before 2000 often favored the details of carved carvings on the costumes. This feature appears very little in the new artifacts. Besides, the group of subjects (2), which were not in their funeral sculptures from past to present, also appear in their artworks nowadays. Some are also colored like Jrai T'buan.

Fig. 2.2 Some very small wooden funeral sculptures of the Jrai Mthur in Krong Pa district, Gia Lai province - Source & time: Ho, 2019

while the demand in reality for this type is coming less and less. About the type of decorative sculptures, it is extremely necessary to promote the formation of identity of sculptures in parallel with catching up with the trend of modern art, and only when the artisans get rid of familiar topics, then they could be free to compose. The reality showed that in the context outside the cemeteries, some artisans are now just "making copies" of sculptures that used to be created by their ancestors rather than "creating/composing the new artworks" (exactly repeat the groups of topics (1) or (2) for example). In other words, those unsuitable-in-space sculptures could be called "fake artworks" which can gradually make the next generations create them as a habit without understanding the core meaning or the birth reasons of the traditional sculptures. If this method of cultural conservation is wrong, "the more we try, the worse it gets," then it may eventually lead to a period of "cultural disruption" once again.

4 Discussion and Conclusion

Nowadays, the proportion of locals adopting new religions is much higher than the proportion of people who still retain traditional beliefs in the Central Highlands of Vietnam. Therefore, the number of wooden funerary statues cannot be restored here because the custom of *pothi* is progressively rare, but it is possible to increase the aesthetic quality of those sculptures. Some solutions, such as the government should set aside a budget to help families who want to retain traditional funeral customs, and provide enough wood so that they could be more creative in sculpting and decorating tombs. The "real" artworks abandoned after the ceremonies should be collected (with the permission of the local people) for displaying in museums or cultural agencies to introduce to visitors from other regions and countries, while also making an increase in the understanding and pride of the next generations. In addition, it is very important to study the core initial significance of funeral statues to overcome the effects of cultural disruption.

It is very important to consider whether or not to promote the restoration projects of tomb sculpture by gathering many folk artists to make statues on demand but without the grave-leaving ceremony. Instead of organizing classes or sculpture competitions to create "fake" funerary statues which have never been used in any ritual, it might be better to focus on teaching skills of traditional wood carving (such as how to use special tools like the way their ancestors did to create their own style and define their sculptural identity).

In some certain aspects, the classes where some good folk artists are responsible for teaching the traditional wood carving skills to the younger generation are still cultural activities that should be encouraged. However, the content of the artworks needs to be expanded beyond topics related to grave-leaving rituals (*pothi*) if their purposes do not match. In other words, outside the context of beliefs or religions, modern artistic thinking which still roots in traditional cultural identity is necessary for the development of local folk sculptures suitable for the new context.

References

Cagan, M. (translator: Phan Ngoc) (2004). *Morphology of Art ("Hình thái học của nghệ thuật")*. Writers Association Publishing House: Hanoi, Vietnam; p.29–34; p.273–324.

Chi, T. (1986). *Decorative patterns of the Jrai and Bahnar ("Hoa văn các dân tộc Gia rai - Ba Na")* (pp. 1–78). Culture and Information Department.

Tu Chi (1995). A visit to the tombs of the Jrai and Bahnar ("Thăm nhà mả Bahnar và Jrai"), In Book Tu Chi & Decorative patterns of the Jrai and Bahnar (trong quyển "Từ Chi & Hoa văn Jrai, Bahnar"), 2nd ed.; Vu Ngoc Binh, Nguyen Quang Tue, Department of Culture and Information of Gia Lai Province, Culture and Information Center: Internal circulation in Gia Lai province, Vietnam, 2006; p.88–92.

Condominas, G. (1957). *Nous avons mangé la forêt*. Musée du Quai Branly Publishers.

Dam Bo (translator: Nguyen Ngoc). (2003). Populations montagnardes du Sud - Indochinois ("Miền đất huyền ảo"), Writers Association Publishing House: ; p.1–439.

Dang, N. V., Cam, T., Tran, M. C., Le, D. D., & Ngo, V. B. (1981). *Ethnic groups of people in Gia Lai - Cong Tum province ("Các dân tộc tỉnh Gia Lai - Công Tum")* (pp. 53–157). Social Science Publishing House.

Dao, H. Q. (2007). *Wooden Sculptures in the Vietnam Central Highlands ("Tượng gỗ Tây Nguyên")* (pp. 1–154). Tong Hop Publishing House.

DCISKT (Department of Culture, Information, and Sports of Kon Tum province) (Sở Văn hóa, Thông tin và Thể thao tỉnh Kon Tum) (2013). Report No. 124 / BC-SVHTTDL on the results of the organization of "Mang Den Culture, Sports and Tourism Week, Kon Plong District, the first time in 2013 ("Báo cáo số 124/BC-SVHTTDL về kết quả tổ chức "Tuần lễ Văn hóa, Thể thao và Du lịch Măng Đen, huyện Kon Plông lần thứ nhất năm 2013″); Internal circulation in Kon Tum province: Kon Tum, Vietnam.

Dounes, J. (2006) (translator: Nguyên Ngọc), Forêt, femme, folie ("Rừng, đàn bà, điên loạn"); Writers Association Publishing House: ; p.1–366.

Dourisboure, P. (1873). *Les Sauvages Bahnars* (2nd ed..; Lecoffre Publishers: France, 1875, pp. 1–338). de Soye Publishers.
Freeland, C. (2001). *An introduction to art theory*. Oxford University Press. p.39-44; p.51-52; p.122-123; p.133-143; p.151-153; p.165-170.
GLPPC (Gia Lai Provincial People's Committee) (Ủy ban nhân dân tỉnh Gia Lai) (2009). Summary Report of the International Gong Festival in 2009 in Gia Lai Province ("Báo cáo tổng kết thực hiện kế hoạch Festival Cồng chiêng quốc tế năm 2009 tại Gia Lai"); Internal circulation in Gia Lai province, .
Guerlach, J. B. (1884). Chez les sauvages Bahnars. *Journal of Les Missions catholiques, 1*, 11–25.
Guerlach, J. B. (1887a). Deux ans de captivité chez les Bahnars. *Journal of Les Missions Catholiques, 12*, 2.
Guerlach, J. B. (1887b). Mœurs et superstitions des sauvages Bahnars. *Journal of Les Missions Catholiques, 6*, 11.
Guilleminet, P. (1942). *Le sacrifice du buffle chez les Banhar de la province de Kontum*. Ecole Francaise D'etreme-Orient.
Guilleminet, P. (1951). *L'aspect du temps et les météores dans une tribu du Kontum*. Ecole Francaise D'etreme-Orient.
Guilleminet, P. (1952). *Coutumier de la tribu Bahnar des, Sedang et des Jarai de la province de Kontum*. Ecole Francaise D'etreme-Orient.
Guilleminet, P. (1960). *Langages spéciaux utilisés dans la tribu bahnar du Kontum (Sud Viêt-nam, Indochine)*. Ecole Francaise D'etreme-Orient.
Ho, T.T.N. (2014). The artistic value of tombs and funeral sculptures of the Jrai and Bahnar in the North Central Highlands of Vietnam (Giá trị nghệ thuật của nhà mồ và tượng mồ các dân tộc Gia Rai, Ba Na ở Bắc Tây Nguyên"). Master's thesis specialized in Criticism and History of Fine Arts; Ho Chi Minh City University of Fine Arts: Ho Chi Minh city, Vietnam; p.1–120.
Ho, T. T. N. (2016). The tomb Statues of Gia Rai and Ba Na: Differences in styles between local groups ("Tượng mồ Gia Rai, Ba Na: Sự khác biệt về phong cách giữa các nhóm địa phương"). *Cultural Heritage, 3*(56), 61–64.
Ho, T.T.N. (2018). Variation of Funerary Figurines in the Central Highlands of Vietnam under the Change of Professional Transmission. The 5th Asian Conference on the Art, Humanities and Sustainability (ACAHS 2018): Hiroshima, Japan; p.46–55.
Ho, T.T.N. (2019). Explaining the symbol of banyan tree in decoration at the Jrai tombs in the North Central Highlands of Vietnam ("Giải mã biểu tượng cây đa trong trang trí tại nhà mồ Gia Rai ở Bắc Tây Nguyên"). Annual Scientific Conference, Ho Chi Minh City University of Fine Arts: Ho Chi Minh city, Vietnam; p.16–32.
Hoang, T.H. (2017). Final Results of Study: Preserve and promote the value of wooden folk sculptures of the Jrai and the Bahnar in Gia Lai province ("Báo cáo tổng kết kết quả nghiên cứu và triển khai đề tài: Bảo tồn và phát huy giá trị tượng gỗ dân gian Bahnar, Jrai trên địa bàn tỉnh Gia Lai"); Gia Lai Provincial People's Committee, Department of Science and Technology: Gia Lai Province, Vietnam; p.101–111; p.132–140; p.1–178.
ISSCH (Institute of Social Sciences of Central Highlands) (Viện Khoa học Xã hội vùng Tây Nguyên) (2017). Conference "Economic Development in the Central Highlands: Potentials and Problems" (Hội thảo khoa học "Phát triển Kinh tế vùng Tây Nguyên: Tiềm năng và những vấn đề"); ISSCH: Hanoi, Vietnam; Page 11.
Maitre, H. (1909). *Les Régions Moi du Sud Indo-Chinois - Le Plateau du Darlac* (pp. 1–277). Plon Publisher.
Maitre, H. (1912). Les Jungle Moi - Expoiration et histoire des hinterlands moi du Cambodge, de la Cochinchine, de l'Annam et du Laos; Emile Larose Publisher: Paris, France; p.1–170.
Ngo, V. D. (1986). Funeral sculptures in the Vietnam central highlands - some thoughts on ethnography ("Tượng nhà mồ Tây Nguyên - Vài suy nghĩ dân tộc học ban đầu"). *Ethnography, 3*, 60–65.
Ngo, V. D. (1990). Tombs in the Vietnam Central Highlands and the Pacific Regional Culture ("Nhà mồ Tây Nguyên và văn hóa khu vực Thái Bình Dương"). *Folklore Vietnam, 2*, 54–59.

Ngo, V. D. (1991a). Styles of the funeral sculptures in the Vietnam central highlands from the perspective of poetics ("Phong cách tượng nhà mồ Tây Nguyên từ góc độ thi pháp học"). *Folklore Vietnam, 3*, 57–61.
Ngo, V. D. (1991b). The art in construction of the tombs in the Vietnam central highlands ("Nghệ thuật xây dựng nhà mồ Tây Nguyên"). *Ethnography, 3*, 16–21.
Ngo, V.D. (1993). Funerary architecture and sculptures of the Jrai and Bahnar ("Nhà mồ và tượng mồ Gia Rai và Bơhnar"); Institute of Southeast Asia, Department of Culture, Information and Sports of Gia Lai Province: Gia Lai, Vietnam; p.1–233.
Nguyen, D. C., & Nguyen, K. C. (1937). *The Bahnar in Kon Tum province ("Mọi Kon Tum")* (pp. 1–514). Maridor Publishing House.
Nguyen, T. C. (1979). Wooden sculptures in the Vietnam central highlands ("Tượng gỗ Tây Nguyên"). *Art and Culture, 2*, 61–68.
Nguyen, T.K.V. (2007). The history and culture of Northern Highlands ("Đến với lịch sử - văn hóa Bắc Tây Nguyên"); Da Nang Publishing House: Da Nang, Vietnam; p.112–140, p.154–158, p.271–274, p.337–340.
Nguyen, T.K.V. (2013). Beliefs and religions of the Bahnar and the Jrai ("Tín ngưỡng và tôn giáo dân tộc Bahnar, Jrai"); National Culture Publishing House: Hanoi, Vietnam; p.117–177, p.211–236.
Ocvirk, Stinson, Wigg, Bone, Cayton Le (translator: Le Thanh) (2006). Art Fundamentals ("Những nền tảng của Mỹ thuật"); Fine Arts Publishing House: ; p.18–296.
Phan, C. T., & Nguyen, T. C. (1995). *Funeral Sculpture in the Vietnam Central Highlands ("Điêu khắc nhà mồ Tây Nguyên")* (pp. 1–159). Fine Arts Publishing House.
PHCNSC (Population and Housing Census National Steering Committee) (Ban chỉ đạo trung ương về điều tra dân số và nhà ở) (2010a). Table 6: Population by urban/rural residence, sex, age group and ethnic group, 1/4/1999″. The 2009 Vietnam population and housing census: Completed Results; Hanoi, Vietnam; p.226–280.
PHCNSC (Population and Housing Census National Steering Committee) (Ban chỉ đạo trung ương về điều tra dân số và nhà ở) (2010b). Table 7: Population by urban/rural residence, sex, religion, socio-economic region and province/city, 1/4/1999″. The 2009 Vietnam population and housing census: Completed Results; Hanoi, Vietnam; p.281–312.
Tran, P. (1995). *Folk wooden sculpture of the Jrai and Bahnar ("Điêu khắc gỗ dân gian Giarai-Bahnar")* (pp. 1–88). National Culture Publishing House, Department of Culture, Information, and Sports of Gia Lai Province.
Tran, P. (2008). *Festivals in the Vietnam central highlands ("Lễ hội Tây Nguyên")* (pp. 36–88). World Publishing House.
Tran, T. (1986). The lifestyle of the minorities in the Vietnam Central Highlands ("Nếp sống Thượng"). In Book Tu Chi & Decorative patterns of the Jrai and Bahnar (trong quyển "Từ Chi & Hoa văn Jrai, Bahnar"), 2nd ed.; Vu Ngoc Binh, Nguyen Quang Tue; Department of Culture and Information of Gia Lai Province, Culture and Information Center: Internal circulation in Gia Lai province, Vietnam, 2006; p.26–32.
VCHSC (The Vietnam Central Highlands Steering Committee) (Ban chỉ đạo vùng Tây Nguyên) (2017). Report on socio-economic data of the Central Highlands from 2001 to 2015 ("Tập tư liệu Kinh tế - Xã hội vùng Tây Nguyên 2001–2015"). Hanoi, Vietnam; pages: 170.

Chapter 3
Traditional Concepts on "Sustainability" in Vietnamese Culture and Its Impacts on Forming Modern Values of Sustainable Development

Le Lan Chi

Abstract Vietnam is an agricultural country characterized by wet-rice civilization, and the livelihood of its people mainly depends on the nature. Due to geopolitical characteristics, Vietnam had to encounter unceasingly wars and assert their identity to resist against the Sinicization and Westernization. Does the aforementioned context lead to the fact that "sustainability" is considered as an objective or a value of the Vietnamese after thousand years of national construction and protection? Does the concept of "sustainability" in Vietnamese culture has the same meaning as "sustainability" in the concept "sustainable development," which was newly brought into Vietnam, how does it impact on the shaping of modern values of "sustainable development"? The article applies generalization and "Yi-ology" as the two main methods to answer these questions; in addition, contextual analysis method and comparison are also employed to concretize conclusions inferred via the methods of "Yi-ology" and generalization. Thereby, the paper discovers certain distinctions between the traditional concept of "sustainability" and modern value of "sustainable development" imported to Vietnam, and the process of shaping modern values of "sustainable development" in the country, on one hand, needs to lean on, on the other hand, has to struggle to be liberated from the ties of tradition.

Keywords Sustainability · Traditional culture · Value · Sustainable development

L. L. Chi (✉)
VNU-School of Law, Hanoi, Vietnam
e-mail: lelanchi@vnu.edu.vn

A. T. Nguyen, L. Hens (eds.), *Global Changes and Sustainable Development in Asian Emerging Market Economies Vol. 2*, https://doi.org/10.1007/978-3-030-81443-4_3

1 Introduction

After decades of constant war and confrontation with the postwar socioeconomic crisis, Vietnam implemented an open-door policy, shifted to a market-based operating economy. The new mechanism unleashed the longing for riches and economic resources which had long been constrained, but the fast-growing economy also contributed to a more or less materialistic society, a society of luxuriation, chasing economic growth, ignoring moral values, and destroying the environment. This is a problem of Vietnam and also the problem went through by many other countries in the world when overheated economy has left seriously social, cultural, and environmental consequences. The concept of "sustainable development" appears in the Brundtland Report and is defined as "*development that meets the needs of the present without compromising the ability of future generations to meet their own needs*" (The World Commission on Environment and Development, 1987). However, when "sustainable development" is introduced into the life of each member state, this concept will be welcomed in various ways, depending on not only the political viewpoint, the policy system, the laws of each country but also on its traditional culture. "Tradition - contains the inherent elements of culture and society, embodying the code of conduct, ideological standards, customs, habits, lifestyles and behaviors of a shaped community of people in history, becoming stable and being passed down from generation to generation and being preserved for a long time" (Nguyen & Nguyen, 2002). Traditional culture with the system of values and negative characters is formed and preserved for a long time, shaping psychological factors, personality, and national identity, which can be easily accepted but can also exclude policies, new values from the outside being brought for the transplant due to the conservativeness, obsolescence in the new context, not being adjusted to adapt to the changes of the times. Traditional culture is not only valuable but also intertwined negative characters, which are two sides of a problem. On the other hand, placed in the movement and development of culture, at this time and some contexts, some conceptions and behaviors are considered to be valuable but at other time, in different contexts, such conceptions and behaviors become worthless. Up to now, there have been many valuable studies about culture, values, and negative characters in Vietnamese culture. Therefore, the article is devoted to discussing these traditional factors, which concept is valuable and negative for Vietnam today, upon the receipt of the policy of "sustainable development." The article defines "counterpart factors" of the Vietnamese to receive "sustainable development" policy to an extent, traditional conception of Vietnamese people of sustainability, and whether the concept of "sustainability" in Vietnamese culture is synonymous with "sustainability" in sustainable development introduced by the United Nations. How do values and negative characters in traditional culture negatively and positively impact the formation of modern values of sustainable development?

2 Methodology

In responding to the aforesaid research questions, the context of Vietnam as a specific nation and country is defined. This context includes the geographical context, conditions of the nature and soil related to the geographical context, which in turn leads to social characteristics, historical problems decided by geopolitical factors. This context has been further referred to the villages namely production community, community of people who make ends meet and cultural communities of the Vietnamese. The contextual method is used to personalize Vietnamese conventional way of thinking, the national consciousness of Vietnamese nationals as well as to detail the habits, personalities, and traditional conceptions that form Vietnamese people's viewpoints in the past. The linguistic approach is also set out to identify traditional concepts through the system of proverbs that express the conclusions, life experiences of people in the country through sentences, and phrases with timbre and rhymes. Through these traditional concepts, the generalization methodology is used to identify groups of values that are close to sustainability, namely stability, harmonization as well as to identify negative characters, such as bad habits of giving preference to appearance; short-term way of thinking, which are contrary to intrinsic value; and the sustainability in Vietnamese culture itself. The conflicts between values and negative characters and the complicated and confusing interchanges and conversions in Vietnamese psychology and, the mutual interaction between the state's social management policies and society and the impacts of socio-culture to the implementation of state policies, etc., are interpreted by Yi-ological method, putting the reviewed subjects in the movement of Yin and Yang concept which is concurrently existing and reciprocal.

After clarifying the concept, values in traditional Vietnamese culture in terms of sustainability and adjacency to sustainability, the article uses comparative methodology to identify the similarities and disparities between the sustainability in traditional Vietnamese culture and sustainable values derived from sustainable development from the outside world which have been introduced to Vietnam recently. However, sustainable development with in-depth professional issues from an economic, sociological, or environmental perspective is not the object of this paper. The article only approaches "sustainable development" with the definition of The Brundtland Commissions[1] as well as the concretization of this policy of Vietnamese Government through specific objectives and tasks to compare it with the conceptions, values in Vietnamese culture and tradition.

The generalization method continues to be used to identify the problems of contemporary Vietnamese society that arise from the reservation and transformation processes of traditional culture and are promoting and hindering the policy of "sustainable development" from being realized in Vietnamese social life.

[1] The definition of The Brundtland Commissions is one of the most commonly used definitions with pillars and criteria for evaluating sustainable development.

3 Results

3.1 *Identifying "Sustainability" in Vietnamese Tradition and Culture*

The current geographical position of Vietnam stretches along the meridian from the North to the South, associated with the history of the march to the South, starting from the North of Dai Viet kingdom, through the Central region, associated with the territory of Champa kingdom (Champa) and down to the south where there was a fertile land - part of the Kingdom of Funan and surrounding areas in the southern tip of the Indochina peninsula whose sovereignty had yet to be claimed; bordering the sea to the east, China to the North, leaning against the Truong Son mountain range to the west like a natural rampart. Over one thousand years since independence in 938, Vietnamese national history was associated with wars with feudal Chinese states from the North and wars with other big countries in the twentieth century. "A remarkable feature of Vietnamese history is that the anti-invasion appeared many times with high frequency, for a long time and in the fierce conditions of a small country, it must resist the aggression of a big country and fighting against foreign aggression is not only their own ends but a means to achieve a higher end than independence and peace and national construction" (Phan, 1998). In addition, Vietnamese history witnessed a nearly 500-year period starting from the end of the sixteenth century with constant civil wars between feudal groups located in geographical areas along the length of the country and with problems of division and reunification.[2] It is such geographic and historical characteristics that created the spirit of a Vietnamese nation who always faced wars and thus always desires peace and stability. The status of Vietnam is a small country in the war fighting against big ones not for competition and territory expansion. The victory of Vietnam in the wars with the major countries was the victory in the battle of self-defense, repelling invasion intention of major countries; therefore, almost after each war, the Vietnamese were humble, tolerant, keeping peace with big countries with the highest goal of long-term peace and stability.[3] On the other hand, in peacetime, the policy of ruling the country of dynasties and political regimes also showed harmony and flexibility with the people, attached importance to the people, took civilians as the foundation,

[2] From the middle of the sixteenth century, the Le dynasty and the Mac forces competed for power, forming the South–North dynasties, followed by a war between the feudal corporations Trinh and Nguyen that took the Gianh river as the dividing line between Dang Trong-Dang Ngoai lasted until the Tay Son dynasty and was unified only when the Nguyen Dynasty entered the political arena at the end of the eighteenth century. When the French came to Vietnam, the French policy of "Divide to govern" established different political regimes in Tonkin, Annam, and Cochinchina; the twentieth century saw the separation of the country under the Geneva agreement into two South–North regions at the 17th parallel until 1975.

[3] The postwar diplomatic behaviors of the Tran Dynasty after defeating the Yuan Mong army, the Le Dynasty after the defeat of the Ming Dynasty, and Quang Trung troops after the victory of the Qing Dynasty showed this flexible and peaceful policy.

decreased the pressure imposed on people, as Hung Dao Dai Vuong Tran Quoc Tuan said in 1300: "In peace time, we have to decrease the pressure imposed on people to make deep-rooted successive plans. That is the best policy to defend the country" (Le, 1967). Compared with the history of neighboring countries such as China or the Khmer, it is clear that even though the strong citizen-friendly policy of Dai Viet's rule having its original purpose of national defense: "...fight against the Chinese colonialists and against the expansion of China, the ruling class of Vietnam - the representative of our nation at that time - must try to escape from the Chinese influence deeply rooted in the people and people to create their own bravery. To do so, we must be close to the people, get to know the people, empower the people, limit the authoritarian tyranny, flexibly combine the state concentration and the village democracy, the State and the society, the official (orthodox) and the folk, the exogenous and the endogenous, the traditional reservation and the innovation..." (Tran, 2000), but it still contributed to the harmonization of Vietnamese society and to the creation of a Vietnamese national identity.

In terms of economy and mode of production, Vietnam's economy is a farming agricultural economy that mainly cultivates wet rice along the river basins. Unlike cereal farming which is not much dependent on water sources and nomadic farming which is associated with the shift from one place to another, the wet-rice economy depends on water and irrigation system, requires the stability in terms of the production location, living place of residents, and requires a particularly high demand for water treatment by community collaboration. This has formed a wet-rice civilization with high stability in terms of residence, production location, mode of production, and daily activities of the residents. Vietnamese society was organized into villages and communes, where groups of residents resided, closely associated with the life of each individual for the fulfillment of the demand for water management, natural disaster prevention,[4] and self-defense. The villages acted as autonomous and self-governing institutions - with the governance apparatus and village conventions and rules - the social norms of the village are highly accepted by the State: "Vietnamese village convention - a legal cultural product bearing the characteristics of the history of social struggle and natural struggle for thousands of years of the Vietnamese nation" (Le, 1998). Because the northern colonial government was unable to intervene too deeply in local community of the Vietnamese during more than a thousand years of domination, this model of community organization was typical and formed "village culture" of Vietnamese people; on the other hand, it is these autonomous villages that help Vietnamese people to preserve their local cultural values against the risk of assimilation.

Regarding psychology and lifestyle, because of the aforesaid natural–social characteristics, Vietnamese people's characteristics are similar to sustainability. These are the preference for stability (unwillingness to change) and harmony.

[4]Vietnamese people have to coexist and gradually improve nature with embankment and dyke maintenance because the main area of residence is next to large rivers flowing into the sea or short, steep rivers, sea embankment to reclaim land.

Vietnamese people tend to prefer stability in the way that their living space remains unchanged and geographically stable and farming, cultivation activities and community cultural activities are conducted within villages. Vietnamese people stick together in their villages and consider them as their own small universe. Not many people leave their village, their homeland for another village, or somewhere else.[5] The stability is also reflected in the preservation of traditional production method, the wet-rice agriculture production method has been lasting since the dawn of forming the nation. Industrial production only started upon the arrival of the French in Vietnam when only a part of the population was literally industrial workers, while before that, only a part of the population was doing handicraft in craft villages and urban areas. In Vietnamese people's view, the concept of stability meant "building a house – getting married with a woman – buying a buffalo" ("lam nha – lay vo – tau trau"), which represented the stability in residence ("building a house"), family life ("get married"), and the way to make ends meet ("buying a buffalo" – because buffalo was a means of production), "Buffalo is the opening of a business" ("con trau la dau co nghiep"). The way to make ends meet was confined in the space of survival and the space of survival is the living space.

The preference for harmony of Vietnamese people comes from the "yin and yang" philosophy which "originates from the Southeast Asia, the only place in the world where wet-rice farming was born and people lived on such farming throughout the history" (Tran, 2016), and reproduction belief which promotes the combination of "yin and yang", male and female, parents, the sun and the moon, etc., phenomena and things are interpreted in the "yin and yang" methodology, which looks for contradictory objects, classifies and groups them into "yin" and "yang", in order to solve life's problems by putting events, things in the "yin" group with those in the "yang" group to form contradictory pairs, to find the interferences, harmony, balance between them; interference and harmony are the basis for proliferation and balance is the basis for stability. From eating, dressing, building houses to reproduction, giving birth, getting married, treating diseases, prescription to funeral, the harmony and balance of "yin" and "yang" are always desired. It is the state of harmony and "yin"–"yang" balance that leads to a moderate, harmonious, and non-extreme attitude toward life.

From the aforesaid analysis of values that resemble sustainability and from (geographical, historical, environmental, social) context, the sustainability in Vietnamese value system and in Vietnamese society can be more specifically identified as follows: First, it is reflected through the spirit, national aspiration for peace, long-term stability, no war, and no destruction caused by natural disasters. In other words, from the geopolitical and geo-economic perspective, it can be seen that sustainability is an ideal value in the minds of Vietnamese people, originating from their own demand for peace and stability. Second, sustainability is reflected through the

[5]Although a significant part of the people followed Lord Nguyen to the South from the end of the sixteenth century, this process of "emigration" was also a process of reclaiming a long and vast realm. Although the South has many adjustments in customs and practices to suit the new living conditions, the inhabitants of new land still has the roots and essence of Vietnamese people.

political ideology of decreasing the pressure imposed on people, the harmonious relationship between the State and the people in which the State did not intervene too deeply into village communities and their domestic policies, which brought about a society without major conflict among communities caused by religious, ethnic, and regional conflicts. Last but not least, the sustainability is reflected through the preference for stability and love of harmony of Vietnamese farmers, those who created Vietnamese cultural identity, stemming from the psychosocial characteristics of residents in the wet-rice agriculture in villages (Fig. 3.1).

However, when looking back at history, there was not much theoretical knowledge about "sustainability" in the history of Vietnamese ideology. Like other countries and ethnicities in the world, Vietnamese nationals also thought of development but the question how "sustainable" is in sustainable "development" was not well defined. In the mind of Vietnamese people, stability is the basis for sustainability and for development. In order to reach sustainability, stability must be achieved. In order to develop, stability must be achieved. In other words, stability is a precondition for the purpose of obtaining sustainability and development. Stability means the safety in terms of security, way of making a living, family emotion, and the engagement with the society. Without stability, the conditions to meet these needs will be threatened. Development goes after stability. For individuals, families, generations, "development" is understood as "prosperity," which means the promotion

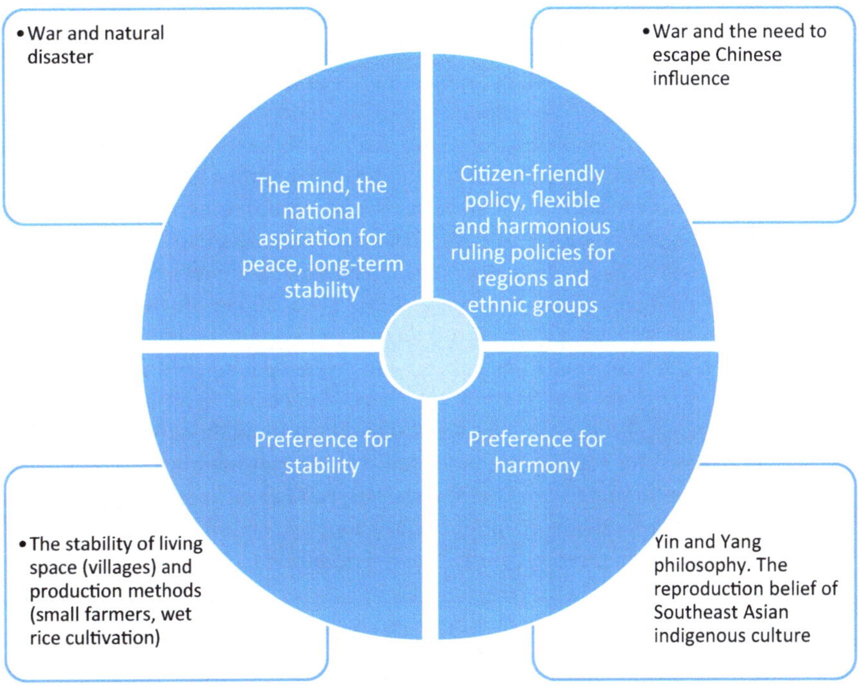

Fig. 3.1 Traditional Vietnamese cultural factors shaping sustainability

in career path as officials, in privileges, the increase in production, land, houses, and better success and prosperity gained by the next generations compared with the previous ones. At a national level, Vietnamese feudal dynasties wanted to build a developed society, but the model of a sustainably developed society was not specific. A peaceful and prosperous society envisioned by Confucianists was even the model of a society that was more or less nostalgic, ambiguous, and mythical as one in the Yao and Shun dynasties in Chinese ancient culture. The lack of a concept of development of sustainability, in particular, and the lack of knowledge about many fields, in general, were a limitation of Vietnamese people. Vietnamese people are "more intuitive than logical" (Dao, 2000), making it unlikely for policies to be long-term, consistent and sustainable.

Lacking the theory of sustainable development, Vietnamese people even lacked the motivation for sustainable development. In traditional Vietnamese culture, individuals and their own development (which was the driving force of community development) have not been properly recognized and only people of the community may be appreciated. "Only as a member of the community, not as an individual, people have little value. That value is associated with his role in the community and in society. Before being a Jack, a Mango, he had to be a farmer, a teacher, a worker or a merchant from one village or another. In the context of the community, people only have the right to choose different shades of the same living spectrum." (Do, 2005). The need to unite individuals to fight against invaders and to conquer the nature has urged people to form community, those who possessed common nature of the community and lived for the community's sake. Individualism was even suppressed and the motivation for community development from a personal perspective was difficult to be activated. Because stability is associated with safety and individual people depend on the community, individuals do not have opportunities for development and their innovation and creativity are eliminated. "Society goes up mostly based on differences with the community. If everyone shares the same lifestyle, way of thinking and feeling, how can social breakthrough and improvement happen? Our country is still poor partly because the Vietnamese tend to be more community-oriented, so individualism tends to be limited. The personality is less appreciated by residents of the wet-rice culture." (Bui, 2006). Small-scale business and small-scale handicraft production industry are also the results of small-scale agricultural production, small businesses, which stably happened in each village with high autonomy and separation from others. The "small and medium" scale is suitable with material resources and weak theoretical thinking of Vietnamese people. The preference for stability is associated with that for harmony, so Vietnamese people do not like to make some breakthroughs, run big business, think big, erase the old, build the new and everything should happen in an order, in the philosophy expressed in the Vietnamese proverb: "nang nhat chat bi" (it means that "many a little makes a mickle").

The mode of small-scale agricultural production also resulted in selfishness, "ruong ai nguoi nay dap bo" (farmers care about their own fields). When people run after their personal benefits, the common good is forgotten and jeopardized, which leads to the way of life similar to an English proverb: "everybody's business is nobody's business" (in Vietnamese ones: "cha chung khong ai khoc" (people do not cry for their common father)), "lam vai khong ai dong cua chua" (no one close his pagoda gate if there are so many monks). For common properties (from common living space to landscape, ecological environment, natural resources), the awareness of preservation and protection is low because people do not consider those properties their own belongings, thus assume no personal responsibility for any consequences, like the meaning of "flood kills all" ("lut thi lut ca lang"). The chase after personal interests is the key to explain why Vietnamese people have no awareness of environmental protection due to their thinking (considering environment is public and movable property and taking advantage of the common to gain personal interests without care about sacrificing their personal benefits for the common good), besides their unawareness. The pursuit of personal small interests will inevitably lead to conflicts with other individuals and the desire to be superior to others. Because people want to be superior to others and chase after personal benefitsconcurrently, the desire of superiority is manifested in external and vain criteria such as title, degree, award, and most dangerously, in the focus on appearance rather than content, leading to the obsession of achievement and the tendency of chasing after trends. When people run after their own benefits, it leads to the conflicts with others and the selfish attitude toward benefits. Such attitude is reflected in various Vietnamese proverbs: "den nha ai nha nay rang" (light is only shed within its owner's house), "trau ta an co dong ta" (buffaloes feed on their own grass fields), "an cay nao rao cay ay" (who fences the tree has the right to eat fruits), "trau buoc ghet trau an" (the chained buffalo hates the fed one), and "con ga tuc nhau tieng gay" (cocks envy each other in terms of their crowing sounds); besides personal interests, there are also selfish group benefits such as "trong lang nao lang ay danh, thanh lang nao lang ay tho" (each village strikes its own drum and worships its own saint).

Therefore, besides positive values and those close to sustainability, there are also many negative characters and negative values. Vietnamese people in the traditional culture uphold community values, the ones that create solidarity among individuals and the strength against invaders, natural disasters, and the stability of the community but that also suppresses personal character, the new, breakthroughs and motivation for development. Personality is suppressed but selfishness arises due to the mode of small-scale production and reliance on the community, due to community responsibility, not personal responsibility. It is the sense of community of Vietnamese people, which dignifies community values, that brings about personal interests, group interests, selfishness, and internal separation. However, when facing common risks (invasion, natural disaster), Vietnamese people give up their personal benefits to unite and to overcome such risks. However, when the threat of common interests no longer exists, people tend to think for their own sake. These are contradictory examples belonging to the two categories of "yin" and "yang," both moving (correlating) and standing still (existing); but there is "yin" in "yang" and there is also "yang" in "yin." When "yin" in "yang" increases, "yang" decreases and "yin" increases, when "yang" decline to a certain level, "yang" in "yin" will increase, leading to a decrease in "yin." The Yi-ological method express values and negative characters approaching sustainability of Vietnamese people in the following figure (Fig. 3.2).

Fig. 3.2 Values and negative characters of sustainability which are both reciprocal and existing in village culture, small agriculture according to "yin and yang" principle

3.2 "Sustainability" in Sustainable Development: The Origin and Acculturation to Vietnam

In modern society with internationalization and globalization, the concept of "sustainable development" from the outside world has also been introduced into Vietnam. In 1980, IUCN – in partnership with the UN Environment Program (UNEP) and the World Wildlife Fund (WWF) released the term "sustainable development" in the *World Conservation Strategy*, despite certain limitations (Seghezzo, 2009), contributed to define the concept and shaped the global conservation and sustainable development agenda with Sect. 20 Toward sustainable development of the Strategy (IUCN et al., 1980). In 1987, "The Brundtland Report focused primarily on the needs and interests of humans, and was concerned with securing a global equity for future generations by redistributing resources toward poorer nations to encourage their economic growth in order to enable all human beings to achieve their basic needs. The report expressed the belief that social equity, economic growth and environmental maintenance were simultaneously possible, thus highlighting the three fundamental components of sustainable development, the environment, the economy, and society, which later became known as the triple bottom line. The report discussed the need to apply integrated, sustainable solutions to a broad range of problems related to population, agriculture and food security, biodiversity, energy choices, industry, and more" (Pisani, 2016). Today, sustainability is a qualitative feature of "sustainable development" recognized with the harmonization and balance of three key factors namely society, economy, and environment. Sustainable development is commonly recognized and simplified as a development model where economic growth is parallel to social justice and environmental protection.

"Sustainability," linguistically, is understood as a sustainable state (Sustainability is the ability to exist continuously, the ability to be maintained at a certain rate or level). In terms of value, if sustainability is understood as a value (value is the nature of the object that is rated by the subject as positive in a comparative relationship between this object and other objects of the same type and is located in a specific space and time), sustainability is understood as a positive nature of development based on the comparison with other unsustainable development models, defined in a broad extent (with three pillars namely economy – society – environment) and in an enduring period. The sustainability in sustainable development directs toward the harmony between individuals and the communities, between people and the nature, the intrinsic equality of distribution, bringing added value to human development including material values as well as spiritual ones. Within each pillar, sustainable development is interpreted and assessed through the following criteria:

Firstly, economic sustainable development. Economic sustainability requires common prosperity for all, not only focusing on profitability for a few, within the permissible limits of the ecosystem nor violating the basic rights of human. The right to use natural resources for economic activities is equally shared. The economy that is considered sustainable is the one with high GDP growth and GDP per capita. The proportion of industry and services is higher than that of agriculture in GDP structure. Economic growth must be a highly efficient growth and it does not accept the achievement of growth at all costs.[6]

Secondly, social sustainable development. Social sustainability focuses on equality and society always needs to create favorable conditions for human development (human development index (HDI) is the highest criteria of social development) and strive to give everyone the opportunity to develop their potential and provide an acceptable living condition. Social sustainability is the guarantee of harmonious social life. There is equality among social strata and gender equality; the gap of living conditions among regions is not large in terms of per capita income, intellectual level, education, health, life expectancy, social welfare, enjoyment of culture, and civilization.[7]

Thirdly, environmentally sustainable development. Environmental sustainability means the proper use of environmental factors, natural conditions, rational exploitation and use of natural resources, protection of environment and quality

[6]The sustainable development in terms of economics is often expressed through the following common criteria: Firstly, gradually reducing energy and other resources consumption through saving and lifestyle changes; Secondly, changing consumption needs in order not to harm biodiversity and the environment; Thirdly, equality in access to resources, living standards, health services, and education; Fourthly, eliminating hunger and reducing absolute poverty; Fifthly, adopting clean technology and industrial ecology (recycling, reusing, and reducing waste, renewing used energy).

[7]Social sustainable development includes a number of main contents quantified through criteria such as: Firstly, population stability, rural development to reduce the pressure of migration into urban areas; Secondly, minimization of negative impact of the environment on urbanization; Thirdly, improvement in education and elimination of illiteracy; Fourthly, protection of cultural diversity; Fifthly, gender equality, giving attention to gender needs and interests; Sixthly, the increase of public participation in decision-making processes.

enhancement of living environment, and a guaranteed human living environment quality. It is to ensure the purity of air, water, land, geographic space, and landscape. Environmental sustainability requires maintaining a balance between protecting the natural environment and the exploitation of natural resources for the benefit of people in order to restrict the exploitation of natural resources at a certain limit which enables the environment to recover and provide long-term support, equalizing the living conditions for humans and other living species on earth.[8]

The sustainability of sustainable development, which originally means the long-term stability, in sustainable development, has shown the harmony in the interests of present and future generations. Sustainable development is divided into three aspects namely economic, social, and environmental aspects with many different evaluation criteria but the core of sustainability in each area, in all three areas, is harmony and more generally speaking, the harmony among the three areas mentioned above. Thus, it can be generalized that the "sustainability" in sustainable development includes two attributes namely long-term (time axis) and harmonious (spatial axis, harmony among communities, geographical areas, between the conduct and relationships between people and people and between people and nature).

When comparing the sustainability in sustainable development policy introduced from outside into Vietnam recently and traditional Vietnamese concepts of sustainability, it can be seen that: firstly, in Vietnamese ideological history, there was neither theory of sustainable development nor policy on sustainable development. As regards policy level and beyond, the Vietnamese political ideology, there was no theory close to the theory of sustainable development accessible on three economic - social - environmental pillars. Secondly, Vietnamese people already have concepts and values close to the two attributes "harmony" and "long-term stability" in the connotation of sustainable development. These are important bases and driving forces, which promote the reception of sustainable development policies and raise the value of sustainable development in Vietnamese culture; Thirdly, it is the traditional culture of Vietnamese people that contains concepts and negative characters that are against the attributes of sustainable development (Tables 3.1 and 3.2).

The "sustainable development" policy was introduced into Vietnam and was received in the course of Vietnam's international integration and learning from international experience. Sustainable development as an exogenous element has step by step been introduced into Vietnam and according to cultural theory, this is a cultural acculturation starting from exposure to reception,[9] and ending with changes

[8] Specific criteria for quantifying environmental sustainability are: Firstly, effective use of resources, especially nonrenewable resources; Secondly, development does not exceed the load-bearing threshold of the ecosystem; Thirdly, protecting biodiversity and protecting the ozone layer; Fourthly, controlling and reducing greenhouse gas emissions; Fifthly, strictly protecting sensitive ecosystems; Sixthly, reducing emissions, solving the problem of (water, gas, soil, and food) pollution, improving and restoring the environment of polluted areas, etc.

[9] The notion of legal acculturation has just emerged for approximately 100 years, indicates a contact, an exchange between two cultures that share close racial, geographical, or historical connections. A number of Vietnamese scholars, of whom Ha Van Tan is a notable representative, has interpreted "acculturation" into Vietnamese by combining two Sino-Vietnamese words, consecu-

Table 3.1 Comparison of concepts and values of Vietnamese people shaping "sustainability" in traditional culture and Attributes of "sustainability" in the modern concept of "sustainable development"

Traditional culture of Vietnamese people	Attributes of "sustainability" in the modern concept of "sustainable development"
Aspiration for peace and for long-term stability	Long-term existence/stability
Citizen-friendly, flexible and harmonious ruling policies	Harmony
The stability of living space and production methods	Long-term existence/stability
Yin and Yang philosophy, the reproduction belief (the preference for harmony in the behavior with the natural environment and social environment)	Harmony

Table 3.2 Comparison of criteria of contemporary modern concept of "sustainable development" and traditional culture of Vietnamese people

Criteria of contemporary concept of "sustainable development"	Traditional culture of Vietnamese people
Economic sustainable development	Lacked theory on economic development; Did not attach much importance on economic development; Highlighted the virtue of hard-working labor and thrift to get further accumulation.
Social sustainable development	Attached much importance on the harmony between individual and individual in community, among communities in society; Did not encourage individual development.
Environmentally sustainable development	Mainly introduced the theory of living in line with rule of nature, rule of "yin-Yang".

of the receiving community/nation. Reception demonstrates the creativity of culture, is the process of creating new values for the receiving subjects and reception is also the process of creating new, added values for the culture of the receiving community/country. This process is hindered by adversities but is also motivated by pros and the problem posed for policymakers is to proactively absorb, transform as well as to take advantage of the pros and limit the cons (Figs. 3.3 and 3.4).

tively "tiếp" (接, contact, receive) and "biến" (变, change, transform), and made up a new compound word: "tiếp biến". "Tiếp biến" reflects the characteristics of acculturation: 1, "contact" =>2, "reception" =>3, "transformation" (as a result of the contact and reception between two or more cultures or culture communities). The process of acculturation could impact two or more cultures through mutual symbioses, exchanges, and communications; however, the impact prevailing is the transition from greater culture with higher level of civilization to the smaller one with lower level of civilization. The smaller culture with lower level of civilization shall receive the values and achievements of the greater culture with higher level of civilization. The nature of this process is receiving and modifying exotic elements to harmonize with the indigenous culture's endogenous elements, which later become exogenous elements, hence increasing the values of the indigenous culture.

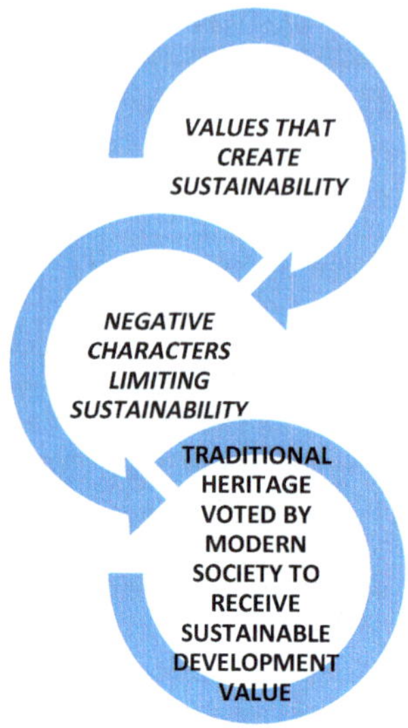

Fig. 3.3 Relationship between values that create sustainability and negative characters that limit sustainability in the traditional heritage of Vietnamese culture

Fig. 3.4 Relationship between exogenous factors (sustainable development value) and endogenous (State policies and traditional heritage preserved by Vietnamese society) in the process of cultural acculturation

The modern Vietnamese society has a myriad of changes. After decades of stabilization of traditional values under feudalism, Vietnamese society applied significant Western cultural values under the influence of French people since 1858 and of the American people in the South of Vietnam in the period of the Vietnam War (1954–1975), and especially the socialist culture in Northern Vietnam since 1954. Ever since the reunification of the country in 1975, the society faced many fluctuations due to the war, which were poverty created by the devastated economy, the value crises among people who contributed to the revolution and postwar emerging people who profited from the war, etc. In many circumstances, the belief crisis was very significant after the socialist system in Eastern Europe collapsed. The downfall of an ideology and of a value that is seemingly sustainable made the Vietnamese people's mindset more fragile. A society with unstable institutions and an incomplete legal framework, a transitional economy to an unstable market mechanism was an opportunity for individuals and businesses to implement business activities involving elements of illegal transfer of assets from state to private ownership, which brought about large sources of capital accumulation, and at the same time enriched and corrupted the section of competent state officials who were entitled to manage public administration. These two groups increased their assets rapidly and the nature of their activity was convulsive, creating a habit of living a fast-paced, luxurious life and contributing to the formation of a more or less materialistic and shallow lifestyle (Vietnamese people consume a lot of alcohol,[10] and buy tons of luxurious cars[11]). This creates social division and dissatisfaction. Thus, from a psychosocial perspective, the contemporary context of Vietnamese society is still a society of the phenomenon of urgent living, short-term way of thinking. It is these characteristics that show a social attitude that has not yet reached the proper state of mind to practice the theory of sustainable development but also needed to further receive the theory of sustainable development. The context of the early years of the twenty-first century when sustainable development was introduced to Vietnam was when Vietnam's economy made significant changes from a commanding economy, focusing on bureaucracy and subsidy to a socialist-oriented market economy. The economy grew fast and people's lives were improved. However, the emerging economy was also caught up in hot demand, with many social problems arising and the ruined environment. Therefore, the goals set by sustainable development are in line with the contemporary context of Vietnamese society and the needs of contemporary Vietnamese society. The state of Vietnam has seen the consequences of hot growth which posed negative environmental problems and social instability.[12] Vietnam also sees the lesson learnt by other hot growth countries, especially the

[10] According to the information of Deputy Minister of Health Nguyen Thanh Long at the Conference to collect comments on the Draft Law on the Prevention of Alcohol and Beer Harm, which was held by the Ministry of Health and the World Health Organization on April 13, 2018, the situation of alcohol and beer consumption in Vietnam is very alarming. "Vietnam ranks second among Southeast Asian countries, tenth in Asia and 29th in the world". See:

https://www.nhandan.com.vn/suckhoe/item/36072902-viet-nam-dung-thu-29-tren-the-gioi-ve-su-dung-ruou-bia.html.

[11] Although Vietnam has the average income per capita in the region, Vietnam is the country that consumes the most luxurious cars in Southeast Asia. See: http://tapchitaichinh.vn/nghien-cuu-trao-doi/viet-nam-la-quoc-gia-tieu-thu-nhieu-xe-hoi-hang-sang-nhat-dong-nam-a-138782.html.

[12] The ninth National Congress of the Communist Party of Vietnam held in Hanoi from April 19 to April 22, 2001, evaluated the process of implementing the Resolution of the Eighth National Congress, which revealed the following weaknesses: The economy did not steadily, efficiently and its competitiveness was low; A number of urgent and harsh socio-cultural issues were slowly resolved; Mechanisms and policies were not synchronized and did not create a strong motivation for development; Corruption, degradation of political ideology, morality, lifestyle in a large number of cadres and party members were very serious.

lesson of China – the neighboring country providing models as great reference for Vietnam - in order to make socioeconomic policies (and later on the Chinese communist government also gave the theory of a harmonious society[13]). On the other hand, the Vietnamese state also takes suggestions of sustainable development as a reference for "socialist orientation" in the realization of a great policy that is not really theoretically clear. That is the policy of building a socialist-oriented market economy.[14] With the shift from the economic theory of communists about income equality for everyone to a market economy operating according to market rules, the equality in terms of income, educational, health and social benefits, cultural enjoyment, equality among different strata in the society and gender equality are close to communist ideals and socialist colors.[15] In 2011, the Executive Committee of the Communist Party of Vietnam identified "The socialist society that our people built is a society of prosperous people, powerful country, fairness, democracy and civilization; possessed by the people; with a developed economy based on modern productive forces and appropriate and advanced production relations; having an advanced culture imbued with national identity; where people enjoy wealth, freedom and happiness with conditions for inclusive development; where all ethnic minority groups in the Vietnamese community are equal, united, respected and helped each other to develop together; a socialist State based on Rule of law, which is of the people, by the people, for the people and under the leadership of the Communist Party; having friendly relations and cooperation with countries around

[13] "In September 2004, the Fourth Plenary Session of the 16th Central Committee of the CPC decided to list "the capability of building a socialist harmonious society" as one of the five governing capabilities that the CPC endeavors to enhance. The general requirements for building a harmonious socialist society are: democracy and law; fairness and justice; integrity and friendship; vigor and vitality; stability and order; and the harmonious coexistence of man and nature". See: http://cpcchina.chinadaily.com.cn/2010-09/16/content_13918117.htm

[14] An unprecedented policy and exception to most countries in the world. However, the head of the Communist Party of Vietnam clarified this policy as follows: "The socialist-oriented market economy is a new economic model in the period of transition to socialism… The development of the socialist-oriented market economy shows the country is very selective about achievements of humankind civilization, in order to develop the active role of the market economy in boosting the production force, labor socialization, technology advances, while improving product quality, contributing to enriching society and improving people's living conditions. However, it is necessary to devise effective measures to limit negative phenomena of the market economy, including pure profits, fierce competition, excessive exploitation and division between the rich and the poor." (Radio Voice of Vietnam, 2003). See: http://vietnamembassy-usa.org/news/2003/11/socialist-oriented-market-economy-concept-and-development-soluti

[15] The institution for the market economy with socialist orientation in the spirit of Resolution of the fifth Plenum of the Central Conference XII "Perfecting the socialist-oriented market economy institution" is perfecting the relationship between the State, markets, and society on the principle of "strong state - effective market - dynamic, creative society". However, the socialist orientation with one of the nuclei is the orientation of distribution relations, whereby the distribution relations ensure fairness and create motivation for development; resources are allocated according to socio-economic development strategies, planning and plans; distribution system is implemented mainly based on labor results, economic efficiency, at the same time as the contribution of capital and other resources as well as the distribution channels of social security system and social welfare.

the world".[16] The sustainable development whose nucleus is the harmony has been favorably accepted into Vietnam due to the compatibility of political factors and political history when Vietnamese people are always finding a political model that harmonizes needs, resolves fairly distribution relations, limits the gap between rich and poor, and stratification. In other words: "Fortunately, socialist ideologies are not unfamiliar with traditional public opinion about a universal society. For us, it is estimated that socialism will create a cultural continuity for our development. For Vietnamese people (most of whom are farmers), socialism is synonymous with social justice, an idea coined by Confucius more than 2500 years ago." (Tran, 2000).

The acculturation of "sustainable development" concept into Vietnam has developed a new component pillar of defense and security besides the three pillars of economy, society, and environment. According to the Viet Nam Sustainable Development Strategy for 2011–2020 (Decision No. 432/QD-TTg of Prime Minister, 2012), sustainable development is a long-term requirement during the national development; an effective and harmonious combination of economic and social development with protecting environment and conserving natural resources, maintaining national defense and social security, aiming for "stable and effective growth along with social equality and progress, environment natural resources conservation, maintenance of stability in politics and society, protection of independence, sovereignty, unity and territorial integrity for national interests". This policy acculturation represents politic ideology of the State of Vietnam as sustainable development in Vietnam is inseparable from the traditional aspiration of the nation, which is sustainability in national defense and security, deriving from the necessity to defense our fatherland. Sustainable development also needs a balanced contribution of resources for defense and security duties.

In recent years, the direction to form a tectonic government that the State of Vietnamese has pursued also establishes a policy foundation for sustainable development, with considerable support between tectonic state model and sustainable development model. "In Vietnam, the term "tectonic government" is a fairly new concept. However, tectonic government can be regarded as the government which orients, sets framework and conditions to develop and serve the public interest. The tectonic government focuses on building an appropriate institutional framework, facilitating and promoting people's capacity and creativity for personal and social benefits". In our opinion, whether learning from governmental models of Japanese, Korean, Singaporean, and Chinese or not, tectonic government is a state model with competent authority (the common feature of the model of State in East Asia influenced by Confucian culture and centralized political tradition) to maintain political stability and social order without over-interfering the market, boost economic

[16] The Executive Committee of the Communist Party of Vietnam, "Cuong linh xay dung dat nuoc trong thoi ki qua do len chu nghia xa hội (bo sung, phat trien nam 2011)" [The Platform of Building the Country in the Transition Time to Socialism (supplemented and advanced in 2011)] See: http://tulieuvankien.dangcongsan.vn/ban-chap-hanh-trung-uong-dang/dai-hoi-dang/lan-thu-xi/cuong-linh-xay-dung-dat-nuoc-trong-thoi-ky-qua-do-len-chu-nghia-xa-hoi-bo-sung-phat-trien-nam-2011-1528

development and harmonize social benefits and conflicts, ensuring a secure society for the present and the future.[17]

In the current context of a market-governed economy, a society in which core values are restored to the norms where citizens have freedom to do business, the Internet, information and communication infrastructure is growing remarkably together with international values and globalization affecting Vietnam via the Internet; the awareness and sociopsychological features of Vietnamese has been dramatically changing, especially in urban areas and among young people. There have been fluctuations in the value scale of Vietnamese people recently when the preference toward stability and harmonization of rural communities has transformed rapidly with considerably high rate of industrialization and urbanization, shifting their occupations from agriculture to industry[18] and their living environment,[19] a more powerful and progressive generation is being established; the number of start-ups founded based on novel and unique ideas[20] are creating independent people instead of community people. This provides a proper basis of awareness, social psychology, created by the new "intelligent" and "elite" population to implement and support constructive policies on sustainable development of the State.

---

[17] However, Prime Minister Nguyen Xuan Phuc, the head of current government, has mentioned this model on many official forums several times and pointed out four primary features of a tectonic government: First, the government needs to construct a strong institutional system for economic development; second, the State only invests in areas which private enterprises cannot invest instead of manipulating the market; third, the government must create a favorable business environment; and forth, the government has to maintain integrity and transparency.

[18] On the socioeconomic situation in the second quarter and the first six months of 2019, according to the General Statistics Office of Vietnam (Tong cuc Thong ke) under the Ministry of Planning and Investment: "In the general growth of the whole economy, the sector of agriculture, forestry and fishery increased by 2.39%, contributing 6% to the general growth; the sector of industry and construction grew by 8.93%, contributing 51.8%; the service sector expanded by 6.69%, contributing 42.2%" (General Statistics Office of Vietnam, 2019a, 2019b). See: https://www.gso.gov.vn/default_en.aspx?tabid=622&ItemID=19251

[19] Based on statistics of the General Statistics Office of Vietnam (Tong cuc Thong ke) under the Ministry of Planning and Investment of Vietnam, the population of urban areas in Vietnam in preliminary 2018 was 33,830,000 over the total population of 94,666,000; that of rural areas was 60,836,000 (Access https://www.gso.gov.vn/default.aspx?tabid=714, on the table "Dân số trung bình phân theo giới tính và thành thị, nông thôn", choose value "Tổng số (Nghìn người)" on variation "Cách tính*" (column 1), choose value "prel2018" on variation "Năm*" (column 2) and choose value "Thành thị", value "Tổng số", value "Nông thôn" on variation "Phân tổ*" (column 3). This means urban areas accounted for 35.74% of the national population, increasing by 16.23% compared with the population 12,880,30 in 1990 (the result of 12,880,30 obtained by the same way excluding value "1990" instead of "prel2018" on variation "Năm*" equivalent to the proportion of 19.51%.

[20] According to the statistics of Echelon Magazine, Singapore is one of the largest online magazines on startup in Southeast Asia, Vietnam currently has approximately 3000 innovative startups, nearly doubling the estimated figure by the end of 2015 (about 1800 enterprises). Topica Founder Institute (TFI) also claimed that Vietnam received 92 investment deals with a total capital of over US$291 million, nearly doubling the number of deals and increasing around 50% of total investment capital, compared with those in 2016 (50 deals with US$205 million). See: https://tinnhanhchungkhoan.vn/thuong-truong/su-troi-day-cua-cac-doanh-nghiep-the-he-moi-245247.html.

Nevertheless, the negative characters in Vietnamese conventional way of thinking, even deeply rooted as national identities, still exist, which are meanness and selfishness in preferring personal short-term benefits to long-term benefits for the community and future generations. Social development makes the aforementioned inferior characters evolve into variations of "tenure way of thinking"[21] and "group interests, interest group" becoming nagging issues in Vietnamese contemporary society. Modern Vietnamese society is still credit-driven,[22] meanwhile the qualifications and qualities are not guaranteed, leading to the "scare and lie," dealing, hiding, law violating way of thinking, public servants lying in reports, interviews, and questions leads to the proliferate phenomenon of lying, skepticism, and collapsed beliefs in society.[23] Virtual values also contradict sustainable development toward the truth and stability.

4 Conclusions and Discussion

Thus, it can be seen that the context of natural and social history of Vietnamese has formed the preference for stability and harmony, together with the desire for peace, prosperity, and longevity of a nation suffered from wars, which has built the foundation for the sustainability of Vietnamese traditions and culture. The history of political ideology, which tended to be citizen-friendly and to decrease the pressure imposed on people, has also formed these favorable attitudes of Vietnamese people. However, the Vietnamese still lack systematic ideologies and theoretical viewpoints on development and sustainable development. In traditions and culture, the Vietnamese have not distinguished between stability and sustainable development;

[21] In negative sense, the notion of "tenure way of thinking" is thinking inappropriately regardless of rules, conditions, and circumstances of an individual or a group of people in an agency, entity, and organization during their term on duty, aiming to pursue short-term objectives and interests for material and immaterial benefits for themselves and their "interest group," ignoring common goals and long-term interests of the community; incompetent qualifications and capacity, lack of information and uncertainty of the circumstances lead to incorrect and inadequate awareness; establishing and implementing guidelines, resolutions, programs, procedures, and plans that are not effective and practical, or even bringing unpredictable results and consequences for the present and the future.

[22] Credit-driven way of thinking has led to misconducts in emulation and reward procedures, bribing for certificates and qualifications. Ministry of Education and Training had to conduct campaigns such as "Say no to misconducts and credit-driven practices in examinations." There is also the "Department of Emulation and Reward" in the administrative system in ministries and ministerial agency which assists Ministers to manage the emulation and reward practice in ministries and departments.

[23] See Nguyen, T. T. N. (2015): Hiện tượng giả dối trong đời sống người Việt từ góc nhìn giá trị học [Deception phenomenon in Vietnamese life from the study of value viewpoint] (based on survey results in 2014), in Tran, N. T. (editor) (2016), *Một số vấn đề về hệ giá trị Việt Nam trong giai đoạn hiện tại* [Issues on Systematic Values of Vietnam in the period of industrialization]; Publishing House, Ho Chi Minh City National University, Vietnam, 412–424.

Vietnamese people still appreciate stability and lack proper knowledge and interpretation about development.

The Western theory of sustainable development has been introduced to Vietnam; on the other hand, the core of this theory is the sustainability in socioeconomic and environmental aspects based on the balance of economic, social, and environmental requirements. Therefore, the implementation of this theory brings numerous advantages from the positive characteristics of traditional culture in achieving sustainability as stated above. Vietnamese government has also acknowledged the positive values of sustainable development theory, realized the similarities with the political direction they follow, thereby accepting and adjusting this theory to suit the context and developmental direction of Vietnam. Regarding superstructure policy, sustainable development is relatively compatible to current guidelines and policies of the State of Vietnam and is favorable to accept the value of traditional and cultural development into Vietnamese society.

Nonetheless, the application and implementation of sustainable development theory are also hindered by Vietnamese deep-rooted traditions and cultures, especially variations of meanness, selfishness, narrow, and short-term visions about individual or group benefits in public sector. These conventional characters remain not only in rural and agricultural areas but also in urban and business areas; in fact, business culture of many enterprises is still influenced by these negative conventions.

Hence, compared with previous research, this paper offers abundant investigation into values and negative characters in Vietnamese traditional culture in comparison with sustainability, especially clarifying the distinction between viewpoints of sustainability in traditional culture and sustainable development policy set in the early twenty-first century. Precedent research has left this matter untouched.

This article has not used sociological survey research methods to assess accurately the variations as well as preservations of values and negative characters related to the sustainability and sustainable development of Vietnamese people. In fact, it is not easy to work out such expected sociological investigations, so the hypothesis on the changes of sustainable values, popular cultural notions of sustainability are mainly justified by generalization method and Yi-ological method, which partially reduces the accuracy of research results.

In order for sustainable development to become a standard for socioeconomic development activities, for codes of conduct between people and people, people and nature in contemporary Vietnamese society, the nation needs to confront the pros and cons of this process. Vietnam also needs to consider their cultural diversity, alternative and multidimensional effects to plan and complete the policies. It is clear that further studies are indispensable to propose solutions that reduce obstacles, promote supports, then internalize the value of sustainable development in Vietnamese culture instead of mere application and imposition of top-down policies so that "sustainable development" actually progress in Vietnamese culture and society permanently.

References

Bui, N. S. (2006). *Những góc nhìn lập pháp [Legislative Perspectives]*. Hanoi, Vietnam: National Political Publishing House; p. 219.

Dao, D. A. (2000). *Việt Nam văn hoá sử cương [An Outline History of Vietnamese Culture]*. Hanoi, Vietnam: Writers Association Publishing House, p. 25.

Decision No. 432/QD-TTg of Prime Minister. (2012). Nguyen Tan Dung on approving the Viet Nam Sustainable Development Strategy for 2011–2020 dated on April 12, 2012. Retrieved March 17, 2020, from http://www.chinhphu.vn/portal/page/portal/English/strategies/strategiesdetails?categoryId=30&articleId=10050825.

Do, L. T. (2005). *Văn hoá Việt Nam – nhìn từ mẫu người văn hoá [Vietnamese culture – looking at civilized people]* (pp. 454–455). Culture and Information Publishing House.

General Statistics Office of Vietnam (2019a). Socio-economic situation in the second quarter and the first six months of 2019. Retrieved March 17, 2020, from https://www.gso.gov.vn/default_en.aspx?tabid=622&ItemID=19251.

General Statistics Office of Vietnam (2019b). Retrieved March 16, 2020, from https://www.gso.gov.vn/default.aspx?tabid=714.

IUCN, UNEP, WWF. (1980). World conservation strategy. Retrieved October 16, 2019, from https://Portals.Iucn.Org/Library/Sites/Library/Files/Documents/Wcs-004.Pdf.

Le, Đ. T. (1998). *Về Hương ước lệ làng [Village Conventions and Rules]* (p. 11). National Political Publishing House.

Le, V. H. (1967). *Đại Việt sử ký toàn thư [Dai Viet history of encyclopedia] Volume II* (p. 80). Social Sciences Publishing House.

Nguyen, T. C., & Nguyen, V. H. (2002). *Giá trị truyền thống trước những thách thức của toàn cầu hóa [Traditional values against the challenges of globalization]* (p. 9). National Political Publishing House.

Nguyen, T. T. N. (2015). Hiện tượng giả dối trong đời sống người Việt từ góc nhìn giá trị học [Deception phenomenon in Vietnamese life from the study of value viewpoint] (based on survey results in 2014), in Tran Ngoc Them (editor) (2015), Một số vấn đề về hệ giá trị Việt Nam trong giai đoạn hiện tại [Issues on Systematic Values of Vietnam in the period of industrialization]; Publishing House, Ho Chi Minh City National University, Vietnam, p. 412–424.

Phan, H. L. (1998). *Tìm về cội nguồn [Back to the origin]* (pp. 545–546). World Publishing House.

Pisani, J., (2016). Sustainable development—Historical roots of the concept. *Environmental Sciences, 3*(2): 83–96. Retrieved March 17, 2020, from https://www.tandfonline.com/doi/full/10.1080/15693430600688831.

Radio Voice of Vietnam. (2003). Socialist-oriented market economy: Concept and development solution. Retrieved March 16, 2020, from http://vietnamembassy-usa.org/news/2003/11/socialist-oriented-market-economy-concept-and-development-soluti.

Seghezzo, L. (2009). The five dimensions of sustainability. *Environmental Politics, 18*(4), 539–556.

The World Commission on Environment and Development. (1987). *Our common future*. Retrieved March 17, 2020, from https://www.un.org/ga/search/view_doc.asp?symbol=A/42/427&Lang=E.

Tran, N.T. (2016). *Hệ giá trị Việt Nam từ truyền thống đến hiện đại và con đường tới tương lai [Vietnam Value System from tradition to modernity and the way to the future]*. Ho Chi Minh City. Vietnam: Culture and Arts Publishing House, p. 205.

Tran, Q. V. (2000). *Văn hoá Việt Nam Tìm tòi và suy ngẫm [Vietnamese Culture: Research and Thoughts]* (p. 746). National Culture Publishing House.

Chapter 4
Educational Renovation in a Restructuring Society: Vietnam's Case Study

Hung Ngoc Le and Phuong Thi Bui

Abstract In Vietnam's society where the economy is under restructuring into a market one, the renovation of education started in 1982 and the comprehensive educational renovation have created huge changes in the national educational system in 2001. The educational renovation has been led by the Vietnam Communist Party's (CPV) policy course, regulated by the country's legislation and has opened up education opportunities, reduced social inequality in education, and changing educational level structure. A number of studies have clarified factors of educational opportunities and their relationship to the human resource and economy, but these studies are relevant mainly to societies with developed and developing market economy. Therefore, based on the systematical theory approach and literature analysis method, this paper concentrates on clarifying the education renovation triangle comprising education policy, education opportunity, and educational level in a society restructuring into a market economy in Vietnam.

Keywords Education · Opportunity · The educational renovation · Education opportunities · Reduce social inequality in education · Restructuring society · Vietnam

1 Introduction

In 1986, Vietnam officially started its economic renovation from a centrally, commanded, administrative economy to a market-guided economy in a society of more than 61.1 million population, among them 19.3% living in urban areas and over 80.7% living in rural areas (GSO Vietnam, 2006, p. 117). In 1986, there were 96

H. N. Le (✉)
VNU University of Education, Vietnam National University, Hanoi, Vietnam

P. T. Bui
Hanoi University of Public Health, Hanoi, Vietnam

A. T. Nguyen, L. Hens (eds.), *Global Changes and Sustainable Development in Asian Emerging Market Economies Vol. 2*, https://doi.org/10.1007/978-3-030-81443-4_4

universities and colleges in the country with 19.2 thousand of teachers and 91.2 thousand of students. On average, there were 15 students among 10 thousand population. In general education including primary and secondary education, there were 13.7 thousand of schools, 426.2 thousands of teachers, and 12.5 millions of pupils making average of 2045 pupils among 10 thousand of population. The annual average GDP (at converted price) was USD 86. Agriculture shared the biggest portion 38.2% to GDP, while industry was of 28.9% and service sector was of 33%. The economic structure by the ownership consisted of mainly two sectors: the state sector, accounting for 46.6% GDP and the collective and individual sector accounting for 62.6%. At that time, there was not almost FDI in Vietnam's economy. Vietnam's economic renewal started with the change of thinking and renovation of economic management and social management. During the 1080s in rural areas, there were production management innovations with agricultural output contract system to farm households and peasants. It made the agricultural production increased and hence farmers' livelihood improved. In urban areas, there were also innovations and implementation of so-called plan 2 and plan 3 of collectives and individuals together with plan 1, the State management organs allocated down to production units. The economic and social thinking was renovated in the process where the people were actively seeking for measures and solutions to improve the living condition, increase productivity in all agriculture, industry, and service. The most obvious reflection open in the direction: the people are allowed to do what was not protected by the State and economic activities that assures the benefit to themselves, the collective, community, and the society are accepted and supported. New way of think and do in the socioeconomic life has been institutionalized in the 1986 CPV Congress documents. According to the CPV's renovation course in 1986, the mechanism of command, centrally and subsidized management was criticized and removed. At the same time, the law and mechanism of market economy was required to study and applied into the economic and social management in order to improve the people's life and progress to realize the target of rich people, strong nation, equal, democratic, and civilized society. The most obvious expression of the renovation of think and policy in the restructure of the economy from the structure of mainly two sectors of the state and collective to a multi-sector including domestic individual, private, and FDI sector. According to the General Statistical Office, in 1989, the FDI sector accounted for 2.1% of GDP (at current price), the state sector accounted for 33.4%, and the non-state sector accounted for 64.5% (GSO Vietnam, 2006, p. 117). The rate of the FDI sector increased very fast and reached almost 16% of the GDP in 2005. Among the non-state sector, the rate of the collective sector including agricultural cooperatives dropped rapidly to less than 1% and the rest made by the individual and private sector.

During twenty renovation years (1986–2005), the growth rate of GDP in Vietnam was of 6.8% annually, of which the state sector was of 6.4% and that of the non-state sector was 5.8% per year; the FDI sector grew at 22.7% annually in 1990–2005. In 2005, the population of Vietnam was more than 83 million people of them about 30% people live in urban areas, the rest of 70% live in rural areas. GDP per capita at converted price was of USD 638 (GSO Vietnam, 2006, p. 117). By 2018, GDP the per capita (at interbank average exchange rate) was estimated as much as USD 2590

(GSO Vietnam, 2019, p. 91, 185, 190). Table 4.1 presents many of the market-trend changes in Vietnam's socioeconomic context, poverty reduction, and improvement in the livelihood of the people. For example, GDP per capita increased from 86 USD in 1986 to 2590 USD in 2018, the general poverty rate by residence reduced from 58% to nearly 7% in the same period (GSO Vietnam, 2006, p. 117; GSO Vietnam, 2019, p. 91, 185, 190).

A question can be raised that in a society of economy restructuring to a market oriented with such changes as presented (see Table 4.1), how education renovation takes place? The answer relates to the relationship triangle of renovation in educational policy, opportunities of the citizen, and education level structure of the labor force in a restructuring society. The research of these three contents of the educational renovation may support available findings in the corelation of education and economy and, at the same time, add new knowledge to present researches that pay less attention on educational policy. The present researches emphasize the role of technical economic factors such as income, expenditure, but neglect social and cultural element including the right for learning and policy on educational reform. The basic conception of this paper can be summarized as follows: First, we can find expressions of educational renovation in educational reform policies and educational renovation is laid down in the CPV course and institutionalized in the State legislation on education. This is because in Vietnam, the CPV plays the leading role, the State manages, and the people own through their participation in the implementation of the provisions in the educational policies and law. Second, the educational renovation is most clearly reflected in the opening up education opportunities and reduction of inequality in education opportunities among social groups. Third, the educational renovation make changes in the educational level structure of the labor force and in turn, the highly educated labor force becomes the motivation of the economy to work by the market mechanism.

Table 4.1 Vietnam's selected socioeconomic indicators, year 1986–2018

Indicators	1986	1995	2005	2010	2015	2018
Population (mil. Persons)	61.1	72.0	83.1	86.9	91.7	94.7
Urban population (%)	19.3	20.7	26.9	30.5	33.9	35.7
Rural population (%)	80.7	79.3	73.1	69.5	66.1	64.3
GDP per capita (USD, in current exchange rate)	86	288	638	1273	2109	2590
The structure of GDP, in which	100	100	100	100	100	100
State (%)	39.7	40.2	38.4	29.3	28.7	27.7
Non-state (%)	60.3	53.5	45.7	43.0	43.2	42.1
Foreign direct investment (%)	–	6.3	15.9	15.2	18.6	20.3
The GDP growth rate compared with previous year (%)	2.8	9.5	8.4	6.4	6.7	7.1
General poverty rate by residence (%)	58.2*	37.4**	19.5	14.2	7.0	6.8***

Note: (*) data in 1993, (**) data in 1997, (***) multidimensional poverty
Source: GSO Vietnam (2006). *Vietnam – 20 Years of Renovation and Development.* Statistical Publishing House, Hanoi, Vietnam, 2006: 117; GSO Vietnam (2019). *Statistical Yearbook 2018,* Statistical Publishing House, Hanoi, Vietnam, 2019: 91, 185, 190

There are many experimental and theoretical researches on the socioeconomic elements of individuals, households, and communities affecting educational opportunities and the role of education for economic growth (Schultz, 1961, p. 1–17; Schultz, 1988; Becker, 1964; Psacharopoulos & Patrinos, 2004, p. 111–134; Sen, 1999; Acemoglu & Robinson, 2013; Zimmerman, 2001, pp. 87–98; Truong, Thai, & Bach, 1999, pp. 115–131; Vo et al., 2001, pp. 177–192; Handa, 2002, pp. 103–128; Iddrisu, 2014, pp. 145–156; Ravallion & Wodon, 2000, pp. C158-C175; Connelly & Zheng, 2003, pp. 379–388; Gharajedaghi, 2005, pp. 81–134, Le, 2015, pp. 235–248; Bui, Cao, Nguyen, Tran, & Haughton, 2001, pp. 91–108; Daren, 2017, pp. 177–192). Education is stemmed depending on macroeconomic elements including growth rates of GDP, national income, and the development of public infrastructure including road, electricity, and microeconomic elements such as income, expenditure levels of household, and the most directive the level of spending by individuals and households on education (Truong et al., 1999, pp. 115–131; Vo et al., 2001, pp. 177–192). However, though economic elements are very important but not the only reasons for educational renovation in a modern society including the restructuring one to market economy like in Vietnam (Truong et al., 1999, pp. 115–131; Gharajedaghi, 2005, pp. 81–134; Le, 2015, pp. 235–248). To clarify, the educational renovation triangle requires to study other noneconomic elements especially institution, policy, and legislation on education. According to the theory of Schultz and Becker, investments in education represent an investment in human capital a combination of educated capacities, trained skills capable to increase labor productivity, and thus increase income of working people and firms (Schultz, 1961, pp. 1–17; Schultz, 1988; Becker, 1964). Therefore, individuals, households, firms, and communities may consider, make choice, and prioritize their investments in education and training to improve the human capital. A study of Psacharopoulos and Patrinos found out that primary education created the largest income in low-income countries with more than 21%, then secondary education with about 16%, and higher education and post upper-secondary education, more than 11% (Psacharopoulos & Patrinos, 2004, pp. 111–134). This may explain why in a restructuring society like in Vietnam many households spend much for their children's education as best as they can so that their children can go for university education. According to Amartya Sen, development as the right for freedom, that is why, to develop the society need to respect, protect, and create conditions for their people to realize the right to work, manage, and access to social services including freedom to access to educational and training (Sen, 1999). To Acemoglu and Robinson, an extractive policy prevents people from participation in a social process is a reason of poverty and slow development. An inclusive policy that may mobilize all people to get involved in the development process is a reason of property and wealthy of a nation (Acemoglu & Robinson, 2013). Therefore, we can consider an inclusive policy that respects the right to learn and creates conditions to widen and realize the right of the people to learn as a direct reason for changes and development of education in reality. This issue opens the requirement of seeking for the root of education renovation in reform, renovation in educational policies in Vietnam.

An important index of educational opportunity is the rate of children going for education of different levels at the right age calculated at percentage of the population of the age X studying at X grade among the population of X grade age. A number of empirical studies found out factors influencing educational opportunities comprising family such as income level (Zimmerman, 2001, pp. 87–98; Truong et al., 1999, pp. 115–131; Vo et al., 2001, pp. 177–192), parental education (Truong et al., 1999, pp. 115–131; Vo et al., 2001, pp. 177–192; Handa, 2002, pp. 103–128; Iddrisu, 2014, pp. 145–1), individual elements of children such as age, sex, child labor (Truong et al., 1999, pp. 115–131; Iddrisu, 2014, pp. 145–156; Ravallion & Wodon, 2000, pp. C158-C175; Connelly & Zheng, 2003, pp. 379–388), and locality (Truong et al., 1999, pp. 115–131; Vo et al., 2001, pp. 177–192;; Connelly & Zheng, 2003, pp. 379–388). Based on regression analysis, for example logistic model (Truong et al., 1999, pp. 115–131; Vo et al., 2001, pp. 177–192), these studies pointed out the impact of the factors like age, sex, family income, and locality where children are living on their schooling opportunities. However, these studies might not do the assessment schooling opportunities in the relation to the whole national educational system where the lower learning grades provide the foundation and inputs for the upper education and learning grades that in turn are the outputs of the lower grades. There needs to apply a systematical theoretical approach to look at educational opportunities in the national educational system and between this system and educational institution, policies, and the labor market system (Gharajedaghi, 2005, pp. 81–134; Le, 2015, pp. 235–248). There possibly schooling opportunities for lower grades like lower-secondary, depend on the opportunity for going to upper-secondary schooling and in turn, the schooling opportunity for upper-secondary depends on the opportunity to go for university education. Because, according to the theory of investment in education by Shultz and Becker (Schultz, 1961, pp. 1–17; Schultz, 1988; Becker, 1964), making decision on schooling always considers expected benefits which are the opportunity to go for further higher education in the future. A lower-secondary student may drop out at any time if he or she finds that there will not be any opportunity to go for grade 10. This may often happen in rural areas where there is tough competition for admission in grade 10 in public schools. This situation may be similar to upper-secondary education: upper-secondary students may easily quit when they see no opportunity to be admitted in a university. While looking at educational opportunities, in the only national educational system from primary to tertiary, one can find out that the outputs of lower grades are inputs of higher grades. That is why, when conditions to enter higher grades are easy, then the schooling opportunities for higher grades will increase. Under the educational renovation in Vietnam, it will be possible that the renewing the method of university recruitment in the direction of increasing the quantity will provide more opportunities for secondary schooling. Some studies found out that education increases income through two levels of impact, at microlevel on health condition improvement, human capital and labor productivity and at macro level, on the level of participation in social labor force, improvement of technical, professional levels of labor force (Bui et al., 2001, pp. 91–108; Daren, 2017, pp. 177–192). In turn, the expectation of multidimensional benefits from education represents an element to increase investments in education and schooling opportunities. In the relation of education to the society, the opening up of university opportunities makes direct change in the structure of professional and technical levels to meet the requirement of the labor market. A society restructuring to a market induces stratification of labor market into types of labor market corresponding to technical levels. Global integration and industrialization and modernization of the country require the structure of the labor market to change accordingly in the direction of increasing the proportion of university, post-university, and college-educated laborers. In short, the empirical and theoretical researches from different corners have clarified every relation between the economy and education, household and education, social community and education, and between education and human capital development and between education and economic growth. However, few researches use systematically theoretical approach to look at the whole triangle of the relations among the renovation of educational policies and renovation of educational opportunities and educational level structure in a market-oriented restructuring society. Therefore, this paper lays down the study objective to clarify this triangle of educational renovation by basing on the analysis of policy data and statistics of population, household, and livelihood consensus in Vietnam.

2 Methodology

This researchuses a qualitative method, namely document analysis method (Browen, 2009, pp. 27–40), where documents have been collected, processed, and analyzed as social facts. The document analysis method have had different advantages such as efficiency, availability, cost-effectiveness, lack of obtrusiveness and reactivity, stability, exactness, and coverage (Browen, 2009, pp. 27–40; Yin, 1994). However, to use this method, one needs to be careful because this method may also have some disadvantages like insufficient details, low retrievability, and biased selectivity (Browen, 2009, pp. 27–40; Yin, 1994). To do a case study of Vietnam on educational renovation in a market-oriented restructuring society, the advantages of document analysis method is prioritized because the study issue requires to collect and process data from two sources. First, a qualitative data source consists of contents of educational institutional renovation of the government and CPV. These documents are (i) Resolutions of CPV's congresses and conferences on the socioeconomic renovation including those on educational renovation (every five years from 1982 to 2016) (Political report presented at the CPV Congress, 1982, 1986, 1991, 1996, 2001, 2006, 2011), (ii) Laws on education including the Law on Children Protection, Care and Education, 1991, The Law on Compulsory Primary Education, 1991, The LOEs in Law on Education, 1998, Law on Education, 2005, The Law on High Education, 2012, and The Law on Vocational Education and Training, 2014. These documents provide the qualitative facts on the right, responsibility, and duty to learning of Vietnamese citizens and the responsibilities of related parties such as the State, Government in the educational renovation in order to realize the objective of education in improving the people knowledge, educating human resources and talents for the country. According to institutionalism, renovating educational policy represents the reason of changes in education consisting of changes in education opportunities and changes in educational levels of the population. That is why a list of Keywords reflecting the renewal of the educational policy is used to analyze the content of these documents. These Keywords are "reform", "renovation", "compulsory education", "diversification", "public school", "semi-public," and "private" school.

Second, the quantitative source presented in the publications of the results of (i) The Population and housing consensus in Vietnam (every 10 years from 1979 to 2019); (ii) The Survey on the people livelihood in Vietnam (every 2 years from 1992 to 2016), and (iii) Surveys and researches relating to educational renovation. These documents provide the quantitative facts on rates of children going to school at the right age for deferent grades from primary to tertiary by sex, nation groups, locality, and the structure of educational levels of the population of age 15 and above. The quantitative facts are processed and analyzed, compared in percentage with the average rates to clarify differences if any, for example by gender, minority groups, locality relating to educational opportunities and levels (Haughton, 2001, pp. 23–24).

3 Results

3.1 Educational Policy Renovation, Opportunity of Going to School and Educational Level

3.1.1 Educational Policy Reform and Renovation

The following is a summary of basic contents of the education renovation process in Vietnam, starting from the educational reform in 1976 to the fundamental and comprehensive education renovation from 2011 until now.

- There are eight political reports presented at the CPV's conferences taking place from 1982 to 2016 (Political report presented at the CPV Congress, 1982, 1986, 1991, 1996, 2001, 2006, 2011, 2016). Among these eight documents, one report was made in 1982 before the start of the renovation (Political report presented at the CPV Congress, 1982) and the other seven were made during the socioeconomic renovation from 1986 to 2016 (Political report presented at the CPV Congress, 1986, 1991, 1996, 2001, 2006, 2011, 2016).
- Education was reformed during the 1976–1981 five-year Plan (Political report presented at the CPV Congress, 1982). The objective of the educational reform was educating new qualified working people to meet the requirements of the economic and cultural development of the country. Reforming education to realize the principle of learning linked to practice, education linked to production, school linked to society. The educational reform consists of reforming the pedagogical system, completing the system of the national education including public education, the system of tertiary and technical and vocational training schools. The reform was continued in 1981–1986. In 1986–2000, the reform was wrapped up, drawn experience, and adjusted and improved the quality of the reform. Educational reforms start with reforming teaching methods, reforming training

teachers, and education management staffs and reaches to the reforming of policies for teachers (Political report presented at the CPV Congress, 1986, 1991, 1996). In 2001–2005, after drawing experience, Vietnam concentrated in reforming the content and method of educating, reforming the system of schools, classes, and the system of education management in the direction of "standardization, modernization and socialization", "education for all", "the whole country becomes a learning society", giving more attention to profession orientation and categorizing secondary pupils (Political report presented at the CPV Congress, 2001). In 2006–2010, education was renovated in a comprehensive manner including organizational structure, management mechanism, educational contents, and methods. At the same time, the renovation was centered in renovating basic grades of the national educational system – preschool education and tertiary education in order to directly realize the objective of educating highly qualified human resources for the country (Political report presented at the CPV Congress, 2006). From 2011 until now, education has been fundamentally and comprehensively renovated including the renovation of the objectives of education from giving more importance to teaching knowledge to more attention given to qualification and ability development of learners; renovating from education contents, forms, methods to examination methods; renovating from organizational, management structure to educating teachers and radically renovating essential elements of the national educational system in the direction of open educational system, life learning, and building up a learning society (Political report presented at the CPV Congress, 2011, 2016).

- The learning right and duty of Vietnamese citizens is recognized and implemented through the compulsory policy of the State and officially has been started in 1981. In the 1981–2000 period, the compulsory policy was concentrated in compulsory primary education for children of 6–14 age in nationally and linked to the target of illiteracy fight and then moved to lower-secondary compulsory in regions of better conditions, especially where it strongly developed the commodity economy. In 2000, Vietnam completed compulsory primary education and in 2011, it completed the compulsory lower-secondary education in all cities and provinces.
- In 1986, before the start of the renovation and during the first years of it, and by the year 1990, there was only one type of public schools in Vietnam and they were owned, invested, and managed by the state. In 1991–2000, schools started to be diversified and as a result, there were other types of schools – semi-state and people-set and private ones. Since 2001, education has been organized with three main types of schools: public, people-run and private ones, and foreign-invested schools.

The educational renovation linked to the learning right of Vietnamese citizens and duty of related parties to ensure this right be institutionalized in the Law on Education (LOE) of Vietnam. The task of making the law was stated in the Resolution of the CPV Congress VI in 1986. In 1991, Vietnam issued two laws: the Law on Children Protection, Care and Education in 1991 and The Law on

Compulsory Primary Education in 1991. According to the provision of these two laws, children under 16 age are Vietnamese citizens; The State guarantees their learning right, encourages them to study well, and creates conditions for their talent development; The State implements the policy on compulsory primary education from grade one to grade 5 for all Vietnamese children of 6 to 14 ages. A child of 6 age shall go to grade 1 from the right beginning of the school year and finish primary school before 15 age. The primary education shall take place in state-owned primary schools, classes, and other types of people-created. The pupils of state-owned school and class do not have to pay fees. So, since 1991, according to these laws, social equality and justice in primary educational opportunities of children of 6 to 15 ages have been recognized by the law and realized by the compulsory primary education policy. However, the provision of two types of state and non-state primary education whereby those children study in state schools do not pay fees may induce the danger of inequality between state and non-state primary education. Both of these laws have been replaced by the LOE promulgated in 1998 and became valid in June 1 of 1999. The LOE in 1998 was replaced by the LOE in 2005. It is noteworthy that many contents of educational renovation on the learning right and renovation of the national educational system stated in the CPV's course have been institutionalized in the provision of LOEs in 1998 and 2005, The Law on Higher Education (LOHE) in 2012, and the Law on Vocational Training in 2014, as specifically as follows.

- Education and training provide the number-one national policy: this conception is written in the foreword of the LOE in 1998 and institutionalized in provisions of the LOE in 2005, Article 9: Education development is the number-one national policy to improve the people's knowledge, educate human resource, and train talent (Law on education, 2005).
- Learning right and equality in learning opportunity of the citizen: Both LOE 1998 and LOE 2005 provide: "Learning is the right and duty of the citizen. All citizen regardless minority group, religion, sex, family root, social status, economic condition are equal in learning opportunity" (Law on Education, 1998, 2005).
- Education compulsory. The LOE 1998 provides that the State decides the plan and the level of public education but does not clearly provide what kind of educational levels. The LOE 2005 clearly provides that the primary and lower-secondary education is compulsory (Law on Education, 2005). However, the State has had the plan of implementing the primary education compulsory and then lower-secondary education where the condition allows.
- A modern national educational system (NES) . Issuing the LOE 1998 allows Vietnam first time to complete the construction of a modern national educational system with fully educational grades and training levels. This has further been institutionalized in the LOE 2005, to which the NES consists of formal and informal education; The NES comprises the following grades and levels: the preschool education includes kindergarten and the preschool, general education comprises the primary and secondary (lower- and upper-secondary), professional

training comprises technical and vocational, and tertiary education is of college, university, master, and doctorate (Law on Education, 1998, 2005).

- Diversifying forms of educational organizations: quite different from before the Renewal when the education was organized in the only form of public schools, the LOE 1998 provides four forms including the public, semi-public, people-set, and private ones. The LOE 2005 provides three forms of education: public schools set up and invested by the State, the people-settings are created and invested by the local people's communities, and the private ones are set up and invested by social, socio-professional associations, economic organizations, or by individuals. Regarding tertiary education before the renewal in 1986, there were only public universities in Vietnam, but during the renewal process, there were some other forms of nonpublic universities. The LOHE 2012 provides three forms of universities (Law on High Education, 2012). They are (i) public universities are owned, invested in the construction of physical utilities; (ii) Private universities are owned, invested in the construction of physical utilities by social, socio-professional associations, private economic organizations by individuals, and (iii) Foreign investment universities are with 100% of foreign investments and the universities with joint stocks by foreign and domestic investors. As one of the grades of the NES, professional education is of four forms similar to those of the university education. The Law on Vocational Education and Training (LVET) provides that national education systems consist of (i) the public, (ii) private, (iii) foreign invested professional units which are with 100% of foreign investments, and (iv) the universities with joint stocks by foreign and domestic investors (Law on Vocational Education and Training, 2014).
- Educational streaming. The LOE 1998 does not mention about "streamline of students", but LOE in 2005 provides educational program creating condition for its streaming. The LVET in 2014 clearly provides that the State has had a policy on streaming students finishing secondary schools to continue vocation education and training relevant to every period of the socioeconomic development (Law on Vocational Education and Training, 2014).

3.1.2 Educational Opportunity: Gender Equality and Difference among Ethnic Groups

The results of the implementation the CPV's policy course, state's legislation on the educational renovation in order to open up more education opportunities and ensure the equality in learning opportunities for all Vietnamese citizen are reflected clearly in the statistic on the rate of schooling at the right age of grades of all general, college, and university. According to the institutional approach, the renovation in the CPV's educational policy in 1986, 1991, 1996 together with the Law on Compulsory Education (1991), The Law on Children Protection, Care and Education (1991), and The Law on Education (1998) provides convenient educational institution for opening more education opportunities. For example, the net primary school enrolment rate increases from 78% in 1993 to 95.5% in 2009. The net lower- secondary school enrolment rate increases from 36% to about 83% and the net upper-secondary school enrolment rate increases from 11% to about 57%. The renovation in educational policy including the Law on University Education promotes the opening up of education opportunities for all social groups from 2009 to 2016.

Table 4.2 indicates the rate of schooling at the right age of the Viet people and other five minority groups – Tay, Thai, Muong, Khmer, and Hmong (these five groups have had the biggest number of population among the 53 minority groups in Vietnam).

Table 4.2 presents the percentage as compared with national average as much as of 100, for easier and more precise comparison (Haughton, 2001, pp. 23–24). In 2009, the national rate of the right-age schooling was 95.5%; the rates of male and female were similar and equal to those of the nation (column 2, Table 4.2). This point proves that in 2009, Vietnam obtained the objectives of education compulsory and gender equality in the primary education. Three ethnic groups, Viet, Tay, and Thai, obtained the objective of compulsory primary education while the Khmer obtained only 90% of the nation rate. The Hmong got the lowest rate, 76% of the national average. Columns 3, 4, 5, and 6 of Table 4.2 show that in 2009, Vietnam got the gender equality in all primary, secondary, and tertiary education; however, as far as it goes to higher levels, the rates of right-age schooling of male and female and ethnic groups decrease strongly. Viet people has had the higher rate than the national average while the rates of other groups are lower than the national average and decreased for upper-secondary, tertiary levels. Especially, this rate for Hmong people was of only 41.3% of the national average for the general level and 11.6% for

Table 4.2 Net school enrolment rates by nation and percentage compared with national average by gender and selected ethnic groups, year 2009–2016

Criteria	2009					2016		
	Primary	Lower secondary	Upper secondary	College	University	Primary	Lower secondary	Upper secondary
Nation	95.5	82.6	56.7	6.7	9.6	97	90.4	68.6
National average	100	100	100	100	100	100	100	100
Gender								
Male	100	98.6	93.7	89.6	94.8	100.1	98.8	92.9
Female	99.9	101.6	106.9	110.5	105.2	100	101.3	107.3
Ethnic groups								
Viet	101.1	105.0	109.0	114.9	115.6	100.5	103.1	109.6
Tay	102.1	106.1	97.9	44.8	33.3	101.1	102.4	101.2
Thai	97.1	88.7	52.7	23.9	11.5	100.8	98.9	66.8
Muong	100.2	100.8	73.0	19.4	17.7	98.1	104.5	95.5
Khmer	90.5	56.1	27.2	13.4	11.5	97.7	78.3	46.4
Hmong	76.0	41.3	11.6	3.0	2.1	93.9	70.0	36.4

Source: The Ministry of Planning and Investment – GSO Vietnam (2011). *Vietnam Census on Population and Housing 2009 - Education in Vietnam: the analysis of basic indicators,* Statistical Publishing House, Hanoi, Vietnam 2011: 36; GSO Vietnam (2018a, 2018b). *Vietnam living standard survey 2016,* Statistical Publishing House, Hanoi, Vietnam, 2018: 118

college and 2–3% for university. This means that the gag between Viet people and Hmong people widens by grade, from preschool to university. The rate of Viet people going to primary school was 2.5 times more than that of Hmong people, and 9 times to secondary, 38 to colleges, and 55 times to universities.

In 2016, the national rates of schooling at the right age of both male and female, all ethnic groups increased for all general grades including primary and secondary. Especially, in 2016, the rate of Hmong people going to general schools increased three times as compared with the rate in 2009 and equal to one of Viet people.

3.1.3 Educational Opportunity: Difference among the Urban, Rural, and Economic Geographical Regions

To clarify the renovation in educational opportunities and the difference in the locality, Table 4.3 presents the rate of schooling at the right age of educational grade in 1998 and 2019 of the whole country. The percentages are also divided by rural and urban areas and ecogeographical regions where by the national average is considered as much as 100.

- Column 2 in Table 4.3 shows that in 1998, there was insignificant differences in the rate of schooling at the right age between rural and urban areas. But, the highland (Tay Nguyen) has had the lowest rate of primary schooling at the right age as compared with the rest of five regions and equal to 90% of the national average.
- Column 3 indicates that the lower-secondary schooling rate in rural area was of 93% of the national average and the gap increased as compared with the urban area. The regions of the Highland and Mekong delta have had the lowest rates of lower-secondary schooling at the right age and were equal to 71–73% of the national average and a half of that of the Red River Delta.
- Column 4 shows the differences in upper secondary net enrollment rate between the urban and rural areas which was more than double (191% in the urban and 76% in the rural areas) and the regional inequality increased: The Highland has had the lowest rate (36.8%), equal to slightly more than one-third of the national average and a half of that of the Northern hill and mountain regions (76%). The inequality in university educational opportunity is very obvious. The rate of the urban area is as high as 4 times that in rural area and the rate of rural people going for the university education at the right age was as much as a half of the national average and equal to 5.4%. The Highland has had the lowest rate of people going to universities at the right age, 17% of the national average and one-tenth of the rate of the Red River Delta.
- In 2019, it was witnessed an increase in the education opportunities of all levels from the kindergarten to the university. Columns 6 and 7 indicate that Vietnam completed the compulsory primary education (98%) and lower-secondary education (over 89%). At the same time, the gap between rural and urban areas and among regions decreased and gradually moved closer to the equity.

Table 4.3 Net school enrolment rates by nation and percentage compared with national average by urban, rural, and economic geographic regions, year 1998–2016

	1998				2019			
National, urban, rural and Regions	Primary	Lower secondary	Upper secondary	Tertiary	Primary	Lower secondary	Upper secondary	Tertiary *
Nation	92.6	61.6	28.8	9.3	98.0	89.2	68.3	23.4
National average	100	100	100	100	100	100	100	100
Urban	103.6	133.0	190.6	229.0	100.3	102.7	111.9	186.3
Rural	99.4	92.9	76.0	58.1	99.9	98.8	94.3	53.4
Economic geographic regions								
Northern	103.1	91.6	77.1	57.0	100.1	101.1	95.3	32.9
Red River delta	103.5	135.6	158.0	174.2	100.8	106.4	122.5	159.8
North central	99.68	101.8	108.0	96.8	100.4	103.6	108.1	83.3
Highland	88.9	70.9	36.8	17.2	98.8	92.8	82.6	28.2
South eastern	101.8	116.4	126.7	149.5	99.7	98.1	94.0	150.4
Mekong delta	94.7	72.9	60.1	59.14	99.1	92.4	81.07	75.6

Note: (*) data 2017
Source: GSO Vietnam (2000). *Vietnam Living Standard Survey 1997–1998*, Statistical Publishing House, Hanoi, Vietnam, 2000: 50, 52; GSO Vietnam (2018a, 2018b). *Survey on population change and family planning on April 1, 2017: Basic results*. Statistical Publishing House, Hanoi, Vietnam, 2018: 35; Steering Committee of Population and Housing Census (2019). *Census on Population and Housing on April 1, 2019: Organization and preliminary results*. Statistical Publishing House, Hanoi, Vietnam, 7. 2019: 55

- Column 8 shows that the rate of upper-secondary schooling at the right age was 68% in 2019 and doubled the rate in 1998. The educational opportunities have been rapidly opened in difficult regions such as doubled in the Highland, and about 30% in the Mekong Delta. But, the opportunities for the upper-secondary education were slowly increased in the Red River Delta and South-Eastern regions. Thus, the difference among regions clearly decreased in 2019 as compared with 1998.
- Column 9 presents facts in 2017 and indicates that the rate of people going for university education at the right age was 23.4%, 2.5 times to that in 1998, a slight increase as compared with the rate of the upper-secondary. However, the gap between the rural and urban areas though narrowed down but is was still large (3.5 times). Similarly, the gap among regions decreased but it was still large: the rate of people going to universities of the Red River Delta was as much as 5.5 times that of the Tay Nguyen (Highland).

3.1.4 Restructuring Education Levels: Gender and Urban – Rural Difference

The opening up of educational opportunities in all educational levels and for all social groups provides the element making changes in the educational level structure of the population in the sense of the output of the national educational system. However, changes in the educational level structure of the population also depend on other elements that need to be interpreted by other theoretical approaches such as the theory of labor market, the theory of labor mobility, and therefore it needs another research. The results of the study should be presented as follows to describe the changes in the levels of education of the population of 15 years and above in a restructuring society to a market economy.

To evaluate the results of education renovation and the gender and local disparities, Table 4.4 presents data on the structure of educational levels of the population over 15 years old. The percentages are compared with the national average considered as 100 by male and female, rural and urban areas in 2006 and 2016.

- Column 2 of Table 4.5 shows that in 2006, there was still 22% of the population of over 15 age "never go to school" or "no certificate" and therefore, resulting in very low human capital; about 53% of them have had compulsory education level and the population having college, university level, and over was 4.5%. So, one person has had university level one or two has had professional level and 19–20 people have no vocational and technical training among them five people have no certificate.
- Column 7 indicates that in 2016, the number of people "never go to school" or having "no certificate" and having "primary education", the rate of "secondary education" increases. The rate of people having "professional level" slightly decreased but being diversified. The rate of population having "university and over level" almost doubled as compared with 2006 and reached 10% in 2016. Thus, on average, one person has had the university and over level one or two has had professional level and nine people have not vocational and technical training. The gender equality gap clearly decreased among the group of university and over level due to the increase of female and the decrease of male.
- In 2016, Vietnam almost obtained the target of gender equality for the university level; however for the secondary level, the gender gap did not decrease but increased because the rate for males increased by 1.5 times while the rate of females increased a little bit.
- A comparison of the urban and rural areas shows that the rural area has had bigger number of the population "never go to school" and "no certificate", doubling the urban area and this disparity did not decrease after 10 years (2006–2016). This is caused by the increased rate of females and slowly decreased rate of male.
- For other levels of education, the difference between the urban and rural area decreased and the fastest for the university level: it was from five times in 2006 down three times in 2016.

Table 4.4 Education levels by the nation's population 15 years old and above and percentage compared with national average by gender and areas, 2006–2016

Education levels	2006					2016				
		% compared with national average 100					% compared with national average 100			
		Gender		Area			Gender		Area	
	Nation	Male	Female	Urban	Rural	Nation	Male	Female	Urban	Rural
Never go to school	8.1	58.0	138.3	59.3	114.8	5.4	64.8	131.5	46.3	125.9
No certificate	14.5	84.8	113.8	69.0	111.7	12.6	81.8	116.7	66.7	115.9
Primary	24.0	101.7	98.7	80.8	107.5	21.3	98.6	100.9	75.6	111.3
Lower secondary	28.7	105.6	94.8	80.5	107.7	28.2	103.9	96.1	79.1	109.9
Upper secondary	12.6	109.5	91.3	142.9	83.3	15.5	157.1	94.2	131.0	85.2
Primary vocational	3.3	142.4	60.6	172.7	69.7	1.9	157.9	52.6	147.4	84.2
Secondary apprentice	–	–		–	–	1.8	150	55.6	155.6	77.8
College vocational	–	–	–	–	–	0.5	160	40	120	80
Professional secondary	4.3	107.0	93.0	176.7	69.8	3.0	96.7	103.3	146.7	76.7
College, university	4.4	113.6	86.4	247.7	43.2	9.4	101.1	100	196.8	54.3
Post-graduate	0.10	200	100	400	–	0.46	123.9	78.3	276.1	15.2
Other	–	–	–	–	–	0.02	150	100	200	50

Source: GSO Vietnam (2018a, 2018b). *Vietnam living standard survey 2016,* Statistical Publishing House, Hanoi, Vietnam, 2018: 118

3.1.5 Restructuring Education Levels: Difference between the Richest and the Poorest Quintile

Table 4.5 presents data on the educational levels of population over 15 age and percent to the national average stemmed 100 of the richest quintile and poorest quintile in terms of annual per capita income in 2006 and 2016.

- Column 4 of Table 4.5 shows that out of people of the richest quintile, there are still people "never go to school". The rate of this group is about 31% as compared with the national average and the rate of people having college and university levels was as much as three times to the national average.
- Column 5 indicates that the rate of "never go to school" in the poor group was 222% to the national average. The rate of people having college and university levels was 4.5% to the national average.
- In 2006, the disparity between the poor and richest quintiles was most clear when the poorest quintile has had the rate of "never go to school" as much as seven times higher than that of the richest quintile. The richest quintile has had the rate

Table 4.5 Education levels by the nation's population 15 years old and above and percentage compared with national average by the richest and poorest quintiles, 2006–2016

	2006				2016			
			% compared with national average 100				% compared with national average 100	
Education levels	Nation (%)	National average as 100	Richest quintile	Poorest quintile	Nation (in %)	National average as 100	Richest quintile	Poorest quintile
Never go to school	*8.1*	100	30.9	222.2	*5.4*	100	20.4	313.0
No certificate	*14.5*	100	55.9	137.9	*12.6*	100	46.0	164.3
Primary	*24.0*	100	70.8	112.9	*21.3*	100	65.3	123.0
Lower secondary	*28.7*	100	81.2	93.0	*28.2*	100	83.7	93.3
Upper secondary	*12.6*	100	150.8	51.6	*15.5*	100	129.7	50.3
Primary vocational	*3.3*	100	190.9	21.2	*1.9*	100	136.8	21.1
Secondary apprentice	–	–	–	–	*1.8*	100	161.1	16.7
College vocational	–	–	–	–	*0.5*	100	140.0	20.0
Professional secondary	*4.3*	100	214.0	16.3	*3.0*	100	170.0	23.3
College, university	*4.4*	100	320.5	4.5	*9.4*	100	241.5	6.4
Post-graduate	*0.10*	100	500	–	*0.46*	100	343.5	–
Other	–	–	–	–	*0.02*	100	300.0	–

Source: GSO Vietnam (2018a, 2018b). *Vietnam living standard survey 2016,* Statistical Publishing House, Hanoi, Vietnam, 2018: 118

of "upper secondary" as much as three times as compared with the poorest quintile. The richest–poorest quintile inequality was of the highest rate for the "college and university level", 71 times.

- After ten years, in 2016, the disparity between the richest and poorest quintiles witnessed a change in different directions and levels. The rate of people "never go to school" decreased but the rate of the poorest people "never go to school" increased, resulting in strongly increasing the inequality between these two groups. The rate of the poorest people "never go to school" was as much as 15 times to the richest quintile and this gap doubled after ten years.
- The rate of the richest people having "upper secondary education" was 2.5 times higher than that of the poorest quintile and this gap decreased as compared with 2006.

- The inequality between the richest and poorest quintiles in terms of college and university level decreased almost a half from 71 times in 2006 down to 38 times in 2016.

3.1.6 Restructuring Education Levels: Differences among Economic and Geographical Regions

Table 4.6 presents the structure of educational levels of the population over 15 age and the percentage of economic geographical regions as compared with the national average considered as 100.

- Row 2 of this table shows that the Highland has had the highest rate of people "never go to school", 168% to the national average and the Red River Delta has had the lowest rate, about a half of the national average. The region of Mekong delta has had the highest rate of "no certificate", 188% to the national average and it is three times higher than the lowest region, i.e. the Red River delta.

Table 4.6 Education levels by the nation's population 15 years old and above and percentage compared with national average by economic geographic regions, year 2008

Education levels	Nation	National average as 100	Northern	Red River delta	North Central	Highland	South Eastern	Mekong delta
Never go to school	7.5	100	161.3	48.0	93.3	168.0	70.7	133.0
No certificate	13.5	100	80.7	49.6	89.6	92.6	94.8	188.0
Primary	23.1	100	91.8	60.2	97.0	113.4	112.1	142.0
Lower secondary	27.9	100	105.0	135.1	110.4	93.6	79.9	60.9
Upper secondary	14.1	100	87.2	122.7	112.1	83.0	119.9	57.4
Primary vocational	3.2	100	103.1	159.4	75.0	84.4	100.0	53.1
Secondary apprentice	1.8	100	127.8	183.3	77.8	77.8	83.3	33.3
College vocational	0.3	100	100.0	133.3	66.7	66.7	133.3	66.7
Professional secondary	3.3	–	130.3	130.3	103.0	97.0	90.9	51.5
College, university	5.1	–	72.6	145.1	84.3	62.8	164.7	47.1
Post-graduate	0.1	–	60.0	280.0	70.0	60.0	190.0	30.0
Other								

Source: GSO Vietnam (2018a, 2018b). *Vietnam living standard survey 2016,* Statistical Publishing House, Hanoi, Vietnam, 2018: 118

- However, the Mekong delta has 57% of the people having "upper secondary", equal to a half of the national average and a half of that of the Red River Delta which has had 123% of people having "upper secondary" to the national average.
- The South-Eastern region and the Red River delta have had the rates of population of "college and university" as high as 145–165% and 1.5 times higher than the national average and three times higher than that of the Mekong delta which is of the lowest rate, 47% to the national average.

Table 4.7 indicates that in 2016, the rate of the population "never go to school" in the Highland was of over 183% as compared with the national average making 6 times higher than the rate of the Red River delta with 31% to the national average. This disparity did not decrease but increased 3.5 times in 2008.

The Mekong delta region has had the rate of people having "upper secondary" as much as 56% to the national average and making only a half of the rate in the Red River delta. The gap between the rich region – the South Eastern and the Red River delta and Mekong delta decreased by 2.6 times as compared with the three-time gap in 2016.

Table 4.7 Education levels by the nation's population 15 years old and above and percentage compared with national average by economic geographic regions, year 2016

Education levels	National	National average	Northern	Red River delta	North Central	Highland	South Eastern	Mekong delta
Never go to school	5.4	100	211.1	31.5	88.9	183.3	57.4	124.1
No certificate	12.6	100	91.3	44.4	92.9	89.7	95.21	194.4
Primary	21.3	100	93	56.8	97.7	116	109.4	148.8
Lower secondary	28.2	100	105	125.9	109.6	98.6	81.2	67.7
Upper secondary	15.5	100	85.2	127.7	107.1	87.1	116.8	56.1
Primary vocational	1.9	100	79	142.1	84.2	73.7	136.8	68.4
Secondary apprentice	1.8	100	105.6	188.9	83.3	50	88.9	38.9
College vocational	0.5	100	80	180	80	60	80	40
Professional secondary	3	100	116.7	123.3	110	86.7	80	63.3
College university	9.4	100	74.5	142.6	87.2	79.8	140.4	54.3
Post-graduate	0.46	100	43.5	243.5	54.5	41.3	102.2	30.4
Other	0.02	100	50	300	50	50	100	50

Source: GSO Vietnam (2018a, 2018b). *Vietnam living standard survey 2016,* Statistical Publishing House, Hanoi, Vietnam, 2018: 118

4 Conclusions and Discussion

This case study of Vietnam is based on the systematical theoretical approach and empirical and theoretical researches on the changes in education in a developed and developing society. It helps to clarify the relation triangle of educational policy, educational opportunity, and educational level in a society restructuring to a market-oriented economy. During the restructuring process from a centrally, subsidized management mechanism to a market-oriented mechanism, there were positive changes in the society with improved living conditions and rapidly reduced poverty. The educational renovation was started with the educational mechanism, policies expressed in the documents of the CPV's on the educational reform before the year 2000 and fundamental and comprehensive renovation of the education since 2001. A number of conceptions about educational renovation such as the learning right and education compulsory together with renewals in the national educational system have been institutionalized in the legislation on the education. This provides the legal foundation for the state organizations to implement policies on compulsory primary, lower-, and upper-secondary education in localities and the conditions allow to development of the economy. This is also the legal foundation allowing households to make more investment in education. The results of the research indicates that educational opportunities measured by the rate of schooling at the right age have been strongly opened for all people regardless their sex, ethnic group, locality, and living standard. Vietnam completed compulsory primary and lower-secondary education and thus achieved gender equality in these two grades. Especially, Vietnam has obtained gender equality in upper-secondary education with the rate of schooling at the right age of girls higher than that of boys and also higher than the national average.

The speed of education opportunity increase may be uneven especially for upper-secondary and university education. The disparity among ethnic groups, urban and rural areas, and economic regions though decreased along with the restructuring process to a market mechanism is large, especially for tertiary education. The university opportunities of the ethnic minority people and people living in the Highland and the Northern hill and mountain regions increased but are still limited as compared with the national average and economically developed regions such as the Red River delta and the South-Eastern region. Within the national education system, the output of general education provides the opportunities for going to colleges and universities. The implementation of the Law on University Education in the direction of respecting the learning right and the need for the development of high-quality human resources might be the factor for opening more opportunities for university education of the people.

The results of research show that the changes in education might induce rapid increase in the proportion of high educational levels and decrease of lower levels including the rate of the people "never go to school" and "no certificate" in the structure of educational levels of the population. The rates of college and university level have doubled in 2006–2016, but the rate of the professional education increased very slowly, though it is diversified from primary vocational to professional secondary levels.

However, in restructuring society educational renovation in all three educational sides - policy, opportunity, and level - always faces with it new issues. One issue going through the whole process of the educational reform and fundamental, comprehensive educational renovation now is how to balance the quality and quantity of education, i.e., what to ensure the opening more education opportunities and, at the same time, improve its quality. For this issue, the policy of educational reform seemingly chose the solution of compulsory education and strengthening occupation-oriented education linked to the renovation of examination, test, performance evaluation methods including the method of upper-secondary final examination. From the systematical theory, the relation between the quantity and quality of education is a kind of dynamic balance between the demand for learning and the freedom to learning of the people, on the one side, and the requirement of development of a market economy. This may require the educational renovation policy to shift the center to renovating educational management staffs and teacher training and renovating research and development work of educational science.

The issue of dynamic balance between the quantity and quality of education directly affects the link to the opening up of educational opportunities, namely how to equally distribute opportunities that are opened up in the society. There may be two most distinguish concepts, first, there need to be more input opportunities of education and narrowing down output in the sense that enrolment by the demand and graduation by the results of study. This concept clearly reflects a center of the educational policy renovation that is renovation of examination method, evaluation of learning results, and renovating forms of graduation examination. However, in the national educational system, the output of this grade is the input of another grade; therefore, giving more importance to the examination, evaluation quality has caused a social problem as it has been in a tradition of an Asian society – "learn to take exam" and the pressure of examination always induces other problems of "additional teaching and learning" and the "success disease". The second concept emphasizes the need to implement at the same time the solutions of renovation of university education and renovation of professional education linked to the renovation of occupation-oriented education, categorizing professional education after lower-secondary school. However, the categorization of professional education may not achieve the expected results when the majority of lower-secondary graduates would continue to go for upper-secondary education. The importance is the post-lower-secondary categorization seems not to be the tendency of the development of education under the impact of globalization and the fourth industrial revolution demands for post-upper-secondary educational level.

Together with the issue of opening educational opportunities is the issue of social inequality in schooling opportunity which is often hidden by another issue causing dispute, issue of "redundant teachers and shortage of workers". Educational renovation in the direction of opening up schooling opportunities for all people has been seen as the reason of creating too many university graduates: redundant teachers means there are too many people having university education level and over and shortage of workers means the shortage of people having professional training. The structure of professional level changes in the direction of increasing the rate of population having university and over education but not at the price of decreasing the rate of people having professional education but reducing the rate of people "never go to school", "no certificate", and of low qualification. It is noteworthy that even when the rate of people having university and over, i.e., teachers doubled in the past decade it is still very low, very "insufficient" to the requirements of the industrialization, modernization of the country, and world integration. Vietnam had achieved the social equality in primary schooling, thanks to the compulsory education and gender equality has been improved in primary and secondary and tertiary levels. However, in upper-secondary level and especially tertiary level, the social inequality in educational opportunities have been decreasing along with renovation educational process but it is still high among ethnic groups, regions, and especially between the rich and poor groups. This again requires the education renovation policy to be integrated with economic renovation policies in difficult economic geographic regions and ethnic minority areas.

In short, this research contributes to clarify a triangle of educational renovation typical for Vietnamese society that is restructuring into a market economy, including the fundamental, comprehensive educational policy with the centers of opening more educational opportunities for citizens and shifting the structure of educational level. The case study of Vietnam may suggest ideas for the development of new conceptions of equality in education opportunities and its relation to educational policy and levels in in a restructuring society.

This research describes the renovation in the educational policy where these elements are seen as institutional elements creating convenient conditions for opening up educational opportunities and reducing social inequality in educational opportunities of all social groups. In turn, the educational opportunity opening provides an element making changes in the educational level structure of the population and being seen as the output of the national educational system. However, conclusions of the research of the educational renovation triangle has just been tested by the research using literature analysis method and, therefore, it needs to be tested by quantitative research with modern statistical models.

References

Acemoglu, D., & Robinson, J. A. (2013). *Why nations fail: The origins of power, prosperity and poverty* (616 pages). Youth Publisher.

Becker, G. S. (1964). *Human capital: A theoretical and empirical analysis with special reference to education* (390 pages). National Bureau of Economic Research (NBER), Columbia University Press.

Browen, G. A. (2009). Document analysis as a qualitative research method. *Qualitative Research Journal, 9*(2), 27–40.

Bui, Q. T., Cao, N. N., Nguyen, T. K. D., Tran, P. B., & Haughton, D. (2001). Education and income. In D. Haughton, J. Haughton, & P. Nguyen (Eds.), *Living standards in Vietnam booming economic period* (pp. 91–108). Statistical Publisher.

Connelly, R., & Zheng, Z. (2003). Determinants of school enrolment and completion of 10 to 18 year olds in China. *Economics of Education Review, 22*(2003), 379–388.

Daren, C. (2017). Education's contribution to economic growth. In *Munich personal RePEc archive*. Retrieved March 8, 2017, from https://mpra.ub.uni-muenchen.de/77365/

Gharajedaghi, J. (2005). *Systems thinking: Managing chaos and complexity: A platform for designing business architecture* (pp. 81–134). Social Science Publisher.

GSO Vietnam. (2000). *Vietnam living standard survey 1997–1998* (448 pages). Statistical Publishing House.

GSO Vietnam. (2006). *Vietnam – 20 years of renovation and development* (124 pages). Statistical Publishing House.

GSO Vietnam. (2018a). *Vietnam living standard survey 2016* (783 pages). Statistical Publishing House.

GSO Vietnam. (2018b). *Survey on population change and family planning on April 1, 2017: Basic results* (307 pages). Statistical Publishing House.

GSO Vietnam. (2019). *Statistical yearbook 2018* (452 pages). Statistical Publishing House.

Handa, S. (2002). Raising primary school enrolment in developing countries: The relative importance of supply and demand. *Journal of Development Economics, 69*, 103–128.

Haughton, J. (2001). Introduction: Miraculous changes. In D. Haughton, J. Haughton, & P. Nguyen (Eds.), *Living standards in Vietnam booming economic period* (pp. 9–38). Statistical Publisher.

Iddrisu, A. M. (2014). The effect of poverty, household structure and child work on school enrolment. *Journal of Education and Practice, 5*(6), 145–156.

Law on Children Protection, Care and Education (1991). Retrieved August 25, 2019, from https://thuvienphapluat.vn/van-ban/Giao-duc/Luat-Bao-ve-cham-soc-va-giao-duc-tre-em-1991-57-LCT-HDNN8-38064.aspx

Law on Compulsory Primary Education. (1991). Retrieved August 25, 2019, from http://vbpl.vn/bogiaoducdaotao/Pages/vbpq-toanvan.aspx?ItemID=11595

Law on Education. (1998). Retrieved August 27, 2019, from https://thuvienphapluat.vn/van-ban/Giao-duc/Luat-giao-duc-1998-11-1998-QH10-44857.aspx

Law on Education. (2005). Retrieved August 27, 2019, from http://vbpl.vn/bogiaoducdaotao/Pages/vbpq-toanvan.aspx?ItemID=18129.

Law on High Education. (2012). Retrieved August 27, 2019, from https://thuvienphapluat.vn/van-ban/Giao-duc/Luat-Giao-duc-dai-hoc-2012-142762.aspx

Law on Vocational Education and Training. (2014). Retrieved August 27, 2019, from https://thuvienphapluat.vn/van-ban/Lao-dong-Tien-luong/Luat-Giao-duc-nghe-nghiep-2014-259733.aspx

Le, N. H. (2015). *System, Structure & Social Differentiation* (306 pages). Vietnam National University Press.

Political report presented at the CPV Congress. (1982). Retrieved August 25, 2019, from http://tulieuvankien.dangcongsan.vn/ban-chap-hanh-trung-uong-dang/dai-hoi-dang/lan-thu-v.

Political report presented at the CPV Congress. (1986). Retrieved August 25, 2019, from http://tulieuvankien.dangcongsan.vn/ban-chap-hanh-trung-uong-dang/dai-hoi-dang/lan-thu-vi.

Political report presented at the CPV Congress. (1991). Retrieved August 25, 2019, from http://tulieuvankien.dangcongsan.vn/ban-chap-hanh-trung-uong-dang/dai-hoi-dang/lan-thu-vii.

Political report presented at the CPV Congress. (1996). Retrieved August 25, 2019, from http://tulieuvankien.dangcongsan.vn/ban-chap-hanh-trung-uong-dang/dai-hoi-dang/lan-thu-viii

Political report presented at the CPV Congress. (2001). Retrieved August 25, 2019, from http://tulieuvankien.dangcongsan.vn/ban-chap-hanh-trung-uong-dang/dai-hoi-dang/lan-thu-ix

Political report presented at the CPV Congress. (2006). Retrieved August 25, 2019, from http://tulieuvankien.dangcongsan.vn/ban-chap-hanh-trung-uong-dang/dai-hoi-dang/lan-thu-x.

Political report presented at the CPV Congress. (2011). Retrieved August 25, 2019, from http://tulieuvankien.dangcongsan.vn/ban-chap-hanh-trung-uong-dang/dai-hoi-dang/lan-thu-xi.

Political report presented at the CPV Congress. (2016). Retrieved August 25, 2019, from http://tulieuvankien.dangcongsan.vn/ban-chap-hanh-trung-uong-dang/dai-hoi-dang/lan-thu-xii.

Psacharopoulos, G., & Patrinos, H. A. (2004). Returns to investment in education: A further update. *Education Economics, 12*(2), 111–134.

Ravallion, M., & Wodon, Q. (2000). Does child labor displace schooling? Evidence on Behavioral responses to an enrolment subsidy. *Economic Journal, 110*(462), C158–C175.

Schultz, T. W. (1961). Investments in human capital. *American Economic Review, 51*(1), 1–17.

Schultz, T. W. (1988). Education investments and returns. In H. Chenery & T. N. Srinivasan (Eds.), *Handbook of development economics* (pp. 543–630). Elsevier Science Publishers, B. V.

Sen, A. (1999). *Development as freedom* (366). Oxford University Press.

Steering Committee of Population and Housing Census. (2019). *Census on population and housing on April 1, 2019: Organization and preliminary results* (p. 124). Statistical Publishing House.

The Ministry of Planning and Investment – GSO Vietnam. (2011). *Vietnam census on population and housing 2009 - education in Vietnam: The analysis of basic indicators* (178). Statistical Publishing House.

Truong, C. T. K., Thai, N. T. D., & Bach, V. H. (1999). Factors influencing the enrolment of low secondary school. In D. Haughton, J. Haughton, S. Bales, T. T. K. Chuyen, & N. N. Nga (Eds.), *Health and wealth in Vietnam: An analysis of household living standards* (pp. 115–131). National Political Publisher.

Vo, T. S., Truong, C. T. K., Doan, H. T., Nguyen, T. T., Do, H. V., Ho, B. T., & Le, C. N. (2001). School going and quitting by students. In D. Haughton, J. Haughton, & P. Nguyen (Eds.), *Living standards in Vietnam booming economic period* (pp. 177–192). Statistical Publisher.

Yin, R. K. (1994). *Case study research: Design and methods* (2nd ed., 171 pages). Sage.

Zimmerman, J. F. (2001). Determinants of school enrolment and performance in Bulgaria: The role of income among the poor and rich. *Contemporary Economic Policy, 19*(1), 87–98.

Chapter 5
Tourism Climate Indicators (TCI) Applied in Moc Chau District (Son La, Vietnam)

Hoa Thu Le, Khanh Nguyen Ngoc, and Nhung Nguyen Thu

Abstract Climate is an important resource of tourism industry; climate indicators can be measured and assessed. The determination of the right time for tourism activities contributes to the orientation of making investment plans for managers and advising vacation time for visitors. To determine the right time, the researched team used 7 climate parameters: total precipitation, monthly average temperature, relative humidity, average maximum temperature, average temperature of dark temperature, minimum average of relative humidity, and daily average of sunshine hours and wind speed to calculate TCI model in Moc Chau. The TCI calculation results based on 10-year climate data (2008–2018) show that Moc Chau is a place with favorable conditions for year-round tourism activities, preferably from October to April next year.

Keywords Tourism climate indicators (TCI) · Tourism industry · Moc Chau · Vietnam

1 Introduction

Tourism is an important economic sector of the global economy. In 2018, the tourism industry achieved 1451 million visitors and achieved a total turnover of $ 1401 billion, creating an average of $ 5 billion per day for the world economy (World Tourism Organization (UNWTO), 2019). Tourism has been improving the quality of life for millions of people, changing the lives of communities in many countries,

H. T. Le
Tay Bac University, Son La, Vietnam
e-mail: hoale@utb.edu.vn.com

K. N. Ngoc
Vietnam Geography Association, Hanoi, Vietnam

N. N. Thu (✉)
VAST - Institute of Geography, Hanoi, Vietnam

A. T. Nguyen, L. Hens (eds.), *Global Changes and Sustainable Development in Asian Emerging Market Economies Vol. 2*, https://doi.org/10.1007/978-3-030-81443-4_5

not only in developed countries but in developing countries also that have benefited from the tourism industry. Therefore, the development of the tourism industry is the object of many scientific studies with many different perspectives, including studies on the relationship between human health and climate, creating an application platform for studying the linkage between climate and tourism industry.

Climate and weather are considered a country's tourism resources, having profound impact on human health and tourism. During the past nearly 100 years, many researches on climate index with human health and resort activities have been conducted (Landsberg, 1972; McMichael, Woodruff, & Hales, 2006). Many studies have even shown that climate affects the tourism industry as deeply as agriculture, namely: climate decides the location selection and tourism experience (Hamilton, Lau, & Lau, 2006; Hu & Ritchie, 1993). From the perspective of organizers, Scott and McBoyle (2001), Matzarakis (2006), Kozak, Uysal, and Birkan (2008), Scott and Lemieux (2010), and Gómez-Martín, Armesto-López, and Martínez Ibarra (2017) point out that climate affects crop length, quality, and profitability of tourism industry. Meanwhile, Matzarakis (2006) and Gómez-Martín et al. (2017) suggest that climate and weather affect tourism dynamics and activities. In line with the above viewpoint, Ayscue, Curtis, Hao, et al. (2015) and Gómez-Martín et al. (2017) have pointed that: The climate influences the decisions of investors, managers, and tour operators. Therefore, climate is an important criterion in assessing the image of a destination for tourism and is considered a tourism development resource of a locality. Climate parameters need to be specifically quantified to assist travelers in planning a vacation, managers in exploiting and making their own decision in investment.

In recent years, there have been many studies on climate index for tourism such as tourism climate index (TCI) (Mieczkowski, 1985), beach climate index (BCI) (Morgan, Gatell, Junyent, et al., 2000), and developing tourism climate index (TNDT) (De Freitas, Scott, & McBoyle, 2008). However, from its birth to the present, the TCI proposed by Mieczkowski is the most widely used indicator to assess the suitability of climate for tourism. The purpose of TCI is to provide a comprehensive measure to quantify and assess the world climate for tourism activities in general by integrating all climate-related variables into a single indicator; this indicator has three advantages: firstly, TCI determines the appropriate time to conduct tourism activities at a specific time in any locality when meteorological parameters are available; secondly, quantifying TCI uses data stored for many years, so it is possible to show the change of TCI at different stages, thereby quantifying the local climate change, this is significant in assessing the impact of climate change on tourism as currently (Amelung, Nicholls, & Viner, 2007; Hein, Metzger, & Moreno, 2009; Nicholls & Amelung, 2008; Scott & McBoyle, 2001); thirdly, identifying TCI helps compare climate potential in different localities, allowing comparison of destination climate differences among different regions. Because of the diversity and integrated climate indicators for tourism, TCI is applied to the research in many countries around the world (Amelung et al., 2007; Cheng & Zhong, 2019; Kozak et al., 2008).

Our study selects the mountainous district of Moc Chau to assess the tourism climate index because of the importance of the district in the Moc Chau national tourist area, in particular, and the Northern Vietnam tourism, in general. According to development planning, Moc Chau has been identified as one of the 10 National Tourist Areas, a driving force for tourism development for the whole region. Although in many reports and studies, it is always confirmed that climate resources played the most important role in the tourism attractions in Moc Chau, no studies have yet quantified the local tourism climate index. Determining tourism climate index in Moc Chau creates scientific basis and creates decision-making tools for tourists, in particular, and the tourism industry, in general, for Moc Chau district.

2 Methodology

2.1 Study Area

Moc Chau is a mountainous plateau district of the limestone plateau in the Northwest region of Vietnam, with an average altitude of more than 1050 m above sea level, with a natural area of 1081.66 km^2 ranked eighth among 12 city districts of Son La province. The whole district has 2 towns (Moc Chau and Moc Chau Farm) and 13 communes (Dong Sang, Muong Sang, Long Sap, Chieng Khua, Chieng Hac, Tan Lap, Chieng Son, Tan Hop, Phieng Luong, Quy Huong, Na Muong, Hua Pang, and Ta Lai).

Overall, Moc Chau has a very cool climate, the average annual temperature is only about 18.7 °C; in the hottest month, the average temperature reaches 23.2 °C and reaches 12.3 °C in the coldest month. Moc Chau has two distinct rainy and dry seasons: the dry season from November to March next year, the average rainfall is about 20-40 mm/month; rainy season is from April to October, the average monthly rainfall ranges from 100 mm/month to 330 mm/month. According to the observed data, the number of sunny hours at Moc Chau station is very moderate, reaching 1665 hours/year (Table 5.1).

Thus, Moc Chau is located in the tropical monsoon climate but due to the impact of the high belt, it brings both the characteristics of the subtropical and temperate highland climate, which is very convenient to develop tourism for whole years.

Table 5.1 Temperature, rainfall, average monthly sunshine hours in Moc Chau district, period 2008–2018

Month	1	2	3	4	5	6	7	8	9	10	11	12	Average in year
Average tem (°C)	12.3	13.6	17.0	20.5	22.4	23.2	23.1	22.6	21.3	19.0	15.9	13.0	18.7
Average rainfall (mm)	21.8	21.4	42.1	103.2	188.2	231.5	286.3	326.7	254.6	132.9	37.9	18.5	1665
Average monthly sunshine hours	118	138	165,9	192,8	153,1	148,0	147,7	148,7	143,4	134,6	146,0	127,7	1665,4

Source: Calculation based on data provided by the National Center for Meteorology and Hydrology

2.2 *Methodology*

For this study, we collect and use monthly average data on climate factors at Moc Chau station, Son La province, for 10 years from 2008 to 2018.

The process of determining climatic indicators is determined as follows:

1. Collect data.
2. Calculate the comfort index daytime (CID) on the basis of average maximum temperature and monthly average minimum humidity (°C).
3. Calculate the average daily comfort index (CIA) (°C) which is a combination of daily average temperature (°C) and average daily relative humidity (%).
4. Determination of monthly rainfall (R)(mm).
5. Determination of monthly sunshine hours (S)(h).
6. Determine the average wind speed W (km/h).
7. Calculate tourism climate index using the following relationship:
   TCI = 8CID + 2CIA + 4R + 4S + 2 W. (Table 5.2)
8. Determine the appropriate time for taking place tourism activities at local area (Table 5.3).

3 Results

Through the research steps, we have the results of the subindex and tourism climate index in Moc Chau district as follows (Table 5.4).

Table 5.2 Additional parameters/indicators in tourism climate index

Parameters	Climate change	Effect on TCI	Contribution level (%)
CID	Daily maximum temperature and daily minimum relative humidity	Description of thermal comfort when tourist activity takes place at its maximum	40
CIA	Average daily temperature and daily relative humidity	A description of thermal comfort within 24 h including when sleeping time	10
P	Amount of rain	Reflecting the negative effects of outdoor activities	20
S	Number of sunny hours	Making a positive contribution to tourism, but there is a potential risk of sunburn and discomfort on extremely hot days	20
W	Wind speed	Being positive in nature because of the cooling effect in hot areas	10

Table 5.3 Classifying tourism climate index according to comfort level

NO	Tourism Climate Index	Comfort level
9	100–90	Ideal
8	90–80	Excellent
7	80–70	Very good
6	70–60	Good
5	60–50	Acceptable
4	50–40	Normal
3	40–30	Unexpected
2	30–20	Very undesirable
1	20–10	Extremely undesirable
0	10–0	Imposed

Table 5.4 Results of sub-indicators and tourism indicators in Moc Chau district, Son La province

Month	1	2	3	4	5	6	7	8	9	10	11	12
CIA	3	4.5	5	5	3.5	3	3.5	3.5	4.5	5	5	4
CID	2.5	2.5	3.5	4.5	5	5	5	5	5	4.5	3	2.5
R	4.5	4.5	4	2	0	0	0	0	0	1	4	4.5
S	1.5	2	2	2.5	3	2.5	2	2	2	2	2	2
W	5	5	5	5	5	5	5	5	5	5	5	5
TCI	**63**	**77**	**81**	**77**	**60**	**54**	**56**	**56**	**64**	**71**	**80**	**73**

In general, Moc Chau district has a favorable climate to develop tourism in whole years. During the year, there are 2 months with excellent tourism climate index (March and November); 4 months have a very good tourist climate index (February, April, October, December); 3 months have a good TCI (January, May, September); 3 months tourism climate index reached the lowest level (May, June, July) but still acceptable in the general classification (Fig. 5.1).

Because Moc Chau is located on a plateau with an average elevation of 1050 m, the daily comfort index (CIA) and the minimum to maximum temperature threshold (CID) show that Moc Chau climate is cool throughout the day and night, favorable for traveling vacation. This is also a district with a moderate number of hours of sunshine and wind speed, which brings a pleasant feeling to visitors, without the feature of being sunburned or too strong wind obstructing the movement so it is very convenient for outdoor activities (hiking, camping, cycling…). Especially, March and November are the best climate months for visitors to Moc Chau, all of the meteorological elements are ideal for human health and safe for visitors' travel (Fig. 5.2).

From May to August, the district's CTI reached the lowest level of the year (54–56%). The reason is that the district is located in the tropical monsoon climate zone; during this period, the district is influenced by the southwest monsoon combined with the tropical convergence band, so the rainfall is much higher than other

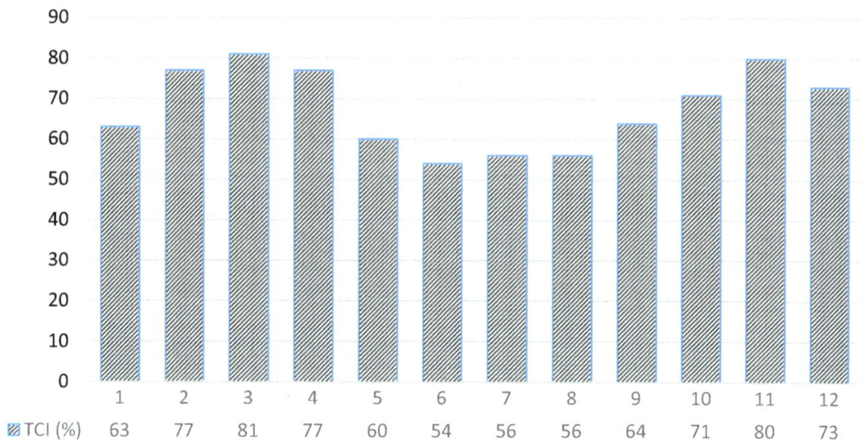

Fig. 5.1 Changes in tourism climate index of Moc Chau district in months of the year

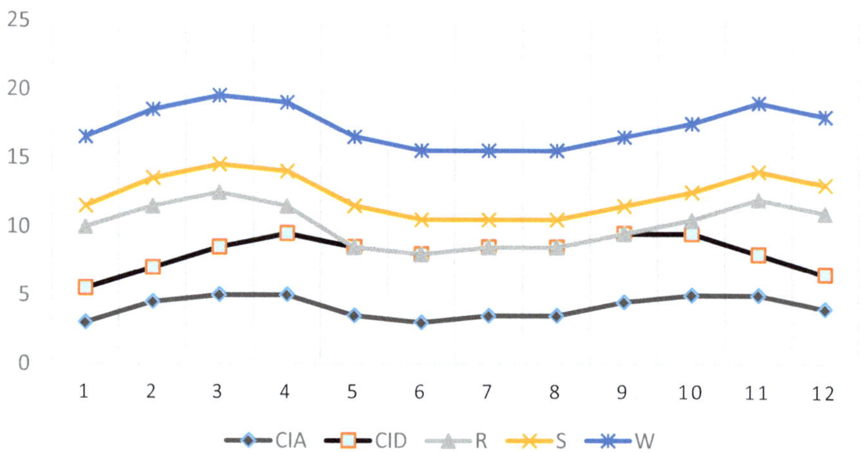

Fig. 5.2 Variation of sub-indicators related to tourism climate index in months of the year in Moc Chau district

months of the year (Table 5.1). It rains making travel activities difficult, especially affecting the travel and outdoor experience activities of visitors.

In addition, CTI in January dropped sharply to 63% compared with December at 70% and February to 77%. The main reason is due to the impact of the polar air block, brought by the northeast monsoon, making the temperature here reach the lowest level of the year; at the same time, the number of sunny hours is also sharply reduced because of heavy fog this month.

4 Conclusions and Discussion

The application model of the tourism climate index (TCI) of Mieczkowski in Moc Chau district, Son La province, Vietnam, once again confirms that tourism activities are not only affected by the heat element of the atmosphere but also by other factors: rain, sun, and wind. Research results show that Moc Chau can organize tourism and recreation activities all year round because all months of the year have an acceptable climate index for tourism activities. The study also provides travelers with the best time to visit Moc Chau from September to April next year. Especially, March and November are the two best climate months to relax and experience here. In addition, the TCI in Moc Chau provides a scientific basis for tourism managers and travel agencies to plan investments and establish appropriate businesses to achieve the highest efficiency.

References

Amelung, B., Nicholls, S., & Viner, D. (2007). Implications of global climate change for tourism flows and seasonality. *Travel Research, 45*(3), 285–296.

Ayscue, E. P., Curtis, S., Hao, H., et al. (2015). Forecast and weather-related information used among coastal tourism businesses. *Tourism Geographies, 17*(4), 603–626.

Cheng, Q., & Zhong, F. (2019). Evaluation of tourism climate comfort in the grand Shangri-La region. *Mt Sci, 16*(6), 1452–1469.

De Freitas, C. R., Scott, D., & McBoyle, G. (2008). A second generation climate index for tourism (CIT): Specification and verification. *International Journal of Biometeorology, 52*(5), 399–407.

Gómez-Martín, M., Armesto-López, X., & Martínez Ibarra, E. (2017). Tourists, weather and climate. *Atmosphere, 8*(12), 255.

Hamilton J.M., Lau M.A., Lau M. A (2006). The role of climate information in tourist destination choice decision making. Tourism and Global Environmental Change.

Hein, L., Metzger, M. J., & Moreno, A. (2009). Potential impacts of climate change on tourism; a case study for Spain. *Current Opinion in Environmental Sustainability, 1*(2), 170–178.

Hu, Y., & Ritchie, J. R. B. (1993). Measuring destination attractiveness: A contextual approach. *Travel Research, 32*(2), 25–34.

Kozak, N., Uysal, M., & Birkan, I. (2008). An analysis of cities based on tourism supply and climatic conditions in Turkey. *Tourism Geographies, 10*(1), 81–97.

Landsberg, P. H. E. (1972). A limited review of physical parameters. *Secrétariat of the World Meteorological Organisation, 123*, 49.

Matzarakis, A. (2006). Weather-and climate-related information for tourism. *Tourism and Hospitality Planning & Development, 3*, 99–115.

McMichael, A. J., Woodruff, R. E., & Hales, S. (2006). Climate change and human health: Present and future risks. *The Lancet, 367*(9513), 859–869.

Mieczkowski, Z. (1985). The tourism climatic index: A method of evaluating world climates for tourism. *The Canadian Geographer/Le Géographe canadien, 29*(3), 220–233.

Morgan, R., Gatell, E., Junyent, R., et al. (2000). An improved user-based beach climate index. *Journal of Coastal Conservation, 6*(1), 41–50.

Nicholls, S., & Amelung, B. (2008). Climate change and tourism in northwestern Europe: Impacts and adaptation. *Tourism Analysis, 13*(1), 21–31.

Scott, D., & Lemieux, C. (2010). Weather and climate information for tourism. *Procedia Environmental Sciences, 1*, 146–183.

Scott D. and McBoyle G. (2001). Using a "tourism climate index" to examine the implications of climate change for climate as a tourism resource. *Proceedings of the First International Workshop on Climate, Tourism and Recreation*.

World Tourism Organization (UNWTO), (2019). Highlights, 2019 Edition. Retrieved September 4, 2019, from https://www.e-unwto.org/doi/book/10.18111/9789284421152

Chapter 6
Carbon Footprint of Vietnam's Small Urban Areas (Ha Dong District, Hanoi)

Nguyen An Thinh

Abstract Increasing urbanization advocates the compact city. This study classifies small urban areas in the Ha Dong district (Hanoi, Vietnam) based on indicators for a compact city and calculates the carbon footprint of urban clusters based on data household of lifestyle surveys. A step use approach combining exploratory factor analysis, hierarchy cluster analysis, and carbon footprint is used. The results sort out four main factors characterizing the compact city such as quality of infrastructure, the density of open space, transportation pattern, public transportation, and urban green space. Small urban areas in Ha Dong are grouped into 4 urban clusters based on the factor of quality of infrastructure. The carbon footprint (CF) of the Ha Dong district is 6.66 ton.year^{-1}.person^{-1}, which is higher than that of the average world, about 3 times over the GHG target, and nearly 6 times than that of GHGs in Vietnam. The urban cluster C3 shows the highest carbon footprint, whereas the C4 has the lowest one. Recommendations based on the study result include the following: raise study area as the number of high-density cities increase (compact city), develop the urban green spaces and public transport system, and improve tools for urban planning based on the criteria of sustainable development and green growth.

Keywords Urban small area · Urban clusters · Compact city · Carbon footprint · Factor analysis · Hanoi city · Vietnam

1 Introduction

Sustainable cities offer a considerable challenge for contemporary land use planning. During recent years, Vietnam urbanized at a rate of approximately 3.4 percent per year (WB (World Bank), 2010); by 2013, about 33.47 percent of the land was urban. Urban planning until now contributed to environmental pollution, which

N. A. Thinh (✉)
VNU University of Economics and Business, Vietnam National University, Hanoi, Vietnam

A. T. Nguyen, L. Hens (eds.), *Global Changes and Sustainable Development in Asian Emerging Market Economies Vol. 2*, https://doi.org/10.1007/978-3-030-81443-4_6

hampered the economic growth and the sustainable development of the country. The urban population is predicted to double in the next 10 years, which necessitates agriculture land transformed to urban land by four times (WB (World Bank), 2010). Urban areas developed without planning could lead to the decrease of infrastructure qualities, impacting environment and urban landscape. During the United Nations Conference on Sustainable Development (Rio 20+) in 2012, decision makers agree for promoting an integrated approach to build sustainable cities and urban settlements. It would not create livable habitats for urban people but deal with other ecological and environmental problems.

Climate change, population growth, and urbanization are considered significant issues that need to be considered in the process of land use planning and urban planning ensuring sustainable development. Vietnam has witnessed urban sprawling faster and faster during different periods of socioeconomic development (WB (World Bank), 2010). Consequences of unplanned urban are inefficient land use and environmental pollution. Sustainable urban planning requires both improving the land quality and decreasing the emission of green house gases (GHGs) . In European country, compact cities have been shown as good examples of sustainable urban planning and climate change responses in terms of creating livable habitats for urban people and promoting relatively high residential density with mixed land uses and based on an efficient public transport system (Dantzig & Saaty, 1973).

In the international literature, there were many urban areas developed from 1950 to 1970, resulting from the increased housing demand after wars. As a result, different problems arising from the process of urbanization: the irrational land use, the overuse and exploitation of natural resources, environment pollution, traffic jam, and green house gas emission. In 1973, the term "compact city" was invented and defined as a way to efficient urban planning, which allowed us to limit the process of urban sprawl and to reduce energy consumption (Dantzig & Saaty, 1973). Brundtland commitment in 1987 first introduced the term "sustainable development" and stated that "transforming the old urban architecture into the compact city was considered as a good practice to archive sustainable development." Compact city was the structure of urban planning, which enabled citizens to cycle and walk easily as well as use public transportation effectively (Elkin, McLaren, & Hillman, 1991). Compact city was regarded as the more sustained urban architecture in comparison with urban sprawl because its structure helped to reduce using private cars and to increase the effect of infrastructure (Williams, Burton, & Jenks, 2000).

While urban sprawl became ineffective and led to inequality society (Bourne, 1992), compact cities offered good conditions for society development (Garcia & Riera, 2003). Different research on the relationship between the compact city and sustainable development showed that the compactness and the sustainability had interactive effects (Neuman, 2005). Transportation was one of the major resources of carbon dioxide emission. One solution that could be implemented was to adjust land use planning (Barton, 1990). Through efficiency urban planning and effective land use, the amount of fossil fuel decreased by 10–15 percent, which means the amount of CO_2 released also dropped by 10–15 percent through the changes in using transportation (Rickaby, Steadman, & Barrett, 1992). Compact cities were developed by mixing land use in a reasonable scale for the purpose of effective energy consumption (Owen, 1992) because the effect is higher in high-density cities (WB (World Bank), 2010). Obviously, the urban structure and energy consumption have close linkage.

This study deals with classifying small urban areas and calculating carbon footprint in a case study of Ha Dong district (Hanoi city). Ha Dong district is selected as a case study area. It is located along two sides of the National Road no. 6, is 13 kilometer away from city center, and covers 4833.66 square kilometers. Ha Dong is an urban district, which is contiguous to 3 rural districts (Thanh Oai district in the south, to Thanh Tri district in the east, and Hoai Duc district in the west). Thus, it becomes the west gate to the suburb of Hanoi. In the history of development, Ha Dong became a district of Hanoi because of the urban sprawl. In addition, the speed of urbanization accelerates, which results in more and more effect showing clearly in the urban landscape of Ha Dong. Meanwhile, Ha Dong is affected by population growth and urban resettlement. In general, Ha Dong's infrastructure is not synchronous. The center areas have good infrastructure, in contrast to the suburb areas. Therefore, practical solutions need to be considered to solve problems related to urban planning.

The rest of the paper consists of four parts: part 1 shows the introduction and literature review; part 2 introduces material and methodology of factor analysis and carbon footprint calculation; part 3 focuses on the results of small urban area classification and calculating carbon footprint for these areas; finally, conclusion and discussion are pointed out in part 4.

2 Methodology

2.1 Conceptual Study Model

Figure 6.1 shows flowchart of steps for studying carbon footprint of urban small areas. Once land use map in the year 2010 had been used to design an urban small area map, the factor analysis and hierarchy cluster analysis were applied to group small urban areas into a reduced number of urban clusters. The analysis is based on data collected by an urban architecture questionnaire. Then responses of citizens' lifestyle questionnaire were used to create input data for carbon footprint estimation for different small urban areas. Finally, an urban planning recommendation was conducted based on analyzed results.

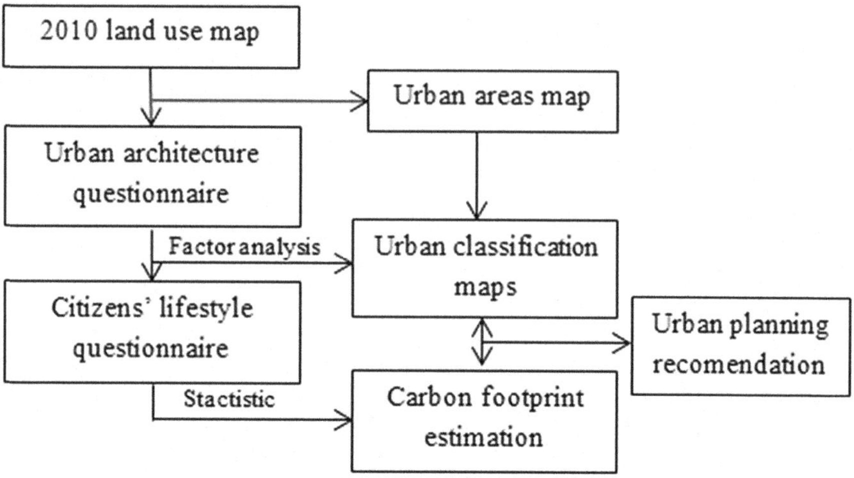

Fig. 6.1 Flowchart of steps for studying carbon footprint of urban small areas

2.2 *Multivariate Analysis*

2.2.1 Factor Analysis

Factor analysis is a method of multicriteria analysis, which allows reducing from a large number of variances to a smaller number of factors that account for the most variance among the original data. Factors are nominated by applying principal component analysis to a standardized correlation matrix. A table of factor loadings shows which variables are grouped together on which common factors and the degree of correlation between individual variables and the factors. The factors are interpreted as axes in state spaces, and the meanings of the axes are inferred from the variables that are most correlated with them.

* Factor matrix:

$$x_i = \sum_{i=1}^{k} \alpha_{ij} f_i + e_j, \tag{6.1}$$

where x_i are observed variables, f_i are the common factors, α_{ij} are factor loading of factor x_i, and e_j is the measurement error for x_i.

For factor analysis, to calculate the covariance of any two observable variables,

$$r_{jk} = \alpha_{j1}\alpha_{k1} + \alpha_{j2}\alpha_{k2} + \ldots + \alpha_{jm}\alpha_{km}. \tag{6.2}$$

* Principal component analysis was used to determine eigenvalue and eigenvector of correlation matrix. Eigenvalue λ_i show the proportion of the variance of x_j. Eigenvector shows the attribute of component I,

$$x_j = \sum_{i=1}^{m} w_{ij} x_j, \tag{6.3}$$

where x_j are observed variables, z_j are matrix components, and w_{ij} are variables j loading of component i.

* Varimax Rotation: Once principal component analysis had been completed, a variable in the n dimensional space specified by the factors involved, and factor loadings are the cosine of the angle formed by a vector from the origin to that coordinate and the factor axis. Varimax rotation function is as follows ($\gamma = 1$):

$$R_{VAR} = agr\max_r \left(\sum_{j=1}^{k} \sum_{k=1}^{p} (AR)_{ij}^4 - \frac{\gamma}{p} \left(\sum_{j=1}^{k} (AR)_{ij}^2 \right)^2 \right). \tag{6.4}$$

2.2.2 Hierarchical Cluster Analysis

A hierarchical cluster analysis based on the Euclidean nearest-neighbor distance attempted at identifying relatively homogeneous groups of small urban areas in Ha Dong district based on factor scores. Euclidean nearest-neighbor distance is defined using simple Euclidean geometry as the shortest straight-line distance between a commune and its nearest neighbor. Theoretically, an algorithm that starts with each small urban area in a separate cluster and combines clusters until only one is left was used for this analysis. Hierarchical clustering, consequently, created a hierarchy of clusters, which may be represented in a tree structure called a dendrogram.

This analysis is based on Euclidean distance metrics as follows:

$$\begin{gathered} a - b_2 = \sqrt{\sum_i (a_i - b_i)^2} \\ \min\{d(a,b) : a \in A, b \in B\} \end{gathered} \tag{6.5}$$

where a and b belong to two sets of observations A and B and d is the chosen metric.

Results of a hierarchical cluster analysis showed that there are four groups of the small urban areas as urban clusters.

2.3 *Carbon Footprint Calculation*

Carbon footprint (CF) is a measurement of total cumulated GHGs emitted directly and indirectly over times resulting from different human activities. There are different methods of GHG estimation depending on the object (individuals, residents, nations or companies, factories, and economic sectors). The result of carbon footprint estimation becomes the foundation of implementing policies and strategies for the purpose of GHG reduction. To quantify the impact of human activities and lifestyle of residents, the study uses the method of GHG estimation (with particular Vietnamese CO_2e index) presented by Carbon Footprint Ltd., US (retrieved from *http://www.carbonfootprint.com* in 2015):

$$\mathrm{CF} = \mathrm{CF_H} + \mathrm{CF_T} + \mathrm{CF_S}. \tag{6.6}$$

2.4 Questionnaires

Two questionnaires were conducted for survey as follows:

1. Urban architecture questionnaire: as shown in Tables 6.1 and 6.2, this questionnaire was designed based on 30 indicators belonging to 7 criteria of compact city. The 3-point Likert scale quantifies the properties of small urban areas. Data collected from this questionnaire were then used in factor analysis and hierarchy cluster analysis. A total of 135 respondents were involved in the survey.
2. Carbon footprint questionnaire: this questionnaire enables us to collect data from 3 sectors for carbon footprint estimation (Table 6.3). A total of 231 respondents agreed to provide information according to the questionaire.

3 Results

3.1 Classifying Urban Small Areas

As shown in Table 6.4, once factor analysis had been completed, 19 out of 30 compact city variables were selected and quantified into 4 factors as follows:

Table 6.1 Criteria of a compact city

Criteria	Definition
Mixed land use (C1)	Density of open spaces; the number of commercial zones/services places
Take advantage of compact building design (C2)	Housing architecture
Provide a variety of transportation choices (C3)	The quality of transportation
Preserve open space, farmland, natural beauty, and critical environmental areas (C4)	The quality of public places (parks, schools, hospitals, and playgrounds)
Strengthen and direct development toward existing communities (C5)	The environmental quality
Make development decisions predictable, fair, and cost effective and create a range of housing opportunities and choices (C6)	The effect of infrastructure planning
Develop ground space (C7)	The number of ground spaces

Table 6.2 The interpretation of criteria and indicators for urban architecture questionnaire

Criteria	Indicators (Variable)	Likert scale
C1.	I1. Open space density	Low (1), average (2), and high (3)
	I2. Water body density	
	I3. Number of commercial sites (markets, shopping malls, etc.)	Poor (1), average (2), and good (3)
	I4. Number of service sites (restaurants, cinemas, etc.)	
C2.	I5. Housing architectural style	Houses (1), houses and high buildings (2), and high building (3)
C3.	I6. Number of public transport	None (1), small (2), and reasonable (3)
	I7. Quality of roads	Low (1), average (2), and good (3)
	I8. Wide of roads	Narrow (1), average (2), and wide (3)
	I9. Wide of footpaths	None (1), narrow (2), and wide (3)
	I10. The number of over bridge	None (1), small (2), and reasonable (3)
	I11. Bus lane	None (1), small (2), and reasonable (3)
	I12. Vehicle lane	
	I13. Type of road network	Radial road network (1), ring road network (2), and both (3)
	I14. Number of lighting	None (1), little (2), and reasonable (3)
	I15. Number of traffic sign	
	I16. Number of lanes (in main roads)	2 lanes (1), 3–4 lanes (2), and >4 lanes (3)
C4.	I17. The quality of public space	Low quality (1), average quality (2), and high quality (3)
	I18. The quality of schools	
	I19. The quality of hospitals	
	I20. The quality of playgrounds	
C5.	I21. Density of green spaces along main roads	Low (1), average (2), and high (3)
	I22. The waste treatment process	None (1), average (2), and good (3)
	I23. The water waste treatment process	
	I24. Water quality (in rivers and lakes)	Heavy polluted (1), slight polluted (2), and good quality (3)
	I25. Air quality	Polluted (1), only polluted in rush hours (2), and good
C6.	I26. Electric capable network	Disordered (1), reasonable (2), and ground capable (3)
	I27. Internet capable network	
	I28. Water consumption	Rain water (1), tap water (3), and both (2)
	I29. Energy consumption	Coal (1), green energy (3), and both (2)
C7.	I30. Number of round space	None (1), low (2), and high (3)

Table 6.3 Varimax rotated component matrix

Indicator	Factor 1	Factor 2	Factor 3	Factor 4
I4	0.833	0.030	0.054	0.185
I8	0.828	0.270	0.256	0.006
I9	0.816	0.260	−0.018	−0.164
I16	0.801	0.276	0.141	−0.075
I24	0.799	−0.108	−0.363	−0.016
I3	0.796	0.185	0.185	−0.029
I30	0.778	0.168	0.012	0.067
I29	0.763	0.267	0.261	−0.053
I20	0.732	0.404	0.176	−0.099
I26	0.723	0.297	0.319	−0.088
I15	0.713	0.340	0.282	−0.112
I7	0.701	0.444	0.251	0.072
I5	0.682	0.186	0.320	0.188
I16	0.643	0.490	0.210	−0.184
I23	0.634	0.486	0.131	−0.066
I2	0.218	0.844	−0.125	0.132
I13	0.185	−0.057	0.890	0.079
I11	0.307	−0.006	0.108	0.858
I21	0.339	−0.067	0.019	−0.670

Table 6.4 Total Variance Explained

	Initial Eigenvalues			Rotation sums of squared loadings		
Factors	Total	Percentage of variance	Cumulative	Total	Percentage of variance	Cumulative
1	10.373	54.592	54.592	8.774	46.177	46.177
2	1.441	7.583	62.176	2.158	11.358	65.858
3	1.119	5.890	68.065	1.581	8.323	65.858
4	0.974	5.124	73.189	1.393	7.331	73.189

- Factor 1: "*the quality of infrastructure*" includes 15 variables correlating and showing the synchronization between infrastructure and urban landscape of individual residential areas.
- Factor 2: "*the density of open space*" describes the appearance of open spaces inside urban.
- Factor 3: "*transportation pattern*" shows the distribution of two main road types inside urban.
- Factor 4: "*public transportation and urban green space*" is a dipole factor showing the inverse correlation of two variables. It indicates that public traffic spaces and urban green spaces are not planned logically.

Based on the mathematical relationship between 4 factors and the 2010 land use map, four urban classification maps were created (as shown in Figs. 6.2, 6.3, 6.4, and 6.5).

The total percentage of variability of factor 1 accounts for 75 percent. Moreover, factor 1 describes a direct correlation. Thus, small urban areas in Ha Dong are grouped into 4 urban clusters based on the quality of infrastructures (As shown in Fig. 6.6).

Urban cluster C1 includes new small urban areas having similar properties in complex building, each of which is a combination of housings, commercial zones, schools, hospitals, sport centers, playground, etc. In addition, there are several urban green spaces and underground spaces in cluster C1. Overall, these areas have good quality infrastructures, creating livable environment for residents and promoting socioeconomic development.

Urban cluster C2 includes old small urban areas with the high diversity in its pattern: complex buildings, containing housings, commercial zones, schools, hospitals, playgrounds, etc. However, the area of green spaces and ground spaces is limited. Overall, cluster C2 offers reasonable infrastructure quality for residents.

Urban cluster C3 includes residential areas that have been developed along two sides of National Road number 6. Thanks to special location, the number of populations accelerates, which results in the highest population density in Ha Dong district. There are many main offices of Ha Dong's Government in cluster C3. In these areas, the development of urban areas is spontaneous because of a scarcity of

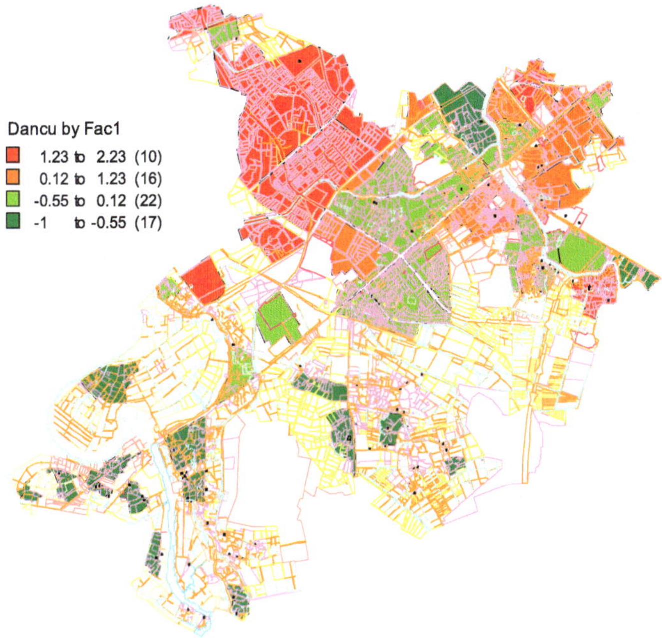

Fig. 6.2 Map of the quality of infrastructure (Factor 1)

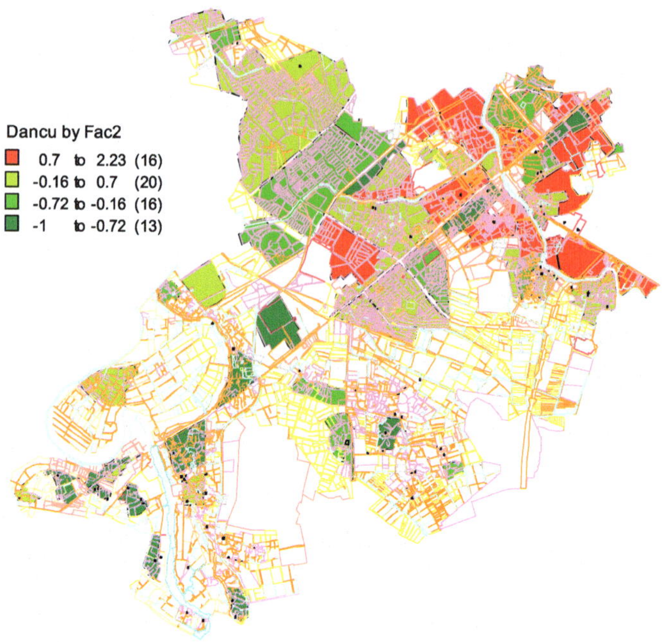

Fig. 6.3 Map of the density of open space (Factor 2)

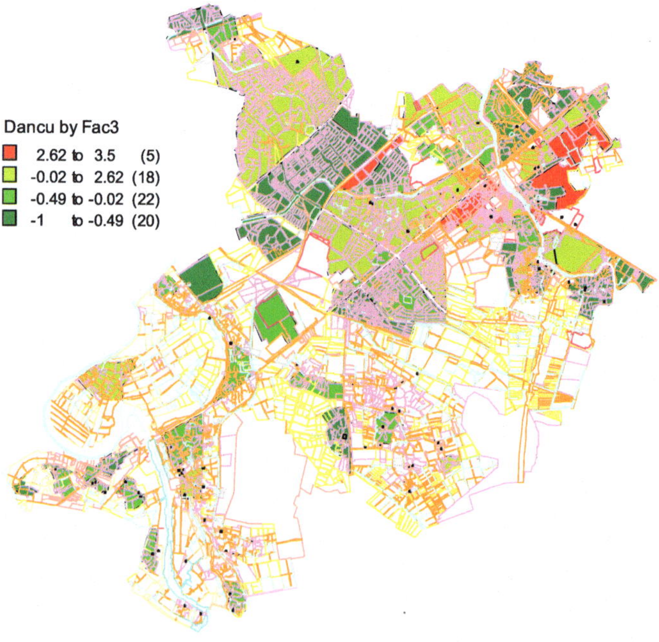

Fig. 6.4 Map of transportation pattern (Factor 3)

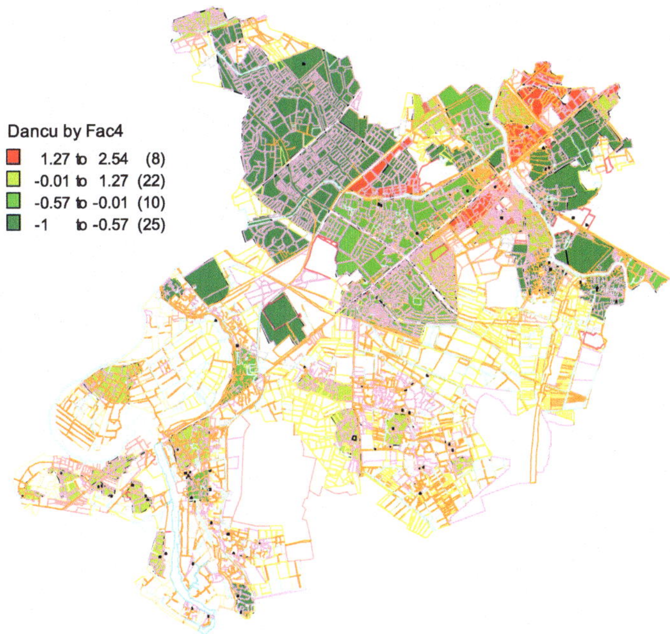

Fig. 6.5 Maps of public transportation and urban green space (Factor 4)

planning. Urban areas chopped into many blocks. Otherwise, the public transport system is developed, whereas urban areas are shot of green spaces.

Urban cluster C4 includes suburbs of Ha Dong district. In the part, some of them were part of Ha Dong district because of the urban sprawl, whereas others were urban areas when Ha Dong became urban district. Urban areas are chopped into many blocks. Most of the houses are built for a long time, which makes them have bad quality. In cluster C4, many people use charcoal stoves instead of gas cookers and electric cookers. Both the infrastructure system and the potential of socioeconomic are limited. The impact of urbanization cannot promote the development of these areas. It may lead to many environmental and social problems.

The results of study show the differences of energy using, transportation, and lifestyle. Hence, the relationship between urban pattern and lifestyle is close.

3.2 Carbon Footprint

Among three indexes of CF, CF_H has the lowest value, accounting only for 13 percent (equivalent to 0.91 ton/year/person). It is followed by that of CF_T is 1.18 ton/year/person (about 18 percent). The highest value belongs to CF_S, with the index being 4.57 ton/year/person (equivalent to 69 percent). In total, that of CF of Ha Dong district is 6.66 ton/year/person, which is higher than that of average CF in the

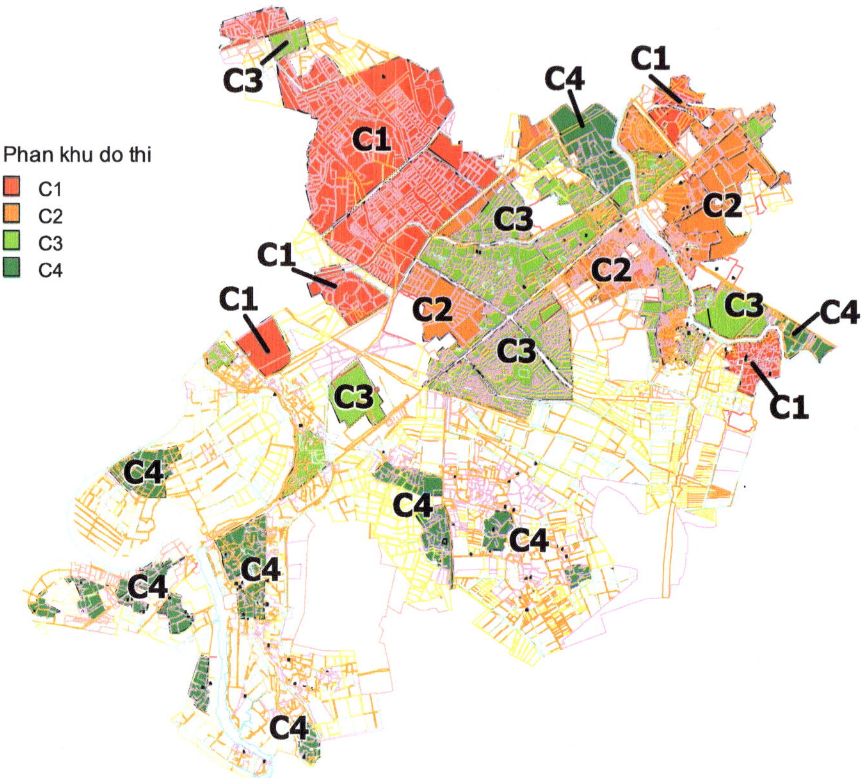

Fig. 6.6 Map of urban clusters in Ha Dong district

world (4 ton/year/person), is 3 times more than that of GHGs target (2 ton/year/person), and is nearly 6 times than that of GHGs in Vietnam (1.18 ton/year/person).

Urban clusters C1 and C2 are high-density urban areas among four clusters. They offer certain benefits from effective urban architecture, providing a variety of choices. As a result, the demand for transportation decreases. However, the figure of CF_H is quite high, resulting from high electricity divided consumption. The figure of CF_S is also considerable due to consuming high GHG emission products as imported products, plastic bags, joining high GHG emission activities, and a high number of transportation means.

Urban cluster C3 witnesses different negative impacts of urbanization such as rapid population growth. The number and the density of population in C3 are highest, which lead to great demand for transportation. Thus, the figure of CF_T of C3 is highest and the difference between that figure of C3 and three other clusters is considerable. In addition, the figure of CF_S of C3 is biggest, resulting from high GHG emission lifestyle as using many imported products, buying many furniture and electric divides, and using plastic bags for shopping.

Cluster C4 is not affected by urbanization. Most of the residents in C4 are rural villages, being located separated and mixed with agriculture land. They remain old

Table 6.5 Carbon footprint of Ha Dong urban (unit: ton/year/person)

Sectors	N	Min	Max	Mean	Std. Deviation
CF_H	56	0.12	5.40	0.91	0.84
CF_T	56	0.12	7.08	1.18	1.23
CF_S	56	2.11	8.21	4.57	1.20
CF	56	2.83	15.40	6.66	2.14

Table 6.6 Carbon footprint of urban cluster C1 (unit: ton/year/person)

Sectors	N	Min	Max	Mean	Std. Deviation
CF_H	10	0.36	1.68	0.93	0.45
CF_T	10	0.36	1.44	0.81	0.46
CF_S	10	2.72	5.66	4.19	0.83
CF	10	4.62	8.18	5.94	1.01

Table 6.7 Carbon footprint of urban cluster C2 (unit: ton/year/person)

Sectors	N	Min	Max	Mean	Std. Deviation
CF_H	13	0.36	1.08	0.55	0.19
CF_T	13	0.36	1.56	0.84	0.36
CF_S	13	2.50	6.44	4.70	1.19
CF	13	3.34	8.48	6.09	1.36

Table 6.8 Carbon footprint of urban cluster C3 (unit: ton/year/person)

Sectors	N	Min	Max	Mean	Std. Deviation
CF_H	24	0.12	5.40	1.01	1.06
CF_T	24	0.36	7.08	1.70	1.70
CF_S	24	3.13	8.21	4.88	1.27
CF	24	4.47	15.40	7.59	2.64

Table 6.9 Carbon footprint of urban cluster C4 (unit: ton/year/person)

	N	Min	Max	Mean	Std. Deviation
CF_H	10	0.24	3.60	1.05	0.97
CF_T	10	0.12	1.20	0.71	0.34
CF_S	10	2.11	5.85	4.04	1.18
CF	10	2.83	7.48	5.80	1.60

habitats such as using charcoal stove instead of gas cookers or electric cooker, which results in the highest figure of CF_H. However, their demand for transportation is low, which brings about the lowest figure of CF_T. In addition, the citizens in C4 have healthy lifestyle such as enjoying organic and seasonal products (they grow by themselves), joining none-GHG emission activities, only using furniture and electric divides if necessary (Tables 6.5, 6.6, 6.7, 6.8 and 6.9).

3.3 Conclusion and Discussion

The compact city is recognized as a good lesson for urban planning in developed countries. Urban analysis and classification based on the criteria of compact city therefore are useful tools for planning. Urban architecture helps to change the way to consume energy inside cities. The results show the relationship between urban landscape and lifestyle of human inside urban landscape. In 2010, World Bank presents report, namely, *"Cities and climate change: An Urgent Agenda,"* and shows results such as increased density can reduce energy consumption; urban design and mobility are crucial in CO_2 emissions; encouraging denser cities and greater reliance on public transportation; change in urban landscape architecture (compact cities are more sustainable than sprawling cities); and change in using energy, toward using alternative resources. This support the favorable solutions for analyzing the relationship between urban pattern and resident's lifestyle in urban planning.

Increasing the number of high-density cities (compact city): Density may concern in terms of building density, population density, or infrastructure density as the ratio of urban green space. From sustainable development viewpoint, high-density cities emit CO_2 lower than low-density cities. For instance, Japan's urban areas are five times denser than Canada's. The consumption of energy per capita in Japan is 40 percent lower than in Canada. In Madrid, city density is 10 times higher than Atlanta, and Madrid's CO_2e emissions per capita are four times lower than in Atlanta (WB (World Bank), 2010). Obviously, the compact city becomes a model of sustainable city that is applied widely over the world.

Cities may have high density for several measures such as mixing land use in planning and developing the public transport system. It not only reduces using energy and natural recourses but also saves the environment and improves the land use effect. For study area, urban clusters C1 and C2 need to apply density cities model because these areas have good-quality infrastructures and high building density.

Developing the urban green spaces and public transport system: Developing urban green space means design in individual spaces, making small parks inside the city. A good example is the green spaces of Singapore, calling "garden of the world". Otherwise, Government needs to implement more policies about transportation such as restricting private cars, improving the public transport system. In many high-density country in European, people travel by public transportation for a long journey and ride bikes in a short distance. This solution brings about many benefits such as saving energy, reducing green house gas emission, and management transportation more effectively. Another good example is transportation in Copenhagen (Denmark); 55 percent of population travel by bicycles, which results in reducing 90,000 ton CO_2 per year. For study area, cluster C3 needs to apply this solution because this zone has good-quality public transportation but lack of urban green spaces.

Improving tools for urban planning based on the criteria of sustainable development and green growth: Urbanization and population growth bring many difficulties such as the shortage of land, environment pollution, and traffic congestion. Yokohama also dealt with these challenges from the 1960s to the 1980s. Government changed in urban planning policies, which helped Yokohama escape from these difficulties. For instance, Government spent more money on improving transport system such as building underground highways, which may decrease air pollution. Not only that but also Government considered about traffic planning the most when they planned the city. As a result, Yokohama is one of the 11 most livable cities in the word, becoming an Ecological Economic City (Eco2 Cities) in the world. It brings many benefits to improving living standard and economical competitive advantages. Yokohama becomes a model of urban planning toward sustainable development and green growth.

In Vietnam, urbanization causes different changes in urban land use, which results in many new challenges and opportunities. Dealing with these problems, urban planning plays a significant role. Not only that but also it is the orientation and determination factors toward sustainable development. Decision-makers should seek scientific indicator system, which becomes the framework for assessing the effect of urban planning and the extent of sustainability in urban management:

- Carbon footprint is the indicator of green urban city;
- The compactness is a quantified indicator for assessing "the sustainability"; and.
- The relationship between the carbon footprint and the compactness should be used for assessing the sustainable development of urban areas.

References

Barton, H. (1990). Local global planning. *The Planner, 76*(42), 12–15.

Bourne, L. (1992). Self-fulfilling prophecies? Decentralization, inner city decline, and the quality of urban life. *American Planning, 58*(4), 509–513.

Dantzig, G., & Saaty, T. L. (1973). *Compact city: A plan for a livable urban environment* (244 pages). W.H. Freeman.

Elkin, T., McLaren, D., & Hillman, M. (1991). *Reviving the city: Towards sustainable urban development* (282 pages). Friends of the Earth.

Garcia, D., & Riera, P. (2003). Expansion versus density in Barcelona: A valuation exercise. *Urban Studies, 40*(10), 1925–1936.

Neuman, M. (2005). The Compact City fallacy. *Planning Education and Research, 25*(1), 11–26.

Owen, S. (1992). Energy, environmental sustainability and land-use planning. In: Breheny, M. (Ed.), *Sustainable development and urban form*. pp. 79–105.

Rickaby, C.A., Steadman, P., Barrett, M. (1992). Patterns of land use in English Towns: Implications for energy use and carbon dioxide emissions. In: Breheny, M. (Ed.), *Sustainable development and urban form*. pp. 182–196.

Williams, K., Burton, E., & Jenks, M. (2000). *Achieving sustainable urban form* (388 pages). Taylor & Francis.

World Bank (WB). (2010). *Cities and climate change: An urgent agenda* (92 pages). The World Bank.

Chapter 7
Agriculture Land Conversion and its Implications for Food Requirements and Farming in Vietnamese Northern Mountains

Nguyen An Thinh and Nguyen Phung Quan

Abstract Vietnamese Northern Mountains have seen urbanization with slow pace in relationship with the effects of the agricultural land conversion on local people's livelihoods, food and farming. This paper deals with the conversion of agricultural to built-up land in a case study of Tran Yen (Yen Bai, Vietnam). Satellite images are used to examine the relationships between land use land cover change (LULCC) and local livelihood diversification strategies during 1994–2016. Ethnographic fieldwork is combined with land-cover change mapping, which reveals drivers of LULCC. The findings provide scientific basis for policy makers in approaching complementary approaches to integrating livelihood policies to stabilize the lives of people in mountains. Support policies should be provided to households directly affected by these planning projects.

Keywords Agriculture Land Conversion · Food requirements · Farming · Object-oriented approach · Vietnamese Northern Mountains

1 Introduction

The conversion of agricultural land to nonagricultural land use supports infrastructure development, urbanization, and industrialization (Tan, Beckmann, Van Den Berg, & Qu, 2009). Not only the area of agricultural land that was used as residential land has declined but also the number of farmers has declined. As a result of reduced livelihoods from agricultural, people have gradually moved to new

N. A. Thinh (✉)
VNU University of Economics and Business, Vietnam National University, Hanoi, Vietnam

N. P. Quan
The Committee on Ethnic Minority Affairs (CEMMA), Hanoi, Vietnam

A. T. Nguyen, L. Hens (eds.), *Global Changes and Sustainable Development in Asian Emerging Market Economies Vol. 2*, https://doi.org/10.1007/978-3-030-81443-4_7

livelihoods (Firdaus & Ahmad, 2011). Over 80% of the smallholder farmers worldwide are food insecure, and their life depends on agricultural land (Cruz, 2010). About 75% of poor people live in rural areas and depend on agriculture. Their demand for food and production patterns changed when their livelihoods changed. In developing countries, challenges of food supply and distribution systems relate to rapid urban population growth (FAO, 2005). Land use land cover change (LULCC) could be detected by multiple means difference method. In the 2000s, many studies applied effective tool for detecting changes in the space-time model of the landscapes changes at various scales. The results of studies indicate that the area of agricultural land is decreasing, but the area of construction land is increasing considerably. However, all studies just carry out the image classification step and finding the variation of the area of the object class (Fazal, 2000; Mohammed, Elhadary, Samat, & Omar, 2015; Mukhopadhyay, Mukherjee, & Ghosh, 2013; Statuto, Cillis, & Picuno, 2016).

Since the start of economic renovation in Vietnam, socioeconomic and LULC has changed more and more significantly. With the emergence and development of industrial zones, economic zones, urban areas, and other nonagricultural projects, a large amount of farmland has disappeared. Studies on agricultural land conversion in major cities of Vietnam are conducted. Applications of GIS for detecting and analyzing spatial changes as well as quantifying results show the urban growth processes and its impact on the land use distribution (Tran, 2008). Landscape metrics are used to analyze agricultural conversion in periurban areas of Hanoi, Vietnam (Pham, Pham, Tong, Nguyen, & Pham, 2015). Remote sensing data and ethnographic fieldwork are used to examine how land use and land-cover change (LULCC) has occurred across three borderland provinces of Vietnam (Lai Chau, Ha Giang and Lao Cai). The relationships between LULCC and local livelihood diversification strategies are analyzed. The loss of agricultural land reduces traditional agricultural livelihoods and threatens food security. It brings opportunities for households to diversify their livelihoods and well-being sources (Turner & Pham, 2015). A few studies argue that the impact is negligible, when people's income is not much difference after the acquisition of agricultural land.

2 Methodology

2.1 Study Area

The research is carried out in Tran Yen district, which is located in the northeast of Yen Bai province, with an area of 62,857.99 ha and geographic coordinates from 21°31′48 N to 104°59′00″E. The North borders of Van Yen district, Ha Hoa district, and Phu Tho province are in the south, Yen Binh district, and Yen Bai city in the east, Van Chan district in the west. Tran Yen district center is Co Phuc town, 13.5 km away from Yen Bai city. Tran Yen also has the Hanoi—Lao Cai railway running through, with the Red River running in the direction of Northwest—South East, and Highway 37 and Highway 32C and 03 Provincial Road 163, 166, and 172 are land roads of district.

Under the industrialization and modernization policy, Tran Yen district has seen changes in agricultural production, in particular, and economic activities, in general. According to the report of 2015, the area of agricultural land decreased 1170.22 hectares; Nonagricultural land increased 1237.86 hectares. Labor in agriculture, although tending to decline, still accounts for a high rate of 78.39% (2015), while labor in industry, construction occupation, and commercial services in spite of the trend is quite slow, 6.25%. The economic structure has shifted sharply in the direction of decreasing the proportion of agriculture and forestry from 52.7% in 2005 to 41.0% in 2010, the industry and construction increased from 27.4% in 2005 to 35% 0% in 2010, and trade services increased from 19.9% in 2005 to 24.0% in 2010.

With the rapid restructuring of the economy toward reducing the share of agriculture and forestry, the area of agricultural land has been transformed into residential land, roads, factories, etc. On the other hand, with the proportion of labor in the agricultural sector still occupying a high proportion, the acquisition of agricultural land has a significant impact on the livelihood of the people here. For these reasons, we have chosen Tran Yen as the area to investigate and analyze the impact of agricultural land transfer on livelihoods of people.

2.2 *Data Collection*

The study is based on both primary and secondary sources of data and has an exploratory design. The primary data were collected through field surveys, whereas the secondary data were obtained from government and institutions through published and unpublished reports, records, and literature.

In order to detect the change of agricultural land and urban growth in the study area, we used two Landsat 5 TM satellite images acquired in October 1994, October 2004, and Landsat 8 October 2016. Because these days have low cloud cover <20%, the object is clearly visible. In addition, we used Google Earth, a map of land use status as a basis for classification.

2.2.1 LULCC Detection

The object-oriented approach uses information such as shape, contextual relationship of the objects, and thematic knowledge to classify remotely sensed images. This approach generates segmentation and classifies objects. The eCognition software provides solutions to select object features such as spectral statistics, texture, shape, and topological features. This approach is applied to obtain LULCC maps from Landsat images. Four types of LULC are classified: agricultural areas, built-up areas, water bodies, and forests. The objects are decoded by eye, and the subject has developed the key to image interpretation (Table 7.1).

Segmentation parameters such as bands, scales, color/shape ratio, and compactness/smoothness ratio are tested at different values. Segmentations and rule-based classifications are undertaken at different scales using the same band composition (bands 1–5, 7 Landsat 5; band 1–7 Landsat 8), color/shape ratio (0.3/0.7), and compactness/smoothness ratio (0.5/0.5). A segmentation value of 50 is chosen because it produces the most homogenous segments for spectral values and texture. The first set of criteria is used to classify water class. Then, three classes of agriculture and forest are classified. Max different is evaluated to assign built-up pixels.

Table 7.1 Image interpretation keys

No.	Objects	Images	Fields
1	Water bodies		
2	Built-up areas	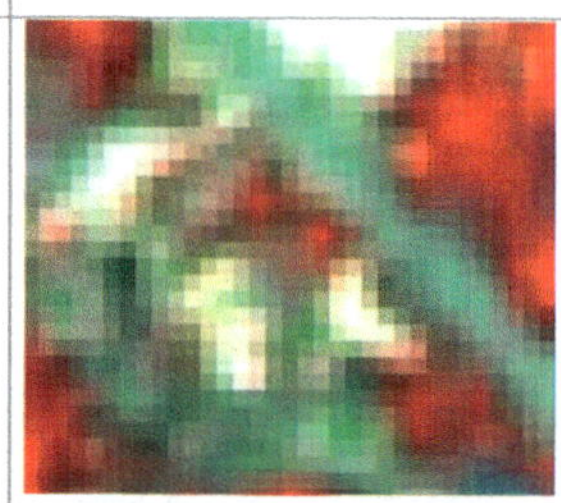	
3	Agricultural areas		
4	Forests	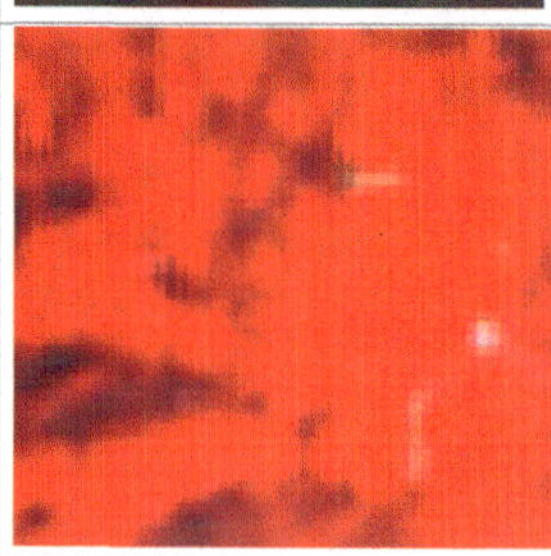	

2.2.2 Food Requirement and Farming Analysis

To analyze the changes in livelihoods under the impact of agricultural land use change, a top-down approach has been applied using structured interviews. According to the list of farmers in the acquired agricultural land and the list of industrial-oriented industrial development projects in the study area provided by the district People's Committee, two survey sites were selected: Bao Phat commune and Co Phuc town. 75 households have been selected in 2 hotspots for land acquisition: 1 is the location of the garment factory and 2 is the location for the resettlement site. The first interviews were conducted in December 2017, to assess whether questions in the questionnaire were appropriate. The second interview was conducted in February 2018. After reviewing the data collected, two sets of responses did not prove to be reliable and were rejected. Finally, 73 sets of feedback can be used as the basis for other data used and analyzed. Structured interviews focused on basic household member information, land conversion, transgenic livelihoods, change in income from livelihoods, and expenditure on households before and after land acquisition.

The whole sample was divided into two groups, corresponding to two different land acquisition areas, namely, land for the construction of the garment factory in 2007–2009, land for the 2016 road in Co Phuc town, and land for construction. Set up resettlement area in 2015 in Bao Dap commune. We have applied econometric methods to determine the degree of impact of agricultural land loss on the choice and the outcomes of household livelihood. Here, we divide the livelihood activities into four main groups: formal workers, informal workers, agricultural workers, and non-farm workers. To analyze livelihood determinants, we apply a sustainable livelihood framework with five sources of capital: natural capital, social capital, human capital, material capital, and financial capital. The change in funding is considered in terms of agricultural land loss. In this research, we use two out of five sources of capital.

Two sources of capital, which are natural capital and human capital, is used to analyze the effects of factors on the livelihood choices of households.

- Natural capital: covers the area of land owned/labor (100 square meter/labor) and lost agricultural area (when the area Cultivation is reduced, leading to increased idle time, the family is forced to move to new livelihoods.)
- Human capital: expressed by household size, dependency ratio (this ratio is calculated by the number of family members under 15 and above 59 years, divided by the total number of members in the age group 15 to 59 years) (affects participation in wage work), age, and level of household head (affect livelihood choices).

We also examine the impact of factors on livelihood outcomes by analyzing the remaining three sources, including the following:

- Physical capital: This is reflected in changes in family facilities between before and after land acquisition (reflecting changes in livelihood outcomes, livelihoods are also improved).
- Social capital: These are all social resources that can help people make a living. Social capital is expressed through participation in social organizations. Here,

we would like to see what kind of support groups is available to people when they lose agricultural land.

- Financial capital: in the form of access to formal and informal loans. Households receiving formal or informal loans may use this resource to generate income or consumption.

In addition, we analyze the variation in household expenditures between before and after land acquisition, to assess the impact of the acquisition of livelihoods and ultimately to influence to the life of the household.

3 Results

3.1 Land Use Land Cover Change

Table 7.2 shows that the most significant changes are an increase in *built – up* (6173.73 ha), followed by a decrease in *Agriculture land* (roughly 667.81 ha), a decrease in *bare soil* (166.72 ha), and a decrease in *forest* (5975.02 ha). Spatial patterns for these changes are visible in the LULC maps in 1994, 2004, and 2016 (Fig. 7.1). The level of urbanization and the variation in agricultural land use are determined according to image classification.

3.2 Changes in Food Requirement and Farming in Agricultural Land Conversion Areas

The research holds an interview in two locations in Bao Dap commune and Co Phuc town, with 61 valid votes at Bao Dap and 12 votes at Co Phuc. The interviewees were 78.1% male and 21.9% female, mostly in the 40–60 years. After synthesizing and processing forms, we have obtained the following results:

Table 7.2 Land use land cover change in the period 1994–2016

Land use land cover	Area in 1994 (ha)	Area in 2004 (ha)	Area in 2016 (ha)	2016–1994 Area change (ha)	2016–1994 Percent Change (%)	Annual Rate of change (%/year)
Agriculture	3282.18	2324.21	2614.37	−667.81	−20.35	−0.93
Built-up	4865.87	4616.68	11039.6	6173.73	126.88	5.77
Forest	53162.42	53879.60	47187.4	−5975.02	−11.24	−0.51
Bare soil	198.50	145.62	31.78	−166.72	−83.99	−3.82
Cloud shadow	0	745.25	418.53	418.53		
Water	1215.21	1005.46	1467.11	251.9	20.73	0.94
Total	62724.18	62716.82	62758.78			

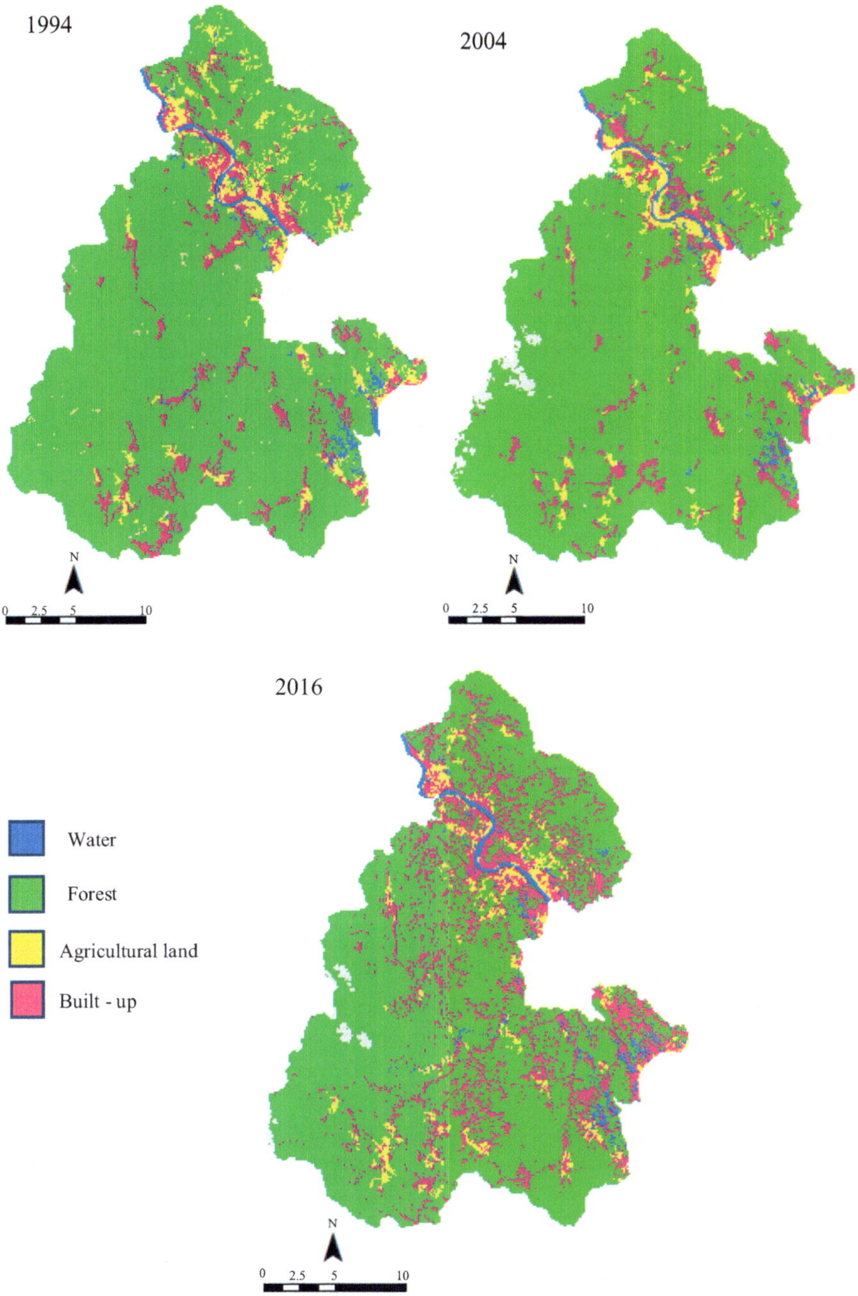

Fig. 7.1 Land use land cover in 1994, 2004, and 2016

Factors influencing the choice of livelihoods (Table 7.3).

* **Human capital:** Prior to land acquisition, 56 households accounted for 76.7% of the total labor force, while the remaining 17 were nonagricultural workers, accounting for 23.3%. But only 47 households have their main source of income based on agricultural activity. After land acquisition, laborers in households mainly work as hired laborers for individuals, factories, etc. The households still have land recovered or not continue to cultivate on that land but transferred to business activities. Nonagricultural activities, or farming to serve the needs of the household, but not agricultural products for sale, earn income. As a result, the number of households with major labor force in agriculture decreased sharply to 16 households accounting for 21.9% and the number of households having the main labor force in nonagricultural sectors increased to 57 households, accounting for 78.1%. Main income generating activities of households are also changing, with only one household having a main source of income based on agricultural activity, which is the conversion of agricultural land after harvesting rice to grow mulberry to feed silkworms. The rest of the main income sources are nonfarm-based activities, accounting for 98.6%.

* **Natural capital:** In the survey areas, there were three times of land acquisition. There were 61 households that have been cleared by 2015 to build resettlement area in Bao Dap commune. In Co Phuc town, there were 8 households that have been recovered land to build garment factory in 2007 and 7 households have been recovered land for road construction in 2016, of which 3 households of five in 2007 have been withdrawn (Table 7.4).

The results of the survey show that the area of agricultural land of households is significantly reduced due to land acquisition. Of which, group 1 has 6.47% average area of agricultural losing land corresponding to 124.1 m^2, group 2 decreased 46.5% and 1054.5m^2, and group 3 of average land area decreased to 92% corresponding to 369.3 m^2.

The average area of agricultural land after acquisition of household groups decreased, leading to a significant decrease in the area of agricultural land on agricultural labor. In which, especially categorized as group 3, the average area of agricultural land is 401.4 m^2/agricultural labor, reduced to only 32.1 m^2/agricultural labor, which indicates that significant means of livelihood of farmer households have been reduced considerably after land acquisition. This is a big shock to them.

Table 7.3 Change the labor force in both agricultural and nonagricultural sectors

	Before losing land				After losing land			
	Number of	% (Percent)	Main income earning	%	Number of employees	%	Main income earning	%
Agricultural labor	56	76.7	47	64.4	16	21.9	1	1.4
Nonagricultural labor	17	23.3	26	35.6	57	78.1	72	98.6

Table 7.4 The average area of agricultural land of the household groups conforming the land loss rate (Unit: square meters)

Group	Before losing land (2006)	After losing land (2017)	Compare 2006/2017 (+/−)	(%)
Group 1 (<30%)	1915.6	1791.5	−124.1	−6.47
Group 2 (30–70%)	2268.4	1213.6	−1054.5	46.5
Group 3 (>70%)	401.4	32.1	−369.3	−92.0

Using the method of monetary compensation, outside compensate damage, support for job change was also made by money. On the other hand, many people wanted to compensate and support by this form because having a large cash flow is each farmer‘s dream.

As such, natural capital source is automatically moved to financial capital source. Previous agriculture land is an important means of livelihood for farmers, now transferred as money accounts. Depending on the purpose of using money, financial capital could be convert into physical capital or human capital.

In addition, the survey showed that 18 households still have land but no further cultivation, of which 11 households have an area of over 1000 square meters but no more cultivation. Some other households, although, still have a lot of land but only partially farming. The reason is that after being recovered, the area of agricultural land is leveled to build factory or resettlement site, causing some sections of the canal system to be filled, affecting irrigation. The dry season is short of water and flood season flooded, leading to crop failure. Therefore, although there is still agricultural land, households can no longer farm and have to work as hired laborers or as workers in factories.

Factors affecting livelihood outcomes.

* **Social capital:** Social capital is considered by the number of groups and organizations that households participate in. Each family can participate in different groups. Of the interviewed households, 11 households did not participate in the group. Farmer association has the largest number of households with 40 households. The Red Cross has the smallest number of households with 0 households. All respondents indicated that they did not receive cash assistance or shift orientation after land acquisition from all stakeholder groups they participated in. This shows that social capital plays no role in the selection of livelihoods after land acquisition (Table 7.5).

* **Physical capital:** Depending on the amount of compensation received and the condition of each household, the difference in using the money varies for each household. It can be seen that compensation for land acquisition is largely used for the purpose of improving living conditions. Accordingly, 24 households built or repaired their house after land acquisition, accounting for 16.44%, and 5 households bought equipment, accounting for 6.85%. It can be seen that the recovery of material conditions of households has improved significantly. There is rotation from

Table 7.5 Purpose of using compensation money

Code	Purposes (1 household may choose multiple purposes)	Number of households
1	Saving	24
2	Built/repair of houses	12
4	Investment in production transformation	4
5	Land purchase	0
6	Buying equipment in the house	5
7	Daily expenses	31
8	Others (for children to school, repayment, medical expenses, etc.)	21

financial capital to physical capital, however, mainly in the means of living, a small amount of investment in means of production.

Using this money for the purpose of building houses, purchasing assets (physical capital), and other purposes, people's livelihood in the long run will be difficult. When they are poor, they may lose the opportunity to develop themselves, their family, and their social life.

With regard to the physical capital used for the community as the roads are being built, the future will create conditions for people to improve the conditions for exchanges with the outside environment.

* **Financial capital:** Table 7.6 illustrates some statistical descriptions of the sources and total value of loans borrowed by households in the last 2 years. About 27.4 percent of the total sample households reported having at least one loan from the rural credit markets (formal and informal). The participation rate in informal credit was higher than that in formal credit. The most important source of borrowing is borrowing from relatives, and the least used loans are loans from credit cooperatives. The most important source of borrowing is borrowing from relatives, and the least used loans are loans from credit cooperatives. With the average loan amount of VND 44,900,000, the largest loan and the smallest loan are VND 200,000,000 and VND 6,000,000, respectively. The purpose of borrowing from households is very different (Table 7.6).

Figure 7.2 shows that in the past 2 years, households have gradually concentrated on using loans to invest in production/employment, with credits from official credit. However, this number is still very small. The vast majority of loans are due to emergencies, unexpected families needing money for medical treatment, paying for schooling, or spending the river every day. So the loans are not too big, and the borrowed funds are mostly from family members or from neighbors. For these reasons, financial resources do not really play an important role in improving the livelihood of people after losing agriculture land (Fig. 7.3).

Table 7.6 Sources and total value of loans for households

Sources of loans	Number of households with at least one loan by source	Credit participation (%)
Formal credit	10	13.69
Commercial banks	7	9.59
Social policy bank	2	2.74
Credit cooperatives	1	1.37
Informal credit	15	20.55
Friends	2	2.74
Relatives	8	10.96
Neighbors	5	6.85
Any of the sources above	20	27.40

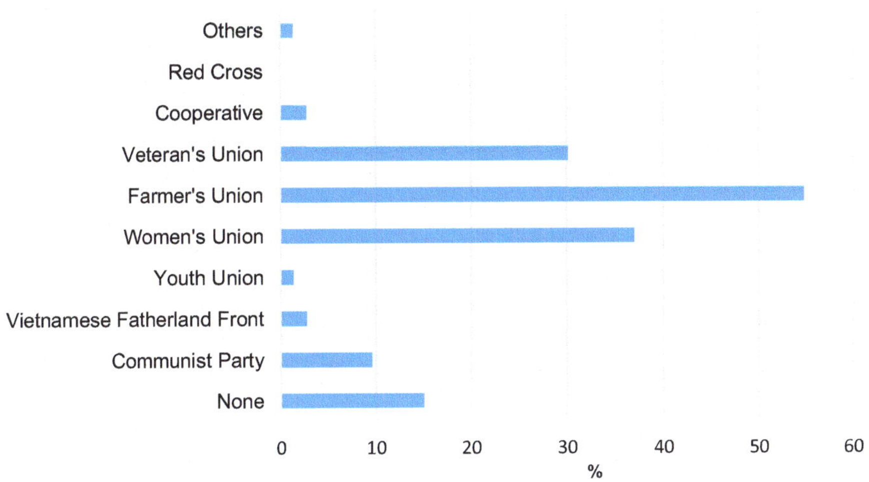

Fig. 7.2 Proportion of households with at least one member participating in associations/groups (Source: Own calculation from author's survey)

4 Conclusions and Discussion

Local development planning projects have led to changes in cover and agricultural land use. The study provided an overview of the life and livelihood of people in Tran Yen district before and after land reclamation. In the past, almost 100% of households were dependent on agriculture for their livelihoods. The land area recovered is proportional to the level of life. The larger the area of agricultural land is in the collection, the more it is affected. Farmers' livelihoods have changed drastically when there are nonfarm types or hired labor in garment factories. Although the factories have solved part of the problems of employment for households losing land, there are many shortcomings. One of them is the garment factory that employs only

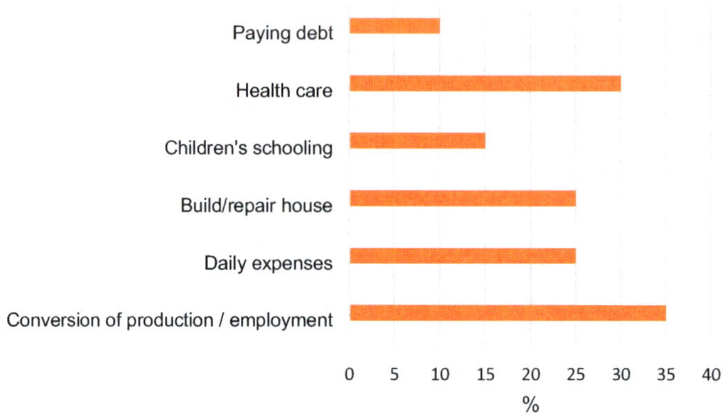

Fig. 7.3 Purpose of using money from loans from 20 households to borrow

women under 50 years old. As a result, most people still lack stable jobs and have to look for seasonal jobs. At the same time, the costs of buying food and foodstuffs of households increase considerably, putting pressure on their livelihoods, so most households think that their life becomes more difficult, The economy is declining compared to before the agricultural land is recovered. In addition, when conducting household interviews, we received some household complaints even though the agricultural land was not recovered but could not be cultivated due to flooding from the planning project. Our findings point to the importance of policy makers in approaching complementary approaches to integrating livelihood policies in the region to stabilize the lives of people. At the same time, support policies should be provided to households directly affected by these planning projects.

Acknowledgements This study was funded by the Vietnam Committee for Ethnic Minority Affairs (National Project code CTDT.39.18/16-20). The authors would like to thank the community members of the study area for participating in the research, who were most collaborative in completing the questionnaires and in providing discussion opportunities on the results.

References

Cruz, L. (2010). Responsible Governance of Land Tenure. An Essential Factor for the Realization of the Right to Food. Land Tenure Working Group Discussion Paper 15.

FAO (2005). Tracey-White, J. Rural-urban marketing linkages – An infrastructure identification and survey guide. FAO Agricultural Services Bulletin 161.

Fazal, S. (2000). Urban expansion and loss of agricultural land – A GIS based study of Saharanpur City, India. *Environment and Urbanization, 12*(2), 133–149.

Firdaus, G., & Ahmad, A. (2011). Impact analysis of urbanization on rural livelihood – An empirical study of an urban Centre of Delhi, India. *International Journal of Urban Sciences, 15*(3), 147–160.

Mohammed, K. S., Elhadary, Y., Samat, N., & Omar, N. Q. (2015). GIS and remote sensing techniques for measuring agriculture land loss in Balik Pulau region of Penang state, Malaysia. *Asian Economic and Social Society, 5*(2), 30–41.

Mukhopadhyay, A., Mukherjee, S., Garg, R.D, Ghosh, T. (2013). Spatio-temporal analysis of land use - land cover changes in Delhi using remote sensing and GIS techniques. International journal of Geomatics and Geosciences, 4(1): 213–223.

Pham, V. C., Pham, T. T. H., Tong, T. H. A., Nguyen, T. T. H., & Pham, N. H. (2015). The conversion of agricultural land in the peri-urban areas of Hanoi (Vietnam): Patterns in space and time. *Land Use Science, 10*(2), 224–242.

Statuto, D., Cillis, G., Picuno, P. (2016). Analysis of the effects of agricultural land use change on rural environment and landscape through historical cartography and GIS tool. Agricultural engineering, XLVII(468): 28–39.

Tan, R., Beckmann, V., Van Den Berg, L., & Qu, F. (2009). Governing farmland conversion: Comparing China with the Netherlands and Germany. *Land Use Policy, 26*(4), 961–974.

Tran, T. V. (2008). Urban expansion and loss of agricultural land in the north of Hochiminh city: A GIS and remote sensing approach. *VNU Journal of Science, Earth Sciences, 24*, 104–111.

Turner, S., & Pham, T. T. H. (2015). "Nothing is like it was before": The dynamics between land-use and land-cover, and livelihood strategies in the northern Vietnam borderlands. *Land, 4*, 1030–1051.

Chapter 8
Urban Exclusion: Theoretical Approaches and Emerging Trajectories for Vietnam

Nguyen Dang Dao

Abstract There are only a few research projects focusing on the effects of urbanization on Vietnamese society, particularly social exclusion and inclusion, while urban exclusion is one of the most pressing matters in Vietnam nowadays. Thereby, it leads to many sustainable development issues that attract the attention of social science researchers and policy-makers. Among sustainable issues related to urbanization such as urban-rural relations, urban conflicts, and environmental justice, urban exclusion has received the least attention, although it is currently one of the research centers of urban studies in anthropology, sociology, economics, and development studies. This paper aims to gather comprehensive theoretical approaches and case studies from the works of Robert Park, Jehoel-Gijsbers, Loïc Wacquant, Philippe Bourgois, and other scholars. After building a theoretical framework, this paper examines studies in Vietnam, explains similarities, differences, or complementary to world studies, identifies research gaps, and suggests suitable research directions for the upcoming research projects. The result suggests that it is worthwhile to strive for high-quality research about the urban exclusion of vulnerable groups regarding Vietnam's evolving efforts toward sustainable development.

Keywords Urban Exclusion · Urbanization · Social Exclusion · Theories · Sustainable Development

1 Introduction

This paper evaluates existing research documents that focus on urban exclusion in various cities in the world. Exclusion (or social exclusion) is a process, a condition, and a result of a combination of social inequality, economic and power disadvantages, relegation and denial of systematic personal or community rights,

N. D. Dao (✉)
Vietnam Academy of Social Sciences, Vietnam National University, Hanoi, Vietnam

A. T. Nguyen, L. Hens (eds.), *Global Changes and Sustainable Development in Asian Emerging Market Economies Vol. 2*, https://doi.org/10.1007/978-3-030-81443-4_8

opportunities, and resources. In developing countries, which are experiencing the rapid growth of urbanization, new social issues have emerged and vulnerable groups have been increasingly excluded from the mainstream groups. This requires social science scholars, including economics, sociology, anthropology, urban studies, and development studies, to develop a theoretical framework that can be applied to field research in the Global South.

1.1 Urban Exclusion: An Overview

In the early years of the twentieth century, French Sociologist Émile Durkheim started to evaluate how social order and stability could be sustained in a society where social distortions and a transition from a peasant society to industrial and modern society happened simultaneously. This concept first appeared in policy discourses in France, and the application of other European countries later had an increasing impact on the analysis of social disadvantages in Europe in the few past decades (Aasland & Fløtten 2001). Social exclusion is then used in policy analysis and planning in developed countries—where the process of urbanization, globalization, and migration was underway (Francis, 1997).

The origin of urban social exclusion, thereby, is considered as a consequence of economic restructuring and globalization in the world. The structure of urban labor, therefore, has changed. As the needs for manual workers reduced, while the demand for service workers soared up, unemployment, social exclusion, and marginalization appeared (Mustered, 1998). There were some dwellers who benefited from these changes; however, the rest had not fully been involved in the cultural, social, and political system (Sassen, 1991).

1.2 Urbanization in Vietnam

In Vietnam, for the past 20 years, not many researchers have been interested in urban exclusion. It was only when the process of industrialization and modernization led to a prominent urbanization trend, the interest of researchers shifted to study the issues of the cities in the cities. Vietnam is currently experiencing one of the most powerful urban transformation processes in the world. This is closely related to the socioeconomic reforms established in the 1980s, which gradually liberalized the economy and accelerated the rapid transition to the urban society through the processes of urban-rural migration combined with the transfer from the agricultural sector to industrial and service sectors. Over the next 25 years, cities and towns are expected to grow at an average rate of 6% per year, soaring the rate of Vietnam's urban population from one third to half (OECD, 2018).

The city, in the twentieth century, was not the center of socioeconomic life as the majority of Vietnamese people were farmers and they primarily lived in rural areas. As such, the process of urbanization has led to the demographic and social change and affected the rural-urban relationship with the trend of civilization and modernization of life. The scale of migration from rural to urban areas has been expanding day by day and thereby attracting the attention of researchers, especially in big cities like Hanoi, Ho Chi Minh City, Da Nang, and Can Tho (Danielle, 2010)

Urban development was initially not a priority in the government agenda, yet gradually narrowing the development gap between rural and urban areas and relocating population from crowded cities to new economic zones in rural and mountainous areas have become an important strategy. In the late 80s of the twentieth century, the rapid development of urbanization caused the rapid flow of rural to urban migrants. Along with the urbanization process, the relationship between rural and urban areas has become inseparable and moving from rural to urban areas has become the survival strategy of many farmer households.

Before the Doi Moi period, the pace of urbanization in Vietnam was slow. Yet, the situation changed from 1999 after the Doi Moi took place. Until 2015, Vietnam has 629 cities, and this number has increased to 755 in 2010, including 2 cities, Hanoi and Ho Chi Minh (Pham, 2013).

According to research named "Facing The Urban Transition in Hanoi: Recent Urban Planning Issues and Initiatives" of the Center for Cultural Studies—Urbanization and the Academy of Social Sciences, the process of urbanization of Vietnam, although taking place in many forms, can be divided into two main categories:

- Communities experience a planned urbanization process;
- Communities experience spontaneous urbanization (often referred to as slums).

Since the 2000s, the government has embarked on a large-scale urbanization plan with the aim of promoting economic development and urbanization. This means that policy-makers tend to revolve around converting agricultural land to urban land for industrial development, and residential building, as well as replanning slums by relocating slum dwellers into public housing quarters. In all cases, people face many challenges, namely, (1) seeking new means of living, such as employment, food access, new family time management habits, and travel patterns; (2) competing on scarce resources between rural households and land use rights; and (3) adapting to rapid change of land use, development models, infrastructure, and legal framework.

In particular, the areas in transition, or periurban areas, are potential sites for researchers to learn about Vietnam's economic, political, social, and cultural changes. Serving as a bridge between the traditional rural environment that once dominated the economic and sociocultural history of the country and the urban environment, this suburban interface allows researchers to discover the urban integration (Luong, 2018).

2 Methodology

Secondary research: summary and collation of existing academic materials to find the research gap and build a theoretical research. Combine and compare the approach of academics in the U.S., the U.K., and Europe.

Source:

Information from the Internet (newspaper, online library).

Books.

Information obtained from government research institutes.

A Theoretical Framework and different approach for urban exclusion.

According to the study of Gerda Jehoel-Gijsbers in the study "Explaining Social Exclusion: A theory model tested in the Netherlands" of the Dutch Institute for Social Studies, social exclusion can be related to nonmaterial characteristics (relationship dimensions) and material aspects (distribution dimension) (Jehoel-Gijsbers, 2007). The causes of social exclusion may lie at the collective level, but individual characteristics and behavior also contribute to the theoretical framework. Social exclusion is often divided into two main dimensions:

Economic-structural exclusion: focus on a distributional dimension with an Anglo-American approach.

Socioculture exclusion: focus on the aspect of the relationship with the French school approach.

As for the first approach, American scholars focus on two factors, which are material (income and goods) and nonmaterial (social rights). This approach stems from material deprivation, especially basic and material needs, debts and payments for housing fees, living expenses, etc. Besides, the social exclusion also involves inadequate access to government services such as health care, education, housing, and legal aid, management, social services, debt subsidies, employment, social security, and some social services such as banking or insurance.

The second approach focuses on two factors, namely, social integration, including social relations and social networks, and cultural integration, including values and norms. Nonsocial inclusion refers to the lack of participation in formal and informal social networks, including recreational activities and inadequate social support (Table 8.1). Combined with the risk factors mentioned above, Jehoel-Gijsbers theorized social exclusion through the following table (Jehoel-Gijsbers & Vrooman, 2004):

Learning from two abovementioned approaches, researchers focusing on urban social exclusion have developed their own theoretical framework for social exclusion in the cities. In the context of overloading of theories, policy analysis, and practice, empirical studies on urban exclusion can be divided into two trajectories. The first one is the Human-based and location-based approach. This human-based approach focuses on the impact of policies on the living conditions of disadvantaged groups, regardless of their geographical location. It concerns multidisciplinary issues such as equal access to everyone and the ability to benefit from equal legal, education, training, health, and market systems. Besides, the location-based

Table 8.1 Factors of Social-Culture exclusion

Microlevel (individual and household)			Macrolevel		Intermediate level
Background (Fixed)	Background (changeable)	Characteristics of social exclusion	Social Development	Government	Government agencies, businesses, and citizens
Gender, age, civil status, family composition, social context, and ethnicity	Adaptability, health, education, position in the labor market, and income	Economic/ structural shortcomings and cultural/ social shortcomings	Economic recession, individualism, bureaucracy, **urbanization**, emigration, and aging of the population	Policy gaps, supply gaps, access gaps, and getting supplies	Ineffective policy implementation, waiting times, financial barriers, risk-taking options by employers or banks, and discrimination

approach also studies the issues mentioned above and considers spatial change. This means that while analyzing any issue, academics have to acknowledge that disadvantaged groups in different parts of the world will be affected in various ways (Atkinson, 2000). Although theoretically, these two approaches are similar in many ways, in fact, the human approach recognizes exclusion from a human-based viewpoint, while the location-based approach emphasizes more on the interaction between exclusion and urban space. The second approach called the Area effects approach. Associated with a location-based approach is the debate around regional and neighborhood influences on urban exclusion. This debate focuses primarily on recognizing the size of the neighborhood as the primary analytical scale and research on how the location affects the conditions and opportunities of residents (Atkinson & Kintrea, 2001). This school argues that living areas can be the main cause or at least a decisive factor that contributes to various forms of exclusion (Galster & Hedman, 2013).

Based on the proposals and evidence contained in various studies, Atkinson has distinguished six types of area effects (Table 8.2). In each of these types of influences, specific mechanisms that link individuals/households to the region have been pointed out (Atkinson, 2000).

3 Result

Research on marginalized and excluded groups in urban Vietnam is still limited as urban researchers currently are focusing on Urban Migration, Urban Poverty, and Marginalizations. Thus, in Vietnam, there is a huge research gap for social sciences and humanities researchers.

Table 8.2 Six types of effects in Area effects approach

Area effects	Mechanisms
Concentration	– Focus on service
	– Many households live together
Location	– Labor market
	– Housing market
	– Public housing allocation and accessibility
	– Geographical isolation
Milieu	– Social network
	– Communication and context leading to deviations and mistakes
	– Associated activities
	– Aspects of daily life
Socialization	– Education
	– Childcare
	– Friendship
	– Isolation
	– Division
Physical	– Environment build
	– Housing quality
	– Quality of public areas, for example, parks
Service	– Education
	– Receive and deal with people who have problems in problem areas

3.1 *Urban Exclusion in the World: A Literature Review*

3.1.1 Urban Migration

In order to understand urban studies and urban migration in specific, conducting a literature review of research of Robert Park is necessary. Robert Park lived in a society where the transformation from traditional to modern, from culture to civilization, and from rural areas to urban density occurs simultaneously. Migration at that time was considered as the primary way to move forward modern and civilization. The origins of urban anthropology, as well as urban sociology, began with Park's observation of the transition from rural to urban areas with the rate of urbanization increasing seven-fold of the US after the Civil War in 1865. Park, when analyzing the two trends of urbanization and migration, said that at that time, the United States had only two types of people: those who went to the city and those who had not (Park, 1937). Nearly a decade after introducing "The Marginal Man", Robert Park emphasized that marginalized people often suffer from negative circumstances, such as inconsistency, insecurity, and emotional instability. However, Goldberg argues that not all marginalized people suffer from the same negative consequences. According to him, they could avoid being excluded by developing and engaging in a marginal culture created by people who share the same situation. In a new subculture, marginalized people will no longer be marginalized, and they will become normal (Goldberg, 2012).

3.1.2 Urban Poverty

Jonathan Greene in "The Contradictory Dynamics of Urban Poverty Management study: Homelessness, the State, and Struggles for Social Inclusion" also affirmed that since the 1990s, urban scholars have begun to focus on analyzing punitive practices introduced by local authorities to eliminate homelessness and other obvious manifestations of poverty and marginalization in city streets.

Although poverty can be observed in many different forms and caused by various factors, all forms of poverty are able to be described through the concept of exclusion. A poor person will always be disadvantaged in one or more dimensions, while the sociocultural context and personal awareness will determine in what circumstances and how many dimensions a person will be excluded (Franz, 2013).

Oscar Lewis introduced an important theory of "poverty culture" in the 1950s. It highlighted the social exclusion situations of the poor people at that time. Also, in this study, he examined the daily activities, behaviors, and thoughts of the poor and argued that the poor have their own culture, poverty culture. Poverty culture emerged from the following series of conditions that Lewis classified according to three levels. (Lewis, 2002)

At the individual level: poverty culture includes a strong sense of marginalization, hopelessness, laziness, impotence, feeling of inferiority, and dependence;

At the family level: poor culture represents the absence of childhood; early start of sex or marriage, high rates of abandonment of wives and children, a tendency for families to lack privacy, competition for limited quantities of goods, and attention from mothers;

At the community level: lack of participation and effective integration with larger social organizations, low wages, chronic unemployment, lack of jobs, no savings, no wage reserves, and real in-house.

3.1.3 Urban Marginalization

In order to comprehensively understand where social exclusion originated from and in what form does social exclusion occur in cities, it is crucial to understand the process of marginalization. Although marginalization is often applied in the field of urban research, researchers have not found a suitable methodology to fully measure and analyze urban marginalization (Dunne, 2005).

The concept of marginalization was also first introduced in 1928 with an article by Robert Park titled "Marginal Man", in which Park describes the struggles of migrants to adapt to dominating cultures. (Park, 1928). While debates initially focused on issues of race and ethnic relations, this concept was later extended to career, gender, and scientific innovation (Goldberg, 2012). Other scholars also sought to expand this concept to include anyone excluded from one or more social groups. For example, Deegan defines a marginalized person as a person who is self-aware of his/her culture, identity, status, and experiences in the world and has access to resources but cannot socialize and become a part of the mainstream. Marginalized individuals or groups are people standing on the sidelines of social, political, economic, ecological, or physiological systems. They are not able to access public services and resources and have freedom of choice (Gatzweiler, 2011).

However, there are also scholars who have pointed out that entering individual groups in similar marginalized situations can share a common culture or identity. Goldberg is the first to argue that individuals who are marginalized or excluded from society can develop a marginal culture/subculture. Therefore, a subculture may be marginalized, but individuals may not feel marginalized in this new subculture. (Goldberg, 2012) In some cases, in fact, a group may choose to stay out of the culture to preserve their own identity and independence (Scott, 2009). A case study of the Haitian community in Guadeloupe, Brodwin pointed out that a group's self-creation is shaped by the experience of marginalization in a specific time and place, such as the culture of urban poverty or culture of slums and neighborhoods (Brodwin, 2003).

3.2 New Trajectories for Urban Exclusion Studies in Vietnam

By evaluating the literature review and the current situation of urban areas in Vietnam, it can be witnessed that the research on urban exclusion in Vietnam still has many gaps to be filled. New research directions that social science researchers include economics, sustainable development, anthropology, and sociology can focus on are Urban Everyday Politics, Ruralization Urban, and Urban Policy.

3.2.1 Urban Policy

The dominant research direction in Vietnam is the study of the impact of urban policies on vulnerable groups, including rural-urban migrants, slum-dwellers, people living in the suburbs, and the urban poor. For example, Erik Harms, after living in the most famous metropolis in Vietnam, Ho Chi Minh City, finally came up with these two books, namely, Saigon's Edge and Luxury and Rubble. For Luxury and Rubble, he analyzed Ho Chi Minh City's development master plan based on 17 months of fieldwork in Phu My Hung and Thu Thiem. If Phu My Hung represents esthetics, wealth, and affluence, Thu Thiem is the area with 14,600 households being dismantled to serve the government's urban development plan. Erik focuses on the impact of how the urbanization process has affected the vulnerable group in Thu Thiem when their houses are removed and replaced by commercial centers and villas. Nevertheless, the main conflict in his book is not about the people at Thu

Thiem who oppose the government's urban development idea, instead, people who are very supportive and eager to see the development of Ho Chi Minh city. Thu Thiem residents felt that they were being controlled and there was an injustice in distributing the compensation, which resulted in a much smaller amount of money they should have been received. Although they support the idea of a civilized, polite city, real estate companies and governments did not engage them in the grand plans and policy (Harms, 2016). Likewise, by evaluating various urban policies in cities of Vietnam, academics like Erik Harms can find the shared pattern and point out the differences.

3.2.2 Ruralization Urban

Another direction for Vietnamese researchers is ruralization urban process. There are many studies on social groups in cities that demonstrate the importance of family networks, kinship, and close and personal links between people living in cities and their hometown (Pocock, 1960). Oscar Lewis noted that when looking at the relationship between rural and urban areas, especially in countries in the Global South like Vietnam, it must be noticed that this relationship is interrelated and interdependent. Furthermore, difference between urban and rural areas, cities, suburbs, and rural areas has been gradually disappearing due to the inner migration flow and the lifestyles of migrants. The line between the rural lifestyle and urban lifestyle is being blurred. This phenomenon possibly leads to excluded situations for rural-to-urban migrants. They can be excluded by the city dwellers who live in the same neighborhood because of differences in cultures and lifestyles.

3.2.3 Urban Everyday Politics

The third research trajectory that is suitable for cities in Vietnam is urban everyday politics. Benedict Kerkvliet, with "Daily Politics in Our Rural and Social Societies", is one of the leading researchers that set the foundation for this trajectory. Politics is the act of control, distribution, production, and use of resources, including power, money, land, and education. Collaboration, cooperation, bargaining, agreement, conflict, or competition are acts related to production, distribution, and use of resources. It recognizes the fact that politics can emerge and exist in all areas and aspects of life. Different from the Official politics that focuses only on a small part of politicians, leaders, and policymakers and the way they organize, implement, and adjust fecal policies, everyday politics refer to the way normal citizens accept, comply, adjust, and protest against standards, regulations, powers, production, or allocation of resources (Benedict, 2008). Everyday politics, thereby, often refers to individual behaviors, which can take place everywhere and moreover, done by people who do not consider their actions to be political. Hence, it also intertwines in the activities of living, producing, and distributing resources of individuals and small groups and in the interaction between them and those of the same class or classes (Ball & Guy, 1993).

Everyday politics can exist in many different forms, but generally scholars classify them into four categories: support, submission, adjustment—avoidance—and resistance. Almost most studies of urban exclusion today focus on daily resistance. Everyday political research will highlight the social exclusion of urban residents with common activities and policy making in Vietnam. Marginalized and excluded group in urban areas in Vietnam often chooses their own forms of resistance in accordance with the law in Vietnam. Therefore, Vietnamese researchers could concentrate on the resistance of city dwellers who are excluded from mainstream groups in Vietnam.

4 Conclusions and Discussion

The study of urban social exclusion has emerged since the nineteenth century in Europe when the process of urbanization began to appear and thrive in the wealthier countries in the Global North. There are many theories and approaches to urban exclusion coming from different schools in England, America, and France, and these theories can be applied to studying cities in the Global South. However, in Vietnam, research on urban exclusion has not yet really developed, as social science researchers, especially scholars in anthropology and sustainable development studies, pay more attention to primitive communities living in remote areas. Therefore, this article suggests three new research directions for researchers in Vietnam, including urban policy, ruralization urban, and urban everyday politics. Those directions are among the most suitable to the context of urbanization in Vietnam.

Given the above analysis and suggestions, it can be seen that the theoretical framework of urban exclusion still has not been constructed in Vietnam. Urban exclusion combines urban studies and social exclusion, and it requires a separate approach and framework. At the same time, Vietnam's urbanization rate is very fast, reaching 40% in 2019, with 50% of the population living in urban areas (Kiet, 2019). New issues in urban areas of Vietnam have been appearing consistently in the past few years and many more will attract the attention of academics. Therefore, the research gap will be increasingly expanded and the author would like to invite more academics to contribute to the debate on urban exclusion in Vietnam.

References

Aasland, A and T. Fløtten (2001). Ethnicity and social exclusion in Estonia and Latvia. *Europe-Asia Studies, 53*(7), 1023–1049.

Atkinson, R. (2000). Combating social exclusion in Europe: The new urban policy challenge. *Urban Studies, 37*(5), 1037–1055.

Atkinson, R., & Kintrea, K. (2001). Disentangling area effects. Evidence from deprived and non-deprived neighbourhoods. *Urban Studies, 38*(12), 2277–2298. https://doi.org/10.1080/00420980120087162
Ball, A., & Guy, B. P. (1993). *Politics, power and authority, modern politics and government* (pp. 19–37). Macmillan.
Brodwin, P. (2003). Marginality and subjectivity in the Haitian diaspora. *Anthropological Quarterly, 76*(3), 383–410.
Benedict, J. (2008). Everyday politics in peasant societies (and ours). *The Journal of Peasant Studies, 36*(1), 227–243. https://doi.org/10.1080/03066150902820487
Danielle, L. (2010). Facing the urban transition in Hanoi: Recent urban planning issues and initiatives., Center of Urbanization Culture Society.
Dunne, J. (2005). Marginality. A conceptual extension. In R. M. Dennis (Ed.), *Marginality, power and social structure: Issues in race, Class and Gender Analysis* (pp. 11–27). Elsevier.
Franz, G. (2013). *Marginality—A framework for analyzing causal complexities of poverty*. Springer. Retrieved October 17, 2019, from http://link.springer.com/chapter/10.1007-94-007-7061-4_2.
Francis, P. (1997). *Social capital, civil society and social exclusion, in development theory and practice: Critical perspectives*. Palgrave.
Kiet, A (2019). *Vietnam's urbanization rate in 2019 to reach 40%, HanoiTimes*. Retrieved August 18, 2019, from http://www.hanoitimes.vn/news/viet-nam/2019/02/81e0d34e/vietnam-s-urbanization-rate-in-2019-to-reach-40/
Galster, G., & Hedman, L. (2013). Measuring neighbourhood effects non-experimentally: How much do alternative methods matter? *Housing Studies, 28*(3), 473–498. https://doi.org/10.1080/02673037.2013.759544
Gatzweiler, F. (2011). *Marginality: Addressing the root causes of extreme poverty, ZEF working paper no 77*. Center for Development Research, University of Bonn, Bonn.
Goldberg, C. (2012). Robert park's marginal man: The career of a concept. *American Sociology Laboratorium, 4*(2), 199–217.
Harms, E. (2016). *Luxury and rubble: Civility and dispossession in the new Saigon* (pp. 154–177). University of California Press.
Jehoel-Gijsbers, G. (2007). *Explaining social Exclusion A theoretical model tested in the Netherlands, The Netherlands*. Institute for Social Research Journal, ISBN No 9789037703252.
Jehoel-Gijsbers, G., & Vrooman, C. J. (2004). *Social exclusion in the Netherlands* (pp. 5–11). Social and Cultural Planning Office.
Lewis, O. (2002). *The culture of poverty* (pp. 269–278). Waveland Press.
Luong, H. (2018). The changing configuration of rural–urban migration and remittance flows in Vietnam. *Sojourn: Journal of Social Issues in Southeast Asia, 33*(3), 602–646.
Mustered, S. (1998). *Urban segregation and the welfare state: Inequality and exclusion in western cities*. Londo: Routledge.
OECD (2018). *OECD urban policy reviews Vietnam 2018: Policy highlights*.
Park, B. (1928). Human migration and the marginal man. *American Journal of Sociology, 33*(6), 881–893.
Park, R. (1937). *Cultural conflict and the marginal man in Everett V Stonequist, The Marginal Man, Introduction.* New York: Charles Scribner's Sons.
Pocock, D. (1960). Sociologies: Urban and rural. *Contributions to Indian Sociology, 4*(3), 63–81.
Pham, T. (2013). Urbanization context: Anthropological research issues in Vietnam. *Journal of Social Sciences, 4*(176), 23–37.
Sassen, S. (1991). *The Global City: New York, London, Tokyo*. Princeton, NJ: Princeton University Press.
Scott, J. (2009). *The art of not being governed: An anarchist history of upland Southeast Asia*. Yale University Press.

Chapter 9
Applying Structural Equation Modeling (SEM) to Analyze Factors Affecting the Entrepreneurial Intention of the Students of Vietnam National University, Hanoi

Nguyen Cam Ngoc and Nguyen The Kien

Abstract This study aims at identifying factors affecting Vietnam National University, Hanoi, students' entrepreneurial intention. The samples are 300 students from the VNU University of Social Sciences and Humanities, VNU University of Economics and Business, and VNU University of Technology, Hanoi. Based on the application of SEM model, the results show six factors affecting students' entrepreneurial intention including confidence, entrepreneurship, education, capital, technological skills, and work skills. The higher the factors, the higher the entrepreneurial intention.

Keywords Structural Equation Modeling (SEM) · entrepreneurial intention · students · Vietnam National University, Hanoi

1 Introduction

Small and Medium Enterprises (SMEs) are considered as an important contributing factor to the development of the national economy and well-being as well (Hoelscher & Elango, 2012; Hatten, 2006). Developing countries confront the poverty, unbalanced increase, and unemployment, in which, unemployment is asserted as the most serious issue (Dev & Mahajan, 2003). Entrepreneurship plays a significant role in fostering the economy and creating much more jobs for workers (Moica et al.,

N. C. Ngoc
VNU University of Social Sciences and Humanities, Hanoi, Vietnam

N. T. Kien (✉)
VNU University of Economics and Business, Hanoi, Vietnam

A. T. Nguyen, L. Hens (eds.), *Global Changes and Sustainable Development in Asian Emerging Market Economies Vol. 2*, https://doi.org/10.1007/978-3-030-81443-4_9

2012). According to the statistics of Vietnam Chamber of Commerce and Industry in 2016, new businesses, especially small and medium enterprises, contribute to nearly 50% of GDP and attract more than 90% of new labor. Therefore, the development of startup activities will be a driving force to help solve employment problems, reduce unemployment, and turn Vietnamese economy to be more dynamic at the same time.

VNU, Hanoi, has also conducted many activities in recent years to provide advice on starting a business for students. According to a study of iPrice group and 500 Startups Vietnam, in 2017, VNU was one of the national leading training units in the number of student startups. In March 2017, the President of VNU Hanoi signed a decision to establish the Center for Enabling Startups and Knowledge Transfer with the mission of promoting the commercialization of science and technology products of VNU and other research institution as a bridge to create value from scientific research products and promote entrepreneurship development in VNU and Hanoi. The startup movement among students has been quite dynamic; one of the successful startups is the case of a student of the VNU University of Economics and Business, Nguyen Ngoc Quan, when starting a business called Wine Co., Ltd., Vietnam, in 2016. Starting from being a CEO of a media company when he was very young, just after 2 years of graduation, Quan's company has now grown into a medium-sized one with about 100 employees. The company operates in marketing and consulting brand strategy and provides communication solutions for many major partners such as: Dairy cow by BOO, Lincup by Lintimate, Vincom Ice Rink, The Elephant Company, Hailecao Studio, etc. Another example is a startup search project of the University of Technology—VNU in 2018 when the student group created WORKSVN in two forms: website and mobile application with the goal of connecting businesses with students. According to statistics from the Google Play app, up to now, the WORKSVN app has had more than 10,000 downloads and an average of about 2000 visits per day with user appreciation.

However, a large part of students still have poor initiative in finding jobs as well as starting a business. The question is what factors affect the VNU students' entrepreneurial intention? In Vietnam, SEM model is currently applied in many fields, but the number of studies into students' entrepreneurial intentions using SEM model is quite limited. Therefore, this study applies SEM model to analyze factors affecting VNU students' entrepreneurial intention, thereby making policy suggestions to further promote the startup movement among VNU students, in particular, and students from the whole country, in general.

2 Methodology

2.1 Literature Review

Intention refers to a cognitive situation that engages personal experience for a specific purpose or behavior (Anderson and Gerbing, 1988). The theory of Planned Behavior (TPB) suggests that entrepreneurial intention is the result of the intent and actions of brave individuals described as modern-day heroes (Ajzen, 1987). Based on the analytical framework of GEM, Micozzi and Lucarelli (2016) analyzed data from 37 economies and concluded that gender and confidence in one's skills affect personal mood and strengthen the individual's entrepreneurial intention. Van Gelderen et al. (2008) showed two important factors that influence an individual's entrepreneurial intentions based on market awareness and self-improvement. Ooi (2015) studied the impact of educational business on Malaysian students with 235 final-year students as samples collected from four schools in northern Malaysia show the important role of universities and colleges in stimulating and nurturing entrepreneurship intention (EI) among graduate students. Sahinidis, Vassiliou, and Hyz (2014) indicated that there is a robust linkage between the Personal Attraction and EI, alike to the discovery with other sample category. They also found a solid correlation between Perceived Behavioral Control or Self-Efficacy and EI. At last, anew as conclusion in the previous studies, Social Norms and Valuations were asserted to influence both directly and indirectly on EI. Patrick et al. (2019) reviewed the EI literature in the research context of startup firms among university and college students. The current research also suggests a model to raise the insight into the factors impeding e students from launching a company while studying. This research model, which focuses on EI among undergraduate students, upgrades the theory of Psychological School of Thought and Social Networking Theory in EI literature.

In Vietnam, Le (2007) studied the startup decision process of young Vietnamese entrepreneurs. The study was conducted with a sample of 159 entrepreneurs who established businesses under the age of 30 years during 2000–2006. The research results indicate that young people tend to make the decision on starting a business when meeting all three groups of factors, namely, personal qualities, accessibility to resources, and business opportunities. In addition, the research also shows that entrepreneurial thinking is strongly formed and develops over time and is influenced by many environmental factors. Do (2016) had the study "Factors affecting EI of Business Administration students at the University of Labor and Social Affairs (Ho Chi Minh City campus)." The study used the entrepreneurial potential model of Krueger Jr and Brazeal (1994) and the TPB of Ajzen (1991). The research gathered 315 students at the University. The research results indicate five factors influencing students' entrepreneurial intention, namely, higher education and training, experience and experiences, family and friends, personality, and capital. Do et al. (2017), in the study "Analysis of factors affecting the start-up intention of students of Binh Duong University," conducted a survey of 250 samples and applied the linear structural equation model (SEM) for analysis. The results show that the intention of students to launch a company is influenced by seven factors, which are directly affected by self-awareness, education and training, family, and social environment. Nguyen and Nguyen (2016), in the study "Factors affecting students' start-up intention in Tra Vinh University," conducted a survey of 405 students in different disciplines; the results show that factors affecting entrepreneurial intention through the factor of confidence in entrepreneurial feasibility reflected in teaching activities, extracurricular activities, people's opinions, and business interests, which directly affect confidence. The higher the confidence about the feasibility of launching a company, the higher the student's intention to launching the company.

2.2 Analysis Methods

In this study, the authors applied the method of structural equation model (SEM), which used SPSS 22.0 software and AMOS version 20 software, via six steps: Cronbach's Alpha reliability analysis, Exploratory Factor Analysis (EFA), Confirmatory Factor Analysis (CFA), Structural Equation Modeling (SEM) analysis, Model estimation test by Bootstrap, and Multigroup structure analysis, specifically as follows:

- Step 1: Scale reliability test.
- Cronbach's Alpha coefficient (CA) is normally used to evaluate the reliability of the scale for each observed variable belonging to the factor groups. Peterson (1994) suggested that any factor with CA less than 0.6 should be excluded from the research model (Peterson, 1994). According to Bernstein and Nunnally (1994), observed variables with a total correlation coefficient less than 0.3 are considered as irrelevant variables and should be excluded from the model (Bernstein & Nunnally, 1994).

Step 2: Exploratory Factor Analysis (EFA).

The EFA allows describing the correlation between the impact variables, called "factors." EFA is used in cases in which the relationship between observed and latent variables is unclear. The EFA analysis is conducted in a discoverable manner to determine the extent and degree of relationship between observed variables and the latent factors, which underlie a set of measurements to reduce the number of observed variables' loading base factors. Meyers, Gamst, and Guarino (2016) reported that in the EFA, the Principal Component Analysis extraction method with Varimax rotation is the most commonly used method. A condition for EFA analysis is the following requirements: factor loading >0.3, $0.5 \leq$ KMO ≤ 1, Bartlett test having statistical significance (Sig. <0.05), and total variance percentage $>50\%$.

Step 3: Confirmatory Factor Analysis (CFA).

CFA can be used when the researcher has some knowledge of the latent variable structure. CFA may be the next step of EFA to test whether there is a theoretical model that underlies a set of observations. Indicators for measuring the relevance of the model with data include Chi squared (CMIN), Chi square adjusted according to degrees of freedom (CMIN/df), Comparative Fit Index (CFI), Tucker & Lewis index (TLI), and Root Mean Square Error of Approximation (RMSEA). According to Hair Jr. et al., if $1<$ CMIN/df <3, the model is considered to be a good fit. Nguyen (2011) states that if the model receives CFI values, TLI ≥ 0.9, RMSEA ≤ 0.08, and $P > 0.5$, the model is relevant to the data.

Step 4: Structural Equation Modeling (SEM).

The structural equation model (SEM) is the following step of EFA and CFA, helping to test a set of regression equations at the same time. In this study, the SEM model was applied with the aim of identifying the determinant factors and the degree of influence of factors on "intention of VNU students to start up".

2.3 Data Collection Methods

Determining the sample size is important. However, although there is no specific standard in sample selection, the model still reaches a high practical significance with a relatively small number of variables (100–150) (Hoyle and Kenny 1999; Marsh and Hau 1999). With the scope of research equivalent to district and interdistrict levels, the practical significance of the data is high and very high at the sample size of about 200 samples (Hoogland and Boomsma 1998; Boomsma and Hoogland 2001; Kline 2005). In some studies analyzing employees' satisfaction with management, the number of samples was 150 (Yu and Muthén 2002). According to Hair et al. (1998), in order to perform the EFA factor analysis, the size of the sample applied in the study must be at least 5 times the total number of observed variables. The study has 29 observed variables, so the minimum number of samples is $29 \times 5 = 145$. Thus, in order to identify the factors affecting entrepreneurial intention, the study conducted in-depth interviews and a structured questionnaire survey for students of the University of Social Sciences and Humanities, University of Economics and Business, and University of Technology at VNU, Hanoi. The number of responses collected was 300, of which 296 provided relevant information that could be used for analysis (98.67%).

The questionnaire was designed with closed-ended questions to ask students to assess and compare the current conditions and status of students related to entrepreneurial intention, in aspects related to confidence, technological skills, work experience, entrepreneurship, capital, and education. The assessments use the Likert scale with 5 levels ranging from 1 = strongly disagree to 5 = strongly agree.

2.4 SEM Model Results

2.4.1 Evaluating the Scale Reliability

The scale reliability evaluation was performed by calculating Cronbach's Alpha coefficients for seven groups of factors through a number of cycles. The results obtained (Table 9.1) show that the CA of all factor groups is greater than 0.6 and the correlation coefficient of the total variable is greater than 0.3. Removing one of the base factors in the group will result in a smaller CA. Specifically, Cronbach's Alpha results of "confidence" through 2 cycles showed that variables NT2, NT3, NT4, NT5, and NT6 satisfied the criteria. CA results of "education "showed that variables GD1, GD2, and GD3 met the requirements. CA results of "capital" showed that variables NVKD1, NVKD2, and NVKD3 met the requirements. CA results of "Entrepreneurship" showed that through two cycles, variables TDKN1, TDKN2, TDKN3, and TDKN5 met the requirements. Cronbach's Alpha results of "work experience" showed that variables KNLV1, KNLV2, and KNLV3 were satisfactory. Cronbach's Alpha results of "technological skills" showed that through two cycles, variables KNCN1, KNCN2, and KNCN4 satistied the criteria. CA results of "entrepreneurial intention" showed that through three cycles, variables YDKN2, YDKN3, YDKN4, and YDKN5 met the requirements and were retained to analyze the discovery factor in the following round. Thus, the results obtained had 25 observable variables and were retained for discovery factor analysis (EFA) in the following round.

2.4.2 Exploratory Factor Analysis (EFA)

In this step, all 25 variables obtained through Cronbach's Alpha coefficient test ran EFA analysis and EFA results through a number of qualifying variables were $0.5 < \text{KMO} = 0.832 < 1$, as shown in Table 9.2; in conclusion, factor analysis is consistent with actual data.

- Testing the correlation between measurement variables (Bartlett's Test).
  - Test of hypothesis H0: The correlation of the variables is zero;
  - Bartlett's Test results are valid Sig. = 0.000 < 0.05; in conclusion, the measurement variables correlate with each other in each factor group;
- Testing of variance extracted from factors (% Cumulative variance):
- In the table, the total variance extracted (Total Variance Explained) from Component 7 and Cumulative% column has the value of cumulative variance of factors of 56.263% > 50%, which meets the standard. In conclusion, 56.263% change of factors is explained by factor's component measurement variables.
- Testing Factor loading.
- The results of EFA analysis for the independent variables of the factor rotation matrix showed that the factor loading of the measurement variables satisfied the condition when Factor loading was ≥ 0.5, so all 23 measurement variables were analyzed in the next step (Table 9.3).

2.4.3 Confirmatory Factor Analysis (CFA)

After the EFA analysis, the scales continued to be tested again, using CFA confirmatory factor analysis. The aim was to ensure greater reliability and validity of the scale based on a defined number of known factors in order to limit errors in identifying factors. The CFA results showed the indicators evaluating the relevance of the theoretical model were GFI = 0.930, TLI = 0.931, CFI = 0.942, and RMSEA = 0.047. Therefore, the framework was relevant to actual data. It can be concluded that components measuring "Student's entrepreneurial intention" and independent factors achieved discriminant validity. Therefore, this model was relevant to actual data.

Table 9.1 Results of the scale reliability evaluation

No	Abb	Observable variables	Correlation coefficient	Cronbach's Alpha for excluding variables
Confidence	Cronbach's alpha = 0.836			
1	NT2	You feel that starting a business is quite easy.	0.372	0.829
2	NT3	You think that maintaining the value of your business is not too difficult.	0.461	0.694
3	NT4	You think you have the ability to control the establishment of a new business.	0.461	0.802
4	NT5	You think starting a business brings you more development opportunities.	0.379	0.784
5	NT6	You think you know what needed to start a business.	0.443	0.751
2. Education	Cronbach's alpha = 0.792			
6	GD1	The university provides the knowledge needed for starting a business.	0.510	0.773
7	GD2	The university has business skills training courses.	0.361	0.698
8	GD3	My university often organizes career guidance activities for students (startup workshops and entrepreneurship contests)	0.476	0.783
3. Capital	Cronbach's alpha = 0.921			
9	NVKD1	I can borrow money from friends and relatives to do business.	0.327	0.749
10	NVKD2	I have the ability to accumulate capital (thanks to saving and doing a part time job)	0.483	0.861
11	NVKD3	I can raise capital from other sources (banks, credit funds, ...)	0.629	0.895
4. Entrepreneurship	Cronbach's alpha = 0.820			
12	TDKN1	I believe in my ability to launch a business.	0.429	0.801
13	TDKN2	I have many social relationships.	0.532	0.785
14	TDKN3	My relationships can help my startup.	0.460	0.699
15	TDKN5	I am not afraid of risks in business.	0.398	0.769
5. Work experience	Cronbach's alpha = 0.739			
16	KNLV1	Employee's experience	0.402	0.694
17	KNLV2	Management experience	0.386	0.723
18	KNLV3	Business experience	0.581	0.728
6. Technological skills	Cronbach's alpha = 0.793			
19	KNCN1	The 4.0 technology revolution gives businesses an opportunity to expand their markets.	0.425	0.738
20	KNCN2	The 4.0 technology revolution will help businesses increase their ability to participate in global and regional value chains.	0.392	0.764

(continued)

Table 9.1 (continued)

No	Abb	Observable variables	Correlation coefficient	Cronbach's Alpha for excluding variables
21	KNCN4	Entrepreneurs will study the advanced technologies of industry 4.0 and apply them to improve their position in the value chain.	0.580	0.691
7. Entrepreneurial intention Cronbach's alpha = 0.839				
22	YDKN2	I tend to open my own business after graduating.	0.542	0.790
23	YDKN3	Starting a business is appealing to me.	0.389	0.827
24	YDKN4	I am a person with many business ambitions.	0.493	0.819
25	YDKN5	I believe in my ability to launch a business.	0.462	0.803

Source: Analyses of collected data

Table 9.2 Results of testing KMO and Bartlett's Test of CFA

Kaiser-Meyer-Olkin Measure of Sampling Adequacy.		0.832
Bartlett's test of sphericity	Approx. Chi-Square	6840.424
	Df	300
	Sig.	0.000

Source: Analyses of collected data

2.4.4 Structural Equation Modeling

In the current work, SEM was utilized to determine the determinants and the degree of effects of factors on "Student's entrepreneurial intention." The original SEM model was adjusted to be more relevant to the study. The model of satisfaction factors is described in Fig. 9.1.

The research results showed that the degree of impact of factors affecting VNU student's startup intentions was confidence, 0.082; entrepreneurial thinking, 0.106; education, 0.192; capital, 0.89; technological skills, 0.147; and work skill, 0.156. These factors have a positive relation with students' intention to start a business, which means that the factors above will strengthen the intention of students to launch a company, thereby promoting students to start a business.

3 Conclusion

The main objective of this research is to clarify the determinants of VNU students' EI. With the structural equation model, the study found a positive relation between the factors of confidence, entrepreneurship, education, technology skills, capital, and working skills and the formation of entrepreneurial intention. The most important affecting factor was education, which reflects the reality because when the university focused on training, organizing seminars, and startup idea competitions

Table 9.3 Exploratory Factor Analysis results (Pattern Matrix)

	Factor						
	1	2	3	4	5	6	7
NT2	0.751						
NT3	0.610						
NT4	0.842						
NT5	0.603						
NT6	0.832						
GD1				0.783			
GD2				0.747			
GD3				0.739			
NVKD1						0.698	
NVKD2						0.744	
NVKD3						0.648	
TDKN1		0.655					
TDKN2		0.768					
TDKN3		0.819					
TDKN5		0.848					
KNLV1							0.524
KNLV2							0.841
KNLV3							0.628
YDKN2			0.686				
YDKN3			0.771				
YDKN4			0.815				
YDKN5			0.799				
KNCN1					0.683		
KNCN2					0.749		
KNCN4					0.759		

Factor loading was at its minimum when >0.3; significant when >0.4, and practically significant when >0.5 (Hair, 1998). Source: Analyses of collected data

(such as the Business Challenges 2019 Contest organized by the University of Economics and Business, VNU), the university provided students with necessary knowledge for starting a business, and created a positive environment for the development of startup ideas. The factor was followed by factors of work skills, technological skills, entrepreneurship, capital, and confidence. The findings from this study are useful references in developing strategies to promote entrepreneurship.

However, the study also reveals some limitations; for example, the student survey sample was not enough and did not cover all the universities in VNU, and the study did not compare the impacts of different factors on university students' entrepreneurial intention. This is the premise for further studies by expanding the scope of research on the factors affecting student entrepreneurial intentions in the future.

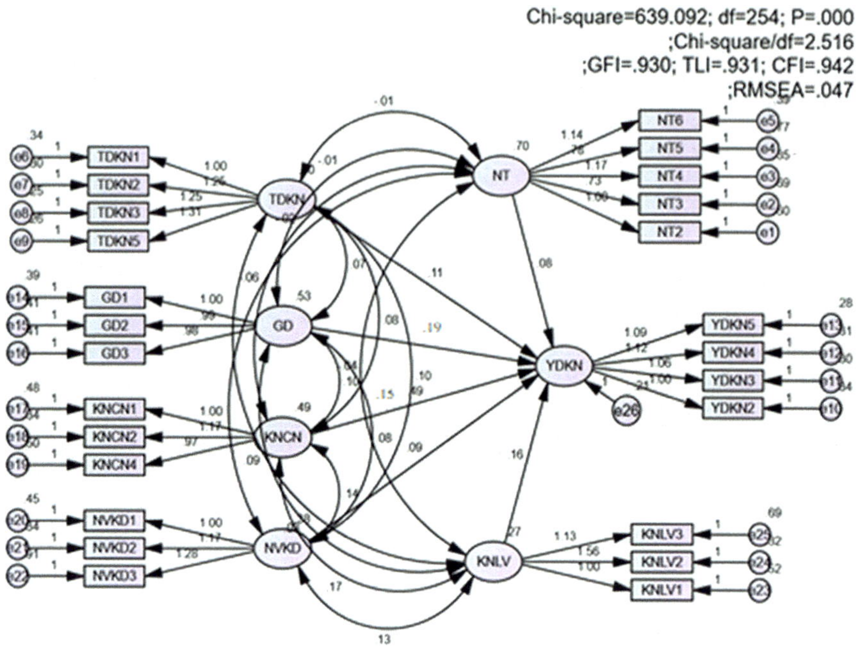

Fig. 9.1 Model to identify factors affecting VNU students' entrepreneurial intention (Source: Analyses of collected data)

References

Ajzen, I. (1987). Attitudes, traits, and actions: Dispositional prediction of behavior in personality and social psychology. In L. Berkowitz (Ed.), *Advances in experimental social psychology*, (Vol. 20, pp. 1–63). Academic Press. https://doi.org/10.1016/S0065-2601(08)60411-6

Ajzen. (1991). The theory of planned behavior. *Organizational Behavior and Human Decision Processes., 50*, 179–211.

Anderson, J., & Gerbing, D. (1988). Structural equation modeling in practice: A review and recommended two-step approach. *Psychological Bulletin, 103*, 411–423.

Bernstein, I. H., & Nunnally, J. C. (1994). *Psychometric theory*. McGraw-Hill.

Boomsma, A., & Hoogland, J. J. (2001). The robustness of LISREL modeling revisited. In R. Cudeck, S. du Toit, & D. Sörbom (Eds.), *Structural equation models: present and future. A Festschrift in Honor of Karl Jöreskog* (pp. 139–168). Lincolnwood, IL: Scientific Software International.

Dev, M. S., & Mahajan, V. (2003). Employment and unemployment. *Economic and Political Weekly, 38*(12), 1252–1261.

Hair, J. F. (1998). *Multivariate Data Analysis* (Vol. 730). Prentice Hall.

Hair Jr., J. F. et al. (1998). *Multivariate data analysis with readings*. Englewood Cliffs, NJ: Prentice-Hall.

Hatten, T. S. (2006). *Small business management, entrepreneurship and beyond (3rd Edition)*. New York: Houghton Mifflin Company.

Hoelscher, M., & Elango, B. (2012). The impact of business climate, foreign population and unemployment on new venture creation. *Developmental Entrepreneurship (JDE), 17*(04), 1–14.

Hoogland, J. J., & Boomsma A. (1998). Robustness studies in covariance structure modeling. *Sociological Methods & Research, 26*(3), 329–367. https://doi.org/10.1177/0049124198026003003

Hoyle, R. H. & Kenny, D. A. (1999). Sample size, reliability, and tests of statistical mediation. In R. H. Hoyle (Ed.), *Statistical strategies for small sample research* (pp. 197–223). Thousand Oaks, CA: Sage Publications.

Kline, R. B. (2005). *Principles and practice of structural equation modeling* (2nd ed.). Guilford Press.

Krueger, N. F., Jr., & Brazeal, D. V. (1994). Entrepreneurial Potential and Potential Entrepreneurs. *Entrepreneurship Theory and Practice, 18*(3), 91–104.

Le, Q. (2007). Study into the startup decision process of young Vietnamese entrepreneurs – HCMC. *Economic Development, 7*, 20–30. (in Vietnamese).

Marsh, H. W., & Hau, K. T. (1999). Confirmatory factor analysis: strategies for small sample sizes. *Statistical Strategies for Small Sample Research, 1*, 251–284.

Meyers, L. S., Gamst, G., & Guarino, A. J. (2016). *Applied multivariate research: Design and interpretation* (1104 pages). Sage.

Micozzi, A., & Lucarelli, C. (2016). Heterogeneity in entrepreneurial intent: the role of gender across countries. *International Journal of Gender and Entrepreneurship, 8*(2), 173–194. https://doi.org/10.1108/IJGE-06-2015-0021

Moica, S, Socaciu, T, & Rădulescu, E. (2012). Model innovation system for economical development using entrepreneurship education, entrepreneurship education, procedia economics and finance, 3, 521–526.

Nhi, D. T. Y., Chuong, H. N., Mai, N. T., Giang, V. T. L., & Thi T. T. (2017). Analysis of factors affecting the start-up intention of students of Binh Duong University. *Scientific Journal of Thu Dau Mot University, 4*(35), 107–114 (in Vietnam).

Nguyen, D. T. (2011). Research methods in business: Design and implementation. *Labor-Society, 539*. (in Vietnamese).

Nguyen, T. H., & Nguyen, T. K. P. (2016). Factors affecting the start-up intention of students at Tra Vinh University. *Socio-Economic Review, 9*(23), 1–9. (in Vietnamese).

Ooi, Y. K. (2015). And Nasiru. A entrepreneurship education as a catalyst of business start-ups: A study on Malaysian community college students. *Asian Social Science, 11*(18), 350–363.

Patrick, C. J., Kramer, M. D., Vaidyanathan, U., Benning, S. D., Hicks, B. M., & Lilienfeld, S. O. (2019). Formulation of a measurement model for the boldness construct of psychopathy. *Psychological Assessment, 31*(5), 643–659. https://doi.org/10.1037/pas0000690

Peterson, R. A. (1994). A meta-analysis of Cronbach's coefficient alpha. *Journal of Consumer Research, 21*(2), 381–391.

Sahinidis, A. G., Vassiliou, E. E., & Hyz, A. (2014). Factors affecting entrepreneurs' intention to start a new venture: An empirical study. *International Journal on Strategic Innovative Marketing, 1*, 148–162.

van Gelderen, M., Brand M., van Praag, M., Bodewes, W., Poutsma, E., & van Gils A. (2008). Explaining entrepreneurial intentions by means of the theory of planned behaviour. *Career Development International, 13*(6), 538–559. https://doi.org/10.1108/13620430810901688

Yu, C.-Y., & Muthén, B. (2002). Evaluating cutoff criteria of model fit indices for latent variable models with binary and continuous outcomes. *Doctoral Dissertation*. http://www.statmodel.com/download/Yudissertation.pdf

Chapter 10
Creating an Added-Value Capital of Forest Based on the Local Knowledge of Tai in Tuong Duong (Nghe An, Vietnam)

Nguyen Hong Anh

Abstract Tuong Duong district is located in the West Nghe, an biosphere reserve, belonging to the world network of biosphere reserve and playing the role of protecting the watershed of Lam river. Among the ethnic groups here, the Tai ethnic group has densest population and holds the most important role in sociocultural development. Moving from Laos to Tuong Duong, the Tai settlements were built along streams and banks of the Lam River (Pao River in Tai's language), and hence, they are also named Tay Pao or their name is known as Tay Hay (swidden Tai); because of the terrain conditions of Tuong Duong, instead of traditional wet rice cultivation, they also exploit forest for slash-and-burn cultivation. Forests give them not only food, clothes, house, etc. but also a spiritual support in traditional culture, and forest is an environment that is creating rational indigenous ecological knowledge on exploiting nature and culture of Tai people. But, in the contemporary context, forest resources have been exploited strongly and become exhausted and cultural space of communities was influenced, which requires measures to create added-value capital of forest. By fieldwork method and approaches of human ecology and cultural ecology, this study deals with how Tai people in Tuong Duong have created the added-value capital of forest by their local knowledge through livelihoods. The value capital of forest was increased, which will be the basis for restoring their lost cultural space and maintaining their religious and spiritual practices.

Keywords Forest · Tai people · Indigenous knowledge · Tuong Duong · Vietnam

N. H. Anh (✉)
Institute of Socio-Economic Geography and Development, Hanoi, Vietnam

A. T. Nguyen, L. Hens (eds.), *Global Changes and Sustainable Development in Asian Emerging Market Economies Vol. 2*, https://doi.org/10.1007/978-3-030-81443-4_10

1 Introduction

Tuong Duong is a mountainous district of Nghe An, where is the confluence of the two rivers Nam Nom and Nam Mo, where begins Lam river (Tai people called Nam Pao) and belongs the Ca river basin, and where there are many streams, waterfalls, rapids, and springs that make great landscapes and bring about potential for tourism with travel routes on river for exploration, commerce, and recreation, but the climate is severe, called "Indochina oven", which makes travel difficult and dangerous. Tuong Duong has the largest forest area of Nghe An province (180,592.1 ha accounting for 18.74% of the total forest area of Nghe An province), of which 115,981.8 ha are natural forests, including special-use forest area (39,289.2 ha) and protection forests (76,692.6 ha).

Located in Pu Mat National Park belonging to the World Biosphere Reserve of the Western Nghe An, the natural forest in Tuong Duong is listed as "special forest ecosystem"(semievergreen and deciduous dry forest). In thousands of hectares of thousand-year-old primitive, there are still a rich variety of species, from subtropical coniferous forests to mixed conifers—broadleaf and closed forests; according to the report of Pu Mat National Park Management Board, there are nearly 2000 species of plants, belonging to 827 genera and 195 families, of which 42 species have been recorded in the Vietnam Red Book and 17 are international species, including Aquilaria crassna, Diospyros mun, Parashorea, Hopea odorata, Shorea roxburghii, Chukrasia, Cloves, Magnolia, Hopea Pierrei Hance, and especially *Lagerstroemiatomentsa* Presl, *Cunninghamia*, and *Fokiena hodginsii* (the current reserve is about 220,000 ha *Cunninghamia* and *Fokiena Hodginsii*). Nontimber forest products such as Bamboos, Lacca-Stick-lac, and Calameae; and 200 kinds of valuable medicinal herbs such as Codonopsis, Amomum Xanthioides Wall, Smilax Glabra, and Reynoutria Multiflora; and 100 kinds of vegetables, forest fruit trees, etc. There are 31 species of endangered medicinal plants that need to be preserved according to the Vietnam Red Book and 13 species of endangered and 15 alarming species (*Eleutherococcus Trifoliatus*, *Rauwolfia*, etc.).

Lagerstroemiatomentsa Presl forest is a pure forest, covering an area of more than 53 ha, located along both sides of National Road 7A in Tam Dinh and Tam Quang communes and in the core area of Pu Mat National Park, most of which are Lagerstroemiatomentsa Presl with 30–40 m height and big circumference. According to Decision No. 4639/QD-UBND approving the planning on conservation and sustainable development of special Lagerstroemiatomentsa Presl forest in Tuong Duong district toward year of 2020, the total of the planning area for conservation and sustainable development of special Lagerstroemiatomentsa Presl forests is 1335.82 ha. Of which, the special core zone is 241.60 ha (including strict protection area of 53.85 ha and ecological restoration area of 187.75 ha); the buffer zone is 1094.22 ha (of which in Tam Dinh commune, 707.66 ha and in Tam Quang commune, 386.96 ha).

In particular, among these species, there are species that are strictly prohibited and restricted to be exploited, such as Fibraurea… but still being "eradicated."

People said that in the past, these plants were easy to look for, so far they could not be found, such as ***paniculata*** tree (a medicinal plant that has the ability to shield and prevent erosion and landslides), Cibotium barometz (hemostasis), Knema tonkinensis, Coix lacryma-jobi, Homalomena occulta, Amomum, Alocasia odora, Fibraurea, Smilax, mushroom of Erythrophleum fordii, even Myrsinaceae, etc. Under the canopy of the forest is 241 animal kinds, belonging to 86 families and 28 orders, of which there are 24 mammals, 25 reptiles, 137 birds, and 15 species of amphibians and there are still precious animals such as tigers, clouded leopards, Indochinese serow, Petauristini, Trachypithecus, Hylobatidae, and Helarctos malayanus (recorded in the Vietnam Red Book), along with many valuable birds such as Phasianidae, pheasants, etc.

In addition, according to the District People's Committee's report in 2018, a total of production forest area is 63,928.6 ha (accounting for 22.73% of the total natural land fund), of which forested forest land in the district accounts for about 63.99%, with 60% of forest coverage, and nonforest land is 681.98 ha (accounting for 0.24% of the total natural land fund) (Nghe An Genenral Statistical Office [NAGSO], 2019).

Tuong Duong is home of 6 ethnic groups Tai, Kh'mu, Tay Poong, Tho, O Du, and H'mong and to approximately 467 thousand people, in which Tai ethnic has densest population (about 324 thousand), resides in the large area of the Western Nghe An, and "plays a very important role to change the mountain area Nghe An in history as well as today, and Tai culture has always played an important role in regional cultural exchanges". The Tai people in Tuong Duong Nghe An were a group from northwestern Tai people, who were moving in along the road through Laos and settling along the banks of the Lam river (Nam Pao) so they were also called Tai Lai Pao (Tay Pao), and therefore, the local knowledge of the Tai people in Tuong Duong based on Tai culture in Vieetnam also contains many similarities with the Lao's culture and also has the unique characteristics of the Tay Pao people, who are living along the Lam River.

In the treasure of Tai proverbs, the phrase "Xa people live by fire, Tai people live by water" exists, when moved to Tuong Duong, Tai people bring with themselves wet rice civilization, but due to the topography and natural conditions and high mountains, they just cultivated wetland fields in combination with shifting cultivation to survive, and even there are villages that cultivate the upland fields, which are also called Tay Hay (swidden Tai) just like calling people Kh'mu also (Vi, 2017).

There are 18 administrative units including 17 communes and 1 town, part of Tuong Duong is located in the areas of two national parks of Pu Mat and Pu Huong, and this research was conducted in three communes of Tam Dinh, Tam Quang, and Tam Hop, because people here have long been accustomed to cultivating paddy fields and exploiting forest products for food, clothing, shelter, and travel, now in the context of conversion to reforestation. The rich system of local knowledge in the traditional culture of Tai's people on the forest exploitation and use is now gradually changing because of the transformation in production, from agriculture to forestry, and new knowledge about afforestation is updating and becoming "recognition" of the people. Under the government's policy, community-based management and afforestation mean that the use of local knowledge of the community and the empowerment of management for the community in protecting and creating added-value of forests is long-term goals to increase income, improve household living standards, protect biodiversity and environmental landscapes, and at the same time, restore lost cultural spaces and ethnic cultures.

2 Methodology

The research used fieldwork method (including PRA, in-depth interviews, group discussions, and expert) to collect field materials and build practical databases to this article. With 75 respondents, the data were collected from the three respondents from each department including Department of Culture and Information of Nghe An, Department of Natural Resources and Environment, Department of Agriculture, Management Board of World Biosphere Reserve in Western Nghe An, and 1 respondent from each division of district, such as Division of Culture, Division of Agriculture, Division of Environment, Division of Forestry, and National Park. They are stakeholders working and directly conducting on agri forestry and understanding ethnic culture. Also, 16 respondents are local people for 2 groups of each commune in three communes Tam Quang, Tam Dinh, and Tam Hop. They are households moving toward producing forestry, so they are updating scientific knowledges of producing forestry. Local knowledge is enriched by combining indigenous and scientific, but beside that, scientific instead of indigenous knowledges. Voice of the people will be the insiders voice that expresses their awareness of forest resources, how they create forest livelihoods, and why they do it aims to (i) identify the role of protecting forests and living spaces of Tai communities in Tuong Duong through the local knowledge system in traditional culture and (ii) identify the role of the Tai community in restoring the living space and creating added-value of forest through local knowledge in the contemporary context. At the same time, the comparative method in the research clarifies the change of local knowledge of the Tai people and the cause of the change.

3 Results

3.1 Forests in the Life and Local Knowledge on Forests in Tai's Traditional Culture

The forest in Tai's language is Pa, and the Tai has a saying: There are trees have fur (meaning old tree), such as bearded old people, the immense green forest, the headwaters forest, worship forest, sacred forest, etc.

The important role of forests in life, not only are confirmed by science, but also have in folklore, the tangible and intangible values, has been recognized by the Tai's community since ancient times. Tai people said "when human live, forests will nourish, when human died, forest will entomb", or "rice in the soil, food in the forest", meaning the forest provides food, clothes, house, medicines, energy … and making the fresh living space, creating a beautiful landscape for Muong village, "there are forests, trees, and fruits. Birds, bees and butterflies will come. If nothing, all butterflies and birds will leave". So they always teach their children to protect the forest today, tomorrow and forever. Elsewhere, forests is a spiritual support for the community.

The forest in the mind of the Tai people is like the heart of the community, which expresses the conventions, customary laws, and traditional cultural values that are worshiped and revered like ancestors. From a scientific perspective, the law of forest protection and development specifies and stipulates forest classification according to functions (i) Special-use forests, protection forests, and watershed forests in order to conserve biodiversity and protect the ecological environment and (ii) production forests to ensure income for people in order to reduce poverty and proceed to get rich. In folklore, the community also has strict regulations in (i) nonencroaching areas, such as sacred forests, all activities of slash and burn are completely banned, in fact, no one dared to violate, because of fear of punishment from nature and condemnation from community, and (ii) the area where everybody is available to exploit, cultivate, graze, get forest products, etc. So, there is a similarity in "planning and territorial organization" between science and folk knowledge; thus, the community's natural cultural space is preserved and ecological environment is balanced.

In traditional culture, the Tai said "where are the bitter bamboo shoots growing, Tai land goes there". The horizontal layer is community's living space that here includes accommodation, play, cultivation, exploitation of natural products, and sacred places, all of it is arranged in an order proposed by the community, which is drawn from life experiences: Houses in the villages and Mueang often are concentrated, followed same side, built along streams and rivers and wet-rice fields in front of villages, and hills, mountains, and upland paddy field behind.

> "Our Tai's village has got stilt houses that stands hillside
> There are stream, where everybody have shower together every afternoon
> In the morning mother goes to work in the fields,
> And in the afternoon father goes to work in the upland field …"
> (Tai's folk songs)

On the vertical layer, the Tai's living space is a vast cosmic world consisting of three storeys: in the highest is heaven (Mueang Pha); In the middle is the soil storey, Tai people use different names to call Mueang Lum (Mueang people)/Mueang Din (Mueang land)/Mueang Pieng; In the ground is Mueang Booc Dai (the Mueang of the tiny ones). In particular, Mueang Pha and Booc Dai are unreal worlds. The real world is Mueang Lum only, and this is a world of people and animals, but at the same time, it is also the abode of spirits and demons. That is also the reason why in the organization of living space of Tai people, the community has prescribed forbidden mountains, forbidden forests, sacred forests, ghost forests, etc. Humans are not only under the influence and control of Mueang Pha but also by many other supernatural powers, such as spirits and demons that accompany the people of Mueang Lum. The land and the forest belong to the gods, the ancestors left behind the common space of the community to be exploited and used in accordance with the regulations of the Mueang, and the community respects that right of possession and inheritance. But if not used, it will be returned to the community and will be divided back to other people in the family or community, not to trade, exchange, or give to the Mueang. The living space of the community going into folk songs is a picture of a harmonious culture and vivid natural environment with a lot of colors.

"High mountains stand on the ground
Small mountains and hill side by side
Mountain winding, and stack up high and higher
Mist wrapped around the rocky mountain
Green tall tree roots stick to roots
Do not know how many animals in the mountains moved to
How many bird species and rats come to reunite?"
(Tai's folk songs)

According to Mr. Vi Van Mieng, before the year 1991, the forest was still abundant, arounding the village, everything was taken from the forest easily, the forest was very close to the village, so every time when the New Year comes near, the older people often go to listen to the "voice" of the forest. If the echoes of the forest are like play folk drum, the rain will be favorable in the year, good harvest. If the echoes of the forest are like the crying of children, then that year will lose crops, people will be sick and the village will suffer, etc. Mr. Luong Tien Phuong, now 90 years old, was a very good hunter when he was young, he made different kinds of traps to trap birds and animals, and he often told his children about biological properties like cries, habits, feeding preferences, breeding season, time and migration places of about more 50 species on the ground, 15 species in the air, 70 species of birds, 20 species of fish, 25 species of insects, etc. He also taught his children how to make traps, set traps to catch birds and animals, or taboo animals that were never caught because they were the ancestors of some Tai families. In the village, besides him, only a few elderly people have such knowledge, but young people cannot know.

Tai's idioms have the saying: "Eating sticky rice, Drinking tube wine, Wearing "xin xua", Living in stilts home". Images of traditional stilt houses are described as "houses have attics, floors on pillars" or "build stilt houses, wood is costly, build mud houses, rope is costly". Mr. Lo Van Kha is famous for building traditional stilt houses. When he was young, he often helped families in the village build houses, he could distinguish between 30 types of wood, where trees can often grow. He said that to make pillars, it is necessary to have big and straight precious woods, hard and strong core, not being eaten by weevil, diameter is only 40 cm; To make walls, it is necessary to have durable, light, decorative carvings, bamboo to make floors, calameae to tie up, and cogongrass to make roofs and some specialized plants to make furniture. Tai people always teach their children not to exploit big trees. *"Do not cut big trees, its Mr. Xan's pillars*[1]*"*. Therefore, when shifting cultivation, with big trees, they just prune branches to get light.

Everyday things that always brought by Tai women when working in this field are basket and knives, and when they go to work in the upland fields, they look for fruits and vegetables to make food. Ms. Vi Thi Hang identified 15 types of tubers, fruits, seeds, or stalks for food and 67 kinds of forest trees to make vegetables in daily meals with cooking methods such as steaming in wooden hips, sausage (mannequin and salad), and sour salt. In particular, bamboo shoots (fresh or sour salt) are always the main food used daily. She can also distinguish nearly 10 types of bamboo shoots with different bitter, sweet, acrid taste, mushrooms (more than 10 types), banana flower, rattan bud, firewood, bitter flower, etc. Tai people think that sour bamboo shoots when used as "spices" for processing meat and insects will eliminate some toxins in meat and insects, fight digestive disorders and food poisoning, reduce the fat of meat, and increase the deliciousness of the dish. Cooking soup or boiling some wild vegetables that have unforgettable bitterness is also a medicine to cure malaria and intestinal diarrhea or women often cook after giving birth for creating milk. In particular, Tai people do not use fish sauce, they combine salt with forest spices to make "Cheo", creating a unique cuisine that the author calls "Cheo culture". Traditional Tai women make drinking wine for daily use and during holidays with enamel making from 22 different species of 18 genera and 14 families, by using root, bark, stem, fruit, leaves, or whole tree.

The research team has collected 39 folk remedies treating 15 different groups of diseases such as diseases of the intestines, respiratory, digestive, fever, sprains, infections, and osteoarthritis, for pregnant women. Ms. Vi Thi Phieng frequently goes to the forest to collect leaves in her remedies, she uses 231 species belonging to 192 genera and 88 plant families, accounting for 5.97% of the total number of medicinal plants in the country, including 72 species of herbaceous plants, 64 species of woody plants, shrubs of 55 species, and vines with 40 species. In particular, leaves, stems, and roots are most used. At least use medicinal plants to have only 7 species.

According to statistics, Tai people can create red, yellow, indigo, brown, and black to dye fabrics from 30 species of 22 plant families, of which 9 types use leaf foliage, 1 species uses stem bark, 6 species use roots, tubers, 2 species use pods, 3 species use seed pods, and 6 species use trunks, combined with ants to create specific colors and keep color stability.

A group of elderly people said that before the year 1991, all forest products were available, every time they went to the upland fields and people took advantage of vegetables, fruits, or trapped small animals to make food. But Tai always take just enough to use, because if taken redundant will have to leave, because no one buys, anyone goes to earn it for themselves. When they are unable to go to the forest in the rainy season, they can take more than vegetables and make dried meat to eat gradually (Tables 10.1 and 10.2).

[1] God, who governs Tai's villages

Table 10.1 Statistics and classification of land and forests in Tuong Duong (ha)

Types of soil and forest	Area	Of which		Rate (%) of Natural area
		Special use	Protective forest	
Agricultural land	11,014.00			3.92
Jungle land	180,592.10			64.24
Natural forest	115,981.60	39,289.10	76,692.50	41.26
Production forests	63,928.60			22.74
Land without forest	681.90			0.24
Natural area	281,129.73			

Source: NAGSO (2018)

Table 10.2 Average household land use area by commune

TT	Commune	housing area (m^2)	Agricultural area (m^2)	Forestry area (ha)
1.	Yên Na	168	4.739	4.4
2.	Yên Hòa	137	4.093	4.5
3.	Nga my	124	4.679	2.3
4.	Tam Quang	156	2.938	3.0
5.	Tam Đình	142	3.280	3.6
6.	Tam Hợp	211	5.786	6.3
7.	Hữu Khuông	200	14.584	2.0
8.	Nhôn Mai	172	7.896	3.0
9.	Mai Sơn	180	6.438	3.5

Source: Report of District Department of Agriculture (2018) processed by research group

Along with that, the community often tells about taboos when going to the forest such as never encroach on forbidden forest areas, the whereabouts of the forest Gods (owner pa), mountain deities (owner pu), the God of rivers and streams (owner huoi), and the land deities (owner din) that are the natural deities, ingrained in the Tai's spiritual realms. Do not go into the murky dense forests, more vines, more vines forest as many ghosts, where people die when going to the forest, places where people are buried, etc. Under the gods are ghosts residing in sacred forests: water ghosts, animal ghosts, and tree ghosts, in which tree ghosts are the scariest, so there are a some of stories often told to children and grandchildren about a few species of tree such as Dipterocarpus (associated with legend of dragon), banyan tree, microcarpa tree, etc. The Witch said that they had to make worships to many people whose soul lost when going to the forest or go to the field, because unknowingly people had committed taboos or ghosts making fun of, or having stories about the harm of trapping a baby animal, or a pregnant animal, or animal of raising child, so that the offspring can never indiscriminately exploit the forest. To avoid risks, if any individual violates the taboo, the teachings of the elderly, they will be punished.

The Tai believe that there is a soul/spirit in each and every life, that is the shadow existing in the body, and everything in this world like hills, rivers, grasses, flowers, birds, and animals have a soul/spirit, of which people have the most trouble 30 souls in the front, 50 souls in the back, small souls wipe each other on the flanks (Vi, 2017), or it is in each human: 30 souls reside in the body, another 900 souls reside on the head (Vi, 2017), of which there is the Lord of souls that resides on the top of the head. In shifting cultivation, the Tai people never clear the forest on the top of the hill but have to protect like the "Lord of souls" of the hill, to make residence for the souls and forest on the top of hill become a place of worship for shifting cultivation cycle. According to scientists, "the method of not slash and burn forest on the top of hill has the effect of maintaining the fields bellow. Forest on the top of hill likes as "a hat" to cover the rain and sun, and keeping the forests at the environmental important point." Thus, in traditional culture, thanks to beliefs, spirituality, and customary law, with the technical indigenous knowledge of upland cultivation, forests are protected.

After 1991, the forest was severely damaged, the main cause was (i) the policy of mechanical migration to develop new economic zones, the population was increase in this place, and people have to slash and burn forests to food cultivation and (ii) the forest management is changed and forest land is managed by teams and forestry farms. Communities, who live long time ago here, are separated from the forest-their natural living space, therefore, the forest has not been managed (Table 10.3) by communities, indigenous knowledge, ethnic culture was changed, and the social structure of swidden farming communities has become loose, etc. This is period of "sacred less" for all beliefs of the community in the management and exploitation of forests.

Since 2000, Vietnam implemented the international convention for the protection of biodiversity and many national parks, and natural reserves have been established to conserve biodiversity. Communities were participated to manage forest by indigenous knowledge and customary law. Forest and land were managed by communities, and at the same time, it is the transition from agroecosystem (shifting cultivation) to agroforestry model, with the aim of ensuring food security and restoring forest ecosystems and biodiversity. Forests have got ecocultural values in traditional culture and now have economic values (poverty reduction) and ecological environment values (erosion control, landslides, and floods are limited) in contemporary context.

Table 10.3 Statistics on forests in Tuong Duong from 2001 to present (ha)

TT	Item	2001–2005	2006–2010	2011–2015	2016–2018
1.	Area of planted forest (ha/year)	10.2	11.7	15.5	25.5
2.	Deforested forest area (ha/year)	24.5	14.2	9.2	8.8
3	Burning forest area (ha/year)	42.4	59.2	43.7	62.8
4.	Proportion of land maintained and protected by biodiversity	14.3	9.7	10.2	8.7
	In which nature reserve area (ha)	234.9	158.3	167.6	165.2

Source: Environment statistics 2001–2015, district statistical yearbook, provincial statistical yearbook, report to Pu Mat Nature Reserve, Tay Nghe An Biosphere Reserve 2018

3.2 Rehabilitation Increases the Forest Capital Base on the Local Knowledge of Tai People

In the past twenty years, after the decision of government to convert shifting cultivation land to afforestation in order to (i) reduce poverty, (ii) ensure production materials for the society, and (iii) restore the lost ecosystem, afforestation has become a new livelihood model for the communities. In traditional culture, Tai people used to only slash and burn forests for cultivation, now it is afforestation. The transfection makes changes in perception of communities, aims to (i) Create added values of forest resources—one of the scientific bases for the rationality in exploiting and using forest resources and (ii) Create a new ecological succession in a positive way. As a result, new knowledge is formed and integrated with traditional knowledge enriching local knowledge and provides an opportunity to restore and preserve the values of ethnic culture.

Since 2000, Tuong Duong began to implementing a forestry livelihood project (afforestation to replace shifting cultivation). Mr. Luong Van Minh has participated in afforestation since the first days of implementing the project, and his family has 3 hectares granted for planting Acacia hybrid. Acacia hybrid is a material plant that has been experimentally deployed in the communes belonging to buffer zone of National Park. According to Mr. Luong Van Minh, Acacia hybrid will remain suitable for two months July and August, because then the weather in Tuong Duong becomes cooler. To plant trees, people dig holes for 30 × 30 × 30 cm, each hole 3 × 3 m apart and planting density varying from 1100 trees to 1660 trees/ha, after preparing seedlings, we choose rainy days, shady or light sun, in moist pits with fertilizer NPK lining and plant trees so the plants will take root quickly. Plants need careful care 2–3 times per year in the first three years on the occasion of October–November, with the weeding, fertilizing, and pruning. After 8 years of testing, the Acacia hybrid was harvested. When harvesting, the family chose to exploit only trees with a diameter of 20 cm. Income of each household is from ten million to 30 million, and income of his family is the highest. Also, according to him, acacia planting techniques and knowledge and the sale of products were supported by the Government. During the planting process, in addition to the agricultural extension training program, he learned through television channels broadcast in ethnic minority languages and also found information on the internet for himself, etc.

In the years 2009–2010, the acacia faced difficulties in price so many households were encouraged to change the hybrid acacia with Melia azedarach and Dendrocalamus barbatus. In fact, the current price of hybrid acacia in Tuong Duong is only over VND 400,000/m^3, while the price of Melia azedarach is 4–5 times higher and the output is easy to consume and cost of Melia azedarach is quite low. In 2012, the Melia azedarach and Chukrasia tabularis were planted, when households were being allocated forests under the decision of 327, up to 300/900 households participated in afforestation and now have planted more than 300 ha of Melia azedarach and about 200 ha of Chukrasia tabularis instead of hybrid acacia. Mr. Kha Van Ngan's family with some households that involved in planting Chukrasia tabularis effectively shared the techniques of planting: "planting 1,600 trees/ha" (Luong Thi My), "the best time to plant Chukrasia tabularis is at the beginning of March, cool spring season "(Kha Van Ngan)," tillage is the most important step, 4 times per year, cut grass regularly, protect the first 5 years from destroying of cattle and buffaloes"(Luong Tien Phuong),"If taken care of after 8 years, they will be harvested for the first time, in 3 to 4 years next, they can be exploited again for the second time, and in the next 5 years, they will be harvested for the third time"(Kha Van Ngan). With The average incomes of planting 100 million VND/ha (Luong Thi Mi) for the first time, after deducting expenses for seeds, fertilizers, pesticides, and other amounts, households have reduced poverty thanks to the Chukrasia tabularis (Luong Tien Phuong). With the planting and care techniques that his family has received from the agroforestry project, along with the dissemination of knowledge in ethnic languages on TV, in addition, he also bought books from Forestry about the biological properties of Chukrasia tabularis and how to plant and care of the tree for an effective harvest.

The Dendrocalamus barbatus (bamboo) tree is an easy-to-grow species that can be used to plant along streams and banks of rivers and can be to harvest bamboo shoots for food and bamboo trees to build houses, furniture, material of paper industry and prevent riverbank erosion when floods come, at the same time, bringing significant income for households. Currently, Tuong Duong district has over 2000 hectares of Dendrocalamus barbatus (bamboo). Mr. Kha Van Toan said: Dendrocalamus barbatus (bamboo) is a tree suitable for wet lands, such as along streams, planting techniques are simple. Take out one-year-old stems, then cut it and plant directly or nursery for 6 months for shoots and plant with density of 250 roots/1 ha. About after 4–5 years, Dendrocalamus barbatus (bamboo) will harvest. Dendrocalamus barbatus (bamboo) is harvested in each 1–2 weeks, both bamboo shoots and trees, the trees with the diameter is and the price as bigger as possible. In Tam Dinh commune, where there were many villages that suffered from landslides during the rainy season, his family experimentally planted 500 roots of Dendrocalamus barbatus (bamboo), resulting in land erosion mitigation, and he shared this experience with households. In the village, the people have planted Dendrocalamus barbatus (bamboo) on hills. They observed that, when the tree is put on a high hill, although it does not grow in wet place, bamboo shoots are not as good as planted along the stream, but the stem is stronger, and it is important to keep the soil and prevent erosion and sliding of soil.

Calamus tetradactylus Hance is a kind of natural product. In Tuong Duong, it is used to grow a lot, and Calamus tetradactylus Hance is used in construction, construction of houses, furniture, and daily-life appliances. Under the situation of fully exhausted Calamus tetradactylus Hance, since 2010, the project of planting Calamus tetradactylus Hance has been implemented in Tuong Duong with about 50 hectares and participating of 400 households. Project 30a supports 50% of seed prices, and the Institute of Rural Industry Research and Development provides technical assistance. Mr. Cao Van Son said that the Calamus tetradactylus Hance is a very easy tree to plant, without care. When the tree grows, Calamus tetradactylus Hance needs to fasten up the forest trunk to grow, after 4 years, the harvest will be each 1–2 times per year, and Calamus tetradactylus Hance is still harvested for 20–30 years. Calamus tetradactylus Hance can be planted as a fence to enclose cattle raising areas, planted in home gardens, and planted at areas where soil is poor. "The time for planting Calamus tetradactylus Hance is in April and May, since the time of harvesting Calamus tetradactylus Hance, we do not have to go for harvest Calamus tetradactylus Hance in natural forest for sale. From harvesting Calamus tetradactylus Hance per year incomes at least VND 5,000,000/3,000 roots, enough money to children's study" (Vi Thi Thom).

The project has built and transferred a nursery of Calamus tetradactylus Hance in Yen Hoa commune with the scale of nearly 2000 m^2, with the production capacity of more than 180,000 roots in 18 months. Currently, people were able the implement all stage of nursery process such soaking, nursery proactively, and provided enough seeds for the locality. Households have reduced costs when buying trees from other places. Mr. Lo Van Kha made "some home appliances for himself (baskets, chairs, rice trays, etc.) and helped restore the traditional craft of the people here, using the labor force of all family members, and promoting restore and develop the traditional handicraft of the commune"(Cao Thi Than).

In addition to mass planting of above, people have started planting perennial native varieties such as Chukrasia tabularis, pterocarpum, Melia azedarach, Syzygium aromaticum, etc. Traditional knowledge on natural trees, such as biological characteristics, the concentrated development location of tree species, the growing season, as well as the harvesting methods, help local people to easily access nurseries, planting techniques, planting times, tending techniques, harvesting, processing, etc. Now, the forest area is growing, increasing the forest coverage in Tuong Duong.

According to the report of Tuong Duong District Department of Agriculture, so far Tuong Duong has over 4500 hectares of mixed forests, of which acacia hybrid area occupies less than 200 hectares, Dendrocalamus barbatus (bamboo) on 1500 hectares, and Melia azedarach is more 2500 hectares, with regeneration of Chukrasia tabularis, Khaya senegalensis, Lagerstroemiatomentsa Presl. Mr. Ho Viet Son, Chairman of Tam Quang Commune People's Committee, said that the biggest breakthrough of afforestation that it has changed perceptions, habits, and customs farming practices on shifting cultivation of people. People are voluntarily converting to afforestation.

However, the planting model of single kind is not yet an effective model to help restore ecosystems, afforestation is a method to simulate the development of nature, and natural forests are mixed species (biodiversity) with multistorey. In ancient silviculture, people also planted by the method of mixed multilayer. In shifting cultivation, the Tai also used the method of intercropping with the multilayer model (traditional knowledge). The multilayer afforestation has brought about special intangible values in the role of watershed protection (concentrated forests with mixed and multistorey structure), protecting the ecological environment. From a single-species forest model people are transferring to a multistorey mixed forest model. Popular model of multistorey mixed species is being implemented here (i)

Melia azedarach-Chukrasia tabularis-Dendrocalamus barbatus (bamboo) and (ii) medicinal plants intercropping the canopy of Lagerstroemiatomentsa Preslforest.

Mrs. Lo Thi Nguyet's family was assigned 6.3 hectares to afforestation, and the family decided to plant multistorey mixed on the basis of intercropping Melia azedarach-Chukrasia tabularis. Dendrocalamus barbatus (bamboo) and Calamus tetradactylus Hance are planted around to keep the land and prevent erosion. She shared her experience, if planting acacia, it needs 1600 stems/1 ha, but for mixed forest planting, her family planted just only need 950/ha of which 600 roots of Melia azedarach and 350 roots of Chukrasia tabularis/1 ha, and alternating Calamus tetradactylus Hance into the poor soil for ultimizing the land, moving Dendrocalamus barbatus (bamboo) up hills to replace acacia. After 3–4 years, Dendrocalamus barbatus (bamboo) and Calamus tetradactylus Hance will harvest, each every 3–4 weeks for Dendrocalamus barbatus (bamboo), twice a year for Calamus tetradactylus Hance. After 5–8 years is Melia azedarach, and after 8–10 years is Chukrasia tabularis. However, in order to succeed, the forest grows well, the family have to plant in the spring crop, because if the community takes advantage of the time to plant the forest at the end of the beginning of winter, the tree will grow poorly, even killed by hoarfrost.

If previously, the Tai people used to exploit and use forest products, now they know how to plant forests. They have realized that afforestation not only brings economic benefits for the family but also contributes to preventing erosion and landslides and limiting flash floods. In 2013, Tuong Duong planted forests exceeding its assigned targets. Due to fragmented and small soil condition, which is not enough 0.5 ha or more for designing concentrated forestation as prescribed, the district has directed the commune to proposal people to plant scattered trees of small trees. In order to encourage people to plant forests, in addition to the support of the province, in five districts, over 2 billion VND has been budgeted to support 70% of the price of oval seed for farmers. The effect from reforestation is very clear. Many farmer households have come out of poverty from forests, and there are hundreds of households earning a minimum income of VND 30–40 million/year from forestry. In 2014, Tuong Duong deployed to plant 1500 ha of forest, but according to the way people register their own areas, choose the right plant variety to proactively output for the product (this is a new feature in the program, afforestation). Based on the aggregated registration data of the communes, most of the people choose Melia azedarach-Chukrasia tabularis-Dendrocalamus barbatus (bamboo). Nga My commune registers to plant an area of 171 ha, mainly Melia azedarach-Chukrasia tabularis, Yen Hoa 186 ha of Melia azedarach-Chukrasia tabularis, and Tam Quang 314 ha (Melia azedarach-Chukrasia tabularis-Dendrocalamus barbatus bamboo) and at the same time, to restore and conserve forest of Lagerstroemiatomentsa Presl and medicinal plants under the Lagerstroemiatomentsa Presl. Since 2015, people have actively shifted from shifting cultivation to afforestation and actively invested in afforestation, they have shared that afforestation is less difficult than shifting cultivation, but their income is many times higher. (Table 10.4).

Since the government allocated land for people to plant and manage and right of forest protection is also given to the community to directly guard rare forests belongs

Table 10.4 Average income per household in 2018 by income source (million VND)

TT	Commune	Agriculture	Forestry	Livestock	Exploiting forest products	Other source
1	Tam Quang	5.5	60	4.5	6	25
2.	Tam Đình	6.9	63	4.5	6.5	16.6
3.	Tam Hợp	8	62	4.4	6.3	12.7

Source: Data of fieldwork in 2018

Pu Huong and Pu Mat protected areas. Communities develop pledges and regulations on forest protection. Regarding plantation forest area, people still intercrop food crops and vegetables in newly planted forest areas or on exploited forest areas in time of regeneration to ensure food security. In addition to intercropping, people rotate the forest plantation, to ensure that when exploiting, there is no clear cutting and no price decrease of forest products.

Regarding the forest area allocated to the community for management, many villages have restored the sacred space, used as a place of worship of the gods, and the sacred forest areas will never be violated by the community.

4 Conclusions and Discussion

In the fact that natural reserves are still being encroached as they are now, the measures to close forests and protect strictly to prevent deforestation are just a situation that lacks its optimal because the following reasons: Forest closure is still unavoidable due to illegal logging, by isolating forests from community responsibilities, affecting forest culture and community livelihoods; The movement of natural ecosystems needs human influence to create new adaptations to the impact of social systems and stimulate the growth of ecosystems; In addition, the lack of management cohesion between localities and biodiversity conservation units is the cause of the unrestrained illegal forest exploitation, which is sporadically spontaneous. The conversion of poor forest ecosystems to plantations is a sign of lax management.

"Forest management based on communities"is understood to (1) regard the community as the legal entity for the exploitation and use of natural resources, (2) associate the responsibility of the people with the national assets, and (3) Use community knowledge in forest protection, management, and development, especially when local knowledge about forests is increasingly lost by deforestation, which means the loss of the most natural cultural space and forest loss. As a watershed area, there will be many environmental consequences.

Preserving and promoting local knowledge are only feasible when applied in practice. Creating livelihoods from forests is a sustainable measure to maintain the ecosystem, reduce poverty from forests, and create an experience system (new knowledge of the community) based on combining economic benefits and traditional knowledge, associated with natural space, at the same time, increasing the value of forests, one of the principles of rational exploitation and use of resources.

Through forest livelihoods, knowledge of intercropping in traditional culture is preserved by being applied to actual production. Many households take advantage of intercropping rice or vegetables in the early years when the forest is not closed or the model of forest regeneration intercropping with medicinal herbs under the forest canopy. Thus, on the scientific basis, the traditional intercropping method has very high scientific value because of the similarity with the scientific basis of the rationality in increasing the value of forest resources. This is like a fertile land for traditional knowledge "intercropping" to be preserved and promoted.

After a period of reforestation, the experience, and knowledge of afforestation, new forest exploitation has been formed and added to the local knowledge treasure of the people here. It is a deeper understanding of crop behavior, acquiring new techniques to suit the new situation—developing forest economy (ecoeconomy).

The conversion of forest livelihoods will not only make the knowledge of wetland and shifting cultivation fading but also enrich the knowledge of forests, because forest livelihoods have helped the community regain it. The forest culture space for the community "swidden", the spiritual beliefs, and the customary laws about the forest do not change, thus deepening the forest culture of the Tai people.

Method of "Rotation and retirement crops"also needs to be understood with a new connotation that can be applied and promoted in the modern context, which will create a new ecological development succession and can make recommendations on land policy issues.

References

Nghe An Genenral Statistical Office [NAGSO]. (2019). *Statistical yearbook Nghe an 2018* (571 pages). Publishing House.

Vi, V. A. (2017). *Tai people in Western Nghe an* (pp. 105–107). The Gioi Publishing House.

Chapter 11
Minimizing the Negative Effects of Irrigation and Hydropower System on Sustainable Development and Environmental Protection in the Huong River Basin

Nguyen Tien Thanh, Nguyen Dinh, and Nguyen Hoang Son

Abstract Recently, the effects of irrigation and hydropower system in Vietnam generally and Hue particularly have been becoming more complex because of the need for considering hydrologic and hydraulic structures in terms of sustainability and environmental protection. In practice, irrigation and hydropower operations may yield several undesirable impacts like unsafe conditions for dam operations, flood risks for areas located at the downstream of dams, drought risks or environmental impacts cause natural habitat loss. All things may lead to the unsustainable development of social-economic and adversely affect natural resources and environment. The goal of this paper, hence, is to propose the solutions for a minimization of the negative impacts of irrigation and hydropower operations in relationship with natural resources and environment, taking the Huong river basin located in Thua Thien Hue province of Vietnam as an example. In this study, the methods of hydrologic and hydraulic modeling, statistical, and geographic information system (GIS) are applied. On the basis of attained results, an increase in flood storage capacity of Binh Dien and Huong Dien reservoirs is proposed. In addition to this, solutions related to the land cover and inter-reservoir operational process in the Huong river basin are given.

Keywords Huong river basin · Reservoirs · Environment · Hydropower · Natural resources

---

N. T. Thanh (✉) · N. H. Son
Thuyloi University, Hanoi, Vietnam
e-mail: thanhnt@tlu.edu.vn

N. Dinh
Thua Thien Hue Provincial Commanding Committee of Natural Disaster Prevention and Control, Search and Rescue, Thua Thien Hue, Vietnam

A. T. Nguyen, L. Hens (eds.), *Global Changes and Sustainable Development in Asian Emerging Market Economies Vol. 2*, https://doi.org/10.1007/978-3-030-81443-4_11

1 Introduction

Hydropower is considered to be one of the largest sources of renewable energy. It is interested as cheap, flexible, and low polluting renewable energy sources (Kumar et al., 2011). Meanwhile, the irrigation reservoirs are considered to be significant role in flood and drought control as well as water supply. According to the report of the Intergovernmental Panel on Climate Change (Kumar et al., 2011) and SREX (SREX, 2015), it is illustrated that hydropower could be vulnerable to extreme weather events (e.g., heavy rainfall). On the other aspect, hydropower could lead to the vulnerability for living-hood in the downstream (e.g., inundations). Therefore, assessment of these constructions mainly relies on water availability assessment as well as flood and drought management and control. These play a central role in the objectives of sustainable development.

The Vietnamese climate has both tropical and monsoon characteristics due to completely covered by the domain of intertropical zone and monsoon. Consequently, climate regime from region to region is unevenly distributed in space and time. Combined with the topographic conditions of mountainous and hilly, a high frequency of extreme events is recorded in the whole country. Flood events and prolonged droughts can both often record in the same year and region. This is a big challenge for the socio-economic development. Located in the coastal area of central Vietnam, Thua Thien Hue province generally and Huong river basin particularly are frequently hit by tropical cyclones, resulting in devastating floods, landslides, and other natural disasters. Importantly, the issues are compound by the inappropriate discharge from the hydropower and irrigation reservoirs and role of dam operation during heavy rainfall events. Under such conditions, the issues of sustainable development and environmental protection have been drawn much attention from scientists and policy-makers, especially in a global warming. So the study concentrates on fully interpreting the effects of hydropower and irrigation system and dam on hydrologic and hydraulic characteristics. Then solutions for sustainable development are proposed to minimize negative effects of these systems. The HEC-HMS and HEC-RAS models are widely applied due to its advantages for estimating the hydrologic and hydraulic characteristics as pointed out several publications (Halwatura & Najim, 2013; Sardoii, Rostami, Sigaroudi, & Taheri, 2012; Sharma & Mujumdar, 2016; Wang, 2014). Hence, the study uses a system of HEC-HMS and HEC-RAS to estimate hydrologic and hydraulic characteristics under different conditions (e.g., dry season or flood season).

2 Methodology

2.1 Study Area

The Huong river basin is the largest basin in the northern part of the Central Plain, Vietnam. It has an area of about 2830 km^2, a length of 104 km, and average slope of 2.85%. Also, this is a concentrated area of culture, social-economic, and politics activities of the Thua Thien Hue province. It has three major tributaries, namely the Ta Trach, Huu Trach, and Bo. The main flow of Huong river is united and formed from the rivers of Ta Trach and Huu Trach where originated from the mountain range of districts Nam Dong and A Luoi, respectively. Specially, the Ta Trach River and Huong River mainstream originate from the more than 1700 m height Mountain on the northwest of the Bach Ma mountain range. Meanwhile, the Bo River merges into Huong river downstream near the Hue City. Generally, the flow of Ta Trach and Huong rivers is to the northwest from the southeast. They pass the City of Hue and discharge into the Tam Giang lagoon and finally flow to the eastern sea at the Thuan An outlet. Annual precipitation could be reached up to nearly 4000 mm at places (e.g., A Luoi) (Fig. 11.1). The Huong river completely lies in the Thua Thien Hue province where has a wet climate. Maximum monthly precipitation in rainy season (from September to December) is measured at Nam Dong and A Luoi (about 1000 mm). The dry season is from January to August, especially hot and dry in June and July. Over the Huong river basin, the average annual rainfall is always higher than 2600 mm. Recently, lots of irrigation and hydropower reservoirs are developed in the region under the demand of growing energy.

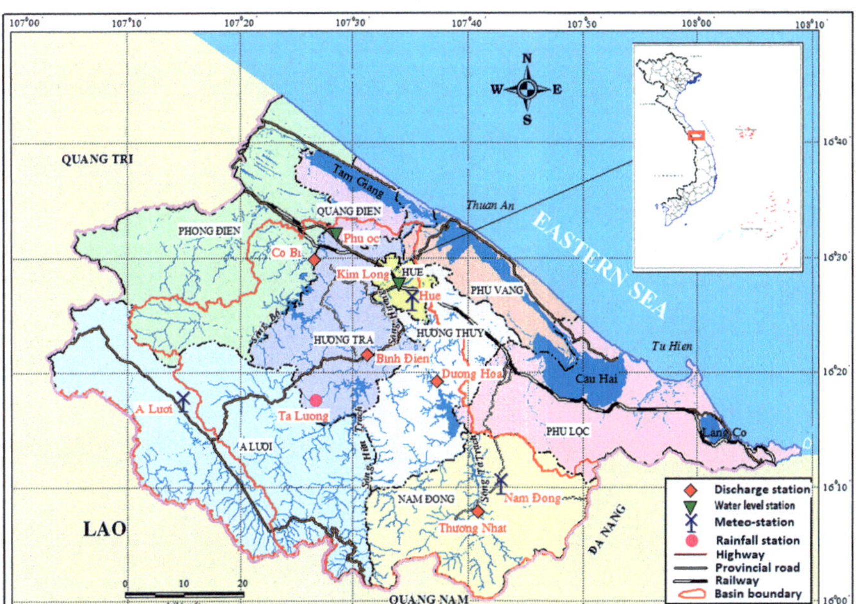

Fig. 11.1 Location of study area

2.2 Data

The information on reservoirs that play a significant role in hydrologic and hydraulic regime is collected from management board of hydropower and irrigation works and Ministry of Industry and Trade of Vietnam. Information on reservoirs in the study is freely available (e.g., https://www.evn.com.vn/d6/news/Thu-tuong-Chinh-phu-ban-hanh-Quy-trinh-van-hanh-lien-ho-chua-tren-luu-vuc-song-Huong-2-10-24660.aspx). It is noteworthy that only reservoirs that have an effective capacity of more than 100 million m^3, over 30 MW for hydropower reservoirs, and over 10 megawatt (MW) for a combined hydropower and irrigation reservoirs are considered. Under these conditions, there are three reservoirs (i.e., Binh Dien, Huong Dien, and Ta Trach) in the Huong River basin. The information on technical documents of Thao Long dam, hydropower, and irrigation works is included. A sample of technical parameters on reservoirs of Ta Trach, Binh Dien, and Huong Dien is shown on Fig. 11.2.

Cross-sections of topographic on the route of river are measured with the information on parameters as presented in Table 11.1. Besides that, hydro-meteorological data are collected from the Vietnam Center of Hydro-meteorological data. Digital elevation model of 30 m resolution and Landsat image are also included.

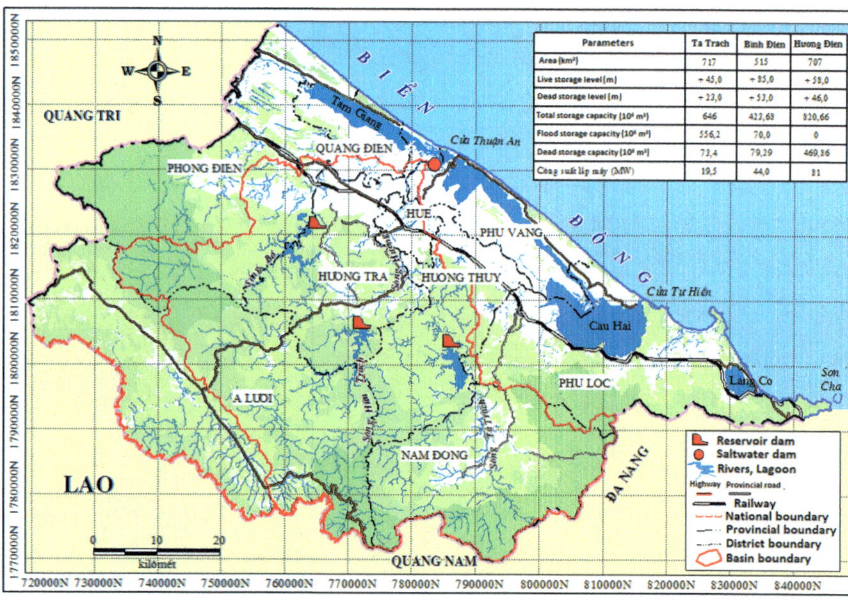

Fig. 11.2 Parameters of reservoirs

Table 11.1 Number of cross-sections and length of rivers

Nr.	River	Length (km)	Number of cross-sections
1	Huu Trach (Binh Dien – Tuan)	7.7	8
2	Ta Trach (Duong Hoa – Tuan)	14.5	24
3	Huong (Tuan – Tam Giang lagoon)	34.3	59
4	Bo (Huong Dien – Sinh)	31.3	37
5	Bo (Bac Vong – Dong Lam)	4.2	8
6	An Xuan Chanel (Dong Lam- An Xuan)	9.64	8
7	Dien Hong Chanel (Dong Lam- Ha Do)	10.2	8
8	Kim Doi (Thanh Ha – Quan Cua)	6.0	7
9	5 xa Chanel (Nham Bieu – Bo river)	14.3	17
10	Dai Giang (Phu Cam – Quan sewer)	27.1	39
11	Nhu Y (Dap Da – Dai Giang river)	15.1	29
12	Pho Loi (La Y - Dien Truong sewer)	5.9	8
13	Other rivers surround Hue city	6.1	16

2.3 *Models*

2.3.1 Hydrologic and Hydraulic Modeling

HEC-HMS, a Hydrologic Modeling System, is designed to describe the physical properties of river basins and the meteorology that occurs on them (US Army Corps of Engineers, 2008). It also describes the result of runoff and streamflow that are produced. It is physically based and conceptually semi-distributed model and easily operates huge tasks in relation to hydrological studies. The components of model include losses, runoff transform, open channel routing, meteorological data, rainfall-runoff simulation, and parameter estimation (Hydrologic Engineering Center [HEC], 2005). The HEC-HMS, plus, uses separate models that compute runoff volume, models of direct runoff, and models of base flow. Lots of methods involved in this model are to estimate the losses. Some of them are primarily designed to simulate the flood events, while others are intended for continuous simulation. Also, seven different transformation methods (e.g., the Snyder Unit Hydrograph Yilma and Moges (Yilma & Moges, 2007) or Clark Unit Hydrograph Banitt (Banitt, 2010) are available in the model. Advantages and disadvantages of some of these methods are clearly documented (Razmkhah, 2016). Version 3.5 of HEC-HMS is used in this study.

In this study, the Hydrologic Engineering Center's River Analysis System (HECRAS, version 4.1.0) is applied. This is a one-dimensional, hydraulic-flow model primarily developed by the US Army Corps of Engineers - Hydrologic Engineering Center (HEC) (US Army Corps of Engineers, 2010a, 2010b).

Specifically, it is designed for the studies in flood-plain management and flood-insurance. By using this model, the floodway encroachment and estimated flood inundation are simulated and evaluated. A number of input parameters that can be changed to get the best results in hydraulic characteristics are included in HEC-RAS. On the basis of these parameters, a series of cross-sections along the stream could be established. At each cross-section, several input parameters could be used to describe the characteristics (i.e., shape, elevation, and relative location) along the stream. Additionally, the main parameters (e.g., the number of cross-sections, lateral, and elevation coordinates for each terrain points) are described. Advantages of HEC-HMS and HEC-RAS are closely connected via the DSS program (Hydrologic Engineering Center [HEC], 2005). This program is incorporated into most of HEC's major application programs. To validate the performance of the model, Nash-Sutcliffe index (Krause, Boyle, & Bäse, 2005) is used.

$$R^2\left(EI\right)=1-\frac{\Sigma\left(Q_{cal}-Q_{obs}\right)^2}{\Sigma\left(Q_{obs}-Q_{obstb}\right)^2}$$

where Q_{cal} is calculated discharge (m^3/s), Q_{obs} is measured discharge (m^3/s), and Q_{obstb} is average measured discharge (m^3/s).

3 Results

3.1 Preprocessing

3.1.1 Sub-Basin and Hydrologic-Hydraulic Network

As the first step, the Huong river basin is divided into 17 sub-basins using the ArcGIS and digital elevation models as shown in Fig. 11.3a. On the basis of sub-basins, hydrologic network is produced for HEC-HMS (Fig. 11.3b, Table 11.2).

The system of reservoirs and river network is simulated as presented in Fig. 11.4. River network for Huong river basin briefly described as Ta Trach (Duong Hoa), Binh Dien, Huong Dien (Co Bi) reservoirs to Thao Long dam and Dien Hong, Kim Doi, Pho Loi, Nhu Y, Dai Giang, and An Xuan tributaries through the sewers of Ha Do, An Xuan, Quan Cua, Dien Truong, Cau Long, and Quan. Then they flow into Tam Dang-Cau Hai lagoon and to the sea at the Thuan An and Tu Hien outlets.

Hydraulic network is shown in Fig. 11.5. Grid cells located in the downstream of Huong have a large area. Location and area of grid cells are defined by Landsat image in combination with digital elevation models of 30 m.

3.1.2 Calibration and Validation of HEC-HMS and HEC-RAS

For the Huong river basin, the flow to reservoirs is calculated on the basis of HEC-HMS at stations (i.e., Thuong Nhat, Nam Dong, Binh Dien, Hue, Kim Long, A Luoi, Ta Luong, Co Bi, and Phu Oc). The outputs of HEC-HMS are automatically connected to HEC-RAS via the HEC-DSS program. Lateral boundary is based on the HEC-HMS output. The stations used to validate are Kim Long on the Huong river and Phu Oc on the Bo river. Lower boundary is the hourly water level data at the outlets of Thuan An and Tu Hien.

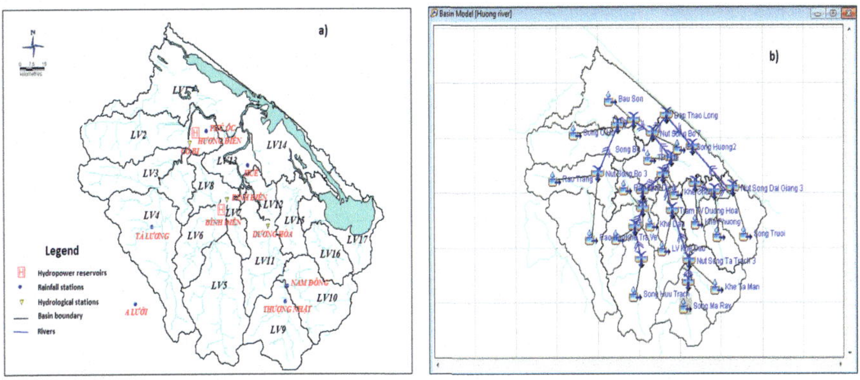

Fig. 11.3 Sub-basin map (**a**) and hydrologic network (**b**) for Huong river basin in HEC-HMS

Table 11.2 Detail information of sub-basin

Name	Restriction zone	Area(km^2)
LV1	Restricted basin by Bau son	454.5
LV2	Restricted basin by O ho river	351.4
LV3	Restricted basin by Rao Trang river	300.5
LV4	Restricted basin by Rao Nho river	409.4
LV5	Restricted basin by Huu trach river	360.9
LV6	Restricted basin by Tra Ve rivulet	157.9
LV7	Restricted basin by day rivulet	58.3
LV8	Restricted basin by Rao Binh Dien river	109.5
LV9	Restricted basin by ma ray river	231.8
LV10	Restricted basin by ta man rivulet	238.8

(continued)

Table 11.2 (continued)

Name	Restriction zone	Area(km^2)
LV11	Restricted basin by Dau rivulet	247.2
LV12	Restricted basin by chon rivulet	89.2
LV13	Restricted basin by Huong river	130.0
LV14	Restricted basin by Huong (downstream) river	293.0
LV15	Restricted basin by Phuong rivulet	161.4
LV16	Restricted basin by Nong river	115.2
LV17	Restricted basin by Truoi river	147.3

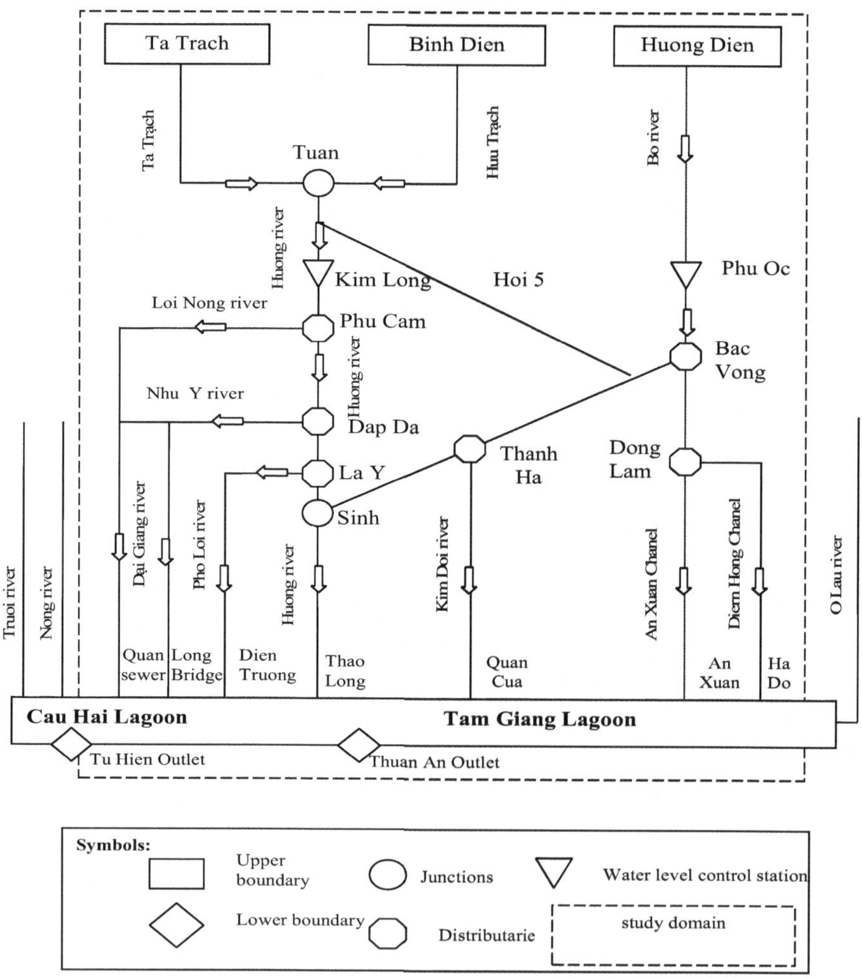

Fig. 11.4 Diagram of river network and irrigation system in Huong river basin

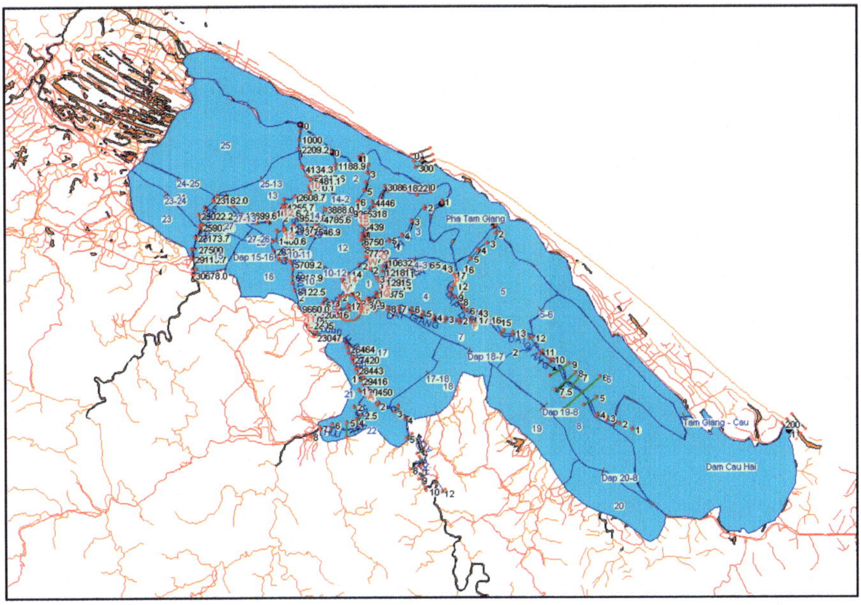

Fig. 11.5 Map of hydraulic network in HEC-RAS

Table 11.3 Parameters of HEC-HMS model

		Parameters								
Sub-basin	Station	CN	Ia	tLag (h)	Cp	Qbq (m^3/s)	Rc	R	x	k (h)
Bo river	Co bi	60	2	12	0.45	91.7	0.97	0.1	0.2	6
Huu trach	Binh Dien	60	2	12	0.52	68.2	0.97	0.05	0.2	6
Ta trach	Duong Hoa	60	2	6	0.42	76.5	0.97	0.1	0.2	6

As first step, the performance of the HEC-HMS model is fully clarified under two cases (i.e., daily and hourly discharge). For daily discharge simulations, daily precipitation and discharge data in 1983 and 1986 are used to calibrate the model. Daily precipitation and discharge data in 1984 and 1987 are used to validate the model. As a result, parameters of HEC-HMS are presented in Table 11.3.

In general, the Nash index for all stations (i.e., Co Bi, Binh Dien, and Duong Hoa) reaches over 0.5. Specifically, at Binh Dien station, the Nash index could be reached up to closely 0.7 for both calibration and validation (Fig. 11.6). It is documented that the performance of model well captures the measured data. It should be noted, however, these values are considered as acceptable values due to uneven distribution of rainfall stations over the basin. The monitoring time is not synchronized. More importantly, rainfall regime is not fully interpreted the discharge regime of the river. The reason for this is come from multiple factors affected the discharge regime of the river like elevation slope, patterns of weather conditions, and vegetation.

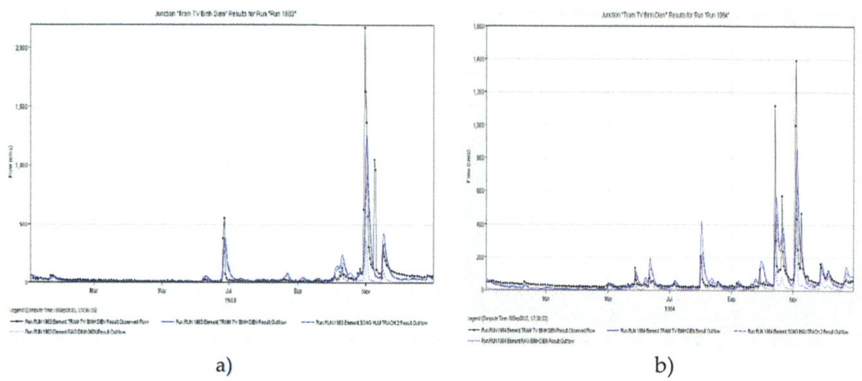

Fig. 11.6 Calibration in 1983 (**a**) and validation in 1984 (**b**) between calculated and measured discharge for Binh Dien

For hourly discharge simulations, measured data at Co Bi (October 14–16, 1981; October 15–19, 1985), Binh Dien (October 13–15, 1984; October 15–18, 1985), and Duong Hoa (October 10–13, 1986; November 17–23, 1987) are used to calibrate and validate, respectively. The simulations closely fit the measured data with the Nash values of 0.9, 0.94, and 0.92 for calibration and 0.78, 0.9, and 0.95 for validation at Co Bi, Binh Dien, and Duong Hoa stations, respectively. It is emphasized that the peak of flood events could be well captured by the model. An example for Duong Hoa station is shown in Fig. 11.7. All outputs of HEC-HMS are used as inputs for HEC-RAS.

As the second step, the performance of the HEC-RAS model is fully interpreted under three cases (i.e., daily water level, hourly water level in water, and dry seasons) at stations Kim Long and Phu Oc. Daily water level in 1984 and 1999 at Kim Long and Phu Oc is used to calibrate and validate the model, respectively. The results are in very good agreement with the Nash indices of calibration (0.56 at Kim Long, 0.57 at Phu Oc) and validation (0.66 at Kim Long and 0.59 at Phu Oc). In case hourly water level in water season, a series of data during the flood event from September 13, 1984 to October 30, 1984 is used to validate. It is illustrated an agreement with the Nash indices of 0.63 and 0.77 between the water level simulations and measurement for Kim Long and Phu Oc, respectively. In dry season, a series of data from June 1, 1984 to August 31, 1984 is used to validate the model. The Nash value of 0.62 is estimated for both Kim Long and Phu Oc. Figure 11.8 shows the results of validation for Kim Long in both flood event and dry season.

As depicted in the Fig. 11.8 that the peak values are underestimated in comparison with measured data, but acceptable simulations due to the uneven distribution of rainfall stations. Consequently, the climate and hydrological regime are not completely clarified for the basin. The parameters of HEC-HMS and HEC-RAS, then, are used to assess the changes in hydrologic and hydraulic features for the downstream of Huong river basin under the impacts of hydropower and irrigation reservoirs.

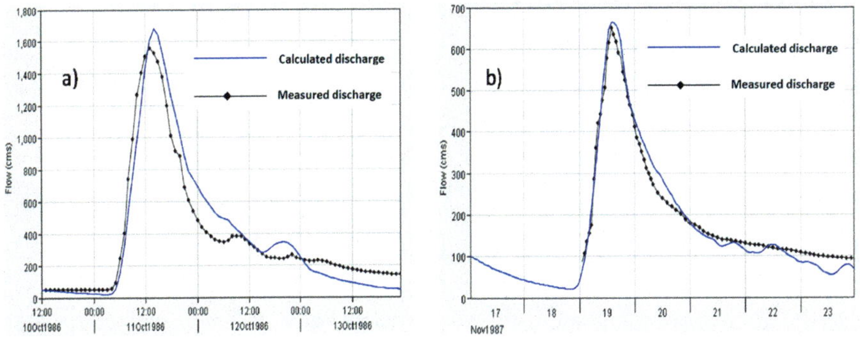

Fig. 11.7 Calibration in 1986 (**a**) and validation in 1987 (**b**) between calculated and measured discharge for Duong Hoa

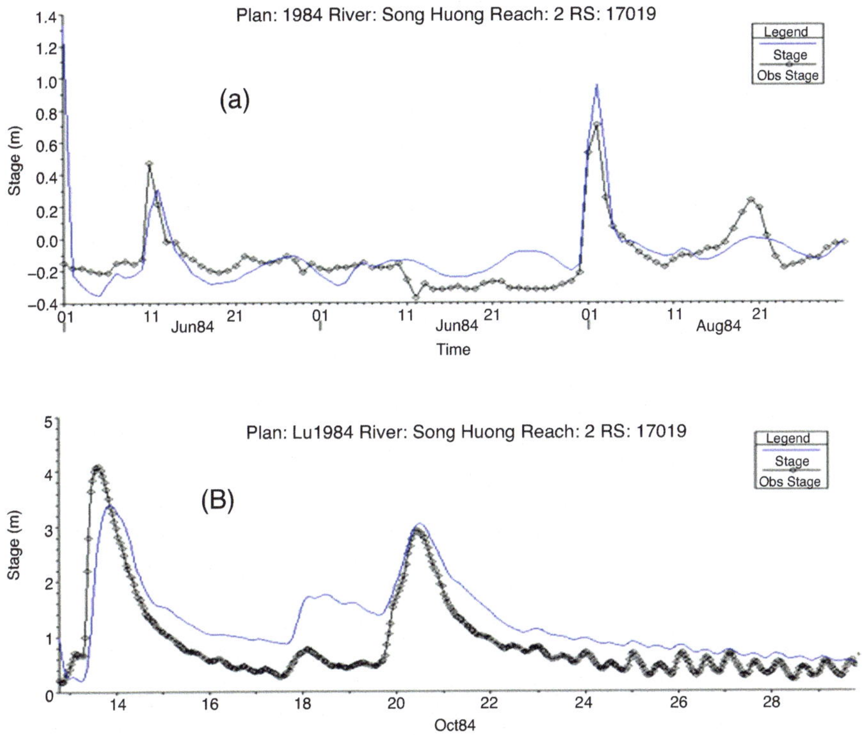

Fig. 11.8 Validation of HEC-RAS for Kim Long in dry season (June 1, 1984 – August 31, 1984) (**a**) and in flood event (October 1984) (**b**)

3.2 *Main Impacts of Hydropower and Irrigation Reservoirs on Hydrologic and Hydraulic Regimes*

The features of hydrologic and hydraulic for the downstream of Huong river basin are considered at water level control stations (Kim Long and PhuOc).

3.2.1 Changed Flows in Flood Season

For the flood season, typical flood events in 1999 and 1983 are selected to interpret the changes in hydrologic and hydraulic characteristics. The cases are mentioned including (1) TH1: Nonreservoirs and dam, (2) TH2-PA1: Reservoirs and dam are operated under each reservoir operation produce, (3) TH2-PA2: Reservoirs and dam are operated in collaboration with the flood controlling under water levels of flood alarms at the downstream, (4) TH2-PA3: Reservoirs and dam are operated with the increasing of flood storage capacity for Huong Dien and Binh Dien.

Table 11.4 shows changes in flood characteristics before and after constructions at Kim Long in corresponding to the historical flood in 1999. It is illustrated that with the constructions of reservoirs and dam, the characteristics of floods are significantly changed in comparison with nonreservoirs and dam. With the reservoirs and dam, the flood peak is pulled down to 5.81 m (TH2-PA1), 5.13 m (TH2-PA2), and 5.1 m (TH2-PA3) from 6.09 m (TH1). More importantly, the time of flood occurrence that water level is higher than 3 m is significantly decreased by 16 to 25 hours. The flood intensity to the maximum water level is reduced by 0.02 to 0.3 (m/hour).

Similarly, Table 11.5 shows changes in flood characteristics before and after constructions at Kim Long in corresponding to the historical flood in 1983. It is observed that with the constructions of reservoirs and dam, the characteristics of floods are relatively changed in comparison with nonreservoirs and dam. With the reservoirs and dam, the flood peak is pulled down to 4.7 m (TH2-PA1), 3.81 m (TH2-PA2), and 3.56 m (TH2-PA3) from 5.0 m (TH1). More importantly, the time of flood occurrence that water level is higher than 3 m is significantly reduced by 8–9 h. The flood intensity to the maximum water level is reduced by 0.02 to 0.2 (m/h).

Table 11.4 Changes in flood characteristics before and after constructions at Kim Long in corresponding to the historical flood in 1999

Flood characteristics at Kim Long	TH1	TH2-PA1	TH2-PA2	TH2- PA3
Maximum water level Hmax (m)	6.09	5.81	5.13	5.1
Flood intensity to the average water level (m/hour)	0.20	0.15	0.09	0.1
Flood intensity to the maximum water level (m/hour)	0.54	0.52	0.24	0.35
Time for water level ≥3 m (hour)	101	97	96	93

Table 11.5 Changes in flood characteristics before and after constructions at Kim Long in corresponding to the historical flood in 1983

Flood characteristics at Kim Long	TH1	TH2-PAI	TH2-PAII	TH2- PAIII
Maximum water level H_{max} (m)	5.00	4.70	3.81	3.56
Flood intensity to the average water level (m/hour)	0.09	0.07	0.06	0.05
Flood intensity to the maximum water level (m/hour)	0.35	0.33	0.15	0.14
Time for water level ≥3 m (hour)	57	48	49	48

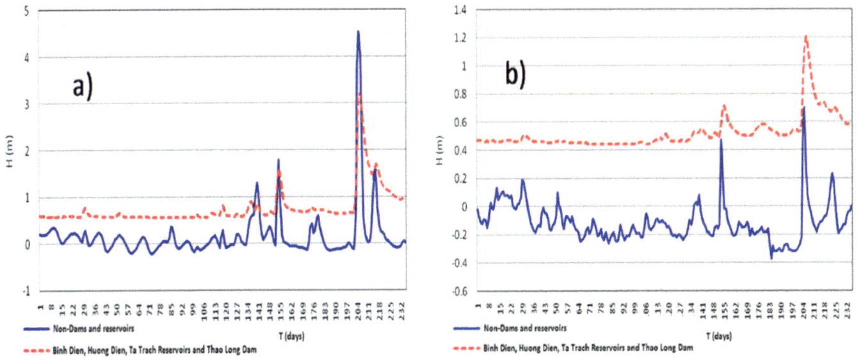

Fig. 11.9 Changed water level in Phu Oc (**a**) and Kim Long (**b**) in dry season for interannual flows with and without dams and reservoirs

3.2.2 Changed Flows in Dry Season

The dry season is defined from January to August. The results in changed flows at the Kim Long and Phu Oc are documented in Fig. 11.9.

As shown in Fig. 11.9, it is illustrated the role of reservoirs and dam is reasonable. The flows are stable with the effects of minimized tidal. Water level is increased by 0.81 m opening the conditions of potential water exploitation for agriculture and water supply for other areas.

3.3 Proposal on Minimizing the Negative Effects of Irrigation and Hydropower System on Sustainable Development

The results show a significant effect of irrigation and hydropower system on the changes in hydrologic and hydraulic features over the Huong river basin. Advantages of these systems are salinity preventing, water level control in dry season, and reduced-flood peaks. The potential disadvantages are mentioned (e.g., potential inundations in the downstream of Huong river when unreasonable reservoir operation or reduced sediments at the Thuan An outlet). Importantly, flood with the frequency of less than 10%, Ta Trach plays a central role in reducing the inundations

Table 11.6 Proposed Flood storage capacity for reservoirs

Reservoir	Present		Proposal	
	Vpl (10^6 m^3)	% Vho	Vpl (10^6 m^3)	% Vho
Ta trach	556.2	86.1	556.2	556.2
Binh Dien	70	16.5	150	35.4
Huong Dien	0	0	200	24.4
Total	626.2	33	906.2	48

Table 11.7 Water level of flood peak in the downstream of Huong river basin under the increasing of flood storage capacity

	Flood event in 1983			Flood event in 1999		
	TH1: Vpl	TH2: Vpl2	Changes in water level (m)	TH1: VPL	TH2: Vpl2	Changes in water level (m)
Kim long	4.70	3.56	−1.14	5.81	5.10	−0.71
Phu Oc	4.43	4.30	−0.13	4.60	4.47	−0.13

in the downstream. Meanwhile, the roles of Binh Dien and Huong Dien reservoirs are negligible under the issued reservoir operation processes of Vietnam. In dry season, all hydropower and irrigation system could be seen from the stabilizing and increasing of flows. Thao Long dam prevents the negative effects of tidal on the downstream of Huong river basin. With the obtained results, proposal on minimizing the negative effects of irrigation and hydropower system is concentrated on the construction solutions. Presently, Ta Trach reservoir has the largest flood storage capacity of 556.2 million m^3 in comparison with a total storage capacity of 646.0 million m^3. Meanwhile, flood storage capacity of Binh Dien is 70 million m^3 out of total storage capacity of 423.68 m^3 (accounting for 16.5%). Notably, there is not a flood storage capacity for Huong Dien reservoir. Table 11.6 gives proposed values for minimizing the negative effects of reservoir system with Vpl is flood storage capacity and Vho is total storage capacity.

The study calculates the efficiency of this solution for specific cases as follows: (a) Binh Dien reservoir with the flood storage capacity of 150 million m^3 and water level before floods of +65.03 m; (b) Huong Dien reservoir with the flood storage capacity of 200 million m^3 and water level before floods of +51.63 m; (c) Ta Trach reservoir with the flood storage capacity of 556.2 million m^3 and water level before floods of +25 m; and (d) Thao Long dam is completely opened. For historical flood events in 1983 and 1999, the results in water level of flood peak are presented in Table 11.7.

Obviously, it can be seen from the Table 11.7 that Binh Dien and Huong Dien reservoirs with the increasing of flood storage capacity lead to the water level reducing at the downstream of Huong river basin in both 1983 and 1999. Specially, the water level of flood peak at Kim Long on the main Huong river falls down to 3.56 m from 4.7 m in 1983 and 5.1 m from 5.81 m in 1999. Besides that, the study also proposes to enhance the vegetation cover in the upstream of reservoirs system. With the forest zone, it prevents the erosion, landslide or jumping sand, and reduces the sediment of reservoirs.

4 Conclusions

The study presents the effects of reservoirs and dam system on the hydrologic and hydraulic characteristics in the downstream of Huong river basin. Importantly, the main impacts of hydropower and irrigation reservoirs on hydrologic and hydraulic regimes are fully interpreted in both flood and dry season. The water level in dry season is increased for exploitation activities in a variety of different fields like agriculture. Meanwhile, flood flows are partly controlled with the reservoir and dam system. More importantly, the solutions of nonconstruction and construction are proposed to minimize the negative effects of these systems on flows and water level in the downstream. The proposal is to increase the flood storage capacity for Binh Dien and Huong Dien from 70 to 150 million m^3 and from 0 to 200 million m^3, respectively.

References

Banitt. A (2010). Simulating a century of hydrographs e Mark Twain reservoir. In *Proceedings of the 2nd Joint Federal Interagency Conference*. Las Vegas, NV, USA.

Halwatura, D., & Najim, M. M. M. (2013). Application of the HEC-HMS model for runoff simulation in a tropical catchment. *Environmental Modelling & Software, 46*, 155–162.

Hydrologic Engineering Center [HEC]. (2005). *HEC-DSS Vue HEC data storage system visual utility engine: User's manual.*

Krause, P., Boyle, D. P., & Bäse, F. (2005). Comparison of different efficiency criteria for hydrological model assessment. *Advances in Geosciences, 5*, 89–97.

Kumar, A. T. S., Ahenkorah, A., Rodriguez, R. C., Devernay, J.-M., Freitas, M., Hall, D., Killingtveit, A., & Liur, Z. (2011). Chapter 5: Hydropower, Report. *IPCC special report on renewable energy sources and climate change.*

Razmkhah, H. (2016). Comparing performance of different loss methods in rainfall-runoff modeling. *Water Resources, 43*(1), 207–224.

Sardoii, E. R., Rostami, N., Sigaroudi, S. K., & Taheri, S. (2012). Calibration of loss estimation methods in HEC-HMS for simulation of surface runoff (case study: Amirkabir dam watershed, Iran). *Advances in Environmental Biology, 6*(1), 343–348.

Sharma, P., & Mujumdar, S. (2016). Dam break analysis using HEC-RAS and HEC-GeoRAS—A case study of Ajwa reservoir. *Journal of Water Resources and Ocean Science, 5*(6), 108–113.

SREX (2015). *Vietnam special report on managing the risks of extreme events and disasters to advance climate change adaptation.* Retrieved September 7, 2019, from http://www.vn.undp.org/content/vietnam/en/home/library/environment_climate/viet_nam_special_report_on_managing_the_risks_of_extreme_events_and_disasters/

US Army Corps of Engineers. (2008). *Hydrologic Modeling System (HEC-HMS) Application Guide Version 3.1.0.* Davis, CA: Institute for Water Resources.

US Army Corps of Engineers. (2010a). *HEC-RAS. User Manual.* Hydrologic Engineering Center. Version 4.1.

US Army Corps of Engineers. (2010b). *HEC-RAS. Hydraulic Reference Manual*. Hydrologic Engineering Center. Version 4.1.

Wang, C.H (2014). Application of HEC-RAS model in simulation of water surface profile of river. In *Applied mechanics and materials* (vol. 641, pp. 232–235), Trans Tech Publications.

Yilma, H.M., & Moges, S.A (2007). Application of semi-distributed conceptual hydrological model for flow forecasting on upland catchments of Blue Nile River Basin, A case study of Gilgel Abbay catchment. Catchment and Lake Research 200.

Chapter 12
Socio-Economic Development toward Sustainable Ecological Model in Vietnam

Pham Cong Nhat

Abstract The paper deals with a general concept of socio-economic development following ecological model. It is a model of socio-economic development based on three important pillars: green economy development, sustainable environmental protection, and building a democratic social institution also known as construction and socio-economic development according to sustainable ecological model. Socio-economic development under the sustainable ecological model is being chosen by many countries in the world as an inevitable trend, especially in the world climate conditions with many changes in the negative direction as currently. Part of the content of this paper initially takes some characteristics of socio-economic development following the sustainable ecological model in Vietnam today. Accordingly, the paper stated that, like many other countries in the world, Vietnam is currently choosing socio-economic development based on a sustainable ecological model. However, besides the favorable factors, Vietnam is facing many difficulties and challenges in choosing this development model. There are many solutions for Vietnam to choose for socio-economic development based on the current sustainable ecological model, but in particular need to pay special attention to solutions related to development green economic development, sustainable environmental protection, and building socio-economic institutions more and more complete and democratic.

Keywords Green economy · environmental protection · sustainable development · democratic social institutions · sustainable ecological models

P. C. Nhat (✉)
VNU University of Social Sciences and Humanities, Hanoi, Vietnam

A. T. Nguyen, L. Hens (eds.), *Global Changes and Sustainable Development in Asian Emerging Market Economies Vol. 2*, https://doi.org/10.1007/978-3-030-81443-4_12

1 Introduction

In the history of its development, it seems that this is the first time humanity has faced to the great threats and challenges in the development caused by the influences of its own uncontrolled actions. The predictions that the development of the humanity is confronted with the disintegration of social and natural ecological structures seem to become a reality for all nations over the world, especially countries that are directly affected by climate change as Vietnam. As a developing country in the context of many impacts from climate change trends in the world, the selection of solutions for sustainable ecological development is considered as a smart choice of Vietnam. According to the Vietnamese Government: Vietnam always determines sustainable development as both a requirement and a goal: "rich, strong, fair and civilized people; contribute to preserving peace, strengthening cooperation, prosperity in the region and the world". At the same time, The Vietnamese Government also affirmed: "Sustainable development is the right choice, if not the only one. This poses many challenges to all countries, especially for developing countries"(Vu, 2018). This paper bases on the analysis of the real situation, then recommends some soulutions to construct and develop the society as well as the economy in the sustainable ecology model – a development model regarded as one essential trend associated with specific conditions in Vietnam today.

2 Methodology

- General analytical methods. We use general analysis methods (including qualitative and quantitative analysis methods) to analyze and describe the socio-economic changes and development of Vietnam according to the model eco sustainable. In particular, in this paper we also pay attention to the use of case analysis methods. Through this method, based on comparative analysis of data from developed and developing countries in the region and around the world to confirm the case of Vietnam - a country that still belongs to a group of developing countries, but boldly selected and initially successful in socio-economic development based on the sustainable ecological model.
- Interdisciplinary approach. Although this is a philosophical study, the research object of this paper is related to the research subjects of many different sciences, so in addition to using general research methods of philosophy, we also use a number of related interdisciplinary approaches such as ecology, ecology economics, ecology politics, ecological ethics…, to analyze and do so. It is clear that the current trend of socio-economic development and development in Vietnam follows the sustainable ecological model.
- Some of the content presented in this paper is also part of the research results of some of the same research topics of my work with colleagues at National University of Hanoi (VNU), Vietnam in recent years, etc.

3 Results

3.1 General Concept of Socio-Economic Development Following Sustainable Ecological Model

Sustainable ecological model is a new term in ecological sciences that describes an artificial ecological structure that includes natural, social, and human factors existing and interacting, assisting each other in order to achieve a stable, long-term (sustainable) development goal. Developing the society and the economy in the sustainable ecology model is also a concept used by many countries in recent years. This concept which refers to the construction and development of a particular country's socio-economy bases on a sustainable ecology model and three basic pillars: green economy, sustainable environmental protection, and development and democratic social institutions.

Green economy. According to the economists, green economy is defined as an economy which mitigates environmental risks and ecological scarcity and aims at the sustainable development without impairing the environment. On the one hand, the green economy is closely related to ecological economics. On the other hand, it also has some connections to politics (Kahle & Gurel-Atay, 2014). Visualizing the green economy is a combination of three elements: economic, social, and environmental. Sustainable green economy is an economy that generates a lot of profit or useful values toward the development of human society's life. Today, the concept of green economy is not only understood in terms of activities in the economic sector alone but also in cultural and social activities, in which environmental friendly factors are considered Core elements in activities to create green economy. The report of the green economy of the United Nations Environment Program (UNEP) in 2011 also states that "to be green, an economy must not only be efficient, but also fair. Fairness implies recognising global and country level equity dimensions, particularly in assuring a just transition to an economy that is low-carbon, resource efficient, and socially inclusive" (UNEP, 2011). A green economy is an economy or model of economic development basing on sustainable development and knowledge of ecological economics. The primary goal of green economy is to create jobs, ensure sustainable economic growth, and prevent environmental pollution, global warming, resource depletion, and environmental degradation. One of the distinguishing features of green economy is the emergence of green labeling and eco-labeling methods that have appreared when the consumers have to face the friendly measurements for the environment and sustainable development. Many industries are beginning to apply these standards as an effective way to promote the green practices in a globalized economy… Green economy development is considered as one of the basic pillars of socio-economic development in the ecological and sustainable model.

Environmental protection and sustainable development. Environment in a narrow sense does not take into account natural resources, but only includes natural and social factors directly related to the quality of human life. The environment in a broad sense is all the natural and social factors necessary for human living and production, such as natural resources, air, land, water, light, landscape, and relations society. The concept of environment, in terms of sustainable development, is understood in a broad sense including all factors of infertility and biology, nature, and society, which have direct or indirect effects on health, human life. Human life whether in the early period or even in the modern development period depends very much on the surrounding environment. Humans are facing many challenges when the environment is decline or degraded environment. Environmental degradation is a change in the quality and quantity of environmental components, adversely affecting the lives of people and nature, destroying ecosystems, and extinction of wildlife. In particular, environmental components are understood as elements that create the environment: air, water, land, sound, light, underground, mountains, forests, rivers, lakes, sea creatures, ecosystems (Johnson et al., 1997).

Sustainable environmental protection and development must first be is to protect the natural environment so that it is always clean and to ensure that every organism can survive there. In the ecological sciences, the concept of sustainability (originated from sustainable and ability) is the property of biological system to maintain the limitless diversity and productivity. Principle of organization for sustainability is the sustainable development including four interconnected sectors: ecology, economics, politics, and culture (Andy et al., 2013; James, Magee, Scerri, & Steger, 2015). By approaching it continually and dynamically, the result of this process can form a sustainable system (Wandemberg, 2015). However, sustainability is first manifested in the natural environment. Therefore, environmental protection and sustainability must first contribute to constructing and preservating the nature sustainably. The natural environment itself is a perfect ecosystem, but since the appearance of human beings and their excessive intervention, the natural environment has become unbalanced, polluted, which threatens the survival of every living organism. Protecting the environment (first of all is natural environment) sustainably is a very important premise of constructing and developing the society and the economy in a sustainable ecological model.

Democratic social institutions. Today, "democracy" has become a fundamental term of popular and dominant political theories in the world. This is reflected not only in a large number of democratic countries in reality, but also in how most countries in the world, regardless of its political system, call themselves a system. Democratic system, or rule on behalf of the people and for the benefit of the people, including Vietnam. A political regime or a government takes the element of "democracy" into the fundamental, fundamental principles for maintaining, managing, and developing society thereby creating a specific institution called democratic society institutions. Democratic social institution is the system of institutions, values, norms of freedom, and democracy that incorporates the principles of organization and operation of democracy that guarantees human rights and the ability to self-protect human rights of each individual. The democratic society is a social framework founded on democracy - a state regime built on positive political ideology whose official goal is to create a system of adequate mechanisms and legal environment through reformed and progressive measures. In a democratic society, all activities of the state and citizens are governed by a democratic system, in which the ownership of the people is respected and ensured by the legal system through the tool that is mainly the rule of law. As the present democratic social institutions apply universal suffrage for their voting system (Rex, 2003), it is necessary to mobilize people's participation, social criticism, and fight for social progress. In fact, now a democracy is based on three main pillars: *first, the market economy* is the economy operated by the influence of market rules with many economic components which exist competitively and equally. A perfect market economy is often regarded as the backbone of the democratic society. *Second, the rule of law*, is the form of the State building the law to manage society and putting itself under the law. In the state, the rule of law is always placed on the top with the principle "People have the right to do all that the law does not prohibit and use the law to protect their legal rights and benefits. State agencies and cadres, civil servants can only do what the law allows. All management decisions of the State must be transparent" (Nguyen, 2014). Third, *democratic social institutions*, it is the society run by democratic social institutions within the framework of a complete rule of law. In democratic social institutions, all thoughts, aspirations, and voices of the people are respected. The state was born not to protect the interests of the faction but to be a tool for the people to implement their mastery. The core of democracy lies in the recognition of the difference. The nature of the difference is each one's own. Human society is always diverse and complex, because it has both a common and a separate one. If it is impossible to find the common, the collective, the community, it is difficult to form society. But if we only acknowledge the existence of the common and deny the existence of the particular, then society cannot develop well. During the period of feudalism, a social regime lasting thousands of years, people only impose common things on everyone. Only in modern times, when the sense of individualism is formed, do people respect the privacy and privacy of each person. In the process of operation and management, only when the State creates the necessary mechanisms and legal systems to respect and protect differences, only then can a real democratic society be created. Democratic social institution is also one of the important objectives in socio-economic development in the sustainable ecology model at the present.

3.2 Some characteristics of Socio-Economic Development Following the Sustainable Ecological Model in Vietnam Today

Developing a sustainable society and economy in the sustainable ecological model is one of the essential trends of many countries in the world, including the ones that are suffering many impacts of global climate change trends such as Vietnam.

In Vietnam, from The eighth Congress (1996), The Communist Party of Vietnam has started to set goals for the strategy of constructing and developing the social-economy in the direction of "rapid economic growth, high efficiency and sustainability gowith solutions for social problems, security and national defence ensured, consolidation of national security and defense, improvement of public life..." (VCP, 1996, p. 82). The viewpoint of Vietnam's socio-economic development has been affirmed in the documents of the ninth, 10th, and 11th Congresses, especially in the 2011–2020 socio-economic development strategy. That is "rapid development associated with sustainable development, sustainable development is required throughout the Strategy" (VCP, 2006, p. 98). At The 12th Congress, the viewpoint of constructing and developing the social-economy in a sustainable ecological model was concretized once again by The Communist Party of Vietnam with the strategy "renew the growth model". Therefore, "Model of growth in the coming time will prefer the combination of intensive and extensive development with more concentrations on intensive development, whose goal is to raise the productivity and competitiveness by effectively exploiting scientific-technological advances, increasing labor productivity, improving the quality of human resources, being active in international integration, developing rapidly but sustainably, harmonizing short-term and long-term goals as well as economic growth and cultural development, exercising social progress and equality, being eco-friendly, and ameliorating the material and spiritual life of the people" (VCP, 2016, p. 87). Particularly, The Politburo, The Central Committee of the Communist Party of Vietnam has recently issued Resolution no 23-NQ/TW on 23rd March 2018 on *Orientation for constructing the national industrial development policy* until 2030 with the vision to 2045. One of the primary guiding viewpoints emphasized by the Resolution was: "Make use of as much as possible the country's advantage in the period of golden population; exploit thoroughly the achievements of the Fourth Industrial Revolution (4.0), the advantages of the trade in order to develop effectively some key industries with the strategies, competitive edge; develop decisively the information technology industry, electronic industry; develop centrally the processing and manufacturing industry; develop considerably the intelligent manufacturing industry; focus on developing the green industry" (VCP, 2018). According to the analysis above, sustainable ecology model is urgent and essential for Vietnam's current society and economy.

Being a developing country carrying out industrialization and modernization, in recent years, Vietnam has been recognized as one of the countries with fast socio-economic development pace in the region and over the world. After more than 30 years since the reform of Vietnam, the country's socio-economic development has generally been improved, the people's material life has been gradually improved, and the spiritual and social life have been developed progressively.The development of Vietnamese society and economy has had development achievements which are worth encouraging. However, due to many objective and subjective reasons, Vietnam's socio-economic development in the past few years also has created many challenges to the demands of socio-economic development in a sustainable ecological model. In terms of economy, due to the limitations of science, technology, and the quality of the labor force; the current level of production in Vietnam is still backward and unstable. It can be seen clearly in the field of agricultural production.

Vietnam is one of the countries with advantages in agricultural production, yet it does not know how to make use of this strength. Although the labor force in the agricultural sector is relatively crowded (accounting for 41.9% of the country's labor force in 2016), (GSO Vietnam, 2016), labor productivity is low due to the limited qualifications of labor and backward technology. Therefore, agricultural contributions to the overall growth of the economy are insufficient and unsustainable. (In 2016, the growth rate of the agricultural sector was only 1.36%, much lower than 2011; the contribution ratio of agriculture accounted for only 16.32% in the general structure of the economy), (Communist Review, 2017). Industrial production and service have made major contributions to the Vietnamese economy. However, in reality, the growth of the Vietnamese economy over the past time mainly relies on exploitation, and export of raw materials. Highly sustainable industrial sectors such as financial services, transportation, tourism, information technology, etc. have developed but not firmly.

In Vietnam, the environmental protection in a sustainable way is facing lots of challenges. Due to limited public awareness of environmental issues and inadequate legal system, especially the law on environmental protection, over the past few years, we have accidentally prioritized the mode of production and the life style which is unfriendly to the nature. This is one of the main causes leading to current difficulties in protecting the ecosystem in Vietnam. Some key environmental indicators even fall below average compared to other countries in the area and in the world. According to the latest study published in early 2016, the overall ranking of Vietnam after measuring five indicators, was at 131st in the world, lower than China at 109th place, Syria at 101st, and the United States at 26th. Finland, a country in Northern Europe, ranked top in the chart. Specifically, in terms of wastewater treatment, Vietnam scored 19.8/100 points, ranking in 124th place among 139 countries. Regarding forest cover, with a score of 23.97/100 points, Vietnam was at 100th in the total of 116 countries. Whereas, this Southern Asian country was near bottom in Climate and Energy rankings, placed at 105th out of 113 countries (*Tien phong* Newspaper. Online version, 2016). Continuous environmental degradation along with natural disasters increasingly linked to climate change have been causing significant human and material damage as well as putting huge pressure on the nation's sustainable development.

During the past years, "the socialist rule of law has been progressively improved and perfected; its effectiveness and efficiency have been increased". However, in order to have a real democracy accelerating the sustainable socio-economic development, many shortcomings and challenges need to be overcome. Viewpoints and perceptions toward social democracy are, in general, still controversial. Moreover, there are great difficulties in implementing people's mastery in economic, political, and social fields due to an adequate system of mechanisms, policies, and laws. "The mastery of people in many areas and fields is still violated. At times, the exercise of the democracy is limited or just a formality in some places; it is also reported that some people distort the democracy and use it as a pretext to disrupt the national unity or disturb national security, social order and safety" (VCP, 2016; p.168). In addition, many pressing social problems are going on, such as fall back into poverty, youth unemployment, increasing social evils, and criminal offences. These issues along with corruption and red tape, which are escalating and become more sophisticated, are leading to the erosion of social trust, hindering the renovation, and the sustainable socio-economic development.

3.3 Propose some Solutions

All these drawbacks and challenges prove that to attain the core objectives of the national strategy for sustainable socio-economic development as analyzed above, Vietnam should take the following measures comprehensively:

First, it is necessary to agree on and understand thoroughly the views of the Party and the State of Vietnam on linking socio-economic development objectives with sustainable ecological models. In the context of market economy and the impact of many factors, there are still many localities, industries, and production and business units (including all state, collective, private or foreign investment sectors – FDI) continue to pursue economic growth, increase profits at all costs, including implicitly violating or taking advantage of the State's legal loopholes, especially the law on environmental protection. The phenomenon of some enterprises, among them private enterprises, enterprises with foreign direct investment for many reasons, attach much importance to the profit growth of enterprises did not seriously implement the requirements. On waste treatment to protect the environment – conditions must be committed to implementation right from the time of licensing; besides, in some localities in the country, for a long time, due to the reason of focusing on the growth achievement only or to attract more enterprises to invest in production development in the locality, the security work has been overlooked environmental protection, etc.... are bottlenecks that are very worrisome in the strategy of environmental protection and sustainable development. Therefore, it is necessary to continue to deeply understand and grasp the views of The Party and State of Vietnam on sustainable development, resolutely eliminate the idea of economic development at all costs. Because if economic development is overlooked or neglected the goal of social development and environmental protection, the price to pay will not be fully measured; even the result will be "anti-development". Sustainable development cannot be just a common way, but more importantly, it needs to be socialized, becoming a concrete awareness and practical action of every subject, of the whole society (Hoang, 2009; p.34).

Second, stimulating Vietnam's "Green Growth Strategy" in every sectors of the economy will be a good way to go. One of the goals of this strategy is to accelerate the economic restructuring, make the best use of natural resources, and reduce greenhouse gas emissions by researching, applying green technology, and developing infrastructure. As a result, Vietnam can improve economic efficiency, respond to climate change, contribute to the poverty alleviation, and create driving force for the sustainable development of the economy. The standpoint is that green growth is by the people and for the people, society will develop in harmony with natural environment in order to reduce the poverty, enhance citizen's material and spiritual life. To do so, it is necessary to restructure the economy and improve the institutions in a way that encourages economic sectors to employ effectively natural resources to make high value added. Besides, the country should restrict and then eliminate the industries exhausting natural resources and polluting the environment. In the short time, the exploitation and production of raw materials (petroleum, coal, rare ores, etc.) need to be slow down if technique and technology cannot meet the requirement of a green economy. Vietnam should also study, apply, and update widely modern technology in the country as well as in the world to be more effective in the use of resources and friendly to the environment. Additionally, it is essential to strengthen the propaganda on the active role of the economic sectors and every single individual in establishing a green economy nationwide. In company with the improvement of the laws and policies, the country needs to build an appropriate machinery of government in charge of managing and running the green economy in every field.

Third, reviewing the system of policies, planning, socio-economic development plans on green economy. Develop and step by step to improve the system of policies and legal documents on green economy, with a focus on low carbon production, greening production in all fields, changing consumption habits, and technological innovation toward environmental friendliness. Research and build a system of criteria of green economy, standards and standards of green economy in accordance with Vietnam's conditions, meeting international cooperation requirements. Issuing necessary sanctions in production, business and consumption to implement green growth to ensure sustainable socio-economic development. Strengthening the relationship in direction to integrate the contents related to green growth of national target programs, existing national key programs and projects such as new rural construction, responding to climate change, economical and efficient use of energy, combating desertification, developing high technology, providing clean water for rural people, ensuring jobs, reducing poverty, and ensuring hygienic and food safety.

Training and developing human resources: Strengthening the organization of training and fostering knowledge about green economy, green growth for cadres, especially for staff working in policy making, developing strategies, planning, and socio-economic development plans. Implement training and development of human resources on green growth in education and training. Study and select subjects and contents on ecological economics, sustainable development, green growth, renewable energy, clean technology, sustainable consumption, environmental auditing, and resource exploitation. Sustainability, ecosystem, and biodiversity protection at all levels and levels of education in the college and university education system.

Fourth, promoting the strategy for durable management and protection of the environment should be integrated into Vietnam's actions. To be precise, departments, branches, administrative agencies, and enterprises have to implement fully and properly viewpoints, guidelines, policies of the Party, and the State on environmental management and protection, which is basis for sustainable development. The perception about the relationship between economic growth and ecological protection must be changed. Economic growth is not always the priority over the measures against potential polluting sources and polluted areas. Importance should, in addition, be attached to the investments in research and application of environmental protection technologies. Regarding international relations, it is a must to intensify global cooperation in training and developing Vietnam's human resources, applying world's advanced technologies, which are adaptable to enterprise's situation, in ecological protection. Entrepreneurs, moreover, should be ready to renovate their production technologies to be more modern, eco-friendly, and resource-saving. Meanwhile, the government has to revise, amend, supplement, and finalize policies on management of investment in science and technology for the good of the environment. As a result, the policies supporting investments at any price must be removed because they can turn Vietnam into a landing field for outdated technologies. In the short term, Vietnam had better concentrate on curbing and minimizing new sources of pollution, especially, on controlling the projects that discharge a great volume of sewage into environment, polluting manufacturing sectors such as steelmaking, mining, thermal power, paper production, and textile dyeing, controlling the discharge of seriously polluting factories as well as those with outdated technologies. For the contaminated areas, there must have methods for amelioration and rehabilitation. Simultaneously, Government should increasingly invest in environmental infrastructure and technology. For the time being, more attention needs to be paid on enhancing capacity to respond to climate change and reducing greenhouse gas emissions; encouraging scientific research, development, and application of ecological technology; intensifying and diversifying investments in environmental protection; strengthening the propaganda in order to raise public's awareness of the importance of environmental protection. In the long term, it is crucial to finalize the legal system for environmental protection. By which, Vietnamese society and economy will have a legal basis for its sustainable socio-economic development.

Fifth, renovating Vietnamese social institutions in the pursuit of democracy and progress which meet the demand for ecologically sustainable socio-economic development certainly play an important role. This should be even considered a key solution to ensure the success in the strategy of sustainable socio-economic development. To realize this, Vietnam's first mission is to reform its leadership style toward political system. On the one hand, the country consolidates Party's leadership; on the other hand, it builds a mechanism for democracy, an insurance for the effectiveness of the political system. Then, Vietnam's socialist model, which is based on democracy, rule of law, and modernity, is continued and perfected as emphasized in the Resolution of 12th National Party Congress: "In its organization and operation, the Government has to promote democracy, conform to the requirements of the rule of law, produce positive changes, and obtain better results. Building the rule of law socialist state must be carried out synchronously in legislative, executive, judicial aspects, and be carried out synchronously with the renovation for a streamlined, effective and efficient political system along with economic, cultural, and social renovation. Continuing to complete the mechanism for legal and constitutional protection" (VCP, 2016; p.175). Additionally, it is necessary to keep renewing the operation of institutions of the state, namely Vietnamese Fatherland Front, Ho Chi Minh Communist Youth Union, associations, and unions. Pursuing the democracy and reality, these organizations create a real democratic society to encourage the active participation of Vietnamese from all classes in the establishing and pushing the socio-economy based on the sustainable socio-economic development model.

Sixth, continue to enhance the implementation of Vietnam Sustainable Development Strategy until 2020 with the vision to 2030.

In order to complete these missions, Vietnam's viewpoints and solutions are:

Viewpoints: Human beings are the center of sustainable development. The government should consider the role of people as the key subject, resources, and targets of sustainable development, aim to fully meet the material and spiritual demand of people of all strata, make the country wealthy and strong, the society democratic, equal, and civilized; develop an independent and self-reliant economy with active international integration for sustainable development (VCP, 2012). Sustainable development is the requirement lasting throughout the process of national development; Sustainable development is the common work of the whole Party, people, authorities at all levels, ministries, agencies, localities, enterprises, social organizations, communities, and individuals; Sparingly and effectively exploit natural resources, particularly nonrenewable resources, preserve, and improve living environment; build a learning society, environmentally-friendly lifestyle, sustainable production, and consumption. Besides, science and technology are the foundation and driving force of national sustainable development. Modern, clean, and environmentally friendly technologies need to be widely applied in production fields.

Solutions: Continue to improve institutional system and national administration quality for sustainable development; enhance investments and effectively use State-funded capital to implement the Strategy for Sustainable Development in Vietnam; Accelerate dissemination and education to raise the whole society's awareness of sustainable development; Intensify the role, responsibility, and participation of businesses, socio-political organizations, social-professional organizations, non-governmental organizations, and residential communities in deploying, consulting, criticizing, and proposing sustainable development policies; Amplify the role and effects of science and technology, push up technological renovation to implement sustainable development; Expand international cooperation in the fields of information exchange, science, and technology for sustainable development, develop environmental economics (waste treatment, energy-saving technology, low-carbon, and waste renewal technology, etc.…) (Pham, 2018a, 2018b; p.731–741).

4 Conclusion

Constructing and promoting sustainable socio-economic development are not only an objective tendency, but also an urgent requirement of Vietnamese economic development. Based on the features of socio-economic development in the country as well as the negative impacts of climate change on a global scale, in the past years, Vietnam has actively prepared and deployed the strategy for sustainable socio-economic development, which has gained enormous encouraging results. However, besides the positive achievements, Vietnam is also facing many difficulties and challenges in its socio-economic development strategy based on a sustainable ecological model, such as re-poverty and increase social inequality, environmental pollution, and the effects of changing global trends, etc. To successfully achieve this socio-economic development strategy and resolutions of the Party and State of Vietnam, it is necessary to synchronously implement the main solutions as proposed in this paper. In particular, it is important to pay attention to some basic solutions such as: promoting green growth strategy, actively protecting the environment, paying more attention to social progress and democracy in order to promote strategic success socio-economic development under the current sustainable ecological model in Vietnam. Of course, the most important factor is the right attitude and determination of all levels, sectors, and individuals in transforming Vietnam's strategy for sustainable socio-economic development from vision to reality in the next year.

Acknowledgments To complete this paper, I would like to express my special thanks to my colleagues and colleagues who collaborated and helped me to complete research projects related to the topic of the paper. I would also like to thank *New-Kku International conference* (9th) allowed me to publish the results of the study for the first time at the Conference before completing this paper.

References

Andy, S., Paul, J., Thom, J. A., Lin, P., Sarah, H., Hepu, D., & Felicity, C. (2013). *Reframing social sustainability reporting: Towards an engaged approach.*

Communist Review. (2017). *Vietnam socio-economic overview in 2016.* Updated on 11th June.

GSO Vietnam. (2016). *Press release announcing socio-economic statistics of 2016.* Online newspaper, updated on 28th December.

Hoang, D.C. (2009). Sustainable development in Vietnam: Some theoretical and practical issues. *Journal of Philosophy, 8,* 219. Retrieved August, from http://citiesprogramme.com/archives/resource/circles-of-sustainability-urban-profile-process

James, P., Magee, L., Scerri, A., & Steger, M. B. (2015). *Urban sustainability in theory and practice.* Routledge.

Johnson, D. L., Ambrose, S. H., Bassett, T. J., Bowen, M. L., Crummey, D. E., Isaacson, J. S., Johnson, D. N., Lamb, P., Saul, M., & Winter-Nelson, A. E. (1997). Meanings of environmental terms. *Journal of Environmental Quality, 26,* 581–589.

Kahle, L. R., & Gurel-Atay, E. (Eds.). (2014). *Communicating sustainability for the green economy.* M.E. Sharpe.

Nguyen, T.D. (2014). Completing institutionalization, promoting people's right to mastery, successfully implementing tasks of 2014 and creating foundation for rapid and sustainable development. Electronic Nhan dan newspaper, updated on January 1.

Pham, C.N. (2018a). Developing the society and the economy in sustainable ecogolical model in Vietnam. *Proceedings of the workshop: 9th NEW-KKU Intenational conference on socio-economic and environmental issues in development.* Labour – Social Publishing House, p. 731–741. ISBN: 978–604–65-3529-4.

Pham, C.N. (2018b). Some issues in ecological philosophy. *Proceedings of 14 th international conference on humanities and social sciences 2018 (IC-HUSO 2018) 22nd -23rd November 2018*, Faculty of Humanities and Social Sciences, Khon Kaen University, Thailand. ISBN (e-book): 978–616–438-332-6. (https://hs.kku.ac.th/ichuso/Proceeding-ichuso-2018.html) paper number: ICHUSO-204, pp. 1324–1334.

Rex, M. (2003). *A system of rights*. Print publication date: 1997. Print ISBN-13: 9780198292937. Published to Oxford Scholarship Online.

UNEP (2011). *Towards a Green Economy: Pathways to sustainable development and poverty eradication*. Retrieved from www.unep.org/greeneconomy

VCP. (1996). *Documents of the 8th National Congress* (p. 82). National Political Publishing House.

VCP. (2006). *Documents of the 11th National Congress* (p. 98). National Political Publishing House – Truth.

VCP. (2012). Decision No. 432/QD-TTg of the Prime Minister. Approval of sustainable development strategy of Vietnam period 2011–2020. Updated on 12nd April.

VCP. (2016). *Documents of the 12th National Congress* (pp. 87–168). National Political Publishing House – Truth.

VCP. (2018). Resolution No. 23-NQ/TW of the Politburo to orient the policies on national industrial development until 2030, with a vision toward 2045. VOV Online Newspaper, updated on 23rd March.

Vu, D.D. (2018). *Phát triển bền vững là sự lựa chọn đúng đắn nhất (Sustainable development is the best choice)*. Retrieved December 17, from http://www.tinmoitruong.vn. Updated on December 17.

Wandemberg, J.C. (2015). *Sustainable by design*. Amazon. p. 122.ISBN 1516901789. Retrieved February 16, 2016.

Chapter 13
Perennial Cropping System Development and Economic Performance of Perennial Cropping System in Dak Lak Province, Vietnam

Thi Thuy PHAN, Thi Thanh Thuy BUI, Thi Minh Hop HO, Duc Niem LE, and Philippe Lebailly

Abstract The paper deals with the perennial crop development in Dak Lak Province by using the mixed data collection of quantity and quality. At the same time, the returns of two major perennial crop systems including monocropping and intercropping systems concentrating on coffee and pepper were analyzed. The findings indicated that development and adjustment of perennial crop systems in Dak Lak province highly related to economic, political, and social changes; and underwent through five main stages as following large-scale perennial crop plantations as early-stage; perennial crop systems as state-owned and cooperative farms; intensified perennial crop systems; mixed perennial crop systems; and specialized and diversified perennial crop systems. Additionally, the study revealed that intercropping systems were more efficient than monocropping systems due to the potential presence of economies of scope. Consequently, the results provided useful information for local households in making sustainable production strategies and policymakers in enacting suitable policies as well as diagnosis of the future evolution of perennial crops for the coming years.

Keywords Perennial crop systems · Historical development · Economic performance · Dak Lak · Vietnam

T. T. PHAN (✉)
Tay Nguyen University, Dak Lak, Vietnam

Gembloux Agro-Bio Tech, University Liège, Liège, Belgium

T. T. T. BUI · D. Niem LE
Tay Nguyen University, Dak Lak, Vietnam

T. M. H. HO · P. Lebailly
Gembloux Agro-Bio Tech, University Liège, Liège, Belgium

A. T. Nguyen, L. Hens (eds.), *Global Changes and Sustainable Development in Asian Emerging Market Economies Vol. 2*, https://doi.org/10.1007/978-3-030-81443-4_13

1 Introduction

There were various definitions of cropping systems and interacting with farming resources such as land, labor, and capital showed by previous authors (Diepart & Allaverdian, 2018; Rana & Rana, 2011). Furthermore, the cropping system is an important component of the farming system of which the perennial crop sector is one of the elements of the cash crop system (FAO, 2019). As compared to annual crops, perennial crops can be harvested multiple times before dying or replacing. On the other side, regarding many authors such as (Alexander, Paustian, & Smith, 2015; Araya et al., 2012; Galati, Crescimanno, Gristina, Keesstra, & Novara, 2016; Hatfield & Walthall, 2014; Nguyen, 2017), the changes of agricultural, in general, and cropping systems, in particular, are results of climate change, agro-technology, socio-economic transformation, liberalization, and globalization. In which, cropping systems not only changed along the time, the local, the regional, and international economic situation but also sustainably adapted to the specific conditions to ensure human needs under population growth (Darnhofer, Bellon, Dedieu, & Milestad, 2010; Hatfield & Walthall, 2014). In other words, these changes require to move to another place or to develop new practices for survival (FAO, 1999; Fresco & Westphal, 1988; Lebailly et al., 2015). Thus, the major trends and trajectories of different farm types were accurately understood to apply appropriate strategies in protecting natural resources as well as securing households' income by enhancing productivity (Jayne, Chamberlin, & Headey, 2014). Subsequently, determining the evolution of cropping systems to support future research and extension might be enclosed to the best service of the rural communities (Herridge et al., 2019). Moreover, economic performance is the key in determining the resilience and sustainability of farming systems because economic change caused the threads to agricultural lands and natural resources (Barbier, Burgess, & Grainger, 2010; Zinnanti, Schimmenti, Borsellino, Paolini, & Severini, 2019). Thus, a study about the historical development of perennial crop systems and its economic efficiency is essential.

The perennial crops such were introduced to Vietnam at the end of the nineteenth century by the French. Until now, they have become the top 10 commodities in the world and play a crucial part in agricultural development and export earnings. Interestingly, Vietnam had become the biggest exporter of black pepper and the second largest for coffee from 2011 to 2013 (ICC, 2019; OECD, 2015). Perennial crops have been growing popularly in Central Highlands of Dak Lak, Dak Nong, Gia Lai, Lam Dong, and Kon Tum thanks to the favorable conditions.

Going beyond, the changes in driving forces such as ecological, technical, socio-economic, political issues resulted in either challenge in perennial crop production or adopting different approaches (Pham et al., 2001; Phan, Le, Ho, Burny, & Lebailly, 2019a). In other words, there was a significant relationship between the changes such as economic reform, government support (by direct and indirect policies), and evolution of perennial crop systems (Lindskog, Dow, Axberg, Miller, & Hancock, 2005; Scherr, Mankad, Jaffee, & Negra, 2015). However, empirical evidence on the analysis of the evolution and profitability of the different perennial crop systems in Dak Lak Province, Central Highlands, Vietnam is limited. In such circumstances, the purpose of this study is to give a comprehensively real picture of the temporal and spatial transformation in perennial crop systems at distinct stages. Furthermore, evolutionary information can be used to explain how farms generate and adapt changes in driving forces at different levels (Han Quang, Azadi, Dogot, Vu, & Lebailly, 2017). Besides, an analysis of the economic efficiency of two representative perennial crop systems provides empirical evidence and opportunities for practices. This can help farmers in choosing perennial crop production systems as well as policymakers in enacting the appropriate strategies to perennial crop system in forthcoming years.

2 Methodology

2.1 *Study Area*

This study was carried out in Dak Lak Province with advantageous conditions for agricultural development like basaltic soil and flat or gently hilly relief, especially for perennial crops such as coffee, pepper, etc. a previous document, Dak Lak Province was classified into six agro-ecological zones by soil types, rainfall patterns, and altitudes based on various sources (Karimov et al., 2016). Additionally, as a result of discussions with head and key informants at the provincial level, agricultural experts, and survey implementation, Dak Lak Province was separated into three perennial crop zones by fertility criterion as below:

Zone 1: Highly favorable for perennial crops growing (including Buon Ma Thuot City, CuM'gar, Cu Kuin, Krong Ana, Krong Buk, EaHleo, Krong Ana, Krong Pak Districts).
Zone 2: Moderately favorable for perennial crop growing.
Zone 3: Unfavorable for perennial crop.

Moreover, based on secondary data, observation, and conducting the survey, two distinct cropping systems chosen in this study are monocropping systems (Mono-Coffee Systems-MCSes and Mono-pepper systems-MPSes) and intercropping systems (coffee and pepper intercropping-CPI).

2.2 *Data Collection*

For reliable findings, the study collected both qualitative and quantitative data from primary and secondary sources (Faruque et al., 2017; Ulmer & Wilson, 2003).

The primary data were gathered from visual observation, key informant interviews (KIs), Focus Group Discussion (FGDs), and household surveys through questionnaires. Furthermore, the economic performance of two major perennial crop systems showed via using simple and stratified random sampling of perennial crop households.

Secondary data were gathered from differently official sources such as annual statistic books published by the General Statistics Office (GSO), the Dak Lak Statistics Office, Provincial People's Committee, maps, historical books, annual reports of local authorities in Villages, Communes, Districts, and Province. Furthermore, research documents and scientific reports were also collected as complementary sources for this study.

KIs and FGDs provided qualitative data on the following subjects: (1) the history of changes of local perennial crop systems regarding coffee and pepper; (2) current cropping patterns and changes of perennial crop systems overtime; (3) key driven forces of change related to adopt the current perennial crop systems (4).

Simultaneously, to collect quantitative data, a list of 86 households who cultivated mono and intercropping systems related to coffee and pepper in the three villages was conducted by the same probability of statistical units (FAO, 2016). It means the study collected 10% of villagers (selected households (N) = 0.1 n (n = populated size)), who are cultivating coffee and pepper crops. The households list provided 90 plot samples for data about perennial crop production because some households owned more than one piece. The farms that were between 0.5 ha and 2 ha are similar to the average area of local farms. The data at given year including (1) profile of perennial crop system such as area, yield, age; (2) the cropping calendar (timing of plot care crops and harvesting); (3) annual cost (inputs, labor); sale of crop and revenues (Table 13.1).

2.3 Data Analysis

Analysis of the evolution of perennial crop systems.

The historical analysis was obtained by hierarchy, chronology, and timeline methods.

Economic analysis.

This study examined the differences in economic returns by using comparative analysis and descriptive statistical analysis like means, percentages, charts, and growth rates. For a detailed explanation, indicators that include production cost, revenue, value-added, and profit show how the best economic performance of

Table 13.1 The distribution of surveyed samples by research sites

Region	Commune	MCSes	MPSes	CPI	Total Samples
BMT	Ea Kao	12	9	9	30
Cu Mgar	Cu Sue	12	8	15	35
Cu Kuin	Ea Ktur	8	11	6	25
Total samples		32	28	30	90

perennial crop systems generates for households (Hill & Bradley, 2015; Newton et al., 2012). Furthermore, a statistical hypothesis testing was requested to find whether or not there is a difference in the nonparametrical distribution of distinct groups based on the Kruskal Wallis-test (Chu Nguyen & Hoang, 2008).

3 Results

3.1 The Development of Perennial Cropping Systems in Dak Lak Province (Fig. 13.1)

3.1.1 Large-Scale plantations of Perennial Crop Systems under the Regime

Dak Lak has exploited the best appropriate conditions for the growth of the perennial crop such as coffee, rubber by the French since the 1920s (ICC, 2019; Pham et al., 2001; Tran & D'haeze, 2005). In this era, two major crops such as coffee and rubber were planted initially for economic purpose. At that time, coffee and rubber trees were only practiced and developed by large-scale plantation in highly suitable fertility for perennial crop growing regions-Zone 1. There was a limitation of a low population density, a dependence of natural conditions, a lack of manpower, know-how knowledge, and simple implemented tools. In this context, the production was inefficient as a result of mainly mono-coffee and rubber systems (Do, 2016).

3.1.2 Existence of State and Cooperative Owned farms of Perennial Crop Systems

The perennial crops recovered after the Revolution of Vietnam. In the past, most plantations operated under State ownership and cooperatives. However, in 1981 the private ownership admitted under Decree 100 (Khoan 100) (Meyfroidt, Vu, & Hoang, 2013). To take advantages of natural conditions, monocropping systems were applied causing absolute exploitation of land area in Zone 1. Furthermore, the high population density thanks to resettlement program (New economic Zones-NEZs) made the perennial crops expanding on moderately favorable for perennial crops' growth-Zone 2 (Lindskog et al., 2005; Maurice, 2004). Although other cash crops like pepper and tea were planted at the beginning of the 1980s, lands were owned by the state in formal management (namely, State ownership) (Muller & Zeller, 2002). This leads to stagnancy in the economy with low productivity of perennial crops (Table 13.2).

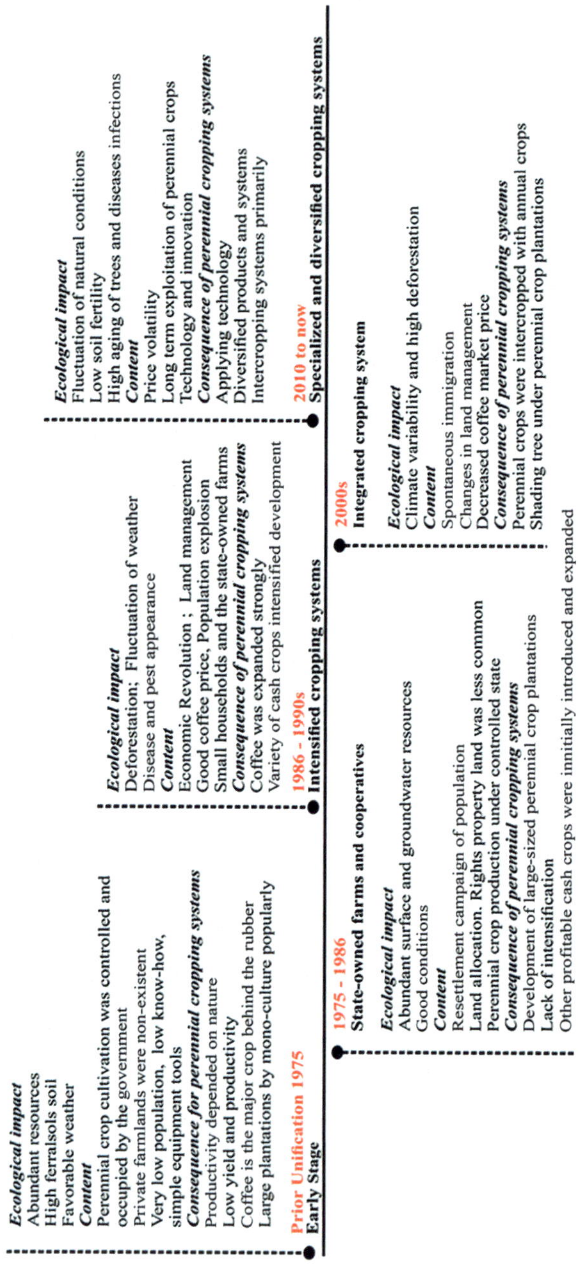

Fig. 13.1 Important historical milestones and consequences of perennial crop systems. Source: Secondary data, KIs, and FGDs

Table 13.2 The number of farms and producers participated in owned state and cooperative production

	State owned Farms		Cooperatives	
Year	Number (Farms)	Producers (People)	Number (Farms)	Producers (People)
1978	12	3930	17	9121
1981	19	23,885	122	40,600
1985	28	43,521	184	58,728

Source: (People's Committee of Dak Lak, 1986)

3.1.3 Intensifed Perennial Crop Systems

From 1986 to the 1990s, the perennial crop sector had a significant increase in Dak Lak, especially in the coffee-growing area. The reasons for rapid growth of the coffee field were the policy privatization, economic liberalization, state-sponsored migration, and price spikes (as the frost in Brazil, a large part of the coffee production damaged led to a sharp reduction in the international coffee supply that then made the high price of coffee) (Dang & Shively, 2008) to lead Vietnam as becoming the world's second-largest exporter of coffee (Pham et al., 2001). Furthermore, labor and capital intensive have been because of introduced fertilizer, improved rural infrastructure, and markets as well as expanded irrigated systems since 1992. Additionally, local governments explained an increase in the coffee-growing area due to demographic resettlement, socio-economic liberalization, and legislation towards land ownership. Perennial crops were planted to Zone 2 and 3 (unsuitable lands for growing perennial crops). Furthermore, spontaneous migration was also a reason for the diversification of perennial crops. Besides, the direct and indirect government-supported funds encouraged the development of perennial crops (Dang & Shively, 2008). By direct support, the government enacted financial, technological, and economic policies. Additionally, the local government stimulated to apply new crops or efficient crop management methods. In such, farmers could access the finance as well as knowledge of crop protection (Cheesman, Son, & Bennett, 2007; D'haeze, Deckers, Raes, Phong, & Loi, 2005). Additionally, indirect support has made a large number of migrants from constrained and crowded regions to Dak Lak Province as well as more investment in rural areas. However, the over-exploitation of perennial crops resulted in some enormous vulnerability such as land degradation and deforestation (Fig. 13.2).

3.1.4 Mixed Perennial Crop Systems

Unfortunately, in 2004, coffee price decreased dramatically. As Ponte S, the global coffee chain changed significantly due to new consumption fashion, the evolution of corporate companies. In other words, technical development resulted in low-cost Robusta processing to parallel inelastic demand which contributed the decrease of international coffee price to below 40 cents per pound in 2004 (Doutriaux, Geisler,

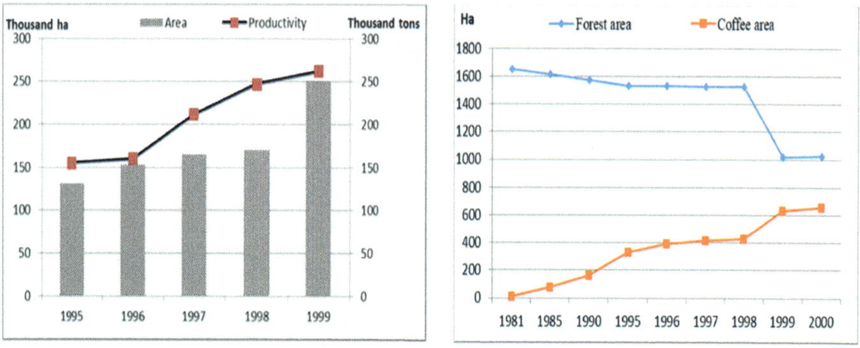

Fig. 13.2 The coffee-growing area during 1986–2000 (Source: Dak Lak PSO, 1986, 2011)

& Shively, 2008; Ponte, 2002). Moreover, frequently natural disadvantages such as drought, pests, and infectious diseases are challenges to farmers. To solve these, the government implemented many policies including a ban of planting more coffee trees through enforcement of land use rights and limitation of bank loans related (Cheesman et al., 2007). Also, the government encouraged convertibility of coffee in the unsuitable areas to more profitable and fewer water crops such as cashew and cassava, cotton, pepper, and corn (Lindskog et al., 2005; Meyfroidt et al., 2013). Apart from coffee, hence, other crops such as rubber, pepper, and cashew in both perennial and annual crop mixed systems developed steadily over this period. However, as the previous period, an expansion of perennial crops is a reason for deforestation resulting in various other obstacles such as degradation and soil erosion.

3.1.5 Specialized and diversified Perennial Crop Systems

From 2010, perennial crops continuously faced challenges. For instance, aging of coffee tree stock that accounts for over one-third of the provincial coffee-growing areas was 15–20 years resulting in low yields and returns (Ho, 2018). Apart from pests and infectious diseases on pepper, a fall in the price of rubber, coffee, and pepper made farmers into difficulties such as bad debts and discouragement (Scherr et al., 2015). Under this condition, different crop systems practiced. In terms of coffee, applying technology and innovation for perennial crops were popular such as irrigation, seedling, and processing. Especially, certified coffee production attracted participants. Although linkages, certified production, and development of specialty coffee trees have given more efficient than conventional models, certificated production has not been applied widely because of strict requirements (Table 13.3).

On the other hand, the 2014 good price of pepper motivated farmers to convert from coffee plantations or intercropped orchards. Unfortunately, pepper price from 2016 to now has dropped significantly causing many difficulties for producers. To

Table 13.3 The area and number of producers by certificate in Dak Lak Province

	2013		2015			2017			2018		
Certificate	(2)	(3)	(1)	(2)	(3)	(1)	(2)	(3)	(1)	(2)	(3)
UTZ	16.8	14.5	29	19.5	14.5	17	15.6	11.0	7	6.2	5.6
4C	36.8	23.0		37.1	25.4	14	25.2	15.5	4	8.7	6.7
RFA	7.4	3.2	3	4.3		3	2.1	–	3	5.1	2.8
FLO	0.2	0.4	1	3.0	0.5	5	0.6	0.3	6	0.7	0.5
GI	–	–	10	–	–	12	–	–	12	–	–

(1): Number of certificate; (2): Area (thousand hectares); (3) Producers (Thousand people)
Source: (Provincial People's Committee of Dak Lak, 2018)

mitigate market risks, intercropping systems such as coffee and pepper and diversified perennial crop systems (coffee, pepper, and fruit) have become more popular. According to People's Committee of Dak Lak, there are over 39,000 ha of the intercropping area with 11,000 ha intercropping models of coffee and pepper (Provincial People's Committee of Dak Lak, 2018). Intercropping systems including coffee trees and others (pepper and cashew) are generating higher profitability than conventional ones (Nguyen & Phan, 2017; Phan, Le, Ho, Burny, & Lebailly, 2019b). Additionally, diversified systems including perennial crops and fruits (avocado and durian) are popular with higher returns and sustainability (Ho, Hoang, Wilson, & Nguyen, 2017). Nonetheless, the intercropped area is still limited. A lack of government intervention and farmers' financial constraints negotiation also led to inefficiency.

3.2 *Current Perennial Crop Systems Comparison*

3.2.1 Characteristics of Perennial Crop Systems

Monocropping and intercropping systems profile.

The brief information of three perennial crop systems is presented in Table 13.4 as following below:

For MCSes, the average cultivated area was at 1.1 ha, in which the density of 958 trees per ha, its density was lower than the technical advice (1100 trees/ha) (ICC, 2019; Phan, 2015) because of many trees dying without replacement at the gap (coffee growers converted other crops instead of a new planting of coffee tree). Also, the yield reached only 2.1 tons per ha which linked to (Amarasinghe, Hoanh, & D'haeze, 2015), because of causing the high proportion of aging tree stock (17 years in this study) and natural disaster like floods and drought. On the other hand, MPSes initially were formed from very small acreage of residential gardens in the last decades. Although under favorable prices recently, they have been developed strongly, the surveyed data showed that the average size of MPSes was the smallest one in comparison to the rest, by 0.8 ha.

Table 13.4 General information of three selected systems

Indicators	MCSes ($n = 32$)	MPSes ($n = 28$)	CPI ($n = 30$)
1. Average plot area (ha)	1.1	0.8	1.0
2. Density (tree/ha)			
Coffee	958	–	964
Pepper	–	1344	914
Average age of system (years)			
Coffee	17	–	13
Pepper	–	7.43	7.3
Yield (tons/ha)			
Coffee	2.1	–	2.3
Pepper	–	2.3	1.8
Crop losses	25	27	22
Distance (km)	1.8	0.5	1.7

Source: Authors' own calculations

Regarding pepper density, it was not high, only 1344 trees/ha as compared to recommendation (about 1700 trees/ha). Pure pepper gardens were converted from renewing an old coffee plantation instead of an initial design for MPSes. Unfortunately, black pepper was sensitive to pests and diseases. As statistical data in 2017, there was about 2000 ha experiencing the plant diseases (equal to 13.2% of all plant diseases in the whole country such as foot rot or quick wilt disease, Pollu disease, slow decline or slow wilt, and stunt disease). Since the 2000s, MPSes have been planted significantly with 7.43 years to create yields at 2.3 tons/ha. In this study, most MPSes were grown by in an unregulated way (in areas not zoned for farming) as well as more wooden and concrete pillars were utilized rather than live plants. Even though rubber trees could be used as pillars, play areas, ponds, and rice fields were used to grow pepper. For diversified systems as CPI, it can obtain higher yields, more returns or mitigate environmental damages such as in Uganda with coffee-banana intercropping (van Asten, Wairegi, Mukasa, & Uringi, 2011), Thai Lan with rubber-based intercropping systems (other perennial crops or fruit) (Romyen, Sausue, & Charenjiratragul, 2018). According to the study, CPI has quite young with 7.3 years of pepper, 13 years of coffee. With the average area was 1.0 ha and the density was 964 coffee trees and 914 pepper trees per ha, CPI primarily includes two intercropping methods: group (a small sub-area of coffee and pepper was planted in the orchard) and intersection (two coffee rows or three coffee rows or five coffee rows to intercrop one pepper row where pepper was designed at the intersection point of coffee holes). Additionally, the yield of CPI was at 2.3 tons of coffee and 1.8 tons of pepper/ha (Table 13.4). For KIs and FGDs, while the density of coffee is often from 950 to 1100 trees/ha, this of crowded pepper depends mainly on types of pillars (concrete, wood or alive trees). This is because high pepper density causes a decrease in yield due to the competition for space and light.

Cropping Calendar

Figure 13.3 illustrated the calendar of coffee and pepper crops in distinct perennial crop systems during the cropping year by various activities.

Under dry season with high sunlight and evapotranspiration during six months, irrigation and pruning of coffee become the main activities in MCSes. The coffee trees, for example, are provided water from January to April by pumping from private wells or drip irrigation systems. Meanwhile, pruning often occurs after harvesting season that significantly attracted by both men and women, sometimes, households hired nonfamily labor by fixed-paid wage (3000 VND/coffee tree), with the average of pruning coffee trees about 40 trees/day/person. Harvesting lengthens from October to December when coffee is harvested totally. Additionally, fertilizer is used at least three times for breaking flower, feeding fruits, and recovering post-harvesting among two seasons. On the other hand, weed control was carried out in mainly wet season. However, weeding that needs to be done at least monthly is costly by requiring more labors as shown. Nowadays, producers considered applying equipment and herbicide to kill grass rather than by hand.

In terms of MPSes, crop care activities are considered the same as MCSes. However, pruning has just carried out with living poles for pepper growing on the vine (alive trees as Cassia siamea and/or Leucaena leucocephala; fruit) during the rainy season. Besides, turning to irrigation, pepper although requires less water than coffee, it needs to be irrigated more regularly. For instance, coffee needs about three rounds per year, pepper is at least ten-day- irrigated cycles. Additionally, on the rainy season, drainage needs to be done as fast as possible to avoid flooding the soil. On the other hand, harvesting is a constraint when farmers have to spend at least two-three month with a high demand of labors, from February to April.

Concerning CPI, farmers considered costing more workdays as compared to others. As responded by interviews, most of the planters implemented concurrently

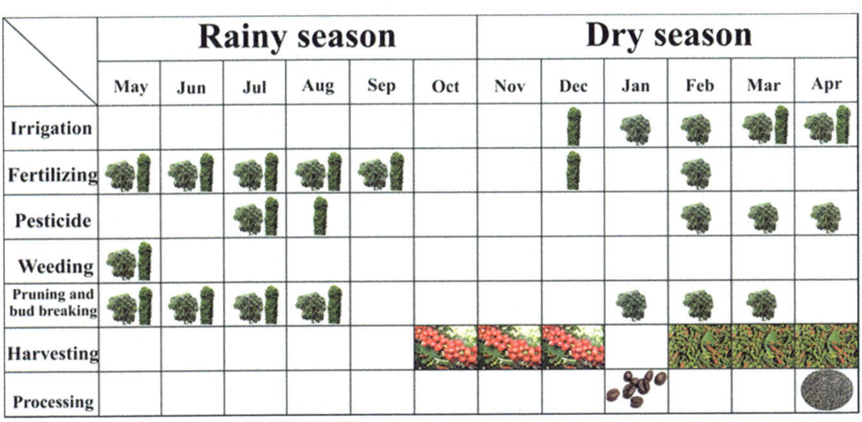

Fig. 13.3 Coffee and pepper crop calendar during a cropping season. (Source: Secondary data, KIs, FGDs, 2017/2018)

some operations such as fertilizer, irrigation, and weeding. However, some farmers separated such acts on crops for growth. Nonetheless, this act requires many labors and technology to secure each other.

3.2.2 Comparison of Economic Performance

Annual Cost

The annual cost of monocropping systems comprised 43.6 million for MCSes and 86.7 million for MPSes while CPI was at 86.3 million VND per ha. These costs were higher intermediate and labor costs, which were the two main components.

For intermediate costs, MPSes were the highest at 38.7 million VND per hectare while CPI was at 28.5 VND per ha (see Table 13.5). As Phan et al., fertilizer and pesticide seem to be overused in production, accounting for over 50% of input costs (Phan et al., 2019b) . This is because (1) price of pepper that is currently higher than that of coffee motivated farmers to invest more; (2) pepper is sensitive to pest and infectious disease making higher usage of pesticide. This is evident in the survey as there was a high incidence of disease (i.e., foot rot, slow decline) on most plantations (i.e., 90% of the surveyed households owned at least ten crops lost by diseases) (Duong, 2019; Nguyen & Bui, 2011). However, due to promising profitability, the 2017 pepper-growing areas exceeded provincial master plans, accounting for 150% of growth (NIAPP, 2017). This led to a glut in the market and a fall of pepper prices by half compared to the previous year.

On the other hand, MCses were at 18.5 million VND per ha which seem to be more suitable for poor or fewer capital farmers.

Table 13.5 The cost and return of perennial crop production in 2017. Unit: Thousands of VND/ha

	Monocropping systems				Intercropping systems		
	MCS (n = 32)		MPS (n = 28)		CPI (n = 30)		
Items	Mean	SD	Mean	SD	Mean	SD	Sig.
1. Annual cost	43.6	11.1	86.7	39.3	86.3	23.3	0.00*
Intermediate cost (IC)	18.5	6.8	38.7	25.4	28.5	12.2	0.00*
Labor cost	21.5	4.7	39.2	13.7	45.8	13.6	0.00*
Loan interest[1]	1.9	3	2.6	4	2.9	4.3	NS
Depreciation[2]	1.8	0.9	6.1	2.4	5.5	4	NS
2. Gross output	80.8	19	253.5	89	285.4	82.7	0.00*
3. Net farm income	55.4	13.3	166.7	59	200	77	0.00*
4. Profit	37	12	135	51.3	165	76	0.00*

[1]Interest rate: 10%
[2]The life expectancy of perennial crops was 25 years
Source: Authors' own calculations
Asymp. sig. of Kruskal–Wallis test: 0.00. NS: Nonsignificant. Most amortization equipment (equipment owned by the farmers) considered to be linearly fixed at 25 years

With labor costs mainly for harvesting, both MPSes (45.5%) and CPI (58% of annual costs) were more labor-intensive than MCSes (see Table 13.5), especially collection of black pepper is costly (i.e., 40 kg fresh pepper/day collected as compared to 100 kg fresh coffee/day). This created an intense demand for labors in the harvesting season because harvest duration could not be lasted longer due to crop characteristics. Interestingly, CPI has the harvest periods are from September to November (for coffee) and February to April (for pepper) facilitating family labor usage. Thus, this system had the greatest number of family labor days comparing the others, about 213 days over the year (a working day = 160,000 VND). In other words, CPI creates more opportunities for family labors than others. To conclude, the available evidence showed that MCSes incurred the lowest production costs, whereas MPSes had the highest.

Returns

The revenues of MCSes, MPSes, and CPI gained about 81, 254, and 286 million VND per ha (where total earnings equal coffee and/or pepper yield multiplied by coffee and/or pepper price). Net farm income was about 37 for MCSes, 167 for MPSes, and 200 million per ha for CPI, respectively (Table 13.5).

MCSes seem to have lower economic efficiency in revenue and profit criteria than the others. Simultaneously, the CPI had the best performance among the three systems due to economic scope of coffee and pepper consistent with many previous studies (Phan et al., 2019a; Phan et al., 2019b). For instance, the profit of CPI accounted for 165 million VND while MCSes and MPSes reached 37 and 135 million VND per ha, respectively.

4 Conclusions and Discussion

There is a great consensus on the growth of perennial crop systems in the Dak Lak Province during different stages. Under this circumstance, appropriate perennial crop systems were applied as an adoption for an institution, society, economy, and environment.

The changes of perennial crop happened through types of crops and systems in which cultivated area increased significantly, even in the unsuitable areas. The perennial crop systems showed the trends including firstly, large-scale coffee and rubber plantations were likely to be low know-how and simple equipment tools under regime depending on natural conditions considerably; secondly, perennial crop systems as state owned-farms and cooperatives paralleled production of individual rights thanks to land allocation; thirdly, to take the most advantages of the market, perennial crop such as coffee was intensified by households' resource utilization; fourthly, reducing risk and maximizing the efficiency of the market, integrated systems such as perennial crops, annual crops, and livestock should be practiced; fifthly, to increase the efficiency of the resources (land, labor, and capital) and resilience, the specialized perennial crops should be applied technology such as the production of specialty products and certificate as well as intercropping systems.

Moreover, perennial crops required different activities for each system. This was carried out by households at different levels depending on their capabilities. Unfortunately, returns are very priced sensitive (Huynh, 2018). Meanwhile, environmental production encountered risks such as pests and infectious diseases. Although the government offered guidance and support like technical advice, funds, and seedling, these limited due to a concentration on the coffee sector.

Currently, two major perennial crop systems have been existing including monocropping and intercropping systems. In this, intercropping systems demonstrated to be likely more returns than monocropping ones. In other words, significant output complementarity between coffee and pepper crops production was discovered thanks to the potential presence of economic scope. However, there is a need to have correct guidance for farmers' on-farm practices (density, preventing pests and infectious disease, the amount of used fertilizer and pesticide). Besides, the successful models should be widened for farmers through the sharing experience at different levels from province to villages. Moreover, the training programs should be organized to give academic knowledge for farmers who are crucial.

In the coming years, the government should enact policies that improve the sustainability of perennial crop systems. For the vulnerable production, reorganization of perennial crop production and marketing activities (building strong cooperatives among 6 main agents like producers, scientists, banks, the government, companies, and investors) would enhance the value chain performance as well as the power of farmers. Likewise, associations such as farmers unions and women associations should operate more efficiently.

References

Alexander, P., Paustian, K., & Smith, P. M. D. (2015). The economics of soil C sequestration and agricultural emissions abatement. *The Soil, 1*, 331–339.

Amarasinghe, U.A., Hoanh, C.T., D'haeze, D., Hung, T.Q (2015). Toward sustainable coffee production in Vietnam: More coffee with less water. Agricultural Systems 136: 96–105.

Araya, A., Stroosnijder, L., Habtu, S., Keesstra, S. D., Berhe, M., & Hadgu, K. M. (2012). Risk assessment by sowing date for barley (Hordeum vulgare) in northern Ethiopia. *Agricultural and Forest Meteorology, 154*, 30–37.

van Asten, P. J. A., Wairegi, L. W. I., Mukasa, D., & Uringi, N. O. (2011). Agronomic and economic benefits of coffee–banana intercropping in Uganda's smallholder farming systems. *Agricultural Systems, 104*(4), 326–334.

Barbier, E. B., Burgess, J. C., & Grainger, A. (2010). The forest transition: Towards a more comprehensive theoretical framework. *Land Use Policy, 27*(2), 98–107.

Cheesman, J., Son, T.V.H., Bennett, J (2007). Valuing irrigation water for coffee production in dak Lak, : A marginal productivity analysis. Research Report. Retrieved September 15, 2019, from http://se.ueh.edu.vn/wp-content/uploads/nghien-cuu/19-Tran-Vo-Hung-Son-Valuing-irrigation-water-for-coffee-production-in-Dal-Lak-Viet-Nam.pdf

Chu Nguyen, M. N., & Hoang, T. (2008). *Analysis data for research with SPSS* (p. 295). Hong Duc publisher.

D'haeze, D., Deckers, J., Raes, D., Phong, T. A., & Loi, H. V. (2005). Environmental and socio-economic impacts of institutional reforms on the agricultural sector of Vietnam: Land suitability assessment for Robusta coffee in the dak Gan region. *Agriculture, Ecosystems and Environment, 105*(1–2), 59–76.

Dak Lak PSO. (1986). *Dak Lak Socio-economic Statistical Data Source (1975–1985)*. Dak Lak.

Dak Lak PSO. (2011). *Dak Lak Socio-economic Statistical Data Source (2000–2010)*.

Dang, T. H., & Shively, G. (2008). Coffee boom, coffee bust and smallholder response in Vietnam's central highlands. *Review of Development Economics, 12*(2), 312–326.

Darnhofer, I., Bellon, S., Dedieu, B., & Milestad, R. (2010). Adaptiveness to enhance the sustainability of farming systems. *A review. Agronomy for Sustainable Development, 30*(3), 545–555.

Diepart, J.C., Allaverdian, C (2018). *Farming systems analysis: A guidebook for researchers and development practitioners in Myanmar*. GRET. Retrieved September 15, 2019, from https://www.gret.org/publication/farming-systems-analysis-guidebook-myanmar/?lang=en

Do, V.D. (2016). Những thành tựu và định hướng phát triển kinh tế Đắk Lắk qua các thời kỳ (Achievements and orientation for Dak Lak province over stages). Editor Le Ngoc Diep; Political theory, 2016, 359.

Doutriaux, S., Geisler, C., & Shively, G. (2008). Competing for coffee space: Development-induced displacement in the central highlands of Vietnam. *Rural Sociology, 73*(4), 528–554.

Duong, T. O. (2019). *Thực trạng sản xuất hồ tiêu tại Tây Nguyên, Vietnam (The sistuation of pepper production in Central Highlands)*. Buon Ma Thuot. Retrieved September 15, 2019, from http://wasi.org.vn/doi-net-ve-thuc-trang-san-xuat-ho-tieu-tai-tay-nguyen-hien-nay

FAO. (1999). *Agricultural and farm systems - concepts and definitions*. Retrieved September 15, 2019, from http://www.fao.org/3/w7365e/w7365e04.htm

FAO. (2016). *Agricultural cost of production statistics guidelines for data collection, compilation and*. FAO. Retrieved September 15, 2019, from http://gsars.org/wp-content/uploads/2016/02/Handbook-on-Agricultural-Cost-of-Production-Statistics-28-02-2016.pdf

FAO. (2019). *Cropping systems diversification to enhance productivity and adaptation to climate change. Zambia*. Retrieved September 15, 2019, from http://www.fao.org/policy-support/resources/resources-details/en/c/1183655/

Faruque, G., Sarwer, R. H., Karim, M., Phillips, M., Collis, W. J., Belton, B., & Kassam, L. (2017). The evolution of aquatic agricultural systems in Southwest Bangladesh in response to salinity and other drivers of change. *International Journal of Agricultural Sustainability, 15*(2), 185–207.

Fresco, L. O., & Westphal, E. (1988). A hierarchical classification of farm systems. *Experimental Agriculture, 24*(4), 399–419.

Galati, A., Crescimanno, M., Gristina, L., Keesstra, S., & Novara, A. (2016). Actual provision as an alternative criterion to improve the efficiency of payments for ecosystem services for C sequestration in semiarid vineyards. *Agricultural Systems, 144*, 58–64.

Han Quang, H., Azadi, H., Dogot, T., Vu, D. T., & Lebailly, P. (2017). Dynamics of agrarian systems and land use change in North Vietnam. *Land Degradation & Development, 28*(3), 799–810.

Hatfield, J. L., & Walthall, C. L. (2014). Climate change: Cropping system changes and adaptations. In K. V. A. Neal (Ed.), *Encyclopedia of agriculture and food systems* (2nd ed., pp. 256–265). Elsevier.

Herridge, D.F., Win, M.M., Nwe, K.M.M., Kyu, K.L., Win, S.S., Shwe, T., Cornish, P.S (2019). The cropping systems of the central dry zone of Myanmar: Productivity constraints and possible solutions. Agricultural Systems, 169, 31–40.

Hill, B., & Bradley, B. D. (2015). *Comparison of farmers' incomes in the eu member states*. University of London. Retrieved from https://www.researchgate.net/publication/280086961_Comparisons_of_Farmers'_Incomes_in_the_EU_Member_States_-_Study_for_the_European_Parliament

Ho, Q.T. (2018). *Economic analysis of sustainable coffee production in Vietnam*. PhD disseration, p. 250.

Ho, T. Q., Hoang, V. N., Wilson, C., & Nguyen, T. T. (2017). Which farming systems are efficient for Vietnamese coffee farmers? *Economic Analysis and Policy, 56*, 114–125.

Huynh, A.C.T. (2018). Vulnerability context: A study on livelihood pathways of the Indigenous people. In *Livelihood pathways of Indigenous people in Vietnam's central highlands* (pp. 1–34). Springer.

ICC (2019). *Country Coffee Profile: Vietnam*. Retrieved from http://www.ico.org/documents/cy2018-19/icc-124-9e-profile-vietnam.pdf

Jayne, T. S., Chamberlin, J., & Headey, D. D. (2014). Land pressures, the evolution of farming systems, and development strategies in Africa: A synthesis. *Food Policy, 48*, 1–17.

Karimov, A.A., Nguyen, T.T., Cadilhon, J.J., Truong, T.Kh., Tran, V.T., Chau, T.M.L., Dinh, T.N.T (2016). *Value chain assessment report for avocado, cattle, pepper and cassava in Dak Lak province of Central Highlands of Vietnam*. ILRI. Retrieved September 15, 2019, from https://pdfs.semanticscholar.org/d2be/5874d7f126f1ef0695e84452b25a686187a1.pdf

Lebailly, P., Peemans, J.-P., Vu, D.T., Diepart, J.-C., Dupuis, D., Mai Lan, P., Han Quang, H (2015). *Développement rural et petite paysannerie en Asie du Sud-Est: Leçons d'expériences au Vietnam et au Cambodge*.

Lindskog, E., Dow, K., Axberg, G. N., Miller, F., & Hancock, A. (2005). *When rapid changes in environmental, social and economic conditions converge: Challenges to sustainable livelihoods in dak Lak*. Vietnam. Retrieved from https://mediamanager.sei.org/documents/Publications/Risk-livelihoods/SEI_Lindskog_Dak_Lak_2005.pdf

Maurice, G. (2004). Concept for sustainable upland farming in dak Lak Province. Vietnam. Retrieved September 15, 2019, from http://www.mekonginfo.org/assets/midocs/0001359-farming-concept-for-sustainable-farming-system.pdf

Meyfroidt, P., Vu, T. P., & Hoang, V. A. (2013). Trajectories of deforestation, coffee expansion and displacement of shifting cultivation in the central highlands of Vietnam. *Global Environmental Change, 23*(5), 1187–1198.

Muller, D., & Zeller, M. (2002). Land use dynamics in the central highlands of Vietnam: A spatial model combining village survey data with satellite imagery interpretation. *Agricultural Economics, 27*(3), 333–354.

Newton, A. C., Hodder, K., Cantarello, E., Perrella, L., Birch, J. C., Robins, J., & Cordingley, J. (2012). Cost–benefit analysis of ecological networks assessed through spatial analysis of ecosystem services. *Journal of Applied Ecology, 49*(3), 571–580.

Nguyen, H. Q. (2017). Analyzing the economies of crop diversification in rural Vietnam using an input distance function. *Agricultural Systems, 153*, 148–156.

Nguyen, T.T., Bui, C.B. (2011). How to prevent the most serious diseases of black pepper (Piper nigrum L.) A case study of Vietnam. In *PC Annual Meeting in Lombok, Indonesia*. Retrieved September 15, 2019, from http://iasvn.org/en/upload/files/WVW96GJAG5VNblackpepper2014.pdf

Nguyen, T. D. N., & Phan, T. T. (2017). Economic performance of coffee and pepper intercropping in Quang Hiep commune, Cumgar district, dak Lak province. *VJAS, 15*(9), 1277–1287.

NIAPP (2017). *Vietnam pepper industry development planning to 2020, and vision 2030*. Retrieved from http://www.niapp.org.vn/info/en/dtcl/giaidoan4/viet-nam-pepper-industry-development-planning-to-2020-and-vision-2030/54161

OECD (2015). *Agricultural Policies in Vietnam*. Retrieved September 15, 2019, from https://doi.org/10.1787/9789264235151-en

People's Committee of Dak Lak. (1986). *Báo cáo tóm tắt kế hoạch phát triển kinh tế-xã hội năm 1986 (preliminary summary report of socio-economic development)*. Province.

Pham, D. H. P., Nguyen, N. T., Le, V. D., Pham, T. H., Espaldon, M. V. O., Magsino, A. O. (2001). *Impacts of changes in policy and market conditions on land use, land management and livelihood among farmers in central highlands of Vietnam*. Retrieved September 15, 2019, from http://citeseerx.ist.psu.edu/viewdoc/download?doi=10.1.1.506.8156&rep=rep1&type=pdf

Phan, H. T. (2015). *Tài liệu đào tạo nông dân kỹ thuật sản xuất cà phê bền vững (Coffee, Technical for sustainable production of development)* (HDI). VNU. Retrieved September 15, 2019, from http://gcpconnect.coffee/sites/gcp-connect.org/files/TOF.pdf

Phan, T. T., Le, D. N., Ho, T., Burny, P., & Lebailly, P. (2019a). Which perennial crop farm approach generates more profitability? A case study in dak Lak province, Vietnam. *Asian Social Science, 15*(9).

Phan, T. T., Le, D. N., Ho, T., Burny, P., & Lebailly, P. (2019b). Economic analysis of perennial crop Systems in dak Lak Province, Vietnam. *Sustainability, 11*(1), 81.

Ponte, S. (2002). The 'latte revolution'? Regulation, markets and consumption in the global coffee chain. *World Development., 30*(7), 1099–1122.

Provincial People's Committee of Dak Lak. (2018). *Coffee annual report 2017–2018*. Retrieved September 15, 2019, from https://daklak.gov.vn/documents/10181/872901/293.pdf

Rana, S. S., & Rana, M. C. (2011). Cropping system. *Department of Agronomy, College of Agriculture, CSK Himachal Pradesh Krishi Vishvavidyalaya, Palampur, 80.*

Romyen, A., Sausue, P., & Charenjiratragul, S. (2018). Investigation of rubber-based intercropping system in southern Thailand. *Kasetsart Journal of Social Sciences, 39*(1), 135–142.

Scherr, S. J., Mankad, K., Jaffee, S., Negra, C (2015). *Steps toward green: policy responses to the environmental footprint of commodity agriculture in East and Southeast Asia*. EcoAgriculture Partners. Retrieved September 15, 2019, from https://ecoagriculture.org/wp-content/uploads/2015/08/Steps-Toward-Green-Book-File-Final-for-Upload.pdf

Tran, T. Q. C., D'haeze, D. (2005). *Assessment of water, fertilizer and pesticide use for coffee production in Dak Lak Province*. IPSARD, Working Paper. Hanoi. Retrieved September 15, 2019, from http://agro.gov.vn/vn/tID2802_Assessment-of-water-fertilizer-and-Pesticide-use-for-coffee-production-in-Daklak-province.html

Ulmer, J. T., & Wilson, M. S. (2003). The potential contributions of quantitative research to symbolic interactionism. *Symbolic Interaction, 26*(4), 531–552.

Zinnanti, C., Schimmenti, E., Borsellino, V., Paolini, G., & Severini, S. (2019). Economic performance and risk of farming systems specialized in perennial crops: An analysis of Italian hazelnut production. *Agricultural Systems, 176*, 102–645.

Chapter 14
Sustainable Agriculture Development in Vietnam

Thi Hong Linh Phi and Thi Thanh Huyen Bui

Abstract In Vietnam, agriculture contributes significantly to economic growth, poverty reduction, food security, and social welfare. Over the last 30 years since the reform in 1986, agricultural sector in Vietnam has still shown many inner limitations and had negative spillover impacts on social and environmental dimensions. Based on the perspective of sustainable development, agricultural development in Vietnam has been far from being sustainable. In this paper, we will propose a group of indicators in order to measure sustainable agricultural development and use them to analyze the case study of Vietnam. The study results offer policy implications to develop a sustainable agricultural system in Vietnam.

Keywords Agriculture · Sustainable development · Sustainable agriculture · Vietnam

1 Introduction

Agriculture plays a critical role in developing countries. A well-developed agricultural sector can ensure food security as the population increases, generate jobs for rural workers, offer more opportunities in foreign trade and create a solid platform for industries (World Bank, 2008). Over the last few decades since the 1986 reform, the agricultural sector has achieved great success that significantly contributes to the socio-economic development of Vietnam. Different to harsh condition of starvation before 1986, Vietnam has not only ensured food security but also boosted the agricultural export turnover over the years. The agricultural sector has experienced a better structural transformation contributing to income generation for farmers and successful pursuit of Vietnam's development goals. However, agricultural development in Vietnam has revealed a number of weaknesses including low growth rate,

T. H. L. Phi (✉) · T. T. H. Bui
National Economics University, Hanoi, Vietnam
e-mail: linhph@neu.edu.vn

A. T. Nguyen, L. Hens (eds.), *Global Changes and Sustainable Development in Asian Emerging Market Economies Vol. 2*, https://doi.org/10.1007/978-3-030-81443-4_14

insignificantly growing labor productivity, and rapidly growing emission in agricultural production. Therefore, these weaknesses will hinder agricultural development in Vietnam. Thus, Vietnam need to focus on offering better policies for sustainable agricultural development, contributing to the achievement of the national development goals in the future.

This study will provide a systemic review of existing theories to clarify the implications of sustainable agricultural development (contents, measurement criteria), use the recommended indicators to produce a situational analysis of Vietnam's agricultural development and thus propose adjustments in the future.

Literature Review Sustainable agricultural development has been much researched by many researchers. However, the concept of sustainable agricultural development has not been completely agreed yet. There are more than 70 definitions of sustainable agricultural development (Zhen & Routray, 2003). Of which are some concepts that the research team can access as follows:

In 1989, on the basis of the Bruntland Commission's definition of sustainable development, the FAO Council defined sustainable agriculture and rural development as:

> "… the management and conservation of the natural resource base, and the orientation of technological and institutional change so as ensuring the attainment and continued satisfaction of human needs for present and future generations. Such sustainable development (in the agriculture, forestry and fishery sectors) conserves land, water, plant and animal genetic resources, is environmentally non-degrading, technically appropriate, economically viable and socially acceptable".

By 1997, FAO specified the criteria to meet the sustainable development process of agriculture and rural areas, including: *(i) Meeting the basic nutritional requirements of present and future generations, qualitatively and quantitatively, while providing a number of other agricultural products; (ii) Providing durable employment, sufficient income, and decent living and working conditions for all those engaged in agricultural production; (iii) Maintaining and, where possible, enhancing the productive capacity of the natural resource base as a whole, and the regenerative capacity of renewable resources, without disrupting the functioning of basic ecological cycles and natural balances, destroying the sociocultural attributes of rural communities, or causing contamination of the environment; (iv) Reducing the vulnerability of the agricultural sector to adverse natural and socioeconomic factors and other risks, and strengthening self-reliance.*

Mollison and Slay (1994) asserted that sustainable agriculture is a system designed to be ecologically stable, economically viable, capable of meeting the human demands without exploiting the land, and polluting the environment.

According to Zhen and Routray (2003), an important principle of sustainable agriculture development is achieving long-term efficiency goals while preserving natural resources, especially land and water.

Swaminathan (2006) suggested that sustainable agriculture development must be stable and long-term efficiency without causing harms to the ecological and social system.

According to Pretty (1995), agricultural development is a learning process, not a goal, so in his study "Agriculture sustainability: concepts, principle and evidence" in 2008, he suggested that sustainable agriculture development should adhere to the following principles: (i) Applying ecological and biological cycles to production process; (ii) Restricting the use of nonrenewable inputs that are harmful to either environment or health of both producers and consumers; (iii) Effectively use of knowledge and skills of producers, thereby improving self-reliance and using human capital to replace external costly production factors; (iv) Effectively use collective capacity of people to solve common problems of agriculture and natural resources such as disease, irrigation.

Vu (2013) employed various approaches in sustainable agriculture development including (i) organic agriculture; (ii) conservation agriculture; (iii) selecting appropriate farming sites; (iv) limiting the use of external inputs; (v) integrated pest management; (vi) integrated nutrition management; (vii) production biogas in farms; (viii) expanding grassland areas; (ix) agroforestry systems, agriculture coping with climate change. In addition, the author also mentioned the concept of sustainable agriculture development based on the value chain, which is about developing a sustainable agricultural value chain from production to commerce, encouraging sustainable production processes.

Luu and Truong (2016), although not giving a specific concept of sustainable agriculture development, when analyzing the current situation of sustainable agriculture development in Vietnam during 1989–2014, approached the following aspects: (i) economic (growth rate, labor productivity, efficiency of capital use, contribution of factors to agricultural sector's growth); (ii) social (quantity and quality of labor, poverty rate, income inequality between rural and urban areas, improvement in the health issues and nutrition of citizens, educational and training level of rural civilians); (iii) environmental (emission from both agricultural production and soil as well as quality of water).

Some studies above showed that existing literature on sustainable agriculture development has approached different perspectives such as agricultural practices or the results of development. From the practical perspective, sustainable agriculture development refers to the principles and ways that agriculture production process must comply to the continuity and expansion of agricultural production to meet the demand that requires the improvement of the citizens' life quality (FAO, 1989, 1997; Pretty, 1995, 2008; Vu, 2013). Using result-based perspective, Zhen and Routray (2003) and Luu and Truong (2016) concluded that sustainable agriculture development deals with aspects of growth, the capacity to use resources, and the spillovers effect of agriculture that can impose the opposite effects to promote the development of agriculture. The division of these approaches is not fully complete because in fact, the results of sustainable agriculture development and agricultural practices are closely related. Sustainable agriculture development depends on the sustainability of the practice in agriculture. Particularly, the approach according to practical perspective is more appropriate for ecological studies. From an economic perspective, this study sees sustainable agriculture development as a result, agricultural practices will be considered factors affecting sustainable development outcomes.

With this approach, based on the overview of some concepts above, "sustainable agriculture development" can be seen as the process of maintaining the increase in scale and improving efficiency. The productivity of the agricultural sector also creates the positive impacts to social and environmental aspects.

With this concept, sustainable agriculture development covers the following basic contents:

Maintaining an increase in scale: maintain a stable and long-term growth rate.

Improving production efficiency: shown through improving the efficiency of using input resources to improve productivity, quality, and the competitiveness of agricultural products.

The positive spillovers effect on society: agricultural growth leads to improvement in social issues such as raising farmers' incomes, contributing to reducing the income gap between rural and urban areas, improving the living standards of farmers (rural areas), fulfilling the living conditions of people in order to increase the quality of the population in the agricultural and rural areas.

The positive spillovers effect on the environment: agricultural production is associated with environmental protection, preservation of natural resources, and biodiversity.

These will be the basis for giving out the indicators applied to analyze the situation of sustainable agriculture development in Vietnam in the following content.

2 Methodology

Based on the concept of sustainable agriculture development proposed above, the authors analyze the situation of sustainable agriculture development in Vietnam according as follows:

Firstly, analyzing situation of sustainable agriculture development in each specific aspect of the content.

Secondly, analyzing overall situations of sustainable agriculture development level through measuring the sustainability of agriculture.

With such an approach, the study is conducted as follows:

2.1 Data Collection

This study uses secondary data collected from reports of agencies and organizations from Vietnam and other foreign countries, including data collected from the Statistical Yearbook of General Statistics Office of Vietnam, data from the Vietnam Household Living Standards Survey of General Statistics Office of Vietnam, statistics of the World Bank and the World Food Organization (FAO). This study uses desk research approach to collect this type of data.

2.2 Data Analysis

After collecting the data, the authors use descriptive statistics approach to analyze data and then evaluate the current situation of agriculture development in Vietnam in the following steps:

1. Measure sustainable agriculture development in each specific aspect of the content as follows:

   Maintaining an increase in scale: analyzing growth rate.

   Improving production efficiency: analyzing the efficiency of the resource use, including labor (labor productivity) and capital (through increase in investment), and dependence on external resource factors.

   Spillovers effect on society: analyzing indicators of per capita income, expenditure structure, qualifications, and physical strength of the people in agricultural sector.

   Spillovers effect on the environment: analyzing indicators of agricultural-generated waste (CO_2) and the use of chemical fertilizers and pesticides.
2. Measure overall sustainable agriculture development: calculating and analyzing Sustainable Agricultural Index (SAI).

   This index was used by Hatai and Sen (2008) to calculate Orissa's sustainable agriculture index in India. The SAI method is similar to the human development index (HDI) released in 1992 by the United Nations. Specifically:

$$\mathrm{SAI} = \left(\mathrm{I_E} + \mathrm{I_S} + \mathrm{I_{EN}}\right)/3$$

   In which:

   Sustainability indicators for each sector is calculated based on the average of the I_{Xi} sub-indices. Where: I_E is the economic index, I_S is the social index, I_{EN} is the environmental index. These components are calculated according to one of the two below formulas according to the principle: if the component index has a positive impact on sustainability, select formula (14.1), whereas if the component index has the opposite effect, choose formula (14.2).

$$\mathrm{I_{Xi}} = \frac{Xi - Xmin}{Xmax - Xmin} \tag{14.1}$$

or

$$\mathrm{I_{Xi}} = \frac{Xmax - Xi}{Xmax - Xmin} \tag{14.2}$$

   To calculate the sub-indices, the authors applied the calculation and selected the indicators based on specific indicators measuring each aspect of the sustainable development content in accordance with the data conditions in Vietnam. In principle: (i) It is feasible to collect calculated data. (ii) Change with space and time. (iii) Ensure sustainability on each pillar: Economic, social, and environmental. (iv) Widely accepted (Zhen & Routray, 2003).

Since then, data used to calculate SAI include (Table 14.1):

However, because the gap between the max and min values of the two groups of economic and environmental indicators is too large, the research team applies the way that UNDP adjusted to calculate the income index in HDI as follows:

$$I_E = \frac{\ln(Xi) - \ln(Xmin)}{\ln(Xmax) - \ln(Xmin)}$$

$$I_{Ei} = \frac{\ln(Xmax) - \ln(Xi)}{\ln(Xmax) - \ln(Xmin)}$$

The value of SAI will be in the range (0÷1), when the SAI value is closer to 1, the agriculture is more sustainable, and the closer to 0, the less sustainable. In more details, based on the Likert scale, it can be divided into five intervals: 0÷0.2 means unsustainable, 0.2÷0.4 means low sustainable, 0.4÷0.6 means medium sustainable, 0.6÷0.8 means relatively sustainable, 0.8÷1 means very sustainable.

3 Results

3.1 Increasing the Scale of the Agricultural Sector

Based on the statistical yearbook's data of the General Statistics Office, the authors calculated the agricultural growth rate of Vietnam over the last decade as follows (Fig. 14.1):

The figure above shows that the agriculture's growth is unstable and tends to decrease.

Table 14.1 Data used to calculate SAI (SAI)

	Data	X_{max}	X_{min}
IE	Agriculture labor productivity (USD constant price 2010)	140.340,96 (Argentina- 2010)	204,15 (Mozambique- 2010)
IS	Rural to urban income ratio (%)	100	0
	Percentage of trained employed agriculture population	100	0
IEN	The volume of chemical fertilizer (kg/ha)	6.225,78 (Qatar-2010)	0,084 (Rwanda- 2010)
	The volume of CO_2 emission by agriculture (ton/ha)	8.156,11 (Turks and Caicos Islands)	0,105 (Saint Pierre and Miquelon)

Source: The authors

Calculating from the statistics of the General Statistics Office, the average growth rate of the agricultural sector was 3.83% in the 2001-2005 period, 3.38% in the 2006-2010 period, 3.06% in the 2011-2015 period, and only 2.46% in the 2016-2018 period. The growth rate of the agricultural sector decreased, especially in the period from 2009 up to now, due to: (i) In the period of 2005-2014, the agricultural sector received little investment attention and necessary support, the proportion of investment in agriculture was low and decreased. At current prices, the percentage of agricultural investment in the total social investment decreased from 7.4% in 2005 to 6.15% in 2010 and 5.04% in 2014. At constant prices, the growth rate of investment capital for agriculture in some years was even negative, such as in 2011 (−8.3%), 2012 (−9.91%), 2014 (− 4.79%); (ii) From 2013 up to now, implementing the project of restructuring agricultural production, the proportion of investment in agriculture has tended to increase (reaching 6.31% in 2018), but the weather is unfavorable, consecutive epidemics make agricultural production fluctuate.

Therefore, the growth rate was not only decreasing, but also unstable. There are several years that agricultural sector had quite good growth, but there are years the growth rate decreased sharply, such as in 2009 and 2016. The reason for the sharp decrease in the growth rate in 2009 (only 1.91%) was due to the continuous epidemics in the Central Highlands and Southern Central Coast. In 2016, growth rate also decreased to 1.36% due to the cold weather, frost in the Northern provinces, drought and saltwater intrusion in the southern provinces (Mekong Delta) at the beginning of the year, incidents of marine environment in the Central region mid-year.

It is shown that Vietnam's agricultural sector is still heavily dependent on natural conditions, reflects its outdated production technology level. At the annual Climate Change Summit in Katowiice (Poland), Germanwatch (2019) shows that Vietnam is one of the 10 countries most affected by climate change. Thus, extreme weather events will tend to increase, directly threaten Vietnam's agricultural production.

Fig. 14.1 Growth rate of GDP and labor productivity of agriculture sector. Source: Calculated from Statistical Yearbook 2018 - GSO Vietnam (2018a)

3.2 *Production Efficiency of the Agricultural Sector*

Labor efficiency, as shown by the labor productivity of the agricultural sector, has improved but still very low.

Although the labor productivity of the agricultural sector in Vietnam increased over time and faster than the GDP growth rate, which were 6.7%, 3.69%, and 2.47% in the periods of 2016-2018, 2011-2015, and 2006-2010, respectively, it still remains low. In 2018, labor productivity of agriculture reached 24.46 million VND/labor (constant 2010 price), only about 38% of the labor productivity of the whole economy.

In addition, Vietnam's agricultural labor productivity is lower than many other countries with similar income levels and Vietnam's labor productivity gaps with these countries are also increasing. According to World Bank data, in 2018, Vietnam's agricultural labor productivity reached 1209.85 USD, approximate 56.7% of the low-middle income countries, 38.2% of middle-income countries, 70.4% compared to India, and 56.8% compared to Uzbekistan (calculated at US $2010 constant price). A low labors productivity once again reflects the outdated production technology level of Vietnam's agricultural sector (Fig. 14.2).

The efficiency of investment capital in the agricultural sector is low, reflected in the high coefficient of increase in investment output in the agricultural sector.

On average, in the period of 2006-2010, the coefficient of increase in investment output in the agricultural sector was 4.16. It was 3.53 when 2009 was excluded (because, agricultural growth decreased sharply due to natural disasters, epidemic in 2009). It was 3.22 in the period of 2011-2015 and 5.72 in the period of 2016-2018 (excluding 2016, it was 4.2) (Table 14.2).

The coefficient of increase in investment output tends to increase in the period of 2016-2018 due to the impact of the agricultural production Restructuring Scheme, whereby, applying science and technology is promoted. However, with low level of

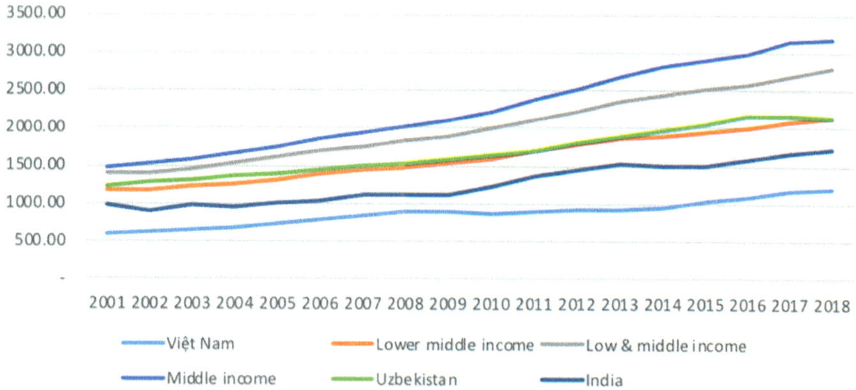

Fig. 14.2 Agricultural labor productivity of Vietnam and some countries. Source: World Development Indicators, Last Updated Date 10/7/2019

Table 14.2 The coefficient of increase in investment output

Period	The coefficient of increase in investment output in agricultural sector	The coefficient of increase in investment output whole economic	The coefficient of increase in investment output in agricultural sector/ The coefficient of increase in investment output whole economic
2006-2010	4.16/3.53[a]	6.29	66.1%/56.1%[a]
2011-2015	3.22	5.39	59.7%
2016-2018	5.72/4.2[a]	4.98	79.33%/66.8%[a]

Source: Calculated from Statistical Yearbook 2018 - GSO Vietnam (2018b)
[a]Calculated data after excluding 2 years with considerable changes in agricultural growth rate due to natural conditions (2009 and 2016)

technology in agricultural production, in addition with comparing the coefficient of increase in agricultural output to the whole economy, it is clear that efficiency of investment in the agricultural sector in Vietnam is low. The ratio of the coefficient of increase in investment output of agricultural sector to the coefficient of increase in investment output of the whole economy is increasing, meanwhile, the coefficient of increase in investment output of the whole economy of Vietnam is very high (1.5 to 2 times higher) compared to other countries with the same period of rapid growth and the same level of technology as Vietnam today (South Korea in the period of 1981-1990 was 3.2, Japan in period of 1961-1970 was 3.2, China in the period 1991-2003 was 4.1).

Agricultural production is a processing nature, dependent on agricultural materials imported from abroad.

The processing nature of agricultural production is reflected in the fact that domestic agricultural production is highly dependent on inputs import both for the crop and livestock industries. Domestic livestock depends on food sources and raw materials for processing food from abroad. According to the Ministry of Agriculture and Rural Development, domestic maize and soybean production only meet 50-55% of the demand for animal feed production and processing, thus, along with the growth of the domestic livestock industry, the import of food and feeding ingredients also increases. In 2018, 70% of total raw materials for animal feeding were imported.

In cultivation, currently, Vietnam can only produce a number of types of fertilizer such as NPK, urea, and phosphorus, some types of domestic materials cannot be produced at all and must be imported. Besides potassium fertilizer, SA fertilizer, Vietnam also has to import a large amount of pesticides. Production dependent on material produced from abroad makes the efficiency of agricultural production low.

3.3 Spillovers Effect on Society

Based on the evaluation criteria proposed above, an analysis of the spillovers effect of agriculture on social aspects through analyzing living standards and quality of population in agriculture sector and rural areas shows the followings:

Firstly, the income disparity between urban and rural areas is increasing.

Increasing in per capita income, the relative gap between rural and urban areas (according to survey data of Household Living Standards (2016) has narrowed, the average monthly per capita income of rural area was 47.83% compared to rural in 2006, increased to 53.2% in 2016) but, the absolute gap is getting bigger, in 2006 the average monthly per capita income of rural area was 883,7 thousand dongs higher than of rural area but in 2016 the gap increased to 1443.5 thousand dongs (constant price, 2010).

Secondly, living standards in rural area are still low.

With a low income level, expenditure in rural areas is also much lower than that in urban areas, according to the results of the Household living standard survey in 2016, the average monthly expenditure per capita of rural areas is only 56.7% compared to urban areas. At the same time, due to the lower income, the ratio of expenditure to income of rural areas is also higher than that of urban areas (in 2016, the average urban population spent 67.2% on expenditures, while rural areas spent 71.6%). Thus, the accumulation of rural people is much lower than in urban.

In addition, analyzing the expenditure structure of rural people shows that: because of the lower income, rural people are spending a larger proportion of income on essential needs (food, drinking, smoking) compared to urban people. In the expenditure structure for life, rural people spend about 52.9% on food, drinking, smoking; only 4.9% on education, while the corresponding figures in urban areas are 48.7%, 6.7%. The share of spending on education is lower but the proportion of spending on health in rural areas is higher than in urban areas (6.4% vs. 4.8%) means that the quality of life rural people is lower and rural people face more difficulties in improving their quality of life.

In addition, the poverty rate in rural areas is still high. According to data released by the General Statistics Office (2019), the income poverty rate of rural areas in 2016 was 7.5%, much higher than in urban areas (2%). Not only income poverty, multidimensional poverty in rural areas is also still quite severe, in 2018, the multidimensional poverty rate in the rural area was 9.6% compared to 1.5% in the urban area. These data show that not only the income is lower but also the living condition in rural areas is worse.

Thirdly, labor quality in agriculture sector is much lower than in other sectors.

Data from Statistical Yearbook of General Statistics Office of Vietnam showed that the percentage of trained employed population of agriculture is much lower than other sectors in the economy, in 2018, percentage of trained employed population at 15 years of age and above of agriculture was only 4.1%, while the average percentage of the whole economy was 21.9%. The process of improving the labor force in agriculture has also been slower than in the whole economy (Table 14.3).

Table 14.3 Percentage of trained employed population at 15 years of age and above

	2010	2013	2014	2015	2016	2017	Pre. 2018
TOTAL	14.6	17.9	18.2	19.9	20.6	21.4	21.9
Agricultural sector	2.4	3.5	3.6	4.2	4.1	4.2	4.1

Source: GSO Vietnam (2018a, 2018b)

In addition, according to the 2016 Household Living Standard Survey, following the highest degree of people, agriculture sector accounted for only 11.9% of the total number of labor with a college degree while industry had 21.9% and service had 66.2%. The agriculture sector only accounted for 4.5% the total number of labor with bachelor degrees, the industry contributed 18.8%, and services had 76.7%. Meanwhile, the agriculture sector accounted for 82.2% of the labor who have not completed the first grade or have never gone to school, 65.1% of the workers without a degree. The low degree of labor is the reason for many difficulties to the agriculture sector in receiving and applying modern production techniques to production. This is also the reason for the unsustainable economic indicators in agriculture development analyzed above.

Beside the low degree of employees, the quality of health of the rural population is also low, reflected by the high rate of malnourished children and the slow rate of reduction. According to data in 2018, in rural areas, the current situation of malnutrition of under-five children is as follows: rate of weight-for-age malnutrition was 16.1%, rate of height-for-age malnutrition was 26.7%, rate of weight-for-height malnutrition was 6.7%, those rates are much higher than in urban areas (corresponding figures are 7%, 10.2%, and 4.8%).

The low quality of health of the agricultural and rural population is a factor that makes it difficult for having sustainable agriculture development not only in the present but also in the future.

3.4 *Spillovers Effect on Environment*

Assessment of sustainability in agricultural development in Vietnam over the years from the perspective of Spillovers effect on the environment shows the following manifestations:

Firstly, agricultural production is still not environment-friendly because of overusing chemical fertilizers and pesticides.

In Vietnam, environment-friendly farming practices such as organic agriculture, climate-resilient agriculture, and low-input agriculture are few. In 2017, organic production land was 57.01 thousand hectares, equivalent 0.53% of total agriculture land. Although the rate of organic production land in Vietnam was higher than in Thailand (91.26 thousand hectares, equivalent 0.41%), Indonesia (208.04 thousand hectares, equal to 0.36%), it is still very small and lower than in Philippines (200,06 thousand hectares, equivalent 1.16%); China (3023 thousand hectares, equivalent 0.59%) (FiBL & IFOAM, 2019).

Because of the low rate of organic production land, agricultural production in Vietnam is still not environmentally friendly because of overusing chemical fertilizers and pesticides:

3.4.1 Chemical Fertilizer Consumption

In 2016, Vietnam's chemical fertilizer consumption was 429.78 kg/ha, reduced by 40.92% comparing to 2002 (304.96 kg/ha). However, Vietnam is still in the group of countries with high chemical fertilizers consumption in the world. We used 1.8 times higher than Uzbekistan, Thailand (2.65 times), the Philippines (2.73 times), India (2.59), and the average of the Asia Pacific region (1.29 times). The continuous increase in using chemical fertilizers has shown that the policies to encourage businesses to produce organic fertilizers and encourage farmers to use organic fertilizer during their cultivation toward a green agriculture of the Ministry of Agriculture and Rural Development have not been effective (Fig. 14.3).

Intensive farming and chemical fertilizers consumption to improve productivity are making agricultural land degrade; soil, water, and environment are polluted. This has negatively affected the sustainability of Vietnam's agriculture.

3.4.2 Pesticide Consumption

In 2016, Vietnam's pesticide consumption was 157.28 kg/km^2 down 30.67 kg compared to 2000. Vietnam's pesticide consumption was lower than China and Malaysia, but still higher than some countries in the region, specifically India used only 28.5 kg/km^2, Thailand used 98.6 kg/km^2. Therefore, in the coming time, Vietnam must replace chemical pesticides with biological pesticides to reduce toxic

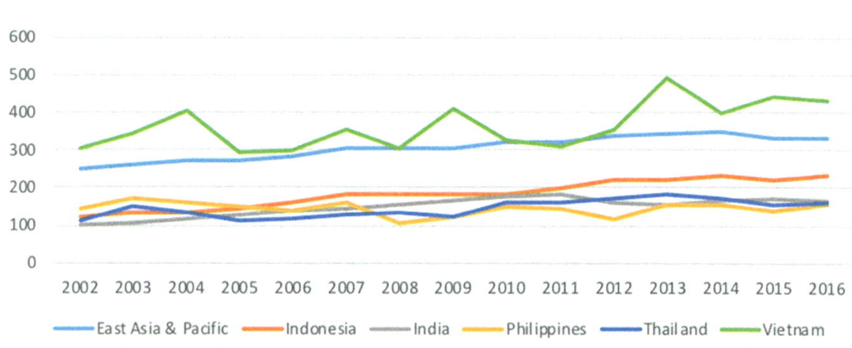

Fig. 14.3 Chemical fertilizer consumption in Vietnam and some selected countries. Source: Calculated from World Bank data, https://data.worldbank.org/indicator

Table 14.4 Pesticides consumption in Vietnam and some selected countries (*Unit: Kg/km²*)

Year	Vietnam	Thailand	India	China	Malaysia
2000	187.95	102.52	24.84	246.86	567.92
2010	178.01	327.57	22.26	343.20	814.63
2016	157.28	98.60	28.05	344.45	574.10

Source: Calculated from World Bank Data, https://data.worldbank.org/indicator

chemicals in pesticides to be released and penetrate air, water, and sediment, thereby limiting the impact on biodiversity, soil, water, and air quality (Table 14.4).

Second, greenhouse gas emissions by agricultural production are increasing rapidly.

The total amount of carbon dioxide (CO_2) released from agricultural production into the air still tends to increase, from 57.43 million tons in 2000 to 65.22 million tons in 2016. On average period 2000-2016, the growth rate amount of CO_2 of Vietnam's agricultural sector released into the air was 2.144%/year, higher than the growth rate of China (1.439%/year), and Thailand (0.339%/year).

The average CO_2 emission per 1 ha of agricultural land in Vietnam was 9.32 tons/ha, which is also higher than Thailand (3.39 tons/ha), Indonesia (7.3 tons/ha), India (4.06 tons/ha) (calculated from FAO data, 2019). As of 2016, Vietnam was still in the category of high average agricultural CO_2 emissions.

In the period 2000-2016, the proportion of CO_2 emissions of the agricultural sector in total CO_2 emission decreased from 39.26% to 29.82%. The reason for the decreasing proportion of CO_2 emission of agricultural sector is the growth rate of emissions of agricultural production is slower than the growth rate of emissions of industrial production, transport, which shows the efforts of Vietnam's agricultural sector to reduce greenhouse gas emission and slow down the drawbacks of climate change (Fig. 14.4).

Considering the structure of CO_2 emission sources in Vietnam, rice cultivation emits the highest amount of gas (44.23%), followed by synthetic fertilizers (17%). The least CO_2 emissions are burning - crop residues (0.67%) and Burning - Savanna (0.18%) (FAO, 2019).

The high amount of chemical fertilizers and pesticides consumption together with greenhouse gas emissions from agricultural production is constantly increasing, reaches high levels and leads to environmental pollution, affecting people's health, especially rural people.

3.5 *Overall Sustainable Agriculture Development Measurement*

Measuring overall sustainable agriculture development by using Sustainable agriculture index (SAI) in the period of 2010-2016 gives the followed result (Table 14.5):

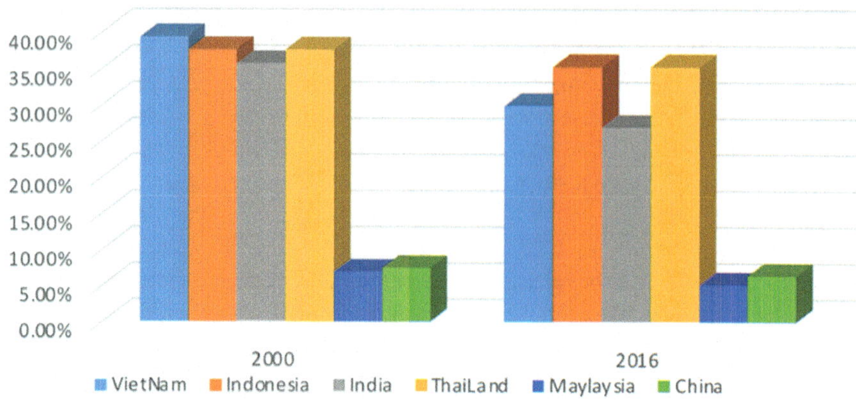

Fig. 14.4 The proportion of CO_2 emission in agricultural sector (%). Source: Calculated from FAO data (http://www.fao.org/faostat/en/?#data/GT)và EDGA (2018)

Table 14.5 Sustainable agriculture index of Vietnam in the period of 2010-2016

Year	Labor productivity (constant price 2010 USD)	I_E	Ratio per capita income between rural and urban (%)	Percentage of trained employed agriculture population (%)	Is	Chemical fertilizer consumption (kg/ha)	CO_2 emissions (tons/ha)	I_{EN}	SAI
2010	864.8	0.2090	50.2	2.4	0.263	323.3	9.432	0.432	0.301
2012	922.4	0.2191	52.8	3.0	0.279	352.3	9.474	0.428	0.309
2014	966.8	0.2264	51.4	3.6	0.275	397.8	8.911	0.425	0.309
2016	1099.6	0.2464	53.2	4.1	0.287	429.8	9.320	0.420	0.318

Source: Calculated from data of FAO (2019), World Bank (2019), GSO Vietnam (2018a, 2018b)

In the period of 2010-2016, the sustainable agriculture index (SAI) tended to increase from 0.301 in 2010 to 0.318 in 2016, the average growth rate was 0.88%/year. This shows a more sustainable trend of Vietnamese agriculture. However, with the SAI was 0.318 in 2016, Vietnam's agricultural sustainability is still low (within the range of 0.2÷0.4).

Analyzes the components of SAI:

Although economic index (I_E) growth rate is the highest (the average growth rate of 2.78%/year), it is the lowest component index of SAI, which is only about 0.2÷0.4. This shows that the economic sustainability of Vietnam's agriculture is very low, similar to the content analyzed above (growth rate is decreasing, agricultural output is unstable, the efficiency of input factors is not high).

The social index (I_S) increased 1.44%/year (slower than the increase of I_E), the value of I_S is little higher than I_E and also in the range of 0.2÷0.4. This means that the impact of economic growth on the social is positive but very little. This is because the growth rate of agriculture is not high enough to deal with the social problem in agriculture sector and rural area. Most of social progress in agriculture and rural areas are not directly due to the development of the agricultural sector but because of the impact of other policies and programs such as the New Rural Program, the Vocational Training Program for Rural Laborers of the Ministry of Agriculture and Rural Development.

The environment index (I_{EN}) is the highest component index but tends to decrease (decline rate is 0.47%/year) and lies at lower bound of 0.4÷0.6, showing that environmental sustainability is low and decreasing. It means that environmentally friendly farming methods have not developed strongly enough to restrict the environmental pollution.

4 Conclusions and Discussion

From the analyzes above, it can be seen that agriculture development in Vietnam is not sustainable: (i) growth rate is decreasing, unstable, the ability to maintain the increase in scale, and the resistance against negative factors is still limited; (ii) The efficiency of input resources (capital and labor) is low, domestic agricultural production is highly dependent on imported materials; (iii) The Spillovers effect of agricultural development on social aspects is limited (living standards in rural area are still low, the quality of the population in agriculture and rural areas is much lower than in urban areas); (iv) Agricultural production do not protect environment (chemical fertilizers and pesticides and greenhouse gas emissions are continuously increasing and reaching a higher level); and (v) The overall assessment shows that the sustainability of Vietnam's agriculture is still low.

From above limitations, there are requirements for innovation in Vietnam's agricultural production in the coming time toward maintaining a high and stable growth rate based on the application of high-tech science and technology to increase productivity, increasing the efficiency of using input resources, applying green agricultural cultivation methods to protect the environment. With such development directions, some key solutions should be implemented, including:

Restructuring the agricultural sector toward sustainable development.

In order to develop agriculture sustainability, increase productivity, increase efficiency of agricultural production, reduce environmental pollution, it is necessary to accelerate the restructuring of the agricultural sector toward sustainable development, i.e., development of industries and production, utilizing comparative advantage of agricultural products, developing high-tech application agriculture, and green agriculture.

Enhancing investment attraction for developing science-applied agriculture, high-tech agriculture, and green agriculture.

To improve production efficiency and labor productivity, reduce the consumption of fertilizers, pesticides, and emissions due to agricultural production; the application of science and technology, development of hi-tech agriculture, and Green agriculture industry are the inevitable. However, in order to attract investment in agricultural development toward this direction, it is necessary to complete policies to attract investment in developing agricultural production, such as land accumulation policy; credit policy; tax policy.

Training and retraining for laborers to improve agricultural labor skills.

The low in quality of labor in agriculture is the main reason for difficulty in sustainable agriculture. In order to develop agriculture sustainability, it requires farmers to have knowledge, awareness of sustainable agricultural production; advanced techniques and farming methods; responsibility and ability to apply the achievements of science and technology in the production process; knowledge of markets, laws, business. Therefore, it is necessary to provide training and retraining courses in order to equip farmers with this knowledge.

Reorganizing the agricultural production model toward modernization.

Building and reorganizing modern agricultural production models is an important content to promote sustainable agriculture development. Accordingly, the agricultural production models must create favorable conditions for the application of science, technology, and advanced agricultural production methods, and perform the function of linking farmers with the market and stakeholders to strengthen the value chain of agricultural products, increase the value of agricultural goods.

Strengthening linkage in agricultural production.

It is necessary to strengthen linkages in agricultural production, including vertical linkages between farmers and stakeholders (enterprises and researchers) and horizontal links (between farmers), not only within a locality but also regional linkages to form agricultural value chains, agricultural clusters to increase the value of agricultural goods.

In conclusions, with about 15% of GDP and 38.1% of labor (as of the end of 2018), the agricultural sector still plays an important role in the economy, so sustainable agricultural development is still a condition that ensures successful implementation of Vietnam's development goals. Although the article still has certain limitations, the comments are drawn after analyzing the status of agricultural development in Vietnam over the pastime and some suggestions mentioned above will be suggested to the development of Vietnam's agriculture in the coming time, contribute to the implementation of the set development goals toward a prosperous Vietnam.

References

EDGA. (2018). Fossil CO_2 emissions of all world countries, 2018 report. Retrieved August 17, 2019, from https://edgar.jrc.ec.europa.eu/overview.php?v=booklet2018&dst=CO2emi

FAO. (1989). The state of food and agriculture - world and regional reviews sustainable development and natural resource management. *Rome, Italia, 1989*, 65.

FAO. (1997). Guidelines for the integration of sustainable agriculture and rural development into agriculture policy. Retrieved August 20, 2019, from http://www.fao.org/3/w7541e/w7541e00.htm#Contents

FAO. (2019). Retrieved August 1, 2019, from http://www.fao.org/faostat/en/?#data/GT

FiBL & IFOAM. (2019). *The world of organic agriculture statistics and emerging trend 2019.* Retrieved August 12, 2019, from https://shop.fibl.org/CHen/mwdownloads/download/link/id/1202/?ref=1

Germanwatch. (2019). Global Climate Risk Index 2019 - Who Suffers Most From Extreme Weather Events? Weather-related Loss Events in 2017 and 1998 to 2017. Retrieved August 19, 2019, from https://germanwatch.org/sites/germanwatch.org/files/Global%20Climate%20Risk%20Index%202019_2.pdf

GSO Vietnam. (2018a). *Household living standards survey 2016* (pp. 219–224). Statistical Publishing House. (in Vietnamese).

GSO Vietnam. (2018b). *Statistical yearbook 2018* (pp. 158–824). Statistical Publishing House. (in Vietnamese).

Hatai, L. D., & Sen, C. (2008). An economic analysis of agricultural sustainability in Orissa. *Agricultural Economics Research Review, 21*, 273–282.

Luu, T.D., Truong, D. (2016). Sustainable agriculture development in Vietnam in the new context of international economic integration. *Vietnam's economy in the integration period: Opportunities and challenges.* University of Commerce, Hanoi, Vietnam, pp. 789-808 (in Vietnamese).

Mollison, B.C., Slay, R.M (1994). (translated by Hoang, V.D). *Outline of Sustainable Agriculture.* Hanoi, Vietnam: Agricultural Publishing House. (in Vietnamese).

Pretty, J. (1995). *Regenerating agriculture: Policies and practice for sustainability and self-reliance* (p. 320). National Academy Press.

Pretty, J. (2008). Agriculture sustainability: Concepts, principle and evidence. *Phil. Trans. R. Soc. B, 363*(1491), 447–465.

Swaminathan, M. S. (2006). An evergreen revolution. *Crop Science, 46*(5), 2293–2303.

Vu, T. B. (2013). Sustainable agricultural development: Theory and practice. *Economics and Development, 196*, 37–45. (in Vietnamese).

World Bank. (2008). *World development report 2008: Agriculture for development* (pp. 1–25). World Bank.

World Bank. (2019). World development indicators. Retrieved from https://data.worldbank.org/indicator/NV.AGR.EMPL.KD accessed on 10/8/2019.

Zhen, L., & Routray, J. K. (2003). Operational indicators for measuring agriculture sustainability in developing countries. *Environmental Management, 32*(1), 34–46.

Chapter 15
Community Participation in Urban Planning in Vietnam Toward Sustainable Development: Prospects and Challenges

Ta Quynh Hoa, Nguyen Quang Minh, Nguyen Cao Lanh, and Nguyen Hai Ninh

Abstract The transition towards the open market economy in Vietnam since "Doi Moi" (Economic Reforms) in 1986 has triggered significant social transformations, especially in megacities, such as Hanoi and Ho Chi Minh City. The increasing diversity of players in the market economy with greater roles exercised by the private sector and the community has put a demand on the urban planning process that must be more open and more community participation. Even though the approach of "community participation" has been applied to many urban planning projects in developed countries since the 1960s and achieved fruitful results for communities, this is still new in Vietnam with many challenges and limited opportunities for application. This paper aims to oversee the concept of community participation in Vietnam context, to review the institutional framework that is expected to pave the way for activities of community, and also to analyze the actual capacity of the community through two case studies in Hanoi city, i.e., "the project of improving living environment in old quarter of Hanoi" and "the detailed planning project of Le Duan – Giai Phong – Bac Linh Dam road crossing six districts of Hanoi". The paper also redefines and highlights the roles of the government, the local authorities, and experts as three key stakeholders in facilitating the community participation in detailed planning projects. The initial findings should then be discussed for effective planning tools and relevant process to enable community involvement in urban development and regeneration toward social sustainability.

Keywords Social transformation · Urban planning · Community participation · Social sustainability · Hanoi

T. Q. Hoa (✉) · N. Q. Minh · N. C. Lanh · N. H. Ninh
Hanoi University of Civil Engineering, Hanoi, Vietnam
e-mail: hoatq@nuce.edu.vn

A. T. Nguyen, L. Hens (eds.), *Global Changes and Sustainable Development in Asian Emerging Market Economies Vol. 2*, https://doi.org/10.1007/978-3-030-81443-4_15

1 Introduction

1.1 The Context of "Reform" in Vietnam and the Demand for "Major Changes" in Urban Development and Management

Reform refers to "economic reform" which was adopted by the Steering Committee of Vietnam Communist Party (VCP) in late 1986 and started to be applied nationwide in early 1987 when the whole country had been shaken for a long time by a social crisis as a consequence of a one-decade economic downturn and inflation. Everyday life in the final years (1984–1986) was a real hardship for the majority of Vietnamese people (Tran, 2015). Until the end of 1986, the whole country was on the verge of collapse, as the Soviet Union and the Eastern Block states fell into total crisis, both socio-economically and politically. The previous state management in economic development – socialist plan economy or centrally subsidized economy – had to be replaced with a more "open-minded" economic policy called "market economy with an orientation towards socialism" or in short "socialist market economy" (Karadjis, 2005). This policy encouraged the private sector to paticipate in (and contribute to) the socio-economic development as the third party along with state-owned companies and cooperatives by making full use of financial capitals and especially the so-called "think-tank" among the qualified economists who had never been able to advise the government or the local authority in policy making, because they could be regarded as "political dissidents" or "counter-revolutionists" for thinking in a different way from the pathway led by the VCP. The positive results of the economic reform can be relected in the economic growth rate (consistently over 7% per year) and the increase in GDP per capita (188 USD in 1991 to 1260 USD in 2011) (Government of Vietnam, 2016).

The vigorous economic development triggered a rapid urbanization in almost every major city in Vietnam, particularly in Hanoi and Ho Chi Minh city. Today, with over eight million and nine million inhabitants respectively (GSO Vietnam, 2019), Hanoi and Ho Chi Minh city have joined the network of Asian Mega-cities and both cities have to face numerous problems in urban planning, design, and management, such as farmland clearance and acquisition on a large scale without a proper livelihood policy for the local community, chaotic cityscape, housing shortage, ghost settlements, flooding, traffic congestion, illegal housing construction, etc. The urban growth rate continues to rise steadily, from 28.5% in 2007 to 31.7% in 2012, 35.2% in 2017 (Statistica, 2017) and has reached 40% in 2019 and is reported to exceed 50% by 2040 (Vietnamnet Website, 2019).

The conventional urban planning concept "top-down", which is highly characteristic and typical for the old-fashioned thinking in state management as well as in urban planning management, turns out to be a huge obstruction on the way to efficiency and sustainability (Danielle, 2010). A new approach – in the opposite direction – called "bottom-up" should be considered and added to the "top-down" process as a balance weight to ensure social equality, as the voice from the most underprivileged social group involved in a planning project will be heard, instead of being largely ignored. The top-down concept is still important in view of city management, and on a smaller scale, planning project management while the bottom-up can offer opportunities for a dialogue between the authority and the community and find a solution to a conflict or disagreement (Hieu, 2007). New challenges continue to arise while old problems remain unsolved. Therefore, it is high time to shift from one-way planning management to a more citizen-based planning model.

1.2 Notions of Community and Community Participation in Vietnam and Worldwide

1.2.1 Notions of Community and Community Participation in Global View

Essentially, community in the world is often closely connected with two dimensions: social (benefit and interest) and spatial (geography and region) ones. Waters stated "Community is a group of people sharing the same interests and living together in the same area" (Waters, 2000) or "Community is a term that consists of two aspects: social and spatial. In general, members of a community often try to reach a common goal, even when they have different views and characteristics" (Wilcox, 1994).

The community participation in urban planning is defined by the World Bank as "Community participation is an active process by which beneficiary/client groups influence the direction and execution of a development project with a view to enhancing their well- being in terms of income, personal growth, self-reliance or other values they cherish" or "...the community participation provides a cooperation process between people living in the same community who work together to strive for a common goal by participating in the decision-making process and finally achieving the results that are worth the common endeavors" (Wilcox, 1994).

Community participation as an issue that can be interpreted as a process by which there are contributions of the community to the making and the implementation of a decision (Healy,1997; Raine, 2006; Sanoff, 2000). However, the community participation in this process may vary in both level and scope.

In urban planning, city is regarded as a physical space, an economic space as well as a community space. Nevertheless, the material factors and economic issues are usually emphasized, rather than "community" and "human being". As Jan Gehl has already pointed out, the way the cities are planned and designed has changed consistently over the past 50 years. Today, urban planners consider the cities a huge machine with numerous functional zones but most of them go without taking good care of human beings and social life. After so many years, people just realized that paying due attention to people – putting people first – is a key to achieving a sustainable city as the most important target to aim at in the twenty-first century

(Gehl, 2010). That is to say, only when community participation taken into account, can the human factor in urban planning be fully respected.

1.2.2 Notion of Community and Community Participation in Vietnam

In Vietnam, community is defined in the Vietnamese language dictionary as "all the people living in a certain territory that are well connected and organized into a society and have similar characteristics". In Vietnam's Encyclopedia, it is interpreted as "a large group of people that shares some common features, such as social classes, occupations, customs of habitation and cultural environment. This notion encompasses also origin and ethnic tradition".

The notion of community in urban planning and building in Vietnam has been mentioned, but not yet clarified. The Planning Act No. 30/2009/QH12 just refers to "community" in Point 6, Section 1 as follows: "urban planning must ensure "publicness, transparency and harmony among different interests of the state, of the community and of an individual" (National Assembly of the Socialist Republic of Vietnam [NASRV], 2009).

Currently, there is only one more notion related to community highlighted in the Revised Land Act – Article 5 of Section 3, as it comes to "land user": "Community comprises Vietnamese people living in the same village, the same ward, the same hamlet, etc. organized into clusters that have similar customs, traditions and origins" (National Assembly of the Socialist Republic of Vietnam [NASRV], 2013).

In addition, the notion of "community" has been used in some teaching documents in urban planning and urban management as follows: "Community is a group of people living in a certain territory, thus they have the same or similar way of thinking, and feeling about the social unity in the local administrative unit, and they are able to participate in the activities or actions that should be taken for their own benefits". Based on this definition, the community is restricted within the local population characterized with geographical features.

But, in reality, the notion "community" in urban planning must be understood in a broader sense, which does not entirely coincide with "local population", because different planning projects require different aims, different scope of influence, different benefits for different groups of residents. In Vietnam, under specific circumstances, the concept of community should be further investigated, and enlightened as well, if necessary.

The community participation in Vietnam has been determined in the 2013 Constitution (Point 2, Article 28) as follows: "people have the right to participate in the state management and societal planning, discussions and recommendations of local development problems to the local authority, as well as the provincial/central government" (National Assembly of the Socialist Republic of Vietnam [NASRV], 2013).

The community participation in Vietnam has been reflected so far in the collection of the public opinions and relevant organizations only for establishing a planning program (Ta, 2015; Dao et al., 2013). At present, there have been no major studies or legal documents systematically or comprehensively analyzing how community participation should be scheduled or implemented in urban planning, so that it can ensure the democracy and the harmony of interests among the stakeholders. Conflicts, if any, will be solved. In the meantime, the efficiency in the implementation of a planning project can be secured and maximized.

1.3 Policy and Institutional Changes to Foster the Community Participation in Urban Development Process of Vietnam

In Vietnam, the implementation of Doi Moi since 1990s has fostering the changes in institutional frameworks toward facilitating community involvement (Hostovsky, MacLaren, & McGrath, 2010) and (Ta, 2015). The important role of local communities in urban planning and development was demonstrated in the Directive CT19 issued on 22 January 1990 of the Chairman of Ministers Council as "... cities belong to people, are shaped by people and operate for people". The Socio-economic Development Strategy (2005–2015) also clearly reflects the change of state in attitude about the role of communities through two major objectives:

Strengthen the decentralization process through Grassroots Democracy Ordinance (2007) to foster and institutionalize the participation of local commune and community- based organizations (CBOs) in development activities at the local level. This is the important factor for the formation of a civil society, in which the relationship between government and local residents is consolidated while the local community can still have chances to participate in decision-making process and dealing with issues that are strongly related to the community's life (René & Thong, 2005; Ta, 2009; Wells-Dang, 2010).

Enhancing the participation of community in development planning, supporting service delivery for the elimination of poverty, and ensuring the community's social safety (National Assembly of the Socialist Republic of Vietnam [NASRV], 2007).

Besides, the perception of the state in urban planning has changed significantly. In the centralized planned economy, urban planning was considered the process of the resources managed by the state and allocated to meet the specific targets while state was a dominating actor. However, in the recent time of market economy, urban planning has been regarded as a tool of controlling spatial development/investment for the public interest, particularly the development projects undertaken by the non-government sectors – other actors than state. As already noted, the increasing diversity of players in the market economy has put a strong demand on the urban planning system and decision-making process to be more open to the community and civil society participation (Hieu, 2007; Ta, 2009; Hostovsky et al., 2010).

Recently, Vietnam has built up an institutional framework for community participation in urban development activities, including: (a) Communities have the right to access urban planning information (National Assembly of the Socialist Republic of Vietnam) in article 32; (b) Communities are consulted and asked for opinions to contribute to urban planning process (National Assembly of the Socialist Republic of Vietnam [NASRV], 2009) in article 9 and article 21; and (c) Communities have the right to monitor the implementation of urban planning and regeneration projects as specified in Decision No. 80/2005/QD-TTg and in Decree No. 29/2007/ND-CP.

According to Vietnam's Urban Planning Law, there are three levels of urban planning relevant to three kinds of spatial planning: City master planning, District planning, and Detailed planning. The detailed planning is performed for the areas according to the urban development and management requirements or investment demands (National Assembly of the Socialist Republic of Vietnam [NASRV], 2009). Detailed planning is conducted at the local level so it reflects the strong relationship with local residents' interests, needs, and expectations. Based on the goals of detailed planning and subjects of the planning process to be affected, the detailed planning can be classified into three types as below:

Type 1: Detailed plan for new construction urban areas. This planning type is implemented to meet the construction investment requirements which are related to the conversion of land use purposes. This detailed planning type often is conducted on vacant land or agricultural land that has a small number of people who had to be relocated. Detailed plan type 1 includes: detailed planning of fun public centers, green parks, commercial areas, etc.

Type 2- Detailed plan for regenerating existing urban areas. This planning type is implemented for the purpose of redeveloping an existing urban area, which is related to the conversion of land use functions and the local residents' relocation. Detailed plan type 2 includes: Detailed planning for residential reconstruction, planning for a functional area in the existing residential area, renovating both sides of urban roads, etc. For this type, the issue is aimed to deal with the relationship and influence between the new planning area and the surrounding areas in order to have a good interaction while mitigating negative impacts on neighboring areas.

Type 3 - Detailed plan for preserving and renovating urban heritage areas: this planning type is a tool to manage and regulate the urban development process, is aimed to preserve and improve historical, cultural, architectural values in the urban cultural heritage zones. This planning type implements on the projects with more considering on the different impacts of development process to the preservation than land use changes. This planning type 3 has to carry out under institutional frameworks, such as the heritage law, the environmental law, regulations on preservation of historical districts, etc.

The detailed planning process as regulated in the Urban Planning Law promulgated No 30/2009/QH12 and the decree No 37/2010/NĐ-CP includes 03 main phases: Phase 1 is to formulate urban detailed planning program and this phase should be conducted for a month; Phase 2 is to formulate, appraisal, and approve of the detailed plan and total time for this phase is 6 months; and Phase 3 is to manage the implementation of the detailed plan. The plan and regulation announcement should be carried out for 30 days.

Main stakeholders involved in the detailed planning process are the investors, consultant units, local management authorities, community, and community-based organization that are defined in the Fig. 15.1 below. The stakeholders, their concerns, and interests are differentiated from different types of detailed planning.

Based on Urban Planning Law and the Decree No. 37/2010/NĐ-CP, the task of collecting community opinions is required in phase 1 and phase 2, including opinions from agencies, organizations, and local residents. This task is carried out before appraisal step in order to submit for approval. Planning information shall be announced by phase 3.

In Vietnam, detailed planning is the final phase in urban planning process to prepare for an area according to "urban development, management or construction investment demands" (National Assembly of the Socialist Republic of Vietnam [NASRV], 2009) - Item c, Article 18. It can be said that a detailed planning affects the community within the planning area directly and enormously. Therefore, the participation of the local community in all planning steps helps ensure the quality of planning that is expected to meet the physical and spiritual requirements for the community as well as to secure the overall sustainable development of the city. In the detailed planning projects, the available tools will be used to accurately assess physical factors such as the current land use, the conditions of technical infrastructure, landscape architecture, environment, population and age structure, labor force, and socio-economic structure, etc. However, within a period of 6 months of making detailed planning, the planners - who often live outside the planning area and are

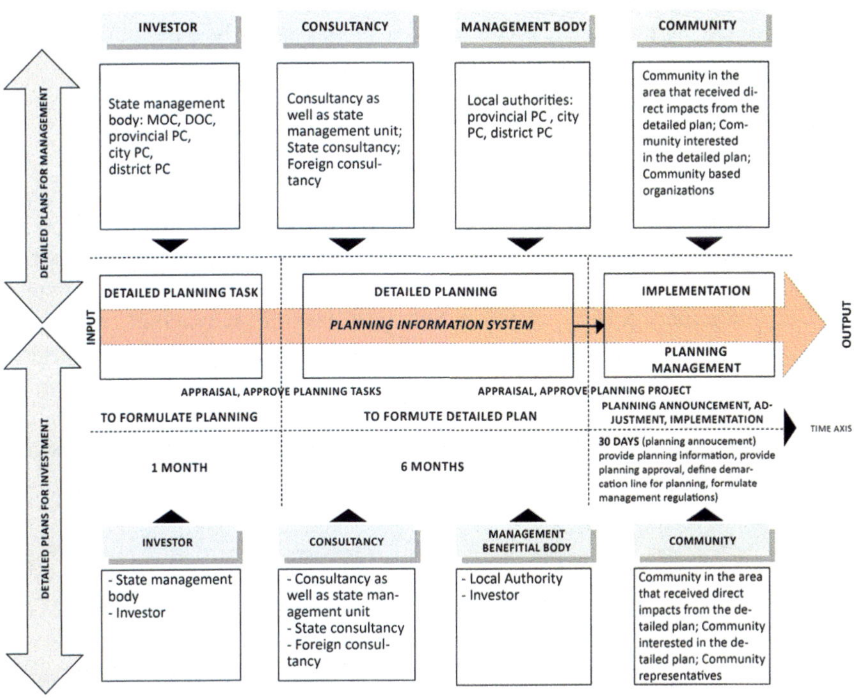

Fig. 15.1 Urban detailed planning process (*Source*: Authors, 2019)

just considered "the outsiders"- find it difficult to identify and assess intangible factors in this area, such as socio-cultural factors, lifestyles, changes of habits, etc. of the community, especially the elements of "sense of places" – the issues that are considered to be the most important in urban planning toward sustainability today. It is obviously impossible to plan and develop urban areas without understanding the material, spiritual, and spiritual life of the local community where people are united by a common way of life to make their communities become places (ACVN and KAS, 2010; Aprodicio, 1995; Thao, 2001; Held, 2006).

However, the process of mobilizing community participation and collecting community opinions is still based on limited technical model, formalistic, and less transparent (Dobertein, 2004). Briffetta, Obbard, and Mackcc (2003) and Obbard, Lai, and Briffet (2002) also assumed that Vietnamese urban planning process with community participation also suffers from inadequate staffing and experience, lack of monitoring and evaluation, poor coverage of baseline information, the failure to integrate competent and relevant ministries as well as agencies, and finally the lack of transparency in the public involvement. Currently, the community participation in urban planning is considered a mandatory administrative procedure to obtain approvals rather than an effective approach for urban planning formulation (Ta, 2009; Ta, 2015). Furthermore, there are no legal documents defining clearly the responsibilities of stakeholders, the mechanism of coordination among the stakeholders to mobilize community participation from the first stages of planning project, such as identifying the goals and objectives of the project, conducting surveys of the existing conditions, actual needs, and resources from the community.

1.4 Research Aims

The key objectives of our research is to answer these four questions: (1) who are considered community and what are the characteristics of community participation in detailed planning in vietnam's context?; (2) what are the roles of community relations and community coherences in preserving the tangible and intangible values of a community in detailed urban planning projects in case of regeneration, refurbishment, and preservation?; (3) how about the public awareness and capacity in participating to address the issues related to detailed planning for urban regeneration and improvement?; and (4) how to mobilize the community's resources when improving the detailed planning process in case of reconstruction, refurbishment, and preservation?

2 Methodology

2.1 Study Areas

The study was primarily conducted in Hanoi, one of the two largest cities in Vietnam, and has the highest rate of urbanization across the country. For over 1000 years, Hanoi history has always been associated with the process of urbanization. Hanoi Citadel, the Old Quarter, and the French Quarter are considered the historical, cultural, and administrative centers of the city. The inner city was home to about 50,000 people in 1902 when it became the capital of Indochina, and there were more than 800,000 in 1975 when the Vietnam War ended.

Since the economic reform in 1986, the city's population has increased significantly by about 3% per year and had reached 2.8 million by 2000. According to the UNDP's World Urbanization Forecast, the expansion of the city's administrative boundaries in 2008 doubled its population to nearly 6.4 million. The metropolitan area is expected to cover 3436 km^2 and accommodate more than 15 million people by 2020 (Divya & Helga, 2017). The Master plan of Hanoi capital city toward 2030 with a vision toward 2050 approved in 2011 would shape a vision Hanoi in the future as follows: *"... the capital of a country of more than 100 million people - a civilized, elegant and sustainable capital city that keeps on growing. After 30 years since the start of the Economic Reform, Hanoi's economy is developing rapidly and people's living standards have been improved significantly"*. However, it is accompanied by the emergence and worsening of "urban problems", such as rapid urbanization leading to population explosion while the construction, transportation, and technical infrastructure fail to keep up with the urbanization process; the quality of the living environment in old urban areas has been seriously affected, and the preservation and conservation of the core values of urban heritage are also negatively influenced (UN-Habitat, 2010). Coping with these huge challenges, a livable city has become a development target toward the sustainability in the coming period to strive for. Putting people at the center of development, emphasizing the importance on the human values and the quality of the spiritual and material life of all people, promoting democracy, enhancing people's participation are absolutely necessary (UN-Habitat, 2010) and (Ta, 2015).

Therefore, the study is conducted with two specific case studies in some central districts of Hanoi city, corresponding to two types of detailed planning projects that require the participation of many stakeholders from the central government and local authority, state agencies, technical groups, professional associations, nongovernmental organizations, and communities. The results of the project have a direct impact on the living environment of the residents, the right of the people, and the interests of stakeholders. The two case studies are:

Case study no. 1: Detailed planning project to preserve and improve the living quality of a street block of Hang Buom street - Hang Giay street – Ta Hien street – Luong Ngoc Quyen street in the Old Quarter of Hanoi city.

Case study no. 2: Detailed planning project to improve and regenerate the cityscape on both sides of a main road axis named Le Duan – Giai Phong – North Linh Dam road running across six districts: Hoan Kiem, Ba Dinh, Dong Da, Hai Ba Trung, Hoang Mai, and Thanh Xuan in Hanoi city (Figs. 15.2, 15.3, and 15.4).

The Case study no. 1 was conducted in the Old Quarter of Hanoi. Covering 91 hectares and located at the heart of the city, the Old Quarter is one of the oldest areas in Hanoi. It was recognized as National Historical Heritage in 2004. With the vision as "The City of Culture - Water Surface and Greenery", successful preservation and redevelopment of historical districts like the Old Quarter would be very crucial for achieving the Vision and sustainable development. Not only regarded as a historical area, the Old Quarter is also ranked among the most thriving commercial and business districts of the city where various types of commodities and services, both whole sale and retail, can be found. It is also a well-known tourism destination. But the Old Quarter must now be facing critical problems: degradation of all of its values. Problems have arisen from the following contradictions: overloaded physical environment versus vital social environment, poor infrastructure versus thriving economic activities, inadequate living environment versus wealthy material conditions of local residents, overcrowded population versus limited spaces, and tourism development versus heritage and identity preservation of the district. Thus,

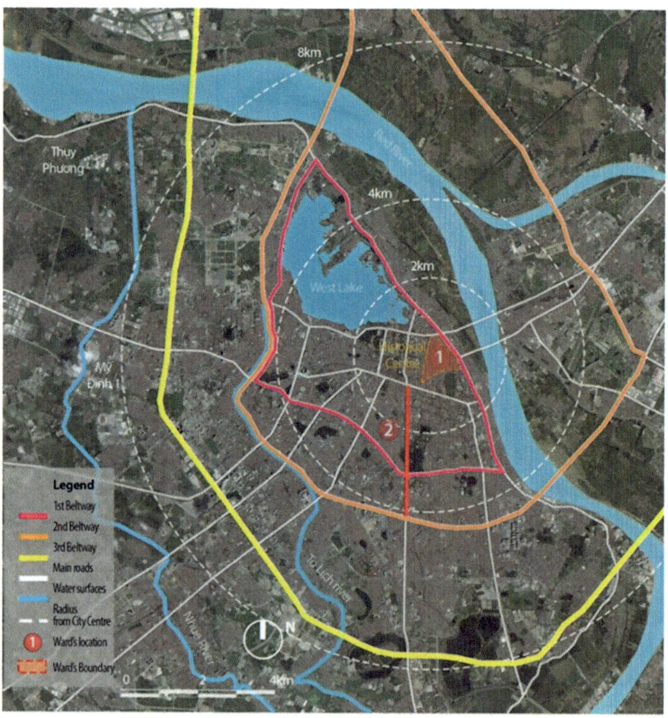

Fig. 15.2 Location map of two case studies (*Source*: Authors, 2015)

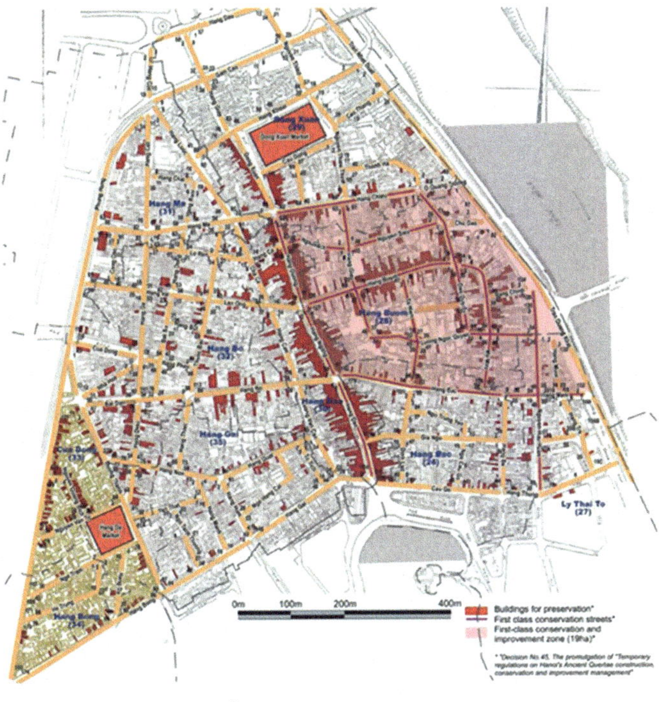

Fig. 15.3 Location map of case 1 – The old quarter of Hanoi (Source: Ta, 2015)

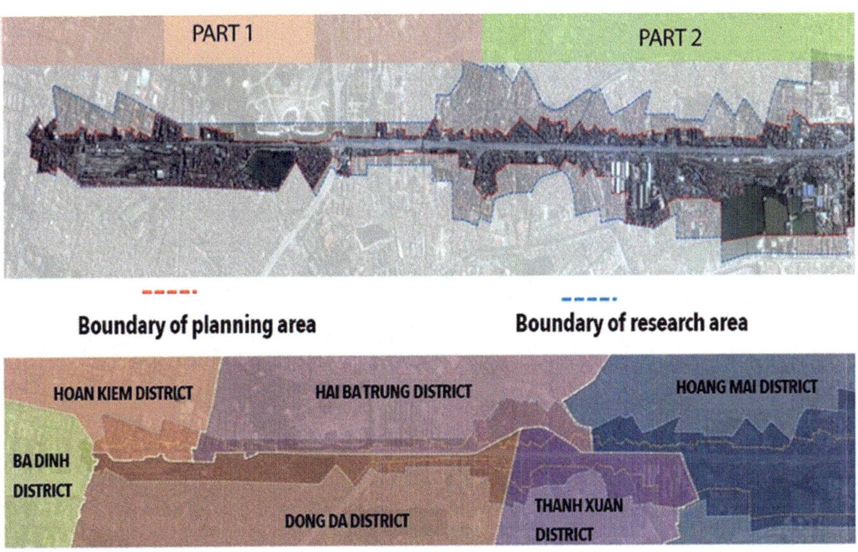

Fig. 15.4 Location map of case 2 – both sides of Le Duan – Giai Phong – Linh Dam road, Hanoi (*Source*: Ta, 2015)

sustainable development of the area has long been an immense challenge to Hanoi city, its Government, its citizens, and professional organizations to deal with.

With the effort to promote sustainable development and conservation for the Old Quarter, an integrated study with the collaboration among sociologists, economic experts, and urban planners was established in the project "Sustainable development for Hanoi Old Quarter" – a biggest pilot project ever undertaken as part of the Program "Comprehensive urban development of Hanoi Capital to 2020 – HAIDEP" sponsored by JICA.

This case study aims to verify the role of the community, the cohesion of the community in the historical urban areas, the awareness of the community about their responsibility in coordinating with stakeholders to preserve the core values of the region and to improve living environment of the Old Quarter. The research also took into consideration the feasibility of mobilizing community resources to improve and upgrade their own living spaces.

The case study no. 2 is a detailed planning project to renovate both sides of a major road located in a large area crossing over 16 wards and 6 districts in the inner city of Hanoi. This case involves many community groups that have different characteristics and relations with the detailed planning area. The research team reviews and assesses the level of community participation in the detailed planning process specified in the legislative documents, considering the role of every stakeholder: the investor (from city government), consultancy, ward, and district authorities. Through this case study, the research also analyzes the mechanism of interaction, information exchange among stakeholders in order to create proposals in improving and supporting the process of community involvement. The research also highlights the challenges of the community participatory approach in that type of detailed planning funded by the state.

2.2 *Research Methods*

The research team includes architectural and urban planning experts from National University of Civil Engineering (Hanoi, Vietnam) who participated in all phases of the two case studies, assessing and verifying the community participation in the two specific types of detailed projects as mentioned above.

2.2.1 Case Study No. 1

The project includes four steps with the integration of community involvement as follows:

Step 1: To conduct surveys and analyze the existing living conditions in the pilot street block with community participation approach methods: In-depth questionnaire survey was conducted with 253 households in the street block, mapping with community involvement, photo voices making – to provide cameras for community to reflect their opinions through photos made by themselves.

Step 2: To establish visions along with development strategies for street block – Discussions among stakeholders and community members.

Step 3: To propose for detailed planning solutions to renovate the pilot street block – Focus group discussions and charrettes with core members of community.

Step 4: To define priority activities for action plan– To conduct three pilot activities with community participation.

2.2.2 Case Study No. 2

The research is conducted in two phases of the detailed planning project:

Phase 1: Formulation – Appraisal – Approval of detailed planning tasks (conducted within one year).

Phase 2: Formulation – Appraisal – Approval of detailed planning project (conducted within 2 years).

In both phases of the project, community consultation was conducted through several community meetings in the districts and wards where the project was carried out. During the community meetings, local residents and community's representatives were required to fill the questionnaires sheets. After collecting questionnaire sheets, the research team analyzed and assessed the roles of community participation in finding appropriate solutions to the detailed road master plan.

3 Results

3.1 Case Study No. 1

3.1.1 Results of Applying Community Participatory Method in Detailed Planning Process

The stakeholders in the detailed planning process consisted of Representatives of the local ward people committee (ward PC), architectural and planning experts, core group of community (including leaders of resident groups, representatives of women union, elderly group, veteran union, etc. in the ward), local residents, and other organizations (nongovernmental organizations, community-based organizations).

The community participated in all four steps of the detailed planning project to improve living spaces of pilot street block while preserving intangible values of the area in the Old Quarter of Hanoi. The way that community involved in the detailed planning process could be illustrated in Diagram 2 below.

The tools used in each step followed exactly step by step the community participatory approach. In the first step of surveying and analyzing existing conditions, the current characteristics of the local areas were defined by local community through the method of mapping with community (See Figs. 15.7 and 15.8). Local residents who joined the research identified the nice architectural and historical buildings that were related to the formation of the location and defined historical houses that are seriously damaged or deteriorated. The "photo voices" tool that provided cameras for local residents helped people to express their opinions, feelings, and expectations for the places where they are living in a very realistic way. The formulating and proposing for vision, goals, and strategies of the development of the street block, choosing long-term and short-term action plans were conducted in several community meetings, focus group discussions, and stakeholders' meetings during planning process (See Figs. 15.1 and 15.4). In the final stage, the core group of community and some local residents proposed three pilot activities to be implemented including: (1) To Install pent-roof system to improve the walking environment and streetscape; (2) To improve common public spaces inside a damaged house at 17 Hang Buom street; and (3) to establish urban design guidelines for the street block surrounded by Hang Buom street - Hang Giay street – Luong Ngoc Quyen street – Ta Hien street. The meeting gained commitment of local community to contribute 40–50% of the renovation cost for those pilot activities instead of making no financial contribution as previous activities that the local authority had done before (Figs. 15.5, 15.6, 15.7, 15.8 and 15.9).

3.1.2 Assessment on Community Capacity when Applying Community Participatory Method

By using some qualitative criteria of standards, awareness, willingness, activeness, ability of taking initiatives, leadership, making decisions, etc. before and after applying the method, the research team tried to find out which factor could be changed or not be changed from each partner, especially of the community (Tables 15.1 and 15.2). The impact of this method on the quality of resources and on the whole results of the project could be clearly demonstrated. After 1.5 years of working with more than 200 local residents and 28 core residents, there were some assessments of the community capacity changes as below:

3.1.3 Mobilizing Financial Sources from Community

Before applying the method: the renovation work and living environmental upgrading were carried out regularly but depended much on different investors whose responsibility were regulated by local authorities and budget were limited. There was no coordination and financial sharing among housing owners. Moreover, the activities of living condition improvement ignored so many channels to mobilize financial such funding organizations, foreign organizations for conservation and restoration, etc.

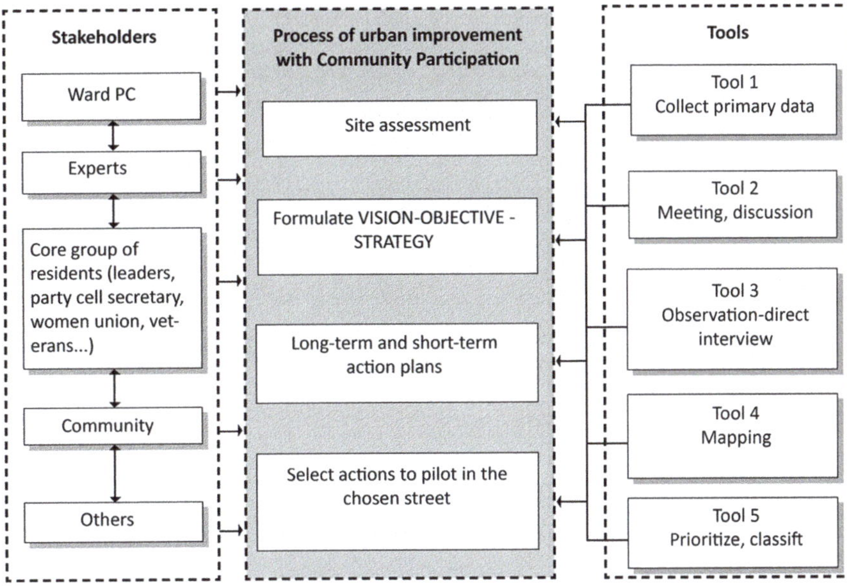

Fig. 15.5 Urban improvement process with community participation in case study no. 1 (*Source*: Author)

Fig. 15.6 Meeting – Discussion (*Source*: Ta, 2015)

Community participation in finance for renovation and living environment upgrading regardless of the fact that the housing ownership would be a new direction for the desire to solve the problems of the area today, particularly the declination in the values and living conditions of the area. Below is a chart summarizing the financial contribution for renovation work before and after the use of community participatory methods (Fig. 15.10):

Fig. 15.7 Site observation and analysis (*Source*: Ta, 2015)

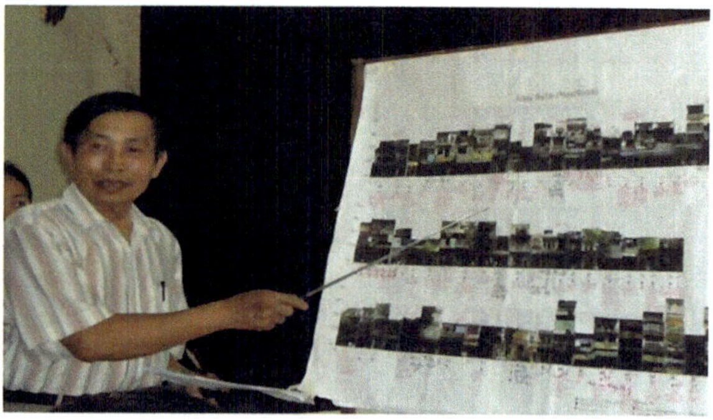

Fig. 15.8 Mapping the existing condition (*Source*: Ta, 2015)

Fig. 15.9 Problem ranking (*Source*: Ta, 2015)

Table 15.1 The assessment of core residents before and after applying the method (Authors)

	Awareness	Skill & education	Willingness	Activeness	Ability of taking initiatives	Leadership, decision making
Before	To be aware of the role and responsibilities of community in public activities Enthusiasm about and responsibility for social work, activities	Different levels Lack of knowledge and ability to rank issues	Willing to participate in activities with the participatory approach and for the benefit of community	No active proposals for activities with community participatory Operate under the direction of the local government	n/a	n/a Not familiar with initiative role Do not really understand their right to do something for their living space
After	More active Flexibility in carrying out problems and issues	Have knowledge and ability to find out, rank, and evaluate local issues	Willing to contribute their money to community activities	Have raised activeness of and creation in activities with the community		
Changed factor	X	X	X	X		
Unchanged factor					X	X

3.2 *Case Study No. 2*

3.2.1 Results of Conducting Community Consultation in Detailed Planning Process

In this case, the research team was an independent unit to examine and verify the level of community involvement in the detailed planning project through analyzing the result of community consultations performed by the consultant group. The consultancy conducted community consultations for detailed planning solutions within 2 years, from August 2012 to July 2014. They carried out six community meetings in six different districts that the main road project runs through with the participation of community representatives of 16 wards. The questionnaire survey forms had been formulated by the consultancy followed the guidelines of the Hanoi Architecture and Urban Planning Authority (HAUPA) that only focused on the matter if the community agreed or did not agree with the detailed planning proposal for the both sides of the main road. The research team conducted questionnaire surveys and

Table 15.2 Criteria to assess community in general before and after applying the method (Authors)

	Awareness	Skill & education	Willingness	Activeness	Ability of taking initiatives	Leadership, decision making
Before	Low awareness, were afraid to participate in the planning process and raise their voicesfeeling helpless, fear of failure Do not trust the local authority Did not want to change anything Pragmatic, always ask benefit when they participate in any activities	Diversified different level Not self-confident about their standard, ability Do not have ability to assess, evaluate, and rank problems	Do not feel willing with common activities Afraid to take the responsibility	Familiar with top-down approachlis-tening and following the direction from government Depending, do not want to make decisions, test, etc.afraid of risk, failure	n/a	n/a Do not really understand and aware of their right and their power
After	Aware their power and ability Self-confident	Have ability to assess, evaluate, and rank problems in small scale (in one house)	Willing to join and contribute to common activities	Find out problems and discuss how to deal with actively Actively mobilize financial sources from community		
Changed factor	X	X	X	X		
Unchanged factor						

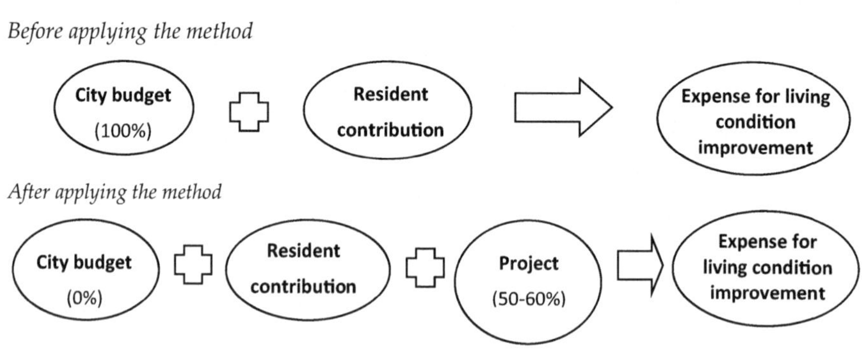

Fig. 15.10 Results of mobilizing community's financial resources by using community participatory method (*Source*: Authors)

Table 15.3 Results of conducting community consultation for the detailed planning project to renovate and improve the main road (*Source*: Authors,

Number of community's representatives	Agreed (%)	Disagreed (%)	Others (%)	Total
292	198 (67.8%)	28 (9.6%)	66 (22.6%)	292 (100%)

interviews to assess the change in awareness and behavior of the community. Results of the consultation meetings are mentioned in Table 15.3.

3.2.2 Assessment of the Change in Awareness and Behavior of community's Representatives

All people who answered the questionnaire survey stated that the assessment of local existing conditions in terms of socio-economy, infrastructure, and land use would be very important before formulating planning programs since it will impact directly the quality of detailed planning.

The local residents admitted that after listening to the consultants' presentation on the survey results of existing conditions, they understood clearly and thoroughly the current conditions and urgent issues that they were facing. More than 70% of the local people who were interviewed by the research team realized that it would be necessary to contribute opinions to the planning consultancy regarding the local current situation of socio-economy, land-use planning, technical, and social infrastructure development. This insisted that the local community would expect to join the detailed planning process from the very first step of analyzing the current situation, defining all the issues and demands for the development of the area. Nearly 90% of the community's representatives stated that the meeting to collect the community's opinions helped them improve the accuracy of the data to assess socio-economic status, land-use situation, infrastructure status, etc. Gathering the community's opinions was a method that people would like to choose the most when asked about the effective ways to mobilize people's participation in making detailed planning programs and finding solutions.

Figure 15.11 illustrates the process of detailed planning of the pilot case study in the practices when applying community participation.

3.2.3 Assessment of the Roles and Involvement of Stakeholders in Performing of Community Consultation

Hanoi People's Committees (Hanoi PC): This was the project owner as well as project investor. This was a highest city administrative agency responsible to approve of such a kind of project for the aim of strengthening the project management (Table 15.4).

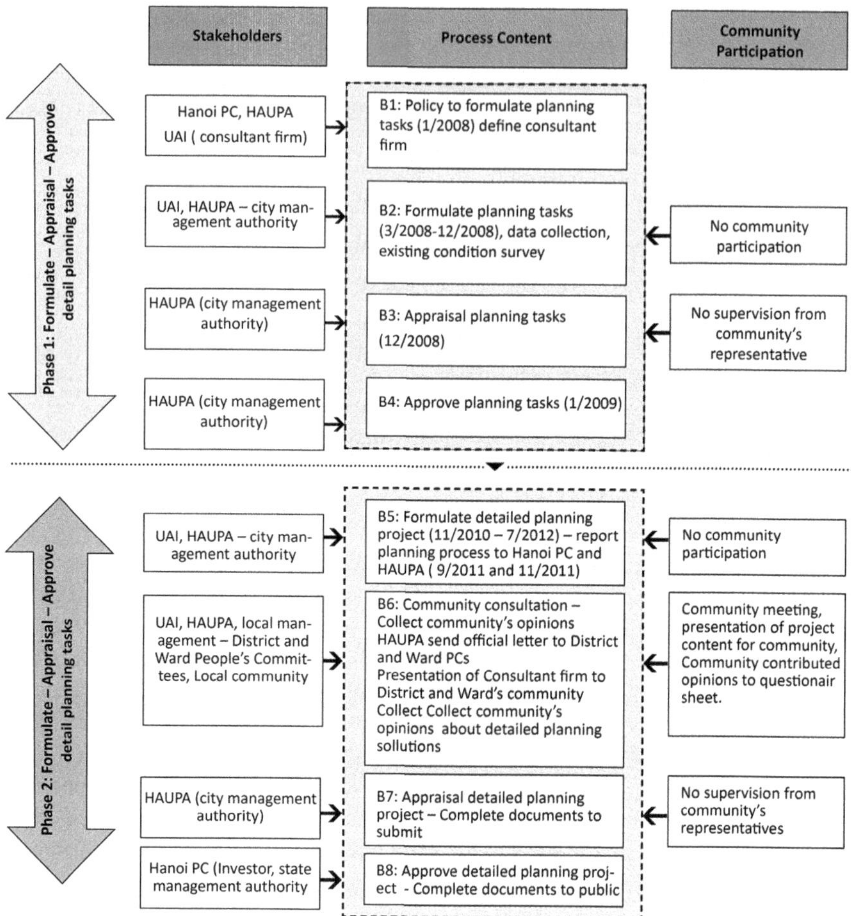

Fig. 15.11 The process of detailed planning for road axis Le Duan – Giai Phong – Linh Dam with community participation in Case study no. 2 (*Source*: Authors)

Hanoi Architecture and Urban Planning Authority (HAUPA): This was a judging body to give requirements for community consultation and receives report from related agencies including a consultant company and District, Ward People's Committees. However, with the responsibility as "to be the judge in one's own case", HAUPA only wanted to organize the fastest consensus from the people without considering the quality of the consulting community's representatives.

People's Committees (PC) of districts and wards related to the project: These were local management agencies responsible for coordinating with the consultancy unit and conveying planning project information to the community for the purpose of collecting their opinions. The effectiveness of the opinion collection depended greatly on the leadership of the district and ward PC. In some districts such as Dong Da and Ba Dinh district PCs, the leaders deeply understood the planning location and made very rightful and thoughtful suggestions to the consultants. However, in

Table 15.4 Roles and responsibility of stakeholders in detailed planning process to perform community consultations (*Source:* Authors)

Stakeholder	Role in the project	Responsibility
Hanoi People's committees (Hanoi PC)	Project owner as well as project investor	Highest city administrative agency responsible for project approval
Hanoi architecture and urban planning authority (HAUPA)	City management body, to do judging for detalied plan	To provide requirements for community consultation and receive reports too
People's committees (PC) of districts and wards	Local management	To coordinate with the consultant unit To deliver project information to the community through different channels
Consultant group	Formulating urban planning tasks and planning solutions	To conduct community consultations To collaborate with local management bodies

other districts, the community's opinions were not so effective and made less contribution to the project planning solutions.

Urban and Architectural Institute (UAI) from National University of Civil Engineering: That was a consultant firm that is responsible for formulating urban planning tasks and developing planning solutions of the project. UAI is also responsible for collaborating with district and ward PCs to conduct community consultations and collect the public opinions. This research team collaborated strongly with the consultant group to assess effectively the role and the involvement of the local community.

Local community including local community's representatives and local residents affected by the project: Local community's representatives were members of socio-political organizations in the project area, for instance veteran association, women union, residential groups, etc. which would act on behalf of the local residents to attend almost every meeting and briefing on detailed planning tasks and solutions with the consultant. However, at some meetings, the local community's representatives did not raise the actual voices for the local community.

4 Conclusions and Discussion

4.1 Discussion

4.1.1 Who Are Urban Community?

Through the above case studies, it can be seen that the urban communities in Hanoi consist of the three main components as: (1) Residents who are living in the research area of the detailed planning project; (2) Organizations and individuals that have properties and interests attached to land and buildings within the boundary of a detailed planning; (3) Socio-professional organizations, nongovernmental organizations (NGOs) that operate and serve the community's interests in the area where the detailed plan is implemented.

However, the components of the "community" are different for different types of detailed planning. For detailed planning of renovation, upgrading or preservation: communities include communities in the planning area and surrounding areas, the organizations, and individuals operating in the area under the impact of the project, and scientists, social-professional organizations, etc. which are interested in all the aspects related to the living environment of the region, whereby the community plays the most important role because they are the most knowledgeable people knowing so much about the local issues as well as understanding which resources they have in hand to help improve the living environment in the areas.

4.1.2 Who Are community's Representatives?

Community representatives in detailed planning are individuals or organizations that legally represent the interests of community groups affected by the detailed planning project. Community representatives are those who are trusted by the community and can represent the desires and expectations of the community. Community representatives include: (1) Representatives of the local residents: Heads of neighborhood groups, or heads of villages and hamlets, representatives of social organizations, and other organizations: members of the Women's Union, the Elderly Association, Veterans Association, youth groups, core community groups, etc.; (2) Representatives of professional organizations such as Vietnam Association of Architects, Vietnam Urban Planning and Development Association, Vietnam Urban Association, etc.; and (3) Representatives of nongovernmental organizations (NGOs), nonprofit organizations (NPOs) who work to support the communities.

4.1.3 Core Group of Community

The core group of community is made up of those who are responsible for the community and trusted by the community. The core group of community is usually elected by the community in the neighborhood or the village. They are considered "leaders" of the local residents in participating in urban planning process conducted within their living area, and they have a significant influence on the community. The core group of community includes the head of local neighborhood areas, or hamlets and usually the most respected people in the community. They also have a good knowledge about the tangible and intangible values of the place where they are living in, and always keep good relationships with all the members in the community. That is the channel to bring planning information, policies, and strategies of the project to the community and gather the opinions to be used in the planning process.

4.1.4 What Is the Current Level of Community Participation in Vietnam?

Through the pilot case studies, it may be useful to know that the level of community participation in detailed planning projects is relatively low. Referring to the Arnstein's ladder of community participation in planning process (Arnstein, 1969) with eight levels (1. Manipulation; 2. Therapy; 3. Informing; 4. Consultation; 5. Placation; 6. Partnership; 7. Delegated power; and 8. Citizen control) and the Kingston's ladder (Kingston, 1998) of six levels ((1). Public right to know; (2). Informing the public; (3). Public right to object; (4). Participation in defining interests, actors, and to determine agenda; (5). Public participation in assessing consequences and recommending solutions; and (6). Public participation in final decision), it is revealed that the community participation in Vietnam is only reaching level 4 or level 5 on Arnstein's ladder and equivalent to level 3 on Kingston's ladder. Community participation in Vietnam has been fixed at the levels of notification and consultation with a limit number of community or community representatives to involve. The community has not actively involved in urban planning processes and made no (or almost no) decisions for the (re)development of their neighborhood. Thus, for a more effective participation of the community, it is necessary to study the processes and tools to mobilize the community engagement effectively from the first step: enhancing information exchange, not only one-way but also establishing a bilateral informing mechanism (exchange, discuss, understand the requirements of the people, and cooperate with the people) and gradually increasing the level of participation. In the future, people will initiate actively and discover the issues as well as the problems by themselves before making their own decisions.

4.1.5 Advantages and Prospects when Applying Community Participatory Method in Detailed Planning Projects at the Local Level

For detailed planning Type 1: The participation of the local community in surrounding areas contributes to ensuring the harmony and coherence among new planning areas and existing developed areas.

For detailed planning Type 2 and Type 3: The community participation contributes to preserve the region's intangible core values, the "sense of places", making place to be environmentally friendlier. Additionally, community participation will help improve the quality of public spaces for people, provide appropriate solutions to increase accessibility, attractiveness, safety, and inclusiveness of the spaces, mobilize resources from the community (logistic conditions, human resources, and financial resources, etc.). Community engagement also contributes to increasing the community liaisons.

Having community consensus in making or preserving the living place is very important to the process of detailed planning. "When the local residents are regarded as a cultural entity in an open and democratic environment, they will have better cognitive capacity and be able to think more freely and creatively" (Nguyen, 2008). This will facilitate the community's consensus in other aspects of detailed planning project.

The participation of the local community in attaching directly to the "place" will be the most effective at the following stages:

1. The stage of evaluating the current situation and identifying problems to be solved: At this stage, members of the local community can help planners identify the characteristics and intangible values of the places related to religion, beliefs, spirituality, habits, lifestyles, etc., even the so-called fun and fear of the community. The community involvement in current situation assessment will help to identify the problems that the community must be facing, as well as the tangible and intangible values that community needs to seek and enrich.
2. The stage of spatial design: At this stage, all the members of the local community can help planners identify the main factors in the process of place making: The spatial elements; social factors (related to the process of formation and development of a place); spiritual factors (feelings, emotions, memories, etc.) (Wilcox, 1994). Especially, for the detailed planning projects of renovating, decorating or preserving historical areas (Type 3), the participation of the community has a very positive effect on identifying and developing elements of the "place" to keep the authenticity of the historical areas.

4.1.6 Challenges when Applying the Community Participatory Approach in Vietnam Context

Lack of legal documents guiding the implementation of community participation: At present, there are no legal documents defining the responsibilities of stakeholders, coordination mechanism between the parties to mobilize more community participation right from the stages of determining the purposes and objectives of the project, surveys on current situation and needs, as well as actual resources from the community. Public consultation is only regulated when the planning project is almost completed, and there is little chance for community opinions to have a great impact on the planning solution. The process of summary, explanation, and receipt of comments is regulated in Article 21 of the Urban Planning Law 2009 but just described in a very general way. The Urban Planning Law has not regulated the specific role that stakeholders are responsible to play for receiving, explaining, and responding to the communities. The planning information channel to the community is only a one-way flow. The community is informed about a planning project when the planning programs and planning solutions have been approved. It is essential that the community needs to be provided with full information about the planning project from the first stage of the planning process to be able to exchange ideas, negotiate, and discuss to gain some signs of consensus and commitment from community. Moreover, there are no specific regulations on financial support to engage community in the planning processes. Currently, there is no stage budget for the activities of mobilizing community participation in urban planning.

Lack of knowledge, awareness, and skills from all stakeholders.

Consultants: the mobilization of community participation at the level of notification and public consultation can be a mandatory step to complete the project and to submit it for approval. For community-based pilot projects, consultants are supported financially and technically to verify the effectiveness of the community participation and they try to apply community participatory methods in some steps of the urban planning process, and then focus on the stage of existing condition surveys to get more information for planning. However, the process of planning information exchange among stakeholders and community has not yet been established. In reality, the consultants consulted with the community in a formalistic and perfunctory manner that has not been able to secure any effective results.

Managing agencies: it is essential to recognize that the mobilization of community participation in urban planning is only a requirement of the administrative procedure to gain an approval of the project. There are currently no comprehensive reports on the results of community consultations as well as no methods to mobilize community involvement in legal documents or regulations promulgated with the project before it could be approved.

Investors: Mainly interested in economic profits (especially private developers). Most investors do not have proper awareness of the role of community and community participation contributing to the effectiveness of planning. Therefore, investors consider the mobilization of the community participation a time-consuming/cumbersome administrative procedure and do not want to invest resources (time, human resources, physical resources) in implementing the community involvement in a practical part of the urban planning. The investors always manage to minimize the cost and time for the implementation of the planning process with the integration of community consultation procedures, so it can only be implemented in a formalistic manner.

Community: People have limited awareness, knowledge, and skills to participate in urban planning processes. In addition, people are not provided with sufficient information about urban planning works as well as the impacts of the planning projects on the living environment, the rights and responsibility of the community when projects are implemented. Community members are not explained clearly what they should benefit from the project, what may be affected when the urban planning project is approved and implemented. At the same time, with the process of prescribing the current community participation, the community is asked in a number of consultations and their ideas can be accepted in the urban planning process but they do not know exactly the mechanism of explanation and feedback of management agencies. They are not certain either if their ideas and comments can be helpful, how much they may help adjust the planning solutions, how the stakeholders have responded to fulfill the wishes and requirements of the community.

4.1.7 Proposal of Community Participation in Urban Detailed Plans

In order to integrate community consultation in all phases of detailed planning process, it is critical to add some supplementary steps and provide a variety of effective tools to enhance the community participation and improve the quality of detailed urban planning projects.

The proposed detailed planning process with supplementary steps of community consultation and supervising organization is presented in the Fig. 15.12.

In the Fig. 15.12, phase 0 is added with the aim to define clearly the policies to perform detailed planning projects. In phase 0, the local residents' representatives and community-based organizations' representatives should attend stakeholders' meetings to have the review of the existing conditions of planning in the proposed planning areas with the direction and policies for planning. Through those meetings, they can contribute their opinions to the decision making of the investors and project management board.

Phase 1 - To formulate, appraise, and approve of detailed planning programs: the process of information transferring to community on detailed planning programs should be added before collecting opinions of the community. The research encourages the community's representatives to contribute to the appraisal process of detailed planning programs.

Phase 2 - To formulate, appraise, and approve of detailed planning projects: The tools to engage the community participation as community meetings should be added with direct and indirect interviews of the local community and focus group meetings and it is advised to facilitate the circular flows of planning information exchange among stakeholders and community. In the step of project appraisal, the community supervision board should review the report of the community's consultation and contribute opinions to the management and appraisal of the authority.

Phase 3 - Implementation and management detailed planning project as approved: This phase should be added with the tools to distribute the information approved in the planning project to the public while setting up the exhibitions or project display events. Apart from that, the information of the planning project should be published at the local authority's information center via local websites, newspapers, etc. to ensure that the delivery of information to the community should go smoothly.

4.2 Conclusions

Through two pilot case studies conducted in Hanoi city, it is noted that the application of community participation method in urban detailed planning process has shown its positive impacts to improve the effectiveness and quality of urban planning, especially for detailed planning projects of preserving or regenerating the urban historical areas in Vietnam. The community can involve effectively in urban planning process if they are provided sufficient information and good collaboration with local authorities, consultants as well as other stakeholders from the first stage

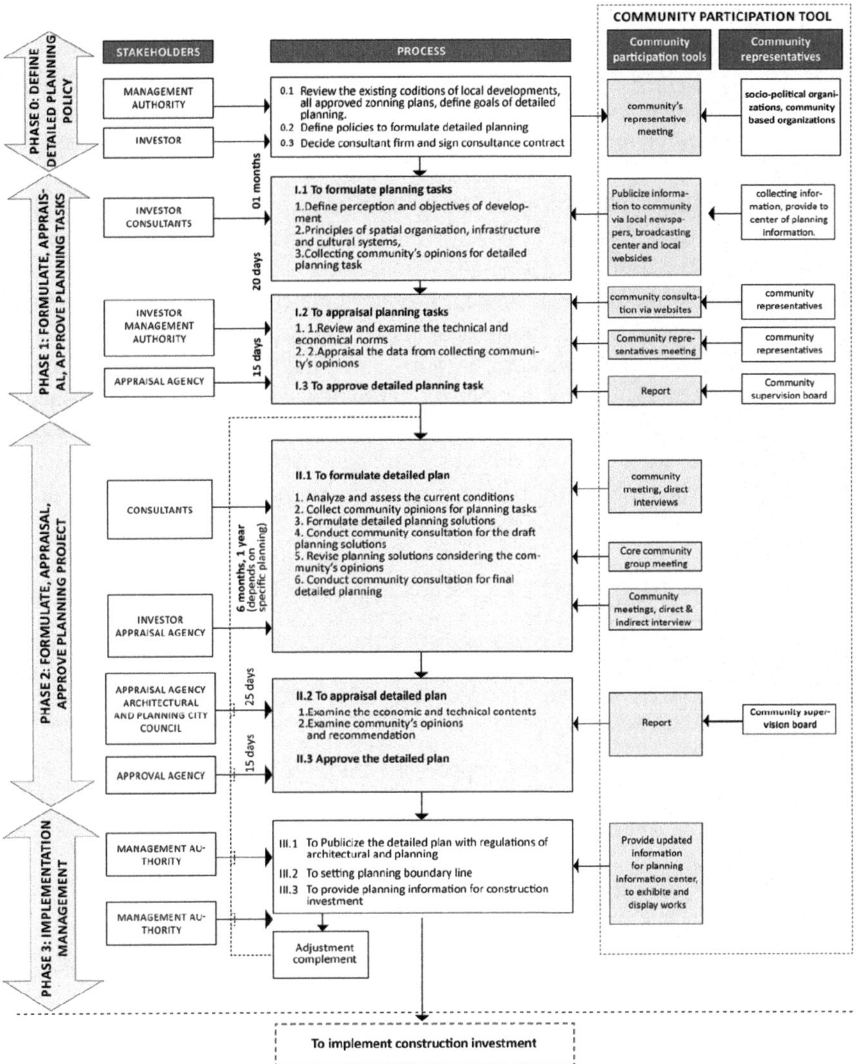

Fig. 15.12 *Detailed planning process with supplementary steps to facilitate community participation.* (*Source*: Authors, 2018)

of such a planning process. With suitable tools to work with community, the local authorities can mobilize a huge amount of resources from community, not only human resources but also financial sources to contribute to the urban regeneration process. Even though the level of current community engagement in Vietnam context is relatively low compared with other developed countries, and the community participation is now facing many challenges, it will be improved significantly if the following important factors will be taken into account: Firstly, improving the community and local authority's awareness of the right and role of the local community so that people will be ready to take part in a planning process, hereby improving the quality of their living environment; Secondly, formulating a more suitable mechanism of cooperation among local authorities, planning experts, investors, and community as well as community-based organizations (COB). As a result, it is possible to have a consensus and agreement for whole process of urban planning that should go on the right track from very first stage of defining planning policies until the final phase of planning implementation and management. The two-way planning information exchange should be facilitated through some channels, such as community meetings, public exhibitions, local newspapers or websites, etc. to enhance the community's accessibility to planning information. Finally, it is necessary to set up financial support regulations to mobilize the highly potential community involvement, not only gaining from technical planning fees as often seen today. If all the aforementioned factors are properly considered and well implemented in reality, the community participation in urban planning process in Vietnam will be a driving force to help us achieve sustainable development.

Acknowledgments The research had received great support from: Hanoi Old Quarter Management Board, Hoan Kiem District people committee, Ba Dinh District people committee, Hai Ba Trung District people committee, Dong Da District people committee, Urban Architecture Institute – UAI at National University of Civil Engineering (Hanoi, Vietnam), Mr. Tran Viet Anh – Director of Hanoi Ancient Quarter Management Board to conduct interviews and questionnaire surveys.

Data Availability The data that support the findings have been received consent from all the mentioned committees, surveyees, and co-authors before presenting in this study. It is available from the corresponding author, Hoa Quynh Ta, upon reasonable request.

References

ACVN and KAS. (2010). Project report *"Promoting people's participation and governance in Vietnamese cities"* through the Association of Cities of Vietnam (ACVN) European Commission and Konrad-Adenauer Stiftung, Hanoi, Vietnam.

Aprodicio, L. (1995). *The mega-urban regions of Southeast Asia* (pp. 215–241). University of British Columbia Press.

Arnstein, S. R. (1969). Ladder of citizen participation. *Journal of the American Institute of Planners, 35*(4), 216–224.

Briffetta, C., Obbard, J. P., & Mackcc, J. (2003). Toward SEA for the developing nations of Asia. *Environmental Impact Assessment Review, 23*, 171–p196.

Danielle, L (2010). *Facing the urban transition in Hanoi: recent urban planning issues and initiatives*, Institut national de la recherche scientifique Centre - Urbanisation Culture Société.

Dao, T. N., Do, T. L., & Nguyen, Q. T. (2013). *The community's participation in urban architectural heritage in Hanoi* (pp. 63–p69). The international Conference on Sustainable Built Environment for Now and Future.

Divya, L., & Helga, J. S. (2017). The new Hanoi: Opportunities and challenges for future urban development. *Cities, 72*, 70–81.

Dobertein, B. (2004). EIA models and capacity building in Vietnam: An analysis of development aid programs. *Environmental Impact Assessment Review, 24*, 283–318.

Government of Vietnam. (2016). Economic reform and international integration in Vietnam – 30 years looking back. Report.

Gehl, J. (2010). *Cities for people*. Island Press, Washington.

GSO Vietnam. (2019). *National Population and Housing Census Report* (pp. 68–69).

Healy, P. (1997). Collaborative planning: Shaping places in fragmented societies.

Held, D. (2006). *Models of democracy*. Stanford University Press.

Hieu, N. N. (2007). *Urban planning and the procedure of making decision on development control – The case of union park in Hanoi*. Builder.

Hostovsky, C., MacLaren, V., McGrath, G (2010). The role of public involvement in environmental impact assessment in Vietnam: Towards a more culturally sensitive approach. Journal of Environmental Planning and Management, 53(3), 405–425. Retrieved from https://www.statista.com/statistics/444882/urbanization-in-vietnam/

Karadjis, M. (2005). Socialism and the market: China and Vietnam compared, International Journal of Socialist Renewal. 27. Retrieved from http://links.org.au/node/14

Kingston, R. (1998) Web Based GIS for Public Participation Decision Making in the UK. National Center for Geographic Information and Analysis, Proceedings of the Workshop on Empowerment, Marginalisation, and Public Participation GIS, Santa Barbara, Ca, Retrieved from http://www.ccg.leeds.ac.uk/vdmisp/publications/sb_paper.html

National Assembly of the Socialist Republic of Vietnam [NASRV] (2007). Ordinance on Democracy at grass root level, No 34/2007/PL-UBTVQH11, Vietnam.

National Assembly of the Socialist Republic of Vietnam [NASRV] (2009). Urban Planning Law, No 30/2009/QH12, Vietnam.

National Assembly of the Socialist Republic of Vietnam [NASRV] (2013). Urban Construction Law (revised), No 16/2003/QH11, Vietnam.

Nguyen, P. N. (2008). *Housing improvement in Hanoi ancient quarter with community participation*. NUCE.

Obbard, J. P., Lai, Y. C., & Briffet, C. (2002). Environmental assessment in Vietnam: Theory and practice. *Journal of Environmental Assessment Policy and Management, 4*(3), 267–295.

Raine, M (2006). Approaches to participation in urban planning theories, University of Oulu, Department of Architecture, Finland.

René, P., & Thong, N. Q. (2005). The role of civil society in urban environmental rehabilitation: A case study (Thanh Xuan district, Hanoi, Vietnam). *Environment & Urbanization, 17*(1), 237–p247.

Sanoff, H. (2000). *Community participation methods in design and planning*. Wiley.

Statistica. (2017). *Urbanization rate in Vietnam.*

Ta, Q. H. (2009). Urban planning with community participation – Possibilities to apply into Vietnam's urban context. *Journal of Science and Technology in Civil Engineering*. National University of Civil Engineering, No 06–12/ 2009. Hanoi, Vietnam.

Ta, Q. H. (2015). *Urban detailed planning with community participation in Vietnam urban context towards sustainable development, PHD dissertation in regional and urban planning – Code 62580105*. National University of Civil Engineering.

Thao, T. Q. (2001). *Introduction to urban studies*. Construction Publishing.

Tran, V. T. (2015). Vietnam – Over the past 40 years and in the years to come – A better oriented market economy is needed, the times journal, issue no. 33.

UN-Habitat. (2010). *International conference: Hanoi millenniums – City past and future, 12–13 October 2010*. Hanoi.

Vietnamnet Website. (2019). Vietnam's Urbanization rate in 2019 to reach 40%. Retrieved from https://english.vietnamnet.vn/fms/society/218251/vietnam-s-urbanization-rate-in-2019-to-reach-40-.html

Waters, N. (2000). *The community-planning handbook*. Earthscan.
Wells-Dang, A. (2010). Political space in Vietnam: A view from the 'rice-roots'. *Pacific Review, 23*(1), 93–112. https://doi.org/10.1080/09512740903398355
Wilcox, D. (1994). *The guide to effective participation*. Delta Press.

Chapter 16
Current Status and Impact of Imbalance of Sex Ratio at Birth in the Son La Province, Vietnam

Tong Thi Quynh Huong

Abstract In Vietnam, imbalance of sex ratio at birth is becoming one of the most difficult problems for population and family planning, which has been attracting the attention of the community. Due to misconceptions, son preference is still evident in social life. In 2018, Son La had the highest sex ratio at birth, approximately 120 boys/100 girls. Notably, this situation occurs everywhere in the province, including in rural and urban areas. Combined with general sources of Son La sex ratio, 2010–2018 of Son La Department of Population and Family Planning, the article analyzes and clarifies the imbalance in sex ratio at birth and negative impacts on family and society in Son La.

Keywords Sex imbalance at birth · Consequences · Son La

1 Introduction

The sex ratio at birth is an important quantity in a population study, understood as the number of boys born compared to 100 girls born in a given time period, usually a year in a country, a region or a province. This indicator shows how many boys will be born for every 100 girls born. Usually for every 100 girls born there will be between 103 and 107 boys, and will generally be very stable over time and space across continents, countries, regions, and ethnic groups.

The imbalance in the sex ratio at birth occurs when the number of live births are higher or lower than the normal threshold for every 100 girls, i.e., when the sex ratio at birth is greater than 107 or less than 103 compared to 100 girls. Any significant change in the sex ratio at birth deviates from the normal biological level to reflect intentional interventions to some extent and will affect the natural imbalance, threatens the stability of the population.

T. T. Q. Huong (✉)
Tay Bac University, Son La, Vietnam

A. T. Nguyen, L. Hens (eds.), *Global Changes and Sustainable Development in Asian Emerging Market Economies Vol. 2*, https://doi.org/10.1007/978-3-030-81443-4_16

Sex imbalance at birth has occurred from the beginning year 80 of the twentieth century, seriously affecting economic and security situation security, politics, society of many countries in Asia. In Vietnam, the sex imbalance at birth appears later, from these the first year of the twenty-first century, which took place at a fast pace, became increasingly widespread and came severity. It can be said that the gender imbalance has been one of the worrying issues, being of top concern in Vietnam in general as well as provinces in particular. According to forecasts of the General Department of Population and Family Planning, with the current situation of gender imbalance at birth, by 2050 Vietnam will lack about 2.3–4.3 million women. An excess of males of marriageable age can lead to a breakdown of the family structure, a part of men will have to marry late, many of whom are unable to marry.

Son La is the province with the average population size in the country. In 2010, the province had 1092.7 thousand people, by 2016 increased to 1208.2 thousand people. The population growth rate tends to decrease over the years but it is still high compared to the national average. In 2010, the natural population growth rate was 1.85%, by 2016 it decreased to 1.45%. Such a decline is a result of population education and family planning in reducing fertility (between 2010 and 2016 the birth rate decreased from 24.3‰ to 18.5‰).

In the population structure, the proportion of men and women is quite different, the structure of the male population has always accounted for a higher proportion than the female population and remained quite stable in recent years, especially when the Son hydroelectric plant project La attracts a large number of male workers to build the project. The percentage of men in the total population is 50.3%, the percentage of women is 49.7%.

The gender imbalance at birth in Son La is at an alarming rate. Statisticing in the whole province, the sex ratio in Son La is currently approximately 120 boys/100 girls. Notably, this situation occurs in all districts and cities in the province, including in rural and urban areas. Some districts that have had high rates of inequality at birth for many years are Phu Yen, Song Ma, Muong La, Moc Chau, Quynh Nhai, and Son La City.

2 Methodology

The author uses professional methods to study the theoretical and practical basis of sex imbalance at birth, lessons learned on reducing the sex imbalance at birth in the world and in Vietnam, applied in research in Son La; using methods of collecting, analyzing, and synthesizing documents and data; field method: expert method; sociological survey method to clarify the current situation of sex imbalance at birth in Son La. On that basis, the author assesses the consequences of the imbalance of sex ratio at birth on family and society in Son La.

3 Results

3.1 Situation of the Problem of Sex Imbalance at Birth

The current gender imbalance at birth in Son La is alarming, according to specialized statistics of Son La Department of Population and Family Planning, the period 2010–2018 tends to increase, especially from 2013 up to now, specifically: In 2010, 109.8 boys for every 100 girls; by 2013 increased to 114.4; 2016 was 118.7; and 2018 was 118.4 - the highest in the country (Fig. 16.1).

Besides, according to incomplete statistics, Son La's sex ratio at birth is high even in the first birth (117.1/100). At the second birth, the pressure to produce a son has been reduced and brought the sex ratio at birth close to the biological balance. However, at the third birth or higher, the pressure to have a son is shown most strongly. The ratio at this time increased very high to 124.3 boys for every 100 girls; especially for couples without sons, this ratio reaches 145.2/100. The reason is that if in the past, a son had only one way to have a child until he had a new son, and the sex ratio at the last birth was very high. In demography, this is called a "stopping rule" or in other words, the sex factor has decided to stop giving birth rather than the number of children already. But now, due to the rapid decline in fertility and the easy access to prenatal sex selection techniques, the "stopping rule" in Son La has also changed so that the sex ratio at birth has a third or higher birth in a group of mothers who have not had such a son.

In addition, the level of sex imbalance at birth in couples with high education, the better-off family economy is often much higher than that of poor couples. Highly educated women actively use contraception and proactively adjust their desired number of children. These women often have better economic conditions to be able to afford prenatal sex selection services and they meet both goals: small family size and having a son. Population and labor statistics of the General Statistics Office

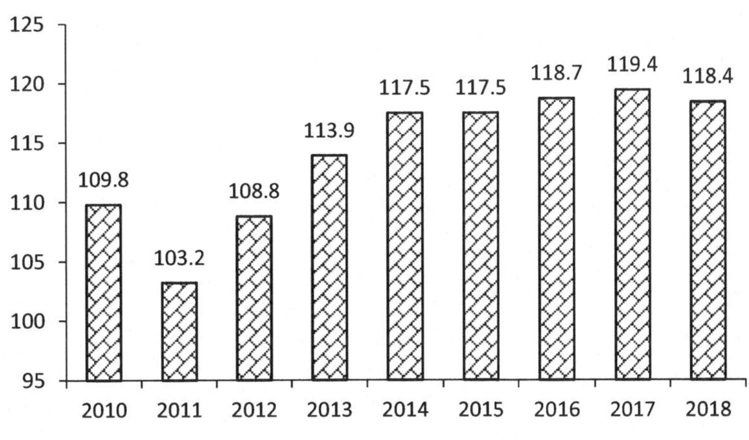

Fig. 16.1 Sex ratio at birth in Son La period 2010–2018

Table 16.1 Sex ratio at birth between urban and rural areas in Son La

Area	2010	2011	2012	2013	2014	2018
The whole province	109.8	103.2	108.8	114.4	117.5	118.4
Rural	127.9	117	108.7	114.6	119.5	121.4
Urban	109	102	109	114.1	116.2	115.4

show that when women's educational attainment increased, the sex ratio at birth also increased, from 109 to 116/100 (in mothers with primary education), up to 119 boys/100 girls (in high school mothers), and up to 124/100 in the group of mothers with university or higher education (Table 16.1).

On a local scale, the sex ratio at birth is also different. Some districts have continuous high fertility ratios for many years, especially in the period of 2013–2017, such as Bac Yen, Moc Chau, Quynh Nhai, and Son La City. Typically, in 2016, Bac Yen had a sex ratio at birth of 151.4 boys for every 100 girls - a serious gender imbalance at birth that agencies here used the word "red alert" to describe. However, by 2018, these localities had significantly improved sex ratio at birth. At the same time, some districts in the previous years that had sex ratio at birth at an average level compared to the whole province increased sharply in 2018 such as Muong La, Thuan Chau, and Yen Chau. According to the statistics of the Department of Population and Family Planning, in 2018, Son La has 5/12 districts with sex ratio at birth higher than the provincial average, 7/12 districts have SBR lower than the provincial average. The districts with the highest sex ratio at birth are Yen Chau 133.9 boys for every 100 girls, Muong La 132.4. The locality with the lowest sex ratio at birth is Quynh Nhai 103.1, achieving a balanced sex ratio at birth (Table 16.2).

Son La, Quynh Nhai, and Bac Yen are the localities that have made great changes in the implementation of population policies and brought about high efficiency, helping to improve the sex ratio at birth.

In Son La City, in recent years, the natural population growth rate has maintained at 1.1%, fertility has decreased from 0.3 to 0.5‰. There are many villages over the past 10 years without people violating the policies on population and family planning. However, like other localities in the province, the gender inequality rate in Son La City tends to increase in the period of 2013–2017, specifically in 2013 was 116.9 boys/100 girls, year 2014 was 127 boys for every 100 girls, by 2015 the city's sex ratio at birth increased to 137.2 boys for every 100 girls; in which, the high ratio communes are Hua La, Chieng Xom commune, Chieng Sinh ward. In the past 3 years, the sex ratio at birth of the city decreased and in 2018 reached 111.7.

Quynh Nhai District is also one of the localities with the highest fertility disparity, especially in 2015 and 2016, if in 2015, the statistics were 116 boys/100 girls; by 2016 this figure will be 128/100, much higher than the provincial average. Muong Chien and Chieng On communes have the highest rates of sex imbalance at birth. According to the average statistics in August 2016, there were 8 boys and 2 girls for every 10 babies born. However, due to the effective implementation of population policies, the district's sex ratio at birth has decreased sharply and reached the lowest level in the province, 103.1 in 2018.

Table 16.2 Sex ratio at birth by locality 2010–2018

District / city	2010	2011	2012	2013	2014	2015	2016	2017	2018
Bac Yen	132.5	117.8	114.4	109.4	102.4	108.8	***151.4***	134.1	114.8
Mai Son	100.4	109.9	106.0	113.7	116.9	123.9	117.6	126.2	115.8
Moc Chau	112.3	117.5	104.6	137.0	113.0	121.6	112.5	115.7	112.5
Muong La	120.4	109.0	109.0	123.1	128.3	118.0	124.9	117.2	***132.4***
Phu Yen	106.0	109.4	112.2	116.7	110.4	112.4	119.3	123.6	123.0
Quynh Nhai	122.5	108.7	105.4	109.7	120.0	116.3	128.4	115.8	103.1
Song Ma	99.1	104.2	106.2	114.7	118.4	114.2	108.2	121.8	110.6
Sop Cop	96.4	157.4	119.7	102.0	111.0	105.2	115.3	107.8	114.9
Thuan Chau	114.8	123.8	111.9	108.7	123.6	116.5	118.3	117.1	126.5
Van Ho	–	–	–	–	133.5	132.0	111.2	115.2	121.0
Yen Chau	107.2	97.8	106.0	105.1	104.2	106.2	111.5	107.1	***133.9***
Son La city	113.3	109.9	107.9	116.9	127.0	***137.2***	121.8	123.6	111.7
The whole province	**109.8**	**103.2**	**108.8**	**113.9**	**117.5**	**117.5**	**118.7**	**119.4**	**118.4**

Meanwhile, Yen Chau and Muong La are the two localities with a sudden increase in the sex ratio at birth only in 2017 and 2018. If in 2017, the sex ratio at birth of Yen Chau is only 107.1 boys. For every 100 girls, a year later reached 133.9 - the highest in the province (in 2018, Yen Chau had 691 boys born compared to 516 girls). In the same period, Muong La's sex ratio at birth increased from 117.2 to 132.4 boys for every 100 girls (in 2018, Muong La had 989 boys being born compared to 747 girls). The reason given is that most of the people still think of men and women as disdain, want a son to continue the family line and shoulder the heavy work of the family.

These are alarming figures showing that Son La province has had a serious gender imbalance at birth and needs timely interventions to control this situation.

3.2 *Consequences of a Sex Imbalance at Birth*

The imbalance in the sex ratio at birth will cause bad effects on the family and society.

For the family: Family happiness will not be complete if it is difficult for an adult male to have a chance of getting the wife he wants, to be single. This causes anxiety, psychological stress, takes more time in finding a wife, for family members, affecting happiness, and the sustainable development of the family economy.

A wife must try to give birth to a son by being forced by the family and husband, or having an abortion because of the sex selection of the fetus which will adversely affect the health of the woman and the sustainable development. To maintain the family economy because of trying to give birth to a son, it means that women have to give birth many times, leading to their health declining. In addition, each birth can cause unpredictable risks to the health of both mother and child. Every abortion causes anxiety and fear for the woman. Abortion can cause catastrophic events such as ice, cervical damage, infection, placenta and can lead to infertility. Abortion and the attempt to have a son all affect the sustainable economic development of the family due to the increased costs for health care for women, reduced income of family members, and must spend a lot of time to take care of women's health.

For society: According to the statistics of Son La Department of Population and Family Planning, in the next 5 years, Son La will have about 5000–7000 men unable to get married; in the next 10–15 years, the number of men who will not get married will reach tens of thousands of people. The situation of excess of men of marriageable age may lead to the breakdown of the family structure, a part of men. The world will have to get married late and many of them will not be able to get married.

Increasing the sex ratio at birth not only does not improve the status of women, but also increases gender inequality: many women have to get married earlier, the rate of divorce and women's remarriage will increase, gender violence, prostitution, and trafficking will increase, increasing the risk of HIV infection and social diseases, causing political, economic, and social instability. According to the statistics of functional agencies, in the province, there are 229 cases of citizens who are women and children absent from the locality suspected of being tricked into selling abroad, mainly China. Particularly in 2015, the number of victims who were tricked into selling abroad in cases of trafficking and trafficking in children was 34 victims, most of them were women and children of Mong and Thai ethnic groups aged between 15 and 30; living in remote and isolated areas, with low educational levels, difficult and light economic circumstances, gullibility, and ignorance. A part of uneducated, underemployed female students is easily seduced and cheated by criminals, who are then tricked into selling abroad as illegal wives, becoming prostitutes at foreign prostitutes, exploited labor. Currently, the province has begun to have areas where prostitutes appear, which are at risk of causing social disorder and disorder, increasing drug addicts and HIV-infected people across the province. Therefore, the sex ratio at birth is considered one of the important indicators to evaluate gender equality.

4 Conclusions and Discussion

In Son La province, the rate of sex imbalance at birth is in the state of "red alert". The sex ratio at birth tends to increase, especially since 2013, here in 2018 was 118.4 boys/100 girls - the highest in the country. It is worth noting that the difference in sex ratio at birth occurs across districts and cities in the province, including in rural and urban areas. Both rural and urban areas in Son La have increased sex ratios at birth during the period 2010–2018, but in urban areas it was much higher than in rural areas, 121.4 and 115, respectively. On a local scale, the sex ratio at birth is also different. Some districts have continuous high fertility ratios for many years, especially in the period of 2013–2017, such as Bac Yen, Moc Chau, Quynh Nhai, and Son La city. Typically, Son La city has a sex ratio at birth of 137.2 boys for every 100 girls in 2015, Bac Yen for 151.4 in 2016. To 2018, Yen Chau and Muong La are the two localities with the ratio. The sex ratio at birth is highest in the province with 133.9 and 132.4 boys for every 100 girls. These are alarming figures showing that Son La province has had a serious gender imbalance at birth.

There are many reasons for the gender imbalance in Son La, but the main reason is that the concept of men and women, having a son to continue the lineage is still quite heavy not only in remote areas, ethnic minority areas, but even in towns, townships, residential groups, State officials, etc. The imbalance in the sex ratio at birth will have many consequences: Impact on The province's population structure in the future will result in male and female surplus in society, delayed marriage among men or an increase in the rate of single living. Gender imbalance leads to an increased risk of gender-based violence. Some industries will suffer from labor shortages such as preschool teachers, primary school, midwives, nurses, etc. Difficulties in marriage, the risk of spreading sexually transmitted diseases, migration, one common fact is that increasing the sex ratio not only does not improve the status of women but also increases gender inequality and violence against women and girls. Social crimes, increased demand for sex trafficking, sexual abuse, and trafficking networks for women, etc.

Therefore, it is necessary to provide specific, practical, and coordinated solutions among levels and sectors, including behavioral change communication solutions including propaganda, advocacy, and education, persuading people to see all the risks, the consequences of the imbalance of the sex ratio at birth so that people can voluntarily implement it, do not participate in the process of prenatal selection to really bring about lasting effects firm.

Reference

Son La Department of Population and Family Planning. (2019). *General statistics on sex ratio at birth in son La period 2010–2018*. Son La.

Chapter 17
Obstacles in the Sustainable Development of Industry in Ethnic Minorities' Areas of Vietnamese Mekong Delta (Kien Giang and Can Tho)

Tham Trinh Chi, Nhuong Le Van, Kha Huynh Hoang, Hieu Le Van, Phuc Nguyen Thi Ngoc, Tran Ngo Ngoc, and Thuy Ha Thi Thu

Abstract Promoting the development of industrial economies sustainably in the areas where ethnic minorities are living is one of key goals in Vietnam as this supports to create a balanced development amongst ethnic groups as well as regions. During industrialization era, combining sustainable industrial engagement with the development of ethnic minority can be considered as indispensable need (Baulch et al, 2007). This study has analyzed some main obstacles in the sustainable development of industry in the localities where ethnic minority people have been living in Can Tho city and Kien Giang province and proposed some solutions. Based on the mixed method, the research data were collected through studying documents, conducting survey, carrying out interview, and taking field trip. The research results have shown three major issues that were concluded from data analysis. Firstly, although it is much potential, the industrial activities in some areas where ethnic minorities have been living in selected localities are still underdeveloped. Secondly, some obstacles including low literacy rate and labor qualification, lack of capital and technology investment, limitation of infrastructure, lack of market information and government policy, and the dispersion of industries were analyzed in order to judge how challenging in developing industrial economies in the localities. Finally, a number of solutions have been proposed to support the development of industry in the local areas consisting of training some appropriate professions to the local situation, giving loan with low interest rates, advising on local industrial development, upgrading transportations system, and disseminating and enforcing government policies.

T. T. Chi (✉) · N. Le Van · K. H. Hoang · H. Le Van · P. N. T. Ngoc · T. N. Ngoc
School of Education, Can Tho University, Can Tho, Vietnam
e-mail: tctham@ctu.edu.vn

T. H. T. Thu
Social and Humanity Research Institute (Thai Nguyen University), Thai Nguyen, Vietnam

A. T. Nguyen, L. Hens (eds.), *Global Changes and Sustainable Development in Asian Emerging Market Economies Vol. 2*, https://doi.org/10.1007/978-3-030-81443-4_17

Keywords Ethnic minority · Industry · Obstacle · Sustainable development · Can Tho · Kien Giang

1 Introduction

Vietnam has 53 different ethnic minorities with 13.4 million people. It accounts for 14.6% of the Vietnamese population in 2018 (GSO Vietnam, 2018). Ethnic minorities in Vietnam mainly live in the Northwest part, Central Highlands, Mekong Delta region, and Central Coast (Phan, 1973). Since 2016, Vietnamese Party and Government have had many policies on socio-economic development in places where ethnic minorities have been living. Specifically, the Prime Minister has issued 41 documents mentioning the priority of socio-economic development for ethnic minority and mountainous areas. Amongst them, there are 15 projects and policies exclusively for ethnic minorities. Currently, there are still 54 effective policies in supporting and prioritizing the development of some areas where ethnic minorities are living (Institute of Ethnology, 1984).

Through the above policies, ethnic minority people in Can Tho city and Kien Giang province are concerned with many specific and preferential policies. However, the socio-economic activities in ethnic minority areas are still undeveloped. This makes a big gap in the development and life quality between ethnic minorities and the remaining population in Vietnam. Especially, the development of industries in the area is quite limited and faces a lot of difficulties. It is believed that to develop the industrial economy in the area where ethnic minorities have been living in Can Tho and Kien Giang sustainably, it is necessary to identify the causes of limitations and difficulties in industrial development in these areas. Based on identifying reasons, solutions that are valuable and appropriate to the actual situation will be proposed to solve the problems.

2 Methodology

2.1 Research Questions

(a) How was the industrial economies developed in some areas where ethnic minorities have been living in Can Tho city and Kien Giang province?
(b) How challenging developing industrial economies sustainably in the localities where ethnic minorities have been living?
(c) What are solutions that can be proposed to support the development of industry in the local areas sustainably?

2.2 Data Collection

Can Tho city and Kien Giang province were selected as research areas because these are two provinces/cities in the Mekong Delta region which are quite crowded with ethnic minorities. Besides, these two areas respond well and more appropriately to specific criteria (different in potentials and economic background, Can Tho is a big city in the region central and Kien Giang is a coastal province), so Can Tho and Kien Giang have been selected as research areas. Additionally, we clearly identify the specific locations in each place to collect data including the districts of Ninh Kieu, Binh Thuy, Cai Rang, O Mon, Thoi Lai, Co Do (Can Tho city); Rach Gia, Giong Rieng, and Go Quao (Kien Giang province).

To have a secondary data for this study, literature review and previously published scientific results in the field were selected. Specifically, the documents on economic activities of ethnic minorities in Vietnam and the Mekong Delta region, planning for socio-economic development in areas where ethnic minorities live, socio-economic development in Can Tho city and Kien Giang province was studied by the authors. Understanding the theoretical background of the research issue helps the authors to have an initial view about the research object. This also supports the researchers to carry out the study easier and more oriented.

Additionally, in order to collect a primary database for this study, specific research methods were selected including survey, interview, and field trip. Specifically, the authors have arrived some district in Can Tho city and Kien Giang province in order to ask for permission on carrying out survey and interview with local residents as well as taking field trip. For instance, the authors came to ask for permission from the People's Committees of districts including Ninh Kieu, Binh Thuy, Cai Rang, O Mon, Thoi Lai, Co Do in Can Tho city and Rach Gia, Giong Rieng, Go Quao in Kien Giang province. After that, the local authorities introduced us to those who could complete the survey and interview. Before asking them to join in survey and interview, the authors have provided them with specific information about the research which helped the participants to understand the research overview and the purpose of the survey as well as interview. For illiterate people, the researchers have asked their parents and local authorities for permission before conducting the survey and interview. It is true that the surveys and interviews were conducted due to reading and analyzing every single question clearly so the illiterate citizens can understand and answer. The authors then selected the answer according to the participants' opinion. It is important that all personal information about the respondents has kept confidential and only used for legitimate research purposes. It was also a mandatory commitment made prior to conducting any interview or survey for the study.

2.2.1 Survey

Survey was considered as a main data collection type which has supported to collect wide range information from 130 households, 150 workers, 50 Entrepreneurs, and 50 local leaders. It is believed that this helps the researchers to gather broad and comprehensive information (Creswell, 2014) as the survey subjects were different in ages, races, incomes, academic background, occupations, etc. Furthermore, a large number of samples as above are enough to ensure the reliability, accuracy, and objectivity for the research results. The survey helps the researchers to know about which industries are developing, what difficulties in industrial development in ethnic minority areas are and how these difficulties impact on the sustainable industrial development.

2.2.2 Interview

To collect profound information as well as increase the reliability for the research, in-depth interviews have been selected. Through this data collecting way, the study have selected ninety different interviewees including 30 households, 30 workers, 15 entrepreneurs, and 15 local leaders. Those subjects were different in occupations, ages, and experience; therefore, they have provided the authors with comprehensive and diverse information. This can support to improve the study results' reliability (Creswell, 2014; Montello & Sutton, 2013). Through the interview, the authors can further understand about the obstacles in industrial development in the selected area as well as the participants' suggestions on how to improve the sustainable development in the local industry.

2.2.3 Field Trip

A fieldwork was conducted to actively explore the research subjects. In particular, we had six field days in two provinces/cities mentioned above. Through fieldwork, we have observed our research object in detail as well as collected a lot of authentic data for the study. For instance, the authors focused their attention on people's housing, transportation system, operation status of factories, and the lives of local residents to better understand the current situation of industry development, and the quality of life in the localities. It can be said that fieldwork helps us proactively evaluate and assess the development of industrial economies in many different localities. Moreover, the information collected from the field trip supports us compare and contrast with what was collected through surveys and interviews (Denscombe, 2010).

2.3 Data Recording

In order to collect, store, and prove the research results, a number of data collection tools such as notebooks, recorders, cameras, and telephones were used in this study. It is true that the questionnaires and interview questions are important to guide the data collection as well as to assist the authors collecting comprehensive information. Notebooks, cameras, voice recorders, and telephones are tools that can help the researchers to store collected data (Creswell, 2014). Data recording helps the researchers not only to prove their research results but also increase the research results' credibility.

In this study, all the information gathered from the survey and interview was required to be stored by completed questionnaires. This means that the respondent has completed the questionnaire they submited to the researchers. For the interview, the researchers have transcribed the respondent's answer in a notebook. In some cases, photographs and videos were conducted in order to store a few other important and necessary information. It should be noted that the storage of information by the above means was carried out with the consent of the participants.

2.4 Data Analysis

In order to analyze the collected information, 23rd version SPSS (Statistical Package for the Social Science) has been used. In particular, SPSS supports the authors to revise information systematically as well as put data into different themes and explain the research results. After that, the authors aggregate data and present it in a tabular form. The analysis of data from SPSS software also provides the researchers with important and necessary information to create a practical basis for this study.

The researchers have based on following process and technique that were suggested by Creswell (2014) in order to analyze collected data:

Step 1: Prepared and organized the data;
Step 2: Read through all data of interviews;
Step 3: Organized the material into segments of text before interpreting the meaning of data;
Step 4: Coded the data based on the meaningful segments and sort them into some different categories;
Step 5: Described the theme with typical meaning of each data sort. The researchers have focused on the main message of the category in each description;
Step 6: Analyzed and concluded on the research results.

3 Results

3.1 Major Industries in the Local Areas

The research results have shown that there are a few industries that have been developed in selected localities; however, most of them are small scale industries. In particular, industrial companies with modern technology and big scale are located in industrial zones. Small scale industries are developed in the rural areas where ethnic minorities are settling. Amongst 50 industrial factories that were surveyed, there are 13 enterprises of confectionery processing and 11 enterprise of rice milling. In addition to this, ice processing is the next most popular industrial activity in the research area with seven factories working in this field. Besides, industry of boat/canoe building is quite common in the selected localities; but its scale is quite small with building small boats. Although seafood processing, wood, and shoe production are industries that bring high profit, they are not popular in Can Tho city and Kien Giang province with the corresponding number of five, three, and two enterprises. Especially, those industries are mainly located in district central where not a lot of ethnic minorities are living. In other words, small scale industrial economies are more popular in the local areas where ethnic minorities are settling (Table 17.1).

It is indicated that basically there are similarities in industrial development in Can Tho and Kien Giang. In general, the types of industries that are developed in these localities are totally the same. However, there are differences in the popularity and scale of industries. Specifically, the confectionery processing industry is most popular in ethnic minority areas in Can Tho while the rice milling industry is the most popular in Kien Giang. The next popular industries in Can Tho are rice milling, seafood, and ice processing. Although the confectionery processing industry is also quite popular in Kien Giang, it is much less common in Can Tho. Following

Table 17.1 Major industries in areas where ethnic minorities have been living

Types of Industry	Can Tho		Kien Giang	
	A number of enterprise	Percentages	A number of enterprise	Percentages
Confectionery processing	8	32.0	5	20.0
Boat/canoe building	2	8.0	4	16.0
Furniture wood processing	2	8.0	1	4.0
Ice processing	3	12.0	4	16.0
Rice milling	4	16.0	7	28.0
Seafood processing	3	12.0	2	8.0
Shoe processing	1	4.0	1	4.0
Others	2	8.0	1	4.0
Total	25	100	25	100

Source: Survey, December 2018, $n = 50$

industrial economies that are quite famous in Kien Giang are boat/canoe building and ice processing. The shoe manufacturing and furniture wood processing industries are not well known in both Can Tho city and Kien Giang province.

It can be seen that industry in both Can Tho city and Kien Giang province needed a lot of labors but it has not require high technology. In specific, only seafood processing and shoe manufacturing industries have been requiring advanced technology, but such industries are less popular in the localities. Based on the research results, it is asserted that some industries with modern technology such as the seafood processing and shoe manufacturing industries are usually developed and located in industrial zones which are planned and invested by government and local authorities. In addition to investment funding, these places have developed infrastructure and abundant labor with relatively high qualifications. Specially, there are few ethnic minorities living in these places. In contrast, in rural areas with underdeveloped infrastructure, low literacy rate, and labor qualification as well as lack of investment capital, only small scale industries can be developed. A leader of a rural commune in Kien Giang province has shared that "Although operating big companies and enterprises will create more jobs as well as increase the local citizens' income, nobody knows how to open more businesses because of lacking of capital, technology and infrastructure. Especially, no one has consulted on such issues yet". This is explained that the national and local government policy and investment directly influence on the socio–economic development in areas where ethnic minorities are living (Braunholz-Speight, 2008).

3.2 *Obstacles in the Sustainable Development of Industry in Ethnic Minority Areas*

Based on surveying 50 industrial entrepreneurs, the research results have shown that there were different factors affecting sustainable development of industry. These factors have affected the sustainable development of the industry differently in terms of level and nature as described in Table 17.2.

In this part, all such factors can be analyzed as below:

Table 17.2 The proportion of entrepreneurs who face difficulties according to seriousness

Difficulties	1	2	3	4	5
Low literacy rate and labor qualification	0.0	4.0	28.0	26.0	42.0
Lack of capital and technology investment	0.0	8.0	32.0	18.0	42.0
Limitation of infrastructure	4.0	4.0	48.0	20.0	24.0
Shortage of information and policy	4.0	4.0	52.0	12.0	28.0
The dispersion of industries	0.0	20.0	42.0	18.0	20.0

Source: Survey, December 2018, $n = 50$

Note: 1 = absolutely not serious; 2 = not very serious; 3 = serious; 4 = very serious; 5 = absolutely serious

3.2.1 Low Literacy Rate and Labor Qualification

The survey result of 150 ethnic minorities in Can Tho city and Kien Giang province has indicated that the educational level of ethnic minorities is very low and lower than that in Vietnam as well as in the Mekong Delta region (Fig. 17.1). Specifically, in the Mekong Delta region only 7.0% of the total population is unlettered. This figure for Vietnam is only 2.0%. Based on Figs. 17.1, 14.0% of the population is ethnic minority who was illiterate. The highest proportion belonged to the ethnic minority group with primary education level, with 47.0%. Besides, there were 38.0% of minority people who have reached secondary school. On the contrary, there was an extremely low rate of minority people with higher education level, with 1.0% only.

In the Table 17.2, it was indicated that low literacy rate and labor qualification were the most influencing factors on the development of industry sustainably. In particular, 68.0% of businessman found that lack of literacy and professional qualifications affects both industrial productivity and quality very serious and absolutely serious. Additionally, there were also 28.0% of entrepreneurs believed that literacy rate and labor capacity can affect industrial development. Obviously, only 4.0% of participants asserted that the development of industry was not very seriously affected by educational level and labor quality. Considerably, educational/intellectual level and professional qualification have made negative effects on the sustainable development in the localities where ethnic minorities have been living in both Can Tho city and Kien Giang province.

Low literacy rate and labor qualification were identified as the most dominant factors which harm on industrial productivity and quality. All participants said that "When workers are low in qualification, they cannot create products that are competitive in terms of design, quality and price". Especially, as stressed by businessman that "Workers with low qualification are not capable of applying advanced

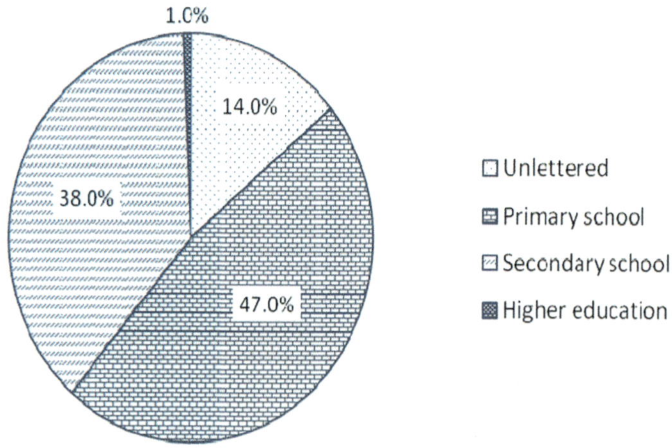

Fig. 17.1 Educational level (Source: Survey, December 2018, $n = 130$)

technology in industry, so the development of industrial economy becomes obsolete with low economic efficiency". This is evident as previous studies have concluded that low literacy and labor qualification are the biggest barriers to the socio–economic development in localities where minority people have been living (Institute of Ethnology, 1984). It is more important that labor quality is a decisive factor for the development of industry in areas with ethnic minorities (Phan, 1973). Hence, in order to develop the industry in the selected areas sustainably, one of the most important priorities is to improve the educational level and employee qualification.

3.2.2 Lack of Investment Capital and Technology Innovation

Being interviewed, 100% of entrepreneurs said that "We do not have enough money to set up our businesses or company; therefore, we must borrow with interest from the government or local authorities". In some cases, households have to borrow money at high interest rates in order to invest in improving their industrial technology (Fig. 17.2). In particular, up to 38.0% of businessman had to borrow from 30.0 to 50.0% of their business capital. In addition, 34.0% of people who are operating industrial enterprises have borrowed from 50.0 to 70.0% of their businesses' capital. It can be said that the proportion of people who have enough money to do their industrial business was very low, with 10.0% only.

The surveying results presented in Table 17.2 have shown that 60.0% of entrepreneurs claimed that capital and technology investment influence on the sustainable development of industry sector in very serious and absolutely serious scales. It also means that such business people strongly believed that lack of funding and technology access affects the production scale and quality of the industry sector. There were 32.0% of such entrepreneurs said that capital and technology were influential factors in average. However, no one thought that the investment of funding and technology did not effect on industrial sectors' development.

Undoubtedly, investment capital and technology innovation were the second most dominant elements that have influenced on the sustainable development in

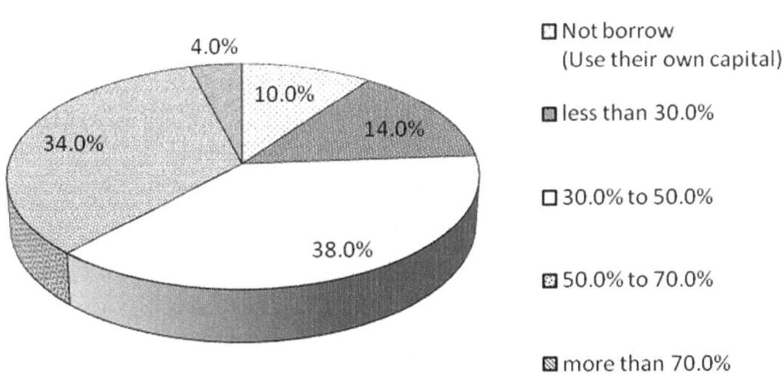

Fig. 17.2 The proportion of capital borrowed for businesses (*Source: Survey, December 2018, n = 50)*

industry sector. Being interviewed, 27 of 30 businessman and local administrators have mentioned that "Lack of investment capital and modern technology not only influences on product quality but also affects natural resources in negative ways". Particularly, one of local leader emphasized that "A lot of industrial entrepreneurs have no investment in wastewater treatment technology, so they made local water polluted seriously". On the other hand, air pollution was other bid problem to the sustainable development in industry in areas where ethnic minorities have been living. Some of business people have admitted that their industrial economy had a great impact on the environment but currently it lacks of investment in technology. It means that they have no solution in this situation. It can be deduced that although local businessman are aware of how to develop industry sustainably, they have to give up because of limitation of capital and technology investment. Thus, it can be seen that capital and technology are factors that have a major impact on the development of industries in the ethnic minority areas (Braunholz-Speight, 2008).

3.2.3 Limitation of Infrastructure

The next important dominant element to the industrial development was infrastructure, especially transportation system. The interviewing results shown that infrastructure in most areas where ethnic minorities have been living in Can Tho city and Kien Giang province is limited compared with actual demands for industrial development. Specifically, some factors including electricity, rural transportation, school, and clinic have been noted by all 50 interviewees. As stressed by all interviewed people, transportation system can be considered as the most affecting factor to the development of industry sector in areas where ethnic minorities have been living. In other words, the local residents found that it is very difficult for them to travel and transport goods because of the limitation of transportation system.

In regard to the impact of infrastructure on the industry sector in Can Tho city and Kien Giang province, 48.0% of interviewees asserted that infrastructure has an average impact on industry development. Subsequently, 24.0% of the 50 respondents claimed that the infrastructure limitation affected local industry absolutely serious. Similarly, 20.0% of participants agreed that this factor has harmed on the sustainable development of industry on scale of very serious. Especially, regarding to traffic limitation 40.0% of respondents have stressed that the local industry was absolutely influenced (Table 17.2). It can be understood that only 8.0% of respondents pointed out that limitation of infrastructure has not impacted on the local industrial development seriously. It can be concluded that infrastructure shortage has impacted on the sustainable development of local industry in negative ways; therefore, improving infrastructure is an indispensable need for the localities where ethnic minorities have been living in Can Tho city and Kien Giang province.

There is no doubt that infrastructure in general and transportation in particular is the third dominant factor to the sustainable development of local industry in this case. The influence of this factor on industry was evidenced by the local residents' typical examples. Firstly, the local children cannot go to school as they are living in areas without traffic. This explains why in areas with minority groups have been living, the literacy rate and labor qualification were very low as analyzed in the previous section. Secondly, due to inadequate shipping, the quality of fruits and seafood is low or agricultural output is significantly reduced. This means that the processing industry lacks raw materials. Thirdly, the health of local people is not protected due to lack of clinics. This is a major obstacle for those areas that are lacking high quality labor. A minority said that "I can only let my children finish primary school so that they can read and write. My wife and I have to earn for living, so we cannot take them to school. Obviously, the local traffic is too bad; therefore, they cannot go to school by themselves". A local leader found that "Although the locality has great potential but due to poor infrastructure, the investors are not interested in this place". It is evident that the infrastructure is a basic condition to exploit the natural potential which supports to operate economic activities in general and industry sector in specific in ethnic minority areas (World Bank, 2009). Hence, the limitation of infrastructure, especially transportation system, adversely affects the sustainable development of the industry in the areas where ethnic minorities are located in Can Tho city and Kien Giang province.

3.2.4 Shortage of Market Information and Government Policy

According to the survey results presented on Table 17.3, a lot of respondents were never or rarely provided with market information and government policies. In particular, the proportion of local households who never or rarely received market information and state policies was 48.4% and 26.2%, respectively. This percentage for workers is slightly lower than that of local residents with the corresponding ratio of 47.3% and 18.0%. The proportion of local entrepreneurs and leaders who have got market information and national policies is quite higher than that of other groups. Specifically, there were 54% of businessman or local leaders sometimes and often have known about market information and related policies. Thus, the proportion of residents and workers who were disseminating about trading information and government policies, especially policies related to minorities, is lower than therate of entrepreneurs and local leaders.

The research results have also indicated that the shortage of information and policies influenced on the development of local industry sector (Table 17.2).

Table 17.3 Regularity on access to information and policy (%)

Objects	1	2	3	4	5	6
Residents (n = 130)	48.4	26.2	16.2	9.2	0.0	0.0
Workers (n = 150)	47.3	18.0	16.7	16.0	2.0	0.0
Entrepreneurs (n = 50)	26.0	14.0	32.0	22.0	2.0	4.0
Leaders (n = 50)	16.0	18.0	30.0	24.0	4.0	8.0

Source: Survey, December 2018; Note: 1 = never; 2 = rarely; 3 = sometimes; 4 = often; 5 = usually; 6 = always

Specifically, 52.0% of respondents claimed that local industrial development can be affected by information and policy limitation seriously. Moreover, as asserted by 28.0% of respondents such as lack of information and policies absolutely serious affected the development of the industry in the minority locality. This is implied that the unsustainable industrial development in the areas where ethnic minorities are located in Can Tho city and Kien Giang province was greatly affected by the lack of market information and government policies.

Although lack of information and policy was not considered as typical dominant factor, it has made a great impact on the sustainable development of industry in this situation. It can be understood that market or trading information and policies support local residents who have been living in Can Tho city and Kien Giang province to understand about the development trend of industry in Vietnam as well as on over the world. In other words, if this lacks of market information and specific policies, it is difficult for both businessman and local people to decide what should they develop and how can they operate for their companies (Hoang, Pham, Tran, & Hansen, 2007). On the other hand, they cannot understand their rights and responsibilities in industrial development.

3.2.5 The Dispersion of Industries

Although being the least dominant factor in industrial development, the dispersion of industry has a certain influence on the sustainable development of this industry. In particular, 42.0% of respondents stated that spatial dispersion affects the industry sector's sustainable development seriously. Especially, 28.0% of respondents said that the spatial dispersion affects the sustainable development of the industry absolutely seriously. In general, spatial dispersion affects the sustainable development of the industry from serious to absolutely serious scale as assessed by the majority of interviewees (Table 17.2).

It is clear that when the industry is distributed in a dispersed manner, the local or national government must invest a lot of money in building roads/streets. Besides, human being has to spend more time and money on shipping raw materials and goods; therefore, both profit and product quality are significantly reduced (Hoang et al., 2007). It should also be noted that the dispersion of industry also makes it difficult to exploit natural resources and labor force. A local leader shared that "We have proposed to build more roads in order to connect some small industrial areas. However, the government claimed that it has to invest too much money on building while the profits will be low". On the other hand, many workers complained that "I do not want to work in industrial zone as it is too far away from my house". The above analysis demonstrates that industrial dispersion has seriously affected the development of this industry sustainably.

3.3 How Local Residents Expect in Order to Develop the Industrial Economies Sustainably?

To further understanding the participants' expectations toward how to develop the local industry more sustainably, 180 people including 60 local residents, 60 workers, 30 entrepreneurs, and 30 leaders in Can Tho city and Kien Giang province were selected for an individual interviewing task. Amongst them, there were 80 participants who are ethnic minorities. The research results have indicated that the participants seriously concerned about a lot of different issues in regarding to developing local industry sustainably. Among all issues that they have mentioned, there were eight typical ones as presented in Table 17.4.

Based on the research results, local people in both Can Tho city and Kien Giang province have concerned most about funds, especially the funds from national government with 92.2% and 96.7%, respectively. In other words, they want to get low-interest loans to carry out various tasks including opening businesses, improving technology, taking part in career training course. The next three typical expectations include developing transportation system, attending to job training without fee, and receiving more local and national policies. As presented in the Table 17.4, the local people in Can Tho city are more interested in getting and understanding local and government policies while the local citizens in Kien Giang province have much worried about developing the transportation system. This is explained by the fact that Can Tho is a developed city located in the center of the Mekong delta region, so the local people are highly educated. In other words, people with good knowledge and awareness are more interested in understanding information and policies. In contrast, because Kien Giang is a small province with difficult travel conditions, people intend to pay their attention to developing transportation system. Although advising on industrial development is the least mentioned expectation by the

Table 17.4 The local people's expectations on development of industry

Local residents' expectations	Can Tho		Kien Giang	
	A number of residents ($n = 90$)	Percentages	A number of residents ($n = 90$)	Percentages
Get more funds	83	92.2	87	96.7
Develop transportation	71	78.9	75	83.3
Train a vocation	71	78.9	72	80.0
Know more state policies	73	81.1	69	76.7
Build more factories	66	73.3	72	80.0
Provide market information	69	76.7	65	72.2
Simplify administrative procedures	63	70.0	67	74.4
Advice on industrial development	59	65.6	51	56.7

Source: Survey, December 2018, $n = 180$

participants, there were 65.6 and 56.7% of local people in Can Tho and Kien Giang paid their attention to this aspect corresponding (Fig. 17.3).

In comparison to the participants' expectations between Can Tho city and Kien Giang province, generally the proportion of participants in Kien Giang have expressed their expectation was higher than that in Can Tho city; however, this also depends on specific situations. More specifically, in regard to government policy, market information, and advice on industrial development, the respondents in Can Tho city have paid more attention to such expectations compared with the participants in Kien Giang province. For instance, if 65.6% of Can Tho participants were interested in consulting for industrial development, this percentage in Kien Giang was only 56.7%. In contrast, for the other five expectations, the respondents in Kien Giang have more concerned. For example, up to 80.0% of Kien Giang respondents wished to have more companies to look for jobs while only 73.3% Can Tho people have expressed on that hope. Similarly, the participants in Kien Giang province have more concerned about investment capital and fund compared with the respondent in Can Tho city, 96.7 and 92.2%, respectively.

3.4 Some Solutions to the Problems

Sustainable development may be defined as the development to meet the needs of the present generation without compromising the needs of the future generations (Sharachchandra, 1999). It is a process whereby the development can be sustained

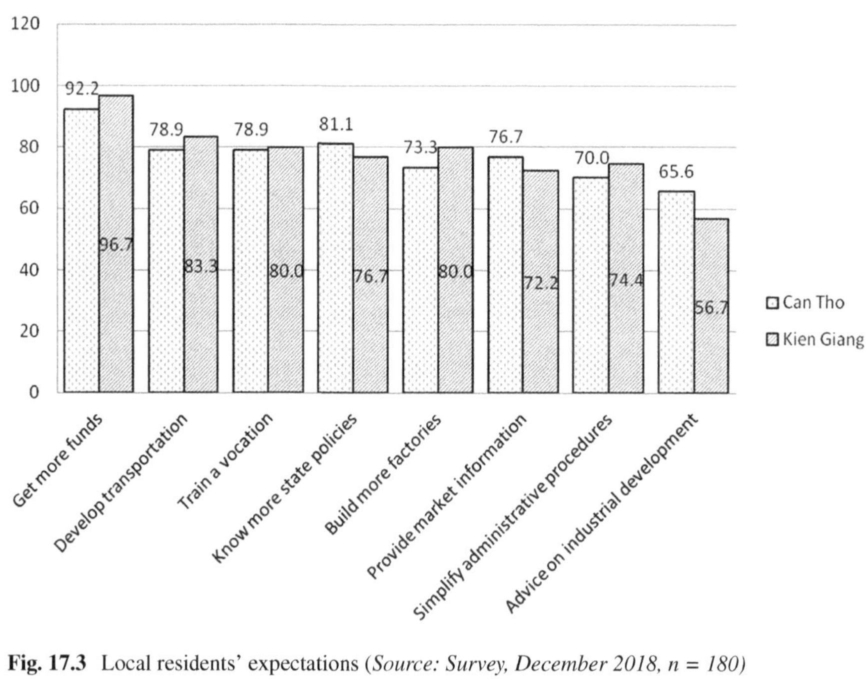

Fig. 17.3 Local residents' expectations (*Source: Survey, December 2018, n = 180)*

for generation. It affords the future generation the same, if not more, capacity to prosper as the present generation. It also means that sustainable development needs to bring high economic efficiency, ensure social security, and protect the environment. According to Erling, Kristin, and David (2014), economic development is not a blessing for the human beings. No doubt, it brings higher material welfare by increasing national output of goods and services on one hand and on the other hand, it pollutes the environment badly by overuse and misuse of natural resources.

The research results on the current situation and obstacles in industrial development in area where ethnic minorities have been living in Can Tho and Kien Giang have indicated the same things as above. It is obvious that low in education, awareness, and qualifications is the cause for the exhaustion of natural resources, environmental pollution, and low economic efficiency in Can Tho city and Kien Giang province. Moreover, the limitation of qualifications, backwardness in technology and market information, lack of investment capital, lack of understanding of state policies have made the economic value of the industry really low. Based on the definition of sustainable development, it is clear that industrial development in areas where ethnic minorities have been living in Can Tho city and Kien Giang province is not sustainable. Firstly, economic efficiency in industry is not high due to shortage of technology, limitation of labor productivity, and low of product quality, so industrial products are unable to be competitive in the international market. Secondly, many local residents are unemployed due to small industry size which cannot create enough jobs for people. This makes the life quality in the local is low. Finally, the environmental pollution and resources exhausting are major problems not only in the present but also in the future. In other words, the industrial development in the study space not only does not meet the needs of the present generation but also threatens the benefits of future generations.

Based on the research results on the current situation and obstacles in industrial development in the area of ethnic minorities living in Can Tho and Kien Giang, a number of solutions were proposed as following:

3.4.1 Training some Appropriate Professions to the Local Situation

Currently, one of the most difficult problems in some places where ethnic minorities have been living in Can Tho city and Kien Giang province is the limitation of labor in both quantity and quality. Especially, the biggest obstacle is that local residents do not have opportunity to take part in service and professional training, so they cannot find good jobs. In some situations, as the local human are still out of work; therefore, they lack of money to pay for such training course/program. To further support them, the local leaders or national government need to provide them with free training courses or low fee training courses which can enable local people to get a job and earn for their living. Another essential thing that can be considered is that local citizens can be lent money with low interest from the local banks which can help them to attend to vocational training courses. However, to enhance such training course to be more effective, the local authorities and governments need to link or cooperate with businessman to introduce suitable jobs to the local employees.

3.4.2 Giving Loan with Low Interest Rates

Not only local people but also entrepreneurs in the area are also sort of capital to improve equipment and technology. Many businessmen have not improved their production technology for a long time so that their products cannot compete with other national and international businesses in trade in terms of design and quality. In other words, those products are only sold to local residents or some neighboring provinces. Thus, in addition to giving local people loans to participate in vocational training courses, the government or local authorities should also allow local businessman to borrow money which can enable them to upgrade technology as well as expand business scale. The most important thing in this case is that the local and national government must issue specific laws or regulations on the use of loans. Besides, the local government needs to support local businesses to understand how to use capital effectively. For example, some new business models need to be introduced to local businesses so that entrepreneurs can understand the international development trend. On the other hand, the government leaders should introduce new technologies in industry that can support local businesses to develop industry more sustainably.

3.4.3 Advising on Local Industrial Development

The research results have shown that many local businesses are still confused in choosing which industry to develop or how to operate it. Many local businesses shared that they only focus on producing traditional goods as they do not know which products are appropriate to the current international market. This means that local governments and large enterprises should support local small businesses in areas where ethnic minorities have been living in Can Tho city and Kien Giang province so that local companies can have better and more reasonable orientations in trading. In addition to advising on choosing industry/products, the local government also needs to help local businesses understanding about potentials/strengths of the locality as well as how to exploit those potentials. Obviously, this helps local entrepreneurs to better exploit natural resources, protect the environment, and develop local industrial sector sustainably.

3.4.4 Upgrading Transportation System

In this study, the limitation of transportation system was one of factors that seriously influence on the development of industry as well as the improvement of life quality for local people in general and for ethnic minorities in specific. So, to better develop the industrial sector in areas where minority people are located in Can Tho city and Kien Giang province, it needs to pay more attention to providing fund as well as high technology in the localities. More importantly, the local government should firstly focus on improving transportation systems in rural areas where has high illiteracy rates. This not only enhances educational standard but also helps to transport agricultural products quickly so that local people can simultaneously improve their education and income. In addition, improving transportation should also be prioritized in areas with potential for economic development as well as in places where many ethnic minorities have been living.

3.4.5 Disseminating and Enforcing Government Policies

It can be said that understanding market information and government policies can help local people, especially ethnic minority people to be clearly aware of their roles and responsibilities in socio-economic development in the locality. In this study, many respondents have stressed that the limitation of information and policies made them difficult to access to new trading markets (entrepreneurs), good job (local residents/minority people and workers), and reasonable orientation for local industrial development (local leaders). It is believed that the local government should build more radio stations or send officials to specific localities to disseminate market information and policies. In this task, it needs to focus on some places that there are a lot of poor citizens and minority people have been living. Such access to information and policies will assist people in achieving economic orientation toward market trends. Certainly, Vietnam's economy in general and the Mekong Delta region in particular are integrating into the world economy, so understanding the market and economic policy plays an important role in the sustainable development of industry sector in selected areas.

4 Conclusions

Although Can Tho city and Kien Giang province are localities with great potential for economic development, many localities are still underdeveloped, especially in areas where ethnic minorities have been living. Specifically, the industry is still rudimentary with a small scale, so there are few goods to export.

There are a lot of obstacles that are affecting the economic development in areas where ethnic minority areas have been living in general, especially influencing on the sustainable development of the local industry. However, all residents, workers, local leaders, and entrepreneurs have claimed that there are five most dominant factors that have impacted on the sustainable development of industry including low literacy rate and labor qualification, lack of capital and technology investment, limitation of infrastructure, shortage of information and policy, and the dispersion of industries. Especially, when being asked about their personal view, the majority of respondents have assessed that the impact of those five dominant factors on the economic development and the sustainable development of industry in areas where ethnic minorities have been living in Can Tho city and Kien Giang province was very serious or absolutely serious.

In general, local people are somewhat knowledgeable about the current situation of local industrial development; therefore, they have expressed their expectations on the emerged issues reasonably. Although they have expected a lot of things, most of them paid more attention to borrowing money, getting vocational training, seeking a good job, and improving the local transportation system.

Based on the current situation of industry development, obstacles in developing industry sustainably and the local residents' expectations, a number of solutions have been proposed in order to promote the development of the local industry more sustainable.

Acknowledgments The data in this paper are under the project "Research the current situation and propose solutions for industrial development in ethnic minority and mountainous areas of Vietnam", coordinated to implement between Social and Humanities Research Institute (Thai Nguyen University) and Can Tho University. During the process of completing this study, the authors have received great deal of help, guidance, and encouragement from a lot of people.

First of all, we would like to express our sincere thanks to Social and Humanities Research Institute (Thai Nguyen University) due to trusting and funding for us to implement this project. It is stressed that all funding to implement this project is provided by the institute.

Secondly, it is our pleasure to thank Can Tho University for allowing and facilitating us to cooperate and conduct this project. The leaders of Can Tho University have consulted and approved administrative formalities for us to carry out the project.

Next, we would like to thank School of Education (Can Tho University) for considering and approving administrative formalities in implementing this project. Especially, the leaders of School of Education have introduced us to some organizations and individuals to collect the study data.

Last but not least, we sincerely thank all organizations and individuals who have supported and provided us with valuable and helpful information and data for this study. Specifically, these are organizations and individuals in the districts including Ninh Kieu, Binh Thuy, Cai Rang, O Mon, Thoi Lai, Co Do, Rach Gia (a city of Kien Giang province), Giong Rieng and Go Quao.

References

Baulch, B., Truong, K. C., Haughton, D., & Haughton, J. (2007). Ethnic minority development in Vietnam. *Journal of Development Studies, 43*(7), 1151–1171.

Braunholz-Speight, T. (2008). *Policies responses to discrimination and their contribution to reducing chronic poverty, background paper to chronic poverty report 2007–2008* (127 pages). Chronic Poverty Research Centre.

Creswell, J. W. (2014). *Research design: Qualitative, quantitative, and mixed methods approaches* (342 pages). Sage.

Denscombe, M. (2010). *The good research guide: For small-scale social research projects* (4th ed., 392 pages). McGraw-Hill/Open University Press.

Erling, H., Kristin, L., & David, B. (2014). Sustainable development: our common future revisited. *Global Environmental Change, 26*, 130–139.

GSO Vietnam. (2018). *Population and labor*. Retrieved April 15, 2019, from https://www.gso.gov.vn/default.aspx?tabid=714.

Hoang, H. T. T., Pham, G. H., Tran, M. B., & Hansen, H. (2007). *Market, policy and poverty reduction in Vietnam*. Vietnam Culture and Information Publishing House. 260 pages.

Institute of Ethnology. (1984). *Ethnic minorities in Vietnam (southern provinces)* (313 pages). Social Science Publishing House.

Montello, D. R., & Sutton, P. C. (2013). *An introduction to scientific research methods in geography* (2nd ed., 314 pages). Sage.

Phan, H. D. (1973). *Ethnographic basis* (263 pages). Professional and University Publishing House.

Sharachchandra, M. L. (1999). Sustainable development: A critical review. *World Development, 19*(6), 607–621.

World Bank. (2009). *Country social analysis: Ethnicity and development in Vietnam (vol. 2): Main report (English)* (277 pages). World Bank.

Chapter 18
Integrating Sustainable Development into National Policy: The Practice of Vietnam

Vo Thanh Son

Abstract By 2015, the United Nations have adopted the 2030 Sustainable Development Agenda with 17 Sustainable Development Goals (SDGs) to further promote sustainable development in the world, with particular emphasis on the importance of integrating sustainable development into each nation's development policy. This paper is an attempt to review the process of sustainable development in Vietnam in terms of institutional arrangement and related policies. By reviewing the framework for integrating sustainable development dimensions into development policies in the world, the paper assesses the initial results of integrating sustainable development into national socio-economic development strategies and plans by the goals and priorities of sustainable development as well as by its dimensions or components in Vietnam. The paper also addresses the main challenges and constraints as the shortage of technical guidance to integrate sustainable development into development policies and the lack of resources needed.

Keywords Sustainable development · Sustainable development goals · Millennium development goals · Integration of sustainable development · Development policy

1 Introduction

The year of 2015 is the time to complete the implementation of Millennium Development Goals (MDGs) and also the time for countries in the world to prepare to implement Agenda 2030 for sustainable development, proposed by United Nations. Therefore, many countries have developed plans to implement the Agenda

V. T. Son (✉)
VNU Central Institute for Natural Resources and Environmental Studies,
Vietnam National University, Hanoi, Vietnam
e-mail: soncres@vnu.edu.vn

A. T. Nguyen, L. Hens (eds.), *Global Changes and Sustainable Development in Asian Emerging Market Economies Vol. 2*, https://doi.org/10.1007/978-3-030-81443-4_18

and specify 17 Sustainable Development Goals (SDGs) into their national development plans. As a responsible country, Vietnam has been pursuing a sustainable development process very early, from participating in the Summits in 1992 (Rio de Janeiro - Brazil), in 2002 (Johannesburg - South Africa) and in 2012 (Rio de Janeiro), to the successful completion of the MDGs and to strong commitment to implementing the 2030 Agenda. In particular, Vietnam has made great efforts in integrating sustainable development and the Sustainable Development Goals into the country's Socio-Economic Development Strategy and Plan and has achieved some remarkable results. These initial results of Vietnam can be shared with the domestic and international community to contribute to the sustainable development process of humankind.

2 Methodology

This paper is one of the results of project "Strengthening the Capacity of Selected Developing Countries to Mainstream Sustainable Development into National Development Strategies" in the framework of the cooperation of United Nation's Department of Economic and Social Affairs (DESA) and Vietnam's Ministry of Planning and Investment (MPI). This report addresses the integration framework applied in the world, the review of Vietnam's sustainable development policies, and the initial results of integrating sustainable development into development policy and some proposed recommendations (Vo, 2016).

Two main methods were used for this study. The first method is an overview of the literature concerning process in the world and in Vietnam, classified by topics and arranged in timeline, to clarify the process of sustainable development, as well as current situation of integrating sustainable development dimensions into Vietnam's development policies. Secondly, the method of consulting experts and stakeholders in the consultation workshop of ministries and nongovernmental organizations was also conducted, in order to clarify and correct the information that is not been shown in the available documents. This workshop was organized by Department of Science, Education, Natural Resources, and Environment (MPI) in cooperation with UN-DESA in Hanoi during 19–20 May 2016 to consult line ministries (such as Natural Resource and Environment-MONRE, Agriculture and Rural Development-MARD, Science and Technology-MOST, Education and Training-MOET, Labor, War-invalids, and Social Affairs-MOLISA, Health-MOH, Transportation-MOT, etc.), several research institutes, I-NGO (UNDP, UNIDO), and local NGO on the objective, approach, contents, and practice of integration of sustainable development into the socio-economic development planning and strategy in Vietnam.

In addition, information and data in the development of Vietnam's Voluntary National Report for 2018 are also used to clarify the results of sustainable development integration in Vietnam (MPI, 2018). Specifically, as a member of the group drafting VNR of Vietnam, the author has an opportunity to consult reports on SDG implementation of ministries, I-NGOs, and NGOs and through this process to synthesize key policies related to SDG implementation as well as those informations related to the integration of sustainable development aspects in Vietnam.

3 Results

3.1 Sustainable Development and Frameworks for Integrating Sustainable Development Dimensions in the World

3.1.1 Context in the World

The United Nations' Sustainable Development Council was established in 1992, and later the High Level Political Forum (HLPF) in 2012, which is the UN organization responsible for promoting sustainable development progress in the world.

The First Summit was held in 1992 (Rio de Janeiro – Brazil) by adopting the Agenda 21 on Sustainable Development, the Second in 2002 (Johannesburg - South Africa) by launching the Program "Implementation of Johannesburg Action Plan" and continuing to affirm the goals set by the Rio Conference in 1992 based on three pillars - economic, social, and environmental, and the Third in 2012 (Rio de Janeiro – Brazil) by emphasizing the trend of sustainable development related to green economy and the contents of sustainable development implemented in the context of climate change. Furthermore, the Outcome Document of the Rio + 20 Conferences of 2012, The Future We Want, demonstrates a strong commitment to mainstreaming economic, social, and environmental dimensions into national development strategies.

In 2001, the United Nations have adopted eight Millennium Development Goals (MDGs) for the period 2000–2015 and in 2015 presented 17 Sustainable Development Goals (SDGs) and 169 specific targets for the period 2016–2030 (United Nations [UN], 2015). This is an important foundation for every country in the world to develop sustainable development goals appropriate to its own context.

The concept of sustainable development has been increasingly clarified in order to build a more practical tool for implementing and evaluating this process in the world and in each country. Several models have been proposed, such as the concept of sustainable development based on three socio-economic-environmental pillars (Bass, 2015), or assessing the interaction between natural and social systems by considering ecosystem services for the benefit of humans (Millennium Ecosystem Assessment (MEA), 2005), or assessing sustainable livelihoods by considering five livelihood capitals proposed by DFID (1999). Thus, the IUCN used the interlocking circles model to demonstrate that the three objectives need to be better integrated, with action to redress the balance between dimensions of sustainability (Adams, 2006).

The concept of "Planetary boundaries"or the life-supporting systems of the planet is proposed by Rockstorm and other scientists in 2009 (Rockstrom et al.,

2009a, 2009b) in order to define the safe operating space for humanity, as premise for sustainable development. This model with nine planetary boundaries can be used to examine the limits of the Earth in the orientation of the world's development strategy and has been interested by policy makers in the world, including governments, United Nations organizations, and international nongovernmental organizations and development organizations, even several Europe countries have the first attempt to apply this concept for their national development policies (Pisano and Berger 2013).

In this context, the "Doughnut model "aims to combine social dimensions with the planetary boundaries, while also institutionalizing the components of sustainable development, and contributing to the process in the world (Raworth, 2014). This model is comprised of three main components such as domains, indicators, and thresholds and the indicators have been selected to measure the current status of the domains. The Doughnut model, therefore, is considered as a model for sustainable development and a conceptual framework combined of the concerns of environmental sustainability and social justice.

Thus, two these models, "Planetary boundaries" and "Doughnut model"can be used as reference to understand the model of sustainable and also be applied while reviewing the implementation of sustainable development in Vietnam.

3.1.2 Frameworks for Integrating Sustainable Development Dimensions into Development Policies

Nature of Integration

The United Nations' International Conference led by UNDESA / UNEP / UNDP on "Sustainable Development Integration in Environmental Policy and Development" (Bass, 2015) emphasized the nature of integration as a process going from separate activity to cooperative activity (synergy) and finally to sustainability. Theory and practice of integrating sustainable development, a consolidated relationship between environment and development, to serve as the basis for the proposed integration of sustainable development strategies into the world's development policies as well as for the country have been proposed (Bass, 2015).

The agenda for nature and environment goes from the philosophy of "Conservation for Preservation", i.e., the conservation works separated from human, to the principle "Conservation for Human", i.e., the nature integrated with the development. In the opposite direction, the development agenda goes from converting and exploiting nature for development to eventually moving toward a harmonious development with nature (win-win). Thus, the level of integration goes from "no integration" to "high integration".

Level and Approach of Integration

Policy integration is a process, neither just as a result, nor the set of coherent policies, not just the sum of policy coordination and policy coherence. Policy integration is a process of strategic and administrative decision-making to achieve a goal, not only including the design and implementation stages but also requiring integrated governmental actions (Cejudo & Cynthia, 2015). Furthermore, policy integration, is not only horizontal integration between ministries, but also vertical integration along different levels of government and even diagonally across different ministries and other stakeholders (ESDN, 2009) (See Fig. 18.1).

In fact, the concept of policy integration has been applied for different sectors. While studying the vertical and horizontal integration of climate change mitigation and adaptation, Di Gregorio et al. (2017) emphasizes that the Indonesian authorities can carry out effective coordination activities to respond to climate change in the land use sector if the sectoral ministries internalize a strong mandate on internal and external climate policy coherence. Furthermore, studies on integrating Australia's sustainable development policy into fisheries also point out the challenges and difficulties in the implementation process although the policy framework with ecologically sustainable development objectives is proved to be coherent (Farmery, Ogier, Gardner, & Jabour, 2019). Integration of sustainable development is applied not only for sector's policies, but also for social sector such as the higher education. A study (Vargas, Lawthom, Prowse, Randles, & Tzoulas, 2019) has shown that vertical policy formulation and integration at international, national, and institutional level has important role to support effective implementation of sustainable development in the education system.

The Organization for Economic Co-operation and Development (OECD) has developed a Technical Guideline to assess sustainability in national development

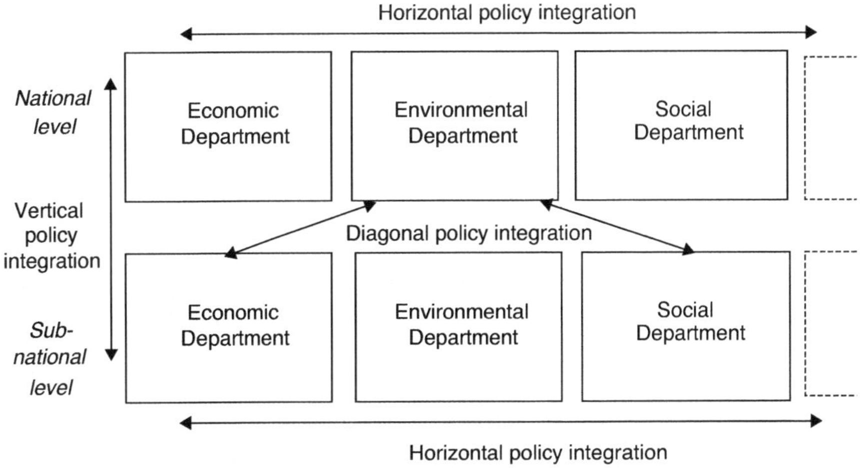

Fig. 18.1 Integrating policies horizontally, vertically, and diagonally (ESDN, 2009, p. 4)

policies as a whole unity, including three dimensions – economic, social, and environmental and applied in sustainability impact assessment (SIA) (OECD, 2010). The Economic and Social Commission for Asia and Pacific (ESCAP) of the United Nations has developed a framework and tools to integrate three economic, social, and environmental dimensions of sustainable development policies into national development policies (UN-ESCAP, 2015). The integration of sustainable development dimensions must be based on the cycle, according to system thinking and while understanding and dealing with trade-offs (UN-ESCAP, 2015, pp. 7–12). Integrated tools including quantitative tools such as development of different scenarios for sustainable development policymaking and quantitative tools such as input-output analysis (UN-ESCAP, 2015, pp. 13–22). This is really good suggestions for other countries to apply integration of sustainable development into national development policies.

Furthermore, the case study of Lake Chad Basin on the interaction between climate change adaptation policy and water resources management also points out the difficulties and challenges in harmonizing the implementation of these policies and shows that "integrated policy thinking" needs to be emphasized in order to achieve sustainable development goals in the short and long term (Okpara, Stringer, & Dougill, 2018). Sector's policy integration has become an approach when forest policy integration and integrated forest management become two constituting elements of integrated forest governance, especially to develop theoretical concepts and analytical dimensions of European policies (Sotirov & Arts, 2018). Under the tendency of more environmentally friendly development, forestry policies in European countries such as France, Germany, the Netherlands, and Sweden have to change in the direction of integrating biodiversity conservation, use of bioenergy, and response of climate change (Sotirov & Storch, 2018). Finally, assessing the practical implementation of 17 sustainable development goals (SDG) in the world, the study has pointed out the limitations of interlinkages and interdependencies between these goals in three areas: across sectors, across societal actors, and between and among low-, medium-, and high-income countries (Stafford-Smith et al., 2017).

3.2 *Initial Efforts to Integrate Sustainable Development into National Development Policy, Strategy, and Plan in Vietnam*

3.2.1 Legal System and Policies to Implement Sustainable Development in Vietnam

Vietnam actively participates in the process of sustainable development in the world, especially in the summits on sustainable development in 1992, 2002, and 2012. Vietnam has also signed many international conventions related to environmental protection and sustainable development.

In 2004, Vietnam promulgated the Strategic Orientation for Sustainable Development in Vietnam (Agenda 21 of Vietnam) (Government of Vietnam, 2004), which emphasized eight principles of sustainable development and 19 priority areas, divided by economic, social, and environmental sectors. The National Sustainable Development Strategy for the period 2011–2020 (Government of Vietnam, 2012a), emphasizes with the overall goal of promoting sustainable, effective growth in line with progress and social justice, conservation of natural resources and environmental protection, and the National Action Plan to implement the 2030 Agenda for Sustainable Development focuses on 17 Sustainable Development Goals (VSDG) up to 2030 and 115 specific targets (Government of Vietnam, 2017).

Vietnam's legal system, especially specialized laws, can also be divided according to economic pillars (such as laws on land, energy, urban, natural resource tax), social (healthcare, education), and environmental (environmental protection, forestry, water, minerals, biodiversity, seas, marine and island, and disaster prevention). This division is also very convenient for promoting activities related to the pillars of sustainable development and therefore, several laws are still in the process of adjusting to adapt to changing reality.

Strategic Orientation for Sustainable Development in Vietnam (Government of Vietnam, 2004) is the foundation for the development and implementation of economic strategies (such as strategies on, green growth, cleaner production, forestry, irrigation), social (such as strategies related to poverty reduction, education, healthcare), and natural resources and environment (such as strategies for environmental protection, forest and water resources, biodiversity, disaster prevention, and response with climate change, and waste management) and is specified in the Socio-Economic Development Strategy 2001–2010 and 2011–2020, and especially in the National Sustainable Development Strategy for period 2011–2020. Furthermore, the priority areas of the Sustainable Development Strategy for the period 2011–2020 are compatible with the 17 UN Sustainable Development Goals (SDGs) and the 17 Development Goals of Vietnam (VSDGs).

Thus, the approach to integrate three sustainable pillars (economic, social, and environmental) into Vietnam's development strategy and policies is relatively systematic and comprehensive.

3.2.2 Institutions and Arrangement to Implement National Sustainable Development Strategy and Sustainable Development Goals (SDGs)

A national organizational structure for sustainable development was formed and developed with the establishment of a National Sustainable Development Council (2005), then renamed National Council on Sustainable Development and Competitiveness Enhancement (2012) which consists of four specialized committees on economically sustainable development, socially sustainable development, sustainable development of natural resources and environment, and on the decade of education for sustainable development (Government of Vietnam, 2012a). The Ministry of Planning and Investment, namely its Sustainable Development Office,

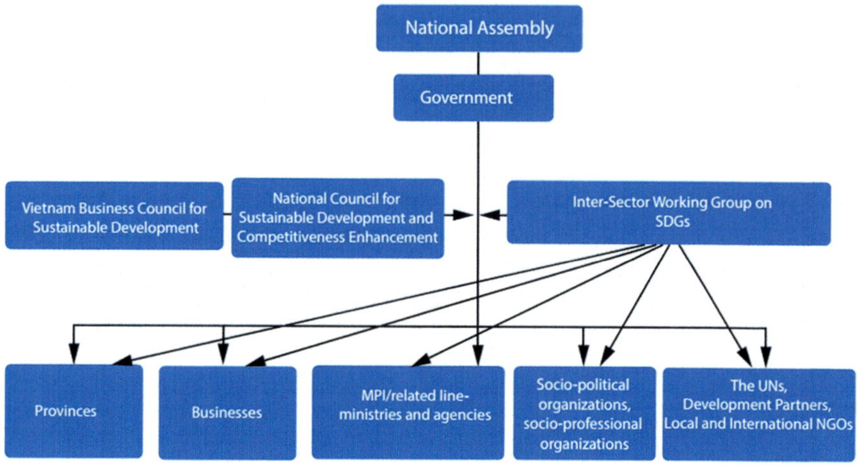

Fig. 18.2 Institutional arrangement for SDG implementation in Vietnam (MPI, 2018)

plays a very important role in developing and coordinating the implementation of the national sustainable development strategy, socio-economic development strategies, and national action plan to implement SDGs in the country. The Institutional arrangement for SDG implementation in Vietnam is demonstrated in Fig. 18.2.

At the sectoral and local level, a Council for Sustainable Development has also been established in several ministries, sectors, and a number of provinces and cities in the country (MPI, 2015, p. 16). At the enterprise level, the Vietnam Business Council for Sustainable Development was established in 2010 by the Vietnam Chamber of Commerce and Industry (VCCI) to develop a sustainable, dynamic, and successful business community, contributing to the national sustainable development course. Based on their functions, ministries or sectors will be responsible for developing, synthesizing, and implementing national sustainable development indicators, as indicated in the Sustainable Development Strategy for the period of 2011–2020 and National Action Plan to implement Agenda 2030 and 17 SDGs. Organizationally, line ministries coordinate with each other through national, ministerial, and local Councils or Steering Committees. In terms of content, the policies of the ministries on sectoral development or local policies on socio-economic development will integrate the national sustainable development goals and are specified in the implementation of the their particular targets. Regarding the planning, the 5-year and annual plans are the basis for ministries and local authorities to realize their goals and targets.

Moreover, the process of integrating sustainable development into the development policy is carried out in the form of vertical and horizontal coordination (See Table 18.1). Vertical coordination is largely about sectoral coordination and management from the central to provincial and district levels. Horizontal coordination is often realized within the same level of government, as between central ministries, or

Table 18.1 The coordination and cooperation between and among the government/authority to implement sustainable development in Vietnam

Administrative level	Agency responsible	Agency responsible	Agency responsible	Coordination and cooperation
Government at central level	Ministry A	Ministry B	Ministry X	Ministries are coordinated by government (prime minister) and ministries have cooperation among them by their main functions and policies concerned
Local authority at provincial level	Department B	Department B	Department X	Departments are coordinated by provincial people' committee (chairman) and departments have cooperation among them by their main functions and policies/tasks concerned
Local authority at district level	Division A	Division B	Division X	Divisions are coordinated by district people' committee (chairman) and divisions have cooperation among them by their main functions and tasks concerned
Local authority at commune level	Unit A	Unit B	Unit X	Units are coordinated by commune people' committee (chairman) and units have cooperation among them by their main functions and tasks concerned
Activities of coordination and cooperation	Sectoral coordination	Sectoral coordination	Sectoral coordination	Sectoral coordination (vertical) and administrative/territory coordination (horizontal)

within provincial and district authorities (i.e., people's committee). At the central governmental level, the Prime Minister coordinates ministries and sectors within the government to achieve the national goal of sustainable development. At the provincial and city governmental level, the Chairman of the People's Committee of the province and city will coordinate relevant departments and agencies to implement the socio-economic development plan and the development of local sectors. This type of coordination is also applied to lower levels, such as for district's and commune's level. Government agencies (ministries, sectors), or local authority agencies at the same level maintain different forms of cooperation to ensure the performance of assigned tasks.

With such an organizational structure, the implementation of the sustainable development process is carried out effectively.

3.2.3 Main Achievements of Sustainable Development Integration into Development Policy in Vietnam

Some Characteristics of Integrating Sustainable Development into National Development Policy

In general, the integration of sustainable development into national development policy is the initial efforts. The process of this policy integration, perhaps, is quite far from achieving full meaning of "policy integration" discussed by Cejudo and Cynthia (2015), in which policy integration is a process, neither just as a result, nor the set of coherent policies, not just the sum of policy coordination and policy coherence.

At present, although there are no formal and institutionalized guidelines on sustainable development integration into socio-economic development policies, the implications for sustainable development are considered and integrated into national and local development policies during the process of policy drafting, approval, and implementation.

1. Integration at the level of the objectives and priority areas of sustainable development.

The integration of sustainable development into the development strategy and policies is clearly demonstrated in integrating objectives and priority areas of sustainable development into the national Socio-Economic Development Strategy for the period 2011–2020 and the implementation of main objectives are measured by the sustainable development indicators (See Table 18.2). Thus, the Green GDP indicator measures the target of rapid and sustainable growth, the Human Development Index HDI - the goal of social development and the Environmental Sustainability Index - the environmental goal.

Table 18.2 The relations between objectives of SEDS 2011–2020 with three dimensions of sustainable development

STT	Objectives by sustainable dimensions	Objective in SEDS 2011–2020	Indicators for monitoring and evaluating sustainable development in NSDS 2011–2020
1	Economic objectives	Rapid and sustainable growth, with GDP growth rate 7–8%/year	Green GDP (methods of calculation still in the process of test)
2	Social objective	Human development index (HDI) achieved at higher average group of the world	Human development index (HDI)
3	Environmental objective	Improvement of environmental quality (Forest, pollution situation, pollution treatment, climate change)	Environmentally sustainable index (methods of calculation still in the process of test)

Source: Communist Party of Vietnam, 2011; Government of Vietnam, 2012a

Furthermore, in the process of implementing sustainable development, perspectives, principles, and objectives of sustainable development are specified and integrated into the strategy, planning, and plans for national and local socio-economic development as well as sectors development policies for the ministries, that link to specific economic, social, and environmental indicators.

This circular 02/2013/TT-BKHDT (MPI, 2013) spent the entire chap. III to guide how to integrate contents of sustainable development into the strategy, planning, and plans of economic and social development of the country, sectors, and ministries while emphasizing *integrating principles* (such as integrating perspectives, objectives, and orientations for sustainable development) and *integrating requirements* (such as strategies, planning, and plans under current regulations). Furthermore, guidelines to conduct the 5-year Socio-economic development period 2016–2020 for sectors and localities (MPI, 2014) have emphasized aspects to ensuring sustainable development, whereas economic development must link closely to the cultural and social development and environmental protection, and pro-active response to climate change and ensuring national defence and security. A study on sustainable impact assessment in planning also proposed a method to formulate socio-economic development strategies, plans, and plans by integrating sustainable development (economic, social, and environmental dimensions) and climate change aspects in the policy making process and these may be the first technical proposals to fully integrate sustainable development into national development policies (MPI, 2011).

Finally, a practical example is that Vietnam has well implemented in-depth integration of the MDGs into important national plans, strategies, and policies (Comprehensive Poverty Reduction and Growth Strategy, Development Strategy Socio-Economic Development 2011–2020), developing specific action plans (5-year and annual socio-economic development plans), and achieved remarkable results (Vietnam Socialist Republic, 2015).

2. Integration of components and dimensions of sustainable development.

   (a) Integration of environment into development policies.

   Strategic environmental assessment (SEA) is one of the important tools aimed at integrating environmental issues into policies, plans, programs, and assessing their relationships with economic and social issues. The Law on Environmental Protection (Vietnam National Assembly, 2014) emphasizes that the SEA is the analysis and forecast of the environmental impacts of a strategic project, planning, and development plan before approval, to ensure sustainable development goals and it becomes important and effective tool for development of strategies, plans, and policies in Vietnam.

   Moreover, the Ministry of Planning and Investment (MPI, 2011) summarized and developed "Guidelines to Implement the Strategic Environmental Assessment (SEA) in setting up strategy, planning and socio-economic development plan", including additional guidance on the integration of climate change issues into strategic environmental assessment.

   (b) Integration of climate change into development policies.

Government's political commitment to integrate climate change into development policy is presented clearly in the National Target Program to Respond to Climate Change, lasting for the period of 2009–2015 and years after that (MONRE, 2008), where all new policies and strategies must consider and integrate climate change in their activities, especially in the formulation and implementation of socio-economic plans of sectors and localities toward sustainability. A number of ministries, such as the Ministry of Agriculture and Rural Development (2011), the Ministry of Natural Resources and Environment (2010), the Ministry of Industry and Trade (2010), and the Ministry of Transport (2011) have developed action plans. Responding to climate change, including emphasizing the consideration and integration of climate change aspects in its development policies (Vo et al., 2015). The Ministry of Natural Resources and Environment has also developed and proposed technical documents to integrate climate change in socio-economic development plans (IMHEN, 2012; Tran, Huynh, & Dao, 2012), but so far, these guidelines have not been institutionalized in planning work.

As an agency of the Government responsible for guiding the formulation and implementation of the country's socio-economic development plans in the short, medium, and long term, the Ministry of Planning and Investment has issued a framework to guide the selection of priority investment projects to adapt to climate change under Decision No. 1485 / QD-BKHDT in 2013 (MPI, World Bank and UNDP, 2015). At the macro level, the Environmental Law (National Assembly of Vietnam, 2014) also has a chapter dealing with climate change, which emphasized response activities integrated into strategies, planning, and social economic development plans. A specific step in the effort to integrate climate issues is that the Ministry of Planning and Investment has issued a circular (No. 05/2016 / TT-BKHDT) guiding the integration of natural disaster prevention and control into socio-economic development planning and plan.

(c) Integration of green growth into socio-economic development plan.

Vietnam's Green Growth Strategy (Decision 1393 / QD-TTg, 25/09/2012) is actually promoting sustainable economic development in the context of climate change by greening production and consumption, and applying modern technology in order to efficiently use natural resources and reduce the intensity of greenhouse gas emissions (Government of Vietnam, 2012b). Furthermore, the National Action Plan for Green Growth period 2014–2020 (Decision 403 / QD-TTg, dated 03.20.2014) specifies the content of sustainable economic development by proposing and implementing nation-wide large-scale programs and projects (Government of Vietnam, 2014).

Ministry of Planning and Investment developed and proposed the methodology for calculating the Marginal Abatement Cost Curve Analysis for the energy sector (MPI and UNDP, 2013a) and the land use, land use change, and forestry sector (MPI and UNDP, 2013b) as a basis for building green growth plan at provincial level. This important tool has been revived from the project "Strengthening Sustainable Development and Climate Change" and used to build the different emission scenarios for socio-economic development and is applied to complete the Provincial Green Growth Action Plans, such as province Bac Ninh (MPI and KOICA, 2015), Quang Nam (MPI and KOICA, 2014a), and Ben Tre (MPI and KOICA, 2014b).

3. Integration of Millennium Development Goals and Sustainable Development Goals into Vietnam's development policies.

   (d) Integration of Millennium Development Goals.

   In the end of MDG implementation period, Vietnam had completed 3/8 targets on poverty reduction, primary education universalization, and gender equality and obtained remarkable results for several goals (MPI, 2018). One of the important lessons learned from Vietnam's success, which is widely recognized by the international community, is that Vietnam has chosen the right ways to "nationalize" MDG, in which, the integration of MDG into national and local socio-economic development strategies and plans is considered an innovative step. Experience of MDG implementation in Vietnam also shows that the 5-year and annual socio-economic development plans are the best mechanism to allow the integration of development goals with resources to realize them.

   (e) Integration of Sustainable Development Goals.

   Successful lessons learned in integrating the Millennium Development Goals are still valid for the implementation of the Sustainable Development Goals on the 2030 Agenda. Specifically, Vietnam has nationalized the global 2030 agenda to set up a National Action Plan to implement the 2030 Agenda with 17 sustainable development goals and 115 specific goals consistent with the country's development conditions and priorities. The National Action Plan sets out the task of "Integrating sustainable development goals in the development of annual Socio-Economic Development Plans and development strategies, policies, planning of ministries, sectors, and local authorities. By 2020, fully integrate Vietnam's sustainable development goals into the content of the Socio-Economic Development Strategy for the period of 2021-2030, the Socio-Economic Development Plan for the period of 2021-2025, and the Development Plans of sectors and localities in the 2021-2030 period" (Government of Vietnam, 2017).

   In fact, the perspective of sustainable development is integrated throughout the Socio-Economic Development Strategy 2011–2020 and the Socio-Economic Development Plan 2016–2020. Many SDGs have been integrated into the national development policy system, from laws, strategies, and socio-economic development plans, action plans of ministries, sectors, and localities. The SDGs will continue to be considered for inclusion in the Socio-Economic Development Strategy 2021–2030, the Socio-Economic Development Plan 2021–2025 as well as the Annual Socio-Economic Development Plan of Vietnam in next time.

   Although the Government has only promulgated National Action Plan in 2017 to implement the 2030 Agenda for Sustainable Development, the Vietnam's

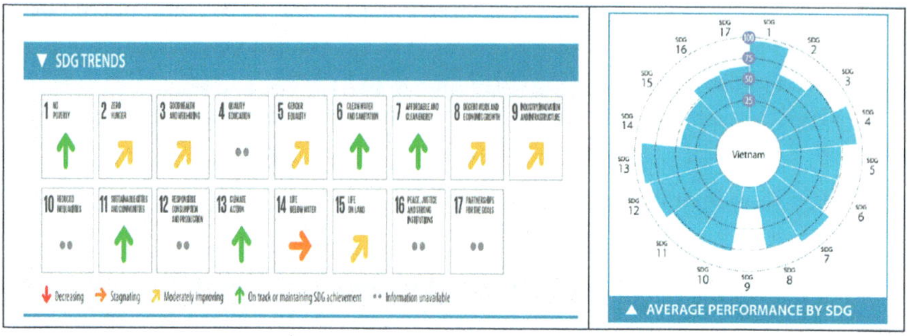

Fig. 18.3 Effectiveness in implementation of Vietnam's SDG Sustainable Development Goals in 2019 (Source: United Nations [UN], 2019, p. 458)

National Voluntary Review on the implementation of the sustainable development goals in 2018 (MPI, 2018) has confirmed five goals with the achieved good results, including SDG 1 (Eradicate poverty), SDG 6 (Clean water and sanitation), SDG 7 (Clean energy and reasonable price), SDG 11 (City and sustainable communities), and SDG 13 (Climate protection action) (See Fig. 18.3). Several other goals also achieved encouraging results such as SDG 1, SDG 2, SDG 5, SDG 8, SDG 9, and SDG 15. The achieved results confirm that these goals have been earlier integrated into the strategy and development plan in order to maintain such positive progress (United Nations [UN], 2019).

Finally, at the request of the Government expressed in the Prime Minister's Directive on sustainable development (No. 13 / CT-TTg, 20/5/2019), the Ministry of Planning and Investment has a responsibility to develop and issue guidelines for integrating sustainable development goals into the Socio-Economic Development Strategy for period 2021–2030 and guidelines for monitoring and evaluating sustainable development goals (SGD) during year of 2019. This demonstrates the urgent need of integrating sustainable development and sustainable development goals into socio-economic development planning and plans in Vietnam.

4 Conclusion and Discussion

Sustainable development in the world is a long-term process, but the important milestones are the United Nations' Summits organized since 1992 up to now and by that time, the concept and contents of sustainable development have been specified, especially the model based on economic, social, and environmental pillars. Vietnam has made great efforts in the socio-economic development of the country toward sustainability with the development and improvement of the institutional system and related policies, especially the strategy of sustainable development, green growth in the context of climate change, and action plan to implement the 2030 agenda.

The Government of Vietnam has obtained initial results in integrating sustainable development and its dimensions into the country's development policies, through the rational integration of goals, principles, and priority areas of sustainable development into Vietnam's entire development policy system, especially socio-economic development strategies and 5-year socio-economic development plans during the period 2000–2020. Furthermore, Vietnam has good success in integrating the Millennium Development Goals (MDGs) for the period 2000–2015 into economic and social policies. Vietnam gives high commitment to implement United Nations' Agenda 2030 by issuing 17 sustainable development goals (SDGs) and 115 targets to implement up to the year of 2030.

However, as the integration is a complex and broad process and there are not many good practices in the world to be consulted, Vietnam is currently facing the challenges and difficulties, as follows:

1. Shortage of technical guidance to integrate sustainable development as a whole in the development policies of Vietnam. In practice, at present, there is no overall comprehensive technical guidelines helping integrating sustainable development into the development policies of Vietnam. Several policies, such as Circular No 02 (MPI, 2013) are recommended only to integrate the principles of sustainable development identified in NSDS into the development policies, i.e., strategy-planning plan but have not yet clarified how to do this integration technically. The policies related to sectoral development, such as Law on Environmental Protection (Vietnam National Assembly, 2014) or Vietnam's Green Growth Strategy (Government of Vietnam, 2012b) or National Target Program to Respond to Climate Change (MONRE, 2008) have only given the principles of integration, but not specific approach and methods to do that. Several technical guidelines of integration (green growth, climate change) have been developed, but not institutionalized officially.
2. Deficient resources to fully integrate sustainable development into development policies. Although the government has great efforts to mobilize all resources possible to build development policies taking into account components and dimensions of sustainable development, such as human resources, technical support, financial resource, and infrastructure but these resources are not adequate to fully meet practical requirements. At the national level as well as at the local level, the shortage of financial resources or lack of financial line or financial source allocated to intersectoral coordinating works in development policy formulation and implementation, may limit the effectiveness and efficiency of integrating objectives and priority actions into development policy system of nation, sectors, ministries, and localities. Many offices for the sustainable development in the provinces do not have specialized staffs responsible for coordinating works, including support and guidance to integrate sustainable development in the policy-making process of local socio-economic development plans. Moreover, sometimes these staffs do not have sufficient knowledge and skills to

perform complex tasks and difficulties in integrating the objectives and dimensions of sustainable development in setting up provincial 5-year or annual socio-economic development plans.

Therefore, a number of recommendations to promote the integration of sustainable development in general, and sustainable development goals (SDGs) in particular, into development policies in Vietnam have also been proposed for implementation as follows:

- Sustainable development must be integrated in the development planning steps to ensure the sustainability of development policies.
- Improving mechanisms and resources suitable for the integration of sustainable development at central and local levels, and relevant ministries.
- Implementation of the Sustainable Development Strategy for the period 2011–2020 and the national and local sustainable development monitoring and evaluation indicators must be properly integrated with the implementation of the United Nations' Sustainable Development Goals (SDGs).
- Continue to consolidate and develop international cooperation in order to use technical, human, and financial support in promoting sustainable development in Vietnam in the context of climate change.

References

Adams, W. M. (2006). *The future of sustainability: Re-thinking environment and development in the twenty-first century*. Report of the IUCN Renowned Thinkers Meeting, 29–31 January 2006: 18 p.

Bass, S. (2015). Conceptual frameworks for integrating sustainable development dimensions paper for UNDESA/UNEP/UNDP workshop on SD integration tools, Geneva, 14–15 October 2015.

Cejudo, G. M., & Cynthia, M. (2015). *Addressing fragmented government action: Coordination, coherence, and integration*. Paper to be presented at the 2nd International Conference in Public Policy, Milan, July 2015. 22 pages.

Communist Party of Vietnam. (2011). *Socio-economic development strategy for the period of 2011–2020*. Communist Party of Vietnam. 26 pages, (in Vietnamese).

Di Gregorio, M., Nurrochmat, D. R., Paavola, J., Sari, I. M., Fatorelli, L., Pramova, E., Locatelli, B., Brockhaus, M., & Kusumadewi, S. D. (2017). Climate policy integration in the land use sector: Mitigation, adaptation and sustainable development linkages. *Environmental Science & Policy, 67*, 35–43.

ESDN. (2009). *Horizontal policy integration and sustainable development: Conceptual remarks and governance examples*. ESDN Quarterly Report.

Farmery, A. K., Ogier, E., Gardner, C., & Jabour, J. (2019). Incorporating ecologically sustainable development policy goals within fisheries management: An assessment of integration and coherence in an Australian context. *Journal of Environmental Management, 249*, 109230. https://doi.org/10.1016/j.jenvman.2019.07.001

Government of Vietnam (2004). *Strategic orientation for sustainable development in Vietnam* (Decision No 153/2004/QĐ-TTg, 17/08/2004), (In Vietnamese).

Government of Vietnam. (2012a). *Vietnam's Sustainable Development Strategy for the period 2011–2020* (Decision No 432/QĐ-TTg, 12/04/2012), (In Vietnamese).

Government of Vietnam. (2012b). *National strategy for green growth* (Decision 1393/QĐ-TTg, 25/09/2012), (In Vietnamese).

Government of Vietnam. (2014). *National Action Plan on Green Growth for the period 2014–2020* (Decision No 403/QĐ-TTg, 20/03/2014), (In Vietnamese).

Government of Vietnam. (2017). *National action plan to implement the 2030 agenda for sustainable development* (Decision No 622/QĐ-TTg, 10/05/2017), (In Vietnamese).

IMHEN. (2012). *Integrating climate change issues into socio-economic development plans* (Institute of Meteorology, hydrology and environment). Environment and Mapping Publishing House. 137 p. (in Vietnamese).

Millennium Ecosystem Assessment (MEA). (2005). *Ecosystems and human Well-being: Synthesis*. Island Press.

MONRE. (2008). *National target program to respond to climate change*. MONRE. 34 p. (In Vietnamese).

MPI. (2011). Guidelines to implement Strategic Environmental Impact Assessment (SEA) in preparing Socio-Economic Development Strategies, Planning and Plans. Vietnam - Denmark environment development cooperation program, component of the MPI' capacity building and environmental planning. Hanoi, 11/2011: 223 p. (in Vietnamese).

MPI. (2013). *Circular Guidelines to implement some contents of the Vietnam Sustainable Development Strategy for the period 2011–2020* (Circular No 02/2013/TT-BKHĐT, 27/03/2013), (In Vietnamese).

MPI. (2014). *Guidelines to develop the 5-year Socio-Economic Development Plan for period 2016–2020* (Decision No 5316/BKHĐT-TH, 15/08/2014), (in Vietnamese).

MPI. (2015). *Assessment of the 10-year progress in the implementation of sustainable development in Vietnam* (Social and natural resources – environmental components). Final report. Sustainable Development Office. Hanoi, 12/2015: 150 p. (in Vietnamese).

MPI. (2018). *Vietnam's Voluntary National Review on the Implementation of the Sustainable Development Goals*. Ministry of Planning and Investment, Hanoi. 6/2018.

MPI and KOICA. (2014a). *Development of a Marginal Abatement Cost Curve to Reduce Greenhouse Emissions for Quang Nam Province*. Project: Supporting the implementation of green growth strategy in Vietnam. Hanoi, 12/2014: 85 p. (In Vietnamese).

MPI and KOICA. (2014b). *Development of a Marginal Abatement Cost Curve to Reduce Greenhouse Emissions for Ben Tre Province*. Project: Supporting the implementation of green growth strategy in Vietnam. Hanoi, 12/2014: 83 p. (in Vietnamese).

MPI and KOICA. (2015). Development of a Marginal Abatement Cost Curve to Reduce Greenhouse Emissions for Bac Ninh Province. Project: Supporting the implementation of green growth strategy in Vietnam. Hanoi, 03/2015: 89 p. (in Vietnamese).

MPI and UNDP. (2013a). A Marginal Abatement Cost Curve Analysis for Energy Sector in Vietnam. Support formulation and Implementation of the National Strategies for Green Growth. Project "Strengthening Sustainable Development and Climate Change". Vietnam Publishing House of Natural Resources, Environmental and Cartography. 71 pages.

MPI and UNDP. (2013b). The Development of a Marginal Abatement Cost Curve Analysis for Land Use, Land Use Change and Forestry in Vietnam. Support formulation and Implementation of the National Strategies for Green Growth. Project "Strengthening Sustainable Development and Climate Change". Vietnam Publishing House of Natural Resources, Environmental and Cartography. 50 pages.

MPI, World Bank and UNDP (2015). Financing Vietnam's Response to Climate Change: Smart Investment for a Sustainable Future. Laying the foundation for Resilient Low-carbon Development through the Climate Public Expenditure and Investment Review (SPEIR). MPI, World Bank and UNDP: 135 p.

National Assembly of Vietnam. (2014). *Environmental protection law*, (No 55/2014/QH13, 23/06/2014), (in Vietnamese).

OECD (2010). *Guidance on sustainability impact assessment*. Organisation for Economic Co-operation and Development.

Okpara, U. T., Stringer, L. C., & Dougill, A. J. (2018). Integrating climate adaptation, water governance and conflict management policies in lake riparian zones: Insights from African drylands. *Environmental Science and Policy, 79*, 36–44.

Pisano, U., & Gerald, B. (2013). Planetary Boundaries for Sustainable Development: From a conceptual perspective to national applications. ESDN Quarterly Report 30 – October 2013. ESDN Quarterly Report N.30. European Sustainable Development Network. 31 pages. http://www.sdnetwork.eu/quarterly%20reports/report%20files/pdf/2013-October-Planetary_Boundaries_for_SD.pdf

Raworth K. (2014). From Will These Sustainable Development Goals Get Us into the Doughnut (Aka a Safe and Just Space for Humanity)? Duncan Green's Discussion on Raworth's Doughnut and SDGs.

Rockstrom, J., et al. (2009a). A safe operating space for humanity. *Nature, 461*(7263), 472–475.

Rockstrom, J., et al. (2009b). Planetary boundaries: Exploring the safe operating space for humanity. *Ecology and Society, 14*(2), 32.

Sotirov, M., & Arts, B. (2018). Integrated Forest governance in Europe: An introduction to the special issue on forest policy integration and integrated forest management. *Land Use Policy, 79*, 960–967.

Sotirov, M., & Storch, S. (2018). Resilience through policy integration in Europe? Domestic forest policy changes as response to absorb pressure to integrate biodiversity conservation, bioenergy use and climate protection in France, Germany, the Netherlands and Sweden. *Land Use Policy, 79*, 977–989.

Stafford-Smith, M., Griggs, D., Gaffney, O., Ullah, F., Reyers, B., Kanie, N., Stigson, B., Shrivastava, P., Leach, M., & O'Connell, D. (2017). Integration: The key to implementing the sustainable development goals. *Sustainability Science, 12*, 911–919. https://doi.org/10.1007/s11625-016-0383-3

Tran, T., Huynh, T. L. H., & Dao, M. T. (2012). *Technical guidance on integrating climate change issues into development plans* (Institute of Meteorology, hydrology and environment (IMHEN)). Environment and Mapping Publishing House. 69 p, (in Vietnamese).

UN-ESCAP. (2015). Integrating the three dimensions of sustainable development: A framework and tools. Greening of Economic Growth Series. United Nations Economic and Social Commission for Asia and the Pacific: 31 p.

United Nations [UN]. (2015). *Global sustainable development report*. 198 p.

United Nations [UN]. (2019). *Sustainable Development Report 2019: Transformations to achieve the SDGs*.

Vargas, V. R., Lawthom, R., Prowse, A., Randles, S., & Tzoulas, K. (2019). Implications of vertical policy integration for sustainable development implementation in higher education institutions. *Journal of Cleaner Production, 235*, 733–740.

Vietnam National Assembly. (2014). *Environmental Protection law No 55/2014/QH13*.

Vietnam Socialist Republic. (2015). *15 years achieving the Vietnam millennium development goals*. Country Report.

Vo, T.S. (2016). Status and options of sustainable development integration into National Development Strategy and policies in Vietnam. Final report for cooperation project between UNDESA/UNDP-MPI "strengthening the capacity of selected developing countries to mainstream sustainable development into National Development Strategies", session on "studying Vietnam's sustainable development integration framework": 53 pages.

Vo, T. S., Nguyen, C. H., Tran, H. N., Bui, C. Q., Nguyen, D. S., Le, V. T., Hoang, V. T., Le, A. T., & Nghiem, P. T. (2015). Chapter 8: Towards sustainable and resilient future. In T. Thục, K. Neefjes, T. T. T. Huong, N. Van Thang, M. T. Nhuan, L. Q. Tri, L. D. Thanh, H. T. L. Huong, V. T. Son, N. T. H. Thuan, & L. N. Tuong (Eds.), *Vietnam special report on managing the risks of extreme events and disasters to advance climate change adaptation* (pp. 305–346). IMHEN and UNDP. Natural Resources and Environment Publishing House.

Chapter 19
Building Human Capital for Sustainable Development: Experience from Some East Asian Countries and Policy Implications for Vietnam

Trong Nguyen Thanh and Ly Pham Thi

Abstract The main objective of this study is to determine the role of human capital and experience in building human capital for sustainable development in some East Asian countries, and providing some policy implications for Vietnam. The paper presents a number of theories and empirical studies on the relationship between human capital and sustainable development to clearly see the role of human capital and sustainable development both theoretically and practically. The study conducted a bivariate correlation analysis, based on Spearman coefficients between Sustainable Development Goals (SDG), Human Development Index (HDI), and Human Capital Index (HCI) to determine the relationship between sustainable development and human capital in East Asia and Southeast Asia.

Keywords Human capital · Sustainable development · Education · East Asia · Vietnam

1 Introduction

The paper reviews theories on the relationship between human capital and sustainable development and empirical studies in three East Asian countries (Singapore, South Korea, and Japan) and in four Southeast Asian countries (Thailand, Malaysia, Indonesia, and the Philippines) to find out if human capital

T. N. Thanh (✉)
University of Economics and Law, Ho Chi Minh City, Vietnam
e-mail: trongnt@uel.edu.vn

L. P. Thi
University of Economics, Ho Chi Minh City, Vietnam

A. T. Nguyen, L. Hens (eds.), *Global Changes and Sustainable Development in Asian Emerging Market Economies Vol. 2*, https://doi.org/10.1007/978-3-030-81443-4_19

is an important factor for the differences in the level of development of these two groups of countries or is not?

Studying experiences of some East Asian countries, the authors found that human capital is considered as a key factor to help resource-poor countries such as Korea and Singapore to quickly convert from the poor countries in the past into the today's developed economy, while other resource-rich countries in Southeast Asia are still among the developing countries. The quality of the human capital is central to promoting and maintaining innovation as well as applying appropriate technologies for sustainable development. Capital and natural resources are passive factors in the production process, people are active agents to accumulate capital; exploitation and conservation of natural resources; building social, economic, and political organizations, etc. If a country cannot build human capital, develop citizen's skills and knowledge, and promote this capital effectively, it will be difficult for that country to develop anything other. Therefore, for sustainable economic development, investing in human resource development should be considered as a top priority in Vietnam.

2 Literature Review

Although the idea of investing in education to improve economic and social efficiency was invented early by economist, it was not until the 1960s that the concept of human capital was proposed fully and systematically accessed. The term "human capital" reflects the value of human capacity. Human capital can be invested through education and training to enhance human skills and knowledge to improve and enhance labor productivity. OECD (2001) introduces a broader definition of human capital as "knowledge, skills, competencies and personal attributes that can facilitate the creation of personal benefits, social and economic". In other words, human capital encompasses the capabilities and abilities of individuals, groups, and societies that enable them to carry out the responsibilities assigned to achieve predetermined goals. Goldin (2014) argues that human capital is "skill that workers own and is considered a resource or an asset". The quality of a human capital is central to promoting and sustaining innovation as well as the application of appropriate technology for sustainable development. Human resources creates the ultimate basis for the wealth of nations. Develop citizens' knowledge, skills, and ability to manage available resources. It is an irreversible fact that human capital is the most valuable asset of any nation. This emphasizes the requirement to build the human capital necessary for sustainable development. The concept of "human capital" became popular in economic theory as a result of recognizing that physical capital alone is not sufficient to bring about long-term development. In fact, many economists later explained that human capital is the most important factor that creates development in the sense that with minimal physical capital, human capital can stimulate or make progress significantly.

The theory of human capital emphasizes the role of education to build human capital in order to increase the productivity and efficiency of workers (Olaniyan and Okemakinde, 2008). In other words, better educated people will be more productive and creative, access, and accelerate technological progress faster. Lall & Kramer-Mbula (2005) point out that developing countries cannot simply import technologies without investing in building "capacity" to learn how to master, acquire, adapt, and improve their existing technology. Pidlisnyuk (2010) also argue that strengthening the role of education helps build the necessary human capital that not only promotes economic growth but is also the best way to raise awareness and skills of employees in order to minimize the harmful effects of society on the environment and protect and preserve the world for future generations.

Sianesi and Van Reenen (2000) note that higher levels of education can correlate not only with a friendly environment, but also with better political and community significance, higher social cohesion, and lower crime rate. Regarding lower crime rates, according to Freeman (2000) there is no clear direct relationship between education level and crime rate. Meanwhile, Kelly (2000) argues that there is an indirect connection between the two variables: while different levels of education lead to income disparities, these inequalities increase crime rates. Green et al. (2003) suggested that educational inequality leads to income inequality, and this affects social outcomes such as crime and health. Therefore, ensuring education equality through access to education will help affect income equality and indirectly affect social cohesion. According to Schuller (2001), social cohesion is a consequence of social capital production. The presence or absence of social capital depends on the level of education. This is argued by Dwiningrum (2013) through the fact that factors related to social capital (institutions, relationships, and values such as trust, rules of conduct) can be had in formal and informal education process. Families, schools, and communities play a very important role in forming a personal personality that will largely affect social capital.

Diaconu & Popescu (2016) focused on studying the relationship between human capital and sustainable development in the 10 countries with the highest sustainable human development index (HSDI) in the European Union in 2010 and 2014. The study concludes that human capital is an important factor for sustainable development in these 10 European countries. The experience of building human capital in East Asian countries shows that investing in education to develop human capital is the main driving force of East Asian economy development; Investment in human resources contributes to high economic growth and also helps in better income distribution in East Asia. Rapid economic growth in Japan is due to early human capital developments; similar to the case of Taiwan, and many other economies. Investment in education has also been found to contribute significantly to the equitable income distribution in East Asia. Moreover, the relatively high social indicators such as life expectancy and quality of life are partly due to education priorities in the national development strategy in general and the investment strategy for education in particular. Some East Asian countries do not have natural resources but have increasingly high-quality human resources thanks to education, training, and technological and technical research, so they have succeeded in economic development.

In summary, theories and empirical studies in several countries have confirmed that human capital is important for growth and sustainable development. So a country with higher human capital will have a higher growth rate. Higher levels of education can correlate not only with a friendly environment, but also for a better political and community sense, higher social cohesion, and lower crime rates.

3 Methodology

This study is conducted through a profile of empirical studies of previous studies in different countries to synthesize and collate whether there is a difference or similarity in the role of human capital in the developed countries to make initial assessments and assessments on the role of human capital.

At the same time, in order to determine the relationship between sustainable development and human capital in some East Asian countries, several composite indicators must be analyzed. First of all, in this study we look at the Sustainable Development Goals index, introduced by Bertelsmann Stiftung in 2015 and since 2016 published annually in collaboration with the Sustainable Development Solutions Network led by Professor Jeffrey Sachs, special advisor to the United Nations Secretary General. Second, we focused on Human Develop Index, which is a measure of average achievement in key aspects of human development: a long and healthy life, understanding and a good standard of living. This index has been reviewed to see if there are any changes between countries, from a development standpoint, when sustainability issues are taken into account. Third, we analyze the Human Capital Index, which has been published by the World Economic Forum (WEF) since 2012 in the Global Human Capital Report. In October 2018, the World Bank (WB) also released the Human Capital Index (HCI) as a measure of national economic success. The index ranks countries by the level of investment in education and health care for young people. The WEF HCI includes four main pillars: health, education, workforce and employment, and favorable environment. The HCI of the World Bank mainly focuses on the education (skills and capacity) of human capital but does not consider the health aspect. Compared to the WEF HCI, the WB HCI has a higher future (potential future human resources) while the WEF looks at the current situation of the labor force (unemployment, complex economy, availability of employees, skilled workers) reflects the current global human resources situation. In this article, we use the HCI index of the WEF published in 2016 and the HCI of the WB published in 2018.

To see if there is a relationship between the ranking positions of countries according to SDG, HDI, and HCI, we conducted a bivariate correlation analysis based on

Spearman's coefficient on SPSS program. Assessing the correlation of two grades of two variables (rank-ordered variables), used when the distribution of the population is assumed to be not the normal distribution or in the case of abnormal observation values (too large or too small). So with the data of seven countries in East Asia and Southeast Asia, the data difference between countries is very large, so we use the correlation according to Spearman. The correlation coefficient (r) is a statistical index measuring the correlation relationship between the two variables, and specifically here we will analyze the correlation coefficient for each pair of indicators: SDG - HDI, SDG - HCI, HDI - HCI (we call this variable x and y). A correlation coefficient has a value between −1 and 1. A correlation coefficient of 0 (or near 0) means that the two variables have no relation to each other; conversely if a coefficient of −1 or 1 means the two variables have an absolute relationship. If the value of the correlation coefficient is negative ($r < 0$), it means that when x increases, y decreases (and vice versa, when x decreases, y increases). If the correlation coefficient value is positive ($r > 0$), it means that when x increases, y increases, and when x decreases, y decreases.

$$r = \frac{\sum_{i=1}^{n}(x_i - \bar{x})(y_i - \bar{y})}{\sqrt{\sum_{i=1}^{n}(x_i - \bar{x})^2 \sum_{i=1}^{n}(y_i - \bar{y})^2}}$$

4 Results

4.1 *Relationship Between Human Capital Development and Sustainable Development*

After World War II, less than 20 years later, since the 1960s, East Asian countries have experienced a period of great development to quickly become the countries with the most developed economies in the world. Although East Asia 3 is not rich in resources, it has successfully transformed from developing countries and quickly ranks among the developed economies in the world. As in the case of Singapore, after gaining independence in 1965, it quickly transformed into a modern developed economy in 1990 with the second highest standard of living in Asia and in 1996 surpassed both Western European countries in terms of GDP per capita. Korea started the era of industrialization in 1961 from a poor, underdeveloped country. In the 1980s, Korea achieved a GDP growth rate of 9.4% per year, followed by a growth rate of 7.2% in period 1990–1995, and in 1996 Korea was admitted to the Organization for Economic Co-operation and Development (OECD). Japan had rapid growth in the past, in 1952 Japan was a low-middle-income country, but only 15 years later advanced to a high-middle-income country and another 13 years to become an economic superpower. Southeast Asian countries, which are rich in natural resources, have lower growth rates and tend to slow down before becoming an economic power. Malaysia has come a long way from 1969 to 1995 with an average annual growth rate of 7%. During this period, there was only a small growth disruption from 1984 to 1986. Similarly, Indonesia also grew rapidly during 1967–1996. During these three decades, Indonesia's average growth rate was 6.8%/year. Thailand also maintained its growth rate of 7.6%/year for nearly four decades. However, the growth rates of these countries have decreased, currently only at 4–6%. The decline in growth rates in these countries occurs when per capita income is still relatively low. Compared to ASEAN 4, East Asia 3 economies have maintained higher growth rates for a longer period of time, and as a result East Asia 3 countries are among the richest countries in the world, Thailand and Malaysia are also currently middle-income countries; Indonesia and the Philippines are the lower middle-income countries (Satoru Kumagai, 2018), (Table 19.1).

East Asia *3* not only achieves high economic growth but is also ahead of ASEAN 4 countries in terms of income distribution and human development. Human development index of East Asia 3 is in the leading group among 174 economies, which is estimated by UNDP, while ASEAN 4 is in the middle and the latter group (Table 19.2).

Although there are many different explanations why East Asian countries have a high growth rate and are quickly converted into developed economies compared to Southeast Asian countries, but the most important reason is investment into human capital. According to Tilak (2002), investing in education to develop human capital is the main driving force of East Asian economic development; Investment in human resources contributes to high economic growth and also helps in better income distribution in East Asia. Tran Van Tho (2016) also emphasized that some East Asian countries do not have natural resources but have increasingly high-quality human resources thanks to education, training, and technological and technical research, so they have succeeded in economic development.

Looking at Table 19.3, we see that while East Asia 3 countries are among the countries with the highest SDG index in the world, ASEAN 4 countries are much lower. Table 19.3 also shows that East Asia 3 countries have high SDGs, while HDI

Table 19.1 Per capita income of East Asia 3 and ASEAN 4 (constant 2010 US$)

Country	GNI per capita	Year
Japan	50210.0	2017
Singapore	54169.4	2018
Korea. Rep	26744.0	2018
Malaysia	11697.1	2018
Thailand	6071.2	2018
Indonesia	4154.8	2018
Philippines	3625.3	2018

Source: World Bank Development Indicators

Table 19.2 HDI indicators in East Asia 3 and ASEAN 4 during 1990–2018

Year Country	1990	2000	2010	2018	HDI ranking 2018
Singapore	0.718	0.819	0.909	0.932	9
Japan	0.816	0.855	0.885	0.909	19
Korea (Rep)	0.728	0.817	0.884	0.903	22
Thailand	0.574	0.649	0.724	0.755	83
Malaysia	0.643	0.725	0.772	0.802	57
Indonesia	0.528	0.606	0.661	0.694	116
Philippines	0.86	0.624	0.665	0.699	113

Source: UNDP (2018)

Table 19.3 Position of East Asia 3 and ASEAN 4 in the world rankings according to the SDG index, HDI and HCI, in 2016

Country	SDG index ranking	HDI world ranking	HCI world ranking
Japan	18	19	4
Singapore	19	8	13
Korea. Rep	27	22	32
Malaysia	63	57	42
Thailand	61	86	48
Indonesia	98	115	72
Philippines	95	111	49

Source: World Economic Forum (2016), The United-Nations (2016), and UNDP (2016)

and HCI are also high, and ASEAN 4 with lower SDG also has low HDI and HCI. In order to see if there is a relationship between the positions occupied by the seven countries presented in Table 19.3 in the three rankings (according to SDG, HDI, and HCI), we conducted a correlation analysis bivariate, based on the Spearman coefficient with the results seen in Table 19.4.

As we can see in Table 19.4, considering the correlation coefficient for each pair, all pairs of SDG-HDI; SDG-HCI; HDI-HCI are strongly correlated because the correlation coefficients are greater than 0.928 with significant level of sig <0.01. Moreover, based on the results presented in Table 19.4, we can say that an important aspect of achieving sustainable development is the accumulation of a high human capital. We can see from our analysis that the leading countries in the SDG index are also among the countries with the highest HCI levels.

To see changes in the SDG, HDI, and HCI rankings of East Asian countries 3 and ASEAN 4, we also conducted an analysis of these indicators for 2018 (see Table 19.5). In 2018, using the HCI of the World Bank, Singapore, Japan, and South Korea were three leading countries in the world, respectively, while Japan's SDG index rose to 15th, South Korea rose to 19th place, while Singapore was downgraded from 19th place to 43rd place.

To see if there is a relationship between the positions occupied by the seven countries from Table 19.5 in the three rankings, the correlation analysis results are

Table 19.4 Correlation between the positions of the seven countries from Table 19.3 in the SDG index, HDI, and HCI rankings, in 2016

			SDG	HDI	HCI
Spearman's rho	SDG	Correlation Coefficient	1.000	0.929[a]	0.964[a]
		Sig. (2-tailed)	0.000	0.003	0.000
		N	7	7	7
	HDI	Correlation Coefficient	0.929[a]	1.000	0.964[a]
		Sig. (2-tailed)	0.003	0.000	0.000
		N	7	7	7
	HCI	Correlation Coefficient	0.964[a]	0.964[a]	1.000
		Sig. (2-tailed)	0.000	0.000	0.000
		N	7	7	7

[a]Correlation is significant at the 0.01 level (2-tailed); Source: Own research

Table 19.5 Position of seven Asia countries in the world ranking according to the HDI, SDG index, and HCI, in 2018

Country	SDG index ranking	HDI world ranking	HCI world ranking
Japan	15	19	2
Singapore	43	9	1
Korea. Rep	19	22	3
Malaysia	55	57	57
Thailand	59	83	68
Indonesia	99	116	87
Philippines	85	113	82

Source: World Bank (2018), The United-Nations (2018), and UNDP (2018)

shown in Table 19.6. There is a fairly strong positive correlation between the seven Asian countries in the HSDI, HDI, and HCI rankings. However, compared to the results obtained for 2016, the correlation is weaker.

4.2 Experience in Developing Human Capital for Sustainable Development in Some East Asian Countries

From the early 1960s to the late 1990s of the last century, East Asian economies to conduct rapid industrialization focused on developing human capital to create a skilled workforce. And well-educated, combined with properly oriented economic policies, is the key to diversifying and upgrading export industries. A good cycle has been created: rising incomes and industrial upgrading have stimulated continued investment in education and skill development, thereby contributing to increased productivity, technological progress, and increased fair growth. The East Asian countries are considered as a good example of successfully transforming from the

Table 19.6 Correlation between the positions of the eleven countries from Table 19.5 in the SDG index, HDI, and HCI rankings, in 2018

			SDG	HDI	HCI
Spearman's rho	SDG	Correlation Coefficient	1.000	0.893[a]	0.893[a]
		Sig. (2-tailed)	0.000	0.007	0.007
		N	7	7	7
	HDI	Correlation Coefficient	0.893[a]	1.000	1.000[a]
		Sig. (2-tailed)	0.007	0.000	0.000
		N	7	7	7
	HCI	Correlation Coefficient	0.893[a]	1.000[a]	1.000
		Sig. (2-tailed)	0.007	0.000	0.000
		N	7	7	7

[a]Correlation is significant at the 0.01 level (2-tailed); Source: Own research

Table 19.7 Annual growth rate of government expenditure for education and GDP growth in the period 1980–1990 (Unit: %)

Country	GDP growth	Growth rate of government expenditure for education
Japan	6.4	1.6
Singapore	6.4	7.1
Korea (Rep)	9.7	9.5

Source: Tilak (2002)

economy underdevelopment to advanced ones thanks to efforts to invest in education to develop human capital.

The aforementioned East Asian economies have firmly believed in the importance of education for economic growth and have rapidly increased public investment in this sector. In the period 1965–1985, the ratio of public spending on education in Korea increased from 15% to over 28%; Japan spent the most on education in the 1960–1970 period, and then gradually decreased in the years 1980–1990; Singapore maintained higher educational spending than the GDP growth rate in the 1980s and 1990s (Table 19.7).

Public spending on education plays an important role in the quality of human capital, but not all countries are positively correlated. Therefore, more important is the effect of these expenditures and this is what makes the difference in the quality of human capital in the States.

The share of public spending on education in total government expenditure in Southeast Asian countries 4 is quite high, even higher than in East Asia countries (Table 19.8), but the educational systems in the East Asia 3 are more developed than in ASEAN 4. For example, according to the World Bank, the percentage of Japanese secondary school graduates was 91.8% in 1975, while this rate in Indonesia was 20%, Malaysia was 45.7%, and Thailand 25.1%. By 2016, this ratio in Southeast Asian countries also improved significantly (Table 19.9). However, compared with

Table 19.8 Government expenditure on Education

Country	% of GDP	% of total government expenditure	Year
Japan	3.5	9.1	2017
Singapore	–	20	2013
Korea (Rep)	5.3	–	2017
Indonesia	3.6	20.5	2017
Malaysia	4.7	21.1	2017
Thailand	-	19.1	2014
Philippines	-	13.2	2009

Source: World Bank Development Indicators

Table 19.9 Secondary Education Enrollment for Select Countries (% Gross)

Country	1975	1980	1985	1990	1995	2016
Japan	91.8	93.2	95.4	97.1	103.4	102.4
Singapore	51.9	59.9	62.0	68.1	73.4	108.1
Korea. Rep	56.3	78.1	91.6	89.8	100.9	99.7
Malaysia	45.7	47.7	52.9	56.3	58.7	85.2
Thailand	25.1	28.8	30.5	30.1	54.1	118.8
Indonesia	20.0	29.0	41.3	44.0	51.5	86.0
Philippines	53.9	64.2	64.4	73.2	77.5	89.1

Source: World Bank Development Indicators

the level of access to higher education of East Asian countries and ASEAN 4 countries, there is still a large gap to date (Table 19.10).

Therefore, an equally important aspect of education spending in East Asia is focusing on the quality of education. East Asian countries have spent a reasonable amount of money on textbooks, learning materials, and training of teachers, and have also spent significant on direct subsidies such as financial and nonfinancial incentives key for students. With a more reasonable and more important level of investment, a reduction in student costs in education can increase significantly in some East Asian economies over the years, contributing to a significant improvement in the quality of education.

Another outstanding feature of the human capital investment model in East Asian economies involves a greater emphasis on natural sciences and an emphasis on engineering training at undergraduate and graduate levels. Improving the quality of labor resources is not only a matter of increasing the education level of the labor force but also increasing the supply of labor to meet the recruitment needs of enterprises. The labor needs of enterprises in these countries are characterized by the fact that university and college graduates are in the fields of science, technology, and engineering. Therefore, the percentage of university graduates in the fields of science, technology, and engineering in East Asian countries is often much higher than the rate of social science graduates, while in ASEAN countries 4 otherwise (Table 19.11). Therefore, the science and technology workforce in East Asia 3 countries is also higher than in ASEAN 4 countries (Table 19.12).

Table 19.10 Tertiary Education Enrollments for Select Countries (% Gross)

Country	1975	1980	1985	1990	1995	2016
Japan	26.3	30.5	27.8	29.6	–	–
Singapore	8.4	7.8	13.6	18.6	33.7	83.9
Korea. Rep	8.8	14.7	34.0	38.6	52.0	93.8
Malaysia	–	4.1	5.9	7.3	11.7	44.1
Thailand	3.3	14.7	19.0	–	20.1	49.3
Indonesia	2.3	3.8	–	9.2	11.3	36.3
Philippines	16.4	24.4	24.9	28.2	29.0	35.3

Source: World Bank Development Indicators

Table 19.11 Tertiary graduates by field in 2016

Country	Engineering (%)	Social sciences (%)
Japan	17.75	8.18
Korea. Rep	22.239	5.625
Indonesia	11.737	31.455

Source: OECD (2019), Tertiary graduates by field (indicator)

Table 19.12 Technical and research workforce per million people

Country	Researchers (per million people)	Technicians (per million people)	Year
Japan	5210	503	2016
Singapore	6730	457	2014
Korea. Rep	7113	320	2016
Indonesia	89	-	2009
Malaysia	2274	130	2015
Thailand	1210	320	2016
Philippines	188	28	2013

Source: World Bank Development Indicators

5 Conclusions and Discussion

The reality of economic development of countries in the world, especially in some East Asian countries, shows that human capital plays an important role in the miraculous development process in these countries; Human capital development brings positive effects on the aspects of sustainable development. Therefore, investing in education to build human capital is an important factor for sustainable economic development.

Vietnam is well aware of the importance of education and training, so Government always regard education and training as a top national policy. Over 30 years of renovation, Vietnam's education and training have achieved significant achievements, making an important contribution to the success of the renovation cause and bringing Vietnam out of a poor country. However, the quality and effectiveness of education and training are still low, especially in tertiary education and vocational education. Vietnam is still poor compared to other Southeast Asian countries and very poor compared to developed countries in East Asia. But with the advantage of a latecomer country, Vietnam can learn from the experience of successes and failures of the former countries, including the experience of developing human capital of East Asian countries. Based on the research on human capital development policies, which are mainly focused on the educational development of East Asian countries, we offer some suggestions for Vietnam as follows:

Firstly, special attention will be paid to the development of education to build human capital. Investing in human capital through education is an important element of development. Education and training are seen as an inseparable aspect of economic development strategies, success in forming achievable human resources, and this capital can be directed toward economic development and education is one of the effective ways to reduce income inequality, promote economic growth and social development. Therefore, the state needs to invest appropriately and prioritize budget spending on education and promote research and development to develop human capital.

Secondly, priority should be given to improving the quality and equity of education by investing in facilities, textbooks, and other teaching and learning materials. Special focus on training capacity building for teachers. Expanding facilities and ensuring teachers' quantity and quality, reducing student-teacher ratio to improve the quality of training.

Thirdly, focus on improving the quality of labor from vocational training and focusing on technical fields for economic growth through formal and informal training channels. Particularly encourage on-the-job training such as vocational training at the production line of the enterprise. The state can create mechanisms to encourage businesses to fund vocational training programs and to coordinate with universities, colleges, and vocational secondary schools in training the workforce and creating linkages between state - school - businesses.

Fourthly, mobilizing resources in society to invest in developing education and training, research and development. In Vietnam, in order to increase investment resources for education and training, in addition to increasing investment spending on education and training from the state budget, the State needs to have policies to mobilize and use effectively resources of all economic sectors, at home and abroad, for human resource development in the context of economic integration and global competition. In particular, attention should be paid to promoting the socialization of educational and training activities, science and technology, encouraging all economic sectors to participate in vocational training, college and university training, and R&D through improving the investment environment, implementing preferential regimes on capital access, tax exemption, reduction of land rent, simplification of administrative procedures.

Human capital development is not just the State's responsibility. The State should have policies and measures to affect the society's awareness of the responsibility of education and equip professional skills to meet the requirements of economic development, seize opportunities career, and personal development in the new era today. In particular, attaching importance to creative capacity, creating a favorable social environment for creative freedom, and equal access to opportunities for the development of workers.

References

Diaconu, L., & Popescu, C. (2016). Human capital - a pillar of sustainable development. Empirical evidences from the EU states. *European Journal of Sustainable Development, 5*(3), 103–112.

Dwiningrum, S. I. A. (2013). Nation's Character Education Based on the Social Capital Theory. *Asian Social Science, 9*(12), 144–155.

Freeman, R.B. (2000). Does the booming economy help explain the drop in crime? *In: U.S. Department of Justice, Perspectives on Crime and Justice: 1999–2000*, Lectures Series.

Goldin, C. (2014). *Human capital.* Retrieved from http://scholar.harvard.edu/files/goldin/files/human_capital_handbook_of_cliometrics_0.pdf

Green, A., Preston, J., Sabates, R. (2003). *Education, equity and social cohesion: A distributional model.* Retrieved from http://discovery.ucl.ac.uk/10005502/1/Green2003Education.pdf

Kelly, M. (2000). Inequality and crime. *Review of Economics and Statistics, 82*(4), 530–539.

Kumagai, S. (2018). The Middle-Income Trap in the ASEAN-4 Countries from the Trade Structure Viewpoint. Retrieved from https://link.springer.com/chapter/10.1007/978-981-13-2859-6_3

Lall, S., Kramer-Mbula, E. (2005). Is African industry competing? *QEH Working Paper No. 122*, University of Oxford.

OECD. (2001). The well-being of Nation. The role of human and social capital. *Centre for Educational Research and Innovation.* Retrieved from http://www.oecd.org/site/worldforum/33703702.pdf

OECD. (2019). *Tertiary graduates by field (indicator).* Retrieved September 9, 2019, from doi: https://doi.org/10.1787/9af26c71-en.

Olaniyan, D. A., & Okemakinde, T. (2008). Human Capital Theory: Implications for Educational Development. *European Journal of Scientific Research, 24*(2), 157–162.

Pidlisnyuk, V. (2010). Education in Sustainable Development: The Role of Universities. *Economic and environment Studies, 10*(1), 125. Retrieved from http://www.ees.uni.opole.pl/content/01_10/ees_10_1_fulltext_05.pdf

Schuller, T. (2001). The Complementary Roles of Human and Social Capital. *Canadian Journal of Policy Research, 2*(1), 18–24. Retrieved from http://www.oecd.org/innovation/research/1825424.pdf

Sianesi, B., & van Reenen, J. (2000). *The returns to education: a review of macro-economic literature.* Centre for the Economics and Education.

The United-Nations. (2016). *Sustainable Development Goals Report.* Retrieved from https://www.un.org/development/desa/publications/sustainable-development-goals-report-2016.html.

The United-Nations. (2018). *Sustainable Development Goals Report 2018.* Retrieved from https://unstats.un.org/sdgs/files/report/2018/TheSustainableDevelopmentGoalsReport2018-EN.pdf.

UNDP. (2016). *Human development report 2016.* Retrieved from http://hdr.undp.org/sites/default/files/2016_human_development_report.pdf.

UNDP. (2018). *Human development indices and indicators.* Retrieved from http://hdr.undp.org/sites/default/files/2018_human_development_statistical_update.pdf.

World Bank. (2018). *Investing in people to build Human Capital.* Retrieved from https://www.worldbank.org/en/news/immersive-story/2018/08/03/investing-in-people-to-build-human-capital.

World Economic Forum. (2016). *The Human Capital Report 2016.* Retrieved from http://www3.weforum.org/docs/HCR2016_Main_Report.pdf

Chapter 20
Mapping Marine Functional Zoning for the Northern Tonkin Coastal Zone, Vietnam

Bui Thi Thanh Huong, Nguyen Van Hong, Nguyen An Thinh, and Nguyen Hieu

Abstract The coastal zone of Northern Tonkin is placed in the Northwest of the Gulf of Tonkin, located on the international maritime road, with rich natural resources, great advantage for developing many marine economic sectors, many seaports and maritime services. Based on geographic approach in coastal space zoning and some law regulations and the legal provisions in Article 33 of the Law on Natural Resources and Environment of the Sea and Islands, Circular No. 74/2017 / TT-BTNMT promulgated December 29, 2017, regulating some techniques on integrated coastal management (ICM) for exploiting and sustainable use of natural resources, statistical data, field data, remote sensing data combined with marine functional zoning methods, maps of marine functional zoning for Northern Tonkin Coastal Zone (Vietnam) was built as follows: (1) protected and conservational area, (2) saving usage natural resources area, (3) active usage natural resources area, (4) security and safety area.

Keywords Marine functional zoning (MFZ) · Tonkin coastal zone · Integrated Coastal Management (ICM) · Vietnam

---

B. T. T. Huong (✉)
VNU University of Education, Vietnam National University, Hanoi, Vietnam
e-mail: buithanhhuong.vn@vnu.edu.vn

N. Van Hong
Institute of Geography, Vietnam Academy of Science and Technology, Hanoi, Vietnam

N. A. Thinh
VNU University of Economics and Business, Vietnam National University, Hanoi, Vietnam

N. Hieu
VNU University of Science, Vietnam National University, Hanoi, Vietnam

A. T. Nguyen, L. Hens (eds.), *Global Changes and Sustainable Development in Asian Emerging Market Economies Vol. 2*, https://doi.org/10.1007/978-3-030-81443-4_20

1 Introduction

The Northern Tonkin coastal zone (NTCZ), located in the western coastal strip of the Gulf of Tonkin, an important Gulf of Vietnam in the marine socio-economic development strategy, ensuring national security and defense, national marine sovereignty, and international integration in the globalization. The NTCZ covers five provinces and cities: Ninh Binh, Nam Dinh, Hai Phong, Thai Binh, and Quang Ninh have a length of about 460 km. The NTCZ is not only a key economic region and has the most vibrant economy in Vietnam but it is also filled with urban areas, industrial parks, and nature reserves of international value. According to the sustainable development strategy of the marine economy to 2030 and vision to 2045, developing basic marine survey, including functional zoning, is one of the prioritized tasks of Vietnam.

Functional zoning for integrated coastal management is basically Marine Spatial Planning (MSP). Currently, it is broadly used to conserve marine environment and sustainably use marine natural resources in order to meet the human's increasing needs. The concept of marine functional zoning (MFZ) originates from the concept of marine spatial planning (Xin et al., 2019), a basic task of the government which divides the marine area into multiple functional zones and provides technical facilities for better development, management, and conservation of the territorial waters (Xin et al., 2019).

Considered to be science-based, MFZ is a decision-making process which prevents unplanned human activities in the managerial process. MFZ is implemented through environmental protection regulation and basic functions for designated marine areas. Before the MFZ system establishment, the intense development of the coastal economy has brought undesirable consequences: coastal resources reduce, coastal erosion increases, coastal reclamation area for urban construction also increase, the linkage between land and sea development is broken, sea pollution and catastrophe are more severe, coastal wetland ecosystems degrade, etc. (Xin et al., 2019).

The result of functional zoning for the NTCZ with four basic functional areas is a new research result contributing to the planning and developmental orientation for the coastal areas of the Tonkin Gulf, serving the general management of the NTCZ.

2 Literature Review

In China, ecological functional zoning was outlined in the 12th five-year plan, marking a change in spatial planning approach from economic orientation to functional orientation. This approach assumes that each region should have a distinct function to focus on promoting its strength according to individual environmental-social requirements. With a function-oriented approach of the region, the Government can monitor regional and local development. Therefore, ecological functional zoning is considered as a tool to guide spatial planning to long-term sustainable development. The zoning is divided into two levels of implementation: national and provincial levels. The zoning of ecological functions at national scale is based on nine quantitative indicators and one qualitative indicator. At the provincial scale, the provincial government will participate in the establishment and zoning. Quantitative indicators include: cultivated land area; water source; environmental load capacity; vulnerability of the ecosystem; the importance of ecosystems; Possible impacts of natural disasters; the level of population concentration; economic development based on GDP; Favorability degree in transportation, with a qualitative indicator as (x) strategic choice. Thus, the results of zoning include four types of zone: optimal development zones; development priority zone; restricted development zone, including ecological functional areas and agricultural production area; prohibited development zone. With this policy, China ensures the goal of both economic development and conservation (Teng, 2019).

Jörg and John (1999) developed trans-disciplinary approaches for integrated spatial planning of coastal areas (ISPCA). The ecological footprint is calculated for coastal areas as a basis for making spatial planning decisions. Research results show that economic development makes the coastal systems exceed resilience threshold (Jörg & John, 1999). Susan and Stewart (2004) studied ISPCA in Central Queensland, Australia. Conflicts between coastal resource users and management-related socio-political issues were considered during the planning implementation. Conflicts that occur between stakeholders in the coastal area are considered in terms of value, trade-offs between conservation and development, access and rights to use of resources; Understanding conflict between stakeholders is essential in the decision-making process to ensure cooperation between social groups (Susan & Stewart, 2004). Counsell, Allmendinger, Haughton, and Vigar (2006) argued that spatial planning is an appropriate solution to integrate policies among different economic sectors in terms of space; 45 interviews were conducted with government officials, stakeholder groups, and scholars to examine the integration degree of spatial planning. The results of the study pointed out the main concerns of stakeholders about the integration level of traditional planning with housing policy, economic development, and transportation. The study also showed that in the process of interdisciplinary integration, several barriers needed to be improved, including traditional planning (Counsell et al., 2006). Davoudi and Strange (2008) conducted research on the role of space and place in strategic planning (Strategic Spatial Planning) in a number of different areas of the United Kingdom. Six case studies on spatial strategy were undertaken. Although there has been broad support for re-orienting the urban planning and location, there is little general understanding of what constitutes the plan. New perspectives, scientific, and practical debates on space and place in spatial planning are presented in this study. Vigar (2008) suggested that planning should be implemented in the direction that integrates with basic contents including: alignment strategies and policies; developing a policy framework; linking policies and actions; negotiate cooperation among stakeholders to resolve conflicts in coastal management. Two types of linkages are considered in policy integration: vertical (national/local planning) and cross-linking (sectoral planning). The interaction between policies and actions at different spatial levels forms the basis for proposing corrective measures and actions. Morphet (2009) examined the role of spatial planning in infrastructure planning and development. The research conclusions are made with consideration of the potential effectiveness of the spatial planning as an integration and future prospect. Macintosh (2013) applied the ISPCA in urban planning adapting to climate change in Victoria, Australia. From the results of the case study, the issues of interest drawn include: risks of damage caused by natural disasters and climate change in coastal areas; opportunity costs and planning equity; encouraging approaches that allow for continued use and development of land but on the basis of protecting the interests of government and the community. The new model is based on a vision that is shared widely with stakeholders and allows identifying priority issues step by step over time.

Five main mechanisms that support the implementation of the ISP methods include environmental impact assessment, planning decentralization, and coast-line delineation. Then, the ISP is applied to solve conflicts/contradiction in coastal Kenya (Tuda, Stevens, & Rodwell, 2014). Successful applied zoning method in a PEMSEA's project (*Partnerships in Environmental Management for the Seas of East Asia*), the coastal region of Xiamen (China) was divided into nine functional zones. These include: Port zone, Tourism zone, Aquaculture zone, Coastal Technology Zone, Maritime Mechanical zone, Mining zone, Nature Reserve zone, Special Functional zone, and resilience zone. Coastal economic activities are prioritized based on the characteristics: development restrictions, limited development, development priorities based on socio-economic benefits, and environmental impacts that the economic activity brings or affects the coast (Mao and Kong, 2018). Thus, environmental zoning is used as an effective tool for the rational use of natural resources in a territorial space. The basis for environmental zoning is the synthesis of natural and socio-economic factors in each zone.

In Vietnam, the functional zoning for integrated coastal management is still quite new and very difficult for managers in Vietnam. First of all, due to the complexity of the sea (sea space), which is reflected in four main aspects: (1) the circulating nature of seawater and the volatility of biological resources types, (2) the interplay of ecological, environmental, and marine resources in three dimensions, (3) the nature of sharing, multisectoral, and often competitive use of marine resource systems always creates demand (sometimes creates conflict) for space which is necessary for human activities, and (4) the interaction between the continent and sea in coastal areas, and between the above sensitive marine resource systems and human intervention. Such issues greatly influence long-term marine zoning plans, current coastal land use planning, as well as the Current policies and institutions governing coastal and marine areas.

According to Decision No. 54/2007/QD-UBND of Da Nang City, management zoning and conservation of coral reefs and marine-related ecosystems from Chao island to Nam Hai Van and Son Tra peninsula include three functional area: the strictly protected zone (core zone), is the zone that includes 36.2 hectares of coral reefs; the zone of ecological restoration; reasonable exploitation zone, enclosing strictly protected areas, and ecological restoration areas. For each of the aforementioned functional zone, the decision also specifies prohibited activities as well as encouraged activities in these zones. Functional zoning for integrated coastal management was first implemented in Vietnam through technical assistance in the Action Plan of Da Nang City for Integrated Coastal Management Action Plan. Accordingly, the coastal area of Da Nang City is divided into 11 zones, including: Conservation zone; restoration zone (corals); Water supply zone (blue lake); restoration zone (sea-grass); Low-intensity use zone; Tourism development zone; zone of industrial activities and seaports; Industrial zone; Fishing zone (inshore); Fishing zone (offshore); Multiuse zone.

The functional zoning of resources and ecosystems of Ha Long Bay area was conducted by Japan International Cooperation Agency, JICA, in 1998. As a result, the coastal area of Ha Long Bay was divided into four main environmental zones: Special conservation zone, including the world heritage site and its buffer zones; Protected zone, including important environmental areas but not enlisted as officially protected; Active management zone, including tidal flats along the shoreline and Bai Chay Bay; and Development Zone, including existing and planned development areas in the Socio-Economic Development Plan of the city and provinces (Hoang, 2019).

Adopting JICA's viewpoint, the Vietnam-US-IUCN's project of Integrated Coastal Management for Ha Long Bay has built a functional zoning map using the coastal area of Ha Long Bay at a scale of 1/25,000 which is feasible and consistent with the multidisciplinary development practices of the coastal areas. The map shows the spatial distribution of 10 different functional sub-zones, belonging to three main zones: (I) Environmental protection zone, including Strict protection zone and Environmental protection and management zone, (II) Marine economic development zone, including the limited marine economic development region and the free sea economic development zone, (III) Coastal economic and social development zone, including Industrial development zone; Tourism economic development zone; Forestry economic development zone; Socio-economic development and urban zone; Agricultural economic development zone; and fisheries economic development zone.

From the above-mentioned studies, some specific types of environmental functional zoning in Vietnam can be listed as follows:

- Functional zoning of integrated environment.
  - Zoning land use by the degree of adaptation to development activities.
  - Zoning by environmental quality.
  - Zoning the environment which receives waste (waste water, emissions).
  - Zoning by environmental sensitivity level.
  - Zoning integrated management of river basins.
  - Zoning integrated management of coastal areas.

Marine spatial zoning serving the orientation of marine spatial planning focuses primarily on Natural Resources and Environment of the marine area, and on the value of marine ecosystems and the rationality of marine exploitation of sectors and localities. In this approach, Vietnam's sea usage zoning is built following the Article 44 of the Vietnam Law of the Sea, the Law of Natural Resources and Environment, Sea and Islands, based on three main groups of criteria for delineating the waters. These are groups of criteria for protecting and preserving marine ecosystems, groups of criteria for economic development that have advantages in natural conditions, natural resources, and positional advantage to develop marine industries such as port services, waterway tourism, aquaculture, energy development, coastal industries, the group of criteria for national security and defense.

Thus, it can be said that environmental zoning has been applied in many planning activities of scheduling, planning, and management of environmental resources in Vietnam in recent years. However, in the above-mentioned programs and projects, the authors have not really focused on researching a complete methodology for functional environmental zoning (Table 20.1).

3 Methodology

The environmental function zone is a part of the territorial hierarchy, has some defining environmental and ecological attributes that can distinguish it from other zones (Dang and Le 2016). Functional zoning (FZ) of coastal areas is the first and important task in the integrated coastal management, conducted on the basis of a comprehensive study of coastal features, based on which the coastal zone is divided into different zone of different roles, depending on its geographical location, natural resources, environmental conditions, etc. FZ of coastal areas is mainly based on the exploitation degree of natural resources, and pollution impacts of major economic activities in coastal areas. The zoning of resources on coastal land can be done in multiple ways, depending on the zoning objectives, the complexity degree of conflicts in resource usage and the conflict status (whether the conflict is being resolved or not), the development level of regions and the scope/boundary of the ongoing zoning plan. In the step-by-step guideline on marine spatial planning, FZ was implemented to divide ownership and usage of land, water resources, and resources associated with their use functions (UNESSCO, 2009). The basic contents of FZ approach in coastal space planning are: (i) Position and design functional areas based on overlaying topographic, oceanographic, and biological factors, development factors in managed coastal areas; (ii) Define a licensing system, legal regulations, and rule of use in each defined functional area; (iii) Establish a mechanism to ensure compliance with the zoning plan and the above-mentioned legal regulations in the process of exploitation and use; and (iv) Create monitoring, evaluation, and adaptation programs for the zoning system (Douvere and Ehler, 2008).

In order to implement functional zoning for integrated coastal management in the northern coastal region, the research team used the methods in Fig. 20.1 below.

Table 20.1 Overview of criteria for development purpose

Purpose of Zoning	Criteria	Source
Ecological zoning serving sustainable development	Land cover; slope; land use Land unit of species under threat; biodiversity Mangrove forest distribution Aquaculture distribution Land unit for tourism development	*Ecological zoning* as a policy tool for sustainable development at the local level (Cabrido)
Environmental zoning	The homogeneity of natural conditions (geology, geomorphology, climate, hydrology, soil, vegetation cover) Specific characteristics of economic development, natural resources exploitation and use (land use, development of industry, agro-forestry, aquaculture) Urgent environmental issues, and natural hazards	Environmental protection planning in the direction of sustainable development at province and district levels
Agro-ecological zoning	Geomorphology Water resource dynamics (salinization map, flooding map, irrigation-regulated regions map of the present and future) Pedagogy; land use type	Agro-ecological zoning in the Mekong Delta under climate change
Criteria for agro-ecological zoning	– territorial integrity (not repeat); homogeneity of a geological architecture forming topographical feature characterized by minor differentiation in climate, soil, vegetation. – there is one or two ecosystems typical for the distribution by latitude and altitude; relatively homogeneous in forestry development planning	Final report of Vietnam Forest ecological zoning (UN-REDD)
Aquaculture ecological zoning	Topography Climate - hydrology Pedology Floral-fauna system Water quality Flooding regime	Study aquaculture ecological zoning in Dong Thap province serving the sustainable exploitation and use of resources
Criteria for zoning by function	Natural function: Existence and transformation of landscape elements such as: Climate, hydrology, soil, topography, geology Function of natural and artificial environment belonging to landscape structure components – function of biological productivity, material transformation, and function in the ecosystem – socio-economic function, esthetic	Law on natural resources and environment of sea and islands

(continued)

Table 20.1 (continued)

Purpose of Zoning	Criteria	Source
Criteria for coastal zoning	(a) having homogeneity and typical characteristics of natural conditions; (b) having important ecosystems with high level of biodiversity, home to many endemic, precious and rare species which are under prioritized protection; (c) having cultural and historical heritages that need to be protected and preserved; (d) having position, potential, and advantages for exploitation and development of economic sectors; (e) having position, potential, and advantages for national defense and security.	Circular no. 74/TT-BTNMT dated April 22, 2011, providing technical guidance for planning exploitation and sustainable use of coastal resources
Criteria for functional zoning of integrated coastal management	(1) optimal development zone; (2) development priority zone; (3) restricted development zone, including ecological functional areas and agricultural production areas; (4) prohibited development areas. Zoning based on quantitative indicators includes: Arable land area; water source; environmental load capacity; vulnerability of the ecosystem; the importance of ecosystems; possible impacts of natural disasters; the level of population concentration; economic development based on GDP; Favorability level in transportation, with a qualitative indicator as (x) strategic choice	Implementing marine functional zoning in China (Teng, 2019)

4 Results

4.1 *Functional Zoning for Integrated Coastal Management, Case Study in Northern Coastal Area of Vietnam*

Based on the methodology of functional zoning for integrated coastal management, the northern coast is divided into four main functional zones:

Group A: Protection and Conservation zone.
Group B: Resource protection and usage zone.
Group C: Active Development zone.
Group D: Safety and Security Protection zone.
Features of each group are detailed as in Table 20.2 and Fig. 20.2.

Functional zone for protection (Zone A) of the northern coastal zone includes three functional zones: (1) function zone of protection Binh Lieu - Mong Cai, (2)

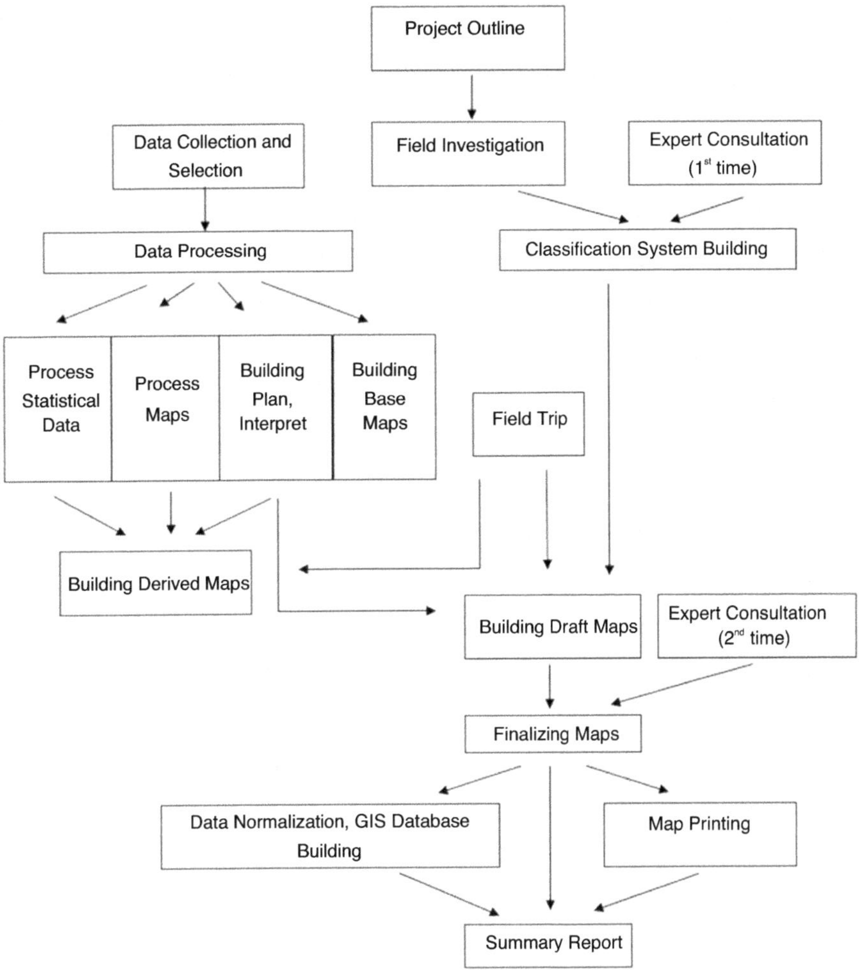

Fig. 20.1 Flowchart of Functional Zoning for integrated coastal management

function zone for conservation Yen Tu, (3) The Ramsa Xuan Thuy zone covers an area of 82473.53 ha, accounting for 6% of the total area of the region, and is a biodiversity and landscape conservation area; riverhead protection, wave break protection, and coastal erosion protection. The functional zone of Binh Lieu - Mong Cai protection zone is in Binh Lieu district and most areas in the north of Tien Yen, Dam Ha, Hai Ha, and Mong Cai cities are the riverhead protective areas of Quang Ninh province in particular and in coastal zone in general. Yen Tu function zone for conservation is a protective forest area in the eastern Yen Tu region located in Cam Pha town and Hoanh Bo district. The Ramsa Xuan Thuy area in Giao Thuy district, Nam Dinh is the largest biosphere reserve area of the Red River Delta, with many roles in protecting the environment and coastal resources.

Table 20.2 Features of functional zoning groups in the Northern Coast

No.	Functional Zoning Group	Extent	Key environmental functions	Characteristics
1	Protection and conservation zone (zone A)	Protective forests of Binh lieu - mong Cai, yen Tu mountain areas, Ramsa Xuan Thuy	Conservation of biodiversity and landscape; riverhead protection, wave break protection, and coastal erosion protection	Ensuring the ecosystem quality of the riverhead area, and the biosphere reserve of the red River Delta
2	Resource protection and usage zone (zone B)	Vang Danh production area, Giao Thuy agricultural and fishery production area, Kim son coastal alluvial area, Hai Phong - Quang Ninh coastal area, Thai Nam Ninh region	Providing food; supply raw materials for production; residence place for rural population communities, small and medium cities	Urgent environmental issues concerning natural resources of vang Danh - cam Pha coal production areas, coastal waters pollution in Quang Ninh - Hai Phong, coastal erosion in Nam Dinh - Kim son
3	Active development zone (zone C)	The mainland coastal area extends from Kom son (Ninh Binh) to mong Cai (Quang Ninh) and the 3 nautical miles to 6 nautical miles area from mong Cai to Ninh Binh.	The residence of the majority of the population; the place to develop multidisciplinary production activities	It is a dynamic development area with diverse economic sectors, located in the national key economic region.
4	Safety and security protection zone (zone D)	Quang Ninh - Hai Phong Sea and island area	Is the area of national defense and security	A prohibited development area, prioritize sustainable conservation combined with coastal safety and security

Functional zone for resource use and protection (Zone B) includes Vang Danh - Cam Pha coal mining area (B1), Giao Thuy development zone (B2), utilization, and protection areas. Kim Son coastal alluvial zone (B3), Hai Phong - Quang Ninh coastal resource use protection zone (B4), Thai Nam Ninh natural resource protection and disaster prevention zone (B5) have an area of 214846.48, which occupies about 16% of the northern coastal zone.

The functional zones of active development (Zone C) have the largest area, 752270.83 ha, accounting for 56.4% of the area of the Northern coastal zone, which is a key economic area with economic development activities, the most dynamic area of the whole Northern region. Coastal localities and cities from Mong Cai, Tien Yen, Ha Long - Cam Pha, Quang Yen, An Lao - Kien Xuong, Hai Phong City, Tien Lang - Kien Thuy, Thai Binh Coastal Area, Hai Hau - Nghia Hung, Kim Son, Quang

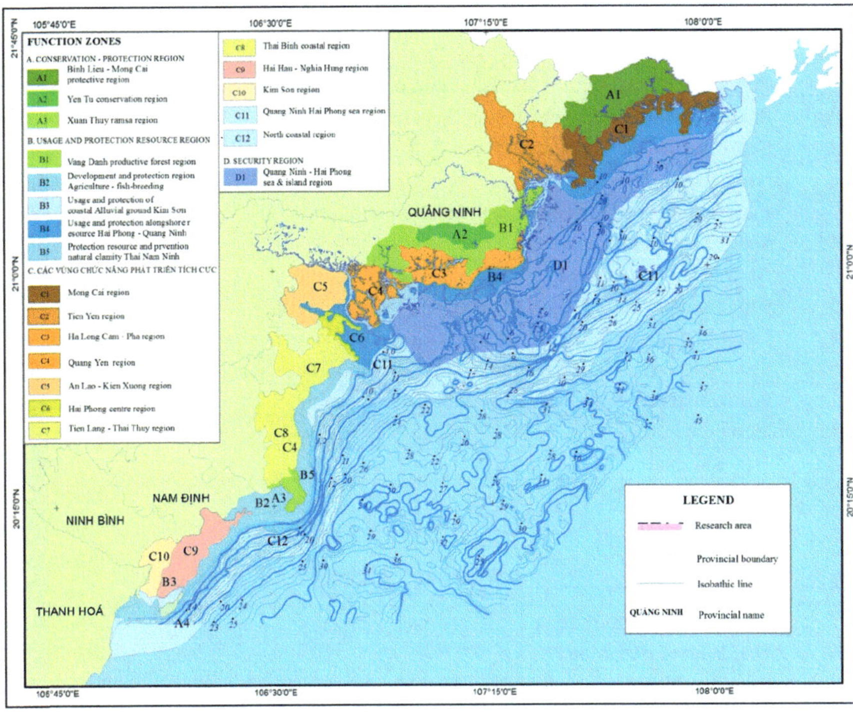

Fig. 20.2 Marine functional zone for the Northern coastal region (Scale 1: 250.000)

Ninh - Hai Phong sea, and island region, the whole coastline the Northern Coastal Region from 1 nautical mile to 6 nautical miles (C12).

Safety and Security Protection zone (Zone D) is an area which prohibits development, prioritized for sustainable conservation in combination with safety and security in Quang Ninh, Hai Phong coastal area with an area of 285,657 ha (accounting for 21.4% of the northern coast region area) concentrating in Cat Ba, Ha Long Bay, Bai Tu Long, Van Don, Co To, and Mong Cai islands. This is an area that needs to conserve and preserve natural ecosystems, ensuring essential security and national defense.

Distribution of northern coastal functional zones in the Functional Zoning map for integrated coastal management is shown in Table 20.3 below:

Table 20.3 Functional zones for integrated coastal management of the Northern Coast

Sign	Name of Zone	Area
A	**Functional zones for conservation and protection**	**82473.53**
A1	Protective functional zone of Binh lieu - mong Cai	56992.57
A2	Conservative functional zone of yen Tu	15286.43
A3	Ramsa Xuan Thuy	10194.53
B	**Functional zones for resource protection and usage**	**214846.48**
B1	Vang Danh productive forest	55826.86
B2	Agro-fishery production zone of Giao Thuy	14549.69
B3	Usage and protection of coastal alluvial ground Kim son	4575.15
B4	Usage and protection of alongshore resource Hai Phong-Quang Ninh	101174.74
B5	Resource protection and natural hazard prevention Thai Nam Ninh	38720.04
C	**Functional zones for active development**	**752270.83**
C1	Mong Cai zone	51483.64
C2	Tien yen zone	59917.58
C3	Ha Long - cam Pha zone	38610.24
C4	Quang yen zone	27967.04
C5	An lao - Kien Xuong zone	33230.63
C6	Hai Phong center zone	7469.37
C7	Tien lang - Kien thuy zone	34306.86
C8	Thai Binh coastal zone	40190.72
C9	Hai Hau - Nghia hung zone	32982.82
C10	Kim son zone	14730.43
C11	Quang Ninh - Hai Phong zone	143492.71
C12	Northern coastal zone	267659.53
D	**Functional zones for security and safety** Quang Ninh - Hai Phong Sea and islands zone	

4.2 Proposing the Economic Development Space Planning Associated with the Rational Use of Natural Resources and Environmental Protection in Coastal Areas of the Northern Coastal Region

Using the results of functional zoning for integrated coastal management of the northern coastal zone, the State's spatial planning orientations for developing the northern coastal spatial region, and principles of properly using natural resources and protecting coastal zone environment, the spatial planning are proposed as in Fig. 20.3 and Table 20.4 below:

Planning zones for conservation development corridors are development axes which are shown in purple pink strip as V sign covering the entire region with development routes such as: coastal route from Tra Co (Mong Cai), Tien Yen, Cam Pha, Ha Long, Do Son, Diem Dien, Tien Hai, Giao Thuy, Hai Hau, and Kim Son. This route is located on the coastal traffic artery with tourist attractions, ports, cultural, economic, and political centers of each locality.

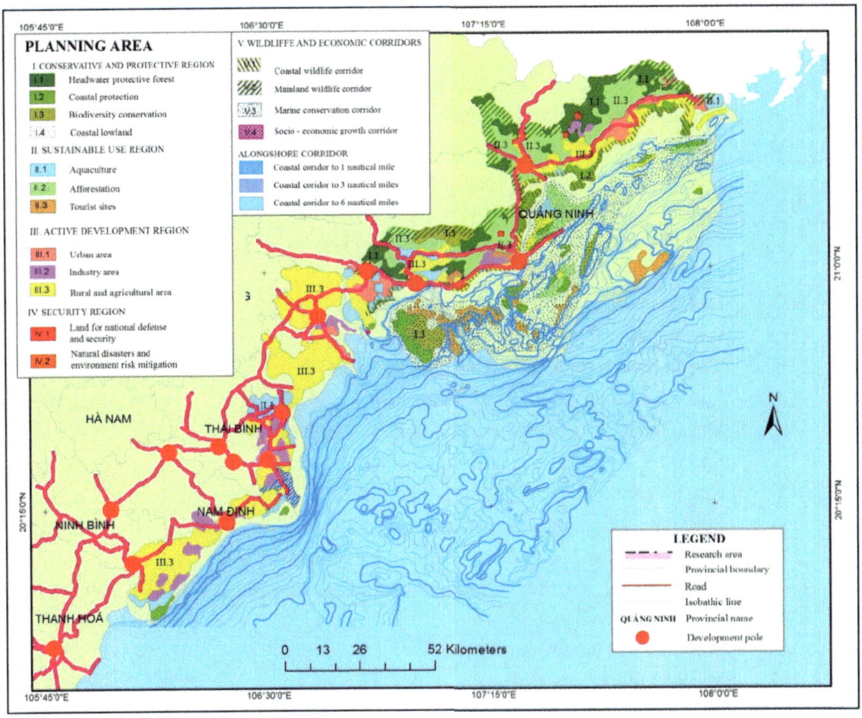

Fig. 20.3 Orientation map of spatial planning for the Northern Coastal region (scale 1:250.000)

Promote economic links along the coastal economic corridor Quang Ninh – Hai Phong – Ninh Binh. This is an important economic corridor, not only in national defense but also in economic significance for the development of the Red River Delta region associated with the Vietnam Sea Strategy to 2030. The economic corridor will be formed on the basis of opening a new route running along the coastal line from Nga Son (Thanh Hoa) through Ninh Binh (in the territory of Kim Son district), connecting Thinh Long to Quat Lam (Nam Dinh) to Thai Binh (from Cong Lau to Tien Hai - Diem Dien) and then to Hai Phong (in Tien Lang District and to the city). This entire route has a length of about 200 km. On this route, in addition to Hai Phong city, there will be many medium-sized cities such as Kim Son, Thinh Long, Quat Lam, Tien Hai, Diem Dien, Tien Lang, and then going to Mong Cai, and associated with them are manufacturing mechanical industrial clusters serving agricultural and fishery production, port industrial complexes, general economic zone (Dinh Vu - Cat Hai and Van Don), and a number of airports (Cat Bi and Van Don). Along this route, there will also be many special aquatic aquaculture zones, many coastal eco-agricultural areas, and sea tourism.

Ha Long-Mong Cai coastal economic route is responsible for international exchange, trade, and tourism development at national and international levels and is also an important axis to counter China. Important economic sectors of this

Table 20.4 Statistical table of zones proposed for planning

No.	Sign	Name of planning area	Area (ha)
1	**I**	**The planning zone for conservation and protection**	**129.472,3**
2	I.1	Priority space for riverhead protection	54.456,9
3	I.2	Priority space for coastal protection	45.945,7
4	I.3	Priority space for biodiversity protection	24.383,7
5	I.4	Space for smart management of coastal wetland	4.686,0
6	**II**	**The planning zone for sustainable use of resource**	**200.441,2**
7	II.1	Priority space for sustainable aquaculture	41.108,3
8	II.2	Development space for productive forest	127.878,5
9	II.3	Priority space for sustainable tourism	31.454,3
10	**III**	**The planning zone for active development**	**264.577,7**
11	III.1	Space for urban environment protection	26.883,7
12	III.2	Space for industrial park environment protection	36.428,3
13	III.3	Space for agricultural and rural protection	201.265,8
14	**IV**	**The planning zone for security and safety**	**6.207,4**
15	IV.1	Space for national defense and security	2.932,4
16	IV.2	Space for natural hazard and environmental incident	3.275,0
17	**V**	**The planning zone for development and conservation corridor**	**395.802,9**
18	V.1	Coastal biodiversity corridor	32.291,5
19	V.2	Mountainous biodiversity corridor	77.859,4
20	V.3	Marine biodiversity corridor	285.652,1

economic route that need to be prioritized for development are: trade, tourism, seaports, aquaculture, shipbuilding, and processing industries.

Along with Ha Long-Mong Cai economic route, to develop coastal economic routes along National Highway 10 from Ninh Binh to Quang Ninh (the sea belt of the whole Red River Delta region) with priority branches as follows: processing, manufacturing, travel, and services. The economic corridors developed in this region are: the economic corridor along National Highway 18: National Highway 18 is one of the arterial roads connecting Hanoi Capital and economic centers of the Northern region to the seaport Cai Lan. The economic sectors prioritized for development on this route are energy industry, major repair and manufacturing mechanics, mineral processing, construction materials production, agro-forestry processing and services. Economic Corridor along National Highway 5: This location focuses on key and spearhead industries with advanced technologies such as: Electrical and electronic engineering, automobile assembly, manufacturing of machinery and equipment, metallurgy, shipbuilding, textile and leather, food industry, and high-end consumer goods manufacturing industry.

Planning area for protection and conservation: The conservation and protection planning area is a priority area for the development of natural ecosystems in order to maintain ecological balance and the sustainability of nature in the industrialization and modernization era. This area encompasses:

The space prioritized for riverhead protection (I.1) with an area of 54456.92 ha is the range of ecosystem located in the West of localities: Mong Cai and Hoanh Bo.

The space prioritized for coastal protection (I.2) with an area of 45945.74 ha is coastal protective forests concentrated in the protective forests and mangrove forests in Van Don, Cam Pha, and Ha Long areas, Cat Ba, coastal Quang Ninh, Hai Phong, Nam Dinh, Kim Son.

The space prioritized for biodiversity conservation (I.3) with an area of 24383.70 ha is the biosphere reserves of Yen Tu, Cat Ba, Xuan Thuy.

The space for smart management of coastal wetland (I.4) has 4685.96 ha, which is a space that needs to be preserved and limited for development. However, each locality can build appropriate models of economic resource development, both ensuring people's livelihoods and ensuring sustainable development of environmental resources. This area (I4) is delineated in coastal areas from Mong Cai to Kim Son.

The planning zone for safety and security: The space for national security and security (IV.1) has an area of 2932.4 hectares concentrated in scattered areas in Mong Cai, Tien Yen, Cam Pha, Ha Long, and offshore islands and sea in the Gulf of Tonkin. Space for natural and environmental disaster (IV.2) with an area of 3275.0 ha is the area that is currently warning many risks of natural resource incidents such as coastal erosion. Nam Dinh and Ninh Binh seas, landslides, and collapses in coal mining areas in Cam Pha, Ha Long or oil spill in Cua Luc Bay area.

The planning zone for sustainable use of natural resources: Different from the areas proposed for planning, the sustainable use planning area allows the optimal use of resources, including minerals, for socio-economic integrated development according to the provincial master plan as well as the relevant districts. This region focuses on space prioritized for sustainable aquaculture (II.1) with an area of 41108.32 ha, space for sustainable development of productive forests (II.2) with an area of 127878.53 ha, and the space prioritized for sustainable tourism development (III.3) with an area of 31454.30 hectares.

The planning zone for active development: This region is concentrated in coastal urban and rural clusters, with densely population and the most dynamic economic development rate in the country. Hai Phong, Do Son, and Hanoi are Vietnam's strong economic growth triangle. The space for urban environmental protection (III.1) with an area of 26,883.7 ha is concentrated in large coastal cities: Mong Cai, Cua Ong, Cam Pha, Ha Long, Bai Chay, Do Son, and Cat Ba, Diem Dien, Tien Hai, Hai Hau, Kim Son. Space for environmental protection in industrial parks (III.2) with an area of 36,428.3 ha is concentrated in industrial parks of Hai Ha Industrial Park, Dam Nha Mac Industrial Park (Quang Ninh), Van Don special economic zone, Ha Long tourist resort, Quang Ninh - Hai Phong, Dinh Vu - Cat Hai.

Development of seaport system: By integrated resources, synchronous and modern development of port system (ports, channels, production equipment, and operation) to ensure the target of 80–100 million tons of commodity going through the ports by 2020.

Shipping development: Build a marine transport fleet, ensuring the transport of about 50 million tons by 2020, meeting over 50% of the shipping demand of the Northern region.

Developing maritime services with important types of services: ship agency and maritime brokerage services; ship-towage services; agency services for freight transport of ships; ship supply services; freight forwarding and tallying services; small ship repair service at the port; marine environmental sanitation services; cargo handling services at seaports.

Development of coastal fishing and aquaculture: Towards the goal of 10,000 tons by 2020 on the basis of restructuring production in the direction of developing off-shore fisheries and aquaculture specialties: Utilizing over 24,000 hectares of ponds and lakes to raise freshwater fish; the intertidal flats are intensified with shrimp, crab, seaweed, etc.; to strongly develop specialty-raising (pearl, sea shrimps, gar-rupa, etc.) at 3000 ha of salt water surface in Cat Ba; develop coastal fishery in combination with development of offshore fishing; associating with domestic and foreign units to build fishing logistics areas of the Tonkin Gulf in Cat Ba and Bach Long Vi in the direction associated with tourism development; build fishing villages.

Tourism development: Taking advantage of the natural landscape in Cat Ba, Do Son islands, and coastal areas to develop large-scale marine tourism in the direction of diversifying tourism products while creating a number of unique and competitive marine tourism products in Cat Ba (underground tourism, mountain climbing, cave exploration, scientific research) to create jobs and income for people on the island and in the fields, and at the same time create motivation to protect ecological, bal-ance resources, and defense-security issues.

Develop regional linkages. The coastal zone of Northern Coastal Area is geo-graphically a transshipment region between the Sea and the Delta, between the Northeast and the Northwest, between the Northern Mountains with the Central and between the Southern provinces with the Northern provinces. Due to its natural position, the Northern Coastal Zone becomes the seaward area (the inlet) of the Northern provinces of Vietnam. The space for agricultural and rural environment protection (III.3) has an area of 201,265.8 ha, concentrated in agricultural areas, intensive wet rice cultivation in Tien Yen, Cam Pha, Thuy Nguyen, Thai Thuy, and Tien Hai, Hai Hau, and Kim Son.

5 Conclusions and Discussion

Based on the methodology of functional zoning for integrated coastal management, the functional zoning map of the Northern Coastal Zone is divided into four main areas: Group A: Functional zone for conservation, protection, Group B: Functional zone for use and protection of resources, Group C: Functional zones for active development, Group D: Functional zones for security and safety protection. Within each group, zones are subdivided based on spatial integrity planning.

Based on the results of the functional zoning map of the Northern Coastal Zone, a planning map for integrated coastal management of the Northern Coastal Zone at 1: 250,000 scale was established with four main groups of zone: (1) Zones for con-servation and protection planning include: priority areas for upstream protection, priority areas for coastal protection, priority areas for biodiversity conservation protection, space for smart management of coastal wetland. (2) The planning zone for sustainable use of resources includes priority space for sustainable aquaculture, space for sustainable development of production forests, and space for prioritizing sustainable tourism development. (3) The planning zone of active development includes space of urban environment protection, space of industrial park environment protection, and space of agricultural and rural protection. (4) The planning zone of security and safety includes defense and security space, space for natural disaster prevention and environmental incidents. (5) Planning zone for corridor development and conservation includes coastal biodiversity corridors, mountain biodiversity corridors, and marine biodiversity corridors.

By using criteria from Teng (2019) combined with the results of natural zoning, socio-economic development planning in climate change and natural disaster contexts, the functional zoning map is built as a basis for integrated coastal zone management planning for the study area. However, some research issues need to be discussed:

1. Does the zoning criteria by Teng (2019) for integrated coastal management functional zoning are appropriate to the natural, socio-economic conditions, and development goals of the zone and field of study?
2. Can the conflicts in socio-economic development associated with exploitation and use of natural resources in the coastal zone of the North be resolved by functional zoning?
3. What is the feasibility of the zoning results if there exists an overlap between the socio-economic development situation and the spatial development planning of coastal areas and functional zoning of the Northern Coastal area?

Acknowledgments This research work was financially supported by the National project (ĐTĐLCN.33/16) “Research and propose solutions for integrated coastal zone for sustainable development of the Northern Coastal Region, case study in Thai Binh province).

References

Counsell, D., Allmendinger, P., Haughton, G., & Vigar, G. (2006). ‘Integrated’ spatial planning - is it living up to expectations? *Town & Country Planning*, 243–246.

Dang, T. T., & Le, T. (2016). *Environmental protection planning: Theoretical basis and practical experience*. Science and Technics Publishing House.

Davoudi, S., & Strange, I. (2008). *Conceptions of space and place in strategic spatial planning* (304 pages). Routledge.

Douvere, F., & Ehler, C. (2008). The Role of Marine Spatial Planning in Implementing Ecosystem-based, Sea Use Management. *Maritime Policy 32*, 759–761.

Hoang, Q. L. (2019). *Research on geographical conditions serving the orientation of general spatial planning of coastal areas in Hai Hau - Nghia hung district* (Nam Dinh province. Doctoral thesis on natural resources and environment management) (150 pages). VNU University of Science, Vietnam National University.

Jörg, K., & John, G. (1999). Coping with complex and dynamic systems. An approach to a transdisciplinary understanding of coastal zone developments. *Coastal Conservation, 5*(2), 163–170.
Macintosh, A. (2013). Coastal climate hazards and urban planning: How planning responses can lead to maladaptation. *Mitigation and Adaptation Strategies for Global Change, 18*(7), 1035–1055.
Mao, Z., & Kong, H. (2018). The evolution of operational modality of ICM in Xiamen. Ocean Coast. Manag (Submitted) in this Special Issue.
Morphet, J. (2009). Local integrated spatial planning - the changing role in England. *Town Planning Review, 80*(4–5), 393–414.
Susan, R., & Stewart, L. (2004). Participatory tools for coastal zone management: Use of stakeholder analysis and social mapping in Australia. *Coastal Conservation, 10*(1–2), 81–92.
Teng, X. (2019). Implementing marine functional zoning in China. Marine Policy. Retrieved from https://www.sciencedirect.com/science/article/pii/S0308597X18304093?via%3Dihub
Tuda, A. O., Stevens, T. F., & Rodwell, L. D. (2014). Resolving coastal conflicts using marine spatial planning. *Environmental Management, 133*, 59.
UNESSCO (2009). *Marine spatial planning: step by step toward management base on ecosystem.*
Vigar, G. (2008). Towards an integrated spatial planning. *European Planning Studies, 17*(11), 1571–1590.
Xin, T. et al. (2019). Implementing marine functional zoning in China, Marine Policy, https://www.sciencedirect.com/science/article/pii/S0308597X18304093?via%3Dihub

Chapter 21
Rainfall Regime and its Impact on Water Resources on Ly Son Island, Central Vietnam

Bui Xuan Thong, Nguyen Van Dan, Nguyen Ngoc Ha, and Van Phu Hung

Abstract Ly Son Island plays a special economic and national security role in Vietnam. However, in recent years, Ly Son is facing a shortage of freshwater and salinity intrusion due to the overexploitation of underground water. This article is the latest research on the rainfall regime and its impact on water resources on Ly Son Island. Based on the time series data of rainfall collected during the period of 1985–2015 of the Ly Son Marine Meteorological Island Station (15°23'N;109°09'E), the authors have calculated all the statistical values of rainfall such as maximum rainfall, maximum of daily, and monthly precipitation and its tendency for the various periods of series data between 1985 and 2015. Using water balance equations and computed results of all maximum values of rainfall, the authors have computed all surface runoff values, respectively. The mean annual precipitation for the period of 1985–2015 was 2278.8 mm. In the rainy months, the rainfall was 1779.6 mm, which occupied 78.1% of the whole year. Over the area of 10.4 km^2 of Ly Son Island, the estimated potential yield of rainfall was 14.17 x 10^6 m^3/year, while for the whole year, Ly Son Island needs only 6.422 x 10^6 m^3 of freshwater. From this, we can suggest that, if there are appropriate measures for collecting rainfall in rainy season and all the days having abnormal rainfall, the scarcity of water resource will be ameliorated on Ly Son Island.

Keywords Water resources · Maximum rainfall · Surface runoff · Ly Son Island

B. X. Thong (✉) · V. P. Hung
Institute of Oceanography and Environment, Hanoi, Vietnam

N. Van Dan
Institute of Resources and Environmental Water, Hanoi, Vietnam

N. N. Ha
National Center for Water Resources Planning and Investigation, Hanoi, Vietnam

A. T. Nguyen, L. Hens (eds.), *Global Changes and Sustainable Development in Asian Emerging Market Economies Vol. 2*, https://doi.org/10.1007/978-3-030-81443-4_21

1 Introduction

In the world, the water resources of islands in the sea and oceans are mostly derived from rainfall. Therefore, we can see that the climate in general and rainfall regimes in particular have a very strong impact on water resources of the island. Especially in the tropical monsoon regions, the rainfall in the rainy season is very large, and in most months of the year, there is rain. There have been many papers published research results on the relationship between rainfalls with the components of the water balance equation according to different spatial scales. Due to the small surface area of the island, the interaction between the island and the atmosphere is limited, although the impact of climate change and rainfall on the island's water resources is still very large, which has been quite carefully described in the works of many authors (Chapman 1985, Falkland 1991).

On the islands, there may not be many hydrographic stations, but most have meteorological observation stations with fairly long datasets of several decades. For this reason, the results of the calculation of climate parameters, frequency distribution as well as rainfall fluctuation trends, have been published in a number of references (World Meteorological Organization [WMO] 1966; Sen 1968; Ramsey 1989; Khaled and Ramachandra 1998; Phan 1999; Yue and Wang 2004; Hennemuth et al. 2013). The assessment results of climate and rain regimes provide the scientific basis for the estimation of surface runoff and the total amount of water from rainfall (Sokolop and Chapman 1974; Vu et al. 2009; World Meteorological Organization [WMO] 1986).

Water balance research results are often very useful for water resources management, for example, facilitating water identification, or the development of new water resources project. Water balance studies can also be used to analyze surface water, groundwater, or combined water resource systems. The detailed components of the water balance equation have been well described in many published literature. Water balance equations can be applied to small islands in two stages, namely, the surface water system and the groundwater system. The reference zone for a surface water system includes the atmosphere above the land and the surface itself and the soil moisture zone. A surface water balance for a given reference zone can be expressed as follows:

Precipitation - evapotranspiration (from all sources) + surface water inflow - surface water outflow + groundwater inflow - groundwater outflow - increases in water storage.

The main input term to the water balance equation of a small island was precipitation, generally in the form of rainfall. Small island water resources are very sensitive to short-term variations in rainfall due to their limited storage capacity within surface and groundwater systems. Rainfall monitoring data on the island are usually used to publish the average annual rainfall only, while to calculate the water resources on small islands, it is necessary to receive rainfall of short rainfall periods. Therefore, using the average annual rainfall to assess the water balance on small islands will be more limited.

This is the reason why we put more attention to the typical durations of rainfall, which would be applicable to very small islands. The duration of rain monitoring was chosen to be one hour, three hours, six hours, twelve hours, and one day (World Meteorological Organization [WMO] 1981, 1994). In the current study, we used daily rainfall.

To understand rainfall regime of Ly Son Island, the authors have used a statistical method applied for meteorological data analyses. Data analysis was performed using the Mann–Kendall method, WMO publication (Wilks 2005).

Climate regime of Ly Son Island is studied well. The present study is on rainfall regime using the dataset of the period of 1985–2015 observed at the Ly Son Marine Meteorological Island Station.

Ly Son Island in Quang Ngai Province is one of Vietnam's major islands with the development of many activities and has the potential to become a logistics center. Ly Son Island is situated between $15^{0}22'00$ and $15^{0}23'00$ N latitude and $109^{0}05'50$ and $109^{0}08'20$E longitude and 30 km far away from Vietnam mainland (Fig. 21.1). Ly Son Island has an area of $10.4km^2$ with the population of 19,307 people (2016). Economic activities are mainly on fishery, garlic cultivation, and tourism. A total of drills and boreholes for water using are 2149. Increasing demand for water on the island has contributed to the degradation and depletion of water resources on the island and especially saline intrusion into the groundwater system (website: lyson.gov.vn, quangngai .gov.vn).

Fig. 21.1 Ly Son Island

Ly Son Island was formed by 5 eruptive volcanoes. Groundwater presents essential resource on Ly Son Island, where the residents and tourists rely on aquifer for domestic supply from 2149 boreholes (2016) (Nguyen 2016; Quang Ngai Statistic Year Book 2016, http://lyson.gov.vn).

Many researchers have considered that water resources on Ly Son Island must focus on surface water resources with rainfall origin (Bui et al. 2017). Rainwater provides an excellent source of freshwater, especially in areas without surface water and groundwater, and is scarce or polluted.

Ly Son Island has only one reservoir Thoi Loi–a supplied source only for agriculture. On the island, there are many projects on groundwater exploitation. There have been only a few studies on surface water resources on Ly Son Island up to now.

There is a Marine Meteorological Station on the island that was established in 1984 and came into operation in 1985 including rainfall recording for the period 1985–2015. The Ly Son Marine Meteorological Island Station is one of the state stations on the system of marine meteorological stations of Vietnam. That is a good condition with 30 years continuing recording data to study the variation of statistical parameters of rainfall and the surface runoff in turn.

1.1 *Characteristics of hydrology and Water Resources of Ly Son Island*

There is no river or stream on the islands. During the rainy season, there are some streams, but it exists for a short time and ends almost after the rain stops. There is only one water reservoir, Thoi Loi, on the island, which has been put in use since 2012. The Thoi Loi water reservoir has a water volume of 27×10^4 m^3 by design, but there is only 60–70% of the water design volume in fact, and this is the only one water resource supplying for agricultural land of 60 ha of Ly Son Island. The water resources used for daily activities are mainly from drilling system, and there are many problems due to salinity intrusions that happen on aquifer qh. This situation requires drilling deeper to bazan aquifer (Bui et al. 2014).

1.2 *Oceanological Characteristics*

Semidiurnal irregular tide regime had dominated all the sea around Ly Son Island. The range of tide was 2.0 m on the spring tide. The sea current had NE direction with the speed of 50–70 cm/s in the summer time and 30– 60 cm/s in the winter period with SW direction. The average sea surface temperature was 26.1 °C, 2–3 °C lower than that of the mainland. The salinity changed from 30–31‰; the highest value was observed at 34%.

1.3 Climate

The climate of Ly Son Island is tropical monsoon and characterized by two seasons, moist and dry; two northeast and southwest wind systems alternately and sometimes strong storms occur, affecting the island. The mean air temperature was 26.7 °C, maximum air temperature was 36.8 °C (August), and minimum air temperature was 15.4 °C (March). The period from March through August is normally the driest season influenced by southwest wind. The wettest season normally lasts from September through February of the next year, being influenced by northeast wind.

The average amount of hours of sunshine was about 2430.3 h/year. The mean relative humidity was about 84.9% of all year around. The mean evaporation was about 964.4 mm. The mean wind speed was 1.5 mps, and the high value of wind speed of 5–10 mps was found during the northeast wind season. The maximal wind speed was 30–40 mps during typhoons. Typhoons occur from September to November, and the average number was 7 per year that strike on Ly Son Island (Bui et al. 2017).

The main purpose of the paper is to access some findings on rainfall regime such as trend of rainfall, maximum values of daily, monthly, and yearly rainfall. From these results, the study comes to estimate the surface runoff and the amount of surface water received from rainfall on Ly Son Island. These results may be useful to conclude that precipitation on Ly Son Island is rather copious in the rainy season and this is the reason why the study comes to propose that, on Ly Son Island, there should be preparation of all kinds of measures for collecting rainfall for supply during the dry season.

Based on the results of the rainfall assessment, the paper identifies the relationship between rainfall and surface runoff. Surface runoff for a particular catchment can be measured using stream gauging techniques or can be estimated. When surface runoff records are unavailable or if records are missing for various reasons, surface runoff can be estimated by some calculation methods. The most common technique available is correlation with rainfall records. Water balance equation is convenient to analyze the water balance of a small island for both surface water system and the groundwater system.

2 Methodology

2.1 Data Collection

The Ly Son Marine Meteorological Island Station was established in 1984 and started collecting the time series data from 1985 to 2015 (Fig. 21.2). Rainfall is measured using a rain gauge, which is read 4 times per day. The set data are examined by experts at the Ly Son Marine Meteorological Island Station.

Fig. 21.2 Ly Son Marine Meteorological Island Station

2.2 Frequency Method and Trend Linear Regression Equation

In statistic computation of rainfall data, the following main formulas are used:

$$\text{Deviation}: \quad \text{Ri} = R_i - R_{tb} \tag{21.1}$$

where R_i is the amount of precipitation in the period i and R_{tb} is the average precipitation of the period computation.

$$\text{Standard deviation}: \quad S = \left[\frac{1}{n}\sum_{t=1}^{n}\left(x_t - \overline{x}\right)^2\right]^{\frac{1}{2}} \tag{21.2}$$

$$\text{Maximum of the dataset}: R_{\max} = MAX\left(R_{1,}R_{2\ldots,}R_N\right) \tag{21.3}$$

$$\text{Minimum of the dataset}: R_{\min} = MIN\left(R_{1,}R_{2\ldots,}R_N\right) \tag{21.4}$$

$$\text{Amplitude of the dataset}: D_{Rx} = R_{\max} - R_{\min} \tag{21.5}$$

where N is the number of the dataset following day, month, season, and years.

One of the methods for analyzing tendency in climate oscillation is the regression method.

The method of regression applied here comprised x variation by time t, x = f(t). f(t) is a linear or nonlinear function. The regression equation for study on the variation of rainfall is as follows:

$$y = ax + b \tag{21.6}$$

where

$$a = \frac{S_y}{S_x} r \tag{21.7}$$

$$b = m_y - am_x \tag{21.8}$$

where m_x and m_y are the average values of x and y and S_x and S_y are standard deviation.

r is the regression coefficient and a has shown a variation of rainfall by time. Sight of value has shown a tendency of increase or decrease in precipitation.

In this study, we use the Mann–Kendall Z test to evaluate trends for a 30-year data series. For this case using a 95% confidence level, a significant trend will be confirmed if the value of Z is greater than 1.96 (Thorsten Pohlert, 2018).

2.3 *Water Balance equation and Surface Runoff*

A simple surface water balance equation applied to a typical small island is described as follows:

Precipitation (*P*) - Evapotranspiration (*ET*) from all sources - surface runoff (*SR*) - groundwater recharge (*R*) - increases in soil water storage (*AV*) = *W*.

This equation may have terms deleted or added wherever appropriate. It can be applied to a single basin or to a large area. Sometimes, a water balance equation is applied to an average situation, using the average values of the hydrological variables and assuming that they represent a long-term average value. In this case, the terms of storage changes are ignored. Long-term reload values are better by averaging the results derived from a long data series.

When storage terms have been omitted in the long-term water balance, the mean value does not change. As usual, annual precipitation is a useful general application for the hydrological study of a specific island. As is often the case, the small island's water resources are sensitive to the short-term changes in rainfall due to their limited storage capacity in surface and groundwater systems. Monthly or weekly time series may be appropriate for the study of water balance, but generally the daily time series and sometimes the hourly time series are appropriate. It is good to analyze rainfall and surface runoff and their relationship to maximum values (Yue et al. 2002).

All of these concepts are useful to study the surface runoff and the regime of rainfall of the Ly Son Island.

We use the following surface water balance equation for islands:

$$P - ET - SR - GWR = 0 \tag{21.9}$$

where.

P: precipitation (mm).

ET: evaporation (mm).

SR: surface runoff (mm).

GWR: groundwater recharge (mm).

And the surface water runoff equation for Ly Son Island is as follows:

$$SR = P - ET - GWR \tag{21.10}$$

Based on the surface runoff (*SR*) results computed, we can estimate the potential surface water for Ly Son Island. The amount of surface water runoff per year over an area is computed using the following formula:

$$\text{Wyear} = \text{Yyear}\left(\text{Flv} * 1000\right) \tag{21.11}$$

where W_{year} (m^3), Y_{year}: surface runoff (mm), F_{lv}: area (km^2).

2.3.1 Precipitation

The main input term for an island's water balance is precipitation, usually in the form of rain. Precipitation is measured using a rain gauge at selected locations. Rain gauges are usually read daily but can also be continuously recorded.

2.3.2 Evapotranspiration

The processes that combine evaporation and transmission are often referred to as total evaporation. Rainfall and evaporation are two important parameters of the water balance of small islands. Brutsaert (1982) recently provided a more detailed document of estimation methods for most practical applications of evapotranspiration. Formulations based on meteorological or evaporation data are often used to estimate evapotranspiration. These formulas have been verified from empirical results.

According to A Falkland (1991) values of hydrologic aridity index (P/E_0) for selected small islands and climatic zone, we can use high values of P/E_0 as humid climates in the Sulu Sea (west side of Philippines) for Ly Son Island of Vietnam.

2.3.3 Groundwater recharge from Precipitation

Recharge from rainfall constitutes all or most of the input to groundwater on small islands. In humid tropics, the average recharge value can reach hundreds of millimeters per year or 25 to 50% of rainfall if it is not lost due to the direct flow to the sea. Recharging values below 10% or even less than 1% of rainfall are common. Several methods have been outlined in many published documents to calculate or estimate these recharge values for islands (Falkland 1991; Nguyen 2016).

3 Results

3.1 Computed Maximal Rainfall of Ly Son Island

The daily rainfall data were used to assess and analyse the rainfall intensity. Table 21.1 shows the results of the computed standard deviation and variation of monthly rainfall on Ly Son Island. The months have been chosen to represent seasons, January for winter, July for summer, and April and October for transition time. From the table, we can see the standard deviation of rainfall in summer and winter is in the range of 81–100 mm. These months have the highest values of standard deviation in the range of 200–350 mm, being coincident with months with the highest rainfall (September to October). Whereas, the standard deviation of yearly rainfall is rather high and it is approximated to be 731 mm (equivalent to nearly 13% of variation). So the monthly rainfall data show a very good pattern.

Figure 21.3 shows the maximum and monthly average rainfall on Ly Son Island in the whole period of 1985–2015. We can see that the maximum rainfall was approximately 1900 mm in September. This is the month of rainy season, and with such large rainfall numbers, the addition of surface water or underground water is worth investigating. The analysis has shown that the maximum daily rainfall in Ly Son was moderated, with the highest value of 418.4 mm occurring on May 18, 1986 (Fig. 21.4). October has a maximum rainfall with an average value of 519.3 mm and a maximum value of 970.7 mm (Table 21.6). The mean annual rainfall on Ly Son Island recorded over 30 years (1985–2015) was 2278.8 mm (Table 21.3). During the recorded period, the rain range was from 1208 mm to 3238 mm (the maximum value of 3238 mm occurred in 2009, and the lowest record value of 1208 mm occurred in 2004). The average number of rainy days is 124 days per year, and the maximum is 160 days (1996). Precipitation can occur in all months of the year. The

Table 21.1 Standard deviation (S) and variation (Sr) of rainfall data (1985–2015) of Ly Son Island

Parameters	Month represented				Yearly
	January	April	July	October	
S (mm)	100	102	81	197	731
Sr (%)	79	163	173	38	13

Fig. 21.3 Maximum and monthly average rainfall of Ly Son (1985–2015)

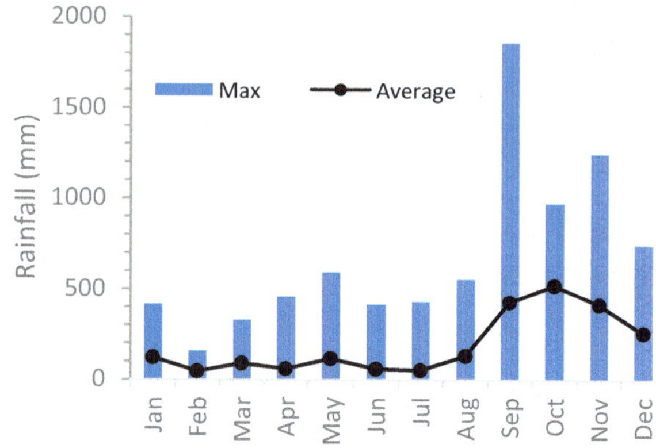

Fig. 21.4 Maximum daily rainfall during various months of Ly Son (1985–2015)

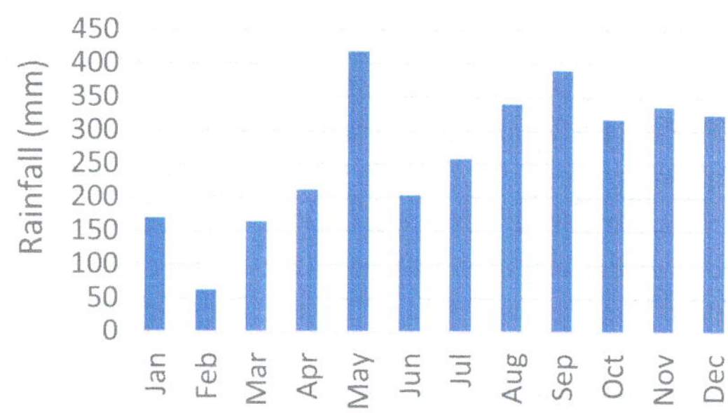

Table 21.2 Maximum daily rainfall frequency (%) in Ly Son in various periods (1985–2015)

	Ranges of rain (mm)				
Periods	<50	50–100	100–200	200–300	> = 300
1985–1990	51.2	26.2	10.7	8.3	3.6
1991–2000	58.3	18.3	15.8	5.8	1.7
2001–2015	61.1	19.4	12.5	4.2	2.8
1985–2015	57.8	20.7	13.2	5.7	2.6

dry season lasts from March until August, meanwhile the rainy season from September until February next year, with the most rain events occurring in the rainy season (78.1%).

The number of days with rainfall in the range of 50–100 mm was highest in October, about 6 days and the number of days of monthly rainfall with range > 100 mm reached the peak of 5 days in September. (Figs. 21.5 and 21.6). Most of the daily rainfall in the study area was in the range of less than 50 mm, accounting for 57.8%, and the rainfall in the range of 50–100 mm with frequency of 30.96% and the rainfall in the range of more than 300 mm with frequency of 2.6% for the whole period of 1985–2015 (Table 21.2). The maximum daily rainfall values changed with each period; from 1985 to 1990, and this value dropped sharply, down by 6 mm/year, but in the next cycle (1991–2000), there was a slight increase, and in

Table 21.3 Average values of rainfall and evapotranspiration of Ly Son meteorological stations (1985–2015)

Month													
Parameters	Jan	Feb	Mar	Apr	May	Jun	Jul	Aug	Sep	Oct	Nov	Dec	Total Annual
Precipitation (mm)	121.2	45.7	88.2	59.3	114.2	57.9	50.2	129.3	424.8	519.3	413.2	255.5	2278.8
Evapotranspiration (mm)	64.4	49.7	45.2	48.7	68.2	90.2	103.9	104.7	81.5	74.4	73.1	73.2	877.2

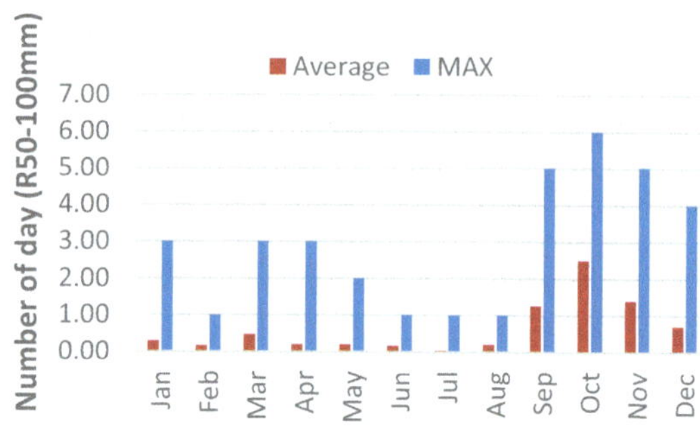

Fig. 21.5 Number of days of monthly rainfall in Ly Son with the range of 50–100 mm

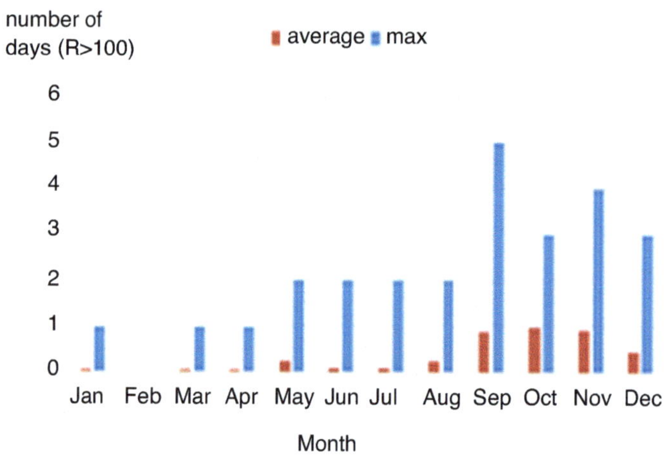

Fig. 21.6 Number of days of monthly rainfall in Ly Son with the range > 100 mm

2001–2015 period, the maximum daily rainfall went up dramatically (Fig. 21.7). But in summary, the trend of maximum daily rainfall was found to be slightly reduced in the whole period of 1985–2015 (Fig. 21.8). These signs may be associated with El Nino activity in the West Pacific.

3.2 *The Trend of Average Rainfall and Maximum Rainfall in Ly Son Island*

3.2.1 Trend of Average Rainfall

The total monthly rainfall, annual rainfall and maximum daily rainfall were used to analyse the trend (Pohlert, 2018). Data are broken down into four phases: before 1991, 1991–2000, 2001–2015, and the whole period from 1985 to 2015. In each period, the rainfall has changed significantly. The average rainfall observed from 1985 to 1990 was about 2363 mm, which is higher than the average rainfall of the

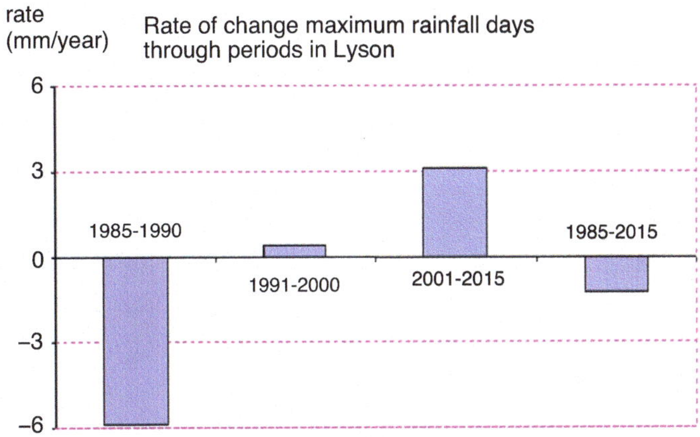

Fig. 21.7 Rate of maximum daily rainfall on Ly Son Island

Fig. 21.8 The trend of maximum daily rainfall on Ly Son Island

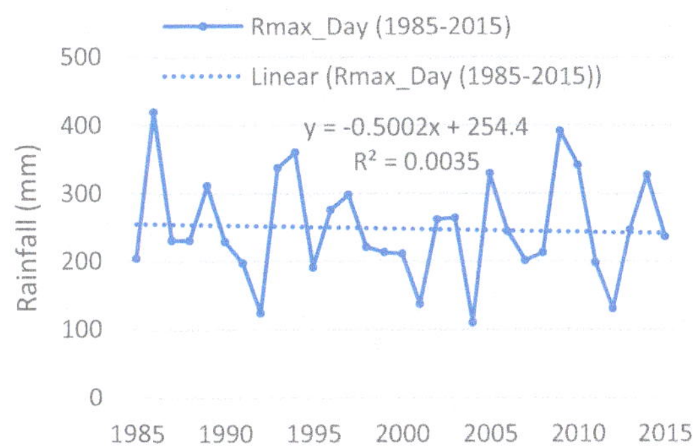

1985–2015 period (2278. 8 mm). The average rainfall in the 1991–2000 period was lower than that in previous years, but was generally greater than that during the annual years, about 61.2 mm. In the recent period (2001–2015), rainfall has tended to decrease by about 100–150 mm compared to the previous two periods, corresponding to a reduction of over 70 mm compared to the all-year average (Fig. 21.9).

The trend of rainfall change in Ly Son with several representative months is shown in Figs. 21.10 and 21.11. The results show that the monthly rainfall in Ly Son has a high rate of change, and there is no clear rule. In the first month of dry season (January), the rainfall is likely to increase, while in the first month of rainy season (July), the total rainfall decreases sharply about 3–5 mm /year (Fig. 21.10 and Fig. 21.11).

The total rainfall has increased in all periods. The highest increase was in the period of 1991–2000 with an increase of 79.9 mm per year (Fig. 21.12). For 30 years from 1985 to 2015, rainfall has increased but not significantly, about 0.1 mm / year (Fig. 21.13).

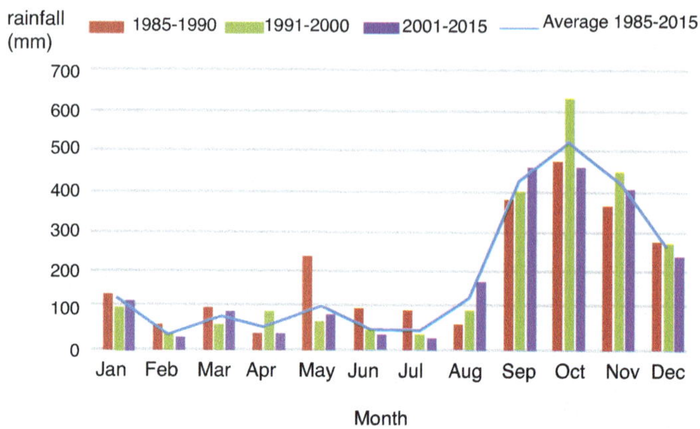

Fig. 21.9 Monthly average rainfall in Ly Son in each period

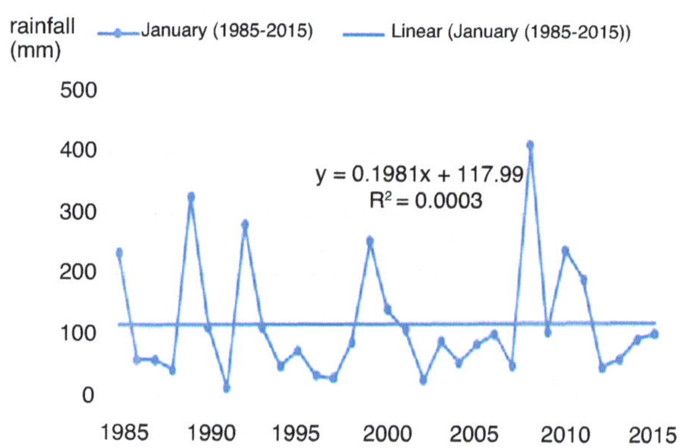

Fig. 21.10 Trend of Ly Son rainfall in January (1985–2015)

3.2.2 Trend of Maximum Daily Rainfall

Concerning the maximum rainfall in a day, the calculated results show that there is no significant change in frequency over time. According to the distribution of rainfall, the majority of daytime rainfall in the range of less than 50 mm had a frequency of more than 50% in all periods, and the frequency of moderate rainfall in the range of 50–300 mm was between 4.2 and 26.2%. Particularly, heavy rainfall in 24 hours (≥300 mm) accounts for a small proportion, ranging from 1.7 to 3.6% and a full-time average of 2.6% (Table 21.2).

While the total rainfall in all periods has increased, the maximum rainfall in the period of 1985–1990 and the whole 1985–2015 period are likely to decrease. However, for the recent years, from 2001 to 2015, the total annual rainfall as well as the maximum daily rainfall is likely to increase (Figs. 21.12 and 21.13). The average annual rainfall measured on Ly Son Island was approximately 2278.8 mm, while evapotranspiration was 877.2 mm (Table 21.3). In the rainy season, from

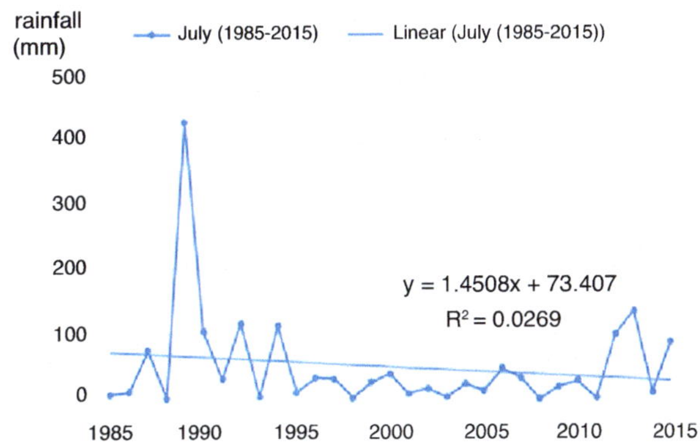

Fig. 21.11 Trend of Ly Son rainfall in July (1985–2015)

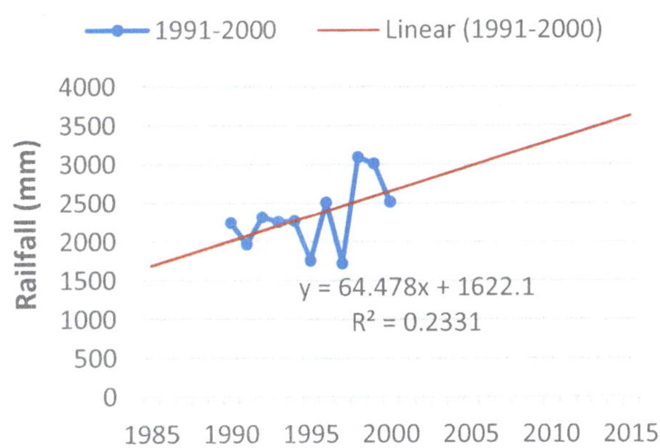

Fig. 21.12 Trend of rainfall in Ly Son in the period of 1991–2000

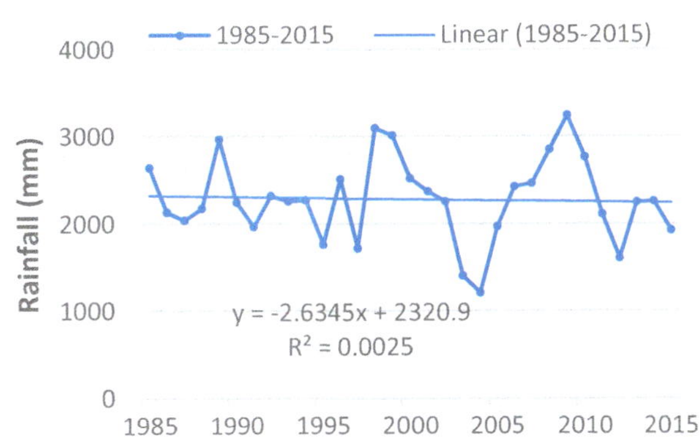

Fig. 21.13 Trend of rainfall in Ly Son in the period of 1985–2015

September to February, rainfall reached 1779.6 mm, accounting for the vast majority of the average rainfall throughout the period (Table 21.4). Reserving this abundant rainfall in the rainy season will solve the lack of water for living and economic development on the island.

3.3 *Calculating the Total Surface Runoff*

Based on formulas (21.9), (21.10), and (21.11) described above, the surface runoff components are computed. With the annual average rainfall of 2278.8 mm/year and the area of Ly Son Island of 10.4 km^2, the total amount of rainfall water on the island was 22.742 x 10^6 m^3 /year or 62,308 m^3 /day.

(−) is the value that does not generate surface flow in that month.

According to the calculation results in the table, the average surface water runoff for the period 1985–2015 was 1419.94 mm, rounded to 1420 mm. With an area of 10.4 $km^2$of Ly Son Island, the total annual surface water flow on the island is 14.17 x10^6 m^3/year. It is noticed that the user demand of water in Ly Son Island was 17,597 m^3/day (1051 m^3 for domestic use, 16,246 m^3 for agriculture, and 300 m^3 for other), and for the whole year, Ly Son Island needed 6.422 x 10^6 m^3. The computed results also show that, if appropriate measures are taken for collecting all kinds of rainfall, enough freshwater will be supplied for the whole Island in a year.

The computed results of surface water runoff and water resources originating from rainfall (Tables 21.5 and 21.6) show the potential yield of rainfall on Ly Son Island. One day with a maximum rainfall of 418.4 mm resulted in a total surface water of 3.2 x 10^6 m^3 on the whole Island, which occupied 22.58% of average annual surface water. One month (October 1998) with a maximum rainfall of 970.7 mm greatly contributed to the average surface flow up to 61.67%, and the average rainfall of October resulted in 4.33 x 10^6 m^3 of surface water flow, which occupied 30.55% of annual surface flow on the island. These figures show that, if there are good solutions chosen for keeping all the waterflow, water shortage on the island will be improved. However, in practice, solutions to store rainwater using boreholes are limited and most of the boreholes are salty at the present time. Furthermore, on Ly Son Island, there is only one reservoir, Thoi Loi, supplying water for agriculture. Therefore, it is necessary to have the solutions of works, non-works, planning and use of water resources, as well as the government concern to soon overcome the water scarcity on Ly Son Island.

Table 21.4 Seasonal rainfall distribution

Station		Annual	Precipitation in dry months							Rainfall in rainy months						
			Mar	Apr	May	Jun	Jul	Aug	Dry season	Sep	Oct	Nov	Dec	Jan	Feb	Rainy season
Ly son	Mm	2278.8	88.2	59.3	114.2	57.9	50.2	129.3	499.2	424.8	519.3	413.2	255.5	121.2	45.7	1779.6
	%	100%	3.9	2.6	5.0	2.5	2.2	5.7	21.9	18.6	22.8	18.1	11.2	5.3	2.0	78.1

Table 21.5 Results of annual average surface runoff calculation for the 1985–2015 period

Month	Average monthly rainfall (P) (mm)	Evaporation (E) (mm)	GWR (groundwater recharge) (%)	GWR (groundwater recharge) (mm)	SR (surface runoff) (mm)
Jan	121.2	64.40	5	6.45	50.31
Feb	45.7	49.70	7	2.98	–
Mar	88.2	45.20	6	5.27	37.75
Apr	59.3	48.70	7	4,40	6.22
May	114.2	68.20	5	6,22	39.81
Jun	57.9	90.20	7	4,12	–
Jul	50.2	103.90	7	3.43	–
Aug	129.3	104.70	6	7.29	17.28
Sep	424.8	81.50	2	10,23	333.12
Oct	519.3	74.40	2	11.01	433.91
Nov	413.2	73.10	3	10.34	329.72
Dec	255.5	73.20	4	10.45	171.82
Total	2278.8	877.20		116.58	1419.94

Table 21.6 The computed results of surface runoff (SR) by maximum rainfall in the period of 1985–2015

Maximum rainfall	Rainfall (P) (mm)	Evaporation (E) (mm)	GWR (%)	GWR (mm)	SR (mm)	W (x10^6 m^3)	%	Average surface flow (x10^6 m^3)
Maximum rainy days (may 18, 1986)	418.4	68.2	0.07	29.29	320.91	3.2	22.58	14.17
Maximum rainfall of October 1998	970.7	74.40	0.02	20.58	875.72	8.74	61.67	
Average rainfall for October (1985–2015)	519.3	74.40	0.02	11.01	433.91	4.33	30.55	

4 Conclusions and Discussion

4.1 Conclusion

The results of the current study show that the rainfall of all periods tends to increase with a value of 0.1 mm / year. The average annual rainfall during the period of 1985–2015 was 2278.8 mm, and the rainfall was distributed throughout the year. In the dry months, the average rainfall was 499.2 mm, while in the rainy months, the rainfall was 1779.6 mm. In addition, it should be noticed that the evapotranspiration in the months had maximum rainfall oscillations in the range of 2–4%. Prevalent rainfall in the range of less than 50 mm had a frequency of 57.8%, the rainfall in the range of 50–100 mm occupied 20.7%, and the rainfall in the range of more than 100 mm had a frequency of 21.5%. The maximum daily rainfall was 418.4 mm (occurring on May 18, 1986), the maximal value of the month with the maximum rainfall was 970.7 mm (October), while the average maximal rainfall value of the month with the maximum rainfall was 519.3 mm.

Based on the water balance equations and the computed results of maximum rainfall as shown above, this study has pointed out that the average surface runoff for the period 1985–2015 was 1420 mm corresponding to a mean annual precipitation of 2278.8 mm. With the area of 10.4 $km^2$of Ly Son Island, the estimated potential yield of surface water by the whole Island was 14.17 x 10^6 m^3 /year.

Furthermore, this study has also shown that the surface runoff of the day with the maximum rainfall (418.4 mm, May 18, 1986) was 320.91 mm and the amount of water flow on the island was 3.2x10^6 m^3, which occupied 22.58% of all year/the whole year. The surface runoff of the month with the maximum rainfall (October, 970.7 mm) was 875.72 mm, and the estimated potential yield of rainfall by the whole Island is 8.74 x 10^6 m^3, occupying 61.67% of all year/the whole year. The average surface runoff of the month with the maximum rainfall (October 1998, 519.3 mm) was 433.91 mm, and the estimated potential yield of surface runoff by the whole Island is 4.33 x 10^6 m^3, occupying 30.55% of all year.

The results of the study on the rain regime and surface runoff on Ly Son Island will be the basis for further implementation for other problems simulating the interaction between rainfall and groundwater replenishment for aquifers of Ly Son Island. On the other hand, the data on the maximum rainfall as well as the intensity of rainfall on Ly Son Island have important reference values to address the problem of exploiting the rainwater potential in the island.

4.2 Discussion

Precipitation is the most important element of water resources on island. Thus, we need to concentrate on researches about rainfall regime. From that, calculating surface runoff then creating a basis for research on underground water recharge from rainfall can be done for the next study. Time series data of rainfall collection in 30 years on Ly Son Island were highly reliable and also applied on water resources research and water resources management in general. The statistical values about rainfall such as daily maximum rainfall and monthly maximum rainfall also have standard deviation and variation within the allowable limits (Table 21.1). The tendency of precipitation increased in most of the periods of observation (Figs. 21.10, 21.11, 21.12 and 21.13). In the present study, rainfall from storm surges that pass through Ly Son Island is not considered abnormal.

According to the annual variable, the rainfall in Ly Son is not evenly distributed among the months of the year and does not show the rainy or dry season. The only difference is that the rainy season and the rainy season are less. Two extreme rainfall events occur in the year, the first extreme coincides with the peak of the rainy season (from September to November), and the second occurs most frequently in May or June. According to the intensity of rainfall, the frequency of maximum rainfall in the day is common at the moderate rainfall level. With heavy rain, Ly Son has a frequency ranging from 19.5 to 23.0%. Prior to the year 2000, the rainfall in the study area was very high during the rainy season. However, in the recent period (2001–2015), very high rainfall during the day sometimes occurs in low rainfall months (from January to March). Rainy days and heavy rains tend to decrease, but the number of days with heavy rains tends to increase. There are no obvious and variable changes in the trend of monthly, seasonal, or yearly rainfall. The total rainfall has increased in all periods, but peak rainfall throughout the period tends to decrease.

Unusual changes in precipitation, rainfall intensity, as well as large differences between the rainy season and dry season on Ly Son Island are related to the anomalies of climate change in the Central Vietnam. Therefore, there is a need for more thorough studies on the impact of climate change on rainfall regime on Ly Son Island.

The method of assessing the surface water potential from rainwater only if the equation of water balance is used is not enough, but the surface runoff distribution on the whole Island must be considered, so as to use other models of simulation (Bui et al. 2019). In the equation of surface water balance, some parameters are still required to accept the unobserved values at the island. The Soil Water Assessment Tool (SWAT) combined with Geographic Information System (GIS) should be recommended for the study of water resources in Ly Son Island.

Acknowledgments This research is part of a National Research Project funded by the Ministry of Science and Technology of Vietnam (The Project Code: KC 09 04/16 - 20, namely "Estimation of potential and variation of surface, underground water resources and proposed appropriate measures for water resource management to support economics - social development on some important islands"). The authors would like to thank the Ministry of Science and Technology of Vietnam for their help.

References

Bui, X. T., Le, T. D., & Truong, V. C. (2017). Estimation of surface water in Ly son island, journal of marine science and technology, 4 (T.17). *Vietnam Academy of Science and Technology, 1859–3079*, 353–363. (In Vietnamese).

Bui, X. T., Nguyen, V. D., Nguyen, M. T., & Nguyen, N. H. (2019). Using the method of soil conservation service curve number (SCS CN) combined with the geographic information system (GIS) to estimate the surface runoff on the co to island, North Vietnam. *International Journal of Advance Engineering Research and Science, 6*(9), 59–75.

Bui, V. D., Stattegger, K., Nguyen, T. T., Phung, V. P., Tran, T. D., & Bui, X. T. (2014). Late Pleistocene - Holocene seismic stratigraphy of Nha Trang shelf. *Marine Petroleum Geology, 2014*, 1–12.

Chapman, T. G. (1985). The use water balance for water resources estimation with special reference to small islands. *Australian Development Assistance Bureau Bull, 4*, 39.

Falkland, A. (1991). Hydrology and water resources of small island: A practical guide. United Nations Educational, Scientific and Cultural Organization 7 (UNESCO).

Hennemuth, B., Bender, S., Billow, K., Preier, N., Keup, E., Krager, O., Mudersback, C., Radermacher, C., Schetter, R. (2013). Statistical methods for the analysis of simulated and observed climate data, applied in projects and institutions dealing with climate change impact and adaptation. CSC Report 13, 41–77, Climate Service Center

Khaled, H. H., & Ramachandra, R. (1998). A modified Mann -Kendall trend test autocorrelated data. *Hydrology, 204*(1–4), 182–196.

Nguyen, V. D. (2016). Ground water in Ly son island and exploitation oriented. *Natural Resources and Environment Journal., 24*(254), 12. (In Vietnamese).

Phan, V. T. (1999). Statistical method in climate. Hanoi National University Publishing, 41–77. (In Vietnamese).

Pohlert, T. (2018). Non - parametric trend tests and change point detection 1–18p. This work is licensed under a creative commons license (CC BY - ND 4.0). Retrieved from http://creativecommons.org/licences/by-nd/4.0.

Quang Ngai Statistic Year Book. (2016). *General Statistics Office of Quang Ngai Publishing*, 377 pages (In Vietnamese).

Ramsey, P. H. (1989). Critical values for Spearman's rank order correlation. *Journal of Educational Statistics, 14*, 245–253.

Sen, P. K. (1968). Estimates of the regression coefficient based on Kendall's tau. *American Statistical Association, 63*(324), 1379–1389.

Sokolop, A. A., Chapman, T. G. (1974). Methods for water balance computation. UNESCO, studies and report in hydrology, N. 17. Paris, 127.

Vu, T. H, Chu, T. H, Phan, V. T (2009). Trend of maximum day rainfall in VietNam in the period of 1961–2007. Journal of nature science and technology. Hanoi National University, Number 25, 3S.

Wilks D. S. (2005) Statistical methods in the atmospheric sciences, Department of Earth and Atmospheric sciences, Cornel University. 26–27, 35–39, 104–109.

World Meteorological Organization [WMO]. (1966). Some Methods in Climatological Analysis (WMO/TN No. 81, WMO-No. 199).

World Meteorological Organization [WMO]. (1981). Selection of Distribution Types for Extremes of Precipitation (OHR-No. 15, WMO/TN-No. 560).

World Meteorological Organization [WMO]. (1986). Manual for Estimation of Probable Maximum Precipitation (OHR-No 1, WMO/TNNo. 332).

World Meteorological Organization [WMO]. (1994). Guide to the Applications of Marine Climatology (WMO-No. 781).

Yue, S., Pilon, P., Phinney, B., & Cavadias, G. (2002). The influence of autocorrelation on the ability to detect trend in hydrological series. *Hydrological Processes, 16*, 1807–1829.

Yue, S., & Wang, C. (2004). The Mann - Kendall test modified by effective sample size to detect trend in serially correlated hydrological series. *Water Resource. Manage., 18*, 201–218.

Chapter 22
Factors Affecting Community Forest Management in Ha Giang Province, Vietnam

Dao Thi Thu Trang

Abstract With the current deforestation in the world, to save the earth and forest resources, residents need to have a long-term strategy for their management and conservation. A form of forest management that has been applied in Vietnam as well as in several other countries in Asia is the model of community forest management. This model was initially promoted owing to its effectiveness in improving the lives of residents and restricting rampant logging, reducing forest over-exploitation. Community forest management is a forest management method based on traditional knowledge and experience of community aspirations, and is aimed at enhancing capacity and strengthening of partnerships to share experience and community stakeholders in order to manage resources sustainably and contribute to improving the material, spiritual, cultural, and ethnic life of communities living in and near forests. This study focuses on the factors affecting community forest management in Ha Giang through the subjective opinions of the respondents. Then, it identifies what factors have been done better and the factors that need to be completed in order to improve the community forest management here. From the data collected, tested scales, and factor analysis, the paper has constructed a linear regression model. The study results are also valid recommendations for organizations looking to develop appropriate strategies for achieving effective community forest management in Ha Giang.

Keywords Community forest management · Likert 5 · ANOVA analysis · Ha Giang · Vietnam

---

D. T. T. Trang (✉)
VNU Economic Business University, Vietnam National University, Hanoi, Vietnam

A. T. Nguyen, L. Hens (eds.), *Global Changes and Sustainable Development in Asian Emerging Market Economies Vol. 2*, https://doi.org/10.1007/978-3-030-81443-4_22

1 Introduction

In recent times, in developing countries in general and in Vietnam in particular, forest resources have been reduced more in terms of quantity as well as quality. Thus, countries face two problems: poverty and the consequences of degradation of natural resources and the environment. Policy and strategy development should be designed and implemented toward two objectives, which are reducing poverty and improving community life, sanitation, and efficiently using natural resources. Therefore, management, protection, and development of forest resources is an urgent task of all residents, as well as the government bodies from the central to the local level. To protect, manage, and develop forest effectively requires the active participation of residents living near the forest.

Village communities are self-governing social units that play a particularly important role in managing and protecting forests, especially community forests. Since the Law on Forest Protection and Development was promulgated in 2004, village communities still have not yet been fully recognized and do not have access to the right to participate in the management, use, protection, and development of community forests, the same as other forest owners. Community forest allocation policies are still inadequate, making community forest management in localities in Vietnam difficult.

Ha Giang is a mountainous province that borders the northern forest; its area for forestry development is 553,138.3 ha. Currently, Ha Giang has six special-use forests, protected species, and habitat areas. With the first national location, the role of forests in Ha Giang is particularly important, not only to preserve the environment and natural resources in Ha Giang and the northern provinces in particular but also for the whole country. Ha Giang Forest is one of the forests that is to be managed, protected, and developed. For a long time, the life of the local people, especially ethnic minorities who have considered a valuable asset for their lives and spirits. In the history of its development, the ethnic communities of Ha Giang province were aware of the benefits of forests and the need to protect the forest, which was formed by the customs, a sense of forest management, which guide and regulate human relations and community forests. However, at present, deforestation and forest degradation in Ha Giang remain, requiring clear strategies to improve the management and protection of forests.

Therefore, it is necessary to study the factors affecting community forest management in Ha Giang so that local authorities have appropriate solutions to improve the effectiveness of forest management here.

2 Literature Review

2.1 Concept

2.1.1 The Concept of Community

In Vietnam, the concept of "community" as used in forest management can be generalized into two categories incorporating the following key points:

First, the "community" is a collection of residents who live close together in a small society where there are similarities in terms of cultural, economic, social traditions, customs, relations in production and life are often intertwined and spatial boundaries in a village. According to this concept, "community" is the "village communities" (hereinafter "villages" are collectively referred to as "rural" in accordance with the Law on Forest Protection and Development (National Assembly of the Socialist Republic of Vietnam, 2004).

Second, the "community" is used in forest management when talking about groups of residents who have intertwined relationships in production and life. Thus, according to this view, "community" is not only the community but also the entire village, including ethnic communities in rural areas and family or community groups in rural households.

Despite these different conceptions of community, most of the comments were that "community" is used in forest management when talking about rural communities.

Article 3 of the Law on the protection and development of forests in 2004 defines "Community is the whole village households and individuals living in the same village, hamlet, village, hamlet or single equivalent" (National Assembly of the Socialist Republic of Vietnam, 2004).

2.1.2 The Concept of Community Forest Management

The concept of community forest management has been discussed for decades, but in fact it has not been fully defined. In the Asia forest network, the views of community forest management generally refer to the activities of the community toward the management and sustainable use of forest resources.

Globally, the concept of community forest management was first defined in 1978 by the Food and Agriculture Organization (FAO) at the world forestry conference. That is "all forestry activities in which the community participates, including works in the garden, harvesting forest products for the needs of human life and planting trees at farms, production and processing of forestry products in households and cooperatives to increase income for the communities living near the forest" (FAO, 1978).

Quach Duong gives a simple concept. That is "process forest management, protection and development based on indigenous knowledge, traditional architecture, festivals and customary law of the community." Community forest management includes activities of individuals and communities relating to forest management, protection, and development (Dong, 2005).

Indeed, it is hardly for a full definition to reflect the reality of community forest management because natural, economic, and social conditions are different in different parts of the world. Thus, the forms of community forest management are also very different. In addition, community forest management is not only framed in community activities but also involves multiple stakeholders such as policy makers, government agencies, nongovernment agencies, sponsorship, and scientists. The involvement of these organizations also impacts on the process of forest management and protection as well as economic and conditions of communities.

Although not a completely accurate definition of community forest management, the process of developing forest communities in fact has not decreased. Community forest management has been done for hundreds of years, and was carried out before the scientists mentioned it. Ecological and social effects of community forests have shown that community forest management is one of the logistic operations that is most efficient at finding out the principles and strategies employed in the management, protection, and development of forest resources (Poffenberger, 2003).

In terms of science, community forest management has only been known since the early 1970s, when droughts in Africa and floods in Asia seriously impaired forest resources. Fuel for rural communities become increasingly scarce. It was at this time the experience of community forest management in India (social forestry model), South Korea (village garden model), Thailand (the village forest model), and in Tanzania (the village plantation forest model) attracted the interest of scientists globally and they are considered a solution to forest development and fuel shortages in rural areas. By the late 1970s, the concept of community forest management was widely recognized worldwide.

In 1978, the World Forestry Congress took the title "community forests" to honor and promote community forest activities (Arnold & Stewart, 1991). In the 1980s, the forest development project expanded to communities around the world, particularly in India and Nepal. The name of the community forest also changed to "Joint Forest Management," "Social Forestry," "Community-Based Forest Management," etc. However, the nature of community forest management has not changed; it is the process in which the residents are centered on forest management, protection, and development. At the end of the 1980s and 1990s, scientists focused more on research institutions in community forest management, including institutional traditions and institutions of the State, in order to create a legal framework for the development of community forests. At this stage, the concept of ownership is given to discuss a wide range, including State ownership, private ownership, community ownership, and free use. For a while the community forest concept was criticized vehemently in the views of Hardin in "The Tragedy of common ownership" that the community ownership of forests is synonymous with liberal use. It is a form of use that every member wants to take, maximizing general benefits for themselves, thus overexploiting the forest (Hardin, 1968).

Contrary to Hardin, Arnold and Jongma suggest that forest communities bring greater efficiency in forest development and community development. They stressed that the community forest is an integral component of rural development, which is mainly aimed at helping poor communities to sustain themselves and grow their lives, etc. Thus, forest community is for the residents and there is residents' participation in management and development (Arnold & Jongma, 1978). With a view like that, they have pointed out three basic goals of the community forest: (1) providing fuel and other vital supplies to serve the basic needs of the community, (2) providing a sustainable source of food and habitat for a process of continuous food production, and (3) having income, resolving employment for local residents. Burda et al. also acknowledged the community forest management: "The people living near the forest understand very well the characteristics of the forest as well as the influence of the forest on their lives. Therefore, they manage the forest more responsibly and effectively than the bureaucratic, inflexible and inconsistent management of the government" (Burda, Curran, & Gale, 1997).

Although community forest management helps residents to live closer to nature and thereby establish institutions, plans to use forest management, and a more efficient manner, community forest management has created a system that is responsive to rapid decision-making and actions to adapt to the changes of specific conditions. The decision is to meet the interests of the entire community, those directly responsible for making those decisions.

Herb also gave arguments in favor of community forest management: "forest management by communities creates opportunities to find solutions to problems responsibly because communities are aware of the realities of everyday life" (Herb, 1991). Despite the debate, the community forest has developed naturally and quickly. Many parts of the world have accepted it and it is seen as an important strategy in the management, protection, and development of forest resources.

Community forest management is not new in Vietnam. Many studies have shown the existence of traditional forest management by mountain communities in different places across the country. With forest allocation policy since the early 1990s and the support of development projects in the country in the forestry sector, the involvement of local communities in forest management has increased.

Community forest management is a traditional form of management in the mountainous region of Vietnam. The Vietnam ethnic communities of different scales: tribe, clan, village, etc., played a vital role both in forest management and in forest protection and residents use forest resources logically within a conventional community. Ethnic communities use forest resources to service their life: wood, bamboo for houses, firewood, herbs for healing, etc.

So far, in Vietnam, in terms of both theory and practice, the community forest management has been recognized. Code management, protection, and development of forests in 2004 confirmed community ownership for forests and there were regulations on forest allocation for village communities. In the highlands, ethnic minorities consider forests as sacred assets, so they should manage and protect them effectively. These forests are managed by residents, with fairly close and effective protection.

From the content of the above definitions, one can see that forest management by the community is shown in two parts.

- The members of the community participate in business and management forests that belong to the general public (State-owned or assigned according to tradition). This management is direct.
- Communities manage forests owned by the State organization through contracts. Participation in the management is directly related to community life such as employment, income, products harvested, or the benefits of the forests and other benefits that cannot be calculated are enjoyed (such as watershed protection, creed, monuments, etc.) Community can join the general management of private forest patches that are delivered to every household in the community to achieve high efficiency in management. This form of management is considered indirect management.

2.2 *The Role of Community Forest Management*

- Community forest management can protect forests more effectively than individuals or households. At the time of the study, community-managed forests should not be subdivided for individual management because when there is a violation, it is more difficult for individuals to solve it than to bring it to the village council. When there is a risk such as a forest fire, a large number of people in the community can work together to cure it instead of an individual or a small family. This is a cost-effective forest management method (requiring little investment from the government), widely accepted.
- Community organizations in the local village work very closely; village leaders are elected democratically and receive State subsidies. Most communities have local conventions (written texts and unwritten) and are highly effective, showing the relationship of social ties of community members in a coherent way. Mechanisms of reward and punishment according to the conventions of the community have proved very effective.
- Managing community forests to meet the needs of protecting water sources for daily life and agricultural production; environmental protection and beliefs; and input for infrastructure construction as well as supplying agricultural products to improve people's living standards. People can change shifts in forest protection and can exchange experiences in using forests for production. People can exchange in guarding to protect the forest and can exchange experiences in using forests for production.

2.3 *Factors Affecting Community Forest Management*

2.3.1 Community Forest Management Policy

- Government Policy.

Based on practical issues in the economy, government policies such as orientation, measures, actions of government to solve problems, toward common development goals, and the effect of economic activity.

Government policies greatly influence the management of community forests. When the government recognized the village community, which means it is an organization that has legal status and is considered to be the object of State land and forests and the management, protection, and local forest (Nguyen, 2004).

Generally, in legal terms, forest allocation is handing local communities the powers of management, and the use of assets attached to forest land. Community residents have the right to manage, protect, develop, and have a common interest in serving the community.

Government policies encourage the community forest management to be done well and vice versa.

- Province Policy.

District policy is from the perspective of policy guidelines and actions of the district authorities in order to achieve the goal of economic, cultral, political, and social development. District policy on forest management are the guidelines, action plans, and methods on how to manage, protect, develop, and use the local forest district to achieve common development. Because the forest is a nature resource, efficient forest management and harvesting will bring great benefit to the community.

The policy of the district on forests also has a great influence on community forest management. Sometimes, the district's policy does not completely coincide with the government's policy. There are government policies that are proposed but not implemented in the district. Conversely, it may not be government policy but if the agency deems it to be necessary locally and efficiently, they can also make specific policy to encourage certain activities to be performed. Therefore, the district's policy on forest management is a new local policy that directly affects the activity in the forest district (Dong, 2005).

District policy encourages land and forest allocation, implementation of community forest management, and brings advantages and effectiveness. Conversely, if the district does not have policies to promote community forest management, it will be difficult to survive.

2.3.2 Understanding of Residents in Community Forest Management

Residents' understanding of forest management is that residents have sufficient information and knowledge needed to manage forests and their role in forest management. Residents become knowledgeable about forest management as they grasp the techniques and principles of exploitation and use of forest resources. It also means knowing how to handle problems of State forest deforestation, forest fires, forest regeneration, etc. (Poffenberger, 2003).

The more residents are knowledgeable about forest management, the more easily they can do it, participating in community forest management and performance management.

2.3.3 The Coordination between State Management Agencies and Residents in Community Forest Management

As forests are national resources and the existence of forests not only affects the lives of residents close to the forest, but also the environment and living conditions of the country. Therefore, the State has always had an important role in forest management. For community forest management, even if the residents are directly managing the forest, close coordination between State management agencies and local communities is still needed. This combination is shown in place: the State agency is entrusted to manage residents. In addition, State agencies also act as a guide, providing technical advice on management, protection, and development of forests, equipped with the certain knowledge of the community. It impacts positively on the management of community forests (Nguyen, 2009).

Because community forest management is a shared responsibility and relies on the collective power to protect common resources, collaboration between residents in forest areas is very important. Residents coordinate forest management and discipline in order to ensure that certain forest management activities are carried out in the community constantly, fully, and effectively.

This combination is that these residents together protect the forest, when problems occur together, including handling the assignments in a clear and fair way. Thus, the close collaboration between residents in forest management has a positive impact on community forest management.

2.3.4 Protective Equipment for Forest Management

Protective equipment for forest management is the equipment needed to help residents to handle problems that occur in forests with ease and convenience. For example, fire-extinguishing equipment when there iswildfire, protection devices, against those who are carrying out deforestation, the media for updating the public on forests and increasing the understanding of the residents, etc. For forest communities, the more fully equipped the community is with forest protection equipment, the more effective the community forest management will be (Swiss Association for Development and Cooperation, 2005).

2.3.5 Traditional Habits

Residents in the nearby forest often have different living habits. The traditional habits, and the concept of forest management and use, affect the management of community forests. The traditional habit of adjusting behavior affects residents and whether they are involved in community forest management or not. Traditional forest management of the various ethnic groups in Vietnam is reflected in the custom of keeping forests, planting trees, building protection regulations, and protection of trees of many villages. Most early forest community recognition and self- managed communities were traditionally well protected in accordance with customs and conventions (Nguyen, 2009). These forests have an important role in production, the meaning of life or spiritual, religious communities, and for almost the entire community to decide on the management, protection, and use of forests as well as benefiting from the forests.

2.3.6 Benefit from Forest Management

Residents see the benefit when they participate in forest management; thus, they will be enthusiastic about the job and promote effective forest management. They benefit from forest management as they are exploit forest use without affecting its long life and other residents. Benefits lead participants to explore creative ways of enhancing the quality of work. Thus, the benefits from forest management influence positive community forest management (Vo, 2012).

3 Methodology

3.1 Research Model and Hypothesis

The research questions are:

- What is the impact of factors on community forest management in Ha Giang?
- What should the Ha Giang provincial authorities adjust to improve the effectiveness of community forest management?

Based on the research purpose, nine following research hypotheses are set out:

- Hypothesis 1 (H1): Government and local forest management policies that encourage forest land allocation and recognition of the role of communities in forest management will positively impact community forest management. This effect is interpreted as the effect in the same direction.

- Hypothesis 2 (H2): Residents' understanding of forest management is that residents have sufficient information and knowledge needed to manage forests and their role in forest management. When residents have an understanding of forest management, they will voluntarily protect and know how to effectively protect forests. This is active promotion of community forest management.
- Hypothesis 3 (H3): State agencies have an important role in forest management orientation. The combination of residents in forest management is very important. Community forest management needs a large number of participants, ongoing management, and all problems being solved quickly and with the resolve of the collective strength to achieve the effect. There must be better coordination between the residents in the community forest areas. The close collaboration between residents will impact positively on community forest management.
- Hypothesis 4 (H4): Protective equipment has a positive impact on overall forest management and community forest management in particular. Protective equipment that is easy and more convenient is needed to help solve the problems that occur in forests.
- Hypothesis 5 (H5): The traditional habits of the residents are a major influence on forest management. If communities have traditionally loved forests and forest protection, they will enthusiastically participate in forest management.
- Hypothesis 6 (H6): Benefits from the forest always attract people. In the process of management, the community protects the forest but can also exploit and use benefits from the forest. The more long-term benefits the community enjoys from the forest, the more they want to participate in forest management.

With the above six hypotheses, the research model is set up as follows (Fig. 22.1):

Fig. 22.1 The conceptual model of the study. Source: Compiled by the author

3.2 Observed Variables

From the theoretical basis, the observed variables of the model will be built as follows (Table 22.1):

3.3 Questionnaires and Evaluation Scales

- Goal-setting questionnaire: validation of opinions of forest management staff, leaders, and residents in Ha Giang province to study the status of community forest management in the district.
- Base construction question: based on an assessment of the variable factors that affect community forest management are shown in the literature review.
- Structure of the questionnaire:
- The content of the questionnaire was divided into two parts:
- Part I: General information about the interviewee's characteristics, including age, gender, place of work, education, etc.
- Part II: Measuring the level of assessment of the factors affecting community forests management in Ha Giang Province.
- Assessment Scale.
- In this study the five-point Likert scale was selected, from totally disagree to totally agree. In addition, the scale represents the name that was chosen to set the questions on additional variables such as age, gender, income, etc.
- Formally, the questionnaire is printed on A4 paper, clearly presented, easy to read, easy to understand, and includes the contents of the above scale.

3.4 Sample Size and Data Analysis

The survey was conducted for forest managers, officers on Residents' Committees, and residents in Ha Giang Province. A total of 400 questionnaires were sent directly to the objects of the study. The total number of questionnaires collected was 355, after entering and cleaning the data, the number of valid questionnaires that were used for analysis is 353 (88.25%).

Research used nonprobability sampling (select elements involved in irregular random patterns). Samples were selected from the forest management staff, the office of the District People's Committee (DPC), and residents in Ha Giang Province.

This test was carried out with the use of SPSS software version 20.0. By using calculation tools, illustration, and test, the study will show the results of linear regression indicating the degree of influence of independent variables on dependent variables in attracting investment with a p value of 5%.

Table 22.1 The observed variables of the research model

	Scale	Code
1	Policy on community forest management	
1.1	The local government recognizes that community is an object of allocated forest land and private forest owners as a way of actually promoting community forest management	CS1
1.2	Local government directs the model building and community forest policy and related activities to promote community forest management	CS2
1.3	Local government building regulations guiding protection and development of forest communities in promoting community forest management	CS3
2	Residents' understanding of community forest management	
2.1	Residents' understanding that their role in protecting the forests will promote community forest management	HB1
2.2	Residents understand that the power of collective forest management will promote community forest management	HB2
2.3	Understanding techniques that use forest resources will impact positively on community forest management	HB3
3	Coordination of communities in community forest management	
3.1	Agency support of state management, residents' management, and the forest will impact positively on community forest management	PH1
3.2	Residents who work together will increase the effectiveness of community forest management	PH2
3.3	The combination of the residents to ensure that forest management is ongoing and timely handling of the work	PH3
4	Protective equipment for the management	
4.1	Protective equipment facilitates community-based forest management	TB1
4.2	Equipment protection reduces the risks of community-based forest management	TB2
4.3	Residents who are equipped with full protective equipment for forest management have a positive impact on the management of community forests	TB3
5	Traditional habits	
5.1	The habit of considering forests as a sacred public property encourages communities to participate in forest management	TQ1
5.2	The traditional customs of the residents to respect and protect the forest has promoted community participation in forest management	TQ2
6	Benefit from forest management	
6.1	Joining forest management can help residents to exploit and receive income and other benefits from forests and will promote community-based forest management	LI1
6.2	Residents are all participating in forest management while promoting community forest management	LI2
6.3	The desire to preserve the green ecological environment promotes community forest management	LI3
7	Community forest management	
7.1	Quan Ba district, Ha Giang province, has been implementing community forest management	QL1
7.2	Community forest management in Quan Ba, Ha Giang, is very effective	QL2
7.3	Residents in Quan Ba are actively engaged in community forest management	QL3

In this study, there are five types of data analysis techniques to be applied, including demographic description, descriptive statistics, reliability test analysis, linear regression, and ANOVA. Demographic and descriptive statistics are two major data analysis techniques whether the author utilizes SPSS to take a summary on how many respondents to choose for each point on the Likert scale. On the other hand, the author has adapted the rule of thumb provided by Nagashima (1970) in which this researcher indicates that the mean value will represent the agreement level of the respondents. In more detail, the mean value of factors higher than 3.5, in between 2.5 and 3.5, and less than 2.5 will indicate the questionnaire's context for which the respondents have high, medium, and low agreement. Furthermore, the author also applies reliability test analysis on organizational commitment and compensation management in the conceptual research model. According to Hair et al. (2006), reliability test analysis will provide the researchers with the level of reliability on the survey scale. This analysis consists of three statistical tests, including Cronbach's alpha value higher than 0.6, Corrected Item-Total Correlation should be higher than 0.3, and Cronbach's alpha if Item Deleted should be less than Cronbach's alpha value overall. Last but not least is linear regression, whether this data analysis technique will provide to the researchers information on how factors in conceptual research model relate to each other. Finally, ANOVA is applied in order to understand differences between respondents who have different characteristics in terms of gender, age, marital status, and education toward organizational commitment, compensation management, and government officials' performances.

4 Results

4.1 Statistical Description

4.1.1 The Independent Variables

The statistical results minimum value (Min) and maximum (Max) of each observed variable in Tables 22.4 and 22.5 shows that the respondents' evaluation of factors differed greatly. The lowest rating for the scale of the variable "policy on community forest management" (CS2), "The coordination of community forest management" (PH2), and was rated as the variable "Benefits from forest management" (LI2).

Overall, the average value (mean) of the independent variables differ (mean = 2.86 to 3.65); this proves that there is a different assessment of the level of importance between the independent variables (Table 22.2).

Considering the frequency of the particular variables assessed, the value is expressed as follows:

Description of variable frequency "Policy on community forests management"

"Policy on community forest management" variable is that those respondents rated pretty high (mean from 2.86 to 3.5). This variable was assessed as a rather large difference between the observed variables. Most respondents agreed that local

Table 22.2 Descriptive statistics of factors affecting community forest management

Descriptive statistics					
	N	Minimum	Maximum	Mean	Standard deviation
CS1	153	1	5	3.50	0.762
CS2	153	1	5	2.86	0.892
CS3	153	1	5	3.19	1.005
HB1	153	2	5	3.56	0.931
HB2	153	1	5	3.61	1.047
HB3	153	1	5	2.98	0.949
PH1	153	2	5	2.93	0.882
PH2	153	1	5	2.86	0.869
PH3	153	1	5	3.02	0.790
TB1	153	1	5	2.96	0.785
TB2	153	2	5	3.11	0.748
TB3	153	1	5	3.04	0.751
TQ1	153	1	5	3.12	0.895
TQ2	153	1	5	3.24	0.903
LI1	153	1	5	3.32	0.893
LI2	153	2	5	3.65	0.963
LI3	153	1	5	3.08	0.977
QLR1	153	1	5	3.37	0.992
QLR2	153	1	5	3.24	0.953
QLR3	153	1	5	3.25	0.813
Valid *N* (listwise)	153				

government recognized the community as forest owners (accounting for 46.4%), only 5.3% did not agree that demonstrated community forest management has been applied in Ha Giang Province. However, 38.5% said that local governments do not have the leadership or clear policy to build a model of community forest management and only guide the convention of forest protection and development in communities (Table 22.3).

Description of variable frequency "Residents' understanding of community forest management"

48.4% of respondents said that residents have an understanding of their role in forest protection and only 11.8% do not agree with that. Therefore, many residents know the power of collective forest management (accounting for 54.2%). However, up to 30.1% on the evaluation of residents' technical knowledge for using forest resources is not high. In general, this can be seen that residents' understanding of community forest management in Quan Ba District, Ha Giang Province, is quite good. This is a favorable factor for the development of community forest management. The evaluation of observed variables HB1 and HB2 have a high mean, 3.56 and 3.61 respectively. Only observed variable HB3 has a mean less than 3. These statistics described reasonably reflect the reality of Ha Giang residents' understanding (Table 22.4).

Table 22.3 Description of the variable frequency "Policy on community forest management"

Code	Content	Variable frequency (%)				
		1	2	3	4	5
CS1	The local government recognizes that community is an object of allocated forest land and private forest owners as a way of actually promoting community forest management	0.7	4.6	48.4	36.6	9.8
CS2	Local government directs the model building and community forest policy and related activities to promote community forest management	2.6	35.9	38.6	19.0	3.9
CS3	Local government building regulations guiding protection and development of forest communities in promoting community forest management	2.6	22.9	39.2	23.5	11.8

Table 22.4 Description of the variable frequency "Residents' understanding of community forest management"

Code	Content	Variable frequency (%)				
		1	2	3	4	5
HB1	Residents' understanding of their role in protecting the forests will promote community forest management		11.8	39.9	29.4	19.0
HB2	Residents' understanding of the power of collective forest management will promote community forest management	2.0	13.1	30.7	30.7	23.5
HB3	Understanding techniques that use forest resources will impact positively on community forest management	5.9	24.2	39.9	26.1	3.9

Description of variable frequency "The combination of community forest management"

Residents think that State management agencies do not support the residents to manage the forests (accounting for 37.3%) and only 24.8% think that they do. In this variable, most of the respondents had neutral answers so they have not clearly commented. Some respondents disagreed more with two observed variables (PH2 and PH3). This shows that generally, residents do not rated the coordination of community forest management highly (Table 22.5).

Description of variable frequency "Protective Equipment for the management"

The most obvious neutral views are shown in the descriptive statistics results for the "protective equipment for the management" variable. The number of respondents who answered "Agree" is equivalent to the number of respondents who disagree. 22.9% of respondents agree that the protective equipment facilitates community forest management and 28.8% do not agree with that. Some argue that protective equipment reduces the risk of community forest management, accounting for 26.1%, slightly higher than 19% of residents who do not agree with this. Specifically, 25.5% said that residents were also equipped with protective equipment for forest management and this is a favorable factor for community forest management in Ha Giang (Table 22.6).

Table 22.5 Description of the variable frequency "The combination of community forest management"

Code	Content	Variable frequency (%)				
		1	2	3	4	5
PH1	State management agencies support communities to manage forests, which will impact positively on community forest management		37.3	37.9	19.6	5.2
PH2	Residents who work together will increase the effectiveness of community forest management	0.7	39.9	36.6	19.0	3.9
PH3	The combination of the residents to ensure that forest management is ongoing and timely handling of the work	0.7	24.8	49.7	21.6	3.3

Table 22.6 Description of the variable frequency "Protective equipment for the management"

Code	Content	Variable frequency (%)				
		1	2	3	4	5
TB1	Protective equipment facilitates community-based forest management	0.7	28.1	48.4	20.3	2.6
TB2	Equipment protection reduces the risk of community-based forest management		19.0	54.9	22.2	3.9
TB3	People are given full protective equipment for forest management and this has a positive impact on the management of community forests	0.7	22.2	51.6	23.5	2.0

Description of variable frequency "Traditional habits"

People in Ha Giang also highly appreciated the traditional habits in forest management. The means of two variables, 2.96 and 3.11, are not very high but show that all activities of the people are related to the forest. 30.7% of respondents said that the local people's habit of sticking to the forest has promoted community participation in forest management. Only 18.3% of respondents said that the people's traditional customs of respecting and protecting the forest will not motivate people to participate in community forest management. That means the people's fine tradition of protecting the forest has a great influence on ideas and willingness to participate in forest management (Table 22.7).

Description of the variable frequency "Benefit from forest management"

The "Benefit from forest management"variable was the highest rated and most homologous among the independent variables. Most respondents agreed that benefits attract everyone involved in forest management. Only 17.7% said that when participating in forest management, people cannot exploit, get income, or benefit from the forest. Similarly, up to 51% of the respondents said that people would be owners when participating in forest management. The only observed variable LI3 was agreed and disagreed on equivalently. Local people are quite confused about whether forest management prevents the forest from being lost or not (Table 22.8).

Table 22.7 Description of the variable frequency "Traditional habits"

Code	Content	Variable frequency (%)				
		1	2	3	4	5
TQ1	The habit of sticking to the forest to promote community participation in forest management	1.3	23.5	44.4	23.5	7.2
TQ2	The traditional customs of the residents of respecting and protecting the forest has promoted community participation in forest management	2.6	15.7	44.4	29.4	7.8

Table 22.8 Description of the variable frequency "Benefit from forest management"

Code	Content	Variable frequency (%)				
		1	2	3	4	5
LI1	Joining forest management can help residents to exploit, earn income, and obtain other benefits from forests and will promote community-based forest management	0.7	17.0	41.8	30.7	9.8
LI2	People are participating in forest management while promoting community forest management		10.5	38.6	26.8	24.2
LI3	Forest management preventing the forest from being lost will promote community forest management	3.9	23.5	41.8	22.2	8.5

4.1.2 The Dependent Variable

According to Table 22.9, the observed variables on the scale "community forest management" have high average values (mean = 3.37, 3.24, 3.25). The average values were similar and there was not much difference in the independent variables. The majority of respondents gave neutral answers to a number of the comments, but agreeing with the statement more than disagreeing. 43.8% of respondents agreed with the statement "Ha Giang province has been implementing community-based forest management" and only about 20% do not agree with that opinion. However, only 34.7% said that community forest management in Ha Giang is very effective; 45.1% could not identify it. Likewise, 34% of respondents said that Ha Giang's people are actively involved in community forest management; the number of people who disagreed was quite low, only 14.4%, and is largely neutral.

Thus, we can see that community forest management in Ha Giang Province has gained quite a lot of interest; however, the results have not really been outstanding.

4.2 Assessment Scale

The survey was carried out to assess the reliability of the scale by Cronbach's alpha coefficient. First of all, Cronbach's alpha coefficient was used for the variable type mismatch. The scale is only accepted when the Cronbach's alpha reliability coefficient is 0.5 or higher.

Table 22.9 Description of the variable frequency "Community forest management"

Code	Content	Variable frequency (%)				
		1	2	3	4	5
QLR1	Ha Giang province has been implementing community forest management	1.3	19.0	35.9	29.4	14.4
QLR2	Community forest management in Ha Giang is very effective	2.0	18.3	45.1	22.9	11.8
QLR3	People in Ha Giang are actively engaged in community forest management here	1.3	13.1	51.6	27.5	6.5

Table 22.10 Cronbach's alpha inspection

Code	Scale	Variables	Cronbach's alpha coefficient	Comment
CS	Policy on community forest management	3	0.734	Good
HB	People's understanding of community forest management	3	0.604	Good
PH	The combination of community forest management	3	0.672	Good
TB	Protective equipment for the management	3	0.688	Good
TQ	Traditional habits	2	0.710	Good
LI	Benefit from forest management	3	0.703	Good
QLR	Community forest management	3	0.603	Good

Table 22.11 Reliability for the scale "Policy on community forest management"

Item-Total Statistics				
	Scale mean if item deleted	Scale variance if item deleted	Corrected item-total correlation	Cronbach's alpha if item deleted
CS1	6.05	2.715	0.547	0.670
CS2	6.69	2.319	0.573	0.629
CS3	6.36	2.021	0.574	0.639

Results are presented in Table 22.10. The results showed that the scales have Cronbach's alpha coefficients greater than 0.6 > 0.5. Thus, scales are built fairly well.

Inspection reliability for the independent variable "Policy on community forest management"

Observing Table 22.11, we can see that on the scale "Policy on community forest management" Cronbach's alpha coefficient is 0.734 > 0.5, which is a good ratio. The Corrected Item-Total Correlations are more than 0.3; thus, the observed variables should be reached to perform further analysis (Table 22.12).

Inspection reliability for the independent variable "People's understanding of community forest management"

With three observed variables, Cronbach's alpha coefficient of the scale "People's understanding of community forest management" is 0.604 > 0.5. This is considered

Table 22.12 Reliability for the scale "People's understanding of community forest management"

Item-total statistics				
	Scale mean if item deleted	Scale variance if item deleted	Corrected item-total correlation	Cronbach's alpha if item deleted
HB1	6.59	2.573	0.454	0.448
HB2	6.54	2.264	0.456	0.438
HB3	7.16	2.822	0.336	0.609

Table 22.13 Reliability for the scale "The combination of community forest management

Item-total statistics				
	Scale mean if item deleted	Scale variance if item deleted	Corrected item-total correlation	Cronbach's alpha if item deleted
PH1	5.88	1.938	0.486	0.576
PH2	5.95	1.931	0.506	0.548
PH3	5.78	2.197	0.464	0.604

Table 22.14 Reliability for the scale "Protective equipment for management"

Item-Total Statistics				
	Scale mean if item deleted	Scale variance if item deleted	Corrected item-total correlation	Cronbach's alpha if item deleted
TB1	6.15	1.563	0.529	0.561
TB2	6.00	1.645	0.528	0.564
TB3	6.07	1.751	0.454	0.656

a good ratio. The correlation coefficient gross, though not as high as on the scale "Policy community forest management," is still more than 0.3, which should meet the requirements. In addition, the HB3 variable is the lowest corrected item-total correlation, which is 0.336. If it is removed, its Cronbach's Alpha coefficient will rise from 0.604 up to 0.609, which is not too big a change. Therefore, all variables are to carry out further analysis.

Inspection reliability for the independent variable "The combination of community forest management"

Observing Table 22.13, we can see the Cronbach's alpha coefficient of the scale "The combination of community forest management" is 0.672. This is a good coefficient. The corrected item-total correlations are more than 0.3 so they should be reached to perform further analysis.

Inspection reliability for the independent variable "Protective equipment for management"

Cronbach's alpha coefficient of the scale "Protective equipment for management" is 0.688. This is a good coefficient. The corrected item-total correlation is greater than 0.3 so that the variable is reliable. Besides, if we exclude the observed variables on the scale the Cronbach's alpha coefficient is lower than the former. Therefore, all variables were retained for subsequent analysis (Table 22.14).

Inspection reliability for the independent variable "Traditional habits"

Cronbach's alpha coefficient of scale "traditional habits"is 0.710. It is the second highest coefficient on the scales. This coefficient is more than 0.5; thus it is a good factor. The correlation coefficients of the total variables are all greater than 0.3, so they are satisfactory and should have been retained to perform further analysis (Table 22.15).

Inspection reliability for the independent variable "Benefit from forest management"

Cronbach's alpha coefficient of scale "Benefits from forest management" was 0.703. This is also a good coefficient. The corrected item-total correlations are greater than 0.3 so that the variable is reliable. Besides, if we exclude the variables, Cronbach's alpha coefficient of the observed scale is even lower than the former. Therefore, all variables were retained to perform further analysis (Table 22.16).

Inspection reliability for the dependent variable "Community forest management"

Cronbach's alpha coefficient of the scale "Community forests management" was 0.603, which is a good coefficient. Its lowest corrected item-total correlation is 0.398 > 0.3, which is satisfactory. If observed variables are removed, Cronbach's alpha coefficient dropped from 0.603 to 0.528, which is not better. The other coefficients are in a similar situation. Therefore, all variables were retained to perform further analysis (Table 22.17).

Table 22.15 Reliability for the scale "Traditional habits"

Item-total statistics				
	Scale mean if item deleted	Scale variance if item deleted	Corrected item-total correlation	Cronbach's alpha if item deleted
TQ1	3.24	0.816	0.550	0
TQ2	3.12	0.802	0.550	0

Table 22.16 Reliability for the scale "Benefit from forest management"

Item-total statistics				
	Scale mean if item deleted	Scale variance if item deleted	Corrected item-total correlation	Cronbach's alpha if item deleted
LI1	6.73	2.753	0.504	0.633
LI2	6.40	2.557	0.506	0.629
LI3	6.97	2.413	0.552	0.570

Table 22.17 Reliability for the scale "Community forest management"

Item-total statistics				
	Scale mean if item deleted	Scale variance if item deleted	Corrected item-total correlation	Cronbach's alpha if item deleted
QLR1	6.49	2.133	0.398	0.528
QLR2	6.61	2.133	0.441	0.458
QLR3	6.61	2.556	0.404	0.520

In conclusion, after testing the scales, we can see that the scale is a good choice and there is no need to adjust.

4.3 Correlation Analysis

After performing exploratory factor analysis, six independent factors (with 17 observed variables) and one independent factor (with three observed variables) were included in the model's test. The value factor is the average of the observed variable's components. Pearson correlation analysis was used to examine the appropriateness of the components included in the regression model. The results of the regression analysis were used to test the hypotheses.

4.3.1 Inspection of the Correlation Coefficient

Inspection of the correlation coefficient is used to test the linear relationship between the independent variables and the dependent variable. The correlation coefficients (r) indicate how tight the linear relationship is. The closer "r" is to 1, the higher the level of strictness and $r = 0$ indicates that the variables do not have a linear relationship.

The inspection was carried out on two sides (two-tailed). The correlation coefficient between variables with itself is 1, and between the variables together is >0. According to the correlation matrix, the correlation coefficient between the independent variables and the dependent variable is significant at the 0.01 level. The corresponding coefficients are:

- "Benefits from forest management" variable and the dependent variable have the correlation coefficient 0.577. This is the average correlation, but it is the highest degree of influence on the dependent variable in the model.
- "The combination of community forest management"variable and the dependent variable have a correlation coefficient of 0.354. This correlation is weak, although it still shows the influence of this variable on the dependent variable.
- "Protective equipment for management" variable and the dependent variable have a correlation coefficient of 0.267. This is the lowest correlation coefficient in all the independent variables, and it also represents a weak influence on "community forest management."
- The correlation coefficient of the "Policy on the management of community forests" variable and the dependent variable is 0.511. Influence of this independent variable is the second strongest in the model. This impact is more than 0.4, which is an average correlation.
- The correlation coefficient of the "Traditional habits"variable and the dependent variable is 0.384. It is also a weak correlation coefficient, which shows that the traditional habits of people in forest protection do not strongly affect "community forest management."

Table 22.18 Correlation matrix between the variables of the model

Correlations

		QLR	LI	PH	TB	CS	TQ	HB
QLR	Pearson correlation	1	0.577[a]	0.354[a]	0.267[a]	0.511[a]	0.384[a]	0.506[a]
	Significance (two-tailed)		0.000	0.000	0.001	0.000	0.000	0.000
	N	153	153	153	153	153	153	153
LI	Pearson correlation	0.577[a]	1	0.257[a]	0.116	0.374[a]	0.265[a]	0.309[a]
	Significance (two-tailed)	0.000		0.001	00.152	0.000	00.001	0.000
	N	153	153	153	153	153	153	153
PH	Pearson correlation	0.354[a]	0.257[a]	1	0.015	0.270[a]	0.021	0.262[a]
	Significance (two-tailed)	0.000	0.001		0.850	0.001	0.801	0.001
	N	153	153	153	153	153	153	153
TB	Pearson correlation	0.267[a]	0.116	0.015	1	0.088	0.201[b]	−0.014
	Significance (two-tailed)	0.001	0.152	0.850		0.281	0.013	0.862
	N	153	153	153	153	153	153	153
CS	Pearson correlation	0.511[a]	0.374[a]	0.270[a]	0.088	1	0.188[b]	0.356[a]
	Significance (two-tailed)	0.000	0.000	0.001	0.281		00.020	0.000
	N	153	153	153	153	153	153	153
TQ	Pearson correlation	0.384[a]	0.265[a]	0.021	0.201[b]	0.188[b]	1	0.148
	Significance (two-tailed)	0.000	0.001	0.801	0.013	0.020		0.068
	N	153	153	153	153	153	153	153
HB	Pearson correlation	0.506[a]	0.309[a]	0.262[a]	−0.014	0.356[a]	0.148	1
	Significance (two-tailed)	0.000	0.000	0.001	0.862	0.000	0.068	
	N	153	153	153	153	153	153	153

[a]Correlation is significant at the 0.01 level (two-tailed)
[b]Correlation is significant at the 0.05 level (two-tailed)

– "Knowing the people on community forest management" variable may affect the dependent variable quite strongly. The correlation coefficient is 0.506. Although it does not show a strong degree of influence, it is also a major factor affecting the variable "Community forest management" in the model.

In addition, Table 22.18 also shows that the matrix of independent variables also has certain relations with each other. This means that independent variables can affect the others, which changed community forest management in Ha Giang Province.

The matrix in Table 22.18 shows that the hypothesis is not rejected and can be included in the model to explain the dependent variable.

4.3.2 Regression Analysis

Regression analysis was performed with six independent variables including: Benefit from managed forests (LI), coordination in community forest management (PH), protective equipment for management (TB), policy on community forest management (CS), traditional habits (TQ), people's understanding of community forest management (HB), and the dependent variable is community forest management (QLR). Values of the independent variables were averaged based on the observed variable component of these independent variables. The value of the dependent variable is the average value of the observed variables on community forest management. Analysis is performed using the Enter method. The variables are taken at the same time to see the relevance of the variables. The results of the regression analysis are as follows.

The results in Table 22.19 show that the regression models were relatively consistent with 0.05 significance. Adjusted coefficient R2 = 0.575 means that there are about 57.5% variance of community forest management, which is explained by the six independent variables. This figure is not too high but still sufficient to meet the requirements of conformity assessment model (adjusted coefficient R2 must be greater than 50%.)

The F-test used in the ANOVA table is allowed to test the hypothesis of fit of linear regression models overall. The audit reviewed a linear relationship between the dependent variable and the independent variables. Significance must be less than 5%; then, the model is considered acceptable. In the ANOVA table (Table 22.20), we see very low signficance (0.00), which means an error of 0%. A regression model is suitable and can be used.

The ANOVA testing results show that the F value = 35.277 (signifcance = 0.000). This represents the fit of the model that was built. This suggests a relationship between the linear regression of independent variables and the dependent variable in the model.

In addition, we consider whether multi-collinearity should be occurring in the model or not. The VIF coefficient is used to examine the multi-collinearity in the model. The rule is that the VIF should be less than 2 signs of multi-collinearity (Hoang & Chu, 2005).

Table 22.19 Assessment of fit of the model

Variables entered/removed

Model	Variables entered	Variables removed	Method
1	HB, TB, TQ, PH, CS, LI	.	Enter
a. All requested variables entered			
b. Dependent variable: QLR			

Model summary

Model	R	R square	Adjusted R square	Standard error of the estimate
1	0.769[a]	0.592	0.575	0.44896
a. Predictors: (constant), HB, TB, TQ, PH, CS, LI				
b. Dependent variable: QLR				

Table 22.20 Testing ANOVA

ANOVA						
Model		Sum of squares	Df	Mean square	F	Significance
1	Regression	42.664	6	7.111	35.277	0.000[a]
	Residual	29.429	146	0.202		
	Total	72.093	152			
a. Predictors: (constant), HB, TB, TQ, PH, CS, LI						
b. Dependent variable: QLR						

Table 22.21 Regression results in the model

Coefficients								
Model		Unstandardized coefficients		Standardized coefficients	t	Significance	Collinearity statistics	
		B	Standard error	Beta			Tolerance	VIF
1	(constant)	−0.737	0.303		−2.434	0.016		
	LI	0.284	0.055	0.308	5.111	0.000	0.769	1.301
	PH	0.145	0.059	0.139	2.458	0.015	0.872	1.147
	TB	0.205	0.063	0.178	3.272	0.001	0.948	1.054
	CS	0.201	0.058	0.211	3.493	0.001	0.770	1.299
	TQ	0.159	0.049	0.183	3.255	0.001	0.884	1.132
	HB	0.260	0.055	0.275	4.681	0.000	0.810	1.235
a. Dependent variable: QLR								

In Table 22.21, the VIF value of the largest variable of 1.301 is quite small (less than 2). Therefore, multi-collinearity does not affect results of the model. Significance of the biggest factors of "The combination of community forest management" variable is 0.01 < 0.05, so that all variables are accepted.

The magnitude of beta of variables is arranged in the following: LI > HB > CS > TQ > TB > PH. In this, the most influential variable is "Benefit from forest management" and the weakest influence is "The coordination of community forest management." Results showed that to promote the effectiveness of community forest management, district leaders need to have the appropriate strategies to impact each factor. They should strongly impact the factors with a large correlation coefficient and do not need to adjust the factors with the correlation coefficient.

4.3.3 Results of Regression Analysis

The regression equation showing the relationship between community forest management in Quan Ba District, Ha Giang Province and the factors: benefit from managed forests (LI), coordination in community forest management (PH), protective equipment for management (TB); policy on community forest management (CS), traditional habits (TQ), people's understanding of community forest management (HB) through the following equation (equation constants removed):

$$Y = 0.308 * \times 1 + 0.139 * \times 2 + 0.178 * \times 3 + 0.211 * \times 4 + 0.183 * \times 5 + 0.275 * \times 6$$

Among them:
Y: Community forest management in Quan Ba District, Ha Giang Province.
X1: Benefit from managed forests.
X2: Coordination in community forest management.
X3: Protective equipment for management.
X4: Policy on community forest management.
X5: Traditional habits.
X6: People's understanding of community forest management.

5 Conclusions and Discussion

Base on the literature review and in consultation with experts, the researchers showed six factors that primarily affect community forest management in Ha Giang. The model has six independent variables and one dependent variable. The data are collected by the survey with Ha Giang provincial managers and people participating in community forest management. The scale has been assessed using Cronbach's alpha test. The results give good coefficients that confirm the correctness of the model.

The research results showed that respondents were mostly well aware of community forest management. They said that the influencing factors built in the model have a strong impact on community forest management in Ha Giang province. However, there are still many ineffective factors that require local governments to have correct views in developing strategies to improve community forest management here.

The research results show that there are many factors affecting the community forest management at different levels. Since then, local governments will have appropriate strategies to promote effective community forest management.

The authors offer the following recommendations:

First, the State as well as local governments need to redefine the legal position of village communities in forest management. It is necessary to address the shortcomings in legal documents related to the legal position, rights and obligations of village communities to promote the role of community forest management. In fact, forest protection and development cannot be based only on administrative, enforcement, and sanctioning measures but also on people (village and hamlet communities) based on protection contracts with a satisfactory benefit mechanism. At the same time, local governments should enact policies to benefit consistent with the principles and techniques of finance or local practices. Benefit-sharing mechanisms are to be based on the highest principles to promote the initiative and creativity of the direct protection of forest development, avoid turning them into employees, so that it is newly sustainable and effective. The current State budget for planting, protecting and developing forests is just a temporary solution, only supportive, inadequate, and not lasting.

Second, the foundation for land and forest allocation is the planning of forest land management and use. From the reality of localities where forests exist, people and grassroots officials play an important role in planning. They have a good understanding of the local natural features. Therefore, planning work needs to be maintained at the commune level, with consultations with the village community.

On the basis of the contents of the Ha Giang Socio-Economic Development Master Plan up to 2025, and orientation up to 2030, the province should continue to build and supplement and adjust in time the allocation of forest for community planning. Particularly, it is necessary to pay due attention to the planning of new rural constructions at the commune level in a way that is suited to the reality associated with land and forest allocation and development and community forest management.

Third, the province needs to support techniques of planting and protecting community forests for people involved in forest management. Updating new science and technology through training helps people to manage and exploit forests more effectively.

Fourth, the province should strengthen the inspection and examination of the implementation of community forest management in the district starting from management planning, management organization, and results of forest exploitation and protection to meet practical requirements. The situation of the illegal exploitation of forests strictly handles cases of illegal or improper forest exploitation.

Appendix

Survey form of community forest management in Quan Ba District, Ha Giang Province.

A. INFORMATION ABOUT THE SURVEY.

1. Purpose of the survey: Only used for the research on community forest management in Quan Ba District, Ha Giang Province.
2. Objects of the survey: forest managers, managers in the District People's Committee and people in Quan Ba District, Ha Giang Province.
3. Survey subjects are **anonymous**.
4. Time survey conducted: 2019.

B. GENERAL INFORMATION ABOUT THE PEOPLE SURVEYED.

(Please fill in the blanks or tick (x) in the appropriate box.)

1. Gender: ☐ Male ☐ Female.
2. Age:

☐ < 30 years old ☐ 30–40 years old.

☐ 41–50 years old ☐ > 50 years old.

3. Qualifications.

☐ Middle school. ☐ Intermediate.

☐ College, University. ☐ Postgraduate.

4. Average monthly income.

☐ < 5 millions VND ☐ 5–10 millions VND.

☐ Từ 1015 millions VND ☐ >15 millions VND.

C. SURVEY CONTENT.

1. How do you evaluate the "Policy on community forest management" in Quan Ba District, Ha Giang Province?

Would you please indicate your agreement with the following statements on a scale of 1 to 5: 1 = strongly disagree, 2 = disagree, 3 = Neutral, no opinion, 4 = agree, 5 = totally agree.

STT	Content	Rating score (1–5)				
		1	2	3	4	5
1.1	The local government recognizes that community is an object of allocated forest land and private forest owners as a way of actually promoting community forest management					
1.2	Local government directs the model building and community forest policy and related activities to promote community forest management					
1.3	Local government building regulations guiding protection and development of forest communities in promoting community forest management					

2. How do you evaluate the "People's understanding of community forest management" in Quan Ba District, Ha Giang Province?

Would you please indicate your agreement with the following statements on a scale of 1 to 5: 1 = strongly disagree, 2 = disagree, 3 = neutral, no opinion, 4 = agree, 5 = totally agree.

STT	Content	Rating score (1–5)				
		1	2	3	4	5
2.1	People's understanding of their role in protecting the forests will promote community forest management					
2.2	People know how to handle deforestation					
2.3	People's understanding the power of collective forest management will promote community forest management					
2.4	Understanding techniques that use forest resources will impact positively on community forest management					

3. How do you evaluate the "The combination of community forest management" in Quan Ba District, Ha Giang Province?

Would you please indicate your agreement with the following statements on a scale of 1 to 5: 1 = strongly disagree, 2 = disagree, 3 = neutral, no opinion, 4 = agree, 5 = totally agree.

STT	Content	Rating score (1–5)				
		1	2	3	4	5
3.1	Agency supports state management and residents' management of the forest, which will impact positively on community forest management					
3.2	People who work together will increase the effectiveness of community forest management					
3.3	The combination of the residents to ensure that forest management is ongoing and timely handling of the work					

4. How do you evaluate the "Protective equipment for management" in Quan Ba District, Ha Giang Province?

Would you please indicate your agreement with the following statements on a scale of 1 to 5: 1 = strongly disagree, 2 = disagree, 3 = neutral, no opinion, 4 = agree, 5 = totally agree.

STT	Content	Rating score (1–5)				
		1	2	3	4	5
4.1	Protective equipment facilitates community-based forest management					
4.2	Equipment protection reduces the risk of community-based forest management					
4.3	Residents who are equipped with full protective equipment for forest management have a positive impact on the management of community forests					

5. How do you evaluate the "Traditional habits" in Quan Ba District, Ha Giang Province?

Would you please indicate your agreement with the following statements on a scale of 1 to 5: 1 = strongly disagree, 2 = disagree, 3 = neutral, no opinion, 4 = agree, 5 = totally agree.

STT	Content	Rating score (1–5)				
		1	2	3	4	5
5.1	The habit of considering forests as a sacred public property encourages communities to participate in forest management					
5.2	The traditional customs of the residents of respecting and protecting the forest has promoted community participation in forest management					

6. How do you evaluate the "Benefit from forest management" in Quan Ba District, Ha Giang Province?

Would you please indicate your agreement with the following statements on a scale of 1 to 5: 1 = strongly disagree, 2 = disagree, 3 = neutral, no opinion, 4 = agree, 5 = totally agree.

STT	Content	Rating score (1–5)				
		1	2	3	4	5
6.1	Joining forest management can help people to exploit, earn income, and obtain other benefits from forests, and will promote community-based forest management					
6.2	People are doing everything to participate in forest management while promoting community forest management					
6.3	Forest management preventing the forest from being lost will promote community forest management					

7. How do you evaluate the "Community forest management" in Quan Ba District, Ha Giang Province?

Would you please indicate your agreement with the following statements on a scale of 1 to 5: 1 = strongly disagree, 2 = disagree, 3 = neutral, no opinion, 4 = agree, 5 = totally agree.

STT	Content	Rating score (1–5)				
		1	2	3	4	5
7.1	Quan Ba district, Ha Giang province has been implementing community forest management					
7.2	Community forest management in Quan Ba, Ha Giang, is very effective					
7.3	People in Quan Ba are actively engaged in community forest management here					

Thank you for your help!

All questionnaires are anonymous.

References

Arnold, J. E. M., & Jongma, J. (1978). Fuelwood and charcoal in developing countries. *Unasylva, 29*(118), 2–9.

Arnold, J.E.M., & Stewart, W.C. (1991). *Common property resource management in India.* Tropical Forestry Papers 24, Oxford Forestry Institute, University of Oxford, England, 15–30.

Burda, D., Curran, F., Gale, M., M'Gonigle (1997), *Forest in Trust: Reforming British Columbia's Forest Tenure System for Ecosystem and Community health.* Report Series R97–2. Victoria, BC: Eco-research Chair in Environmental Law and Policy, Faculty of Law and Environmental, University of Victory, US, 2.

Dong, Q. (2005). *Learn new regulations on forest protection and forest development* (pp. 11–43). Labor Publish.

FAO. (1978). *Forestry for local community development, forestry paper 7* (Vol. 26). FAO.

Hair, M., et al. (2006). The application of structural equation modelling in information systems research. *Information Systems Frontiers, 6*, 325–340.
Hardin, G. (1968). The tragedy of the commons. *Science, 162*.
Herb. (1991). *Seeing the forest among the trees: The case for wholistic forest use* (p. 77). Canada Polestar Calendars.
Hoang, T., & Chu, N. M. N. (2005). *Analysis of research data with SPSS* (pp. 25–60). Thong Ke Publishing House.
Nagashima. (1970). Country-of-origin image: Measurement and cross-national testing. *Journal of Business Research, 58*(1), 103–106.
National Assembly of the Socialist Republic of Vietnam. (2004). *Law on Forest protection and development on 1 April 2004* (pp. 3–4).
Nguyen, B. N. (2009). Community forest management in Vietnam: Current situation, problems and solutions. Proceedings of the National Workshop on Community Forest Management. Community forest management in Vietnam: Policies and practices, FGLG project, Hanoi, 4–20.
Nguyen, V. T. (2004). *General forestry curriculum* (pp. 54–82). Agriculture Publishing House.
Poffenberger, M. (2003). *Communities and Forest management in Southeast Asia* (pp. 147–154). WG – CFM.
Swiss Association for Development and Cooperation. (2005). *Guidance on community forest management techniques* (pp. 27–42). Helvetia Vietnam.
Vo, D. T. (2012). Benefit mechanism in community forest management in Vietnam, Workshop on natural forest management based on people's rights, Center for Community and Environment Initiative (C&E), 7–11.

Chapter 23
Impacts of Urban Expansion on Landscape Pattern Changes: A Case Study of Da Nang City, Vietnam

Do Thi Viet Huong, Bui Thi Thu, Nguyen Bac Giang, and Nguyen Hoang Khanh Linh

Abstract The paper deals with an integration of remote sensing, GIS, and landscape metric indices to employ the spatiotemporal characteristics of urban expansion and to explore the impacts of urban expansion on landscape pattern changes in Da Nang city, Vietnam, from 1996 to 2015. Key landscape change indices were selected to characterize the urban landscape patterns at the landscape and class levels. Several critical urbanization indicators were being developed: urban resident's ratio, urban resident's density, nonagriculture GDP proportion, and nonagriculture labor ratio. The impacts of urbanization on urban landscape changes were determined through analyzing the correlation between the urbanization indicators and landscape change indicators. The results indicate that the built-up area has been increased by 8187.18 ha in an average expansion area of 430.90 ha per year. The urban landscape has undertaken a complicated transformation in landscape structure and composition of which there was the conversion mainly from agriculture land to built-up land. Spatial distribution of different patches became more separated, complex, and irregular, and the patch types became more fragmented. The significant relationship between urbanization indicators and landscape change indicators indicated that the intensity of human activities were decisive factors for the urban development.

Keywords Urban expansion · Landscape pattern changes · Landscape metrics · Da Nang · Vietnam

D. T. V. Huong (✉) · B. T. Thu · N. B. Giang
University of Sciences—Hue University, Thua Thien Hue, Vietnam
e-mail: dtvhuong@hueuni.edu.vn

N. H. K. Linh
International School—Hue University, Thua Thien Hue, Vietnam

A. T. Nguyen, L. Hens (eds.), *Global Changes and Sustainable Development in Asian Emerging Market Economies Vol. 2*, https://doi.org/10.1007/978-3-030-81443-4_23

1 Introduction

Urbanization is considered to be one of the most dramatic land transformations and their ecological consequences (Luck & Wu, 2002; Zhang, Wu, Zhen, & Shu, 2004). Globally, 55% of the world population resides in urban areas as of 2018, and it is projected that by 2050, there will be more than two-third urban population (68%) in the world. Especially, Africa and Asia are urbanizing more rapidly than other regions all over the world (United Nations, 2019). In Vietnam, the urban population was 35.9% in 2018, and it is expected that this proportion will have reached 57.3% by 2050 (World Bank, 2011). The urban expansion causes loss of arable land, devastation in vegetation cover, and rapid increase in the impervious surface and artificial structure (Dewan & Yamaguchi, 2009; Zhang, Tu, & Li, 2016). Moreover, the accelerated urbanization leads to environmental changes and affects ecosystem services (Dai, Wu, & Du, 2017). Humans have the ability to greatly modify their environment, which tends to profoundly alter the pattern and structure of urban landscapes by making more and more patches smaller and leads to the exacerbated spatial heterogeneity and fragmentation of the landscape (Dai et al., 2017; Luck & Wu, 2002; Zhang et al., 2004). Therefore, quantifying its change is essential to monitor and assess the ecological and artificial consequences of urban land use/land cover (LULC) change, as well as to have a proper land use planning and sustainable development policies.

Some recent research has proved the effectiveness of integrating remote sensing, GIS, and landscape pattern metrics for detecting urban sprawl processes and quantifying the pattern features of LULC in the context of urbanization (Dai et al., 2017; Giordano & Marini, 2008; Zhou, Shi, Wu, Ma, & Yu, 2014). A large collection of landscape metrics has been developed to describe landscape patterns and its spatial–temporal dynamics for each LULC from the satellite classification proposed by FRAGSATS (McGarigal, Marks, & Barbara, 1995). In addition, the impact of urbanization on landscape pattern change has also been studied by analyzing correlation coefficients between landscape patterns and urbanization indicators (Yang, Huang, & Li, 2017; Yi et al., 2016; Zhou et al., 2014).

Da Nang is a coastal city in the key economic region of the Central of Vietnam. Since becoming a type-I city under the management of Central Government (in 1997), Da Nang has experienced a rapid development and considered to be one of the cities with a relatively fast and strong urbanization speed (World Bank, 2011). The development of its commercial port, international airport, industrial zones, and new urban areas, as well as the expanding tourism activities along the coastal area, has led to the huge developments in the socioeconomic aspects and spatial structure of the city. In addition to the achievements in the urbanization process, Da Nang is facing pressing issues that deteriorate the living environment quality (Tien, Sy, & Hang, 2006). The change of land use types and the expansion of urban land have resulted in the changes in the natural environment, socioeconomy, and culture of the study area. Previous studies in Da Nang city were mainly focused on environmental quality issues (Tien et al., 2006), LULC change and spatial environmental index

(Tu, Hai, Ha, & Ngan, 2015), urbanization and climate change (The, Cu, Nhuan, & Vuong, 2015), and urban expansion and flood risk change (Huong, Ryota, & Tsutsui, 2013), but studies on the urban expansion and landscape pattern changes have been poorly documented (Linh & Erasmi, 2012). Therefore, in this paper, Da Nang city was selected to study the impacts of urban expansion on the landscape pattern changes. The objectives were to (i) obtain LULC data from the remote sensing images and detect LULC changes in 1996, 2003, 2010, and 2015; (ii) quantify and visualize the urban sprawl; (iii) characterize landscape pattern changes by using landscape metrics; and (iv) explore the impact of the urban expansion on landscape pattern changes.

2 Methodology

2.1 Description of the Study Area

Da Nang city is located in the middle of Vietnam, between the range of $15^{0}55'15''$–$16^{0}13'15''$ north latitude and $107^{0}49'05''$–$108^{0}20'18''$ east longitude. Da Nang is a dynamic city of the key economic zone in the Central of Vietnam with its international airport, deep-water seaport, and National Highway 1. The topography is very diverse, combining mountains and a coastal plain, where the mountainous area dominates with a high range between 700 and 1500 m. The average annual precipitation is 2504 mm, and the mean annual temperature is 25.8 °C (Da Nang Construction Planning Institute, 2014) (Fig. 23.1).

The city has an area of 1283.42 km^2 with a population of 1,064,070 people (2017). It consists of six urban districts, including Hai Chau, Thanh Khe, Lien Chieu, Son Tra, Ngu Hanh Son, and Cam Le, one rural district (Hoa Vang) and one island district (Hoang Sa). Over the past two decades, Da Nang has experienced a rapid urbanization, which is clearly reflected in the increasing population concentration in the inner city. The proportion of the urban population in Da Nang is the highest in the country. In 2017, the average urban population proportion of Da Nang is 87.1%, 2.6 times higher than the national average ratio (33.9%), and higher than that of Ho Chi Minh city (80.1%), the crowdest city of Vietnam.

2.2 Data Sources and Processing

In this study, time series satellite images and demographic statistical data are collected for assessing the temporal and spatial characteristics of the urban expansion from 1996 to 2015 and determining the relationship between urbanization indicators and landscape change indicators. The urban expansion process and LULC changes were investigated by using the image classification of Landsat 5 TM (1996),

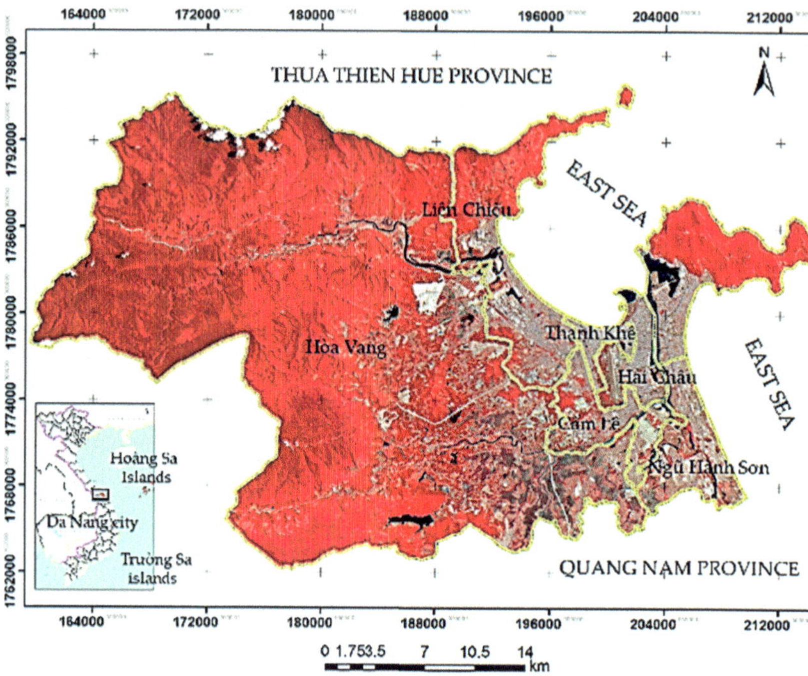

Fig. 23.1 The location of Da Nang city in Vietnam

Table 23.1 Satellite data for image interpretation

Year	Satellite	Acquired date	Cloud cover	Significant period	Sources
1996	Landsat 5 TM	07/07/1996 14/07/1996	0% 9%	Before establishing the city	USGS
2003	Landsat 7 ETM+	14/04/2003 21/04/2003	2% 4%	Recognized as class I city under central government	USGS
2010	ALOS Avnir-2	16/05/2010	3%	Land inventory time	METI & JAXA
2015	Landsat 8 OLI	10/06/2015 01/06/2015	1.,73% 20.5%	Land inventory time	USGS

Landsat 7 ETM+ (2003), ALOS Avnir-2 images (2010), and Landsat 8 OLI (2015) (Table 23.1). The four-period remote sensing images of 1996, 2003, 2010, and 2015 were used to study the spatial–temporal evolution characteristics of the urbanization expansion in Da Nang City. All satellite images were geo-rectified with a topographic map and then masked by the boundary of Da Nang City by using ArcGIS 10.4. The error was controlled within 0.5 pixels. These created a temporal dataset that allowed analysis of the changes in the urban expansion and LULC in a nearly 20-year period. In addition, administrative division maps, topographic map (1:

25,000) in 2000, land use (1:25.000) in 2010 and 2015, and adjustment master planning (1:25.000) of Da Nang City were used for secondary data.

2.3 Methodology

2.3.1 Land Use/Land Cover Classification and Change Detection

Six LULC classes were defined for image classification based on the modified Anderson LULC scheme level I (Anderson, Hardy, Roach, & Witmer, 1976), Vietnam's regulation on land use, and the existing condition of study area, including built-up land, water body, agricultural land, forest land, shrubs, and grassland and bare land.

Object-based image analysis has been applied more frequently for remote sensing image classification compared to pixel-based analysis due to its strength, which is the ability to combine spectral information and spatial information for extracting target objects (Kindu, Schneider, Teketay, & Knoke, 2013; Tamta, Bhadauria, & Bhadauria, 2015). Therefore, in this study, all images were classified into the object-oriented classification based on the class hierarchy by defining the threshold of the indices such as calculated indices—Normalized Difference Built-up Index (NDBI), Normalized Difference Water Index (NDWI), Soil Adjusted Vegetation Index (SAVI), and default indices (NIR, Brightness) using eCognition software 9.1. The formulations for calculating those indices are presented as follows:

$$\text{NDBI} = (\text{SWIR1} - \text{NIR}) / (\text{SWIR1} + \text{NIR}) \tag{23.1}$$

$$\text{SAVI} = (\text{NIR} - \text{RED}) / (\text{NIR} + \text{RED} + 0.5) \times (1 + 0.5) \tag{23.2}$$

$$\text{NDWI} = (\text{GREEN} - \text{NIR}) / (\text{GREEN} + \text{NIR}) \tag{23.3}$$

Finally, the LULC classification results were resampled at a spatial resolution of 30 m. The accuracy of the satellite image classification was assessed using "ground truth" data, land use map, and high-resolution images from Google Earth at the same time as reference data. For evaluation, a grid point with 1 km grid spacing was created and converted into a .kml file that included 955 points. Subsequently, each individual point was trained by visual interpretation of the Google Earth image/previous land use maps. The coded grid points were then overlaid by the Landsat and ALOS satellite images classification in order to compare the accuracy of the results. Four LULC maps of the year 1996, 2003, 2010, and 2015 were produced with six categories.

2.3.2 Urbanization Expansion and Land Use/Land Cover Change Analysis

The urban expansion can be detected by comparing two classified images between the two periods, as 1996–2003, 2003–2010, 2010–2015, and 1996–2015. The urbanization intensity index (UII) was used to analyze the urbanization expansion from 1996 to 2015 (Zhou et al., 2014).

The equation is as follows:

$$UII = \frac{EAB_i \times 100}{TA \times \Delta t_i} \tag{23.4}$$

where UII is the urbanization intensity index; EAB_i is the expansion area of built-up land during a certain period, i; TA is the total area of the study area; and Δt_i is a time span of a certain duration, i.

Spatial and temporal LULC changes were analyzed with GIS by comparing two classified images, 1996–2003, 2003–2010, and 2010–2015. Besides, the indispensability of urbanization expansion was also clarified via analyzing several critical urbanization indicators, in 1997, 2003, 2010, and 2015, such as urban resident's ratio, urban resident's density, nonagriculture GDP proportion, and nonagriculture labor ratio.

2.3.3 Urban Landscape Pattern Metrics Analysis

The LULC map extraction from satellite images was applied for analyzing the urban LULC landscape pattern characteristics. The changes of urban landscape pattern can be detected/defined and measured using landscape metrics, which are used to quantify and categorize complex landscapes into identifiable patterns and reveal some ecosystem properties such as composition, fragmentation, and configuration (Weng, 2007).

Landscape indices for measuring the urban landscape change are performed at two levels, namely class level and landscape level. Six landscape metrics such as percent of landscape (PLAND), largest patch index (LPI), area-weighted mean patch fractal dimension (AWMPFD), interspersion and juxtaposition index (IJI), Contagion (CONTAG), Shannon diversity index (SHDI), and Shannon evenness index (SHEI) were selected for quantifying the urban landscape pattern analysis.

The raster version of FRAGSTATS 4.2 (McGarigal et al., 1995) developed by the Forest Science of Oregon state university was adopted for calculating some landscape and class-level metrics (Table 23.2).

Table 23.2 Landscape metrics utilized to quantify the spatial pattern of the urban landscape in Da Nang city (based on McGarial et al. 1995)

Landscape Metric index	Unit	Significance	Level calculation
Percent of landscape (PLAND)	%	Indicate the proportional abundance of each patch type in the landscape	Class
Largest patch index (LPI)	%	Indicate ratio of the largest patch area to the total landscape area	Landscape/ class
Area-weighted mean patch fractal dimension (AWMPFD)	#	Reflect the complexity of self-similarity of a patch	Landscape/ class
Contagion (CONTAG)	%	Express the agglomeration degree among different landscape types	Landscape
Shannon diversity index (SHDI)	#	Reflect landscape heterogeneity	Landscape
Shannon evenness index (SHEI)	#	Indicate even degree of different landscape types	Landscape

2.3.4 Statistical Analysis

Statistical correlations were calculated between the significant landscape pattern change metrics and critical urbanization indicators. Pearson's correlation coefficient (r) between the urbanization indicators and landscape metrics was applied to quantify the relationship between urbanization and urban landscape patterns. A p-value (less than 0.05) was considered a significant correlation (Field, 2013).

The correlations were performed in such a way that a higher absolute value of the correlation coefficient represented a stronger correlation; positive values indicated positive correlations, and negative values meant that the correlation was negative. All statistical analyses were performed using the IBM SPSS version 26. The impact of urbanization on landscape pattern changes was analyzed in the period of 1996–2015.

3 Results

3.1 Land Use/Land Cover Changes

The image segmentation was done by applying multiresolution segmentation (MS) in eCognition Developer 9.01 software. The MS algorithm is also an optimization procedure that minimizes the average heterogeneity for a given number of objects and maximizes their homogeneity based on defined parameters such as scale, shape, and compactness. Through trial and error to successfully segment objects in an image, four segmentation levels were defined differently depending on the types of satellite image (Landsat 5 TM, Landsat 7 ETM+, ALOS Avnir-2, Landsat 8 OLI) and the nature of LULC to be detected for the analysis (Table 23.3).

Table 23.3 Segmentation levels of the classified objects

Segmentation level	Parameter (scale, shape, compactness)			
	1996	2003	2010	2015
	Landsat 5 TM	Landsat 7 ETM+	ALOS Avnir-2	Landsat 8 OLI
Level 1: Water body and land	10; 0.2; 0.5	10; 0.2; 0.5	15; 0.2; 0.5	10; 0.2; 0.5
Level 2: Vegetation/no-vegetation (built-up land, bare land)	5; 0.2; 0.5	5; 0.2; 0.5	5; 0.2; 0.5	5; 0.2; 0.5
Level 3: Forest land/other vegetation land	30; 0.1; 0.5	30; 0.1; 0.5	20; 0.1; 0.5	30; 0.2; 0.5
Level 4: *Agricultural land and shrub and grass land*	5; 0.2; 0.5	5; 0.2; 0.5	3; 0.1; 0.5	5; 0.2; 0.5

Table 23.4 Rule set classification

Level	Target class	Calculated parameter threshold			
		1996	2003	2010	2015
Level 1	Wa	Mean NIR ≤ 60	LWM ≤ 120 Mean NIR ≤ 60 mean NIR > 0	NDWI ≥0,45	NIR ≤ 600
	La	–	Mean blue >0	–	NIR > 600, Mean blue >0
Level 2	NoVe	Mean blue ≥20	NDVI ≤0.36	Mean blue ≥145	Mean blue ≥700
	VeLa	Mean blue <20	NDVI >0.36	Mean blue <145	Mean blue <700
	BaL	Brightness ≥95	Brightness >95	Brightness ≥165	Brightness ≥2500
	BuL	Brightness <95	Brightness ≤95	Brightness <165	Brightness <2500
Level 3	FoL	SAVI ≥0.80	SAVI ≥0.85	SAVI ≥ −0.15	SAVI ≥0.98
	NoFo	SAVI <0.80	SAVI <0.85	SAVI < −0.15	SAVI <0.98
Level 4	AgL	SAVII ≥0.78	SAVI <0.70	SAVI ≤ −0.35	Brightness ≥1240
	ShGrL	SAVI <0.78	SAVI >0.70	SAVI > −0.35	Brightness <1240

The hierarchical scheme object-based classification of four levels in each image was implemented by approaching fuzzy membership functions. The classification firstly started from the whole landscape into water body (Wa) and land (La) (level 1). Secondly, the land class was further subdivided into more specific class: vegetation (Ve)/no-vegetation (NoVe), in which the no-vegetation was classified into built-up land (BuL) and bare land (BaL) (level 2). The vegetation class was used to extract the forest land (FoL) and no-forest land (NoFoL) (Level 3). And finally, level 4 was used to extract the remaining target class LULC type: agricultural land (AgL), shrub and grassland (ShGrL). The classification of target class was extracted following the defined rule set classification (Table 23.4), in which mainly threshold of default indices (blue, NIR, brightness), Normalized Difference Built-up Index (NDBI), Normalized Difference Water Index (NDWI), and Soil Adjusted Vegetation Index (SAVI) were utilized for achieving LULC classes (Table 23.4).

The series of LULC classification maps over the past 20 years by time are shown in Fig. 23.2, and the trend change of some major LULC area is shown in Fig. 23.2 and Table 23.5.

Table 23.5 and Fig. 23.3 show that the area of built-up areas increased steadily from 1996 to 2003, 2010, and 2015 from 4183.73 ha to 5211.38, 10,140.07, and 12,370.91 ha, respectively. In contrast, the area of agricultural land from 1996 to 2003, 2010, and 2015 has continuously decreased from 8089.31 ha to 8089.31, 5069.31, and 4332.12 ha, respectively.

The decline in agricultural land during this period is suitable for the development strategy of Da Nang City as follows: "service, industry, and agriculture," of which the proportion of service and industry has been increasing and the proportion of agriculture has been decreasing. The city authorities have invested actively in building Da Nang into a modern city with strong industrialization, modernization, and high services. Therefore, most bare lands were reclaimed and covered with industrial zone, infrastructure, and newly built-up areas, which showed a rate of decline from 4805.52 ha to 2563.62 ha in 1996 and 2015, respectively. The area of shrub and grassland increased from 4881.92 ha to 6814.53 ha due to deforestation; forest fires in the West of the City and many urban areas were "hanging planning" to be abandoned along coastal roads.

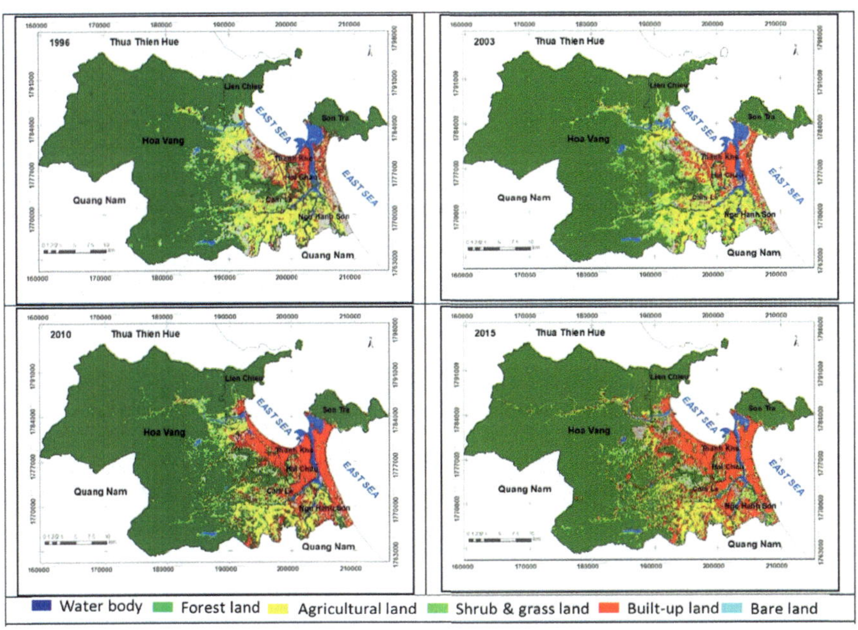

Fig. 23.2 LULC map in 1996, 2003, 2010, and 2015 of Da Nang City

Table 23.5 Area and percentage of LULC types in Da Nang City from 1996 to 2015

	1996		2003		2010		2015	
LULC type	Area (ha)	(%)	Area (ha)	%	Area (ha)	%	Area (ha)	%
AgL	8089.31	8.25	7532.84	7.68	5069.31	5.17	4332.12	4.42
BaL	4805.52	4.90	3254.29	3.32	1540.71	1.57	2563.62	2.61
BuL	4183.73	4.27	5211.38	5.32	10,140.07	10.34	12,370.91	12.62
FoL	74,043.69	74.81	71,024.37	72.44	73,784.11	75.26	69,785.02	71.18
ShGrL	4881.92	4.98	7947.33	8.11	5074.75	5.18	6814.53	6.95
Wa	2731.02	2.79	3069.79	3.13	2431.05	2.48	2173.80	2.22
Total	98,040.00	100.00	98,040.00	100.00	98,040.00	100.00	98,040.00	100.00

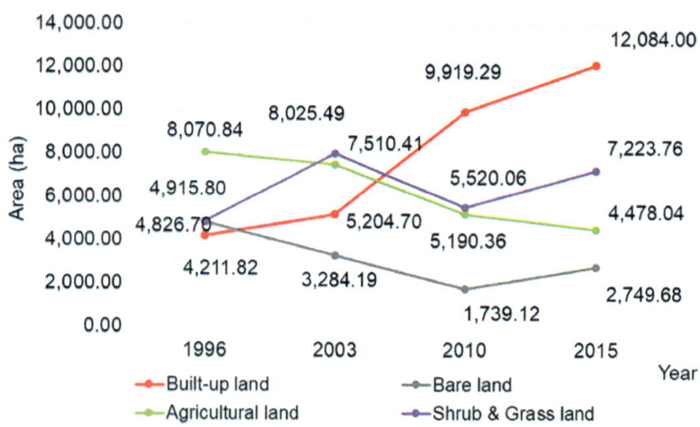

Fig. 23.3 Area of some main Da Nang LULC in the period of 1996–2015

The overall classification accuracy of the LULC map for 1996, 2003, 2010, and 2015 was determined to be 93.51%, 91.31%, 91.20%, and 92.88%, respectively. The overall Kappa coefficient for the four times was over 0.8, which was considered to indicate acceptable or good agreement with the optical data. For the built-up areas, the Kappa coefficient was extracted with good agreement, over 0.85. Therefore, these data were available for continuous study (Table 23.6).

3.2 *Evolution characteristic of Urbanization Expansion*

From the analysis of the LULC changes, Da Nang City has clearly undergone a rapid urban expansion over the past two decades, amounting to 12,370.91 in 2015 as compared to only 4183.73 ha in 1996. The built-up area grew by 8187.18 ha between 1996 and 2015 (2.9 times) and nearly 430.9 ha per year on average. The evolution of urbanization expansion in Da Nang City in the period of 1996–2015 is clearly shown in Table 23.5 and Fig. 23.4.

Table 23.6 LULC classification accuracy assessment

Land use/land cover type	Kappa coefficient			
	1996	2003	2010	2015
Built-up land	0.86	0.85	0.93	0.85
Bare land	0.80	0.70	0.83	0.87
Water body	0.92	0.95	0.83	0.96
Agricultural land	0.69	0.73	0.69	0.88
Forest land	0.89	0.92	0.85	0.85
Shrub and grass	0.67	0.52	0.53	0.68
Overall kappa coefficient	**0.86**	**0.82**	**0.81**	**0.85**
Overall classification accuracy	**93.51%**	**91.31%**	**91.20%**	**92.88%**

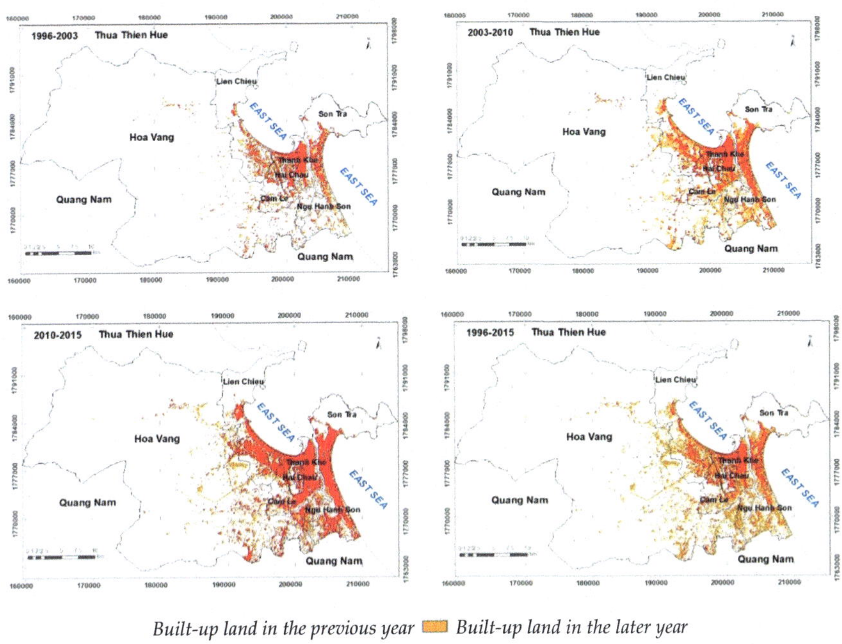

Fig. 23.4 Evolution of urbanization expansion in Da Nang City in the period of 1996–2015

The urban area increased 1027.65 ha from 4183.73 ha in 1996 to 5204.70 ha in 2003 with the average expansion urban area of 146.81 ha per year. The area of built-up area was 4183.73 ha, mainly distributed in the center of Hai Chau and Thanh Khe district in 1996 (which previously belonged to the core center of Quang Nam-Da Nang Province: District 1, 2, and 3). The urbanization process in Da Nang speeded up sharply since the separation of Da Nang city from the Quang Nam-Da Nang Province (in 1997) and the recognition of Da Nang as a type-I city under Central Government (in 2003). In the next seven years (2003–2010), the area of urban land expanded up to 9919.29 ha in 2010 with the average expansion urban area of 704.1 ha per year (nearly 5 times higher in comparison with that of the seven-year period of 1996–2010). Corresponding to that stage, there was the development of emerging pork industry and major infrastructure as well as the road and bridge networks (such as Han Bridge River in 2000, Thuan Phuoc Bridge in 2009) creating connection with other districts (Lien Chieu, Son Tra). After 2010, the city authority accelerated the development of sea park and tourism zones/resorts on the coastal road from Son Tra district to Hoi An city (Quang Nam Province) in the southeast. Therefore, the urban area has not only expanded to the west and northwest but also largely expanded beyond the central city to the south and southeast and along the coastal line, connecting Ngu Hanh Son to Cam Le, demonstrating a complete interconnection of urban areas (Fig. 23.4). During the five-year period of 2010–2015, the expansion area of built-up land was 2230.84 ha. The annual increase in the area of construction land was 446.17 ha in this period, which was lower than that in the 2003–2010 year period, because the urban land fund has gradually stabilized.

The urbanization progress of Da Nang could be divided into three clear stages, which were in the initial construction stage (1996–2003), the rapid development stage (2003–2010), and stable development (2010–2015). The built-up area grew 8187.18 ha within 19 years (1996–2015), and the values of urbanization intensity index were 0.15, 0.72, 0.46, and 0.44 in 4 stages of 1996–2003, 2003–2010, 2010–2015, and 1996–2015, respectively. The high value of urbanization intensity index in the period of 2003–2010 indicated that the urban expansion was great (Table 23.7).

Over the past two decades, Da Nang has experienced a strong urbanization process, which is reflected in the indicators of urbanization such as urban resident's ratio, urban resident's density, nonagriculture GDP proportion, and nonagriculture labor ratio (Table 23.8). The proportion of urban population in Da Nang is the highest in the country. Compared to the national urban population of 33.9%, the urban population of Da Nang is 2.6 times higher than that of Ho Chi Minh City (81.6%). According to the result of a population census, just within 20 years from 1996 to 2015, the population of the city has increased by nearly 1.5 times, from 672,468 to 1,028,838 persons, respectively, in which, the urban population occupied the high

Table 23.7 Built-up land expansion from 1996–2015 in Da Nang City

Duration	The time span of a certain duration	The expansion area of built-up land (ha)	Annual expansion area (ha per year)	Urbanization intensity index
1996–2003	7	1027.65	146.81	0.15
2003–2010	7	4928.69	704.10	0.72
2010–2015	5	2230.84	446.17	0.46
1996–2015	19	8187.18	430.90	0.44

Table 23.8 Some indicators of urbanization in Da Nang City from 1996 to 2015

			Year			
Urbanization indicators	Abbreviation	Unit	1996	2003	2010	2015
Urban population rate	PoUrban	%	79.04	79.59	86.97	87.28
Urban population density	PoDensity	Per/km^2	2581.87	2841.76	3268.08	3657.17
Nonagriculture GDP proportion	GDPNAgri	%	93.34	96.57	97.03	97.04
Average income per person	Income	USD/per	27.13	35.92	102.31	166.66
Nonagriculture employment rate	EmpNAgri	%	67.00	74.18	90.62	92.49

Da Nang Statistical Office [DNGSO] (1998, 2004, 2011, 2016).

rate and increased steadily from 1996 to 2015 Da Nang Statistical Office [DNGSO] (1998, 2004, 2011, 2016).

As shown in Table 23.8, the urban population density has increased by 1.4 times from 1996 to 2015, reflecting the increasing concentration of the population in the inner city. The average monthly income of people has improved dramatically, increasing more than 6 times, from only 27.13 USD in 1996 to 166.66 USD/a person in 2015. Corresponding to the increase in the average income per capita, Da Nang has had no poor households according to the national poverty line (2015). It can be said that the expansion of construction land area in Da Nang is inevitable due to a number of main reasons as follows: (1) coastal location advantage for forming a key economic city, (2) changes in boundary and upgrading urban administrative classification unit, and (3) economic transition.

3.3 *Urban Landscape Pattern Fragmentation Changes from 1996 to 2015*

The urban landscape pattern became more heterogeneous mainly due to the fragmentation. The temporal changes in the landscape pattern metrics at the landscape level of built-up land are depicted in Fig. 23.5.

3.3.1 Largest Patch Index (LPI)

LPI is defined as the percentage of the largest patch in the total landscape (%). It refers to the dominance of one type of patch (McGarigal et al., 1995). In Fig. 23.5a, in landscape level, the value of LPI decreased from 50.77 to 47.43 within nearly 20 years, in which, its value sharply decreased to 35.53 in 2003, corresponding the period of speed urbanization. This decrease of LPI indicated that the dominant landscape of agricultural land had declined according to the orientation development of "service, industry, and agriculture." In contrast, the LPI of built-up land showed a

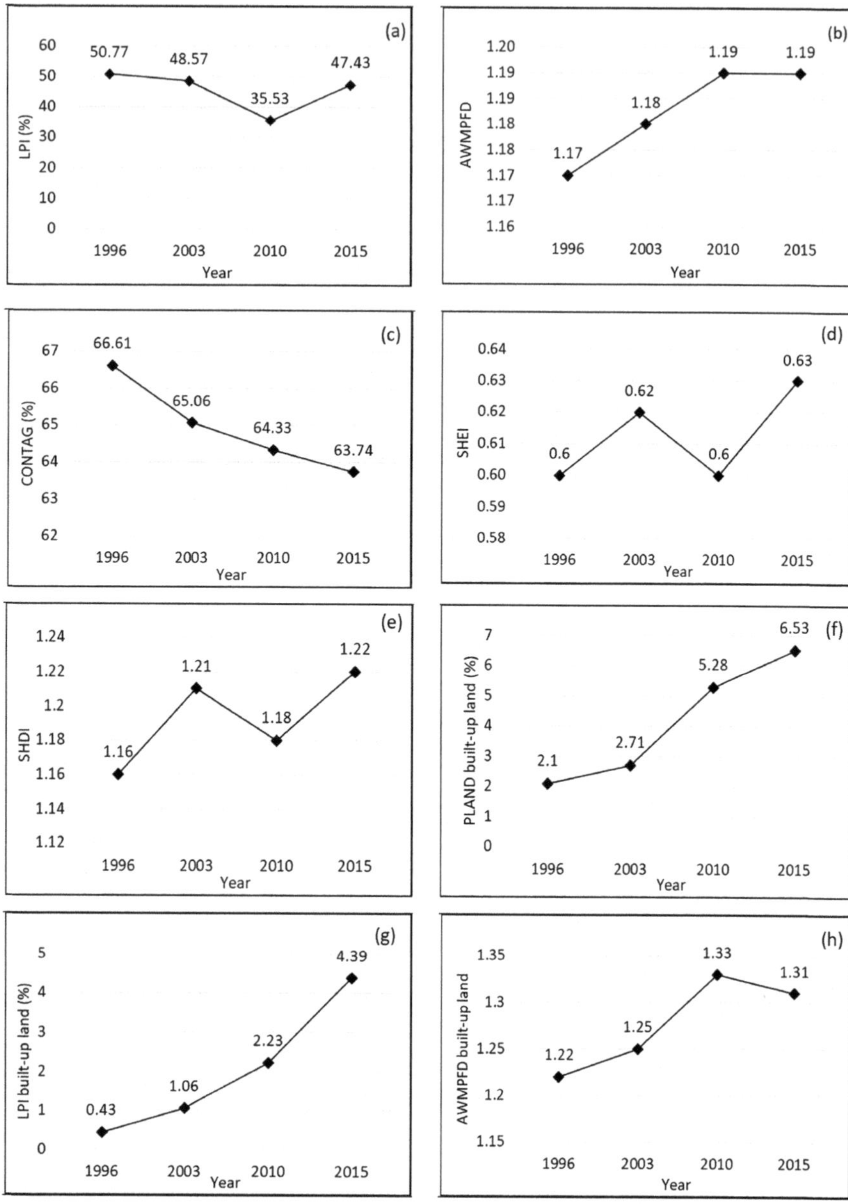

Fig. 23.5 Temporal changes in the landscape pattern metrics at the landscape level (**a**) LPI, (**b**) AWMPFD, (**c**) CONTAG, (**d**) SHEI, (**e**) SHDI, and in the class level (**f**) PLAN built-up land, (**g**) LPI built-up land, and (**h**) AWMPFD built-up land

trend of a gradual increase from 0.43% to 4.29% within the period of 1996–2015 (equivalent to nearly 10 times) (Fig. 23.5g). This indicated the built-up land expanded and became larger in the urbanization process.

3.3.2 Area-Weighted Mean Patch Fractal Dimension (AWMPFD)

In the landscape shape dimension, AWMPDF metric reflects the complexity of a patch. The value of the AWMPFD index is within the threshold as follows: 1 ≤ AWMPFD ≤2. If this index goes to 2, the patch will have a more complex and fragmented shape (McGarigal et al., 1995). From calculation in the landscape level, the value of AWMPDF metric indicated a slight gradual increase from 1.17, 1.18, 1.19, and 1.19 in 1996, 2003, 2010, and 2015, respectively (Fig. 23.5b). This increase represented that the shape of the landscape became more irregular and complex within 19 years of urbanization. This trend was also depicted clearly in the class level for the built-up land and indicated an increase from 1.22 to 1.31 from 1996 to 2015. Especially, this index showed the period of 2003–2010 with the high-value increase from 1.25 to 1.33, which indicated that the built-up land became more irregular and complex in the rapid development stage (Fig. 23.5h).

3.3.3 Contagion (CONTAG)

CONTAG is a metric used to characterize scattered and concentrated landscape types (He, Gap, Niu, & Zhao, 2011). It is affected by the dispersion and interspersion of patch types and regarded as one of the significant indicators for reflection of connectivity and fragmentation over the whole landscape [6, 9]. In this study, the CONTAG was four times smaller, and decreased from 66.61 to 63.74 in the period of 1996–2015 (Fig. 23.5c). This decrease of the CONTAG metric indicated that the spatial distribution of different patches became more separated and the patch types became more disaggregated due to the strength of human interferences.

3.3.4 Shannon Diversity Index (SHDI) and Shannon Evenness Index (SHEI)

The SHDI indicates landscape heterogeneity and is sensitive to less-occupied patch types, while the SHEI expresses such an even distribution of areas among patch type's results in maximum evenness (Dai et al., 2017). As a measure of landscape heterogeneity, SHDI is especially sensitive to the nonbalanced distribution of all patch types in the landscape. SHEI is applied to indicate the diversity of different landscapes or a certain landscape in different periods, which results in maximum evenness (Zhou et al., 2014). Both SHDI and SHEI with a noticeable increase during the period of 1996–2015 indicated that the landscape of study was more fragmented and heterogeneous (Fig. 23.5 d, e).

3.3.5 Percent of Landscape (PLAND)

PLAND is defined as the percentage of the landscape comprising a particular patch type (McGarigal et al., 1995). The spatial analysis of the urban area showed a significant increase in the percentage of landscape index (PLAND) from 2.10% to 6.53% in 1996 and 2015, respectively (approximately 3 times higher). Especially, this trend showed a sharp increase from 2003 to 2010 with a value of 2.71% to 5.28%, respectively (Fig. 23.5f). This corresponded to the decline in other types of land use/land cover (agricultural land and bare land) for expanding the urban area.

3.4 Impact of Urbanization on Urban Landscape Pattern Changes

The impact of the process of urbanization on the landscape pattern change of Da Nang city was defined by analyzing the correlation between the significant landscape pattern change metrics and urbanization indicators. From the calculation of the urban landscape pattern fragmentation changes from 1996 to 2015, five landscape pattern metrics, which were found to be relatively changed much over 19 years, were chosen for the analysis including the percent of landscape for built-up land (PLANDbu), the largest patch index for built-up land (LPI^{bu}), the area-weighted mean patch fractal dimension for built-up land ($AWMPFD^{bu}$), CONTAG, and SHEI. The indicators were utilized for quantifying the relationship between urbanization and urban landscape patterns, which reflect the critical urbanization in the study area highlighting the Urban population rate (PoUrban), Urban population density (PoDensity), Non-agriculture GDP proportion (GDPNAgri), and Non-agriculture employment rate (EmpNAgr). The bivariate correlation between the urbanization variables and urban landscape pattern metric is shown in Table 23.9 with the Pearson' correlation coefficient (r), significant p-value (p), and the determination coefficient (R^2).

The result of Pearson's correlation analysis showed that the sign of the correlation had changed with different urbanization indicators.

Table 23.9 The correlation coefficient between the landscape patterns metrics and the urbanization indicators

	PoUrban			PoDensity			GDPNAgri			EmpNAgr		
	r	*p*	R^2	r	*P*	R^2	r	*p*	R^2	r	*p*	R^2
$PLAND^{bu}$	**0.97**	**0.03**	**0.95**	**0.99**	**0.01**	**0.98**	0.74	0.26	0.54	**0.98**	**0.02**	**0.93**
LPI^{bu}	0.87	0.13	0.76	**0.98**	**0.02**	**0.96**	0.68	0.32	0.47	0.89	0.11	0.79
$AWMPFD^{bu}$	**0.96**	**0.04**	**0.93**	0.98	0.12	0.77	0.82	0.18	0.68	**0.98**	**0.02**	**0.95**
CONTAG	−0.91	0.09	0.83	**−0.98**	**0.02**	**0.96**	−0.89	0.11	0.79	**−0.97**	**0.03**	**0.93**
SHEI	0.24	0.76	0.06	0.57	0.43	0.33	0.53	0.47	0.29	0.37	0.64	0.13

Although SHEI had a significant change during the 19 years of urbanization, there was no significant correlation between the urbanization indicators and Shannon evenness index (SHEI).

The urbanization indicators were significantly positively correlated with $PLAND^{bu}$, LPI^{bu}, $AWMPFD^{bu}$, and significantly negatively correlated with CONTAG.

For the $PLAND^{bu}$, the urban population (PoUrban), population density (PoDensity), and nonagriculture employment rate (EmpNAgr) had a high correlation coefficient with $PLAND^{bu}$. The highest correlation was with PoDensity (0.99), and the lowest was with the urban population (PoUrban) (0.97). The p-value of these three urbanization indicators and $PLAND^{bu}$ was less than 0.05. The determination coefficient (R^2) of these indicators had a high value, ranging from 0.93 to 0.98. The GDPNAgri has not affected the landscape pattern changes during 19 years of the urbanization. This indicated that the urbanization had been significantly affected the urban landscape pattern changes during the 19 years of Da Nang City's urbanization.

For the LPI^{bu}, only the population density (PoDensity) had a high correlation coefficient with LPI^{bu} with the value of 0.98. The p-value of this urbanization indicator was less than 0.05, and the coefficient of determination (R^2) was 0.96.

For the $AWMPFD^{bu}$, only the population density (PoDensity) and nonagriculture employment rate (EmpNAgr) had a high correlation coefficient with $AWMPFD^{bu}$ with the values of 0.96 and 0.98, respectively. The p-value of the two urbanization indicators was less than 0.05, and the coefficient of determination (R^2) was 0.93 and 0.95, respectively.

For the CONTAG, PoDensity and EmpNAgr have had a high correlation coefficient with CONTAG, with the values of −0.98 and − 0.97, respectively. The p-value of these three urbanization indicators and $PLAND^{bu}$ was less than 0.05. The determination coefficient (R^2) of these indicators had a high value, ranging from 0.93 to 0.96.

Thus, the analysis result showed that the landscape pattern in Da Nang had changed significantly for nearly two decades. These changes were due to the effects of policy change of the transition of administrative shift, economic shift, and demographic shift. The expansion of the built-up land in the research area was mainly due to the conversion of agricultural land in accordance with the development strategy policy of Da Nang City in the orientation of "service, industry, and agriculture." The city authorities invested actively in building Da Nang into a modern city with strong industrialization, modernization, and high services. The administrative shift and growth of urban population in urban area were a strong driving force on the landscape pattern change in the study period.

4 Conclusion

The process of urbanization and socioeconomic development has changed Da Nang urban space rapidly. From the research results, the LULC maps of 1996, 2003, 2010, and 2015 have been established by approaching the object-based (oriented) classification with the high accuracy. It is shown that the LULC has been changed over time. The area of built-up land increased about 8187.18 ha during the period of 1996–2015. Da Nang City experienced the rapid urbanization from 1996 to 2015, in which the expansion of urban land occurred quickly in the period of 2003–2010. The area of built-up land expanded, mainly through the conversion of agricultural land. The area of built-up land has increased about 8187.18 ha and about 3757.19 ha agricultural land which was lost during the period of 1996–2015. Through the spatial analysis of the LULC, the expansion of the construction land in Da Nang from the center radiated in the different directions of west–north, south, west–south, and southeast.

Da Nang urban spatial landscape distribution became more separated, complex, and irregular. It has resulted from the significant relationship between the urbanization indicators and landscape change indicators in combination with human activities—decisive factors for urban development. This is the significant reality basis for urban planning, proposing policy, and long-term development strategy to ensure sustainable urban development in future.

Acknowledgments The authors are grateful for the support from the project assigned by Hue University and Hue University of Sciences (ĐHH 2016-01-94). The authors thank Professor Nagasawa Ryota, from Tottori University, Japan, for providing ALOS Anvir-2 image and valuable ideas for this research.

References

Anderson, J. R., Hardy, E. E., Roach, J. T., & Witmer, R. E. (1976). A land use and land cover classification system for use with remote sensor data. Geological survey professional paper, U.S. government printing office. *Washington DC, 964*, 1–28.

Da Nang Construction Planning Institute. (2014). *Explanation adjusting the general planning of Da Nang city to 2030, vision to 2050* (111 pages). Nang.

Da Nang Statistical Office [DNGSO]. (1998). *Da Nang city statistical year book 1997* (159p). Statistical Publishing House.

Da Nang Statistical Office [DNGSO]. (2004). *Da Nang city statistical year book 2005* (289p). Statistical Publishing House.

Da Nang Statistical Office [DNGSO]. (2011). *Da Nang city statistical year book 2010* (315p). Statistical Publishing House.

Da Nang Statistical Office [DNGSO]. (2016). *Da Nang city statistical year book 2015* (407p). Statistical Publishing House.

Dai, E., Wu, Z., & Du, X. (2017). A gradient analysis on urban sprawl and urban landscape pattern between 1985 and 2000 in the Pearl River Delta. *China. Front. Earth Science, 12*, 791–807.

Dewan, A. M., & Yamaguchi, Y. (2009). Land use and land cover change in greater Dhaka, Bangladesh: Using remote sensing to promote sustainable urbanization. *Applied Geography, 29*, 390–401.

Field, A. (2013). *Discovering statistics using SPSS - IBM SPSS statistics* (4th ed., 1079p). Sage.

Giordano, F., & Marini, A. (2008). A landscape approach for detecting and assessing changes in an area prone to desertification in Sardinia (Italy). *International Journal of Navigation and Observation, 2008*, 1–5.

He, X., Gap, Y., Niu, J., & Zhao, Y. (2011). Landscape pattern changes under the impacts of urbanization in the Yellow River wetland taking Zhengzhou as an example. *Procedia Environmental Sciences, 10*, 2165–2169.

Huong, D. T. V., Ryota, N., & Tsutsui, K. (2013). Analysis of urban expansion and flood risk change in Da Nang city in Central Vietnam. *Japan Journal of Agriculture Society System, 29*, 123–134.

Kindu, M., Schneider, T., Teketay, D., & Knoke, T. (2013). Land use/land cover change analysis using object-based classification approach in Munessa - Shashemene landscape of the Ethiopian highlands. *Remote Sensing, 5*, 2411–2435.

Linh, N. H. K., & Erasmi, K. (2012). Quantifying land use/cover change and landscape fragmentation in Danang city, Vietnam: 1979-2009. *International Archives of the Photogrammetry, Remote Sensing and Spatial Information Sciences, Melbourne, Australia, 22*, 501–506.

Luck, M., & Wu, J. (2002). A gradient analysis of urban landscape pattern: A case study from the Phoenix metropolitan region, Arizona, USA. *Landscape Ecology, 17*, 327–339.

McGarigal, K., Marks, A., & Barbara, J. (1995). FRAGSTATS: Spatial pattern analysis program for quantifying landscape structure. Gen. Tech. Rep. PNW-GTR-351. Portland, OR: U.S. Department of Agriculture, Forest Service. *Pacific Northwest Research Station*, 122p.

Tamta, K., Bhadauria, H. S., & Bhadauria, A. S. (2015). Object-oriented approach of information extraction from high resolution satellite imagery. *IOSR Journal of Computer Engineering (IOSR-JCE), 17*, 47–52.

The, D.T., Cu, P.C, Nhuan, M.T., Lieu, T.M, Vuong, D.V. (2015). Application of remote sensing and GIS to study urbanization in Da Nang city and the relationship of urbanization in building a city capable of coping with climate change, GIS National Conference of the year 2015, Ha Noi Construction Publishing House, 2015: 570–576.

Tien, N. T., Sy, P. C., & Hang, H. T. M. (2006). Environmental issues in the process of urbanization - industrialization in Da Nang City. *Science and Technology Development Journal - Environment & Resources, 9*, 75–85.

Tu, D. T., Hai, T. Q., Ha, N. T. T., & Ngan, N. T. M. (2015). Study on Da Nang urbanization using multi -time Landsat imageries for environment protection planning and urban development. *Environmental Journal, 8*, 67–71.

United Nations. (2019). *World urbanization prospects: The 2018 revision (ST/ESA/SER.A/420)* (103p). United Nations.

Weng, Y. (2007). Spatiotemporal changes of landscape pattern in response to urbanization. *Landscape and Urban Planning, 81*, 341–353.

World Bank. (2011). *Vietnam urbanization review -technical assistance report* (238 p). The World Bank in Vietnam.

Yang, Q., Huang, X., & Li, J. (2017). Assessing the relationship between surface urban heat islands and landscape patterns across climatic zones in China. *Scientific Reports, 7*, 1–11.

Yi, Y., Zhao, Y., Ding, G., Gao, G., Shi, M., & Cao, Y. (2016). Effects of urbanization on landscape patterns in a mountainous area: A case study in the Mentougou District, Beijing, China. *Sustainability, 8*, 1–14.

Zhang, Z., Tu, Y., & Li, X. (2016). Quantifying the spatiotemporal patterns of urbanization along urban-rural gradient with a roadscape transect approach: A case study in Shanghai, China. *Sustainability, 8*, 1–19.

Zhang, L., Wu, J., Zhen, Y., & Shu, J. (2004). A GIS-based gradient analysis of urban landscape pattern of Shanghai metropolitan area, China. *Landscape and Urban Planning, 69*, 1–16.

Zhou, D., Shi, P., Wu, X., Ma, J., & Yu, J. (2014). Effects of urbanization expansion on landscape pattern and region ecological risk in Chinese Coastal City: A case study of Yantai City. *The Scientific World Journal, 2014*, 1–9.

Chapter 24
An Assessment of the Pollution Load Capacity of Son La Hydropower Reservoir in the Northwest Mountains of Vietnam

Duc Do Xuan, Hai Luu Duc, and Tuan Do Huu

Abstract Son La hydropower reservoir is the largest reservoir in the Northwest region of Vietnam on the Da River; its water surface area, length, capacity, and average water level are 225 km^2, 120 km, 9,26 billion m^3, and 215 m, respectively. The basin which provides water for Son La hydropower reservoir has an area of 11,075 km^2, including 16 districts and cities and over 100 communes and wards in provinces of Son La, Dien Bien, Lai Chau, and Yen Bai. Water resources within the basin play a decisive role in the water environment composition and biological structure of the wetland ecosystem in Son La hydropower reservoir. Economic, social, and humanitarian activities in the above basin area have direct and indirect impacts on the reservoir water environment. This study calculates the pollution load capacity of Son La hydropower reservoir with nine parameters of pollutants from the basin including DO, BOD_5, COD, TSS, Fe, NH_4^+, NO_2^-, NO_3^-, and PO_4^{3-}, based on the water levels and reservoir capacity when operating Son La hydropower factory.

Keywords Pollution load · Hydropower reservoir · Son La · Northwest Mountains of Vietnam

1 Introduction

The concept of environmental capacity was first used by American scientists when determining the appropriate grazing density in the Kaibab plateau in the early twentieth century (Edwards RY et al. 1974). According to Group of Experts on the Scientific Aspects of Marine Pollution (GESAMP) environmental capacity (also

D. D. Xuan (✉) · H. L. Duc · T. D. Huu
Tay Bac University, Vietnam, Son La
e-mail: doxuanduc@utb.edu.vn

A. T. Nguyen, L. Hens (eds.), *Global Changes and Sustainable Development in Asian Emerging Market Economies Vol. 2*, https://doi.org/10.1007/978-3-030-81443-4_24

known as "absorptive capacity) or" assimilative capacity "is defined as" environmental properties. "field and its capabilities in regulating a certain activity without causing unacceptable environmental impacts" (GESAMP 1986). According to calculations, it takes 21.7 ha of mangroves to absorb phosphorus from the wastewater of 1 ha of intensive shrimp farming and 7.2 ha of mangroves for nitrogen waste of 1 ha of shrimp farming. Intensively, for semi-intensive shrimp farming, the area is 2.4 ha (for nitrogen) and 2.8 ha (for phosphorus), respectively (Robertson and Phillips 1995). P-limited wetlands also have the ability to assimilate low P loads without significant changes in ecosystem structure and function (Richardson C J et al. 1999). Results indicated that 3863.75 tons of COD, 410.24 tons of NH_3-N, and 51.63 tons of P were discharged into Taiping Lake in 2011. While Taiping Lake reservoir could received pollutant according to the current water quality management target (Li X et al. 2014). Phosphor that can be self-cleaningin hydroelectric dams (David G S et al. 2015). In reservoirs, hydrology was an important and additional factor modulating phytoplankton in these tropical reservoirs, directly removing phytoplankton populations and their potential zooplankton grazers by washout, and also affecting nutrient availability (Rangel L M et al. 2012). The reservoir has the ability to load short-term pollution with nutrients from the input source for contaminated rural areas (Ryding S O et al. 1980). The filtration effect of flowing water according to the reservoir structure can reduce pollutant concentration by about 64% for suspended solids and 79% for lead (Legret M et al. 1996). Agricultural land use is directly related to basin pollution; these pollutants are observed to have a negative impact on water quality in freshwater reservoirs in terms of eutrophication and toxicity. The pollution load related to pesticides and fertilizers is calculated from the unit load based on the area used (Tanik A et al. 1999). The loading reservoir is polluted with nitrogen (N) and phosphorus (P) from the seasonal basin, thereby identifying the major polluting sources for the reservoir (Shen et al. 2014). Wetlands have the ability to clean up polluted water contaminated with NH_4^+, BOD_5, COD, and PO_4^{3-} obtained from the Buriganga River in Bangladesh (Saeed T et al. 2016). This study involves calculation of the pollutant loading capacity of the Son La hydropower reservoir in the Northwest region of Vietnam with nine categories of pollutants from the basin, including DO, BOD_5, COD, TSS, Fe, NH_4^+, NO_2^-, NO_3^-, and PO_4^{3-}, based on water level and reservoir capacity when operating Son La hydropower plant.

Estimation of nitrogen load with multiple pollution sources using the SWAT model was done in a case study in Cau River basin in Northern Vietnam, analyzing pollution load depending on rainfall (Tran V B et al. 2017). The activities of basin residents change the nitrogen cycle and phosphor cycle in Nui Coc Lake (Vietnam); the study demonstrates about 66% of total annual nitrogen and 79% of total annual phosphorus removal in the reservoir (Le T P Q et al. 2014). The formation of dams and reservoirs in the Red River Basin of Vietnam, including the Da River system (Hoa Binh hydroelectric dam), has been shown to have self-cleaning for metals (Dang et al. 2010). Calculating the economic value of forest hydrological services for Hoa Binh Hydroelectric Plant in Vietnam indicates that the economic value of forest hydrological services for electricity production ranges from 26.3 million

USD to 85.5 million USD per year, and that the longevity of the hydroelectric plant can be prolonged by about 35–80 years, depending on the state of forest cover in the watershed (Nguyen et al. 2013). In this study, we would like to assess how, after the formation of the Son La hydropower reservoir on the Da River (Northwest of Vietnam), is this reservoir capable of loading pollutants from the basin. Pollution load according to operating water level is calculated and discussed.

2 Methodology

2.1 *Study Area*

Son La hydropower reservoir is a reservoir on the Da River; this is the largest reservoir in the Northwest region of Vietnam. The water surface area is 225 km^2; the length is 120 km, connecting 3 provinces of Northwest Vietnam, namely Son La, Dien Bien, and Lai Chau. With the capacity of 9,26 billion m^3, the average water level is 215 m. The lake is located in the basin of Son La hydroelectric lake, with magnetic coordinates $21^0 15'15''$ to $22^0 45'10''$ north latitude, to $102^0 50'10''$ from $104^0 35'15''$ east longitude.

2.2 *Data Collection and Analysis*

Monitoring data of Son La hydropower reservoir inlet and outlet are a source of data, which is the basis for analyzing the pollutant loading capacity of Son La hydroelectricity reservoir.

Data: the correlation between flood elevation, water level and reservoir capacity during the operation of Son La hydropower plant, 2017, to calculate the pollution load capacity according to the reservoir operating water level.

2.3 *Formula for Calculating the Pollutant Load capacity of Son La Hydroelectricity Reservoir*

The load capacity of hydropower reservoirs is estimated through formulas, using the circular: 76/2017/MONRE, Vietnam Ministry of Natural Resources and Environment.

$$M_{rd} = (C_{sd} - C_{i.out}) \times V_h \times 10^{-3} \times F_S \tag{24.1}$$

In which:

- M_{rd}: wastewater receiving capacity, load capacity for each pollution parameter of the lake; its unit is kg.
- C_c: C_{sd}: the limit value of surface water quality parameters according to the technical standards of surface water quality for usage purpose of the lake; its unit is (mg/l), in that case, that lake is at the level A_2 of National technical regulation on surface water quality 08:2015/MONRE.
- $C_{i.out}$: the analysis results of the output water quality parameters of the lake and separate areas; the unit of calculation is mg/l.
- V_h: the volume of the lake; it is determined on the basis of the volume of the Son La hydroelectric lake on the correlation between elevation level, water level, and reservoir capacity during the operation of Son La hydroelectric plant.
- F_S: safety factor, taken as 0.7.

3 Results

3.1 *Pollution Loading of Son La Hydroelectric Lake Reservoir at 175 M Water Level*

Based on the output water quality data in the water sample at the foot of the Son La hydroelectric dam (Table 24.1) and the lake capacity data (Table 24.2), the possibility of pollution load was calculated for 09 parameters (DO, BOD_5, COD, TSS, Fe, NH_4^+, NO_2^-, NO_3^-, and PO_4^{3-}) of the lake at 175 m, with a volume of 2756 (million m^3) (Fig. 24.1). The numbers were plugged into the formula (24.2) and the following results were derived.

$$M_{rd} = \left(C_{sd} - C_{i.out}\right) \times V_h \times 10^{-3} \times F_S \quad (24.2)$$

M_{DO}	=	$(C_7 - C_{5,4}) \times 2.756 \times 0.7$	=	3086	(tons)
M_{TSS}	=	$(C_{30} - C_{15}) \times 2.756 \times 0.7$	=	28,938	(tons)
M_{BOD5}	=	$(C_6 - C_4) \times 2.756 \times 0.7$	=	3858	(tons)
M_{COD}	=	$(C_{15} - C_9) \times 2.756 \times 0.7$	=	11,575	(tons)
M_{Fe}	=	$(C_1 - C_{0,3}) \times 2.756 \times 0.7$	=	1350	(tons)
$M_{NH4}+$	=	$(C_{0,3} - C_{0,029}) \times 2.756 \times 0.7$	=	522	(tons)
$M_{NO2}-$	=	$(C_{0,05} - C_{0,049}) \times 2.756 \times 0.7$	=	1929	(tons)
$M_{NO3}-$	=	$(C_5 - C_{0,85}) \times 2.756 \times 0.7$	=	8006	(tons)
$M_{PO4}3-$	=	$(C_{0,2} - C_{0,091}) \times 2.756 \times 0.7$	=	210	(tons)

With a water level of 175 m and a volume of 2756 million m^3, Son La hydroelectric reservoir is capable of withstanding 59,475 (tons) of pollutants, including 3086 tons (DO); 28,938 tons (TSS); 3858 tons (BOD_5); 11,575 tons (COD); 1350 tons (Fe); 522 tons (NH_4^+); 1929 tons (NO_2^-); 8006 tons (NO_3^-); 210 tons (PO_4^{3-}).

Table 24.1 Quality of the inlet and outlet water of Son La hydropower reservoir, 2017

List	Parameter	Unit	Input Lai Chau hydroelectric dams	Output Son La hydroelectric dam	National technical regulation on surface water quality 08:2015/MONRE, Column A2
1	Dissolved oxygen(DO)	Mg/L	5.2	5.4	≥ 5
2	Total suspended solids(TSS)	Mg/L	25	15	30
3	Biochemical oxygen demand BOD_5 (20 °C)	Mg/L	4.2	4.0	6
4	Chemical oxygen demand COD	Mg/L	< 9	< 9	15
5	Ammonium (NH_4^+) (calculated by N)	Mg/L	0.024	0.029	0.3
6	Nitrite (NO_2^-) (calculated by N)	Mg/L	0.037	0.049	0.05
7	Nitrate (NO_3^-) (calculated by N)	Mg/L	0.78	0.85	5
8	Phosphate (PO_4^{3-}) (calculated by P)	Mg/L	0.076	0.091	0.2
9	Crude iron (Fe)	Mg/L	0.4	< 0.3	1.0

Source: Son La hydroelectric company, 2017

Table 24.2 Water level and operating capacity of Son La hydropower reservoir in 2017

Month	Elevation of submerged Son La hydroelectric lake in the year	Lake water level (meters)	Lake capacity (million m^3)	Noted
April–august	The water level is shallow	175	2756	Shallow capacity
January–march	Average water level	190	6504	Useful capacity
September–December	Highest water level	215	9260	Overall capacity

Source: Son La hydroelectric company, 2017

3.2 *Pollution Loading of Son La Hydroelectric Lake Reservoir at 190 M Water Level*

Based on the data of the output water quality in the water sample at the foot of the Son La hydroelectric dam (Table 24.1) and the volume of the lake (Table 24.2), it is possible to calculate the pollution load capacity for 09 parameters (DO, BOD_5, COD, TSS, Fe, NH_4^+, NO_2^-, NO_3^-, and PO_4^{3-}) of Son La hydroelectric lake at a water level of 190 m and a volume of 6504 (million m^3). The following results were derived after plugging the numbers into the formula (24.3) (Fig. 24.2).

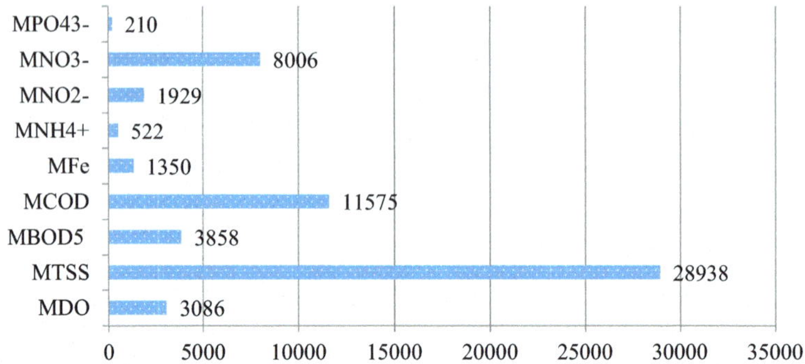

Fig. 24.1 Pollution loading capacity of Son La hydropower reservoir at 175 m water level

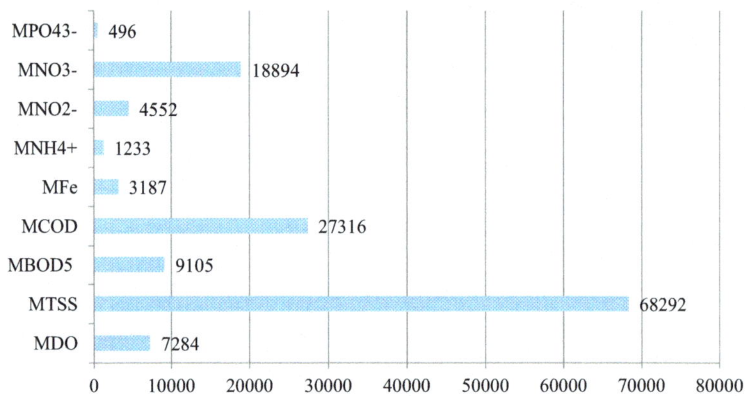

Fig. 24.2 Pollution loading capacity of Son La hydroelectric reservoir at 190 m water level

$$M_{rd} = \left(C_{sd} - C_{i.out}\right) \times V_h \times 10^{-3} \times F_S \tag{24.3}$$

M_{DO}	=	$(C_7 - C_{5,4}) \times 6.504 \times 0.7$	=	7284	(tons)
M_{TSS}	=	$(C_{30} - C_{15}) \times 6.504 \times 0.7$	=	68,292	(tons)
M_{BOD5}	=	$(C_6 - C_4) \times 6.504 \times 0.7$	=	9105	(tons)
M_{COD}	=	$(C_{15} - C_9) \times 6.504 \times 0.7$	=	27,316	(tons)
M_{Fe}	=	$(C_1 - C_{0,3}) \times 6.504 \times 0.7$	=	3187	(tons)
$M_{NH4}+$	=	$(C_{0,3} - C_{0,029}) \times 6.504 \times 0.7$	=	1233	(tons)
$M_{NO2}-$	=	$(C_{0,05} - C_{0,049}) \times 6.504 \times 0.7$	=	4552	(tons)
$M_{NO3}-$	=	$(C_5 - C_{0,85}) \times 6.504 \times 0.7$	=	18,894	(tons)
$M_{PO4}3-$	=	$(C_{0,2} - C_{0,091}) \times 6.504 \times 0.7$	=	496	(tons)

With a water level of 190 m and a volume of 6504 million m^3, Son La hydroelectric reservoir has a load capacity of 140,359 (tons) of pollutants, including: 7284 tons (DO); 68,292 tons (TSS); 9105 tons (BOD_5); 27,316 tons (COD); 3187 tons (Fe); 1233 tons (NH_4^+); 4552 tons (NO_2^-); 18,894 tons (NO_3^-); 496.2 tons (PO_4^{3-}).

3.3 *Pollution Loading of Son La Hydroelectric Lake Reservoir at 215 M Water Level*

Based on the data of the output water quality in the water sample at the foot of the Son La hydroelectric dam (Table 24.1) and the volume of the lake (Table 24.2), it is possible to calculate the pollution load capacity for 09 parameters (DO, BOD_5, COD, TSS, Fe, NH_4^+, NO_2^-, NO_3^-, and PO_4^{3-}) of Son La hydroelectric lake at a water level of 215 m and a volume of 9260 (million m^3). The following results were derived after plugging the numbers into the formula (24.4) (Fig. 24.3).

$$M_{rd} = \left(C_{sd} - C_{i.out}\right) \times V_h \times 10^{-3} \times F_S \quad (24.4)$$

M_{DO}	=	$(C_7 - C_{5,4}) \times 9.260 \times 0.7$	=	10,371	(tons)
M_{TSS}	=	$(C_{30} - C_{15}) \times 9.260 \times 0.7$	=	97,230	(tons)
M_{BOD5}	=	$(C_6 - C_4) \times 9.260 \times 0.7$	=	12,964	(tons)
M_{COD}	=	$(C_{15} - C_9) \times 9.260 \times 0.7$	=	38,892	(tons)
M_{Fe}	=	$(C_1 - C_{0,3}) \times 9.260 \times 0.7$	=	4537	(tons)
$M_{NH4}+$	=	$(C_{0,3} - C_{0,029}) \times 9.260 \times 0.7$	=	1756	(tons)
M_{NO2}^-	=	$(C_{0,05} - C_{0,049}) \times 9.260 \times 0.7$	=	6482	(tons)
M_{NO3}^-	=	$(C_5 - C_{0,85}) \times 9.260 \times 0.7$	=	26,900	(tons)
$M_{PO4}3-$	=	$(C_{0,2} - C_{0,091}) \times 9.260 \times 0.7$	=	706	(tons)

With a water level of 215 m, the total volume of the lake reaches 9260 million m^3; Son La hydroelectric lake has a load capacity of 199,838 (tons) of pollutants, including 10,371 tons (DO); 97,230 tons (TSS); 12,964 tons (BOD_5); 38,892 tons (COD); 4537 tons (Fe); 1756 tons (NH_4^+); 6482 tons (NO_2^-); 26,900 tons (NO_3^-); 706 tons (PO_4^{3-}).

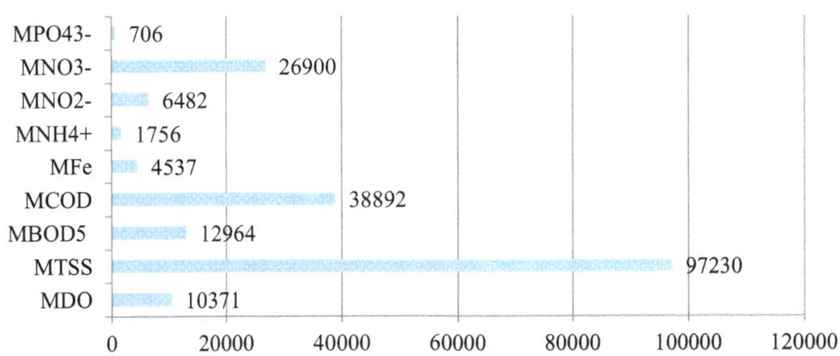

Fig. 24.3 Pollution loading capacity of Son La hydroelectric reservoir at 215 m water level

4 Conclusions and Discussion

The study identified the monitoring data of water quality of Son La hydroelectric reservoir with input and output parameters. In addition, certification of the water level and operating volume of the Son La hydroelectric plant by year was done.

The calculation results demonstrate the ability of Son La hydropower reservoir to load about 399,672 tons of pollutants/year with 09 parameters including DO, BOD_5, COD, TSS, Fe, NH_4^+, NO_2^-, NO_3^-, and PO_4^{3-}.

From January to March, the water level is 190 m, reaching a useful volume of 6504 (million m^3); the lake is capable of loading 140,359 (tons) of pollutants.

From April to August, the water level is 175 m, reaching a shallow volume of 2756 (million m^3); the lake is capable of loading 59,475 (tons) of pollutants.

From September to December, the water level is 215 m, with a volume of 9260 (million m^3), and the lake is capable of loading 199,838 (tons) of pollutants.

The study results have proved that the bigger the water level and the volume of the Son La hydroelectric reservoir will be able to load pollutants higher than the water level and the lower volume.

The results of calculating the pollutant loading capacity of the Son La hydroelectric reservoir are the basis for controlling the sources of pollutants from the seasonal basin of the year, ensuring the capacity of receiving and assimilating pollutants of the lake while maintaining water environment functions.

Acknowledgments The authors would like to thank the Son La Hydropower Company for allowing the use of the environmental monitoring data source of Son La hydroelectric lake.

References

Dang, T. H., Coynel, A., Orange, A., Blanc, D., Etcheber, G. H., & Le, L. A. (2010). Long-term monitoring (1960-2008) of the river-sediment transport in the Red River watershed (Vietnam): Temporal variability and dam-reservoir impact. *Science of the Total Environment, 408*(20), 4654–4664.

David, G. S., Carvalho, E. D., Lemos, D., Silveira, A. N., & Dall'Aglio-Sobrinho, M. (2015). Ecological carrying capacity for intensive tilapia (Oreochromis niloticus) cage aquaculture in a large hydroelectrical reservoir in southeastern Brazil. *Aquacultural Engineering, 66*, 30–40.

Edwards RY, Fowle CD. (1974). The concept of carrying capacity. In readings in wildlife management. In Bailey, J.A, Elder.W, McKinney.T.D. (eds.), *The wildlife society*, Washington DC, USA.

GESAMP. (1986). Organization, IMO/FAO/UNESCO/WMO/WHO/IAEA/UN/UNEP Joint Group of Experts on the Scientific Aspects of Marine Pollution (GESAMP: report of the sixteenth session, London, 17–21 March 1986. Retrieved October 16, 2019, from https://apps.who.int/iris/handle/10665/62801,2019/.

Le, T. P. Q., Ho, Q. C., Duong, T. T., Rochelle-Newall, A., Dang, D. K., & Hoang, T. S. (2014). Nutrient budgets (N and P) for the Nui Coc reservoir catchment (North Vietnam). *Agricultural Water Management, 142*, 152–161.

Legret, M., Colandini, V., & Le, M. (1996). Effects of a porous pavement with reservoir structure on the quality of runoff water and soil. *Science of the Total Environment, 189*, 335–340.

Li X, Lu J, Qian M, Wang X, Fan Z, Wang S (2014). Study on pollution loading and water environmental capacity in watershed case study of Taiping Lake Basin, Anhui Province, China. Department of Environmental Science and Engineering, Fudan University, 1–8.

Nguyen, T. T., Pham, V. D., & Tenhunen, J. (2013). Linking regional land use and payments for forest hydrological services: A case study of Hoa Binh reservoir in Vietnam. *Land Use Policy, 33*, 130–140.

Rangel, L. M., Silva, L. H. S., Rosa, P., Roland, F., & Huszar, V. L. M. (2012). Phytoplankton biomass is mainly controlled by hydrology and phosphorus concentrations in tropical hydroelectric reservoirs. *Hydrobiologia, 693*(1), 13–28.

Richardson, C. J., & Qian, S. S. (1999). Long-term phosphorus assimilative capacity in freshwater wetlands: A new paradigm for sustaining ecosystem structure and function. *Environmental Science & Technology, 33*(10), 1545–1551.

Robertson, A. I., & Phillips, M. J. (1995). Mangroves as filters of shrimp pond effluent: Predictions and biogeochemical research needs. *Hydrobiologia, 295*(1–3), 311–321.

Ryding, S. O., & Forsberg, C. (1980). Short-term load-response relationships in shallow. *Polluted Lakes, Hypertrophic Ecosystems, 2*, 95–103.

Saeed, T., Paul, B., Afrin, R., Al-Muyeed, A., & Sun, G. (2016). Floating constructed wetland for the treatment of polluted river water: A pilot scale study on seasonal variation and shock load. *Chemical Engineering Journal, 287*, 62–73.

Shen, Z., Qiu, J., Hong, Q., & Chen, L. (2014). Simulation of spatial and temporal distributions of non-point source pollution load in the three gorges reservoir region. *Science of the Total Environment, 493*, 138–146.

Tanik, A., Beler, B. B., BaykalI, A., & Gonenc, E. (1999). The impact of agricultural pollutants in six drinking water reservoirs. *Water Science and Technology, 40*(2), 11–17.

Tran, V. B., Ishidaira, H., Nakamura, T., Do, T. N., & Nishida, K. (2017). Estimation of nitrogen load with multi-pollution sources using the SWAT model: A case study in the Cau River basin in northern Vietnam. Journal of water and environment. *Technology, 15*(3), 106–119.

Chapter 25
An Approach for Prioritizing Climate Change Mitigation Measures in Ho Chi Minh City

Doan Quang Tri, Pham Thanh Long, Vuong Xuan Hoa, and Ngo Thanh Tam

Abstract Prioritizing climate change mitigation measures could help identifying the most feasible or most nationally appropriated mitigation actions. This process can also provide important inputs for the development of national climate change strategies or policies. The paper applies the Delphi method to prioritize criteria for potential climate change mitigation technology in Ho Chi Minh City, Vietnam. The consultation process has been done with 10 experts in only two cycles to reach a Kendall (W) value of over 0.5. Then, 21 selected criteria have been used for multi-criteria decision analysis (MCDA) in prioritizing mitigation options in energy, industrial processes, road transport, residential, commercial, aviation, and waste sectors in Ho Chi Minh City. Mitigation options with highest scores will be proposed in Ho Chi Minh City' Action Plan to implement Paris Agreement on Climate Change.

Keywords Climate change mitigation · Delphi method · MRV · Ho Chi Minh City

1 Introduction

Climate change is one of the biggest global challenges for humankind in the twenty-first century. It is an existential risk to the sustainable development goals of all countries and regions. Monitoring Reporting Verification (MRV) systems are designed to increase transparency and check the progress and effectiveness of

D. Q. Tri (✉) · N. T. Tam
Vietnam Journal of Hydrometeorology, Vietnam Meteorological and Hydrological Administration, Hanoi, Vietnam

P. T. Long
Sub-Institute of Hydrometeorology and Climate Change, Ho Chi Minh City, Vietnam

V. X. Hoa
Institute of Hydrometeorology and Climate Change, Hanoi, Vietnam

A. T. Nguyen, L. Hens (eds.), *Global Changes and Sustainable Development in Asian Emerging Market Economies Vol. 2*, https://doi.org/10.1007/978-3-030-81443-4_25

adaptation and mitigation actions. A fully established MRV system will involve the participation of climate management levels, increase the gained benefits, and reduce the cost of information transmission. To develop the plan to implement the Paris Agreement, it is necessary to assess the climate change mitigation actions that are appropriate to the circumstances, development priorities, and resources in Ho Chi Minh City (HCM) . Previously, studies on the evaluation and prioritization of mitigation activities in Vietnam were used to use reference indicators from international reports and were based on the subjective assessments of experts. Therefore, this paper involves research by applying the Delphi method and multi-criteria decision analysis (MCDA) to establish and select criteria for prioritizing mitigation options in HCM. Yedla and Shrestha (2003) studied and examined the impact of including various qualitative criteria for the selection of alternative transportation options in Delhi. The application of the Delphi method will include a three-phase analysis process, which will be used before, during, and after consultation. The consultation process is carried out in two stages. In the first stage, a series of open-ended questions on important issues of mitigation options were developed and sent to the experts to find out the criteria that could help prioritize nationally appropriate mitigation actions. Because this study has been prevalent in the world, the study will obtain the results of the previous studies and summarize general evaluation criteria. Besides, the resulting evaluation criteria, which are specific and appropriate to the national conditions, will be consulted with experts on climate change mitigation. After reviewing the evaluation criteria, the questionnaire on the importance and suitability of the criteria will be sent to the experts for evaluation until the consensus index is reached (Kendall (W) ≥ 0.5) (Musa, Yacob, Abdullah, & Ishak, 2015). Apart from the main goal of climate change mitigation, the climate change mitigation actions should ensure the sustainability of the environment, economy, and social development. The set of evaluation criteria, therefore, needs to fully work towards these goals. Currently, there are no official documents that specify a globally accepted standard for selecting the actions that are appropriate to national conditions (NAMAs) to receive international assistance. Based on the theoretical basis together with the results and experience of international scientific research experiences (Massa, Kim, Wheeler, & Mitchell, 2008; Posada, 2009; Asahi Glass Co. Ltd, 2013; Malaysia Gas Association (MGA), 2014; EX Research Institute Ltd, 2012; Pham, Doan, Vuong, & Ngo, 2019) as well as the uniqueness, importance and independence to ensure the outcome are legitimate. The research team has collected, synthesized, and drawn out the criteria that play a significant role in identifying and developing climate change mitigation actions by following the national conditions (NAMA) that are approved and implemented by different developing member countries of the United Nations Framework Convention on Climate Change (UNFCCC) .

The objective of assessing climate change mitigation measures is to identify activities that reduce greenhouse gas emissions yet still in line with the national conditions. These activities will be meaningful to contribute to reducing greenhouse gas emissions while ensuring economic efficiency, improving the environment, social security, and sustainable development of nations.

2 Methodology

2.1 *Study Area*

Ho Chi Minh City (HCM) is a city with the biggest economy and population size; the city itself has a large amount of GHG emissions, but it also contributes significantly to GHG emissions reduction nationwide (Fig. 25.1). However, the management of GHG emissions in the city has been ineffective so far; the officials' understanding of HG is still limited to some extent. From 2015 to 2017, HCM has actively participated in supporting planning and action implementations to reduce greenhouse gas emissions in line with the national conditions (SPI-NAMA Project), which are launched by the Ministry of Natural Resources and Environment (MoNRE) and the Japan International Cooperation Agency (JICA). The project has achieved prominent results such as guidance on GHG emissions inventory and measurement, reporting, and verification (MRV) of GHG emission reduction. As for GHG emission reduction activities, by the end of 2018, the industry and trade of HCM have primarily focused on dealing with emission reduction in the energy sector. These activities have contributed greatly to the reduction of GHG emissions from industrial activities in the city and the country as a whole. Recent GHG inventory results from HCM indicate that the main sources of emissions include: (i) road transport; (ii) industrial production and construction; (iii) residential buildings; (iv) commercial buildings, public administration buildings, and infrastructure; (v) aviation; (vi) solid waste dumped into landfills; and (vii) domestic wastewater. Therefore, the mitigation activities of HCM will focus on these main causes.

2.2 *Assessing Mitigation Actions*

The assessment of climate change mitigation measures will provide important input for developing a national climate change strategy in developing countries like Vietnam. This assessment can also contribute to capacity building and awareness-raising in developing countries on environmentally sustainable technologies, and increase resilience, and reduce vulnerability to the effects of climate change. The assessment also creates a chance to understand the advantages, disadvantages, and barriers when deploying these technologies. The assessment will provide an overview of solutions or technologies that are suitable for a region or a country, facilitating programs or initiatives support of international assistance on climate change mitigation. When doing the assessment, the selected technologies must be appropriate to the conditions and development strategies of the countries. Factors consistent with national development priorities will ensure that transferred technology and national resources will be used efficiently. Therefore, the description of national development needs and priorities is necessary before moving on to the technology assessment. These priorities will also be shaped by the long-term social and

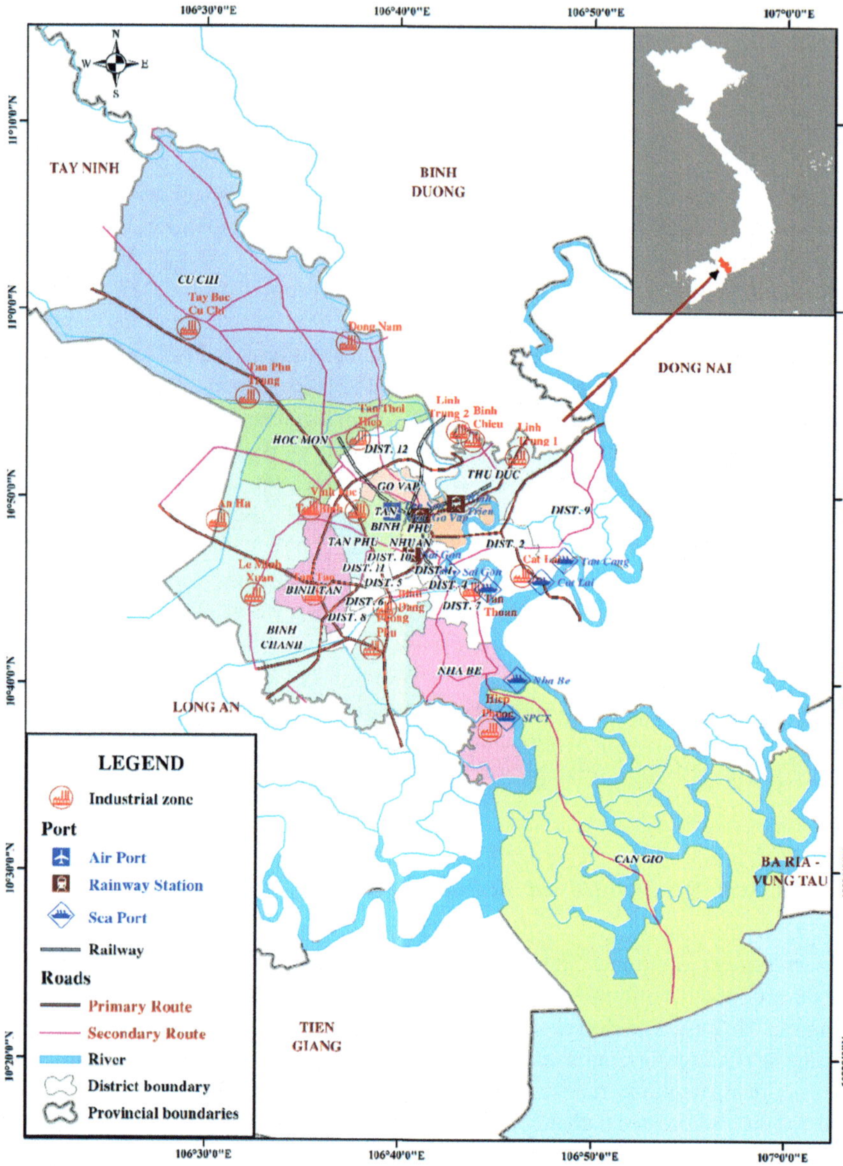

Fig. 25.1 The location of the study area in Vietnam

economic trends in Vietnam, such as increased industrialization and urbanization, because these priorities will have an impact on final technology options.

An important principle when developing climate change mitigation policies is to predict or forecast changes in future climate conditions. From this, it is possible to determine the impact of climate change on energy demand and greenhouse gas

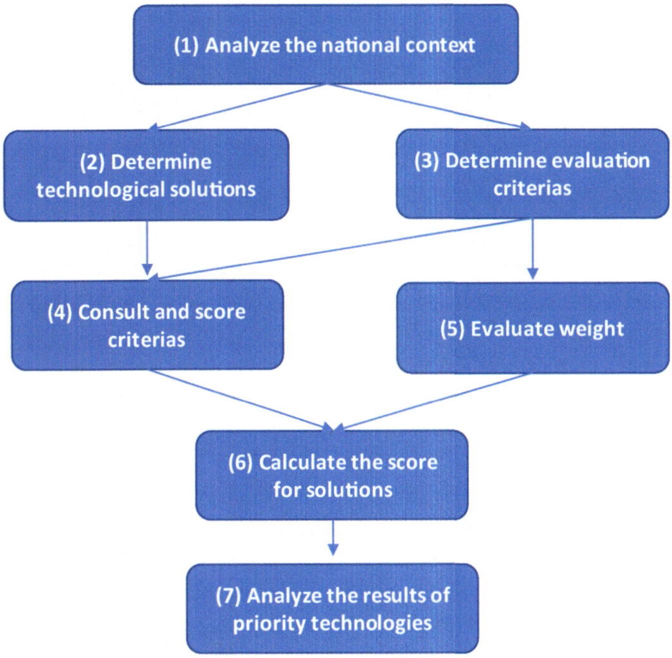

Fig. 25.2 Logical framework for prioritizing climate change mitigation technologies

emissions. To assess mitigation techniques, information on climate change scenarios should be collected and analyzed before conducting a technology assessment. This is one of the fundamental factors for prioritizing climate change mitigation technologies by following the national conditions. The priority evaluation process of climate change mitigation technologies in the study follows the multi-criteria analysis approach described by Dodgson, Spackman, Pearman, and Phillips (2009) and the Guidelines for Prioritizing Mitigation Technology of UNEP DTU Partnership (Ministry of Natural Resources and Environment (MoNRE), 2011; URC, 2011; Wickramasinghe, 2011; URC, 2012a, 2012b). The relevant steps are shown in Fig. 25.2. Typically, in the report on the assessment of climate change mitigation technology in Vietnam, the identification of evaluation criteria is based on the guidance documents of international. However, the criteria in the guidance are relatively common and general. To assess the technology of climate change mitigation that is appropriate for Vietnam, specific criteria need to be developed with respect to Vietnam. Therefore, the study involves the application of the Delphi method to formulate evaluation criteria to prioritize climate change mitigation technology for Vietnam.

1. National context analysis

   An analysis of the national context is aimed at identifying national development goals and priorities, such as socioeconomic, energy, and development goals and plans, as well as response to climate change. Also, this analysis also reviews and analyzes development plans of different economic sectors. Besides, information concerning scenarios on climate change, sea-level rise, and the impacts of climate change on industries has also been collected to provide the foundation for choosing suitable mitigation technology on a national scale.

2. Identify technology solutions to climate change mitigation

   The identification of technology solutions for climate change mitigation can be referred to reports of greenhouse gas reduction scenarios in the world or in the country where the assessment is conducted. If these reports are not available, consult sources such as Climate TechWiki and the guidebook published by UNEP DTU including transport sector, construction, and agriculture (in terms of mitigation) and coastal areas, water, and agriculture (in terms of adaptation) or IPCC evaluation reports (AR3, AR4, AR5) should be taken into consideration. Countries that have developed mitigation scenarios can be referred to reports such as the National Communication on Climate Change and Nationally Determined Contribution. The assessment of mitigation technology solutions mentioned above will help policymakers have a foundation to build a roadmap to implement commitments on reducing greenhouse gas emissions. Moreover, this also provides the private sector with information about potential, relevant, and feasible solutions in that country.

3. Determining evaluation criteria

   To identify the criteria, the study has applied the Delphi method. Accordingly, the criteria will be developed through consultation with experts in the field of GHG mitigation until a consensus is reached. Through the consultation, the main issues or factors that affect the selected technology have been clarified. Besides, the experts have also proposed criteria that satisfy national development priorities. These criteria can be organized into subcriteria in different levels to display the link between development priorities.

4. Consulting and Marking Criteria

   The technology will be implemented based on the selected criteria. Firstly, a performance matrix is developed, in which the rating scale can vary for each criterion. For example, capital costs can be listed directly in monetary units, reducing greenhouse gas emissions in tons of CO_2, and qualitative criteria can be assessed on the Likert scale (or a similar scale). The data of quantitative criteria will be standardized into a mark for each type of technology. Meanwhile, qualitative criteria will be assessed on a scale of 1–5 with increasing importance. The information for evaluation and scoring of the criteria is compiled from national research reports on reducing greenhouse gas emissions and consultations from experts in this field.

5. Weight Evaluation

   The chosen criteria to evaluate the priority of each technology may not be equally important to reach the common goal. Therefore, the given weights to each criterion will reflect their relative importance in the choice of technology options. For example, whether the cost factor is more important than greenhouse gas emissions reduction when choosing technology in the energy sector? If it is, then how much the cost will be? This step aims to assign quantitative values to the relative importance of the criteria. There are many different ways to assign weights, such as using consultations or statistics. However, within the scope of the study, the weights of the criteria are determined by standardizing the results of expert consultations in the "determining evaluation criteria" step.

6. Marking Technology Solutions

   All the information and views collected in the above steps are merged, and several technologies are selected for more detailed analysis. Marking these technology solutions can be done using the multi-criteria analysis model (MCDA) provided by UDP. The technology options are then sorted according to their total score, and two or three technologies with the best score will be selected for further analysis by the SWOT method to further clarify the difficulties and advantages as implementation.

2.3 Development Criteria for Evaluating Priority Options

2.3.1 The Delphi Method

The Delphi method has been applied in many fields ranging from economy, environment, sustainable development, land use, agriculture, transportation, nursing, tourism to climate change (Chan & Lee, 2019; Fefer, Stone, Daigle, & Silka, 2016; Hsueh, Sun, & Yan, 2019; Musa et al., 2015; Nguyen, Vu, Dang, Hoang, & Hens, 2018). There are two ways to apply the method: It will be implemented radiationally or implemented in 4 stages (Linstone & Turoff, 2002). Bunting (2008) used the Delphi method to facilitate interactive participation and reach a consensus on sustainable aquaculture development (Bunting, 2008). For the transport sector, a combination of the Delphi method and the Bayesian Network Model is used to predict highway accidents in developing countries (Mbakwe, Saka, Choi, & Lee, 2016). Delbari, Siew, Aziz, and Ho (2016) used the 2-stage Delphi method together with the hierarchical analysis model (AHP) to identify and prioritize important indicators for aviation services. The future of the 3D printing industry is also developed by using the Delphi method; 18 forecasts have been developed to provide future scenarios for the 3D printing industry (Jiang, Kleer, & Piller, 2017). Le et al. (2015) used the Delphi method in combination with the DPSIR (Divers–Pressures–States–Impacts–Response) (Elliott, Cutts, & Trono, 2014; EU, 1999; Gari, Newton, & Icely, 2015; Martins, Camanho, & Gaspar, 2012; Smith et al., 2016) to evaluate climate change response in local communities in coastal areas (Fig. 25.3). Questions were asked to assess the level of consensus among members of the consultative group. The Kendall value is then calculated after the consensus reached 0.681, the high level of consensus among the members. The study indicates that sustainable ecosystem development and new rural planning are considered appropriate local adaptation measures in the study area. Nguyen et al. (2018) and the research team used the Delphi method to develop a set of sustainable development indicators focusing on environmental and health fields, applied to Quang Tri province.

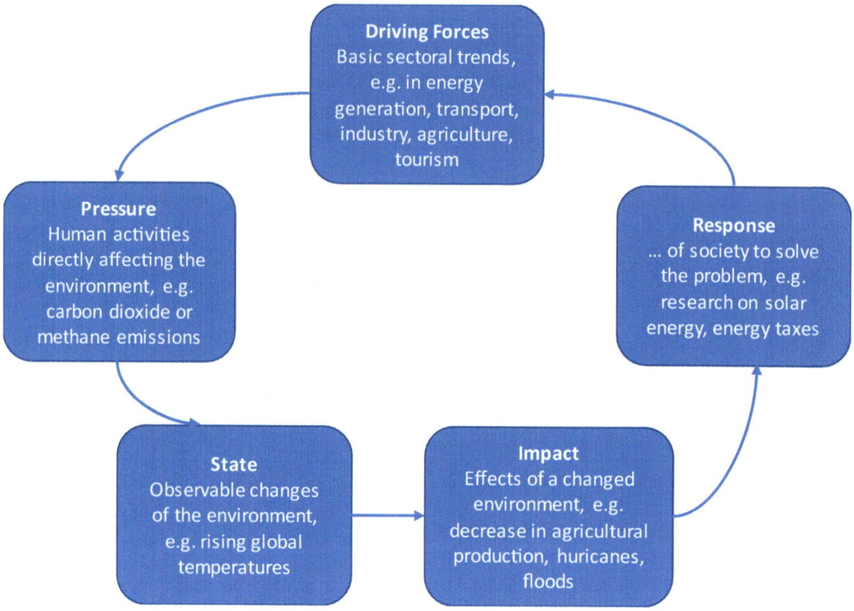

Fig. 25.3 Driving forces–Pressure–State–Impact–Response modified from the original EU framework (EU, 1999)

1. Steps before consultation:

Step 1. Selection of expert groups related to the Delphi method.

The number of experts selected to participate in the consultation for using the Delphi method is 10 people in the field of climate change mitigation from the Department of Climate Change, MoNRE, Vietnam Institute of Meteorology, Hydrology and Climate Change (IMHEN), Energy Institute, Energy and Environment Consultancy Joint Stock Company, Institute of Agricultural Environment, Ministry of Transport, Ministry of Construction, Ministry of Industry and Trade, and so on.

Step 2. Summary of evaluation criteria for prioritizing climate change mitigation technology worldwide according to the technical guidelines of UNEP DTU, IPCC, and UNDP; guidelines on assessing and selecting solutions to reduce greenhouse gas emissions in line with the national conditions (NAMAs); and consultation with experts on some specific conditions in Vietnam.

The indicators include 04 groups

(a) Climate indicators (reducing GHG emissions): The main objective of these indicators is to reduce GHG emissions in each sector. The criteria to consider are:

- GPT1: GHG emissions intensity (GHG emissions per product unit).
- GPT2: Potential to reduce GHG emissions.

(b) Benefit indicators (environmental index, social index, economic index):

- Objective 1 (for the environment): reduce air pollution.
- Objective 2 (for the environment): reduce soil pollution.
- Objective 3 (for the environment): reduce water pollution.
- Objective 4 (for the environment): sustainable natural resource management.
- Objective 1 (for the society): more job opportunities.
- Objective 2 (for the society): improve the quality of life and the health of workers.
- Objective 3(for the society): raise public awareness of environmental protection.
- Objective 1 (for the economy): contribute to a green and sustainable economic transformation.
- Objective 2 (for the economy): scale of investment capital.
- Objective 3 (for the economy): effective investment.
- Objective 4 (for the economy): payback period.
- Objective 5 (for the economy): infrastructure development.

(c) Indicators of the national/local context:

- PT1: have been able to be implemented in the country/locality.
- PT2: consistent with national/local development policies.
- PT3: policies and support mechanisms.

(d) Measurement, reporting, and verification (MRV) indicators:

- MRV1: A system of GHG emission monitoring and sustainable development can be established.
- MRV2: Basic calculations can be developed to quantify GHG emissions reduction and sustainable development.
- MRV3: Reports on the progress of GHG emissions reduction and sustainable development.
- MRV4: Ensure the accuracy and quality of the information reported.

Step 3. Development criteria table and evaluation criteria matrix according to the Delphi method.

2. Consultation:

Step 4. Applying the Delphi method at round 1—working with experts to consult and thereby assess the level of agreement with the construction index. The criteria will be ranked from low to high with increasing importance. Table 25.1 shows an example of a matrix that evaluates the importance and appropriateness of criteria.

Step 5. Data analysis at round 1.

After collecting data using the Delphi Method, the Kendal coefficient can be used to assess the suitability of the indicator. The level of consensus is scored according to the thresholds of 0.0–0.1; 0.1–0.3; 0.3–0.5; 0.5–0.7; and 0.7–1.0, which are equivalent to a very weak level of consensus; weak; medium; strong; and very strong. The Kendall index is calculated as follows:

Table 25.1 Matrix of criteria evaluation according to the Delphi method

	Criterial 1	Criterial 2	Criterial 3	Criterial 4	Criterial 5	Criterial 6	…
Expert 1	1	3	4	5	2	6	
Expert 2	3	5	4	1	6	2	
Expert 3	5	4	6	3	1	2	
…							

When there are two values, x and y, whose value correspond to a set of values of the other sign in the form of statistical distribution, the Kendall coefficient can be used to assess the correlation and consensus. Therefore, experts are the independent variable x and the variable y is rated according to increasing importance.

The Kendall coefficient (W) is calculated by using the following formula

$$W = \frac{S^2}{\frac{1}{12}k^2\left(N^3 - N\right)} \tag{25.1}$$

where N is the number of elements y (the index number); k is the number of elements x (the number of experts); and W is in the range.

$$S^2 = \sum_{j=1}^{N}\left(R_j - \overline{R_j}\right)^2 \tag{25.2}$$

$$\overline{R_j} = \frac{\sum_{1}^{N} R_j}{N} \tag{25.3}$$

where R is the sum of each element of y and R_j is the average of the sum of these elements.

Step 6. Applying the Delphi method at round 2.

In case the Kendall coefficient (W) is greater than 0.5 in the first round, the evaluation process ends immediately in step 5. This means that the experts agreed with the proposed index group. However, if the Kendall coefficient (W) is less than 0.5 in the first round, there will be disagreements between experts. The evaluation process will be repeated until the Kendall coefficient (W) is greater than 0.5.

3. After consultation:

After the data are collected in the final round, the result is analyzed. The weighted value of the criteria will be determined based on the results of rankings evaluated by experts. The results are summarized and analyzed, and a summary report with conclusions is then sent to the experts for storing and reference.

After analyzing and consulting domestic experts on 04 groups of the indicators mentioned above, the author will find 21 important indicators of these 04 groups and set up a questionnaire to conduct consultations with 10 experts.

Through round 1 of consultation, the importance level of criteria set by experts from 1–21 is exactly equal to the number of aggregated indicators.

With the data collected from the table above, the Kendall coefficient is calculated as stated in chap. 2 to be 0.57. With a relatively strong coefficient of more than 0.5, the expert group has a high consensus with the proposed set of indicators, so there is no need to conduct round 2.

This result also helps identifying the indicators of the four selected groups that reach the highest consensus, thereby reducing the set of indicators and increasing the accuracy in prioritizing climate change mitigation solutions in the energy sector related to Vietnam's NDC in the next step of applying the MCDA method. The expert group unanimously agreed on the MRV index group, and all the indicators in this group achieved the highest score compared to the indicators in the remaining groups, so all the indicators in this group were selected as the indicators. And, a mitigation solution is considered feasible and has the opportunity to implement successfully and effectively when it can measure, report, and verify.

2.3.2 Applying Criteria to Assess and Select Priority Mitigation Solutions in Ho Chi Minh City

Based on the built-in index framework, the calculation of the GHG emission reduction indicator is applied based on the experience of calculating the technological assessment of climate change according to instructions of the IPCC and the combined results of the component indicators, and in each component, the index is a set of indicators.

$$MCDAI = a.\mathrm{GPT} + b.KT + c.XH + d.MT + e.PT + g.MRV \qquad (25.4)$$

where MCDAI is an index to select and prioritize activities to reduce greenhouse gas emissions; GPT is the component index—reduction of greenhouse gas emissions; KT is the component index—economic benefits; social is the component index—social benefits; MT is a component index—environmental benefits; PT is the component index—development benefits; MRV is the component index—Possibility of MRV; a, b, c, d, e, and g are the weights of component indexes and indicators in each component index: a + b + c + d + e + g = 1.

The value of indexes is then determined by experts. Accordingly, experts in the field of GHG emission reduction in the energy sector will score component indicators (GPT, XH, KT, MT, PT, and MRV) with a scale from 0 to 5 with increasing relevance and priority. The average score of the component indicators will be used to evaluate the priority.

3 Results of Applying the Criteria to the Evaluation and Priority over Mitigation Options in Ho chi Minh City

The weights of the component indicators were determined during the implementation of the Delphi method. The results of the expert evaluation of the importance and priority of each component indicator will be used to calculate the weight presented in Tables 25.2 and 25.3.

Thus, according to the results of experts in the transport sector, there are 05 solutions with the highest scores, namely T1—petrol used. Biology in transportation (3.72), T9—LED lights for urban roads (3.38), T12—electric/hybrid vehicles in road transport (3.30), T11—vehicles, convenient use of natural compressed air in road traffic (3.12), and T2—conversion of mode of passenger transport from individuals to urban railways (3.11); in industry, there are 08 solutions that are most appreciated, namely I22—recovering biogas for boiler in the beverage industry (2.89), I14—waste heat recovery (WHR) from the drying process of paper in pulp and paper production (2.52), I7—use of vertical continuous kilns (VSBK) in brick production (2.38), I1—waste heat recovery (WHR) from cement production (2.37), I20—using floor-based cooling systems in the beverage industry (2.37), I5—installation of inverter motors (VFD) (2.36), I4—reducing furnace heat loss (2.25), and I10—low pressure central cleaning system for pumping plants in pulp and paper production (2.07); in the field of civil buildings, there are 03 solutions assessed on 03 points, namely R4—using solar water heaters (3.43), R3—using high-performance civil lighting (lights) light-emitting diodes (LEDs) and compact fluorescent lamps (3.31), and R1—using inverter air-conditioners (IAC) (3.05); in the field of commercial buildings, there are 02 solutions and both are evaluated at a fairly good level with C2—applying green building solutions (multi-air-conditioner system, LED lights, 2-layer glass, high-efficiency heat insulation kit) (3.41), and C1—use of multi-air-conditioning systems in commercial buildings (2.93); the aviation sector has 01 solution and is rated at an average level, which is A1—switching from the use of auxiliary engines to the airport ground source (1.95); waste sector has 05 solutions that are highly rated at more than 3 points, namely W4—anaerobic treatment of organic waste, recovering methane for heat and electricity production (4.04), W1—organic fertilizer production (4.02), W2—landfill gas collection for electricity and heat generation (3.59), W6—electricity production from garbage (3.38), and W3—solid waste recycling (3.21). These are also 24 priority solutions and recommendations for implementation in the period of 2020–2030 for Ho Chi Minh City.

Table 25.2 Assessing the importance of the evaluation indicators to prioritize climate change mitigation solutions for Ho Chi Minh City

Criterial / Expert	MT1	MT2	MT3	MT4	XH1	XH2	XH3	KT1	KT2	KT3	KT4	KT5	GPT1	GPT2	PT1	PT2	PT3	MRV1	MRV2	MRV3	MRV4
1	14	4	5	1	15	6	7	16	17	8	9	10	18	11	19	12	20	21	13	2	3
2	7	1	3	2	8	4	9	19	20	10	11	12	21	13	14	5	6	15	16	17	18
3	6	1	2	7	8	4	14	15	9	5	10	3	16	11	17	18	19	20	12	21	13
4	1	2	3	6	4	7	10	13	14	8	15	5	16	17	9	11	12	18	19	20	21
5	5	1	2	6	7	3	8	9	19	4	11	12	13	14	15	10	16	20	17	21	18
6	8	2	3	9	1	4	5	10	11	6	12	7	16	17	18	13	14	19	20	21	15
7	3	1	4	2	8	5	6	14	15	7	9	10	16	17	11	12	13	18	19	20	21
8	5	6	15	1	7	16	8	9	17	2	18	10	19	3	11	4	20	21	12	13	14
9	13	3	4	9	6	7	5	14	10	1	15	11	16	2	12	8	17	18	19	20	21
10	18	12	13	1	2	5	14	6	19	3	7	8	4	9	15	20	16	21	17	10	11
Total score	80	33	54	44	66	61	86	125	151	54	117	88	155	114	141	113	153	191	164	165	155
Variance	900	5929	3136	4356	1936	2401	576	225	1681	3136	49	484	2025	16	961	9	1849	6561	2916	3025	2025
Weight	0.035	0.014	0.023	0.019	0.029	0.026	0.037	0.054	0.065	0.023	0.051	0.038	0.067	0.049	0.061	0.049	0.066	0.083	0.071	0.071	0.067

Table 25.3 Results of evaluation of priority options for potential mitigation measures for Ho Chi Minh City

Criterial Solution	MT1	MT2	MT3	MT4	XH1	XH2	XH3	KT1	KT2	KT3	KT4	KT5	GPT1	GPT2	PT1	PT2	PT3	MRV1	MRV2	MRV3	MRV4	Total score
T1	4	0	0	3	4	3	4	4	1	2	2	0	4	3	5	5	5	5	5	5	5	3.72
T2	4	0	0	0	1	4	4	4	1	2	1	4	4	4	5	5	4	3	3	3	3	3.11
T3	2	0	0	0	1	2	2	3	1	2	2	4	4	4	0	4	2	3	3	3	3	2.42
T4	3	0	0	0	1	2	2	3	3	2	2	2	2	2	5	4	3	3	3	3	3	2.65
T5	3	0	0	0	0	0	0	2	4	3	3	0	2	2	5	4	3	3	3	3	3	2.50
T6	3	0	0	0	0	0	0	2	4	3	3	0	2	2	5	4	3	3	3	3	3	2.50
T7	3	0	0	0	0	3	3	3	5	3	2	0	2	2	5	5	3	2	2	2	2	2.52
T8	2	0	0	0	0	0	1	3	2	1	1	3	4	1	0	4	3	1	1	1	1	1.59
T9	1	0	0	0	0	0	2	3	4	5	5	4	4	4	5	4	3	4	4	4	4	3.38
T10	5	0	0	0	0	4	5	4	4	0	0	4	5	5	5	4	4	1	1	1	1	2.73
T11	3	0	0	0	1	2	2	3	3	3	3	0	3	3	5	4	4	4	4	4	4	3.12
T12	4	0	0	0	1	2	3	4	3	3	3	0	4	4	4	4	4	4	4	4	4	3.30
I1	1	0	0	0	0	1	0	2	3	2	2	0	3	3	5	4	2	3	3	3	3	2.37
I2	0	0	0	0	0	0	0	2	2	2	2	0	3	2	2	3	2	2	2	2	2	1.67
I3	0	0	0	0	0	0	0	2	3	2	2	0	4	2	5	3	2	2	2	2	2	1.99
I4	0	0	0	0	0	0	0	3	3	3	3	0	5	2	5	3	3	2	2	2	2	2.25
I5	0	0	0	0	0	0	0	2	3	4	4	0	5	3	5	4	3	2	2	2	2	2.36
I6	2	0	0	0	0	0	0	2	2	4	3	0	5	3	3	4	2	1	1	1	1	1.84

4 Conclusions and Discussion

The results of the research show that the application of the Delphi and MCDA methods will provide the evaluation process of prioritizing options to reduce greenhouse gas emissions with a more scientific basis. The research involved using the Delphi method to develop a set of priority evaluation indicators; the process of implementing the Delphi method is described in 6 steps with the aim of increasing the accuracy in evaluating solutions and shortening the set of indicators. There are 21 important indicators in 6 groups: (1) reducing emissions; (2) economy; (3) society; (4) environment; (5) development; and (6) measurement, reporting, and verification (MRV).

The application of the MCDA method in the next step helps in prioritizing the selection of 5 solutions from the evaluation results to reduce GHG emissions for the transport sector; 8 solutions in the field of industry; 3 solutions in the field of civil buildings; 2 solutions in the field of commercial buildings; 1 solution in the field of aviation; and 5 solutions in the field of solid waste treatment.

Comparing the research results with the results of Vietnam's National Notice 2, it is shown that the technological priority is different at each time. Based on the practical situation and experience, technologies will be prioritized for development in accordance with the conditions, potential, and national orientation.

From the results of the paper, a number of recommendations are made for further improvement in the future: (1) Firstly, in terms of approach, the set of indexes built in the framework of the article should be further studied and updated in the coming time to suit the practical situation of Vietnam; (2) Secondly, in terms of research methodology: Due to the limited time and conditions, the process of implementing the Delphi method has not been consulted with more experts in different fields related to climate change and energy sustainability; the more opinions it gathered, the more accurate the results of the consultations were confirmed; as well as in the development of a set of priority indicators for GHG emission reduction solutions for different sectors, if the first round consultation is conducted with open questions, the results may be more accurate instead of using results of previous international studies.

Acknowledgments This study is supported by the funding of the program of science and technology in response to climate change, natural resources, and environment management for the period 2016–2020, grant code: BDKH/16-20 with the title "The study and establishment of Monitoring-Reporting-Verification (MRV) system for climate change response activities in Vietnam," grant number: BDKH.32/16-20. The authors would like to appreciate experts from Institute of Energy (MOIT), Industry Safety and Environment Agency (MOIT), Department of Climate Change (MONRE), Institute of Meteorology, Hydrology and Climate Change (MONRE), Institute of Environment and Ecology, Viet Nam Energy and Environment Company, GIZ Viet Nam who contributed to the interview of this research. The authors also confirm that each named authors have substantially contributed in conducting the underlying research and original this manuscript, and to the best of our knowledge, the named authors have no conflict of interest regarding the publication of this manuscript.

Name	Title	Working agency	Expertise
Nguyen minh Bao	Master	Institute of Energy (MOIT)	Energy, energy modeling
Nguyen Quoc Khanh	Dr	MOIT	Energy, energy modeling
Truong Viet Truong	Master	Industry safety and environment agency (MOIT)	Energy
Nguyen Phuong Nam	Dr	Department of Climate Change (MONRE)	Climate change, GHG inventory
Tran Ha Ninh	Master	Department of Climate Change (MONRE)	Climate change
Vuong Xuan Hoa	Master	Institute of Environment and Ecology	Climate change, GHG inventory, MRV, sustainable development
Hoang Tung	Bachelor	Institute of Meteorology, hydrology and climate change (MONRE)	Climate change, GHG inventory
Hoang Thanh Ha	Master		GHG inventory, MRV
Dang Hong Hanh	Dr	Viet Nam energy and environment company (VNEEC)	Climate change, energy, environment
Le hi Hieu	Master	GIZ Viet Nam	Energy

References

Asahi Glass Co. Ltd. (2013). *Energy saving glass windows for buildings (report of JCM feasibility study)*. Airport Ground Power Corporation.

Bunting, S. W. (2008). Horizontally integrated aquaculture development: Exploring consensus on constraints and opportunities with a stakeholder Delphi. *Aquaculture International, 16*(2), 153–169.

Chan, P., & Lee, M. H. (2019). Developing Sustainable City indicators for Cambodia through Delphi processes of panel surveys. *Sustainability, 11*, 3166. https://doi.org/10.3390/su11113166

Delbari, S. A., Siew, I. N., Aziz, Y. A., & Ho, J. A. (2016). An investigation of key competitiveness indicators and drivers of full-service airlines using Delphi and AHP techniques. *Journal of Air Transport Management, 52*, 23–34.

Dodgson, J. S., Spackman, M., Pearman, A., & Phillips, L. D. (2009). *Multi-criteria analysis: A manual*. Department for Communities and Local Government. Retrieved from http://eprints.lse.ac.uk/12761/1/Multi-criteria_Analysis.pdf

Elliott, M., Cutts, N. D., & Trono, A. (2014). A typology of marine and estuarine hazards and risks as vectors of change: A review for vulnerable coasts and their management. *Ocean & Coastal Management, 93*, 88–99. https://doi.org/10.1016/j.ocecoaman.2014.03.014

EU. (1999). *Towards environmental pressure indicators for the EU. First edition 1999* (p. 181). European Union.

EX Research Institute Ltd. (2012). Introduction of mechanical biological treatment (MBT) of municipal solid waste and landfill gas capture, flaring and utilization (Lao PDR), (reports of JCM/BOCM FS).

Fefer, J. P., Stone, S. D. U., Daigle, J., & Silka, L. (2016). Using the Delphi technique to identify key elements for effective and sustainable visitor use planning frameworks. *SAGE Open, 6*(2), 215824401664314. https://doi.org/10.1177/2158244016643141

Gari, S. R., Newton, A., & Icely, J. D. (2015). A review of the application and evolution of the DPSIR framework with an emphasis on coastal social-ecological systems. *Ocean & Coastal Management, 103*, 63–77. https://doi.org/10.1016/j.ocecoaman.2014.11.013

Hsueh, S. L., Sun, Y., & Yan, M. R. (2019). Conceptualization and development of a Dfuzzy model for low-carbon Ecocities. *Sustainability, 11*, 5833. https://doi.org/10.3390/su11205833

Jiang, R., Kleer, R., & Piller, F. T. (2017). Predicting the future of additive manufacturing: A Delphi study on economic and societal implications of 3D printing for 2030. *Technological Forecasting and Social Change, 117*, 84–97.

Le, T. H., Nguyen, A. T., Tran, A. T., Dao, D. C., Luu, T. A., Hoang, L. T. T., Nguyen, M. H., Tran, Q. B., Le, V. H., Uong, D. K., Bui, T. M., Tong, P. T., Hoang, H., & Truong, Q. H. (2015). Impacts of climate change on agro-ecological landscapes in the coastal area of the Thai Binh province (Vietnam) using the Delphi technique. *Climate Change Strategies and Management, 7*(2), 222–239.

Linstone, H. A., & Turoff, M. (2002). The Delphi Method: Techniques and Applications. Addison-Wesley, Mass, pp. 618.

Malaysia Gas Association (MGA). (2014). *Natural Gas Industry Annual Review.* 2014 Edition.

Martins, J. H., Camanho, A. S., & Gaspar, M. B. (2012). A review of the application of driving forces-pressure-state-impact-response framework to fisheries management. *Ocean & Coastal Management, 69*, 273–281.

Massa, G. D., Kim, H. H., Wheeler, R. M., & Mitchell, C. A. (2008). Plant productivity in response to LED lighting, HORTSCIENCE 43(7), 1951–1956. https://doi.org/10.21273/HORTSCI.43.7.1951.

Mbakwe, A. C., Saka, A. A., Choi, K., & Lee, Y. J. (2016). Alternative method of highway traffic safety analysis for developing countries using Delphi technique and Bayesian network. *Accident Analysis and Prevention, 93*, 135–146. https://doi.org/10.1016/j.aap.2016.04.020

Ministry of Natural Resources and Environment (MoNRE) (2011). *National Strategy to respond to climate change of Vietnam.*

Musa, H. D., Yacob, M. R., Abdullah, A. M., & Ishak, M. Y. (2015). Delphi method of developing environmental Well-being indicators for the evaluation of urban sustainability in Malaysia. *Procedia Environmental Sciences, 30*, 244–249.

Nguyen, A. T., Vu, A. D., Dang, G. T. H., Hoang, A. H., & Hens, L. (2018). How do local community adapt to climate change along heavily damage coasts? A stake holder Delphi study in Ky Anh central (Vietnam). *Environment, Development and Sustainability, 20*(2), 749–767. https://doi.org/10.1007/s10668-017-9908-x

Pham, T. L., Doan, Q. T., Vuong, X. H., & Ngo, T. T. (2019). Research and propose a legal framework for stakeholders in the national MRV system in Vietnam. Proceeding of 10th international conference on socio-economic and environmental issues in development, 2019 at National Economics University, Hanoi, Vietnam, 1098–1111.

Posada, F. (2009). CNG Bus Emissions Roadmap: from Euro III to Euro VI. Retrieved from http://www.theicct.org/sites/default/files/publications/CNGbuses_dec09.pdf

Smith, C. J., Papadopoulou, K. N., Barnard, S., Mazik, K., Elliott, M., Patrício, J., Solaun, O., Little, S., Bhatia, N., & Borja, A. (2016). Managing the marine environment, conceptual models and assessment considerations for the European marine strategy framework directive. *Frontiers in Marine Science, 3*, 144. https://doi.org/10.3389/fmars.2016.00144

URC. (2011). *Technologies for Climate Change Mitigation - transport sector.* TNA Guidebook Series.

URC. (2012a). *Technologies for Climate Change Mitigation - agriculture sector.* TNA Guidebook Series.

URC. (2012b). *Technologies for Climate Change Mitigation - buildings sector.* TNA Guidebook Series.

Wickramasinghe, W. R. M. S. (2011). *Technology needs assessment and technology action plans for climate change mitigation.* TNA Report.

Yedla, S., & Shrestha, R. M. (2003). Multi-criteria approach for the selection of alternative options for environmentally sustainable transport system in Delhi. *Transportation Research Part A: Policy and Practice, 37*(8), 717–729. https://doi.org/10.1016/S0965-8564(03)00027-2

Chapter 26
Developing Agricultural Production of Ethnic Minority Households in the Context of Climate Change (Lak District, Dak Lak Province, Central Highland of Vietnam)

Thi Ai Nhi Duong and Van Thao Truong

Abstract Climate change has made agricultural production of ethnic minority households in Lak District, Dak Lak province facing many difficulties and challenges. Qualitative and quantitative research methods were used for analysis. The results of the research showed that developing agricultural production of ethnic minority households is affected by natural conditions, infrastructure economy, socioculture, and customs and government policies, as well as internal resource factors. The article also proposed some solutions for developing agricultural production of ethnic minority households in the study site focusing on strengthening linkages in production and consumption, and changing crop structure in accordance with the climate change context on the basis of promoting local availability.

Keywords Agricultural production · Climate change · Ethnic minority · Lak District · Dak Lak province

1 Introduction

Lak District is located in the Southeast of Dak Lak province, with a natural area of 125,604 ha, and a population of 64,644 people (General Statistical Office of Lak District [GSOL], 2018), with 16 ethnic groups, in which ethnic minorities account for 65%, and there are ethnic groups such as M'Nong, Tay, Thai, Ede, Muong, and Nung. For ethnic minorities, agricultural production is mainly self-sufficient,

T. A. N. Duong (✉) · V. T. Truong
Tay Nguyen University, Dak Lak, Vietnam
e-mail: ainhi@ttn.edu.vn

A. T. Nguyen, L. Hens (eds.), *Global Changes and Sustainable Development in Asian Emerging Market Economies Vol. 2*, https://doi.org/10.1007/978-3-030-81443-4_26

fragmented, small, and attached to mountains and forests based on exploiting elements of natural potential. The agricultural production of this group has difficulties due to the weak economic potential, low education level, and low self-esteem. Developing agricultural production of ethnic minorities contributes to making a part of ethnic minority household richer, creating jobs and increasing household incomes, and contributing to poverty reduction. However, due to major obstacles both in awareness and in customs, their residential areas are often in mountainous, highland, and remote areas, limited in terms of investment resources, market access of inputs and outputs, and supportive policies, which have made the life of ethnic minority households difficult. In addition, climate change in recent years has led to more frequent droughts, natural disasters, and epidemics, which has made the development of agricultural production inefficient and unsustainable. Therefore, based on the analysis of the current situation, the factors affecting the agricultural production development of ethnic minority households are identified; from that, some solutions are proposed to develop agricultural production of ethnic minority households in Lak District in the context of climate change.

Agricultural development, especially in developing countries in the context of climate change, has been a major topic and has received much attention from international scholars in recent decades. For most countries, especially developing countries with majority of rural population, it is important to develop agriculture so that poverty alleviation and ensuring sustainable livelihoods are achieved. Only on this basis, we can overcome self-sufficiency, promote the exchange of goods and social division of labor, form, expand, and perfect various types of markets, and improve living standards and quality of life of the population. However, choosing and prioritizing the type of development is a matter of much wider controversy due to its dependence on the natural and socioeconomic characteristics of the country or the region as well as from time to time.

1. Overview of research on status and factors affecting agricultural economic development in the context of the climate change

   Agricultural development in developing countries faces a lot of constraints, and this is part of the cause of poverty if not overcome.

   First of all, most agricultural households lack resources, especially the lack of land, and the capital needed to expand production, change production structures, and change jobs (World Bank, 2007). In addition, in rural areas, the quality of labor is also poor due to the lack of health and education (Hall & Harry, 2006; World Bank, 2003a). Demographic conditions, such as high population density and fertility rates, in many cases exacerbate these difficulties (World Bank, 2003). These have resulted in low labor productivity and low income (World Bank, 2007). In addition, agricultural production activities also face many risks in terms of natural disasters due to the climate change and markets. It takes a long time for businesses to recover from shocks (World Bank, 2003).

   Market deficiencies in developing countries, especially in rural areas, are also often severe. This results in high transaction costs, asymmetric information, imperfect competition, and many other negative externalities. However, overcoming these defects—the role of the state—is often not well implemented. This can be considered as a failure of the state in regulating markets and providing public services (World Bank, 2007).

Inequality in resource ownership of farm households, especially ethnic minorities in developing countries, is common. They often do not have access to land or the amount of land that is too small to cultivate (Vollrath, 2007; Zezza et al., 2007). Inequalities in access to public services such as health care and training (Jayne, Villarreal, Prabhu, & Guenter, 2006) or insurance and credit services are also important factors that cause poverty and backwardness. In addition, local social institutions such as norms of behavior, beliefs, gender equality, social caste, or the lack of social networks often curb household production (Fafchamps & Bart, 2002; World Bank, 2007).

2 Overview of the situation of researching an agricultural development solution in the context of the climate change

A group of solutions on agricultural development policies in the context of the climate change are discussed here. There are many effective policies and systems that can be implemented; here, we summarize a few key groups:

- The first group of policies can be mentioned as subsidies, subsidies, by product, by household group, or by developments of natural conditions in the context of the climate change and markets (Brooks, 2010; Jones & Andrzej, 2010). Tax and fee reductions also apply to similar conditions (OECD, 2009; World Bank, 2007). In addition, for many types of agricultural products, it is possible to control prices, and apply output floor prices or ceiling prices for input services to ensure the interests of producers. These measures in developing countries today, for many reasons, are generally applied at a low level (OECD, 2009).
- Providing credit with various forms and interest rates for agricultural development especially in the climate change context is also a remarkable measure. Credit provision that is properly applied in parallel with the development of the capital market can create many positive effects.
- Investing in human capital through education, training, and technical assistance to improve productivity in the climate change context and facilitate non-agricultural employment is a long-term policy group (Haggblade, Peter, & Thomas, 2010; Sherraden, Trina, Amanda, & Fred, 2004; World Bank, 2007).

2 Methodology

The communes selected as the research sites are the communes with the highest number of ethnic minority households engaged in agricultural production, including Yang Tao, Dak Lieng, and Krong No Communes.

Both the secondary and primary data are collected for solving the research purposes. Primary data were collected through interviews with 99 ethnic minority households (60 local ethnic minority households and 39 other ethnic minority households) participating in agricultural production in 3 chosen communes by the random stratification method.

The study uses descriptive and comparative statistical methods to analyze the status of the agricultural production development of ethnic minority groups including indigenous ethnic minorities and other ethnic minorities.

3 Results

3.1 The Situation of Agricultural Production of Ethnic Minority Households

3.1.1 Area of some Major Crops of Ethnic Minority Households

The area of annual crops is larger than that of perennial trees, and the average area of annual and perennial crops of the two groups of households is 0.86 ha per household and 0.15 ha per household, respectively (Table 26.1). Due to the climate, weather, and soil conditions suitable for annual crops, annual agricultural production is the strength of the district in general and of ethnic minorities in particular. The main production area is wet rice with the two average groups of households being 0.7 ha per household, and due to the narrowing of the land area, the ethnic minority people have the custom of dividing the land to their descendants after making it. As a result, the family production area that each household has got is getting smaller, and other ethnic minority people are immigrants, so they have to rent and buy land for production. Therefore, the area of agricultural land is zero. In addition, due to climate change, natural disasters, floods, and droughts, a part of agricultural land has been lost and poor-quality land cannot be used for agricultural production.

Table 26.1 Area of some major crops of ethnic minority groups in Lak District

Criteria	Average (Ha/Household)	The indigenous ethnic minority households		The rest ethnic minority households	
		Area (ha/ household)	%	Area (ha/ household)	%
1. Annual crops	**0.86**	0.867	**100**	**0.857**	**100**
– **Rice**	**0.70**	0.700	80.74	0.700	81.68
– **Maize**	**0.08**	0.070	8.07	0.100	11.67
– **Sweet potato**	**0.02**	0.017	**1.96**	**0.017**	**1.98**
– **Cassava**	**0.07**	0.080	9.23	0.040	4.67
2. Perennial plants	**0.15**	0.142	**100**	**0.174**	**100**
– **The coffee**	**0.13**	0.11	**77.45**	**0.16**	**91.95**
– **Cashew**	**0.03**	0.032	**22.54**	**0.014**	**8.05**

Source: Calculated from household survey, 2018

Table 26.2 Productivity and output of some main crops of ethnic minority groups in Lak District

Type of tree	Criteria	The indigenous ethnic minority households	The rest ethnic minority households	Average
Paddy	Productivity (ton/ha)	3.45	3.5	3.5
	Output (ton/household)	2.15	2.29	2.19
Maize	Productivity (ton/ha)	3.05	3.3	3.13
	Output (ton/household)	0.21	0.34	0.25
Sweet potato	Productivity (ton/ha)	11.75	11.2	11.58
	Output (ton/household)	0.03	0.03	0.03
Cassava	Productivity (ton/ha)	19.65	17.9	19.12
	Output (ton/household)	0.66	0.17	0.51
Coffee	Productivity (ton/ha)	2.8	2.9	2.8
	Output (ton/household)	0.165	0.352	0.22
Cashew	Productivity (ton/ha)	1.2	1.2	1.2
	Output (ton/household)	0.04	0.02	0.03

Source: Calculated from household survey, 2018

3.1.2 Productivity and Output of some Main Crops of Ethnic Minority Households

There is a difference in productivity and yield of some main crops of ethnic minority in Lak District. However, the difference between the two groups of households is not large. Other ethnic minority households have greater productivity and output than the ethnic minority. The yield and yield of rice of other ethnic minority households and other ethnic minorities are, respectively, 3.45 ton/ha, 2.15 ton/household, and 3.5 ton/ha, 2.29 ton/household. The cause of this discrepancy is due to the tradition that ethnic minority households produce using improper fertilizer and pesticide content, poor soil, and lack of knowledge and capital. The production process is not focused, so the productivity and output are quite low (Table 26.2).

3.1.3 Quantity and Productivity of some Main Animals of Ethnic Minority Households

Poultry is the livestock with the largest number of animals raised in each household, 22 heads/household and 33 heads/household for ethnic minority and other ethnic minorities, respectively. For the ethnic minority, since they are mainly very easy to raise, many farmers also raised. The number of livestock at least is the buffalo and also the animal with the heaviest weight with 0.42 quintals/head (Table 26.3).

Table 26.3 Quantity and productivity of some main livestock of ethnic minority households

Type of pets	Criteria	Average	The indigenous ethnic minority households	The rest ethnic minority households
Buffalo	Quantity (head/household)	1	1	–
	Productivity (ton/head)	0.42	0.42	–
Cow	Quantity (head/household)	5	6	3
	Productivity (ton/head)	0.32	0.30	0.35
Pig	Quantity (head/household)	3	3	4
	Productivity (ton/head)	0.05	0.04	0.06
Poultry	Quantity (head/household)	25	22	33
	Productivity (ton/head)	0.0009	0.001	0.0009

Source: Calculated from household survey, 2018

3.1.4 Agricultural Production Results of Ethnic Minority Households

The total value of land use efficiency for the group of households has 21.13 million VND / ha, and for other ethnic minorities is 21.68 million VND/ha. The production efficiency per labor of the indigenous ethnic minority households is 3.90 million VND, which is greater than that of the other ethnic minorities. The average added value gained per unit of labor reached 2.25 million VND/labor, and the group of ethnic minority households reached 2.51 million VND/labor higher than the other ethnic minorities, which is only 1.65 million VND/labor. The average mixed income earned per employee reached 1.92 million VND/labor. The capital use efficiency of indigenous ethnic minority groups is higher than that of the other ethnic minority groups. The increase in the cost index indicates that, when investing an additional cost, the value added of the ethnic minority group is 1.85 times, and that of the other ethnic group is 1.65 times. The mixed income on cost ratio shows that, when investing an additional expense, the mixed income of indigenous ethnic minority households is 1.57 times and that of the other ethnic households is 1.35 times (Table 26.4).

3.1.5 Situation of Linkage in Agricultural Production

Linkage in production is considered an indispensable trend of modern agriculture. With the form of linking production households—production households are the form of organizing production between households. Therefore, due to the lack of orientation in production, farmers often look at each other and imitate the same production, which inadvertently forms a form of horizontal linkage between production households. They often use the situation of the previous crop as a basis for the next crop, if the previous crop is high in prices and high in profits, then the number of households participating in the crop will greatly increase. The direction of production in the district cannot guide them on how to produce.

According to the result of the household survey, the proportion of household groups is not high, and the average rate is 38.39%, of which the ethnic minority group is 44.95% and the other ethnic minority group is 23.33% of the rate associated

Table 26.4 Agricultural production results of the ethnic minority groups in Lak District.

Unit: Million VND

Criteria	The indigenous ethnic minority households	The rest ethnic minority households	Average
1. Efficiency of land use			
GO (million VND/ha)	21.13	21.68	21.3
VA (million VND /ha)	13.61	13.50	13.58
MI (million VND /ha)	11.82	11.03	11.58
2. Efficiency of labor			
GO (million VND /ha)	3.90	2.66	3.52
VA (million VND /ha)	2.51	1.65	2.25
MI (million VND /ha)	2.18	1.35	1.92
3. Efficiency of capital			
GO/IC	2.81	2.65	2.76
VA/IC	1.81	1.65	1.76
MI/IC	1.57	1.35	1.50

Source: Calculated from household survey, 2018

with people in the same commune. For those outside the commune, this rate is lower than the average rate of 8.08%, and all groups of households have links with the same lineage for agricultural production. Through this, it was found that the surveyed households were still not aware of the effectiveness of the association, so the participation rate was not high, and not focused, still in the form of self-production, and independence was mainly passive in the consumption of products. Households link with each other in many stages in which the exchange and use of labor make up the highest proportion, with the ethnic minority group reaching 97.10% and the other ethnic minority group reaching 93.33%. In addition, difficulties in the process of linking between still encounter many problems of capacity, lack of information about partners, prices account for a high proportion in the process of linking.

3.2 *Factors Affecting the Development of the Agricultural Production of Ethnic Minority Households*

3.2.1 Biophysical Conditions

The groups of ethnic minority households in Lak District live in steep slopes and remote areas, which are very far from centers of communes, towns, and districts. Land is fragmented. At the same time, the soil quality is bad due to leaching and other physical and chemical properties.

The climate is characterized by the humid tropical highland climate of the valley, with two distinct seasons each year, the rainy and dry seasons. The average rainfall ranges from 1800 to 1900 mm, which is suitable for planting a variety of agricultural crops. Climate change in recent years caused droughts, water shortages in the dry season and floods, and inundation in the rainy season often taking place in some localities in Lak District, especially the ethnic minorities having a significant impact on the agricultural production results of households.

3.2.2 Resource factors for Agricultural Production

1. Labor.

The average number of employees/household is 5 people, in which the indigenous ethnic minority household groups are with 5 laborers/household and other ethnic minority groups are with 4 labors/household, lower than the average (household interview results, 2018). This indicates that the labor force of the households in the district is quite plentiful. In agricultural production, labor and demographics greatly impact the production and business situation of households. Farmers can take advantage of local labor resources in tending, managing, and harvesting. Therefore, it contributes to reducing production costs and raising incomes.

2. Land

Lak District is a purely agricultural district with advantages of soil and climate for agricultural development but still faces many difficulties in the agricultural production process. Because of the lack of the capital as well as fragmented and small land, production costs are still high.

Agricultural production not only meets difficulties in the size of production land, but also the quality of land is increasingly reduced, leading to inefficient agricultural products and low productivity. The main reason is that farmers are increasingly abusing the use of chemical fertilizers and pesticides in the production process. Therefore, soil quality is increasingly degraded and it affects crop yields.

3. Financial Capital

The district has introduced many policies on capital combining banks to support people to borrow money in the production process with low interest rates. But, since the people's intellectual level is not high, the policy access is still not high and not widely available. Therefore, people still produce with traditional means of production, the varieties are not effective, and the productivity is not high.

Capital is one of the important factors to help maintain and expand agricultural production. Especially in the field of agricultural production, a large amount of investment is required. Farmers use their capital to invest in seed sources, expand and take care of plants during growth and development. In addition to their own capital, farmers can mobilize additional capital from other sources such as banks, mass organizations, and associations.

Demand for loans of households for production is relatively high in 2018, with an average of 82.83% of households borrowing with an average loan amount of nearly 11.84 million VND / household. While the amount of loans in ethnic minority households is VND 10.85 million / household, the other group of ethnic minority households is VND 14.11 million/household higher than that of ethnic minority households (interview results from households survey, 2018).

In general, most of the loan sources are assessed to have relatively quick and simple loan procedures. However, the amount of loans is limited, making households not really brave to invest in production. Capital needs for production activities vary depending on the time and type of production and business households, and the loan term of banks is different. Therefore, the rate of payment from time to time also varies.

4. Technical, scientific and technological factors

Agricultural production is inseparable from scientific and technical advances because it has created high-quality and good-quality animal crops. Currently, science and technology are getting more and more attention. Implementing S&T tasks with the topic "Building a model of raising the Central Highlands hybrid sows with wild boars and F1 hybrid breeding techniques in Lak District, Dak Lak province" by the Department of Agriculture and Rural Development presides over the implementation, and the implementation period is from December 2017 to November 2018. Currently, 02 models have 05 Central Highlands sow pigs that have reproduced 30 crosses. In general, the model of raising the Central Highlands hybrid sows with wild boars and the F1 hybrid breeding technique in the district has initially achieved certain results, and the F1 hybrid is adaptable to the environment and growth and development.

5. Equipment and means of production

The equipment and facilities in the households are not adequate, and the level of equipment is still low. The percentage of households equipped with facilities for agricultural production is not much different between the two groups of households. As for other equipment, the percentage of households with special equipment is lower than that of other ethnic minorities, because other ethnic minority groups come from other places, so they are not familiar with or have many relatives, so all means must be purchased by themselves, while the indigenous ethnic minority households have links with people in their lineages, and they have a tradition of solidarity and mutual assistance, so they can borrow from each other for agricultural production. From this, it can be seen that other ethnic minority groups will have higher production costs than ethnic minority households, leading to reduced economic efficiency.

3.2.3 Market of Inputs and Outputs

In the agricultural production process, market price factors play an important role in affecting people's income and life. On the same unit of high price, output will increase the income of the people, but in the case of low market prices, except for those families with capital capabilities, agricultural products are retained. When prices are high, they will consume, and families who are in difficulties at some time need capital for investment and life, they have to accept consumption even at low prices. In addition, the characteristics of the agricultural products are quite bulky and difficult to preserve, so waiting for a reasonable price to consume is quite difficult for people. Therefore, people are often passive in the fluctuations of market prices, which greatly affect the income and life of agricultural production people. In addition, agricultural production in the district has small-scale household economy, lack of linkage, and is limited in productivity, quality, and product competitiveness; the number of establishments purchasing agricultural products is still limited and small in scale so they do not stand out to provide sufficient input materials for people; and the number of outside facilities supplying input materials is still low, not diverse, and rich in products. Therefore, it has not fully met the needs of the people here. Therefore, for farmers, in production, they need to explore the market through mass media, via the Internet. They need to practice the habit of analyzing, judging, and forecasting the market of agricultural products in order to make a rational production plan according to market demands; at the same time, it is necessary to link production in cooperatives in order to proactively coordinate production and actively look for markets for product consumption.

3.2.4 Rural Infrastructure

Rural infrastructure also has an impact on the development of agricultural production. In ethnic minority people areas, rural transport systems, irrigation systems, electricity systems, markets, trading centers, and information systems have not been completed and are not capable to meet the demand, due to this, the development of commodity production and agricultural production of ethnic minority people have not really been promoted. The weakness of rural infrastructure hampers the development of agricultural production as supply of inputs is limited and product consumption becomes more difficult.

Currently, in Lak District, there are 39 irrigation works including 7 pumping stations, 18 spillways, and 15 reservoirs that can serve irrigation for agricultural production. However, at present, many works have been damaged, degraded, and unable to water according to the designed capacity. Particularly for the reservoir system with 15 works, 5 projects are Khe Mon Reservoir (Buon Triet Commune), Nam Kar Lake (Nam Ka Commune), Lieng Krak Lake (Krong No Commune), Hoc Mon Lake (Lien Son Town), and Lien Son Reservoir (Lien Son Town), which has been degraded, damaged, or deposited, and only 60–90% of its capacity is irrigated (People's Committee of Lak district [PCL], 2018).

3.2.5 Public Policy

Policy factors play a very important role in promoting or limiting market economy development in the district. Therefore, the role of the State and localities is very important in timely promulgating of policies and guiding documents for timely guidance of development. In recent years, the State has issued guidelines and policies to encourage the development of agricultural production and agricultural production for ethnic minority households, including policies to encourage agricultural economic development; policies to encourage enterprises to invest in agriculture and rural areas; credit policies for agricultural and rural development; and specific policies to support socioeconomic development in ethnic minority and mountainous areas in the period of 2017–2020.

Lak District has had a number of programs and plans on agricultural production processes such as Plan No. expressed in the agricultural restructuring project toward increasing value added and sustainable development associated with building new rural areas. Thus, these guidelines and policies have created more solid beliefs for production entities in general, ethnic minorities in particular, promoting production in the fields of cultivation, husbandry, and production associated with processing and consuming agricultural products. All things contributed to improve the product quality, develop agriculture sustainably, create jobs for workers, increase incomes, decrease to eliminate hunger, and reduce poverty, step by step renewing the face of the commune in the rural association.

These factors have greatly influenced the development of agricultural production in the district, including the development of agricultural production of ethnic minorities in recent years. The biggest difficulty that constrains the development of agricultural production of this group is the limitation of resource factors including land, capital, capacity of production entities, and labor. Due to a low starting point, and limited resources of the production entity, ability to access policies was limited.

3.3 Some Solutions to Develop the Agricultural Production of Ethnic Minority Households

3.3.1 Promoting the Available Potentials of Natural Conditions

At the same time, there are specific plans to limit the impact of climate change on agricultural production. Continuing to rely on available advantages of natural conditions such as topography and weather to produce agriculture, making the most of the water and soil sources, finding suitable animals and plant crops in the area to diversify the crop, more quality agricultural products can be created to improve the living standards for farmers. There are specific plans to prevent the effects of natural conditions on agricultural production.

3.3.2 Enhancing Resource Factors for Agricultural Production to Limit the Impact of Climate Change

1. Linking expansion of agricultural land area and strengthening measures to improve the quality of land for agricultural production.

- Propagating, mobilizing, and supporting ethnic minority farmers in their villages and in the same lineages to consolidate, exchange plots, convert, transfer, or lease land to accumulate land in accordance with the Land Law, create favorable conditions in production, and increase investment efficiency to develop agricultural production. At the same time, it is easier to deal with the effects of climate change.
- Paying attention to the protection and improvement of land to improve the quality of land, thereby contributing to increasing the productivity of plants and animals.
- Encouraging links between ethnic minority households in the same villages and clans to expand the area of land for agricultural production development.

2. Training and improving the quality of labor.

- Enhancing the training and opening of training courses to improve the intensive level of production in the people, especially the classes in the garden for people to see and practice hands-on to provide full knowledge, planting, and tending skills. Encourage all strata to actively learn to raise awareness to create a foundation for applying scientific and technical advances to the production process.
- There are policies to encourage agricultural officials to work locally to serve production for people.
- Diversify agricultural extension types in training courses, providing useful forms to attract people's participation.

3. Expand rural forms of credit, and increase the amount of loans and the loan term.

Arousing people's capital and attracting investment from outside to build infrastructure to create favorable conditions for agricultural production development can be done. It is necessary to implement preferential policies on investment capital for production households as this is a form of production and business, including

- Based on the planning on the development of agriculture, forestry, and fishery production, there will be a policy to invest in transport infrastructure, electricity, daily-life water, irrigation, information, and processing facilities. Encourage households and individuals to develop agricultural, forestry, and fishery production in the form of production.
- Increase loan size and loan term for production households that need a large capital, a long payback period, especially forestry and for a period of 3 years or more (medium and long-term loans), to have enough time to recover capital and pay principal and interest and continue to invest in reproduction and expanding production.
- Risks in agriculture are quite high, especially droughts, pests, diseases, instability in the market, and price fluctuations that are beyond the control of farmers. In order to help production households limit and overcome risks, soon stabilize production after damage, it is necessary to have policies and solutions on insurance for plants and animals. This is a voluntary means of self-protection that has been popularized in developed countries.

4. Enhancing the applicability of science and technology in agricultural production to reduce the impact of the climate change.

In order for agricultural products to be consumed and highly competitive, there is a need for more active and satisfactory support from the State in high investment in science and technology and effective measures to encourage the maximum mobilization of the participation of all economic sectors, organizations, and scientists in the research, transfer, and application of rural agricultural science and technology.

Today, science and technology have become a direct force to improve efficiency and productivity, science, and technology here are understood both in production and in supply of production materials (livestock breeds, crop) and after harvest. If the producer does not have good plant and animal breeds, the productivity and quality of the product will be poor, and the product will be difficult to consume or even not be consumed. In order for the production and business activities of the production entity to achieve higher economic efficiency, there are a number of solutions to be applied in science and technology:

- Enhancing the import of foreign advanced technologies, especially plant varieties, domestic animals, and machinery and equipment of high technology, suitable to the conditions of each region and each locality.
- Focusing on renovating plant and animal breeds, processing technology, strengthening the care and protection of plants and animals, applying postharvest technology, protection measures and increasing fertility of soil, and protecting water resources.
- Implementing the planning and construction of irrigation works to create water sources for production, and production subjects shall invest by themselves or borrow from credit capital to build water supply systems for production and daily life.
- Planning investment and development of nurseries for industrial plants, forest trees, and seed production establishments or supporting them with conditions for seed production in order to ensure good and high-quality seeds to be supplied to production subject.
- Encouraging production entities to contribute capital to the scientific and technical development support fund to apply it to production and provide technical services to farmers in the region.
- Organizing technical services such as seed services, plant protection services, veterinary services in various forms, contracting for protection, and contracting for service stages.
- Strengthening the system of agriculture and forestry extension on the basis of socialization, and helping farms and farmers improve farming methods and techniques. The agricultural extension system has an active role in disseminating, training, and applying into practice scientific advances such as bringing quality seedlings with high productivity, and training for farmers.
- Planning to change the plant structure conversion is more adaptable to climate change conditions taking place in the study area.

5. Investing in modernizing and strengthening equipment and means of production in service of agricultural production.

Means of production are tools to support agricultural production. However, there are still insufficient facilities for production in households in Lak District, mainly only sprayers and water pumps, poorly constructed barns, and substandard agricultural facilities. As for rice mills, most households rent them. The transportation of products after harvesting is also contracted by households so very few households are equipped with cars, plows, and agricultural machines to transport this, resulting in high production costs. Due to high investment costs and low knowledge of the people, many households are still hesitant to invest in buying or renting machines. This makes losses and postharvest costs still high. In order to equip households with adequate and modern means of production, they need to boldly invest and borrow capital from financial and credit institutions to equip machines for production. In addition, local budgets should only focus on supporting part of the investment in the purchase of machinery and equipment for mechanization development, training support, training, and technical training. Enterprises should be encouraged to research and invest in the development of the mechanical engineering industry to produce agricultural machines with high economic and technical efficiency suitable to the district's socioeconomic situation and soil characteristics. A network of distribution, supply, warranty, and repair facilities in localities should be consolidated and established. Enterprises manufacturing and trading agricultural machines in the district should be encouraged to provide incentives for buyers of agricultural machinery such as deferred payment, maintenance, warranty, repair, replacement, and damaged equipment.

3.3.3 Strengthening the Construction of Rural Infrastructure

- Continue to direct relevant units, People's Committees of communes and towns to conduct monitoring and monitoring of irrigated areas of irrigation works; make plans for natural disaster prevention and control.
- Build and expand the system of canals to ensure irrigation water, and field transport routes.
- Carry out remodeling of degraded works.
- Urge contractors to speed up the construction of irrigation works, ensuring that works are constructed in time; actively store water for production.
- Expand trading markets, rural electricity systems, and information transaction systems to help people have more favorable conditions in agricultural production.

3.3.4 Expanding and Consolidating the System of Supply of Input Materials and Purchasing of Output Agricultural Products

- Establishing purchasing organizations and systems, and forming linkages between production households and companies purchasing enterprises so that after-production products are covered, people do not have to worry much about output for products, people can sell agricultural products to the purchasing company or enterprise or the company will send people to buy at home at market prices or at higher prices than the market from time to time and product quality. Encourage and create a healthy competition between forces engaged in commercial service activities, providing materials and machinery for the production and consumption of products.

- Forming the market forecast organization system. Through extension stations, authorities, mass media, timely, complete, and accurate provision of domestic and foreign markets to farmers in a public and widely manner in order to ensure that farmers have the opportunity to choose the appropriate items and services provided to the market. In addition to making forecasts about the demand for agricultural materials, farmers need to be proactive in updating market information so that the consumption of agricultural products is timely and most effective.
- Through organizations such as the Farmers' Union, the Chamber of Agriculture has formed a service network, providing market information to farmers in a complete, accurate, and timely manner. Avoid the disadvantaged people while consuming agricultural products.

3.3.5 Continuing to Implement policies on Developing Agricultural Production in the Context of Climate Change

- Promoting the propagation and mobilization of economic organizations, professional social organizations, union members, members, and the masses to grasp the party's guidelines and policies and mechanisms, policies of the State, the province, and the district on agricultural and rural development, and consumption of agricultural products. Besides, directing communes to step up the land consolidation and exchange to form a large sample field, which is the basis for developing agricultural production with higher yields and productivity, saving production costs.
- Balancing the budget and having policies to ensure adequate capital for support policies issued from the central to local levels. In other words, adjustments in the Budget Law are needed, or short-term solutions are needed so that adequate resources can be devoted to rural development in general and the agricultural sector in ethnic minority households in particular. In fact, many supporting policies, in recent years, after being formulated and promulgated by the central government, have assigned all implementation responsibilities to the local authorities, but it was not feasible because the district could not allocate resources to perform. This is a fairly common practice for many State policies—when it was first promulgated, it was warmly welcomed by public opinion and localities throughout the country, but after putting it into practice, it does not work.

4 Conclusions

Agricultural production of ethnic minority households in Lak District is mainly short-term crops, in which wet rice is the largest crop with an average area of 0,7 ha/household. Productivity and yield of crops are not high due to many reasons such as lack of resources to invest in production, and the deterioration of soil quality makes crop productivity low. Livestock of great values contributes to the increase in income of each household group. The production value of the ethnic minority group is higher than that of other ethnic minority groups because of the lower expense and higher economic efficiency. Agricultural production development of ethnic minority households is influenced by factors of natural conditions, infrastructure, socio-culture, and customs and policies, and groups of factors, in which the group of policy elements plays the most important role affecting the formation and development of agricultural production. In addition, internal factors are also important, affecting the development of agricultural production both in quantity and in quality. In order to develop agricultural production for ethnic minority people in the area of Dak Lak province in the coming time, some proposed solutions include (i) promoting the available potentials of natural conditions; (ii) strengthening resource factors; (iii) strengthening the construction of rural infrastructure; (iv) expanding and strengthening the system of input supply and purchase of output agricultural products; and (v) continuing to implement policies on agricultural economic development.

References

Brooks, J. R. (2010). *Agricultural policy choices in developing countries* (pp. 132–139). OECD.

Fafchamps, M., & Bart, M. (2002). Returns to social network capital among traders. *Oxford Economic Papers, 54*(2), 173–206.

General Statistical Office of Lak District [GSOL]. (2018). *Year Book of Lak District*. pages 1–8 (in Vietnamese).

Haggblade, S., Peter, H., & Thomas, R. (2010). The rural non-farm economy: Prospects for growth and poverty reduction. *World Development, 38*(10), 1429–1441.

Hall, G., & Harry, P. (Eds.). (2006). *Indigenous peoples, poverty, and human development in Latin America* (pp. 112–120). Palgrave Macmillan. Retrieved from http://www.oecd.org/dataoecd/50/31/46340461.pdf

Jayne, T. S., Villarreal, M., Prabhu, P., & Guenter, H. (2006). HIV/AIDS and the Agricultural Sector in Eastern and Southern Africa: Anticipating the Consequences. In S. Gillespie (Ed.), *AIDS, Poverty, and Hunger: Challenges and Responses* (pp. 203–212). International Food Policy Research Institute (IFPRI).

Jones, D., & Andrzej, K. (2010). Policy responses in emerging economies to international agricultural commodity Price surges. *OECD Food, Agriculture and Fisheries Working Papers., 34*, 23–31.

OECD (2009). Agricultural policies in emerging economies: Monitoring and evaluation (pp. 226-230). Retrieved from http://www.oecd.org/dataoecd/54/30/42347206.pdf

People's Committee of Lak district [PCL]. (2018). Report on socio-economic development situation of Lak district, pages 5–7 (in Vietnamese).

Sherraden, M., Trina W., Amanda M., Fred S. (2004). Overcoming Poverty: Supported Saving as a Household Development Strategy. Working Paper No. 04–13, Center for Social Development, George Warren Brown School of Social Work Washington University, pages 119–128.

Vollrath, D. (2007). Land distribution and international agricultural productivity. *American Journal of Agricultural Economics, 89*(1), 202–216.

World Bank. (2003). Reaching the rural poor - a renew strategy for rural development. *World Bank Publications*, 89–85.

World Bank. (2007). *World development report 2004: Agricultural for development* (pp. 115–121). Oxford University Press.

Zezza, A., Paul, W., Benjamin, D., Gero, C., Katia, C., Esteban, Q., Kostas, S., Panayotis, K., Luca, T., Stefania, D., & Genny, B. (2007). Rural household access to assets and agrarian institutions: A cross country comparison. *Background paper for the WDR, 2008*, 145–158.

Chapter 27
Detecting Flash Flood Susceptible Areas Using a Multicriteria Decision-Making Model: A Case Study of Thai Nguyen Province, Vietnam

Duong Thi Loi, Phạm Anh Tuan, and Nguyen Van Manh

Abstract In this study, Flash Flood Potential Index (FFPI) was integrated with Geographic Information System (GIS) to determine the susceptible areas based on the pre-event characteristics in the study area. Five different physical characteristics that relate to flash flood potentials (slope, land use land cover, soil texture, forest canopy density, and drainage density) were selected to calculate the index maps. Each index was classified from 1 to 10 by assessing their influence levels with the presence of flash flood. As a result, the most susceptible areas were given a value of 10, while the least susceptible areas were assigned a value of 1. These indices were then mapped and integrated into a weighted linear model. The Analytic Hierarchy Process (AHP) was used to determine the weighted correlation among elements based on their importance in this phenomenon. Weighted Flash Flood Potential Index (WFFPI) was generated based on the individual indices from slope index, soil index, land use cover index, forest canopy density index, and drainage density index. The final results described visually the spatial distribution of flash flood potential in Thai Nguyen Province. Accordingly, the susceptible areas with this phenomenon were divided into four levels including very high, high, moderate, and low.

Keywords Flash flood · Flash Flood Potential Index (FFPI) · GIS · Runoff

---

D. T. Loi (✉)
Hanoi National University of Education, Hanoi, Vietnam

P. A. Tuan
Tay Bac University, Son La, Vietnam

N. Van Manh
National Cheng Kung University, Tainan, Taiwan

VAST Institute of Geography, Hanoi, Vietnam

A. T. Nguyen, L. Hens (eds.), *Global Changes and Sustainable Development in Asian Emerging Market Economies Vol. 2*, https://doi.org/10.1007/978-3-030-81443-4_27

1 Introduction

Floods are known as one of the most common types of natural disasters in Vietnam. Like many countries in the world, the number of flood events has increased significantly in the last few decades and has caused many negative effects on the environment and society (Gioti et al., 2013), in which flash flood is the top weather-related killer. Flash floods are defined as rapid-onset hydrologic events of short duration; hence forecasting is difficult. Moreover, in most of the developing countries, the serious shortage of flood warning system limits the ability of flood prevention. The complex variations in climate, land use, and other anthropogenic interventions also lead to the changes in flood risk and make more complicated problems. Along with the development of science and technology, the knowledge about causes of flash flood has developed in recent decades, but most of these researchers concentrate on simulating the process of flash floods based on the recorded data from the meteorological stations. In addition, the need for specific detailed data of flash flood from the stations for such models is a big challenge for many developing countries, where there is a shortage of warning monitoring systems, especially in mountainous areas. Flash floods are known as a natural phenomenon; however, they are directly affected by on-site hydrologic factors such as soil component, slope, and land cover. These factors could promote or inhibit the runoff process, a main cause of the flash flood. Besides, physical factors are often stable and less variable. Therefore, assessing the flash flood potential areas based on the pre-event characteristics is a reliable basis for disaster research and flash flood prediction. It contributes importantly to natural disaster prevention and environmental protection.

One of the previous methods for considering the runoff potential is using the Flash Flood Potential Index (FFPI). This index was made and developed by Greg Smith (2003), and he applied this method at the Colorado River basin, in the National Weather Service (USA). To calculate the FFPI values at Colorado River basin, four input factors (slope, vegetation, soil type, and land use) were selected as parameters and processed with the support of the GIS tool. Some authors modified the original Smith version of the FFPI for application in other study areas, typically, Kruzdlo (2010) and Zogg and Deitsch (2013). These studies improved by considering and changing the weighing input factors in the final equation of FFPI. Obviously, the selection of input factors and weight estimation play a crucial role in this methodology.

According to conventional reasoning, rainfall is often chosen as the closest factor to assess the potential of floods; however, when it is combined with the conditions on the different surfaces, it can cause different effects (positive, negative), which is much more necessary. In addition, a flash flood can even occur with the drought condition and when ground is not saturated. Therefore, in some cases, runoff production processes were selected to study instead of rainfall characteristics. In this paper, the modified-FFPI model was used to assess the risk of a flash flood. The geographic factors at the study area have a serious domination on the timing of runoff, amount of infiltration, and severity of flash flooding. The major factors selected as the index to assess the flash flood potential areas include slope, soil, land use, forest canopy density, and drainage density. Each of these indices contains the value from 1 to 10, corresponding to the probability of flash flood from least to most and then added in a weighted linear model to create the FFPI map. Weights are evaluated based on their influence on the runoff—a direct expression of flash floods. With the support of ArcGIS and ENVI software, digital maps and a satellite image are collected to build the database for studying. Moreover, GIS is a powerful tool for processing spatial data and it also has the ability to combine flexibly with other software in processing the multi-information. This research focuses on identifying the areas with high flash flood potential in the study area, and it can be used as a useful material for helping decision-makers and research about the natural disaster in the local area.

2 Methodology

2.1 Study Area

Thai Nguyen is located in the Northeast region of Vietnam with an area of 3541.5 sq. km (Fig. 27.1). The topography is characterized by a mountainous area running from the north to the south. The study area has a humid tropical monsoon climate with an annual rainfall of more than 2000 mm / year. Thai Nguyen Province is considered as one of the economic centers of the North of Vietnam. However, Thai Nguyen has been affected by flash floods in recent years with serious damages to humans and properties. From 1994 till now, there are about four floods each year in the average and affected areas around 10–40 sq. km (Nguyen and Pham 2009). Thai Nguyen Province becomes one of the most hard-hit with many deaths and reported injuries caused by flash floods in the North of Vietnam.

2.2 Data Collection

In this research, the spatial data such as administrative, hydrology, soil, and land use cover data were collected from different sources to calculate the index maps corresponding with input geographic factors. More specifically, a topographic map and soil map in 2010 provided by the Thai Nguyen Department of Natural Resources and Environment were used to calculate the slope index, soil texture index, and drainage density index. The land use land cover map in 2010 was also provided by the Thai Nguyen Department of Natural Resources and Environment, but the attributes were updated based on fieldwork and compared to Landsat 8 taken in 2016. These satellite data were also used to calculate the Forest Canopy Density. Although the data are derived from different sources and time, this does not affect much the

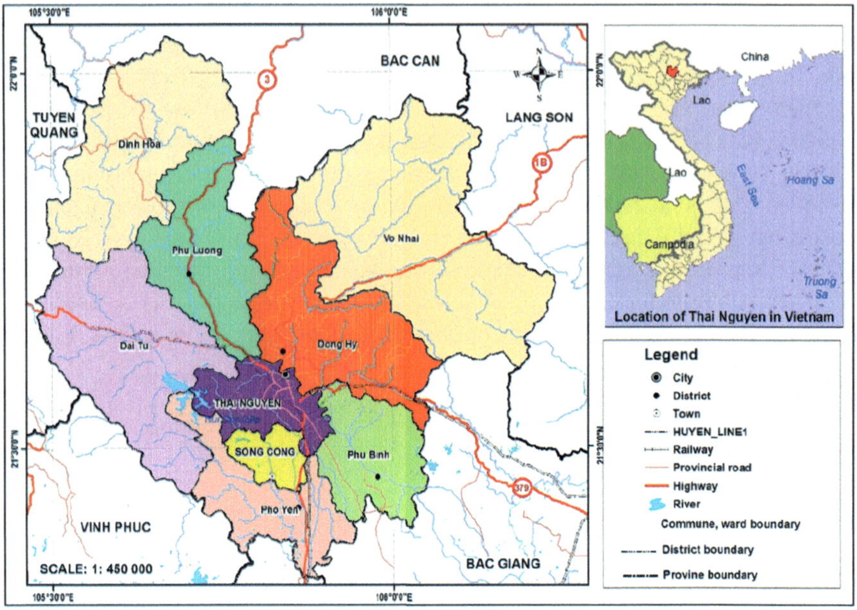

Fig. 27.1 Thai Nguyen administrative map

accuracy of result, because both topography and soil texture are little changed elements, especially topography being a relatively inherent factor. Therefore, there is almost no difference in these factors in a short time. Moreover, land use/land cover is the element that easily changes in a short time under human activities, so these data were updated up to 2016 from the remote sensing data and information from fieldwork to ensure the update of research results.

Five factors were considered to select including slope, soil, land use cover, forest canopy density, and drainage density. Each index map was assigned values from 1 to 10, corresponding to its influence on flash flood risk. The value of 10 corresponds to the highest potential for flash flood, while the value of 1 is the opposite (Zeng et al. 2015). To derive Flash Flood Potential Index maps, all maps were overlaid and weighted based on the AHP (Analytic Hierarchy Process) method. Based on the outcome, the FFPI values were classified and mapped. The final result was divided into four classes of very high, high, moderate, and low potential flash flood. Low potential flash flood areas had the value from 0 to 2.5. Similarly, moderate potential areas, high potential areas, and very high potential areas were given the value from 2.5 to 5, from 5 to 7.5, and more than 7.5, respectively. The methodology is illustrated in the flow chart (Fig. 27.2).

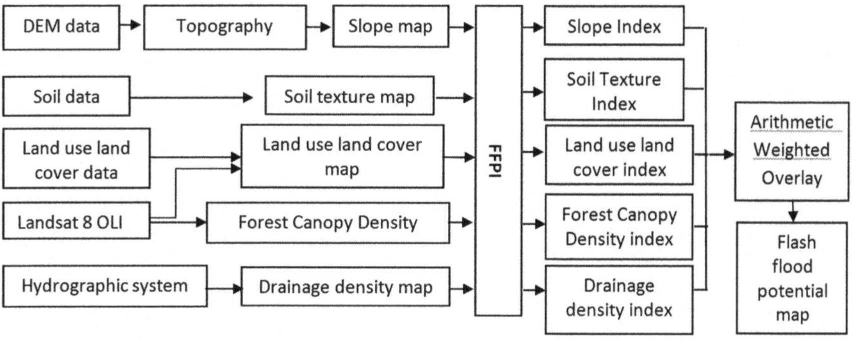

Fig. 27.2 Methodological flow chart

Table 27.1 Classification of slope by FFPI value

Slope (%)	FFPI value	Slope (%)	FFPI value
Above 30	10	From 12 to 15	4
From 27 to 30	9	From 9 to 12	3
From 24 to 27	8	From 6 to 9	2
From 21 to 24	7	From 3 to 6	1
From 18 to 21	6	Under 3	0
From 15 to 18	5		

2.3 Data Processing

- Identification of the slope index.
- The slope was generated from the Digital Elevation Model (DEM) at a resolution of 30 meters by interpolating from the Thai Nguyen topographic map provided by the Thai Nguyen Department of Natural Resources and Environment on a scale of 1: 50.000. Many scientists indicated that the variation of the slope is a crucial factor affecting the runoff time and the amount of infiltration and infiltration rate decreases with the increase in the slope angle in the experimental projects such as Fox et al., 1997 and Akbarimehr and Naghdi 2012. In general, the rain with a high intensity at a slope greater than 30 percent results in a seriously fast runoff in local creeks and stream (Wenbin et al. 2015). Slope values were classified from 1 to 10. Percent slope values greater than 30% were given a 10 on the scale, and the slope values in the range from 0 to 30% were divided from 1 to 9 (Smith, 2003). The classification of the slope is described in Table 27.1. The processing was done with the support of ArcGIS software, and the result is shown in Fig. 27.3.
- Identification of the soil texture index.

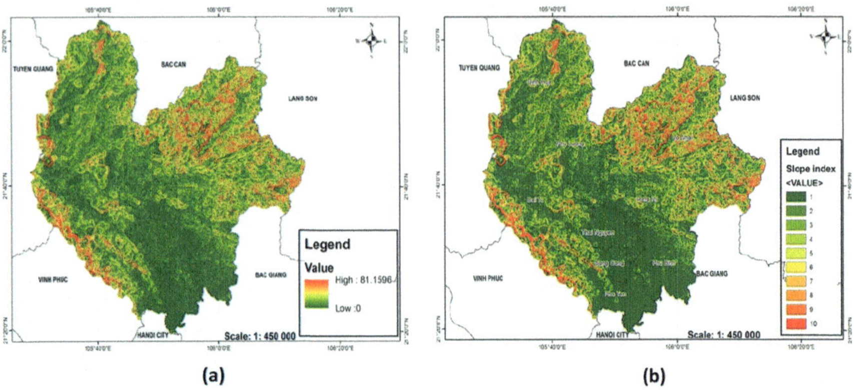

Fig. 27.3 Slope map (**a**) and slope index (**b**)

- The soil map at a scale of 1: 250000 provided by Thai Nguyen Department of Natural Resources and Environment was used to determine the soil texture index. The basis for dividing the soil map by the FFPI value is based on the soil texture, which is determined by the proportions of weight of clay, silt, and sand. A soil texture containing more sand will have better infiltration, but it is hard to create the surface runoff, while more clay soils will restrict the infiltration, but it can promote the runoff. According to the soil map, Thai Nguyen Province was covered by five soil texture types, such as sand, sandy clay loam, silt loam, clay loam, water body, and rocky mountain. In addition, the soil textures with different thicknesses affect infiltration significantly. Accordingly, a soil texture with a thick sand layer will infiltrate better than others, but it will be the opposite of thick clay soils. In the study area, soil textures were distributed with different thicknesses and divided into three levels: from 0 to 70 centimeter (cm), from 70 to 100 cm, and above 100 cm. As a result, the higher the infiltration the smaller the FFPI value by, the FFPI values corresponding to the soil textures are shown in Table 27.2. The combination of soil texture and soil layer thickness was applied to build the soil index, and it is described in Fig. 27.4.

- Identification of the land use land cover index.

- The spatial data from the map of land use in 2010 provided by the Thai Nguyen Department of Natural Resources and Environment were used to generate the land use land cover index map. Accordingly, Thai Nguyen Province was divided into 10 types of land (Fig. 27.5a). The FFPI values were given for the different types of land use land cover based on the storage ability on the foliage and roof of vegetation (Pei-Jun et al., 2007; Ebrahimian et al., 2009) and then classified as shown in Table 27.3. The residential lands, especially the urban areas, have a high flash flood potential due to the domination of impermeable surfaces and compacted soils as well as the shortage of natural vegetation. This speeds up the

Table 27.2 Classification of soil texture by FFPI value

Soil texture	Soil layer thickness (centimeter)	FFPI
Sand	From 0 to 70	1
Sandy clay loam	From 0 to 70	4
	From 70 to 100	3
	Above 100	2
Silt loam	From 0 to 70	5
	From 70 to 100	6
	Above 100	7
Clay loam	From 0 to 70	8
	From 70 to 100	9
	Above 100	10

runoff and causes the flood. In the opposite manner, in areas with better storage capacities such as forest and paddy, the risk of flash floods will be limited. The result of land use land cover index was mapped as shown in Fig. 27.5b.

- Identification of the forest canopy density index.
- Forest canopy density (FCD) is considered as an important index to identify the flood potential because it can intercept rainfall, slowing its fall to the ground. Therefore, it has a positive impact on preventing floods. Besides, it also helps in regulating the interchange of heat, water vapor, and atmospheric gases—the main factors that lead to weather variation. The FCD index was calculated based on the reflective value from bands of Landsat 8 OLI (Operational Land Imager). Landsat 8 taken in October 2016 by United States Geological Survey (USGS) was selected and processed in this paper. In the first step, the bands of Landsat images were converted from DN (digital number) to TOA (top of atmosphere) reflectance. The purpose of this work is to eliminate the negative effects of the atmosphere on image quality. This process has been done by using eq. (27.1).

$$L_\lambda = M_L Q_{cal} + A_L \tag{27.1}$$

where L_λ = TOA spectral radiance (Watts/(m^2 * srad * μm)).
M_L = band-specific multiplicative rescaling factor from the metadata
A_L = band-specific additive rescaling factor from the metadata.
Q_{cal} = quantized and calibrated standard product pixel values (DN)

In the second step, FCD was generated based on the indices including advanced vegetation index (AVI), bare soil index (BI), and canopy shadow index (SI). These indices were calculated as eqs. (27.2), (27.3), and (27.4). The last step, VD (vegetation density)—which was derived from the combination of AVI and BI, and SSI—which was derived from SI by using a linear transformation—were used to calculate the FCD. This process was carried out as eq. (27.5). This process was done with the support of ArcGIS and ENVI software.

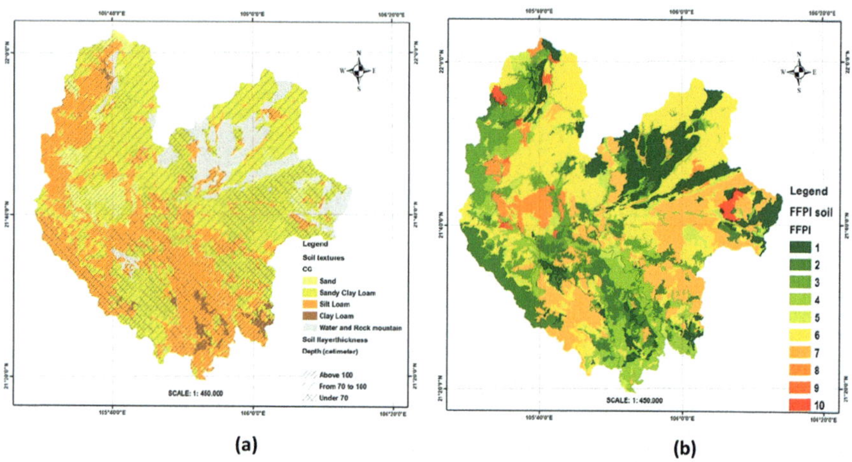

Fig. 27.4 Soil texture map (**a**) and soil texture index (**b**)

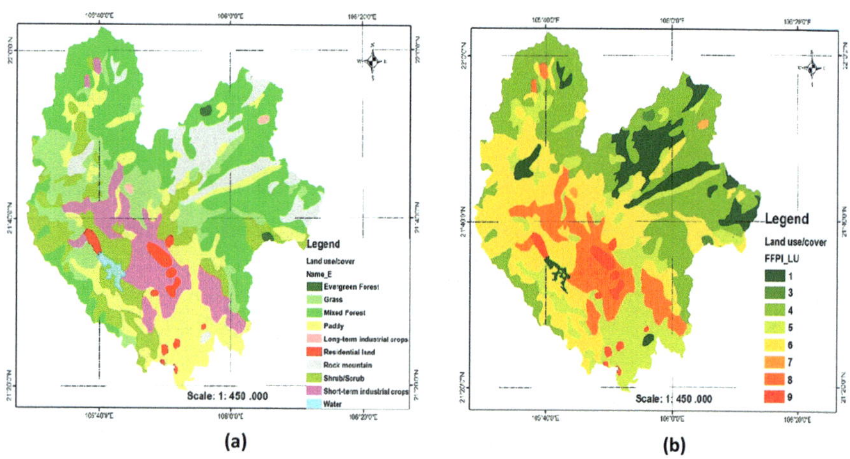

Fig. 27.5 Land use land cover map (**a**) and land use land cover index (**b**)

$$AVI = \sqrt[3]{(B5+1)*(65536-B4)*(B5-B4)} \tag{27.2}$$

$$BI = \frac{(B6+B4)-(B5+B2)}{(B6+B4)+(B5+B2)}*100+100 \tag{27.3}$$

$$SI = \sqrt[3]{(65536-B2)*(65536-B3)*(65536-B4)} \tag{27.4}$$

where B2 is the green band, B3 is the blue band, B4 is the red band, B5 is the near-infrared band, and B6 is the shortwave infrared band.

Table 27.3 Classification of land use land cover by FFPI value

Types of land	FFPI	Types of land	FFPI
Water	1	Evergreen Forest	3
Residential land	9	Mixed Forest	4
Short-term industrial crop	8	Shrub/scrub	6
Long-term industrial crop	7	Grass	6
Paddy	5	Rock Mountain	1

$$FCD = \sqrt{VD * SSI + 1} - 1 \tag{27.5}$$

In this case, the values of FCD were given in the range from 1.44625% to 90.7236%. The FCD Index was then reclassified by the FFPI value from the FCD map with the values ranging from 1 to 10 and is mapped in Fig. 27.6. As a result, the high flash flood potential corresponds with the low FCD and vice versa (Table 27.4).

- Identification of the drainage density index.

- The drainage density was determined based on the total length of stream per basin area by Roberte Horton in 1945. DEM data at a resolution of 30 meters were used first to determine the boundary of river basins by the interpolation method with the support of ArcGIS software. According to the interpolation result, the study area was divided into five river basins, and those are Cau River basin, the Cong River basin, Cho Chu, Nghing Tuong, and Du River basin. In the next step, hydrographic system maps taken from Thai Nguyen Department of Natural Resources and Environment and river basin map were combined to take out the drainage density map. As the result, the drainage density values corresponding to the Nghing Tuong River basin, Du River basin, Cau River basin, Cong River basin, and Cho Chu River basin were 0.68, 0.82, 1.43, 2.0, and 2.75, respectively (Fig. 27.7). After that, this result was classified into the FFPI value and mapped. It can be seen that the number of tributaries affects the flow density, thereby directly impacting the time of flood concentration in the basin (Pallard et al., 2009; Gregogy and Walling 2010), and therefore, the FFPI value increased corresponding with the increase in the drainage density.

- Generation of Weighted Flash Flood Potential Index (WFFPI).

- The Analytic Hierarchy Process (AHP) was applied to evaluate the weights of each index. This method was developed by Thomas Saaty in 1990 and become one of the most famous methods for making multicriteria decisions. Saaty's method describes the level of importance of parameters and their relationship on a scale of 1 to 9. After calculating the weights using the pairwise comparison method of Saaty, the consistency ratio (CR) in this case was 0.018992. As the result, the slope was assessed as the most critical index and assigned the value of 0.47 and the drainage density was determined as the least important index and weighted at 0.04. The soil index, land use cover index, and forest

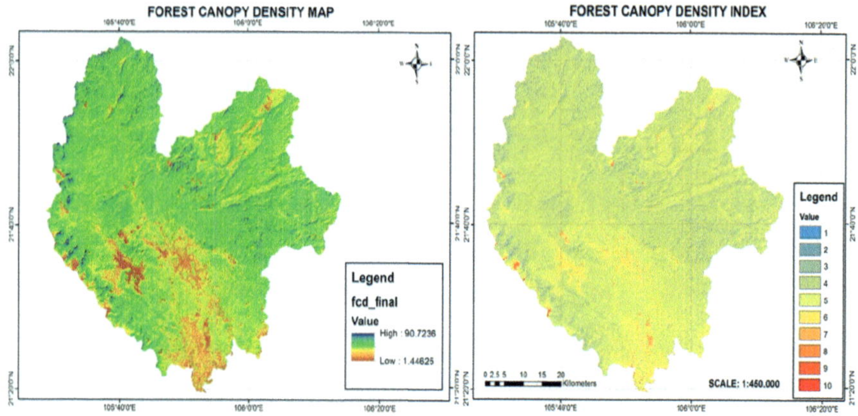

(a) (b)

Fig. 27.6 Forest canopy density (**a**) and forest canopy density Index (**b**)

canopy density index were weighted 0.09, 0.3, and 0.1, respectively. Therefore, the Weighted Flash Flood Potential Index (WFFPI) was generated based on the individual indices from the slope index, soil index, land use cover index, forest canopy density index, and drainage density index and computed as given (27.6):

$$WFFPI = \frac{0.47(M)+0.3(L)+0.09(S)+0.1(F)+0.04(D)}{N} \quad (27.6)$$

where *WUGSI:* Weighted Flash Flood Potential Index.
M: Slope index
L: Land use land cover index
S: Soil texture index
F: Forest canopy density index
D: Drainage density index
N: Sum of weightings

3 Result

Following the methodology described above, the WFFPI was the result of the synthesis from five indices (slope index, soil index, land use cover index, forest canopy density index, and drainage density index). The WFFPI values in the final result ranged from 2 to 8 and were arranged into four classes, in which, a very high flash flood potential corresponds with the WFFPI above 7. In a similar manner, the

Table 27.4 Classification of forest canopy density by FFPI value

Forest canopy (%)	FFPI	Forest canopy (%)	FFPI
From 1 to 10	1	From 50 to 60	6
From 10 to 20	2	From 60 to 70	7
From 20 to 30	3	From 70 to 80	8
From 30 to 40	4	From 80 to 90	9
From 40 to 50	5	Above 90	10

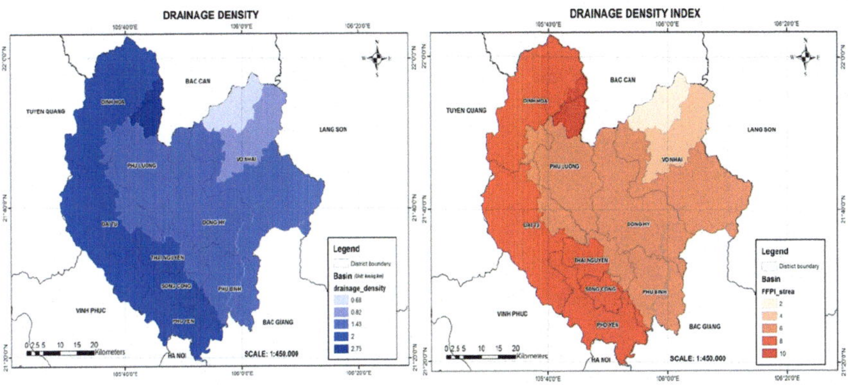

(a) (b)

Fig. 27.7 Drainage density (**a**) and Drainage Density Index (**b**)

WFFPI values from 5 to 7, from 3 to 5, and less than 3 are given the high, medium, and low flash flood potential, and it is shown in Table 27.5.

The result from the WFFPI map indicated the areas susceptible to flash floods. As a result, most of the areas in the study area were at the high and medium flash flood potential with the percentage at 17.54% and 81.06%, respectively. Especially, the areas with a very high flash flood potential occupied 0.27% and was found distributed along the upper streams of the high mountain (Fig. 27.8).

The statistics from the result map showed that the locations with great flash flood potential were distributed in Vo Nhai, Dai Tu, and Dong Hy districts. Vo Nhai was considered as the most serious susceptible district because the areas were at levels of high and very high potential with the flood reaching 24237.7 ha and 635.4 ha, respectively. In addition, Dai Tu and Dong Hy were worrisome districts for flash floods, in which, Dai Tu had 12855.44 ha at the high level and 78.5 ha at a very high level with flash flood potential and Dong Hy was 7992.2 ha and 131.6 ha, respectively. These districts were located at the upstream of the big rivers such as Nghing Tuong River and Cong River. The other districts are Phu Binh, Thai Nguyen, and Song Cong, and their flash flood potentials were less serious than others. The statistics are described in Fig. 27.9.

These results were then verified based on the statistics about flash flood events in the study area from 1969 to 2013 collected by authors. According to these statistics,

Table 27.5 Area and percentage of WFFPI distribution

Class	Area (hectare)	Percentage (%)
Low	3954.31	1.13
Moderate	284141.43	81.06
High	61498.05	17.54
Very high	946.0	0.27

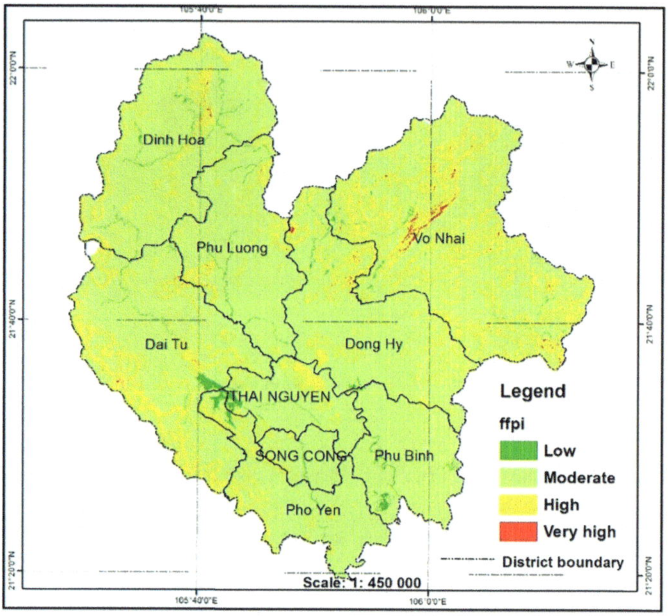

Fig. 27.8 Weight Flash Flood Potential Index (WFFPI) map

the areas having the high flash flood frequency happened to be at some points belonging to Dai Tu District, Dinh Hoa District, Dong Hy District, and Vo Nhai District. It coincided with the results of the classification map. The results proved that the integration of GIS and the FFPI model brings a reliable tool for evaluating the flash flood potential area.

4 Conclusions

Flash flood is one of the natural disasters of special interest, not only because its severity with respect to the environment, society, and economy, but also because of its complex and multifaceted nature. Studying the flash flood is an important aspect of the water resource management. Identifying the areas susceptible to flash flood contributes to reduction in the damages related to this type of disaster and decision making. This study introduces a methodology to evaluate the flash flood potential areas by the combination of WFFPI and GIS environment. By determining weights

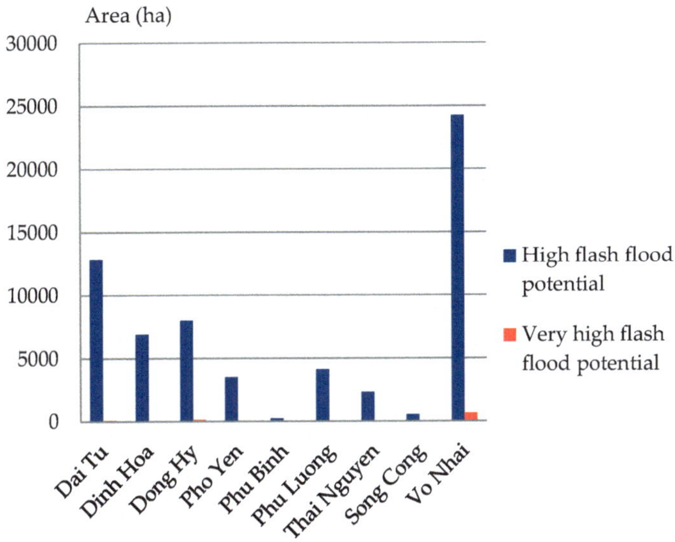

Fig. 27.9 Distribution of Flash Flood Potential Index (FFPI) by districts in Thai Nguyen

of five physical factors (slope, soil texture, forest canopy density, land use land cover, and drainage density), this research also indicates the close relationship between the physical characteristic of study area and flash flood potential. There are some previous projects using the FFPI methods in researching about flash flood, and with this study, the FFPI was modified with the determining weights of factors by the consultation of experts and using the AHP method to fit with features of the study area in this paper.

Moreover, selecting the weights for each factor keeps an important role in this research and changing it will affect significantly the results of the study. Therefore, this work needs to consider carefully and to be based on the actual characteristics of the study area and the consultation of experts.

Besides being dependent on physical elements such as slope, soil, forest canopy, land use, and drainage density, flash flood risk is also affected by the various human activities such as building structures: roads, highways, and hydroelectric, changing the land use structure. This will be added in the next research of the author. It is very necessary to help the research results to be more comprehensive. In addition, it should be noted that the results may not be detailed and absolutely accurate because of the errors of spatial data, the complicated variation of the phenomenon, and limitations on the actual database at the measurement stations. Hence, it is important to raise the level of detail of data from the measurement station to verify the results of the study and evaluate the accuracy.

Acknowledgments This research was supported by the Hanoi National University of Education.

References

Akbarimehr, M., & Naghdi, R. (2012). Assessing the relationship of slope and runoff volume on skid trails (case study: Nav 3 district). *Forest Science, 58*(8), 357–362.

Ebrahimian, M., Lai, F. S., & Abdul, M. I. (2009). Application of Natural Resources Conservation Service curve number method for runoff estimation with GIS in the Kardeh watershed, Iran. *European Journal of Scientific Research, 34*(4), 575–590.

Fox, D. M., Bryan, R. B., & Price, A. G. (1997). The influence of slope angle on infiltration rate for interrill conditions. *Geoderma, 80*, 181–194.

Gioti, E., C. Riga, K. Kalogeropoulos, C. Chalkias (2013). A GIS-based flash flood runoff model using high resolution DEM and meteorological data. *European Association of Remote Sensing Laboratories (EARSeL) eProceedings, 12*, 33–43.

Gregogy, K. J., & Walling, D. E. (2010). The variation of drainage density within a catchment. *International Association of Scientific Hydrology, Bulletin, 13*, 61–68.

Kruzdlo, R. (2010). *Flash Flood Potential Index for the Mount Holly Hydrologic Service 31 Area*. Retrieved from http://www.state.nj.us/drbc/library/documents/Flood_Website/flood-warning/user forums/Krudzlo_NWS.pdf.

Nguyen, T. M., & Pham, H. G. (2009). Several natural disasters in Thai Nguyen province: Situation, causes and solutions. *Journal of Science and Technology, Thai Nguyen, 109*, 155–160.

Pallard, B., Castellarin, A., & Montanari, A. (2009). A look at the links between drainage density and flood statistics. *Hydrology and Earth System Sciences, 13*, 1019–1029.

Pei-Jun, S., Yuan, Y., Zheng, J., Wang, J.-A., Ge, Y., & Qui, G.-Y. (2007). The effect of land use/cover change on surface runoff in Shenzhen region, China. *Catena, 69*, 31–35.

Smith G. (2003). *Flash flood potential: Determining the hydrologic response of FFMP basins to heavy rain by analyzing their physiographic characteristics*. Retrieved from http://www.cbrfc.noaa.gov/papers/ffpwpap.pdf.

Wenbin, M., Yu, F., Li, C., Xie, Y., Tian, J., Liu, J., & Zhao, N. (2015). Effects of rainfall intensity and slope gradient on runoff and soil moisture content on different growing stages of spring maize. *Water, 7*, 2990–3008.

Zeng, Z., Tang, G., Long, D., Xu, H., Chen, Y., & Yang, H. (2015). Development of GIS-based FFPI for China's flash flood forecasting. In *23rd international conference on Geoinformatics*. IEEE. Retrieved from https://ieeexplore.ieee.org/document/7378697

Zogg, J., K. Deitsch (2013). The flash flood potential index at WFO Des Moines. Project at National weather Service, . Retrieved from https://www.weather.gov/media/dmx/SigEvents/2010_Summer_Floods.pdf

Chapter 28
Implementing Agricultural Land Use Solutions to Adapt Climate Change in the Vietnamese Mekong Delta

Huong Thi Hoang and Nam Phuong Pham

Abstract The paper presents theoretical issues related to climate change, agricultural land use, and the impact of climate change on agricultural land use in the Mekong Delta. The research used comparative method and assessment method to point out the difficulties and challenges of climate change for agricultural land use as a basis for proposing solutions for effective and sustainable agricultural land use with climate change adaptation. As of June 2016, 13/13 provinces and cities in the region had been declared as affected by natural disasters, drought, and saline intrusion. Drought and saline intrusion caused about 139,000 hectares of paddy rice to be damaged with more than 50% of the area lost, causing about 215 billion VND loss and making about 400,000 households (1.5 million people) devoid of water with an estimated total loss of nearly 7520 billion VND. The proposed solutions included perfecting policies and laws; changing varieties, crop structure, and economic structure; applying achievements of Industrial Revolution 4.0 in agricultural land use; limiting climate change-induced factors; enhancing the role of communities in the use of agricultural land adapted to climate change; completing the planning; and building synchronous infrastructure and financial guarantee to implement solutions to adapt to climate change.

Keywords Agricultural land use · Adaptation solutions · Climate change · Mekong Delta · Vietnam

---

H. T. Hoang (✉)
VNU University of Economics and Business, Vietnam National University, Hanoi, Vietnam
e-mail: huong.ht@vnu.edu.vn

N. P. Pham
Vietnam National University of Agriculture, Hanoi, Vietnam

A. T. Nguyen, L. Hens (eds.), *Global Changes and Sustainable Development in Asian Emerging Market Economies Vol. 2*, https://doi.org/10.1007/978-3-030-81443-4_28

1 Introduction

The Delta has always proven a difficult environment to manipulate. However, because of population pressures, increasing acidification of soils, and changes in the Mekong's flow, environmental problems have intensified (Stewart & Coclanis, 2011). The Mekong Delta in the Vietnam territory consists of 13 provinces and cities directly under the Central Government with a total area of 3.94 million hectares and a population of 17.5 million people (accounting for 12% of the natural area and 19% of the population of Vietnam). The Mekong Delta is always the largest rice bowl of Vietnam, making important contributions to national food security and export. As of April 2017, the Mekong Delta contributed 50% of rice production, 65% of aquaculture, and 70% of fruit, provided 95% of export rice and 60% of aquatic production for exports of the whole country (Vietnam Government, 2017). Many agricultural products of the Mekong Delta had been present and favored in many places in the world, such as mangoes, rambutan, and star apple. However, the Mekong Delta has been facing many challenges, the most serious of which is the impact of climate change such as sea level rise, saline intrusion, landslides, and erratic floods. This had adversely affected agricultural land use and affected farmers' lives and jobs, as well as the socioeconomic development of the Mekong Delta. Therefore, this paper aimed to answer the questions: What is climate change, what are the causes of climate change, how it affected agricultural land use in the Mekong Delta, and what solutions are needed? Also, this paper aimed to adapt to climate change in order to use agricultural land effectively and sustainably, contributing to socioeconomic development and stabilizing farmers' life.

2 Methodology

Climate change is understood to be the change in the earth's climate system, including the atmosphere, hydrosphere, biosphere, and lithosphere due to various natural and artificial causes. Causes of climate change include two main causes (subjective and objective causes). The subjective reason is due to human impacts such as changing land and water use purposes and the increasing emissions of greenhouse gases from human economic activities. The objective reason is due to the change in nature such as the change in the activities of the sun, the earth changes its orbit, the process of creating mountains and creating continental shelves, the change in many maritime currents, and internal flow of the atmospheric system (FAO, 2011; Peter & Michael, 2011). In addition, changes that increase global temperatures and make rising sea levels are one of the top challenges that mankind has to solve (International Centre for Environmental Management [CEM], 2010). The Mekong Delta in particular and Vietnam in general were also seriously affected by climate change as the sea level rises, especially the situation of invasive seawater took place strongly in coastal areas. This made the area of agricultural land gradually reduced, and many areas were difficult to cultivate, or cultivated with low economic efficiency (Ministry of Agriculture and Rural Development [MARD], 2017).

Agricultural land is defined as land used for production, research, and experiments on agriculture, forestry, aquaculture, salt making, and forest protection and development purposes (Cao & Pham, 2018). Specifically, according to Article 10 of the 2013 Land Law, agricultural land includes agricultural land, forestry land, aquaculture land, salt-making land, and other agricultural lands. Other agricultural lands include land used to build greenhouses and other types of houses for cultivation purposes, including forms of cultivation not directly on land; building stables for cattle and poultry breeding and other animals permitted by law; land for cultivation, husbandry, and aquaculture for study and research purposes; and nursery land for seedlings, land for flowers, and ornamental plants. Agricultural land use is to exploit the useful properties of agricultural land for economic purposes and human life. In nature, agricultural land use is the practice of cultivating, animal raising, salt making, and other activities on agricultural land to create products that serve the needs of the land users and needs of the society.

Data related to research are collected from scientific works published in journals, Internet, and other documents such as legal documents, books related to climate change, climate change impacts on agricultural land use, and climate change adaptation measures. In addition, the study used comparative method and assessment method to point out the difficulties and challenges of climate change for agricultural land use as a basis for proposing solutions for effective and stable agricultural land use, and adaptation to climate change in the Mekong Delta.

3 Results

3.1 The Impact of Climate Change on Agricultural Land Use

According to the scenario of climate change and sea level rise in 2016, if the sea level rises by 1 m, the Mekong Delta is the area most at the risk of flooding (38.9% of the area), in which the two provinces of Hau Giang and Kien Giang have the largest flooded area (80.6% and 76.9%) (Ministry of Agriculture and Rural Development [MARD], 2016). This will greatly affect the production and employment of most rural workers. From the end of 2015 to June 2016, localities in the Mekong Delta suffered from severe drought and saline intrusion. As of June 2016, 13/13 provinces and cities had issued decisions to announce natural disasters, droughts, and saline intrusion in the area. Drought and saltwater intrusion had caused damage for about 139,000 ha of paddy rice, of which more than 50% of the areas were lost, resulted in loss of about 215 billion VND. At the same time, drought and saline intrusion made about 400,000 households (1.5 million people) devoid of water for domestic use (Ministry of Natural Resources and Environment [MONRE], 2016). It was estimated that the total damage during the saline–drought period of 2015–2016 was nearly 7520 billion VND, of which Kien Giang, Ca Mau, and Bac Lieu provinces were the worst-affected areas (about 6050 billion VND) (Central Steering Committee on Natural Disaster Prevention and Control [CSCNDPC], 2017). In 2017 and 2018, the Mekong Delta continued to suffer from unprecedented large-scale landslides and riverbank erosion in the region's history of more than 300 years of development.

Landslides directly threatened people's lives and property, seriously affected the safety of coastal natural disaster prevention and infrastructure works, and degraded coastal mangroves. The total damage caused by landslides on the riverbanks of the Mekong Delta provinces by the end of 2018 was about VND 2766.6 billion. Damages caused by natural disasters and climate change in the Mekong Delta were increasing rapidly with increasing severity (the heaviest damage was from 2017 to 2018 with 7990 billion VND). Generally, in the period of 2010–2018, the total damage caused by natural disasters and climate change in the Mekong Delta was about VND 20,945 billion. According to statistics, from 2013 to now, coastal erosion in Thanh Hai Commune, Thanh Phu District, and Ben Tre Province has cleared 110 hectares of productive land of farmers, of which the directly affected area was about 9.5 km with 97 households were living, caused many households to lose land and houses. This had a direct impact on agricultural land use and residential land and life of Mekong Delta farmers (Nguyen, Ha, & Pham, 2019). Therefore, it is necessary to have comprehensive solutions on efficient, sustainable, and effective use of agricultural land to adapt to climate change and turn the Mekong Delta into a prosperous development area, contributing more and more for the country's economic growth and development.

3.2 Solutions to Use Agricultural Land Adaptable to Climate Change

3.2.1 Completing Policies and Laws

Facing the fact that the area of agricultural land was declining, especially paddy land due to the impact of climate change, it was necessary to amend and supplement the land legislation to suit the current situation. Specifically, in order to use the land with decreasing area, it was necessary to improve the productivity per unit area through the application of technical advances to production or to increase the income from an area unit by changing the income structure, specifically the structure of crops and livestock appropriately and sustainably. One of the first issues that need to be addressed was the quota of land allocation and management of paddy land. But currently, the land law is still inadequate, and the limit of land allocation to individual households directly engaged in agricultural production is limited, affecting the mechanization of production and expanding production following to big scale. This requires revising the current Land Law in the direction of removing the land allocation quota and regulating the area of land allocation based on the agricultural land fund of each locality and the feasibility of production and business plan of land users. In order to promote the accumulation and concentration of agricultural land to meet the requirements of high-tech production, according to the global value chain, the state should issue regulations to encourage farmers to contribute land to produce in-large farm models, avoiding fragmented, scattered, and inefficient use of land. In addition, it also promotes regulations on recovering agricultural land which has been allocated to households and individuals but cannot be used or used ineffectively to assign subjects who use land more efficiently.

In order for the people in the Mekong Delta to be proactive in changing their crop structure, especially converting paddy land into fruit production or aquaculture with higher economic efficiency, the Decree 35/2015/ND-CP about paddy land management and use should be revised in the direction of allowing landowners to convert paddy land into other lands with higher economic efficiency with clear grounds and approved by competent authorities with simple administrative procedures based on the approved land use plan. At the same time, it is not necessary to keep a fixed area of paddy land to ensure food security as at present because in the context of internationalization, it is possible to import rice and export other high-value agricultural products as long as higher profits, stability, and environmental protection. In addition, the state should issue policies to encourage farmers in general and farmers in the Mekong Delta in particular to switch from pure agricultural production thinking to agricultural economic thinking, from quantity to quality associated with the value chain, and from chemical agricultural production to organic and high-tech agriculture. At the same time, the state should issue regulations encouraging enterprises to implement processing industries and supporting industries associated with the development of agricultural economy. Along with that, the state needs to develop mechanisms and policies to promote investment, science and technology, train, develop human resources, and support trade promotion to bring commodity products of the Mekong Delta into international distribution chains.

3.2.2 Changing Varieties, Crop Structure, and Economic Structure

In order to use agricultural land in the context of climate change, it is necessary to research and apply into practice the rice lines which are tolerant with salinity, flood, acid sulfate soil, drought, etc. Besides, it is necessary to grow short-term, high-value cash crops to replace summer–autumn rice in the rice structure. Short-term cash crops increase farmers' income and avoid saline intrusion in the dry season (Nguyen, 2016). At the same time, to build an agricultural production structure based on three focal points of aquaculture–fruit tree–rice, linked with ecological subregions, develop green industry, low emissions, not harming natural ecosystems, focusing on developing renewable energy associated with forest development and coastal protection in the whole region; and develop services—tourism into a strong economic industry based on effective exploitation of potentials, advantages of natural, ecological characteristics, cultural characteristics, and people of the Mekong Delta.

3.2.3 Applying Achievements of Industry Revolution 4.0 in Agricultural Land Use

In order to use agricultural land with high efficiency and stability to adapt to climate change, the use of agricultural land must be associated with the achievements of the Industrial Revolution 4.0 in agriculture, called agriculture 4.0 for short. Specifically, the use of agricultural land must be associated with the sensor application that connects everything to most agricultural farms (IoT Sensors); sensors and smart devices are connected and controlled automatically during agricultural production to help cope with climate change, and improve microclimate in the greenhouse; LED technology is used synchronously in high-tech cultivation to optimize the growth process and applied in countries with little agricultural land or urban agriculture; cultivating in greenhouses, net houses, using hydroponic and aeroponic technologies to isolate the natural environment and actively apply synchronous technology; photoelectric cells (solar cells) are used for efficient use of space and reduced energy costs, and most farm/business equipment is powered by solar power and solar cell batteries; using robots instead of human in tending plants and animals is becoming more and more popular, applied in aging countries and large production scale; drones and satellites are used to survey the status of data collection of farms, thereby analyzing recommendations on the basis of updated database to manage farm exactly; and financial technology serves the farm in all activities from externally connected farms, in order to devise the most effective farm management formula.

It is necessary to apply the achievements of Industry 4.0 and fields such as dairy, pig, and chicken husbandry; farming shrimp; and catfish on an industrial scale. These industries do not require large area scale, and there are models of high-tech applications in chains from production to export, so it is convenient for automation and use of robots (Pham, 2017). In aquaculture, it is possible to apply a farming system combining aquatic and vegetable/flower (aquaponic); producing flowers and fruits is convenient for automating seedling production, mechanizing soil preparation, planting, tending, and harvesting; combined fertilizing and watering; preparations for off-season production; and advanced preservation technology (controlled climate, cold drying, etc.). With flowers, needs more technology to keep the flower fresh. To prioritize the selection of fruit trees with concentrated scale, technology, and market such as dragon fruit, orange, and pineapple; edible mushrooms, medicinal mushrooms/medicinal plants can produce on an industrial scale with high added value in production systems controlled in terms of both climate and farming techniques, occupying a small area of land. Prioritizing the technology of extracting active ingredients with high pharmaceutical properties such as nanocurcumin or baby jackfruit oil, ginseng, etc., proceeds to search for active ingredients with healing and beauty functions. In rice production, proven overseas technologies such as remote-sensing applications in production management and pest control, and crop management tools can be applied on smartphones.

3.2.4 Limiting the Factors that Cause Climate Change

Renovating and upgrading infrastructure are because of the infrastructure accounts for nearly 1/3 of the emissions of greenhouse gases on earth; so improving the environment-friendly infrastructure will help improve the climate change situation. In addition, the convenient transportation system will also play a part in reducing the amount of vehicle exhaust emissions into the environment. Industrial zones need scientific planning and exhaust gas treatment to reduce environmental pollution. Limiting the use of fossil fuels such as coal, oil, and natural gas causes great greenhouse effect.

Preventing indiscriminate deforestation is to protect forest resources in particular and natural resources in general. Besides, every citizen needs to raise awareness, plant trees, and do not discharge waste in the environment; economical use of electricity and water helps reduce environmental pollution; and working close to home limits the use of vehicles in traffic. From there, it increases a certain amount of waste in the environment. Eat smart, enhance vegetables and fruits, plant clean vegetables, and do not use chemical fertilizers and pesticides to limit the amount of toxic substances in the environment. In particular, it is necessary to exploit new environmentally safe energy sources such as energy from the sun, wind, heat, and ocean waves, ethanol from plants, and hydrogen from water hydrolysis.

3.2.5 Enhancing Propaganda, Raising people's Awareness and Training, and Developing Human Resources

Propagate and raise people's awareness to change their thinking, and shift from pure agricultural production thinking to develop agricultural economy in a new and diversified model to meet market requirements in climate change context. The economy of the region must change basically, from production models, production practices, livelihoods, lifestyles, population networks to each household; developing a diversified and quality agricultural economy; building a new countryside associated with the development and application of hi-tech, clean, and organic technologies; actively living with floods and adapting to the region's natural features; efficiently exploiting and using brackish and saltwater; and responding to natural disasters such as storms, floods, droughts, and saline intrusion in the Mekong Delta adapt to climate change.

Training and developing human resources: Develop and implement projects on career change and job creation for agricultural and rural workers toward specialization and professionalization; provide market information; support to attract business participation; vocational training, human resource development, especially high-quality human resources; renovating the training work; and develop human resources for collective and cooperative economy.

3.2.6 Completion of Planning

Completing the planning of irrigation systems and natural disaster prevention is an urgent requirement, ensuring mitigation of damages in the event of natural disasters and actively responding to the most adverse climate change scenarios. A new model in land use planning with a multiobjective approach was implemented. The model integrates many factors: land, hydrology and land use, productivity, and finance in specific local conditions. The model will help to formulate adaptive strategies for the study area in the context of climate change and sea level rise.

The new plan should change from "living with floods" to "actively living with floods"; at the same time, the organization of territorial space in the direction toward forming ecological subregions serves as an orientation for economic, agricultural, and infrastructure development; organize and develop the urban system and rural resident points suitable to the characteristics of natural ecosystems; and specific conditions of the region and each ecological subregion. The criteria set out are models of conversion based on ecosystems, suitable to natural conditions, biodiversity, culture, people, and natural laws. The process of changing production models with long-term vision, improving economic efficiency, reducing vulnerability to risks, and attaching importance to preserving cultural, historical values, biodiversity, and ecological environment of the region.

Implementing the planning of orientations for sustainable agricultural transformation for the subregions in the Mekong Delta; irrigation orientations for sustainable development of the Mekong Delta adapt to climate change; develop a sustainable livelihood conversion plan for the people; plans to overcome river and coastal erosion in the Mekong River delta; master plan toward integration of Mekong Delta development and planning of network of monitoring, warning, and forecasting systems on natural resources and environment, climate change, and database of integrated interfield, and connecting with the Mekong subregion and other regions.

3.2.7 Building a Synchronous Infrastructure System

Infrastructure investment and development projects should be carried out to ensure consistency, and inter-regional and intersectoral development, with focus, key, and reasonable road. Priority is given to urgent works, works of motivational nature, promoting socioeconomic development of the whole region, and essential works for people's life; both structural and nonstructural solutions should be focused. Urgent tasks to prevent landslides and serious subsidence in some coastal areas and riverbanks should be immediately performed. The investigation, survey, construction, and approval of investment projects for implementation should be promoted.

Investing in embankments to create mangrove planting areas; develop mangrove forests to protect the sea dyke and river dyke systems; and develop ecological livelihoods associated with forests. Develop urban systems and rural residential areas suitable to the characteristics of natural ecosystems, specific conditions of the region, and each ecological subregion. Upgrading and modernizing irrigation systems for natural disaster prevention, livelihood development, conversion, development, and sustainable agricultural restructuring in ecological subregions, in which promoting formality public–private partners and ensuring safe houses in the conditions of floods, droughts, storms, floods, cyclones, and sea level rise. In the immediate future, to invest in and build residential clusters, residential lines, and houses in flooded areas; piloting models of houses and buildings; and avoiding storms and cyclones. Invest in complete transportation infrastructure system, prioritize traffic works in areas prone to flooding; work in service of connecting and transiting to promote multimodal transport in the provinces in the Mekong Delta region; and invest in, build, and synchronously develop information and communication infrastructure, electricity, water supply, water drainage, and health networks. Investment in water supply, drainage and wastewater treatment systems, solid waste disposal complexes; and promote recycling, reuse, and energy production from rubbish for the Mekong Delta region in the direction of sustainable development. Implementing and reviewing experiences from projects on sustainable development of subregions in the Mekong Delta region.

3.2.8 Financial Solutions

In order to mobilize financial resources, it is necessary to apply the form of public–private partnership in new construction, upgrading irrigation systems, and supporting the consolidation and completion of in-field irrigation systems to improve resilience to drought and saline intrusion in the Mekong Delta. Stepping up activities to mobilize financial resources for activities of sustainable development, adaptation to climate change in the Mekong Delta; strengthening the efficiency of capital use and speeding up the disbursement progress, and ensuring the efficient use of medium-term public investment capital. Building a flexible mechanism to mobilize financial resources for sustainable development and response to climate change. The government should consider establishing a sustainable development and responding climate change fund in the Mekong Delta. The fund develops a clear operational management mechanism, to mobilize urgent funds, dedicated to each purpose, in accordance with the general principles of sustainable management and regional adaptation.

Select priority and urgent projects for implementation, on the basis of integration, inter-regional, interprovincial, and intersectoral, and at the same time call for donors, international organizations, and further development partners continuously pay attention and appropriate support both technically and financially for the Mekong Delta. Further promote the role of the International Mekong Commission in financial support to share information, manage, and sustainably use water resources among countries in the Mekong basin for the common prosperity of the whole region; and financial contributions from the business community and people for the sustainable and prosperous development of the Mekong Delta to adapt to climate change.

4 Conclusions

The Mekong Delta is the nation's largest granary, making an important contribution to national food security and export. Many agricultural products of the Mekong Delta have been present and favored in many parts of the world, such as mangoes, rambutan, and star apple. However, the Mekong Delta has been facing many challenges, the most serious of which is the impact of climate changes such as sea level rise, saline water intrusion, landslides, and erratic floods. This has adversely affected agricultural land use and affected farmers' lives and jobs, as well as the socioeconomic development of the Mekong Delta. To effectively and sustainably use agricultural land adapting to climate change, it is necessary to synchronously implement policies and laws; changing varieties, crop structure, and economic structure; applying achievements of Industrial Revolution 4.0 in agricultural land use; limiting climate change-causing agents; enhancing the role of communities in the use of agricultural land adapted to climate change; completing the planning; and building synchronous infrastructure and financial guarantee to implement solutions to adapt to climate change.

References

Cao, V.H., Pham, P.N. (2018). Viewpoints on amending and finalizing regulations on the rights and responsibilities of the State in the role of representatives of the entire people's ownership of land and being a public power organization performing the state management function on land, in the context of market-oriented economy. National scientific conference "Proposing to amend and supplement a number of provisions in the land law 2013", November 7, 2018, 12–21.

Central Steering Committee on Natural Disaster Prevention and Control [CSCNDPC]. (2017). Report on the situation of natural disasters and damage caused by natural disasters in 2017, Hanoi, Vietnam. pp.4.

FAO. (2011). *Climate change impacts on agriculture in Vietnam*. pp.12–13.

Government. (2017). Resolution No. 120 / NQ-CP dated November 17, 2017 on sustainable development of the Mekong Delta to adapt to climate change.

International Centre for Environmental Management [CEM]. (2010). Strategic environmental assessment of hydropower on the Mekong mainstream: Summary of the final report, Hanoi. *Vietnam. pp., 13*.

Ministry of Agriculture and Rural Development [MARD]. (2016). Report on damage caused by drought and saline intrusion in the Central highlands, Mekong Delta, Hanoi, Vietnam. pp.4-5.

Ministry of Agriculture and Rural Development [MARD]. (2017). Solutions for sustainable agricultural transformation for the sub-regions in the Mekong Delta. Conference on Sustainable Development of the Mekong Delta in response to climate change. Vietnam. pp.5–6.

Ministry of Natural Resources and Environment [MONRE]. (2016). Climate change and sea level rise scenarios for Vietnam, Natural resources, environment and cartography publishing house, Hanoi, Vietnam. pp.63–70.

Nguyen, T. Q. (2016). Farmers' adaptive measures to climate change induced natural shocks through past climate experiences in the Mekong River Delta, Vietnam. *African Journal of Agricultural Research, 11*(15), 1361–1372.

Nguyen, Q. T., Ha, H. N., & Pham, S. A. (2019). Solutions to adapt to climate change in the Mekong Delta in a new context. *Vietnam Journal of Social Science, 3*, 214–224.

Peter, M., Michael, R. (2011). *Socialist Republic of Vietnam: Climate change impact and adaptation study in the Mekong Delta*. Melbourne, Australia. pp. 1–7.

Pham, P. N. (2017). *Some problems of high-tech agriculture development in Vietnam today*. National Scientific Conference "Investing in developing high-tech agriculture efficiently towards industrialization, August 2017, National Economics University Publishing House, Hanoi, Vietnam. pp. 231–240.

Stewart, M. A., Coclanis, P. A. (2011). Environmental change and agricultural sustainability in the Mekong Delta. *Springer*, 1–3.

Chapter 29
Geothermal Renewable Energy in Vietnam: A Current Status Overview and Proposing Solutions for Development

Van Hiep Hoang, Van Tich Vu, Trong Thang Tran, Xuan Anh Pham, and Thanh Tung Phan

Abstract Climate change and energy are two of the biggest challenges in the world and Vietnam today. To resolve the above-mentioned problem, especially the energy shortage, studying renewable energy systems is becoming necessary. This paper investigates the current status of renewable geothermal energy in Vietnam. This study focuses on determining the heat source based on geothermometry calculations of K^+ and Na^+ cations in the most 18 potential geothermal sources (in the northern part of Vietnam) in association with the assessment and examination of groundwater potential by the hydrogeological survey. With the deep temperatures of reservoirs that varied from 136 to 170 °C, these geothermal reservoirs can be explored for the electricity power plan of the capacities between 4.2 MWe and 17.4 MWe. And as a result, these 18 geothermal resources were estimated to be around 170MWe for the capacity prospects of electricity generation. Furthermore, the comparison of the policy toward renewable energy and people's feedback is illustrated.

Keywords geothermal energy · renewable energy · natural resources · Vietnam · development barriers

---

V. H. Hoang (✉)
VNU School of Interdisciplinary Studies, Vietnam National University, Hanoi, Vietnam
e-mail: hoanghiep412@vnu.edu.vn

V. T. Vu · T. T. Phan
VNU University of Science, Vietnam National University, Hanoi, Vietnam

T. T. Tran
Vietnam Institute of Geosciences and Mineral Resources, Hanoi, Vietnam

X. A. Pham
Vietnam Oil and Gas Group (Petrovietnam), Hanoi, Vietnam

A. T. Nguyen, L. Hens (eds.), *Global Changes and Sustainable Development in Asian Emerging Market Economies Vol. 2*, https://doi.org/10.1007/978-3-030-81443-4_29

1 Introduction

Geothermal energy—clean and renewable energy—was considered to be the natural heat of the earth. The heat source is generated mainly by the decomposition of the naturally radioactive elements as thorium, uranium, and potassium (sometimes caused by geological processes or geological phenomenon). Normally, the temperature increases with increasing depth into earth's core, about 25–30 °C/km, and the temperature at the depth of 10 km would be over 300 °C. In the bottom of the continental crust, temperatures are determined in the range from 200 to 1000 °C, and the heat is transferred to the surface mainly by conduction. In some areas, the heat flows more easily move to the surface due to intrusion of molten magma from depth, the high surface heat flow is created due to a thin crust, or the groundwater has been heated.

Generally, geothermal resources gained above 150 °C are produced on electric power generation, and the resources that have temperature below 150 °C are mainly used for heating and cooling purpose. The resources with temperatures from 5 to 30 °C will be used for heat pump purpose provided both heating and cooling.

The geothermal energy is the immense amount in the world. Geothermal resources are created in areas where geological settings permit a carrier (such as water) to transmit the heat from deep hot zones to or near the surface. Almost geothermal exploration activities concentrate on the areas where the average of resource temperature is higher than 300 °C and drilling is shallower and less costly (Vo 1987). About 2 km geothermal production wells are the most common. Geothermal resources are usually divided into various types (dry vapor or hot water) based on their mean temperature.

Geothermal being a renewable resource does not emit any greenhouse gas emission and particularly the binary technology which is most suitable for geothermal power development in Vietnam, is one of the most environmentally benign energy production technologies. Further geothermal power provides baseload electricity at 90% or above load factor and thus adds to the stability of the electricity grid.

Renewable energy in Vietnam has been officially studied for several decades, with significant research on the assessment of geothermal energy potential being funded by the state budget.

The study from 1983 by the Geological Department produced a map of national hot water resources, with their respective geothermal potential, allowing the identification and prioritization of the regions which warrant further investigation. The study from 1993 by RIGMR was also supported by international organizations (from New Zealand, Italy, and the USA) and identified six locations of geothermal potential (Bang, Tu Bong, Hoi Van, Danh Thanh, Mo Duc, and Nghia Thang).

In 1995, ORMAT, along with PECC1, conducted prefeasibility studies to select six locations for geothermal power plants in the central region. The total capacity was 112.7 MW, consisting of three power plants with a capacity of 55.2 MW planned for the first stage. However, due to several difficulties, the research was not continued.

The study in 2004 by RIGMR concluded that the geothermal energy potential in Vietnam was 340 MW, but that the price of electricity generated was not competitive compared to other sources of energy.

Because of all these factors, many countries despite having modest geothermal resources like Vietnam strive in harnessing this highly valuable and sustainable source of energy.

In this paper, group of authors are going to show an overview observation about the characteristics of geothermal resources in Vietnam, especially in the northern part of our country. Furthermore, we also propose some solutions for potential geothermal power development and management.

2 Methodology

2.1 Study Potential Geothermal Sources (in the Northwest Region of Vietnam)

Active fault zones commonly have great effects on fluid transportation in geothermal reservoirs in the northern part of Vietnam. During fault movement, all the pores and small fractures that meet with the fault plane become interconnected so that the inner part of the fault, the fault core, consisting of breccia or gouge, may suddenly develop a very high permeability. This is evidenced, for example, by networks of mineral veins in deeply eroded fault zones in geothermal fields. Inactive faults, however, may have low permeability and even act as flow barriers. In geothermal reservoirs, the orientation of fault zones in relation to the current stress field and their internal structure needs to be known as accurately as possible. One reason is that the activity of the fault zone depends on its angle to the principal stress directions. Another reason is that the outer part of a fault zone, the deformation zone, comprises numerous fractures of various sizes.

Here, we present field examples of faults, and associated joints in geothermal fields, and potential host rocks for geothermal reservoirs, respectively. We studied several localities of different stratigraphy, lithology, and tectonic settings: fault zones in outcrops from some basins.

Geothermal activity in some basins has long been associated with transport and deposition of sediments, hot mineral spring appearance, making the zone become an important setting for investigating thermal processes. Most studies have focused on fluid chemistry, the applicability of geothermal fluid (Cao 2004, Tran et al. 2016). Epithermal deposits nonetheless form in zones of high permeability in relatively shallow parts of geothermal systems, along steeply dipping faults, fractures that were clear channels of strongly focused hydrothermal fluid flow (Nguyen et al. 2005).

2.2 Analysis of the Inquiry Result on Barriers

Active fault zones play an important role in affecting fluid transportation in geothermal reservoirs in the northwest region of Vietnam. During the movement of faults, pores and small fractures which meet with the fault plane are going to become interconnected, and therefore, the inner part of the fault includes the fault core, consisting of breccia or gouge, which may suddenly become a very high permeability. This is evidenced, for example, by networks of mineral veins in deeply eroded fault zones in geothermal fields. In geothermal basins, the orientation of faults in the relationship with the current stress field and their inside structure may need to be known as exactly as possible. On one hand, the activities of the fault zones depend on its angle to the principal stress directions. In another hand, the outside part of fault zones, the deformation zone, includes the numerous fractures of various sizes.

Group author describes field examples of faults, and associated joints in geothermal fields, and potential host rocks for geothermal basins, respectively. We researched several sites of different stratigraphies, lithology, and tectonic settings: fault zones in outcrops from some basins related to geothermal resources.

Geothermal resource activities in some geothermal reservoirs had long been connected with transport and deposition of sediments, hot mineral spring apparitions, creating the geothermal zone becomes an important setting for investigating thermal processes. Most studies tend to have focused on fluid chemistry, the applicability of geothermal fluid (Cao 2004, Tran et al. 2016). Epithermal deposits nonetheless were created in zones of high permeability in shallow parts of geothermal reservoir systems, along steeply dipping faults that were clear channels of strongly focused hydrothermal fluid flows (Nguyen et al. 2005).

3 Results

3.1 Current Status of Geothermal Energy Potential in Vietnam

According to the result of surveying a total of 269 water sources covered all the regions of Vietnam carried out by RIGMR in 2005 (Table 29.1), we can list water sources as follows: 140 warm water (52.1%), 84 moderately hot water (31.2%), 41 very hot water (15.2%), and four extremely hot water (1.5%). In conclusion, the geothermal resources have been distributed in most regions of the territory of Vietnam (Nguyen et al. 2006).

The distribution of hot springs was as follows: 39.8% in the northern part, 39.5% in the central part, and 19.7% in the southern part of Vietnam.

The geothermal usage in Vietnam has been started from the year of 80s of the end of the last century for the main purpose as drying agricultural products such as tea, manioc, coconut palm, etc., in Hoi Van and My Lam geothermal fields with very good result. However, because of the economy and technique difficulties, the

Table 29.1 Hot springs are classified based on the water temperatures and located areas (Nguyen et al. 2006)

Temperature grades	Areas						Summated by temperature grades	%
	Northwest	*Northeast*	*North plain*	*North-central*	*South-central*	*South*		
Warm (30–40 °C)	35	6	9	11	27	52	140	52.1
Moderate (41–60 °C)	38	3	3	19	20	1	84	31.2
Very hot (61–100 °C)	6	2	2	11	20	0	41	15.2
So hot (>100 °C)	0	0	3	1	0	0	4	1.5
Summated by area	**79**	**11**	**17**	**42**	**67**	**53**	**269**	
%	**29.4**	**4.1**	**6.3**	**15.6**	**24.9**	**19.7**		**100**

geothermal usage is quite limited in that period. Besides, the geothermal resources were also applied and researched for crayfish farming during the winter.

Until now, almost all geothermal sources in Vietnam today are only used for direct utilization such as spa, bathing, hot water swimming pools, and tourist resorts and are as follows: Hot springs My Lam (Tuyen Quang Province), Binh Chau (Ba Ria-Vung Tau Province), Quang Hanh (Quang Ninh Province), Hoi Van (Binh Dinh Province), etc., and recently, people in Quynh Phu and Hung Ha districts of Thai Binh province have used hot water for warming the fish breeding ponds and warming the chickens and pigs farm in the winter.

According to the authors' prediction (topographic conditions, economic development potential, investment capacities), geothermal energy in Vietnam can develop for geothermal power plant (GPP) direction, which will be presented in the next section (in the northern part of Vietnam).

3.2 *The most Potential Geothermal Sources in the Northern Part of Vietnam*

The reservoir temperature information is calculated by the geothermometers through geochemical survey results for geothermal fields. Parameters of geothermal resources are measured based on their characteristic appearance in geological maps, the relationship connecting with tectonic structures, and international references. The thickness of each reservoir system is measured to be 2 m, and the reservoir areas are nearly 2.5 km^2. According to the computing method applications for the potential of geothermal energy (Metcalfe 1986), in over 164 geothermal resources, there are 18 geothermal systems of 18 different places that can be directly applied to renewable energy applications, and as a result, they can be also exploited for electric generation (Figs. 29.1 and 29.2). The surface temperature and flow rate of hot mineral water in these hot springs are used to calculate the natural thermal power that these 18 potential geothermal resources are wasting without applying to human demand. There are 8960 tons of waste heat annually (described in Table 29.2). The deep temperatures of reservoirs are calculated and described in varying from 136 to 170 °C. These geothermal systems can be exploited for electricity generation with the capacities between 4.2 MWe and 17.4 MWe. The total electric generation capacity prospects are measured in nearly 170 MWe (Fig. 29.2) from these 18 geothermal resources.

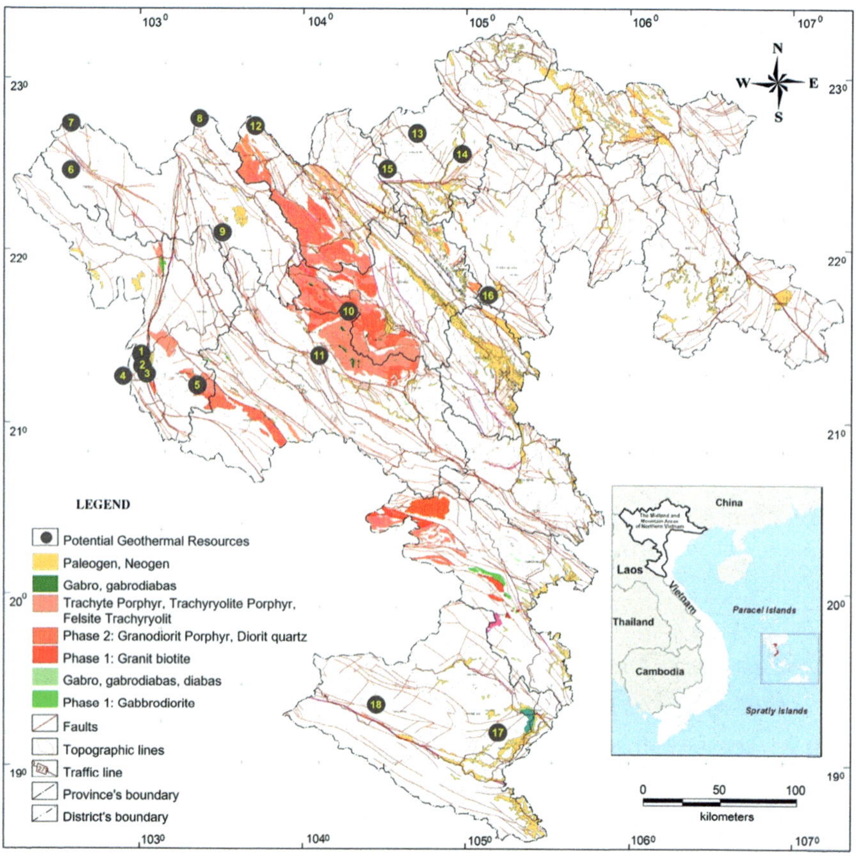

Fig. 29.1 Map describes the most 18 potential geothermal resources in the northwest region of Vietnam. Numbers in the cycles describe the information of these source names including (1) Pe Luong, (2) Na Hai, (3) Uva, (4) Pa Thom, (5) Pa Bat, (6) Pac Ma, (7) La Si, (8) Sin Chai, (9) Nam Cai, (10) Lang Sang, (11) Nam Pam, (12) Lung Po, (13) Bo Duot, (14) Quang Ngan, (15) Quang Nguyen, (16) My Lam, (17) Nam Ron, and (18) Kim Da (Tran et al. 2016). This map was created and modified based on data of the geological map of the Vietnam Scale of 1/1.000.000 (Tran 1973) (Tich et al. 2019)

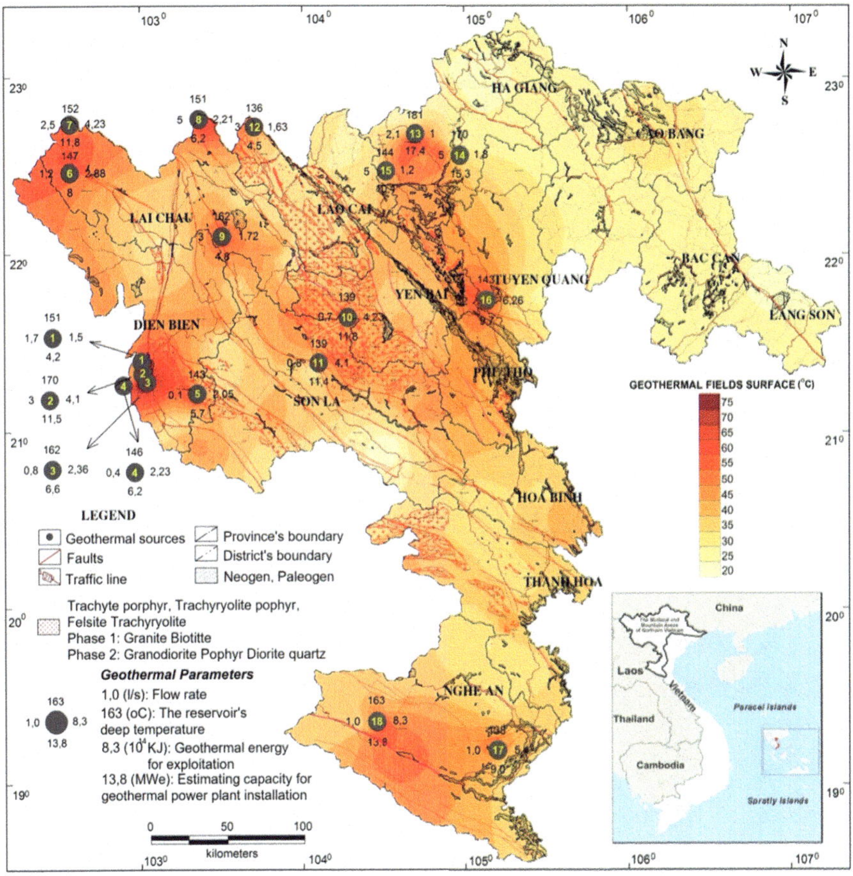

Fig. 29.2 Map illustrates the potential of 18 geothermal resources that related to the surface geothermal field (numbers symbolize the same sources refer to Fig. 29.1) (Tran et al. 2016). This map was created and modified based on data of the geological map of the Vietnam Scale of 1/1.000.000 (Tran 1973) (Tich et al. 2019)

Vietnamese people directly used the geothermal sources (Fig. 29.2). There are two systems for using geothermal energy. The rudimentary traditional systems were applied by the communities living in the northern part of Vietnam for various uses. These various uses can help local inhabitants in tourism development as follows: drying of crops, bathing, concentration of the steam to collect freshwater, and boiling of eggs. The complex systems are using geothermal energy for greenhouse heating, district heating, and refrigeration. In the northern part of Vietnam, traditional systems are evidenced at Uva, My Lam, Thanh Thuy, and Kim Quang Ngan where the steam has been tapped and used for bathing, spas, and ecotourism development.

A large number of geothermal resources are located in the mountain areas in the northern part of Vietnam, usually closely related to volcanic activity. Annual tourist numbers are quite significant, but the domestic travelers mainly want to visit the

Table 29.2 Characteristics of potential geothermal sources in the northern part of Vietnam (Tran 2014)

No.	Name	Location (District, province)	Surface T(*) (°C)	Deep T (°C)	Flow rate (l/s)	Waste heat (KJ/s)	Achievement temperature with 30 °C - output T (KJ/s)	Criterion energy with 30 °C - output T (ton/yr)	Geothermal energy potential for exploration (10^{14} KJ)	Electric generation capacity estimation (MW)
1	Pe Luong	Thanh Luong, Dien Bien	53.80	151.00	1.70	205.14	169.53	182.28	1.50	4.20
2	Pa Thom	Pa Thom, Dien Bien	57.00	146.00	0.40	53.63	45.25	48.66	2.23	6.20
3	Uva	Dien Bien, Dien Bien	74.00	161.00	2.53	164.25	147.49	158.58	2.36	6.60
4	Na Hai	Sam Mun, Dien Bien	78.00	170.00	3.00	666.21	603.36	648.74	4.10	11.40
5	Pa bat	Muong Luan, Dien Bien	61.50	143.00	0.10	21.41	18.48	19.87	2.05	5.70
6	Pac ma	Muong Te, Lai Chau	62.50	147.00	1.20	188.55	163.41	175.70	2.88	8.00
7	La Si	Muong Te, Lai Chau	54.00	152.00	25.00	3037.75	2514.00	2703.09	4.23	11.80
8	Sin chai	Phong Tho, Lai Chau	74.00	151.00	5.00	1026.55	921.80	991.13	2.21	6.20
9	Nam Cai	Sin ho, Lai Chau	62.00	162.00	3.00	465.09	402.24	432.49	1.72	4.80
10	Lang sang	Mu Cang chai, Yen Bai	53.00	139.00	0.70	82.12	67.46	72.53	4.23	11.80
11	Nam pam	Muong La, son La	55.50	139.00	0.80	102.24	85.48	91.90	4.10	11.40

No.	Name	Location (District, province)	Surface T(*) (°C)	Deep T (°C)	Flow rate (l/s)	Waste heat (KJ/s)	Achievement temperature with 30 °C - output T (KJ/s)	Criterion energy with 30 °C - output T (ton/yr)	Geothermal energy potential for exploration (10^{14} KJ)	Electric generation capacity estimation (MW)
12	Lung Po	Bat Xat, Lao Cai	53.00	136.00	3.00	351.96	289.11	310.85	1.63	4.50
13	Bo Duot	Vi Xuyen, Ha Giang	71.00	181.00	1.00	192.70	171.80	207.80	2.10	17.40
14	Quang Ngan	Vi Xuyen, Ha Giang	62.00	170.00	5.00	775.20	670.40	810.90	1.80	15.30
15	Quang Nguyen	Xin man, Ha Giang	56.00	144.00	5.00	649.50	544.70	658.90	1.20	10.40
16	My lam	Yen son, Tuyen Quang	65.50	143.00	6.30	1065.70	934.10	1129.90	1.20	9.70
17	Nam Ron	Tan Ky, Nghe an	57.00	138.00	1.00	134.08	113.13	121.64	5.40	9.00
18	Kim Da	Tuong Duong, Nghe an	73.50	163.00	1.00	203.22	182.27	195.97	8.30	13.80
	Total					**9385.30**	**8044.01**	**8960.93**	**51.01**	**168.10**

T(*): temperature (degrees Celsius-C).

geothermal centers on their journey called "hot spring pilgrimage." Many Vietnamese people want to try "Olsen" and spend their time to experience and discover the exciting hot springs in their whole country.

With the calculated temperature in the northern part of Vietnam, the exploitation of regional energy for power generation purposes, group author proposes that these geothermal resources' governments may use the binary cycle technology to exploit energy for electricity.

3.3 The Inquiry Result on Barriers

3.3.1 Geothermal Power Plan (GPP)

According to our survey results, both groups (geology and nongeology) agreed that policy and technique issues are the two biggest categories of the barrier. For the nongeology group, the biggest category of the barrier is technical issue, while for the geology group the biggest one is the policy issue. In policy issues, the two groups agreed that national energy policy is the most important, and then to the lack of incentives and the lack of investment in research and development. The technical category of barriers that stand up, the two groups agreed that the biggest barriers are lack of information or experience, the data exploration technology, integration, or interpretation (Table 29.3 and Fig. 29.3).

While the geology group responded that the financial issues are major barriers accounted for 21%, the nongeology group responded that they are only accounted for 14%. For the legal issues, the geology group's response is the smallest barrier (13%), but the nongeology group's response is not the smallest barrier (15%).

In general, both groups' responses are not so different and can serve together by taking the average values. Table 29.2 shows that the order of the categories of barriers that stand for geothermal development is technical problems (25%), policy issues (24%), social issues (20), financial issues (18%), and finally the legal issues (13%).

From the survey, results may be temporarily assessed that the biggest category of barriers that stand with the development of geothermal energy in Vietnam is still technical issues in which the lack of information and experience are the most dominant factors (Table 29.3 and Fig. 29.4). This is understandable because there has never been any geothermal resource exploration and any geothermal power generation in Vietnam so that can allow people to get complete information on the geothermal source and geothermal power plants. The policy issue is the second important factor hindering the development of geothermal power. In particular, national energy policy and economic incentives are the most important. The third biggest category of barriers stands on social issues with special emphasis on the lack of specialists. The fourth biggest category of barriers is financial, and the category of legal barriers is evaluated as the smallest.

Table 29.3 Degrees of barriers hindering the geothermal power generation development in Vietnam

Category of	% in total	Barriers	% in total
Policy	24	National energy policy	9.4
		Lack of economic incentives	9.7
		Lack of R&D funding	4.8
		Domestic business protection	0.0
		Other policy matters	0.0
Social	20	Expert deficiency	9.8
		Awareness deficiency	3.1
		Knowledge deficiency	2.2
		Business model deficiency	3.8
		Other land uses	1.2
		Public acceptance (PA)	0.1
		Other social matters	0.0
Legal	13	Environmental matters	1.7
		Legislation/business mechanism	4.1
		Lack of legal incentives	7.3
		Red tape in government	0.9
		Other legal matters	0.0
Fiscal	18	High exploration cost	11.5
		Low selling price	5.0
		Neither loan nor support	0.8
		Other fiscal matters	0.6
Technical	25	Information/experience deficiency	9.1
		Exploration technology	5.3
		Data integration or interpretation	5.0
		Drilling	2.6
		Scaling, erosion, corrosion	0.3
		Reservoir management	1.5
		Other technical matters	0.1
Total (%)	**100**		**100**

In terms of barriers, here we see the barrier to exploration costs being the biggest (11.5%). The second biggest barriers are the lack of experts (9.8%) and the incentive policy (9.7%) to geothermal development.

From a total of 28 barriers that stand investigated, 15 barriers stand noteworthy because of a much higher percentage than these others. Some barriers stand very low to 0%, suggest that interviewees may not be aware of it. The change in the barrier rate from 0% to 11.5% indicates a fairly uniform assessment without focusing on certain barriers. This further confirms that there is no geothermal power plant in Vietnam, so it is impossible to have a clearer assessment of certain barriers (Table 29.3 and Fig. 29.4).

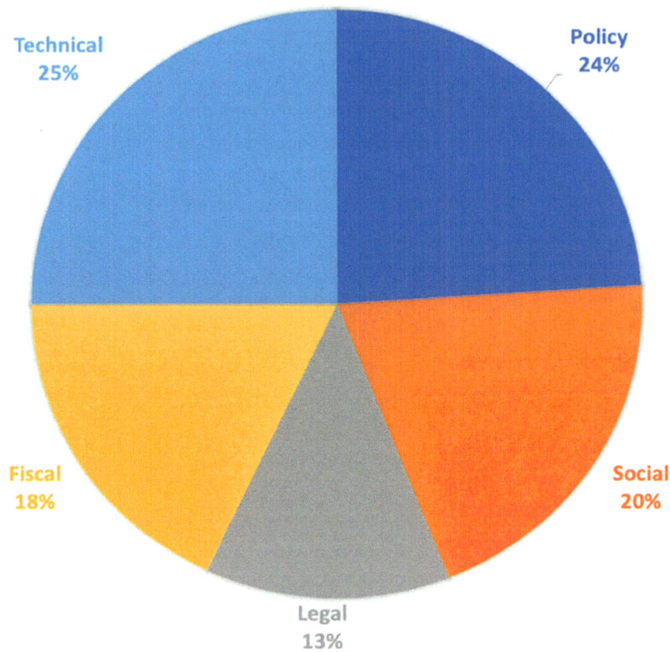

Fig. 29.3 Categories of the barrier to geothermal power generation development in Vietnam

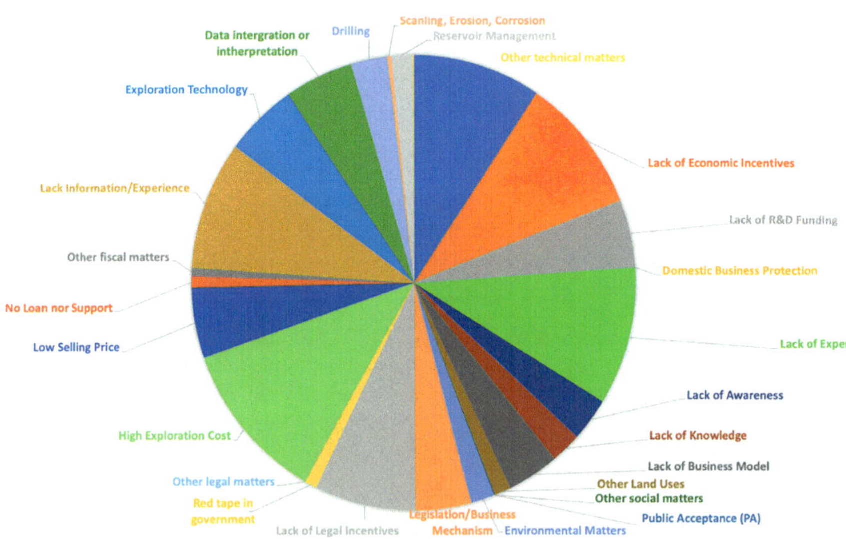

Fig. 29.4 Barriers to geothermal power generation development in Vietnam

3.3.2 The Ground Source Heat Pump (GSHP)

Since there is no GSHP application in Vietnam, almost everyone just understands this system. Therefore, the technical category of barriers is considered the largest, accounting for 27%, of which 12% belongs to lack of installation experience. The financial category of barriers, though accounting for only 18%, of which installation cost is accounted for 14.7%, is the highest among the barriers. Similar to GPP, the social issues here are also considered to be the second largest category of barriers in which the barrier due to national energy policy accounted for 10.6%. In general, as with GPP, there are many barriers to the development of GSHP. This reflects that there is no GSHP application in Vietnam until now. If there were some available GSHP applications, the evaluation of respondents may be another review (Table 29.4, Figs. 29.5 and 29.6).

3.3.3 Innovative Ideas to Remove Barriers

Preliminary assessment shows that geothermal potential in Vietnam could allow the development. Although the intention to build a geothermal power plant has been in

Table 29.4 Degrees of barriers hindering the ground source heat pump installation in Vietnam

Category of	% in total	Barriers	% in total
Policy	24	National energy policy	10.6
		Lack of economic incentives	6.6
		Lack of R&D funding	4.4
		Other factors	2.5
Social	20	Expert deficiency	4.8
		Awareness deficiency	1.8
		Knowledge deficiency	3.6
		Business model deficiency	6.8
		Other factors	0.3
Legal	13	Environmental matters	4.3
		Legislation/business mechanism	3.1
		Lack of legal incentives	4.8
		Other factors	0.3
Fiscal	18	High installation cost	14.7
		Neither loan nor support	3.3
		Other factors	0.3
Technical	25	Information/experience deficiency	12.2
		Hydrogeological information deficiency	4.4
		Installation technology deficiency	7.2
		Heat pump maker deficiency	4.0
		Other factors	0.0
Total (%)	**100**		**100**

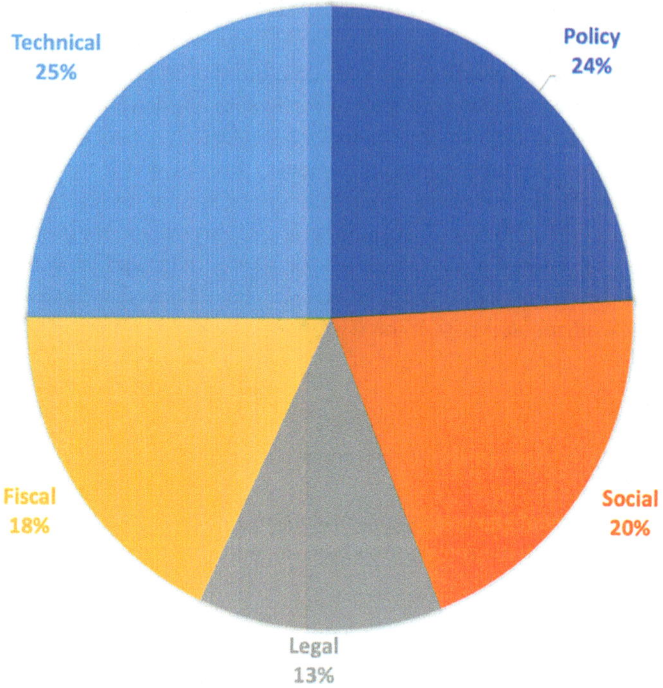

Fig. 29.5 Categories of barriers to ground source heat pump installation in Vietnam

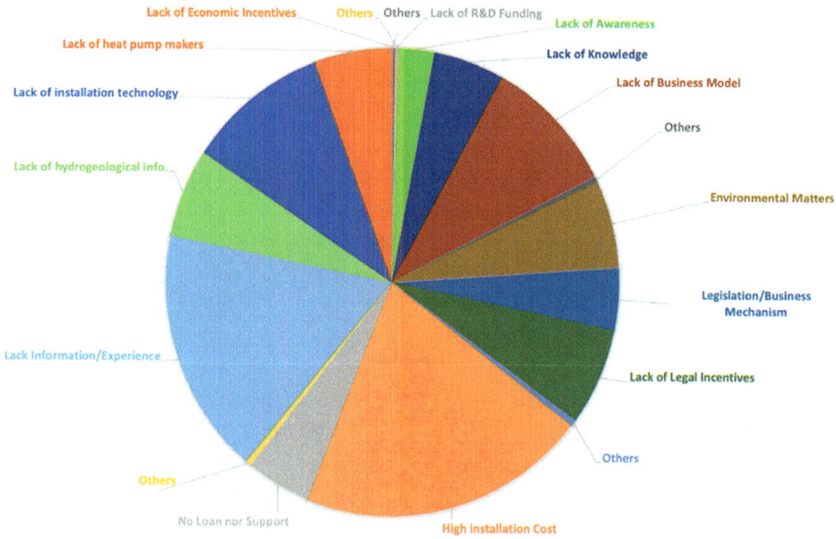

Fig. 29.6 Barriers to ground source heat pump installation in Vietnam

place since 1995, so far there have been no factories. Why? The barriers that are pointed out in the above investigation may be the answer to this question.

Because geothermal energy is still quite new, the community and business do not pay much interest in it, so the government should issue the priority for it first. But to make the policymaker understand geothermal and the benefit from the geothermal application, the role of scientists and specialists is very big. To do so, the geothermal projects should be funded because based on that the scientists and specialists will have proof to prove the effectiveness of the geothermal energy development. Emphasizing that geothermal research must work in partnership with scientists from the countries with geothermal development in the world, otherwise investing in geothermal will not be effective and may waste money.

Brave companies are encouraged to invest in geothermal energy development, especially for pioneering companies. Priority policy must be documented in detail, and this policy must be applied effectively through the management levels from central to a local level.

Parallel with the preferential policies, the government must pay attention to the training of human resources. In the immediate future, it is possible to put selected subjects into the teaching program for undergraduate students in geology and techniques.

The electricity pricing policy for geothermal energy should also be issued early to make it as a basis for the investors to determine their investment in geothermal energy. At the same time, the government also needs to prioritize the purchase of geothermal power by its inherent benefits in comparison with other types of energy including other renewable energy.

If geothermal law cannot be enacted in the current period, geothermal resources need to be included in the mineral law soon, with the same regulations as for other minerals. Of course, it is necessary to refer to geothermal law in countries that already have a geothermal law to make detailed provisions to match geothermal minerals.

Integrating geothermal programs into other socioeconomic research programs is tried if possible. Allocating the favored capital and credit from the development fund, ODA fund, and other bilateral lending funds for geothermal as well as reinforcing the human capacity is prioritized.

The law on renewable energy needs to be issued soon, because this is a breakthrough step and will make favorable conditions to push up the development and utilization of renewable energy as well as creating the market for renewable energy in general and geothermal in particular to develop.

The manpower is reinforced by scientific and technological training for geothermal energy through the cooperation projects. The international cooperation in science and technology on geothermal energy is pushed up.

4 Conclusions and Discussion

1. The geothermal usage in Vietnam has been started from the year of 80s of the end of the last century for the main purpose as drying agricultural products such as tea, manioc, coconut palm, etc., in Hoi Van and My Lam geothermal fields

with very good result. However, because of the economy and technique difficulties, the geothermal usage is quite limited in that period. Until now, almost all geothermal sources in Vietnam today are only used for direct utilization such as spa, bathing, hot water swimming pools, and tourist resorts and are as follows: Hot springs My Lam (Tuyen Quang Province), Binh Chau (Ba Ria-Vung Tau Province), Quang Hanh (Quang Ninh Province), Hoi Van (Binh Dinh Province), etc., and recently, people in Quynh Phu and Hung Ha districts of Thai Binh province have used hot water for warming the fish breeding ponds and warming the chickens and pigs farm in the winter.

2. The total generation capacity of the most potential of 18 geothermal sources in the northwest region of Vietnam is 170 MWe, which is calculated by the method of calculating the generating capacity based on duality technology. The lowest capacity geothermal source is 4.2 MWe (Pe Luong), and the highest capacity geothermal source is 17.4 MWe (Bo Duot). However, in order to conduct visits due to the exploitation and construction of geothermal power plants, it is necessary to assess the ability according to the geological, technical, and technological design and installation conditions associated with the location of each geothermal source. The results of this study open up opportunities for investors and managers on clean energy sources in the future, and it also opens opportunities for the application of geothermal power generation technology currently being popularized.
3. Innovative ideas to develop geothermal energy (electricity production) based on policies current status:

At present, the current energy policy framework in Vietnam does not provide adequate support for developing commercially viable geothermal power projects. The unique risk profile of geothermal power plants requires the implementation of some special policy package. To promote geothermal power plants in Vietnam, it is recommended to implement two policy instruments. The first one is a hybrid scheme of capital subsidy and soft loans. This scheme will help to alleviate the early high-risk profile of the exploration stage of geothermal power development.

Certain fiscal benefits such as tax break, import duty exemption, and land levy exemption of which geothermal power plants are already eligible, are recommended being continued. However, the effect of these policies will be minimal with respect to the overall economics of the geothermal power plant operation. However, they will bring some additional benefits to the potential project developer.

In addition to the recommendation of the policy package, another important recommendation is to develop a detailed geothermal atlas of Vietnam. This atlas when fully completed will be immensely valuable in overcoming the information barrier related to geothermal resources in Vietnam. This, in turn, will greatly facilitate raising the interest among the geothermal power developers around the world.

Finally, the most crucial aspect of realizing the potential of geothermal power in Vietnam is its effective implementation. As Vietnam is now embarking on its journey toward geothermal power development, it is recommended approaching a gradual step-by-step approach to fully utilize the geothermal power potential.

Acknowledgments This work was supported by the Domestic Master/Ph.D. Scholarship Program of Vingroup Innovation Foundation and VNU University of Science—Hanoi through project number KHCN-TB.01 T/13-18 (Northwest Program).

References

Cao, D. G. (2004). Final report of project "research and evaluation of the geothermal potential in the northwestern plain of Vietnam and it is a prospect for economic and social development". *Hanoi: Institute of Geosciences and Mineral Resources, 1*, 179.

Metcalfe, I. (1986). Late Palaeozoic paleogeography of Southeast Asia: Some stratigraphical, palaeontological and palaeomagnetic constraints. *Bulletin of the Geological Society of Malaysia, 19*, 153–164.

Nguyen, T. M., Tsujimori, T., & Itaya, T. (2006). Honvang serpentinite body of the song ma fault zone, northern Vietnam: A remnant of oceanic lithosphere within the Indochina-South China suture. *Gondwana Research, 9*(1–2), 225–230.

Nguyen, T.C., Cao, D.G., Tran, T.T (2005). *General evaluation of the geothermal potential in Vietnam and the prospect of development in the near future*. Proceedings World Geothermal Congress 2005 Antalya, Turkey, 24–29 April 2005, 1–8.

Tran, V. T. (1973). *Geological map of Vietnam Scale 1/1.000.000*. General Department of Geology and Minerals of Vietnam Publication.

Tran, V. T. (2014). *Northern part of Vietnam* (Vol. 1, pp. 908–918). Geology of Vietnam; Vietnam National University Press.

Tran, T. T., Vu, V. T., Dang, M., Hoang, V. H., Pham, H. T., & Pham, X. A. (2016). Assessing the power generation potential of geothermal prospects in the Midland and mountain areas of northern Vietnam. *VNU Journal of Science: Earth and Environmental Sciences, 32*(2S), 225–235.

Vo, C. N. (1987). *Final report of project "evaluating geothermal sources serving as the basis for their exploitation and utilization for energy purposes in some prospective areas, Vietnam"* (245 pages). Institute of Geosciences and Mineral Resources.

Chapter 30
Study on Stand Structure of Secondary Mangrove Forest: *Sonneratia caseolaris-Aegiceras corniculatum* Stand for Introducing Silvofishery Systems to Shrimp Culture Ponds

Kazuya Takahashi and Tuyen Thi Tran

Abstract Secondary mangrove forest, *Sonneratia caseolaris-Aegiceras corniculatum* stand, in the canal network adjacent to the shrimp culture ponds at the Lam River estuary was surveyed. Main objective of this study is to elucidate the relationship between species distribution and flooded duration by brackish water to apply for designing silvofishery systems to the shrimp culture ponds. Land height from datum line and mean inundated duration range from 1.1 m to 1.9 m and from 9.0 hours day^{-1} to 19.6 hours day^{-1}, respectively. The stand mainly contains three or more herbaceous species (associate species: *Cyperus malaccensis* and *Acanthus* spp., and one nonmangrove species: *Phragmites australis*) other than two woody species. Zones of *C. malaccensis* and *Acanthus* spp. distributing are overlapped in height and mean inundated duration, ranging from 1.1 m to 1.4 m, from 16.7 hours day^{-1} to 19.6 hours day^{-1}, from 1.2 m to 1.4 m, and from 16.7 hours day^{-1} to 18.7 hours day^{-1}, respectively. Since it is reported that *Acanthus* has more potential to retain nutrients than woody mangrove species, and *C. malaccensis* called mat grass is an economically valued plant, mangrove buffers with herbaceous species to silvofishery systems in this area are recommended.

Keywords Silvofishery system · Mangrove forest · Herbaceous species

K. Takahashi (✉) · T. T. Tran
School of Agriculture and Natural Resources, Vinh University, Nghe An, Vietnam

A. T. Nguyen, L. Hens (eds.), *Global Changes and Sustainable Development in Asian Emerging Market Economies Vol. 2*, https://doi.org/10.1007/978-3-030-81443-4_30

1 Introduction

Study site named Hung Hoa Commune, Vinh City, Nghe An Province, is located at the estuary of the Lam River, flowing through north-central Vietnam, and shrimp culture is one of the main industries in this area. With rapid expansion of shrimp culture ponds in the beginning of the 2000s, lagoon with mangrove forests had been decreased, which degraded regional ecosystem services. From 1982 to 2000, ca. 47,000 ha of salted area had been reduced especially due to changing the shrimp culture ponds, and ca. 1000 ha of shrimp culture ponds exist in Nghe An Province (as of 2001), accounting for 0.4% of total shrimp culture ponds in Vietnam (Phuong, 2014a). Considering this matter, the central government of Vietnam has launched strategy of Green Economic Development in the Lam River watershed including the estuary (Vietnam Socioeconomic Development Strategy 2011–2020); however, few concrete measures have been conducted.

Silvofishery systems in the shrimp culture ponds, i.e., mixture of mangrove silviculture and shrimp culture, have recently been adopted in many regions, e.g., Ngoc Hien District, Ca Mau Province at Mekong Delta (Clough et al., 2002), Semarang City, Indonesia (Hastuti & Budihastuti, 2017), and it is evaluated as a community-based shrimp culture model contributing for both regional economy and mangrove conservation (Syaiful, Mohammad, Iskhaq, & Andy, 2016). Moreover, mangrove forests provide several ecological services acting as filters for nutrients, sediments, and carbon sink (e.g., Department of Environment and Energy [DOEE], Australia, 2016; Mitra, 2020a); thus, silvofishery systems with mangrove buffers are expected to contribute to environmental restoration in the regions.

For introducing silvofishery systems to the shrimp culture ponds in Hung Hoa Commune, understanding mangrove stand structure in this region is essential to design the mangrove buffer and to determine water operation. Along the Lam River estuary, *Sonneratia caseolaris* is mainly dominated by *Acanthus* spp. and *Cyperus malaccensis* at low layers (Phuong, 2014b). Hung Hoa Commune has a canal network system flooded by brackish water adjacent to the shrimp culture area, and secondary mangrove forests of *S. caseolaris* with *Aegiceras corniculatum* are distributed.

With background mentioned above, this study aims to understand stand structure of secondary mangrove forest, *S. caseolaris-A. corniculatum* stand, especially elucidating relationship between mangrove/associate species distribution in the stand and inundated duration of brackish water in order to introduce and develop silvofishery systems to shrimp culture ponds.

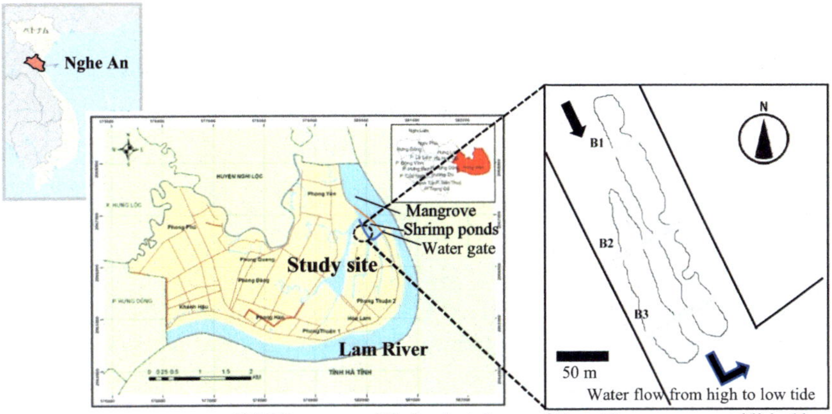

*Middle map: modified cadastral map of Hung Hoa commune published by Hung Hoa commune people's committee (2018).

Fig. 30.1 Study area and location of belt transects

2 Methodology

2.1 Study Area

Secondary mangrove forest in the study site is dominated by *S. caseolaris* and *A. corniculatum* in the canal network adjacent to the shrimp culture ponds at the Lam River estuary in Hung Hoa Commune, Nghe An Province, Vietnam (Fig. 30.1). It is regularly flooded by brackish water; estimated mean highest spring tide and mean lowest spring tide in 2019 are 2.7 m and 0.5 m, respectively (Center for Oceanography, 2018). This canal network has four gates to control water input and output from/to the Lam River for irrigation and flood prevention. The Lam River is an international river, originated in Laos flowing mainly through Nghe An Province, Vietnam, into the Gulf of Tonkin (website; background of key river basins, World Bank document). Climate of Vinh City, Nghe An Province, is affected by northeast monsoon; annual average temperature is 24.6 °C; the warmest month is June, 30.1 °C on average; the coldest month is January, 18.3 °C on average, and annual average rainfall is 1753 mm year^{-1}; minimum precipitation month is March, 48 mm on average; and maximum precipitation month is September, 445 mm on average (website; climate Vinh) (Fig. 30.2).

2.2 Belt Transects

To understand vegetation gradients corresponding to the land height from datum line and inundation duration, three belts, B1–B3 (B1: 18° 41′ 27″, N 105° 45′ 32″ E - 18° 41′ 26″ N, 105° 45′ 31″ E; B2: 18° 41′ 25″ N, 105° 45′ 33″ E - 18° 41′ 23″ N, 105°

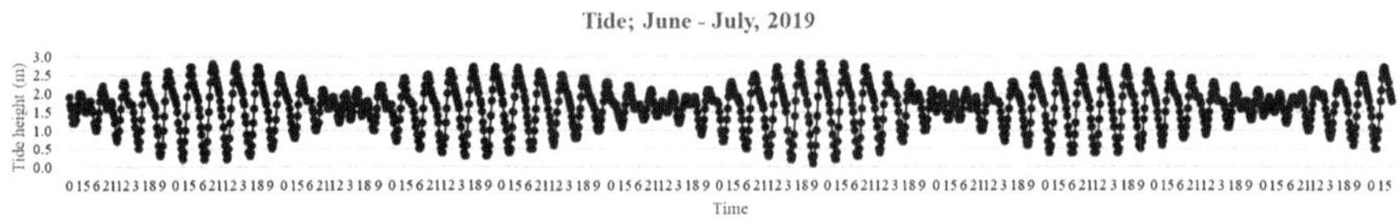

Fig. 30.2 Tidal change at the Lam River estuary (Cua Hoi; 18° 48′ N, 105° 46′ E): June–July 2019, cited from Center for Oceanography, 2018

45′ 32″ E; B3: 18° 41′ 23″ N, 105° 45′ 34″ E - 18° 41′ 21″ N, 105° 45′ 32″ E), were selected in the mangrove stands (Fig. 30.1). At each belt, center line was set up horizontally, and height from the center line to the land surface was measured basically every 2.5 m, and if water body existed, water depth was also recorded. After measuring height, it was converted to the land height based on the relationship between real water depth measured and tide height (Center for Oceanography, 2018). For understanding vegetation gradients, 2-m perpendicular lines to the center line on both sides were set up every 5 m, and all the species occurring and their land coverages by the crowns of trees or plants body of herbs were recorded in the 4 x 5 m rectangular. Belt transect surveys were conducted on June 9 and July 12–14, 2019.

2.3 *Inundated Duration*

Since the vegetation occurs raging approximately from 1.0 m to 2.0 m in land height, average inundated duration per day of this range in unit of 0.1 m was estimated by comparing the land height with change in tide level at Cua Hoi: 18° 48′ N, 105° 46′ E (Center for Oceanography, 2018), i.e., cumulative inundated time for each land height in year 2019 divided by 365 days.

3 Results

3.1 *Vegetation Gradients on the Belt Transects: B1–B3*

Figure 30.3 displays vegetation gradients on the three belts with cross sections of the lands.

B1 is located at the most upstream among three belts (Fig. 30.3a), and vegetation length from the left side to the right side is 31.3 m. It includes 0.1–0.2 m higher spots, but mostly flat at 1.4–1.5 m in land height. *A. corniculatum* and *P. australis* are dominant through the belt, except for *C. malaccensis* being dominant at around 1.2 m in land height on the right side.

B2 is at middle of three belts (Fig. 30.3b), and vegetation length from the left side to the right side is 45.0 m. It has creek-like shallow watercourse having less

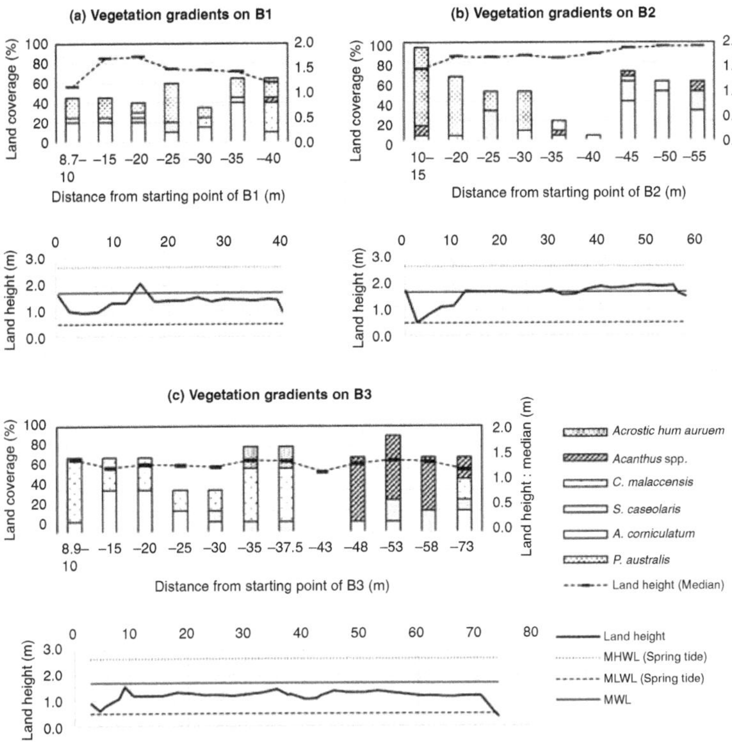

Fig. 30.3 Vegetation gradients on the belts: B1–B3 with cross sections

vegetation. It is 0.2–0.3 m higher than B1, and it is almost flat, but the left side land is 0.1–0.2 m lower than the right side. On the left side, *P. australis* is dominant, while on the right side, *A. corniculatum* is dominant. *S. caseolaris* is scattered through the belt.

B3 is located at the most downstream among three belts (Fig. 30.3c), and vegetation length from the left side to the right side is 64.1 m including no vegetation zone at creek-like shallow watercourse. Through the belt, land height is 1.1–1.4 m lower than upstream belts: B1 and B2. On the left side, *A. corniculatume* with *C. malaccensis* is dominant, while on the right side *S. caseolaris* with *Acanthus* spp. is dominant. In the middle part, there is a slightly higher spot where *Acrostichum aureum* is inhabiting.

3.2 Relationship between Species Distribution and Land Height, and Inundated Duration

Land height data on *A. corniculatum*, *S. caseolaris*, *C. malaccensis*, *Acanthus* spp., and *P. australis* distributing are plotted in Fig. 30.4, and distribution pattern of each species with inundated duration (Fig. 30.5) is summarized in Table 30.1.

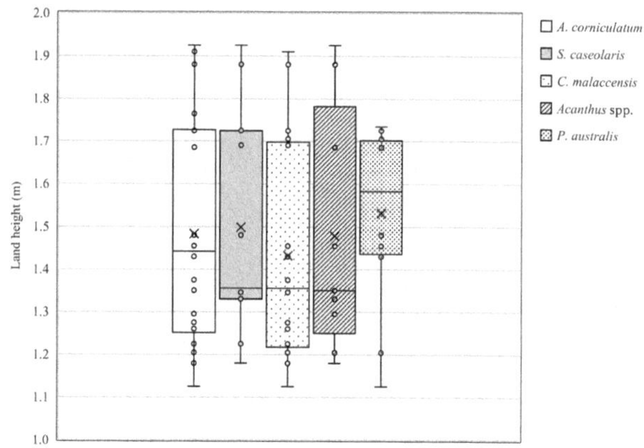

Fig. 30.4 Vertical distribution pattern of five species

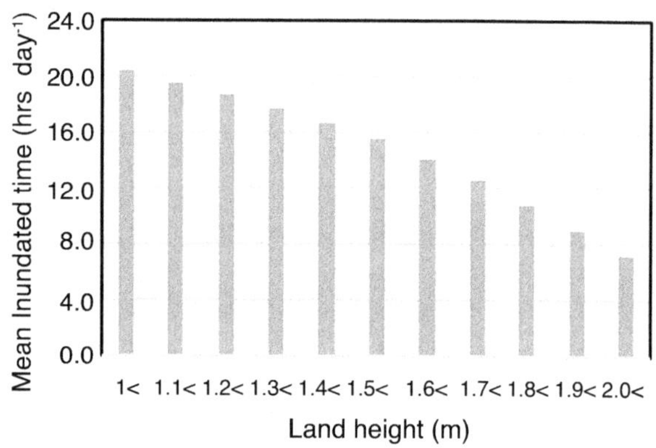

Fig. 30.5 Inundated duration at each land height

Table 30.1 Summary of distribution pattern of five species

Species	Land height from DL (m)				Mean inundated duration (hours day^{-1})
	Min	Max	Mean	Intensive distributed Zone	
A. corniculatum	1.1	1.9	1.5	Relatively homogeneous	9.0–19.6
S. caseolaris	1.2	1.9	1.5	1.3–1.4	16.7–17.7
C. malaccensis	1.1	1.9	1.4	1.1–1.4	16.7–19.6
Acanthus spp.	1.2	1.9	1.5	1.2–1.4	16.7–18.7
P. australis	1.1	1.5	1.5	1.4–1.7	12.7–16.7

Distribution range in land height of these five species is not significantly different, but *P. australis* which is neither mangrove nor associate species (Phuong, 2014b), tends to distribute intensively at 1.4–1.7 m in land height, with inundation duration of 12.7–16.7 hours day^{-1}. Land heights of other four species inhabiting range wider than one of *P. australis*, from 1.1 to 1.9 m in land height, with inundation duration of 9.0–19.6 hours day^{-1}, those species except for *A. corniculatum*, intensively distribute rather lower zones, especially *S. caseolaris* intensively distributes from 1.3 m to 1.4 m in land height, inundation duration of 16.7–17.7 hours day^{-1}. Herbaceous species, *C. malaccensis* and *Acanthus* spp., tend to intensively distribute on the lower zones than two woody mangrove species, from 1.1 m to 1.4 m in land height, with inundation duration of 16.7–19.6 hours day^{-1}, and from 1.2 m to 1.4 m in height, with inundation duration of 16.7–18.7 hours day^{-1}, respectively. The bottom line of intensive distribution zone of *Acanthus* spp. is shifted 0.1 m higher than one of *C. malaccensis*.

In the zone of 1.1–1.4 m in land height equivalent to B3 land height, land cover by *C. malaccensis* is correlated with land height, $r = 0.91$, $p < 0.01$, but one by *Acanthus* spp. is not correlated with land height, $p = 0.09$ (Fig. 30.6).

4 Conclusions and Discussion

4.1 Stand Structure

S. caseolaris and *A. corniculatum* stand in the study site is tide-dominated type of mangrove forest (Woodroffe, 1992), regularly flooded by brackish water except for the gate being operated. Its land height ranges from 1.1 m to 1.9 m, mean inundated duration is 9.0–19.6 hours day^{-1}, and mean water depth at spring tide is 0.8–1.6 m. It includes nonmangrove/associate species, *P. australis* being distributed intensively at MWL, 1.7 m. *S. caseolaris* is typical mangrove species at river estuaries (Khoon et al., 2004) in southeast Asia often with *A. corniculatum* (Phuong, 2014b) in north-central Vietnam.

The stand includes two or more herbaceous mangrove/associate species, *C. malaccensis,* associate species, and *Acanthus* spp., mangrove spp. (Phuong, 2014b). Their intensive distribution zones are overlapped. Mean inundation duration and mean water depth at spring tide of *C. malaccensis* are 16.7–19.6 hours day^{-1} and 1.3–1.6 m, respectively, while ones of *Acanthus* spp. are 16.7–18.7 hours day^{-1} and 1.3–1.5 m, respectively. Previous study from Can Gio, Vietnam, reported that *Acanthus ebracteatus* forms community with *C. malaccensis* and other species at the zone flooded by tides 1.5–2.0 m in height (Hong, 2000), corresponding to the observation in the study site. In the intensive distribution zone of *C. malaccensis,* 1.1–1.4 m in land height, expansion of the habitat area along with increasing of land height was observed, $r = 0.91$, $p < 0.01$.

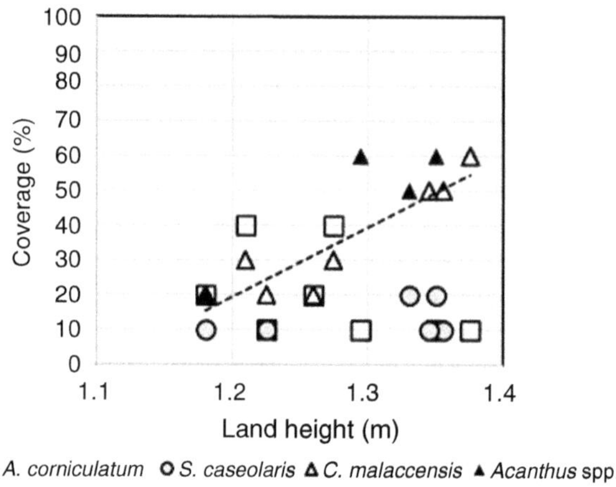

Fig. 30.6 Land height versus coverage

4.2 *Toward Introduction of Silvofishery Systems to Shrimp Culture Ponds*

Mangrove buffers in silvofishery systems are expected to provide ecological functioning of filter for nutrients and sediments. In fact, nitrogen and phosphorus retention effect by mangrove species was reported from silvofishery systems in Indonesia (Hastuti & Budihastuti, 2017).

4.2.1 Filter for Nutrients

Focusing on nutrient retention ability of each species, few studies have been conducted, but Mitra (2020b) reported concentrations of nutrients in senesced leaves of *Acanthus* spp. are higher than *A. corniculatum* and *S. caseolaris* in both nitrogen and phosphorus concentration, indicating that *Acanthus* spp. have higher potential of nutrient retention. Nutrient retention abilities depend on various environmental factors, but mixture of woody mangrove species and *Acanthus* spp. could have more advantage in terms of nutrient retention.

4.2.2 Filter for Sediments

In contrary to nutrient retention, plenty of studies on sediment retention effect of mangrove forests have been reported not only for elucidating filtering effect but also for evaluating erosion prevention effect. Previous study from Ba Lat estuary of the Red River, Vietnam reported that sediments trapped by *A. corniculatum* with *Acanthus ilicifolius* in dry season and wet season are 2.94 g cm^{-2} year and

3.46 g cm^{-2} year, respectively (Santen, Augustinus, Janssen-Stelder, Quartel, & Tri, 2017). Regarding relationship between mangrove/associate species inhabiting in the study site and sedimentation rate, few studies were reported, but it was reported that herbaceous species have rather higher potential of sediment trapping (Willemsen, Horstman, Borsje, Friess, & Dohmen-Janssen, 2016).

4.2.3 Economic Aspects

In order to operate mangrove buffers sustainably, it would be better for buffers themselves should produce some economically valued goods. In this area, *C. malaccensis* called mat grass has traditionally been cultivated as weaving materials; even in year 2018, 17 households still cultivate it and cultivation area of 1.6 ha provides income of 20–30 USD per person month (hearing from Hung Hoa Commune People's Committee 2019). Thus, *C. malaccensis* is appropriate species for mangrove buffer component considering regional economy. It is expected to process it as eco-friendly commodities to circulate in markets.

With reasons mentioned above, buffers of *S. caseolaris and A. corniculatum* with herbaceous species are recommended to set up to the shrimp culture ponds in Hung Hoa Commune.

Acknowledgments The authors specially thank Huong Hoa Commune People's Committee, Nghe An Province, for providing valuable information and kind support.

References

Center for Oceanography (2018). Cua Hoi. In Tide tables 2019 vol.1; ed Quang D.N.; Publishing house of natural science and technology: 18 Hoang Quoc Viet, Cau Giay, Hanoi, Vietnam; 93–117.

Climate Vinh. (2019). Retrieved August 29, 2019, from en.climate-data.org.

Clough, B., Johnston, D., Xuan, T. T., Phillips, M., Pednekar, S. S., Thien, N. H., Dan, T. H., & Thong, P. L. (2002). Silvofishery farming systems in Ca Mau Province, Vietnam. In NACA, WWF, & FAO Consortium Program on Shrimp Farming and the Environment (Eds.), *Report prepared under the World Bank*. Published by the Consortium. 73 pp.

Department of Environment and Energy [DOEE], Australia. (2016). *Coastal wetlands—Mangroves and saltmarshes*. Retrieved June 19, 2019, from environment.gov.au.

Hastuti, E. D., & Budihastuti, R. (2017). Nutrient accumulation in the sediment of silvofishery ponds in Semarang. *Makara Journal of Science, 21*(4), 195–201.

Hong, P. N. (2000). *Effects of mangrove restoration and conservation on the biodiversity and environment in Can Gio district, Ho Chi Minh City Phan Nguyen Hong*, International Workshop Asia-Pacific Cooperation on Research for Conservation of Mangroves, 26–30 March, 2000 -Okinawa, Japan; 14 pp.

Khoon, G. W., Aksornkoae, S., Tri, N. H., Vongwattana, K., Fan, H., Santoso, N., Barangan, F., Havanond, S., Sam, D. D., Pernetta, J. (2004). Inundation regime. In mangrove in the South China Sea; UNEP/GEF project co-ordinating unit, United Nations environment programme: UN building, 9th floor block a, Rajdamnern avenue, Bangkok 10200, Thailand; 6.

Mitra, A. (2020a). Ecosystem services of mangroves: An overview. In *Mangrove forests in India* (pp. 1–30). Cham: Springer Nature.
Mitra, A. (2020b). Mangroves: Nutrient retention box. In *Mangrove forests in India* (pp. 87–114). Cham: Springer Nature.
Phuong, V. T. (2014a). Geographical distribution of mangrove forest in Vietnam. In *National report on mangroves in South China Sea Vietnam*; Research Centre for Forest Ecology and Environment: Dong Ngac, Tu Liem, Hanoi, Vietnam; 1–4.
Phuong, V. T. (2014b). Species distribution and formation. In *National report on mangroves in South China Sea Vietnam*; Research Centre for Forest Ecology and Environment: Dong Ngac, Tu Liem, Hanoi, Vietnam; 5–14.
Santen, P. V., Augustinus, P. G. E. F., Janssen-Stelder, B. M., Quartel, S., & Tri, N. H. (2017). Sedimentation in an estuarine mangrove system. *Journal of Asian Earth Science, 29*, 566–575.
Syaiful, E., Mohammad, R. R., Iskhaq, I., & Andy, M. (2016). Community-based mangrove forests conservation for sustainable fisheries. *Journal Silvikultur Tropika, 7*(3), 42–47.
Willemsen, P. W. J. M., Horstman, E. M., Borsje, B. W., Friess, D. A., & Dohmen-Janssen, C. M. (2016). Sensitivity of the sediment trapping capacity of an estuarine mangrove forest. *Geomorphology, 273*, 189–201.
Woodroffe, C. (1992). Mangrove sediments and geomorphology. In A. I. Robertson & D. M. Alongi (Eds.), *Tropical mangrove ecosystems* (pp. 7–41). American Geophysical Union.
World Bank Document. (2019). Background of key river basins, Vietnam natural hazards project.. Retrieved October 15, 2019, from http://documents.worldbank.org/curated/en/323821468127184247/pdf/E29930REVISED00ronmental0Background.pdf.

Chapter 31
Evaluation of the Effectiveness of Controlling Sheath Blight Disease in Rice Caused by *Rhizoctonia Solani* under Greenhouse Conditions by Applying Biofungicides

Ha Thi Thanh Tuyen and Luu Ba Hoa

Abstract This research was conducted in the greenhouse of the Plant Protection Department, College of Agriculture and Applied Biology, Can Tho University. The aim of this study was to find out the timing application of biofungicides (Validan 5SL and Trico-DHCT) for best effectiveness in controlling sheath blight disease (*Rhizoctonia solani*) on rice. The results found that all of the biofungicide treatments were capable of controlling sheath blight disease on rice under greenhouse conditions with different levels of effectiveness. Among these bio-agents, Trico-DHCT treatment sprayed before pathogen inoculation had higher disease controlling effectiveness than the other treatments.

Keywords Sheath blight disease · Controlling disease · Biofungicide

1 Introduction

Agriculture plays a very important role in the Vietnamese economy in which food production is the most important sector to contribute to the Vietnamese GDP. Vietnam is the second largest rice exporter in the world and also plays an important role in world food security. It is located in a tropical climate, which has a lot of conditions for paddy cultivation and has also cultivated favorable conditions for the development of many pests. Sheath blight disease (SBD) is one of main factors in the damage of rice-growing areas across Vietnam. This disease can cause high yield losses,

H. T. T. Tuyen · L. B. Hoa (✉)
Kien Giang University, Kien Giang, Vietnam
e-mail: lbhoa@vnkgu.edu.vn

A. T. Nguyen, L. Hens (eds.), *Global Changes and Sustainable Development in Asian Emerging Market Economies Vol. 2*, https://doi.org/10.1007/978-3-030-81443-4_31

a massive 25–50% (Agarwal, Mortensen, & Mather, 1989; Gnanamanickam, 2009; Ou, 1985).

In recent years, the trend in rice production of following monocultural production, increasing the number of crops, and using high yielding varieties, are causes of more and more diseases. Besides, SBD is one of the most signficant ones, destroying the rice-growing areas of states, affecteding crop yield, environment, and human health when pesticides are applied indiscriminately. In the countryside, farmers always apply pesticides in excess and continuously to put high pressure on the pathogen resistance in the crop. In order to control SBD, many methods had been researched, until 2010, when the American Institute of Agriculture discovered the some genes have ability of disease resistance to SBD. However, that was applied in research and not in the wider rice-planting areas.

Currently, there are many biofungicides for controlling SBD on rice, some of which were made from biologically active ingredients consisting of Validan 5SL (Validamycin A) and Trico-DHCT, which are two important active ingredients in plant disease control. This chapter focuses on the effect and timing of the application of these biofungicides on control SBD in rice..

2 Methodology

2.1 Objective

The objective was to determine the timing of the application of biofungicides for effectiveness on controlling SBD in greenhouse conditions.

2.2 Design Experiment

Experiments followed the completely randomized design with two factors including 12 treatments and 4 replications. Factor A consists of three biofungicides such as Validan 5SL (Validamycin A), Trico-DHCT, and Evitin 50SC (Hexaconazole 50 g/l). Factor B includes three periodic treatments (spraying biofungicides at 3 days before pathogenetic inoculation, spraying biofungicides at 3 days after pathogenetic inoculation, and combined spraying before and 3 days after inoculation with the pathogen).

Inoculating method: the Jasmine 8 variety of paddy was sown in each pot under greenhouse conditions and inoculated at 40 days after sowing. The fungus *Rhizoctonia solani* cultured in a PDA medium, multiplied the microbial masses on rice husk medium, then 5 g of medium containing *Rhizoctonia solani* were added to each paddy bust and were transferred to a greenhouse under ambient environmental conditions at high humidity.

Application of biofungicides and chemical fungicide: following the recommendations of the manufacturer.

Data recording:

- Percentage of infection (%) = (the number of infected tillers/the total of tillers observed) × 100%.
- Relative lesion height (RLH) = (lesion length/plant length) × 100%.
- Disease index (DI)

$$DI(\%) = \frac{\left[(1\times n1)+(3\times n3)+(5\times n5)+(7\times n7)+(9\times n9)\right]}{tn\times 9}\times 100\%$$

where: n(1), n(3), n(5), n(7), and n(9) = numbers of plants showing a reaction in a scale (1), (3), (5), (7) and (9) respectively (IRRI, 1996); tn = total number of plants scored.

2.3 *Data Analysis*

The SPSS software was applied for calculations using Duncan's test.

3 Results

3.1 *Percentage of Infection*

3.1.1 Percentage of Infection at 5 Days after Inoculation

The ability of SBD prevention and control on Jasmine 85 variety by using biofungicides (Validan 5SL and Trico-DHCT) and chemical fungicide (Evitin 50SC) under greenhouse condition was evaluated by percentage infected, disease index (DI), and the yield components at harvest (Table 31.1).

The result showed that all three treatments reduced SBD through the percentage of infected tiller and significant difference compared with the control at the 0.01 level.

Spraying was analyzed, with combined before and after pathogen inoculation (averaging 34.86% of tiller infected) lowered the percentage of infected tiller significantly more than the other treatments.

As for the interaction between fungicides and the timing of application, there were two biofungicide treatments (Validan 5SL and Trico-DHCT), with spraying combined before and after pathogen inoculation lowering the percentage of infected tiller significantly, though nonsignificantly with Evitin 50SC.

Table 31.1 Percentage of infected tillers (%) caused by sheath blight disease at 5 days after inoculation

Timing application (B) Factor (A)	Percentage of infected tiller (%)			Average factor (A)
	Spraying before inoculation	Spraying before and after inoculation	Spraying after inoculation	
Validan 5SL	45.72 abc	31.78 cd	56.58 ab	44.69 BC
Trico-DHCT	54.68 ab	34.30 bcd	64.44 a	51.14 B
Evitin 50SC	43.49 abc	13.10 de	44.98 abc	33.86 C
Control	60.26 a	68.18 a	67.10 a	65.18 A
Average factor (B)	51.04 A	36.84 B	58.28 A	
Level of significance	F(A)[a]	F(B)[a]	F(AB) [ns]	
Coefficient of variation (%)	43.51			

Means in the same column followed by the same letter are not significantly different according to Duncan's test
Data were converted by arcsine $\sqrt{x}$ before analysis
ns nonsignificant difference
[a]Significant difference at the 0.01 level

Table 31.2 Percentage of infected tiller (%) caused by sheath blight disease at 7 days after inoculation

Timing application (B) Factor (A)	Percentage of infected tiller (%)			Average factor (A)
	Spraying before inoculation	Spraying before and after inoculation	Spraying after inoculation	
Validan 5SL	54.31 ab	32.43 bc	54.77 ab	47.17 B
Trico-DHCT	59.30 a	40.79 ab	65.30 a	55.13 B
Evitin 50SC	49.14 ab	19.33 cd	45.28 ab	37.92 C
Control	73.04 a	79.90 a	78.97 a	77.30 A
Average factor (B)	58.95 A	43.11 B	61.08 A	
Level of significance	F(A)[a]	F(B)[a]	F(AB)[a]	
Coefficient of variation (%)	46.91			

Means in the same column followed by the same letter are not significantly different according to Duncan's test
Data were converted by arcsin $\sqrt{x}$ before analysis
[a]Significant difference at the 0.01 level

3.1.2 Percentage of Infection at 7 Days after Inoculation

All treatments gave a percentage of infected tiller ranging from 37.92% to 77.30%. The treatment applied different fungicides that were lower than and significantly different by the percentage of infected tiller compared with the control (Table 31.2).

At 5 days after inoculation, the method of using different fungicides by spraying with combined before and after pathogen inoculation was lower than the spraying applied before pathogen inoculation or spraying after pathogen inoculation by the percentage of infected tiller.

The interaction between the different fungicides and the timing of application was considered.

Only the treatment that was applied with Validan 5SL by spraying with a combination before and after (32.43% infection) was better than the control (infection 61.97%) and was equivalent to treatment with Evitin 50SC (infection 19.33%).

3.1.3 Percentage of Infection at 9 Days after Inoculation

The result shown in Table 31.3 is of treatment with Trico-DHCT and Validan 5SL, which reduced SBD by 64.99% and 60.52% infected tiller respectively. However, these treatments were higher than the treatment with Evitin 50SC (49,23%), but lower than the control (98.09%).

With regard to the timing of application, the treatment of spraying with combined before and after (average of 53.96% infection) continued to show a higher percentage of infected tiller and a significant difference compared with spraying before pathogen inoculation or spraying after pathogen inoculation (a massive 74.53% and 76.13% respectively).

Analyzing the interaction between some fungicides and the timing of application, the best treatment consisted of Trico-DHCT and Validan 5SL, which had a lower percentage of infection than the other timing of application by spraying a combination of before and after pathogen inoculation the same at 5 days after inoculation.

Percentage of infection at 14 days after inoculation.

Table 31.3 Percentage of infected tiller (%) caused by sheath blight disease at 9 days after inoculation

Timing application (B) Factor (A)	Percentage of infected tillers (%)			
	Spraying before inoculation	Spraying before and after inoculation	Spraying after inoculation	Average factor (A)
Validan 5SL	71.90 ab	38.54 c	71.11 ab	60.52 B
Trico-DHCT	72.72 ab	49.56 b	72.70 ab	64.99 B
Evitin 50SC	57.89 cd	27.77 d	62.03 bc	49.23 C
Control	95,60 a	100.00 a	98.68 a	98.09 A
Average factor (B)	74.53 A	53.96 B	76.13 A	
Level of significance	F(A)[a]	F(B)[a]	F(AB)[a]	
Coefficient of variation (%)	16.41			

Means in the same column followed by the same letter are not significantly different according to Duncan's test

Data were converted by arcsine $\sqrt{x}$ before analysis

[a]Significant difference at the 0.01 level

Table 31.4 Percentage of infected tiller (%) caused by sheath blight disease at 14 days after inoculation

Timing application (B) Factor (A)	Percentage of infected tiller (%)			Average factor (A)
	Spraying before inoculation	Spraying before and after inoculation	Spraying after inoculation	
Validan 5SL	78.16 b	47.02 c	72.91 bc	66.04 BC
Trico-DHCT	80.63 b	58.43 b	76.65 b	71.90 B
Evitin 50SC	67.57 c	37.90 d	63.80 d	56.42 CD
Control	95.60 a	100.00 a	98.68 a	98.09 A
Average factor (B)	80.49 A	60.84 B	78.01 A	
Level of significance	F(A)[a]	F(B)[a]	F(AB)[a]	
Coefficient of variation (%)	12.20			

Means in the same column followed by the same letter are not significantly different according to Duncan's test
Data were converted by arcsine $\sqrt{x}$ before analysis
[a]Significant difference at the 0.01 level

The figures for two treatments such as Trico-DHCT and Validan 5SL (percentage inoculation of 71.90% and 66.04%, respectively) were lower than for the control (98.09%), and there was a significant difference at the 0.01 level. Besides, the treatment with Validan 5SL showed a nonsignificant difference in percentage of infected tiller (Table 31.4).

3.2 *Disease Index*

3.2.1 Disease index at 5 Days after Inoculation

The results in Table 31.5 showed that, using two biofungicides and a chemical fungicide reduced the DI of SBD in rice through the evaluation of the DI between treatments.

At this time, the average DI of all treatments such as spraying before, spraying after, and spraying before and after pathogen inoculation (DI of 5.40%, 6.22%, and 5.30% respectively) showed no significant difference. There was clearly no effect of treatment according to the timing of application.

In general, the majority of treatments showed that reducing ability was expressed by the DI and was different from the control. In the timing of application, the DI of Trico-DHCT (5.83%) was equivalent to that of the Validan 5SL (4.00%) but higher and different from that of the Evitin 50SC (3.39%). At the next spraying after pathogen inoculation, Validan 5SL had a DI of 4.61, equivalent to Evitin 50SC (3.23%). In the spraying before and after pathogen inoculation, the DI of Validan 5SL treatment was low (4.50%) and there was no significant difference compared with Evidin

Table 31.5 Disease index (%) of sheath blight disease at 5 days after inoculation

Timing application (B) Factor (A)	Disease index			
	Spraying before inoculation	Spraying before and after inoculation	Spraying after inoculation	Average factor (A)
Validan 5SL	4.00 cd	4.50 de	4.61 cd	4.17 C
Trico-DHCT	5.83 bc	5.00 bcd	6.25 abc	5.69 B
Evitin 50SC	3.39 de	3.37e	3.23 de	3.33C
Control	8.37 a	8.32 a	8.30 a	8.33 A
Average factor (B)	5.40	5.30	6.22	
Level of significance	F(A)[a], F(B)[ns], F(AxB)[a]			
Coefficient of variation (%)	14.22			

Means in the same column followed by the same letter are not significantly different according to Duncan's test
Data were converted by arcsin $\sqrt{x}$ before analysis
ns nonsignificant difference
[a]Significant difference at the 0.01 level

50SC (3.37%) and Trico-DHCT; however, Trico-DHCT treatment gave a higher DI and was significantly different from the Evitin 50SC treatment.

3.2.2 Disease Index at 7 Days after Inoculation

The treatments were lower with regard to the average DI than the control. In particular, the treatment with Validan 5SL (DI: 12.41%) showed a non-significant difference compared with the treatment with Evitin 50SC (DI: 12.08%) and signficantly lower than the treatment with Trico-DHCT (DI: 26.25%; Table 31.6).

At this time, there was a difference between the average timing of application and the average DI at spraying before pathogen inoculation (16.11%), and spraying before and after pathogen inoculation (15.68%) showed that effective treatment of the disease was not significantly different from the others, that is, it was higher and different from the treatment of spraying after pathogen inoculation (20.15%).

Similar to the previous time, the treatments showed that the effect of different fungicides was at different levels and higher than the control. Specifically, the spraying before pathogen inoculation, where all three treatments showed that the effect on controlling SBD was similar (DI: 11.25%, 14.58%, and 12.28% respectively) and significantly different compared with the control. In spraying after inoculation, only Validan 5SL treatment with DI was 14.58%, a nonsignificant difference compared with Evitin 50SC treatment (DI: 12.88%). In the spraying combination before and after pathogen inoculation, Validan 5SL treatment (DI: 11.42%) showed a high effect, as well as Tricho-ĐHCT (DI: 14.17%) and Evitin 50SC treatment (DI: 11.08%).

Table 31.6 Disease index (%) of sheath blight disease at 7 days after inoculation

Timing application (B) Factor (A)	Disease index (%) Spraying before inoculation	Spraying before and after inoculation	Spraying after inoculation	Average factor (A)
Validan 5SL	11.42 bc	11.25 bc	14.58 b	12.41 C
Trico-DHCT	14.58 b	14.17 b	26.25 a	18.33 B
Evitin 50SC	12.28 bc	11.08 c	12.88 bc	12.08 C
Control	26.14 a	26.25 a	26.36 a	26.25 A
Average factor (B)	16.11 B	15.68 B	20.15 A	
Level of significance	F(A)[a], F(B)[a], F(AxB)[a]			
Coefficient of variation(%)	7.59			

Means in the same column followed by the same letter are not significantly different according to Duncan's test
Data were converted by arcsin $\sqrt{x}$ before analysis
[a]Significant difference at the 0.01 level

Table 31.7 Disease index (%) of the affected sheath blight disease at 9 days after inoculation

Timing application (B) Factor (A)	Disease index (%) Spraying before inoculation	Spraying before and after inoculation	Spraying after inoculation	Average factor (A)
Validan 5SL	22.80b	21.50 b	23.20 b	22.50 B
Trico-DHCT	21.77b	21.57 b	33.33a	25.56 B
Evitin 50SC	18.75 bc	15.42 c	24.17b	19.45 C
Control	37.95 a	37.81 a	37.99 a	37.92 A
Average factor (B)	25.32 B	24.08 B	29.67 A	
Level of significance	F(A)[a], F(B)[a], F(AxB)[a]			
Coefficient of variation (%)	8.73			

Means in the same column followed by the same letter are not significantly different according to Duncan's test
Data were converted by arcsin $\sqrt{x}$ before analysis
[a]Significant difference at the 0.01 level

3.2.3 Disease Index at 9 Days after Inoculation

At this time, the Validan 5SL treatment and Trico-DHCT treatment had an effect on controlling the fungi by an average DI of 22.50% and 25.56% respectively, and was lower and showed a significant difference compared with the control (37.92%), but this treatment was higher than Evitin 50SC treatment (19.45%) (Table 31.7).

About the average of the timing application, all the treatments showed efficiency compared with the control. Spraying of the combination before and after pathogen inoculation and spraying before pathogen inoculation were equally effective in managing the disease; the DI was 25.32% and 24.08% respectively and was higher than spraying after pathogen inoculation (29.67%).

The treatments in the interaction between the applied fungicides and the timing of application were analyzed. Three treatments of applied fungicides had a high average DI of 22.80%, 21.77%, and 18.75% respectively; they were lower and significantly different from the control at 37.95%. In the spraying after pathogen inoculation, Validan 5SL treatment was lower with an average DI of 24.17% and was not significantly different from the Evitin 50SC treatment (23.20%). In the spraying of the combination before with after pathogen inoculation, Validan 5SL treatment and Tricco-DHCT treatment continued to maintain the ability to control the disease only after Evitin 50SC treatment, with the average DI of 21.50% and 21.57% respectively, which was lower than and significantly different from the control (37.81%).

3.2.4 Disease Index at 14 Days after Inoculation

The inhibitory effect of these biofungicide and chemical fungicide treatments was shown to be lower on average DI (Disease index) than the control treatment 51.25%; Table 31.8).

At the time of treatment, the average timing of application affected the inhibition of fungal pathogens. Among them, the spraying of a combination before with after pathogen inoculation has the highest effectiveness (36.50%). However, a nonsignificant difference compared with the spraying before pathogen inoculation (38.26%) was found, but it was significantly different from spraying after pathogen inoculation (41.88%).

Similar to the previous time, analysis of the interaction between the applied fungicide and the timing of application showed that three treatments reduced SBD with

Table 31.8 Disease index (%) of the affected sheath blight disease at 14 days after inoculation

Timing application (B) Factor (A)	Disease index (%)			
	Spraying before inoculation	Spraying before and after inoculation	Spraying after inoculation	Average factor (A)
Validan 5SL	34.07 b	33.27 b	35.17 b	34.17 CD
Trico-DHCT	37.50 b	35.83 b	47.08 a	40.14 B
Evitin 50SC	30.83 b	25.75 c	34.75 b	30.44D
Control	51.00 a	50.25 a	52.50 a	51.25 A
Average factor (B)	38.26 AB	36.50B	41.88 A	
Level of significance	F(A)[a], F(B)[a], F(AxB)[a]			
Coefficient of variation (%)	6.81			

Means in the same column followed by the same letter are not significantly different according to Duncan's test

Data were converted by arcsine $\sqrt{x}$ before analysis

[a]Significant difference at the 0.01 level

the average DI ranging from 30.83 to 34.17%. Besides, this treatment was lower than and significantly different compared with the control (51.00%). In spraying after pathogen inoculation, the Validan 5SL treatment DI (34.75%) was not significantly differeny compared with the Evitin 50SC treatment (34.75%). In spraying of the combination before with after pathogen inoculation, the Evitin 50SC treatment had the best effect on controlling SBD, the second treatment of Validan 5SL and Trico-DHCT maintained the effect well and was lower than the control (50.25%).

4 Conclusions and Discussion

Validan 5SL and Trico-DHCT had affected the management of SBD significantly under greenhouse conditions with different timing of applications. Among them, spraying of Validan 5SL before pathogen inoculation or before and after pathogen inoculation and spraying of Trico-DHCT before and after pathogen inoculation showed higher effectiveness than the other treatments. In addition, the effect of Validan 5SL showed no significant difference Evitin 50SC with regard to the time observation.

In short, the evaluation by percentage of infection and DI at the observable time illustrated that the biofungicide affected inhibition of *Rhizoctonia solani* at different levels.

At 5 and 14 days after inoculation, the Validan 5SL treatment reduced the affected tiller caused by SBD and was not significantly different from the Evitin 50SC treatment. In addition, at 5, 7, and 14 days after inoculation, the Validan 5SL treatment also showed the controlling effect was high compared with Evitin 50SC according to evaluation of the DI. That is, two fungicides *(Rhizoctonia solani and Trichoderma spp.)* contain the antifungal agent *Rhizoctonia solani*, which is commonly used in production today. Vailidan 5SL fungicide contains Validamycin A, which is produced from the *Streptomyces* spp., that is an antibiotic for controlling Validamycin A on rice and other crops (Vu Trieu Man, 2007a, b; Nguyen Manh Chinh et al., 2016). In addition, the Evitin 50SC fungicide contains active hexaconazole, a fungicidal antibiotic that prevents and treats many diseases, including SBD in rice (Nguyen Manh Chinh et al., 2016). Thus, in this study two fungicides were used, Validan 5SL and Evitin 50SC, which had an effect on treating SBD caused by *Rhizoctonia solani*.

Moreover, using Trico-DHCT also reduced the tiller affected by SBD compared with the control and a nonsignificant difference compared with Vailidan 5SL fungicide, although this treatment had a lower effect than Evitin 50SC fungicide. Prasad and Kumar (2011) reported spraying with *Trichoderma* spp. to control SBD on rice. The result was that the percentage of tiller that was affected by *Rhizoctonia solani* was lower than the untreated control and was higher than the control treated by chemical fungicide.

In addition, the investigation of the timing of application has been shown to be effective against pathogen suppression, but there was an interaction with each fungicide at different levels. Specifically, spraying before pathogenic inoculation had more of an effect on disease management than spraying at 9 days after pathogen inoculation, the spraying of the combination before with after pathogen inoculation to achieve control of SBD on rice by 14 days after inoculation.

Treatment with Validan 5SL or Trico-DHCT before pathogen inoculation showed a decrease in the percentage of infection of the affected tiller at 5, 7, and 9 days after inoculation, but the DI was not significantly different from the Evitin 50SC fungicide at 7, 9, and 14 days after inoculation. Fungal *Trichoderma* spp. grow much faster than other fungi (Saksena, 1960). So when they are treated before pathogen inoculation, *Trichoderma* ssp. have much time as well, as they are easy to grow and increase the titer rapidly enough compared with *Rhizoctonia solani*, which was inoculated later. Klein and Eveleigh (1998) reported that *Trichoderma* spp. are directly attacked, and the harmful fungi curl up and release chitinase enzymes that destroy the cell walls of the harmful fungi, forming small molecules that are easily absorbed, which helps plants to resist disease.

Validan 5SL fungicide and Tricho-DHCT fungicide in the spraying combination before with after pathogen inoculation showed the ability to limit the percentage of affected tiller. This result is the same as that in the study by Nguyen Van Hung (2010) in which the biological control of R. *solani* was caused by using *Trichoderma viride*. The treatments (experiments) with *T. viride* before and at the same time with *R. solani* showed a lower infection rate than when using only *T. viride* after *R. solani*

Fig. 31.1 The level of sheath blight disease on rice at 14 days after inoculation. A Control; B Spraying with Trico-DHCT before pathogen inoculation; C Spraying with Validan 5SL before inoculation; D Spraying with Validan 5SL combining before with after pathogen inoculation; E Spraying with Evitin 50SC combining before with after pathogen inoculation; F Spraying of Trico-DHCT combining before and after pathogen inoculation

inoculation. On the other hand, evaluation of the DI of the two bio-fungicides Validan 5SL and Tricco-DHCT with the spraying combination before with after pathogen inoculation showed that the ability to control is maintained until 14 days after inoculation; Validan 5SL has reduced the DI and was not significantly different compared with Evitin 50SC at 5 and 7 days after inoculation. This reason may be explained by the fact that both Validan 5SL and Trico-DHCT had a good ability to prevent when treated before inoculation as well as when analyzed as discussed above so that applying fungicide at 3 days after pathogen inoculation contributed to the increased ability to manage SBD (Fig. 31.1).

References

Agarwal, P. C., Mortensen, C. N. and Mather, S. B. (1989). Seed-borne diseases and seed health testing of rice. Psychopathological papers.

Gnanamanickam, S. S. (2009). *Biological control of rice diseases*, 8, 108 pp.

Klein, D. and Eveleigh, D. E. (1998). Ecology of Trichoderma in Trichoderma and Gliocladium volume 1 (edited by Kabicek Christian. P and Harman Gary. E). Taylor & Francis.

Ou, S. H. (1985). *Rice diseases* (380 pp). IRRI.

Prasad, B. N., & Kumar, M. R. (2011). Comparative efficacy of different isolates of Trichoderma spp. against rhizoctonia solani, incitant of sheath blight of rice. *Indian Journal of Fundamental and Applied Life Sciences, 1*(3), 107–111.

Saksena, S. B. (1960). Effect of carbon disulfide fumigation on Trichoderma viride and other soil fungi. *Transactions of the British Mycological Society, 43*, 111–116.

Chapter 32
Consideration on the Use of Sentinel-1 Radar Image and GIS for Flood Mapping in the Lai Giang River Basin of Binh Dinh Province (Central Coast Vietnam)

Ngo Anh Tu, Grivel Stéphane, Nguyen Huu Xuan, and Phan Van Tho

Abstract The research on building a flood mapping in Lai Giang Basin of Binh Dinh Province in 2016 on the use of Sentinel-1 radar image and GIS initially determines the construction of flooded square in the studied area based on the results of the process of scattering water value threshold, extracting flooded areas on the radar image. This flood mapping is combined with the depth measured from the flood in 2016 traces to apply GIS techniques to create of flood depth map. The results of the study provide a new way of quickly creating flood maps from radar satellite images.

Keywords Flood mapping · Sentinel-1 radar image · GIS · Lai Giang River Basin · Central Coast Vietnam

1 Introduction

Recently, global climate change and flood disasters have caused major impacts that directly impact of flooding include loss of human life, damage to property, destruction of crops, and loss of livestock (Gautam, Hiroyuki, & Mukand, 2019). In fact, the situation of floods has become more and more complicated, making management and forecasting difficult, causing serious losses to the development in Vietnam. This has set the hydrometeorological forecasting agencies the need for practical and timely solutions in order to make rapid and timely flooding forecasts. From there, it contributes to minimize the damage caused by floods.

N. A. Tu (✉) · N. H. Xuan · P. Van Tho
Quy Nhon University, Binh Dinh, Vietnam
e-mail: ngoanhtu@qnu.edu.vn

G. Stéphane
Ministère de la Transition écologique et solidaire, Paris, France

A. T. Nguyen, L. Hens (eds.), *Global Changes and Sustainable Development in Asian Emerging Market Economies Vol. 2*, https://doi.org/10.1007/978-3-030-81443-4_32

The Lai Giang River Basin of Binh Dinh Province in the South Central Coast has a tropical monsoon climate. This basin is proven that this area is more prone to natural disasters than any other region in the country. First, due to the local unfavorable geographical, topographical, and meteorological conditions along with the global climate change (Vo, Vo, & Nguyen, 2019), the region experiences an increase in annual rainfall about 1.750–2.400 mm (Binh Dinh Provincial People's Committee [BDPPC], 2018). However, the rainfall is not evenly distributed over time, concentrated from September to December. In addition, The Lai Giang River Basin is frequently affected by typhoons and tropical low hamlets. On average each year, this basin has experienced an average of 3 to 4 floods, and it has caused many losses of lives and people's wealth.

According to the statistics of the Provincial Steering Committee for Disaster Prevention and Search and Rescue, during the historic flood from November 14 to 18, 2013, there were severe losses: 19 deaths and 14 deaths. More than 101,900 houses were flooded with 510,000 people affected, of which 292 houses collapsed, 418 houses were seriously damaged, and transport and irrigation infrastructure was destroyed by 2125 billion dongs. The December 2016 flood has caused the most severe floods in recent years; it has caused many damages to the people and the transport, irrigation, and educational infrastructure. In the three districts and the battlefield, it was flooded from 0.5 to 1.5 meters, and some places were over 1.5 meters (Binh Dinh Provincial People's Committee [BDPPC], 2018).

Due to the serious nature of floods in the river basins of Binh Dinh Province in general and the Lai Giang River Basin in particular, there have been many typical studies on the application of hydrological and hydraulic models to build flood maps such as the following: Trinh, Nguyen, and Dang (2009) studied "Current situation of flood drainage in Lai Giang estuary"; in 2013, Nguyen Van Hieu et al. studied the application MIKE FLOOD software to calculate flood and simulation plans building dykes in Lai Giang River Basin; Tran Huu Tuyen (2017) has established the program for risk of attachment of river flow in Lai Giang River Basin of Binh Dinh Province in 2017; and Bui (2018) built a flood map of Lai Giang River Basin of Binh Dinh Province. These researches have the common feature of using hydrological and hydraulic models to identify flooded areas.

However, the limitation of this method is the input data that depend on the topographic map, lack of rain gauges, flow measurement, water level, etc. However, in reality, the characteristics of the river basin are always fluctuating in surface cover, amount of sediment, and morphology of the river bed (Ali, Jie, Xuan, & Ram, 2019). Therefore, the use of topographic sections and roughness coefficients for hydrological and hydraulic models will have many errors. In particular, the creation of fast flood maps is a disadvantage of flooding models. Over the years, the development and application of radar satellite images to help quickly identify flooded areas and has been used by many countries to respond to floods (Perrou, Garioud, & Parcharidis, 2018).

In recent times, the satellites image have been increasingly improving the mapping methods from space also under difficult circumstances, such as thick clouds, rugged terrain or other inflexible geographical or meteorological conditions, and capability to acquiring imagery both day and night (Gautam et al., 2019; Tavus, Kocaman, Gokceoglu, & Nefeslioglu, 2018). Inside, the Sentinel-1 is a SAR mission of the Copernicus Program of European Space Agency (ESA). This satellite has been used extensively in mapping flood inundation and has brought about practical benefits for disaster monitoring (Bipinchandra, Romeji, & Loukrakpam, 2019; Borah, Sivasankar, Ramya, et al., 2018; Gautam et al., 2019; Holtgrave, Förster, Greifeneder, Notarnicola, & Kleinschmit, 2018; Kumar, 2019; Jakovljević & Govedarica, 2019).

2 Methodology

2.1 Datasets

This study focuses on collecting data related to the historic flood event in the Laigiang River basin in 2016. The Sentinel-1 SAR data in C-band sensor have been used in this study (Table 32.1). In specialty, DEM (Digital Elevation Model) is a very important role, this DEM has a resolution of 12.5m, provided by ALOS PALSAR, and the DEM is used to delineate watershed boundaries, supporting inundation depth determination.

2.2 Flowchart and Methodology

The paper proposes a research process to develop flood maps from satellite images as shown in Fig. 32.1.

The satellite image data are downloaded from the website for free https://scihub.copernicus.eu/. In this case, the Sentinel-1 satellite constellation consists of Sentinel-1A and Sentinel-1B, which were used. Sentinel-1 data are then processed on specialized software combining DEM data to image the georeferencing in the study area. The paper uses algorithms to extract data on the water layer to produce flooded and nonflooded areas. Finally, these data are combined with the depth of flooding to provide a flood depth map of the study area.

Table 32.1 Dataset to build flood map

No	Data	Type/resolution	Year	Sources
1	S1A-IW-GRDH	VV + VH/10	2016	Copernicus open access hub
2	S1A-IW-GRDH	VV + VH/10	2018	Copernicus open access hub
3	ALOS PALSAR	Raster/12.5	2011	Alaska satellite facility
5	GPS	GPX	2018	Field survey
6	Flood trace	GPX	2016	Investigate flood marks

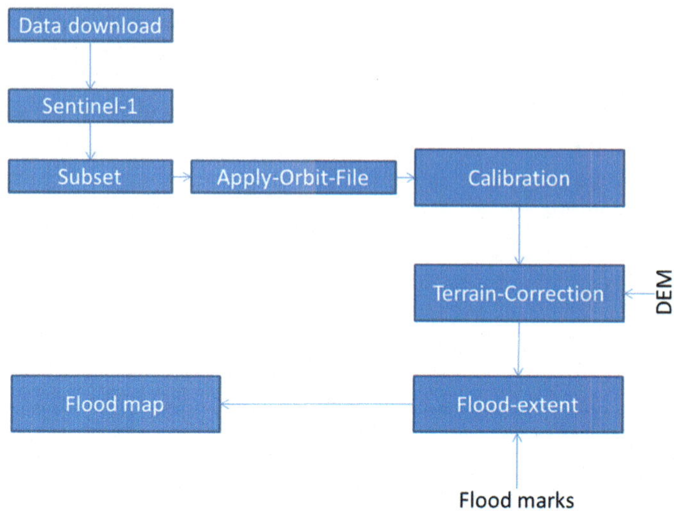

Fig. 32.1 Logical flowchart of study

Remote sensing image processing: The study has used the satellite image of Sentinel-1 dual polarization VV and VH with 10 m spatial resolution on December 16, 2016. The satellite data used in the present study are shown in Fig. 32.2. Preflood imagery acquired on December 16, 2016, and postflood imagery acquired on December 5, 2018, were used to identify changes in inundation over study area.

The data are processed in SNAP toolbox (open source), which can be downloaded from the Copernicus Services Access Hub (https://scihub.copernicus.eu/). To analyze flood mapping, all Sentinel-1 images are first preprocessed with orbit correction, thermal noise removal, calibration to sigma naught, range Doppler terrain correction, and de-speckling (Lee 7x7 speckle filter) by the SNAP toolbox (Jüssi, 2016). Details of processing results are shown in the following Fig. 32.3.

Field survey method: The study conducted random allocation of surveys for the entire basin. For each commune, 30 points were surveyed through GPS measuring equipment to collect flood trace information (Fig. 32.4).

There were 200 sites surveyed in 2018, and then, the survey data are processed on GIS, exported to the map as shown in Fig. 32.5.

GIS application method: This method is used after extracting the flooded area from satellite imagery. Later, the GIS technique uses algorithmic stacking to determine the depth of flooding from the results of surveying flooded areas identified by satellite images. In this study, QGIS software was used by us to build a flood depth map.

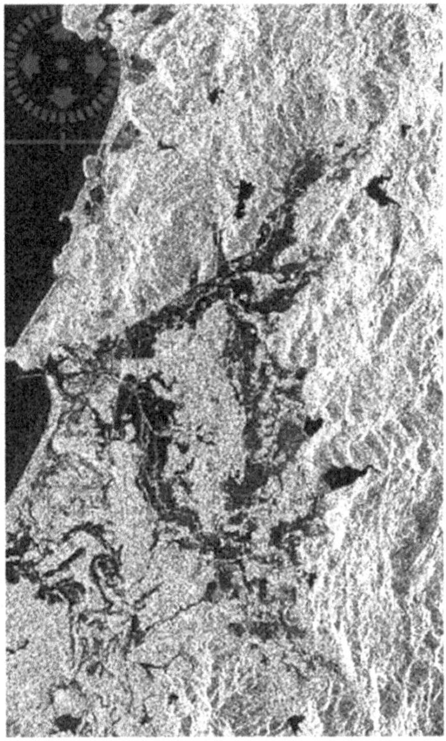

Fig. 32.2 Sentinel-1 SAR data acquired on November 04, 2016

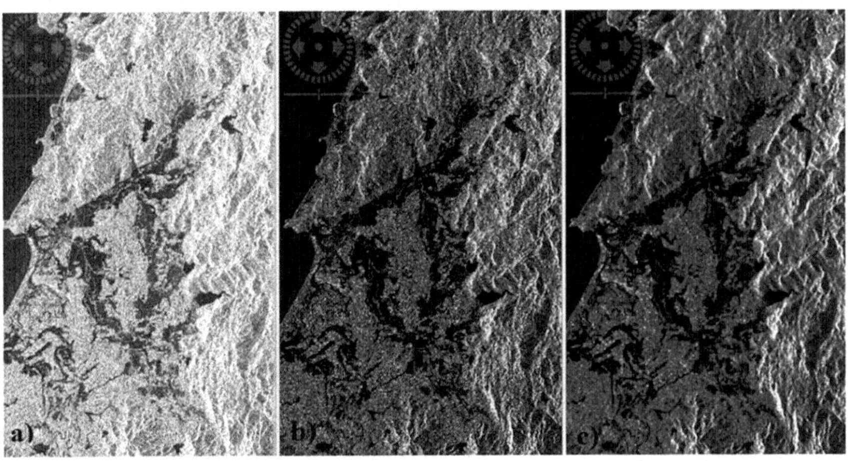

Fig. 32.3 a) Original image; b) after calibration: c) after filtering Lee

Fig. 32.4 Survey points on flood marks (photograph: AnhTu)

Fig. 32.5 Survey points in the Lai Giang River Basin

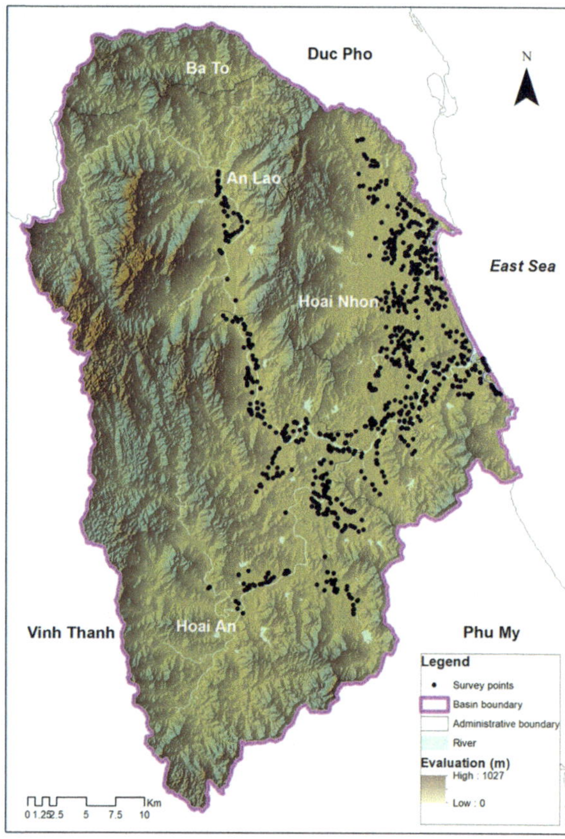

3 Results

3.1 Study Area and the Flood Event

The Lai Giang River is a basin of 1272km^2 (Binh Dinh Province, Central Vietnam). This fluvial system consists of two tributaries (An Lao: 75 km and Kim Son: 64 km). This basin has relatively complex topography. Characteristics of the coastal terrain of the South Central Coast region are as follows: This area has short topography, the west borders are on the high mountains, and the east is a narrow coastal plain. Therefore, this type of terrain has affected the river and high slope; flood concentration time is very fast (Fig. 32.6).

The flood event that occurred in Lai Giang River Basin on December 16, 2016, has caused damages in infrastructure, houses, and offices and also caused the injury to 3 persons (Fig. 32.7).

3.2 The Mapping Procedure

After the completion of SAR preprocessing steps, the study used hydrological maps of the Lai Giang River to extract the flooded area from satellite imagery through GIS technology; the produced images have been preprocessed prior to the final flood map generation (Fig. 32.8).

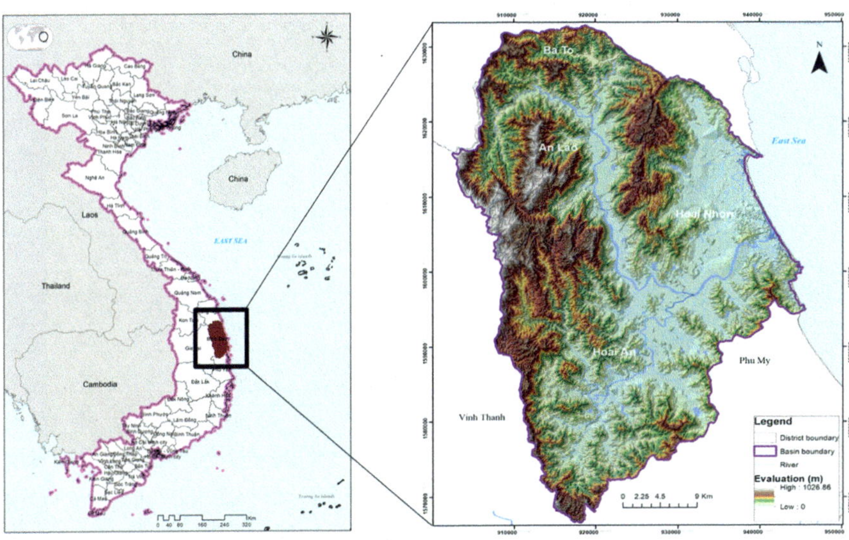

Fig. 32.6 Study area map

Fig. 32.7 Examples of damages after December 16, 2016, flood in Lai Giang River Basin (photograph: AnhTu)

Fig. 32.8 Results of change detection using thresholding

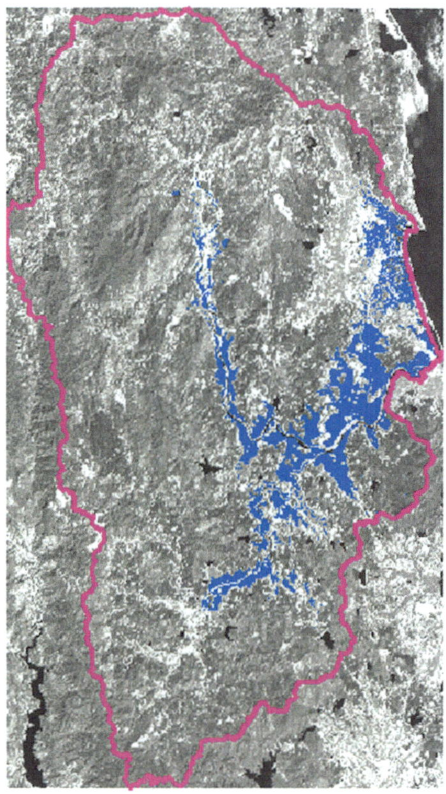

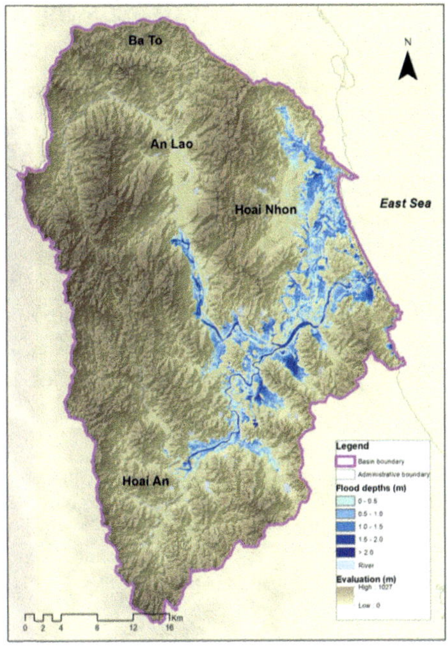

Fig. 32.9 Flood map on December 16, 2016, in Lai Giang River Basin

Table 32.2 Statistics of flood depth in Lai Giang River Basin

Depth (m)	An Lao District	%	Hoai An District	%	Hoai Nhon District	%
0–0.5	55.44	43.39	1977.46	30.27	2746.50	30.16
0.5–1.0	40.34	31.57	2142.89	32.80	2660.26	29.21
1.0–1.5	25.24	19.75	1680.07	25.72	2739.74	30.08
1.5–2.0	6.76	5.29	732.78	11.22	961.30	10.55
Total	127.78	100.00	6533.20	100.00	9107.80	100.0

Flood status map from satellite image is integrated into GIS to combine the results of flood depth measurement points in 2016 from GPS, and then, Kriging interpolation algorithm is used to calculate the interpolation of the space to determine the depth of flooding. Parts of the results are shown in Fig. 32.9.

Through flood data (Fig. 32.9), the study has identified flooding areas in the Lai Giang River Basin (Table 32.2). In Table 32.2, the largest flooded area is in the downstream area of the Lai Giang River Basin. The flooded area is from 1.5 to 2.0 m mainly in Hoai An and Hoai Nhon districts.

4 Conclusions and Discussion

The Sentinel-1 radar sensing satellite image system has the advantage of providing data and helps to view the flooded area compared to optical satellite imagery, because floods occur mainly on cloudy days.

Through research results, it shows the important role of satellite image data in the management of natural resources and environment today, especially monitoring and evaluating natural disasters, with the advantage of being observed in a cloudy weather environment during storms and floods. In addition, the fieldwork combined with the construction and zoning of flood depth plays an important role. This helps managers and citizens to have a visual look. At the same time, it is an effective tool to help prevent and minimize risks of human and property damage during the rainy season.

Acknowledgments The authors would like to thank Quy Nhon University, which is the host agency for ministerial-level research projects "Researching on application of rainfall data from radar satellite images and mathematical models in rapid forecasting of flood risks. Case study of Lai Giang River Basin, Binh Dinh Province, Central Coast Vietnam." They provided data for us to successfully implement this paper.

References

Ali, P. Y., Jie, D., Xuan, S., & Ram, A. (2019). Improved bathymetric mapping of coastal and Lake environments using Sentinel-2 and Landsat-8 images. *Multidisciplinary Digital Publishing Institute., 19*(12), 27888.

Binh Dinh Provincial People's Committee [BDPPC]. (2018). Promulgating flood coping plans in 2018 in Binh Dinh province, No. 3704 / QD-UBND. 30 pages.

Bipinchandra, M., Romeji, N., & Loukrakpam, C. (2019). Flood Hazard risk assessment and mapping of a sub-watershed of Imphal River basin, Manipur, India. Applications of geomatics in civil engineering. *LNCE, 33*, 155–163.

Borah, S. B., Sivasankar, T., Ramya, M. N. S., et al. (2018). Flood inundation mapping and monitoring in Kaziranga National Park, Assam using Sentinel-1 SAR data. *Environment Monitorial Assessment., 190*(9), 520. https://doi.org/10.1007/s10661-018-6893-y

Bui Anh, K. (2018). *Built a flood map of Lai Giang river basin of Binh Dinh province*. Da Nang University of Technology.

Gautam D., Hiroyuki M., Mukand B. (2019). Applications of Sentinel-1 synthetic aperture radar imagery for floods damage assessment: a case study of Nakhon Si Thammarat, Thailand. The International Archives of the Photogrammetry, Remote Sensing and Spatial Information Sciences, Volume XLII-2/W13, 2019 ISPRS geospatial week 2019, 10–14 June 2019, Enschede, The Netherlands.

Holtgrave, A. K., Förster, M., Greifeneder, F., Notarnicola, C., & Kleinschmit, B. (2018). Estimation of soil moisture in vegetation-covered floodplains with Sentinel-1 SAR data using support vector regression. *Journal of Photogrammetry, Remote Sensing and Geoinformation Science, 86*(2), 85–101.

Jakovljević, G., & Govedarica, M. (2019). Water body extraction and flood risk assessment using Lidar and open data. *Climate Change Adaptation in Eastern Europe, 8*, 93–111.

Jüssi, M. (2016). *Synthetic aperture radar based flood mapping in the Alam-Pedja nature reserve in years 2005-2011*. University of Tartu, Faculty of Science and Technology, Institute of Ecology and Earth Sciences, Department of Geography, 1–41.

Kumar, R. (2019). Flood inundation and Hazard mapping of 2017 floods in the Rapti River basin using sentinel-1A synthetic aperture radar images. *Applications and Challenges of Geospatial Technology, 12*, 77–98.

Perrou, T., Garioud, A., & Parcharidis, I. (2018). Use of Sentinel-1 imagery for flood management in a reservoir-regulated river basin. *Earth Science, 12*(3), 506–520.

Tavus, B., Kocaman, S., Gokceoglu, C., & Nefeslioglu, H. A. (2018). Considerations on the use of Sentinel-1 data in flood mapping in urban areas: Ankara (Turkey). *Remote Sensing and Spatial Information Sciences. Vol, XLII-5*, 575–581.

Tran, H. T. (2017). Established the program for risk of attachment of river flow in Lai Giang river basin of Binh Dinh province. *Hue University Journal of Science., 126*(7), 118.

Trinh, V. A., Nguyen, T. T. H., & Dang, H. T. (2009). Current situation of flood drainage in Lai Giang estuary. In *Science and Technology of Water Resources Institute of Vietnam*. Retrieved September 10, 2019, from http://www.vawr.org.vn/index.aspx?aac=CLICK&aid=ARTICLE_DETAIL&ari=1695&lang=1&menu=khoa-hoc-cong-nghe&mid=995&parentmid=982&pid=2&storeid=0&title=hien-trang-thoat-lu-vung-cua-song-lai-giang

Vo N.D., Vo T.D., Nguyen C.C. (2019). *Landsat image processing application for Binhdinh Shoreline Change*. International Conference on Asian and Pacific Coasts APAC 2019: APAC 2019. pp 325–329.

Chapter 33
Studying Shoreline Change in Ky Anh Coastal Area of Ha Tinh Province during 1989–2013 Based on the Digital Shoreline Analysis System (DSAS)

Nguyen An Thinh

Abstract Studying shoreline changes is aimed at determining the important engineering techniques and dealing with challenges in multidisciplinary data integration. It becomes the interests of many researchers, local government, and local stakeholders. Various effective models and systems are developed to identify the trend of changes. The Digital Shoreline Analysis System (DSAS) combined with a Geographical Information System (GIS) is applied to monitor the shoreline changes along the Ky Anh Coast (Ha Tinh, Vietnam) in the period 1989–2013. The shoreline positions are determined using Landsat images; the quantification of the erosion and accretion relationship between 1989 and 2013 in different topographical segments; and a map of affected areas where the shoreline is equal to the high tide water line. The findings offer the decision makers, researchers, and local communities to be benefits of monitoring shoreline change and to help have quickly response to the abrupt change in the area.

Keywords Digital Shoreline Analysis System (DSAS) · Shoreline change · Erosion · Accretion · Ky Anh Coastal Area

1 Introduction

Vietnam is one of the most vulnerable countries damaged badly by hazards and risks of climate change impacts, which affect strongly on over 3000 kilometer long coastal line. The coastal area of Vietnam is mainly influenced by three types of hazards that the integrated effect could be associated with climate changes including drought, sea level rise, and extreme weather conditions. The turns of these primary

N. A. Thinh (✉)
VNU University of Economics and Business, Vietnam National University, Hanoi, Vietnam

A. T. Nguyen, L. Hens (eds.), *Global Changes and Sustainable Development in Asian Emerging Market Economies Vol. 2*, https://doi.org/10.1007/978-3-030-81443-4_33

effects cause floods and erosion. The impacts of both sea level rise and storms result in the erosion of beaches and dune ridges. As well as through the world, from 1990 up to now, the coastal erosion in Vietnam has witnessed an increase in both length and intensity, especially in the low-lying coastal land made by unconsolidated sediments (sand, silt, and clay).

Digital Shoreline Analysis System (DSAS henceforth) has its own advantages for examining historical shorelines compared with traditional ground survey techniques (Thieler, Himmelstoss, Zichichi, & Ergul, 2009). It is considered as a practical application for decision making in coastal management. Most recently, the combination of remote sensing, GIS, and DSAS was applied in coastal studies, which can be found in notable research on determining the rate of shoreline change along the Kenitra Coast, Morocco, during 1969–2009 (Moussaid, Fora, Zourarah, Maanan, & Maanan, 2015), along the Mangalore Coast, West Coast of India (2005–2013) (Aedla, Dwarakish, & Reddy, 2015), in the north of the Coromandel Coast (1972–2013) (Salghuna & Bharathvaj, 2015), along the Tamil Nadu Coast (1978 to 2014) (Natesan, Parthasarathy, Vishnunath, Kumar, & Ferrer, 2015), along the Karnataka Coast (1991–2014) (Hegde & Akshaya, 2015). In Vietnam, several shoreline studies that applied DSAS are listed below: calculating shoreline change in Nam Dinh Coastal Area (To & Thao, 2008); examining the rates of coastal shoreline change in Kien Giang Coast during 1995–2009 (Nguyen, McAlpine, Pullar, Leisz, & Galina, 2015); and detecting long-term mangrove shoreline changes in Mui Ca Mau (Tran, Tien, Phan, Dahdouh-Guebas, & Koedam, 2014). This paper aims at identifying historical shoreline changes during the period 1989–2014 in the Ky Anh Coastal Area of Ha Tinh Province (in central Vietnam). The rate of erosion and accretion was calculated based on the data derived from satellite images and the result of DSAS analysis.

The rest of the paper is organized as follows: Section 2 presents the DSAS methodology; case study results are indicated in Sect. 3; and finally, conclusion and recommendation are determined in Sect. 4.

2 DSAS Methodology

DSAS, or USGS DSAS (United States Geological Survey Digital Shoreline Analysis System), is a GIS tool (a free available ArcGIS extension) designed by the United States Geological Survey (USGS). Its Web-based version (DSAS Web) was published in 2013. DSAS has been used to analyze coastal change based on detecting shoreline movements and calculating the rate of change as follows: (i) mapping historical shoreline position by using available spatial data; (ii) evaluating historical changes and trends of selected transects; (iii) analyzing shoreline geometry; and (iv) predicting shoreline patterns.

The DSAS approach calculates rates of shoreline change through measuring gaps between the shoreline positions in specific periods of time. According to

Thieler et al. (2009), it is possible for DSAS to measure statistical data. Table 33.1 lists measures, which can show the spatial patterns of shoreline change statistics.

3 The Case Analysis

As shown in Fig. 33.1, Ky Anh is the furthest southwestern district of the Ha Tinh Province. The total district area is 105,428 hectares, entailing 7 coastal communes from Ky Xuan in the north to Ky Nam in the south. The coastline of 7 communes spans 63 kilometers, mainly covering beaches, sandy, and the lesser extent of rocky. Behind the dune ridge, delta zones are covered with irrigated rice fields. A small area, which is dryer, is used for growing peanuts and vegetables. There are low hills, which are covered by conifers, mixed leafy trees, and scrub vegetation. Mangroves, which are merely planted, surround rivers. Lakes provide water for the irrigation of rice.

The most significant driving forces (from both nature and humans) of coastal erosion in Ky Anh are storms, floods, and sandy collection (in Vung Ang Economic Zone). The Ky Anh Coastal Area is frequently affected by tropical storms and induced flooding. Storms happened from the southeast to the northwest or from the southeast east to the northwest west over the area. The frequency increases from January to August, decreasing and dipping in December. During the last 50 years, the Ha Tinh Province was affected by 47 storms, 18 of which impacted directly on the Ky Anh Coastal Area. Ky Anh witnessed on average 0.9 storm per year, and there was an increase in the frequency during recent years. In Vietnam, there were 285 tropical storms between 1951 and 2010 (on average 4.75 tropical storms per year) and 38 tropical storms from 2011 to 2013. However, during recent two half decades (1996–2005), there were only 4 storms (on average 0.4 storms per year). This figure was lower than that in the long-term trend analysis, shown. In the study area, a progression from 20 to 200 m, depending on the inclination of the beach, is reported for the period 2003–2010.

Table 33.1 Statistical measures in DSAS

Statistical measures	Definition
Shoreline change envelope (SCE)	A measure of the total change in shoreline movement that considers all available shoreline positions and reports their distances without reference to their specific dates.
Net shoreline movement (NSM)	A measure reports the distance between the initial position and the latest position of shoreline.
End point rate (EPR)	A measure derived from dividing the distance of shoreline movement during the period of time between the initial positions and the latest position of shoreline.
Linear regression rate (LRR)	A measure that determines a rate of change statistic by fitting a least square regression to all shorelines at specific transects.

Source: Thieler et al. (2009)

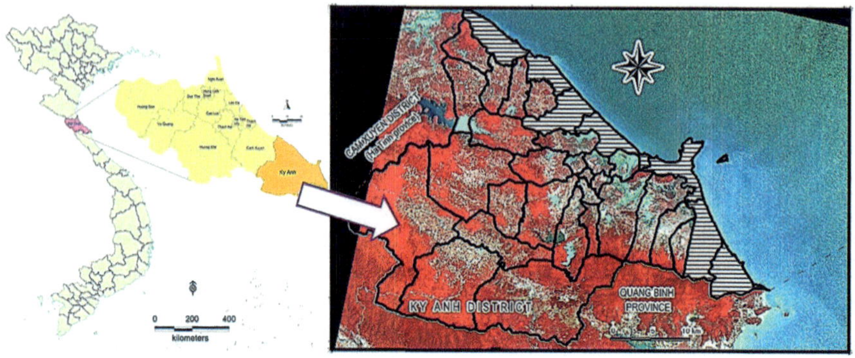

Fig. 33.1 Selected study area in Ky Anh District (Ha Tinh, Vietnam)

Shoreline changes happening in the study area were investigated based on five satellite images, which are available in the period 1989–2013. Movements of both the mean low water (MLW) and mean high water (MHW) are observed through GIS based on DSAS extension developed by the USGS (Thieler et al., 2009). Shorelines were digitized from individual map. Net shoreline movement (NSM) and end point rate (EPR) were calculated. NSM showed the distance between the initial position (1989) and the latest position of shoreline (2013), which indicated the overall change in shorelines position over the 24 years. EPR converted net shoreline movement into an annual rate of shoreline change by dividing the distance of shoreline movement from the initial position to latest position of shoreline passed during a particular period of time period. The 200 m of shoreline in 2013 was chosen as baseline, and 961 transects were created (about the 50 meters of a range of distance) and were numbered in order (Fig. 33.2).

As shown in Fig. 33.3, net erosion and accretion rates were calculated for five sections of coast, using the same boundaries during the periods 1989–2013. During 1989–1996, the rapid accretion of more than 25.25 meters per year (average values are approximately 6.6 meters per year) was observed in the southern Ky Anh District with low erosion of 13.86 meters per year (on average − 3.4 meters per year). The results show that the accretion trend in this period was dominant, and it changed quickly. During 1996–2006, the main trend of shoreline changes was the erosion with more than 5.3 meters per year, which was higher than the accretion rate (3 meters per year) in the study area. Beginning with the erosion, this process happened continuously in whole shorelines with the range featuring from 24 meters per year to 92.5 meters per year during 2006–2013.

Net erosion and accretion rates were calculated for each commune as follows:

- *Ky Xuan Commune:* The coastal zone was observed at 202 transects, which showed that accretion is more than erosion in the period of 1989–2006. The Net Shoreline Movement (NSM) changed from −60 meters at the nearest point to 110 meters at the farthest point. The changing trend in this period was the accretion, with the average highest rates reducing from approximately 14 meters per

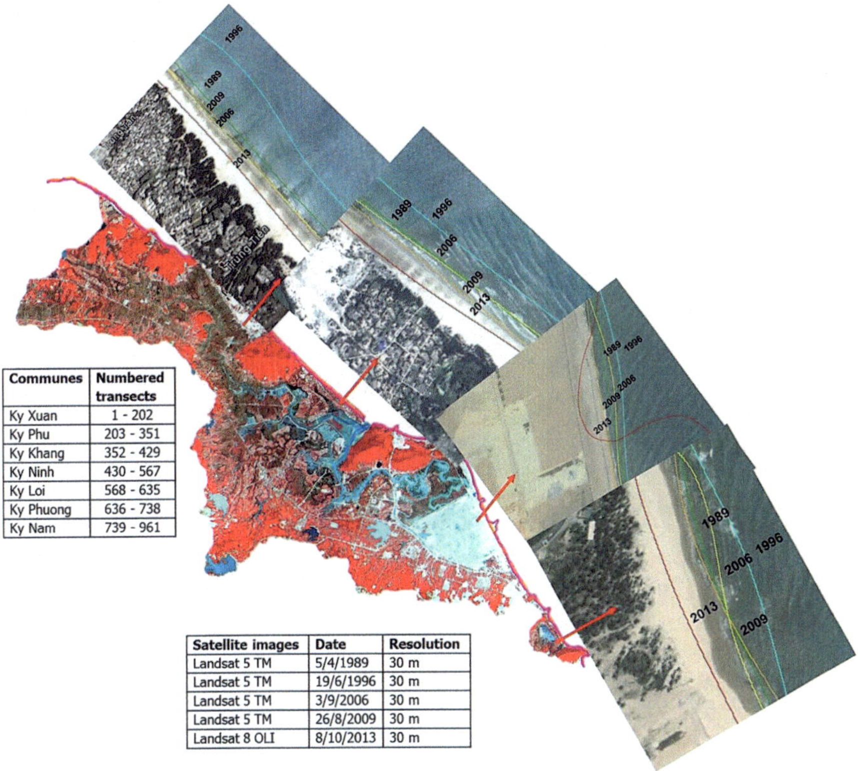

Communes	Numbered transects
Ky Xuan	1 - 202
Ky Phu	203 - 351
Ky Khang	352 - 429
Ky Ninh	430 - 567
Ky Loi	568 - 635
Ky Phuong	636 - 738
Ky Nam	739 - 961

Satellite images	Date	Resolution
Landsat 5 TM	5/4/1989	30 m
Landsat 5 TM	19/6/1996	30 m
Landsat 5 TM	3/9/2006	30 m
Landsat 5 TM	26/8/2009	30 m
Landsat 8 OLI	8/10/2013	30 m

Fig. 33.2 Numbered transects were classified by administrative boundaries

year at the period of 1989–1996 to 5 meters per year at the period of 1996–2006. However, in the year of 2006–2013, the changing trend was transformed into the erosion. The highest rate was 10 meters per year, with the highest movement change of erosion increasing to 160 meters.

- *Ky Phuong, Ky Khang, and Ky Ninh communes:* The shoreline status experienced a change from the accretion during 1989–1996 to the erosion during the period 1996–2013. The shoreline movement changed from 5–20 meters in 1989–1996 to −17.27–0.0 meters in 1996–2006 and to −14–0 meters during 2006–2013. The highest accretion rate was 140 meters per year during 1989–1996. The dominant processes were the erosion, with the highest rate reducing from 170 meters per year in 1996–2006 to 140 meters per year during 2006–2013.
- *Ky Loi Commune:* In Ky Loi Commune, the shoreline movement in 1989–1996 was at the change from −2.5 to 12.3 meters, with the highest accretion rate reaching approximately 88 meters per year and highest erosion rate being −20 meters per year. Therefore, the process of accretion was faster than that of erosion during this period. During 1996–2013, the highest erosion rate increased

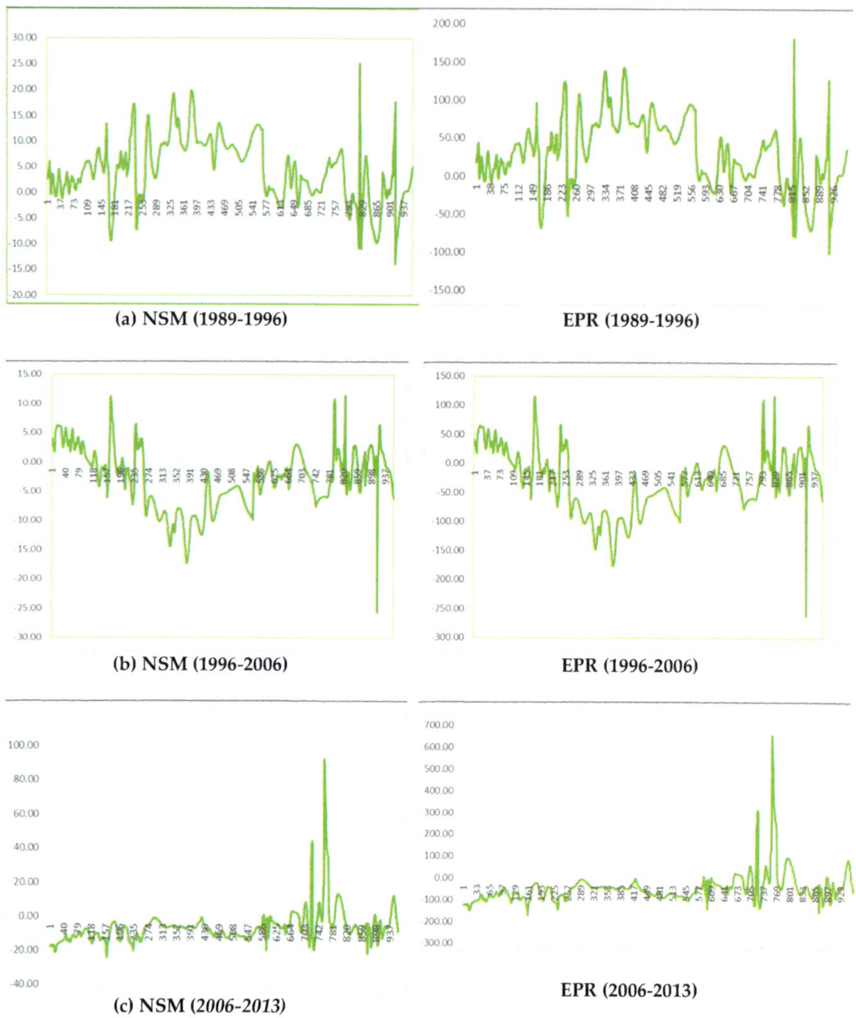

Fig. 33.3 Change in NSM and EPR values during 1989–2013

from −100 meters per year in 1996–2006 to −140 meters per year in 2006–2013. The shoreline movement in 1996–2013 was at the range from −10 to −20 meters.

- *Ky Nam Commune:* The coastal zone in Ky Nam witnessed the erosion more than the accretion during 1996–2013. The highest shoreline movement changed from −13 to 17 meters in 1989–1996; reducing the range from −25 to 6.3 meters in 1996–2006; and maintaining the range between −21 and 10 meters in 2006–2013. Thus, during the period of 1989–1996, the erosion rate was 100 meters per year, whereas the accretion rate was 128 meters per year. The accretion was the dominant process in this period. During remaining years, the highest erosion rate was 260 meters per year, while the highest accretion was 66 meters per year. The domination processes were transformed into erosion, which featured high rates.

During the 24-year period (1989–2013), Ky Anh District witnessed many changes in shoreline. A total of 961 transects were established, where there were 881 transect records of the erosion and 80 transect records of the accretion. There was no transect record of no change (no accretion or no erosion). The movement of erosion ranged from −1.39 to −223.89 meters, the rate ranged from −0.06 to −9.13 meters per year, and the average rate of the whole period was −2.27 meters per year. The erosion point was located in Ky Khang and Ky Ninh communes. The accretion movement fluctuated from 1.99 to 669.43 meters, and the rate fluctuated from 0.08 to 37.31 meters per year. The average accretion rate was 4.81 meters per year, distributing in port construction area in Ky Loi Commune.

In comparison with the changes between three periods, it can conclude that the erosion trend saw an increase in both the movement and the rate during a whole period. The rate of erosion increases from 0.02 to 13.86 meters per year during 1989–1996, to 0.01–25.66 meters per year during 1996–2006; and to 0.02–24 meters per year during 2006–2013. The average erosion rates grow from 3.34 to 5.33 meters per year. The accretion process experienced downward trend in the rate and the movement (Fig. 33.4).

4 Conclusions

Due to significant factors such as storm, flooding, and sandy collection, Ky Anh is regarded as one of the coastal areas experiencing the strongest shoreline changes among Central Coast in Vietnam. The results of DSAS show that during 1989–1996, low erosion was on average 3.4 meters per year. However, from 2006 to 2013, the

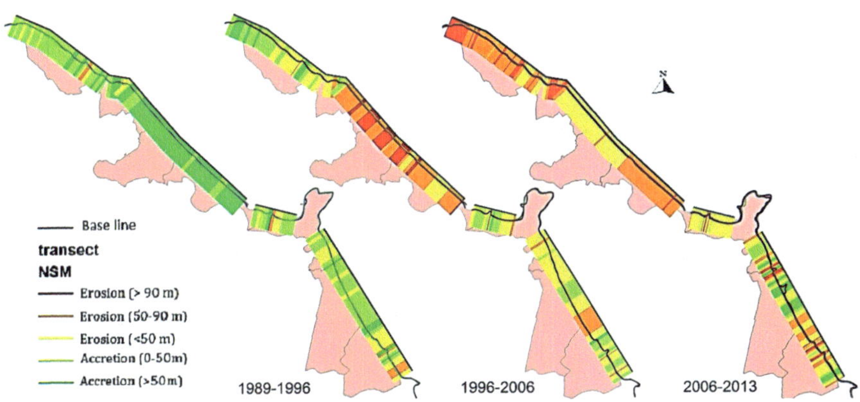

Fig. 33.4 DSAS shoreline change results in the Ky Anh District

erosion witnessed an rapid increase, going up from 24 to 92.5 meters per year. It is similar to recent research on coastal erosion in this area.

Ky Anh is an example of Vietnam's wide and international importance of coastal protection and regional planning. Vietnam has many experiences of coastal protection and flood management; however, the tropical storms and monsoon conditions frequently offer challenges. The local government builds dykes, plants protecting tree ridges, and restores and extent mangroves. The shoreline change extraction and change detection analysis using DSAS could be applied in several fields such as setback planning, hazard zoning, the erosion and accretion studies, regional sediment budgets, and conceptual or predictive modeling of coastal morphodynamics (Aedla et al., 2015). Especially, a setback planning has been designated as an effective spatial planning for coastal erosion mitigation in Ky Anh District.

References

Aedla, R., Dwarakish, G. S., & Reddy, D. V. (2015). Automatic shoreline detection and change detection analysis of Netravati-Gurpur River mouth using histogram equalization and adaptive thresholding techniques. *Aquatic Procedia, 4*, 563–570.

Hegde, A. V., & Akshaya, B. J. (2015). Shoreline transformation study of Karnataka coast: Geospatial approach. *Aquatic Procedia, 4*, 151–156.

Moussaid, J., Fora, A. A., Zourarah, B., Maanan, M., & Maanan, M. (2015). Using automatic computation to analyze the rate of shoreline change on the Kenitra coast, Morocco. *Ocean Engineering, 102*(1), 71–77.

Natesan, U., Parthasarathy, A., Vishnunath, R., Kumar, G. E. J., & Ferrer, V. A. (2015). Monitoring Longterm shoreline changes along Tamil Nadu, India using geospatial techniques. *Aquatic Procedia, 4*, 325–332.

Nguyen, H. H., McAlpine, C., Pullar, D., Leisz, S. J., & Galina, G. (2015). Drivers of coastal shoreline change: Case study of hon Dat coast, Kien Giang, Vietnam. *Environmental Management, 55*(5), 1093–1108.

Salghuna, N. N., & Bharathvaj, S. A. (2015). Shoreline change analysis for northern part of the Coromandel Coast. *Aquatic Procedia, 4*, 317–324.

Thieler, E.R., Himmelstoss E.A., Zichichi J.L., Ergul A. (2009). *Digital Shoreline Analysis System (DSAS) version 4.0—An ArcGIS extension for calculating shoreline change: U.S. Geological Survey Open-File Report 2008-1278.*

To, D. V., & Thao, P. T. P. (2008). A shoreline analysis using DSAS in Nam Dinh coastal area. *GeoInformatics, 4*(1), 37–42.

Tran, T. V., Tien, T. X., Phan, N. H., Dahdouh-Guebas, F., & Koedam, N. (2014). Application of remote sensing and GIS for detection of long-term mangrove shoreline changes in Mui Ca Mau, Vietnam. *Biogeosciences, 11*, 3781–3795.

Chapter 34
How Do Yao Farmers Cope with Extreme Weather Events with their Indigenous Knowledge? A Case Study of Mo Vang Mountains (Yen Bai, Vietnam)

Nguyen An Thinh

Abstract This paper addresses the role of Yao indigenous knowledge in extreme weather event management and focuses on heavy rainfall and drought. In the Mo Vang Mountain (Yen Bai, Vietnam), Yao farmers use diverse information sources to cope with climate risks. Data were obtained from both secondary sources and primary sources. The secondary sources are mainly from the reports and the previous researches, and the primary sources are from focus group discussion (FGD), key informants through interview, and questionnaire survey. Yao indigenous knowledge in extreme weather event management is necessary for the climate risk management and health of the natural environment and its inhabitants. The Yao also summary adaptation strategies to extreme weather events of IK in farming production. The indigenous knowledge is transferred from one generation to another. However, Yao people recently have applied the scientific knowledge in extreme weather event management and farming production.

Keywords Yao people · Extreme weather event · Resource management · Indigenous knowledge

1 Introduction

Indigenous knowledge (IK) is "the knowledge of an indigenous community, accumulated over generations of living in a particular environment" (Masinde and Bagulạ, 2011) or "a body of knowledge built up by a group of people through generations of living in close contact with nature" (Ncube, 2018). IK consists of intellectual, technological, ecological, and medical information. It is traditional and

N. A. Thinh (✉)
VNU University of Economics and Business, Vietnam National University, Hanoi, Vietnam

A. T. Nguyen, L. Hens (eds.), *Global Changes and Sustainable Development in Asian Emerging Market Economies Vol. 2*, https://doi.org/10.1007/978-3-030-81443-4_34

cultural knowledge, which is established where people have survived for a long time and have become familiarized with their surroundings. IK shows its valuation in climate change adaptation and disaster management at community level. Disaster management deals with preparedness, response, and recovery activities to reduce disaster impacts (Mitiku and Hailu, 2017). Disaster management is an isolated, compartmentalized concept and an integrated part of the lives of indigenous people (Sobrevila, 2008).

In Vietnam, for the past 50 years, the average annual temperature has increased about 0.5–0.7 degree Celsius, and sea level has risen about 20 cm (Ministry of Natural Resources and Environment [MONRE] 2016). Climate change has made the disaster, especially storms, floods, and droughts increasingly. The consequences of climate change for Vietnam are serious and present a threat to poverty reduction, the achievement of the MDGs, and the sustainable development of the country. Vietnam is estimated as one of the five developing countries affected by climate change (Asian Development Bank [ADB] 2013). Climate change adaptation is an emergency management issue for sustainable development for any government, organization, community, and individual.

IK system applications vary from region to region over Vietnam. The effects of climate change recently cause decrease in economic development of community in this area. Because this area did not have identical ecosystem, it is easy to change, and the deforestation occurs serious and degradation in land sources, the weather constituents, soil, water, topology, and other biological factors and vegetation cover and many animals are affected by unusual changes such as the increasing or decreasing temperature, the unusual storms, and many other heavy rainfall. The majority of the rural populations in Vietnam Northern Mountains are engaged in farming. Therefore, understanding of IK systems plays an important role to develop a sustainable agriculture to climate change adaptation.

Northern mountainous areas of Vietnam generally have traditional knowledge and experience in agricultural production that helps them to be more flexible with extreme changes in their living environment. Among ethnic minorities living in the area, the Yao ethnic group is quite large, representing the ethnic groups living in upland areas with a tradition of farming on sloping land cultivation. The life of this community mainly depends on nature. Therefore, the people have accumulated a lot of experience and knowledge in agricultural production, resource management, and extreme climatic events over time and from generation to generation. They are very knowledgeable about the local environment. Thus, the study of understanding indigenous knowledge, especially in climate change management of the Yao ethnic group is essential, from which to combine the promotion of the positive nature of indigenous knowledge with scientific knowledge, in order to contribute significantly to the economic development, improve the quality of life for people, and preserve and promote the ethnic cultural identity in mountainous areas.

The findings of this paper show that how does a Yao population in the Vietnamese Northern Mountains use their IK to weather forecast and cultivate crops in the contexts of new context of global climate change and emerged natural disasters such as heavy rains, flash floods, floods, hot extreme weathers, and drought, which are serious climate risks in this area nowadays.

2 Methodology

2.1 Study Area

The selected study area is Mo Vang, which is a remote mountainous commune in Van Yen District (Yen Bai Province, Vietnam). The commune is located in the western part of the Van Yen District and 30 km from the district center. The commune includes 11 hamlets with 852 households and 4426 inhabitants. Five ethnic groups that lived in this commune include Kinh, Thai, Hmong, Tay, and Yao groups (61.5%). The total area of the whole commune is 9986.7 ha. In the commune, there is a district road Mo Vang - An Luong cross over the commune with a length of 18.0 km. The study site is area where agroforestry production is main economic field. Growing cinnamon are the majority in Mo Vang with 1338 ha. Rice cultivation is barely providing trading market but for subsistence mostly due to limited agricultural land (162 hectares for 832 households). In addition, farmers also produce different kinds of livestock such as buffalo, pig, and cattle.

Mo Vang Commune is often affected by extreme weather events such as drought and heavy rains. These impacts affect different dimensions, from human life to agricultural production systems such as altering growing seasons, planting and harvesting calendars, wet, and weed and disease population, and agroclimatic conditions.

2.2 Questionnaires

The data sources in this study were qualitative primary and secondary sources. Primary qualitative data were obtained from focus group discussions (FGDs) with Yao farmers in the study areas, key informant interview, and questionnaire survey.

The phases of the research undertaken in the Mo Vang are detailed below:

* Focus group discussion (FGD).

FGDs were conducted with two groups containing 6–8 persons who have experience concerning topic under investigation. One of the formed groups contains farmers, whereas the other contains expert and collect qualitative information about the subject under study. The number of males and females is relatively equal. During course FGD, the researchers facilitated the discussion and taking notes, recording the discussion session. In addition, activities and practices that Yao Community performed to manage climate change are observed.

* In-depth interviews.

In-depth interviews: The method to collect information from people is referred to as interview. Another precise definition is that any person-to-person interaction between two or more individuals with a specific purpose in mind is called an interview (Kumar, 2001). The interviews with farmers, key informants, and stakeholders were undertaken in collaboration with extension officers, agricultural scientists, and local community members of the Mo Vang Commune. Methods for capturing IK were also reviewed, and those appropriate were selected for use within the study. Participatory research methods were adopted throughout the study, particularly during interactions with the farmers.

* Questionnaire interviewing.

One of the most important techniques of data collection is questionnaire. A questionnaire to get more information from 60 Yao farmers was made. In that questionnaire, a list of 14 written questions was to be answered by 60 respondents of a particular study. Most of the questions are clear and easy for Yao farmers to understand because respondents need to understand the question clearly before answering. There are some open-ended questions for people to express their views on the issues that are researched in their locality.

The advantage of the questionnaire is that it offers great anonymity between researcher and respondents, and these increase the likelihood of obtaining accurate information. On the other hand, it is less expensive to be deployed in research as you can save time, human, and financial resources, and particularly for population study, it is an inexpensive technique of data collection.

* Sampling: Quota Sampling.

Characteristics of quota sampling are as follows: (i) Divide the sample population into subgroups; (ii) figure out the weightage of subgroups; (iii) select an appropriate sample size; and (iv) survey while adhering to subgroup population proportions.

3 Results

3.1 Indigenous Knowledge of Yao People in Daily Life and Farming Production

According to the survey results, it shows that 86.7% farmers of studied Yao farmers in Mo Vang Commune think using domestic animal behaviors as the best heavy rainfall-monitoring tool, cloud formation, and characteristic of plant also are good indications to predict the possible heavy rainfall occurrence. When the clouds are clustered and dark, farmers know that rainfall is coming. A half of the total farmers interviewed agreed that wind formation helps them to point heavy rainfall in next few hours; 46.7% of interviewed farmers said that if it is very hot at night, and this means that a good rain will come to next day and last for some days.

It is same to case of heavy rainfall management that the major of Yao farmer using cloud formation indicator like a forecast method for hot extreme weather in their region when they see in the sky, white or blue cloud are sign of hot extreme weather. The use of plant and tree characteristics as indicators within indigenous climate forecasting systems was indicated by 40% of respondents in the Mo Vang (Tables 34.1 & 34.2).

Nevertheless, wind formation was perceived as the best parameter by 80% of farmers in the study sites to monitor possible drought occurrence. Laos wind with dryness is sign of drought season. Seedlings and leaves that become withered and drooping downs are also indicator of predicting drought. And during drought periods, farmers may adjust fertilizer inputs and adopt drought-tolerant crop varieties. This knowledge was inherited from their ancestors and grandparents; however, some Yao people raised concerns about IK becoming less effective due to the increasing frequency and intensity of extreme climatic events (Figs. 34.1 and 34.2).

Table 34.3 shows that the use of traditional weather/climate prediction indicators and their use in interpreting extreme weather events (heavy rainfall/drought condition).

Table 34.1 Evaluation of some extreme weather phenomena and their effects in recent years

Extreme weather	Time		Total area (ha)	Notes
Drought	30/6/2017	Rice		
		Forest tree (mainly cinnamon)	125.305	3.5 ha of seedlings damaged during the nursery period
Heavy rainfall	2/8/2015	Rice	1.851	Wet rice: 1.661 ha Upland rice: 0.19 ha
		Forest tree (mainly cinnamon)		1044 trees (4 years old)
	10–13/7/2017	Rice	0.566	
	19–20/7/2017	Rice	2.491	Storm no. 2
	5–6/8/2017	Rice	11.881	
		Forest tree (mainly cinnamon)	34.675	
		Houses		10
	4/10/2017	Rice	1.946	
		Forest tree (mainly cinnamon)	0.58	

Source: Report of Mo Vang Commune

Table 34.2 Traditional methods used by Yao to forecast extreme weather

	Number of respondents	Percentage (%)
Extreme climatic events		
(1) heavy rainfall		
Cloud formation	50	83.3
Wind formation	30	50
Characteristic of plants	42	70
Insect/animal behavior	52	86.7
Temperature change	28	46.7
(2) drought		
Cloud formation	30	50
Wind formation	48	80
Characteristic of plants	48	80
Insect/animal behavior	24	40

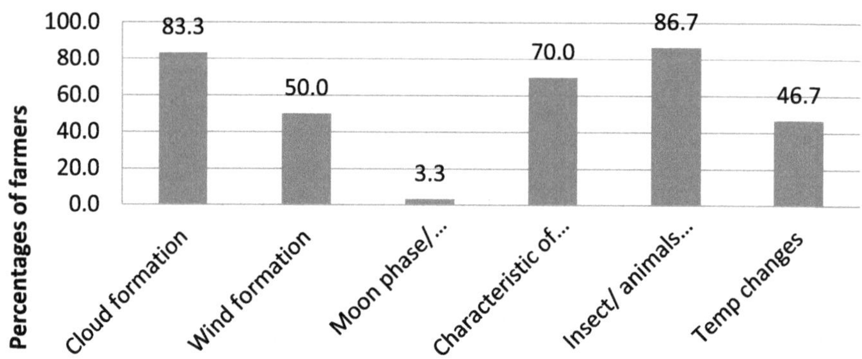

Fig. 34.1 Traditional forecast methods used by Yao farmers for heavy rainfall management

3.2 IK and Scientific Knowledge in Managing Extreme Weather Events

* Use of forecast information sources of Yao farmers in managing extreme weather events.

The result that is got from the research survey indicated that 36.7% of farmers in Yao Community prefer using external indicators only to read the weather and climate to manage disasters. They are mainly young people, people who have much knowledge and experience. They usually use external forecast to get information in managing risks and farming production like weather forecast records on the TV, radio, and other from agricultural officers. While 50% of farmers prefer combining both traditional forecast method and scientific forecast method, only 6.7% of farmers still prefer traditional forecast methods to monitor climate risks. They said that IK system is easier to understand than scientific knowledge and it is still accurate to

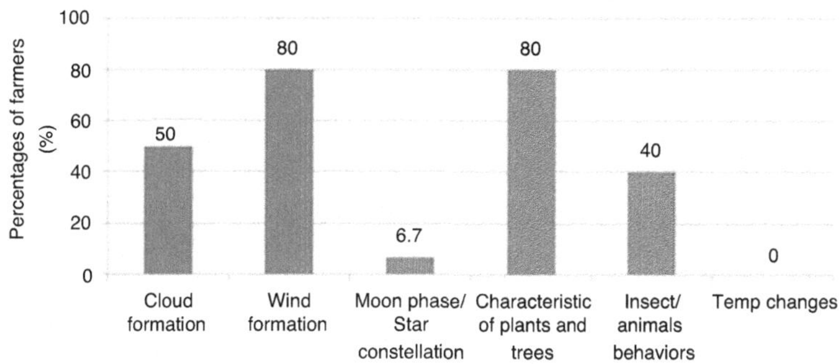

Fig. 34.2 Traditional forecast methods used by Yao farmers for drought management

Table 34.3 Traditional climate prediction indicators and their use in interpreting heavy rainfall condition

Indicator	Indication for weather occurrence
Cloud formation	Dark clouds indicate rainfall/heavy rainfall to occur within a few hours Cloud flies very fast
Wind formation	High frequency in occurrence of wind swirls indicates heavy rain
Insect/animal behaviors	Appearance of red ants and rapidly increasing size are used to predict heavy rain/flood Worm Cranes fly along the river Domestic animal cries out in the forest Dragonfly flies lowly.
Sky	Dark color is a sign of heavy rainfall
Other	On the first day of lunar year, a bottle of water is taken to compare with the same bottle of the last day in the last year. If it decreases in mass indicating good rain in that year. The wet floor predicted good rain last for long time.

farmer. Besides, recent scientific knowledge is usually complicated and they cannot understand in the short time due to lack of knowledge (Fig. 34.3).

* Comparison between IK and scientific knowledge in weather forecast and farming production.

This summary provides a good overview on how farmers in the Mo Vang commune using, conserving, and also maintaining their IK and the trends that still remain even with the introduction of scientific knowledge. In Mo Vang commune, Yao Community still believe in the relevance of IK in some situations, but generally nowadays they use scientific knowledge to cope with climate change and farm production. The table below shows a comparison of IK systems with scientific knowledge using information collected from farmers and extension officers in 11 hamlets in Mo Vang commune (Table 34.4).

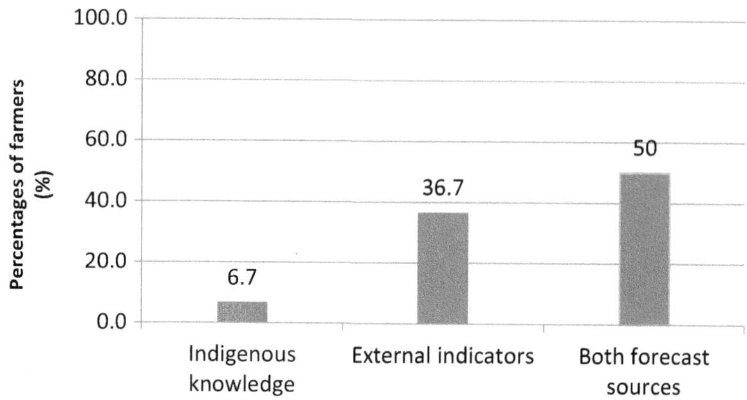

Fig. 34.3 Use of various forecast information sources for farm management

Table 34.4 IK and scientific knowledge

Typology	Indigenous knowledge	Scientific knowledge
Drought indicators/early warning indicators	Wind direction, cloud formation, temp change, Animal behavior, change in plant characteristics	Weather forecast records Rainfall records
Pest management	Getting by hand with sampling/young trees Liming into the stump of tree to kill the pest (for the sampling/young trees)	Add the medication to the rag and wipe it on the tree
Animal diseases management	Make the roof to cover the barn of buffalobuild the cages for livestock and cattle to avoid hot extreme weather	Use of modern medicines to vaccinate and treat animals
Soil conservation	Do not clear weeds to cinnamon forest for keeping water and nutrient richness and into the soil use organic manure and compost to maintain soil fertility in paddy fieldsmixed cropping like corn and cassava to keep water and soil	Applying fertilizer, protein, etc.
Cinnamon forest preservation	Select the indigenous varieties.	

3.3 *Summary of the Drought/Heavy Rainfall Coping and Adaptation Strategies of IK in Farming Used by Yao Farmers*

In contrast with developed countries, the majority of rural populations in Northern Mountains in Vietnam are engaged in farming. The crops of these farmers are based on their local experiences and knowledge developed over the years. The strategies of these farmers and their perceptions of risk vary across the region. Fujisaka (1997) argues that it is necessary to pay attention to local knowledge systems and to analyze their relationship with seasonal predictions. Faced with the uncertainties of climate, economy, and politics, these farmers have been rendered highly risk-averse due to their adaptation strategies, which include indigenous knowledge.

- Drought coping strategies: planting of crops that require less water, crop diversification, adjustment of fertilizer inputs, and use of local climate indicators, for example, cloud formation and wind formation.
- Heavy rainfall coping strategies: Diversity of crops, preparing sandbags, drainage supplies, harvesting soon, and use of local indicators, for example, cloud formation; some of certainly animal behaviors include white storks, hawks, frog, and red ants.

From survey results in Mo Vang Commune, many farmers demonstrated that strategies of cultivating a diversity of crops will be able to increase crop yield. This method helps to reduce total crop failure due to the varying tolerances of crops to environmental stresses. Mixed cropping proved to stabilize yields and conserve soils. There are several strategies used by Yao Community to minimize disaster including growing drought-resistant indigenous crop varieties and use of early warning systems. For example, farmer listens for the cry of certain bird species to predict drought (hawks). Other IK used by Yao Community to predict natural hazards includes cloud formation, shape of moon, and the behavior of some certain animals such as ant and frog.

4 Conclusion and Discussion

Indigenous knowledge plays a very important role in climate adaptation strategy development for farmers in Northern Mountains in Vietnam, including Yao Community in the Mo Vang Commune in Yen Bai Province. People in Mo Vang Commune find that it is difficult to use IK alone in their agricultural practices, and this is because indigenous knowledge is losing value because of increasing climatic variability and change; 50% of the farmers interviewed argued that it is better to combine IK with scientific knowledge in order to counteract the threats posed by climate change.

In the Mo Vang Commune, farmers were found to use insect/animals behaviors to monitor possible rainfall occurrence. It was noted that 86.7% of surveyed farmers in the Mo Vang perceived insects'/animals' behaviors as the best tool. When the crabs come out from the ravine, the appearance of red ants on the ground is used to, for example, farmers can predict good rains. More than 60% of interviewed people use cloud formation as the best tool to predict hot extreme weather. Almost 80% of respondents in the Mo Vang used wind formation and plant characteristics as an indicator for possible drought.

This study noted that several other IK indicators are being used by Yao Community to predict weather patterns, such as the moon phase/star constellation and the temperature; 6.7% of the farmers in Mo Vang are using only indigenous knowledge as a tool to manage climatic risks.

Mixed cropping, growing drought-resistant indigenous crop varieties, and use of early warning systems are strategies that reduce total crop failure due to the varying tolerances of crops to environmental stresses.

Acknowledgements This study was funded by the Vietnam Committee for Ethnic Minority Affairs (National Project Code CTDT.39.18/16-20). The authors would like to thank the community members of the study area for participating in the research, who were most collaborative in completing the questionnaires and in providing discussion opportunities on the results.

References

Asian Development Bank [ADB]. (2013). Vietnam: Environment and climate change assessment. Technique Report. 58 pages.

Fujisaka, S. (1997). Research: Help of hindrance to good farmers in high- risk systems? *Agricultural Systems, 54*, 137–152.

Kumar, R. (2001). *Research methodology: A step-by-step guide for beginners* (415 pages). Sage.

Masinde, M., & Bagula, A. (2011). ITIKI: Bridge between African indigenous knowledge and modern science of drought prediction. *Knowledge Management for Development Journal, 7*(3), 274–290.

Ministry of Natural Resources and Environment [MONRE]. (2016). *Climate change and sea level rise scenarios for Vietnam* (170 pages). NARECA Publisher.

Mitiku, T., & Hailu, W. (2017). The role of Oromo indigenous knowledge in disaster management and protection: The case of Kuttaayee in Ambo District. *International Journal of Multicultural and Multireligious Understanding, 4*(5), 16–28.

Ncube, B. (2018). Insights into indigenous knowledge strategies for coping and adapting to drought in agriculture: A Karoo scenario, South Africa - research. *Indilinga African Journal of Indigenous Knowledge Systems, 17*(1), 92–108.

Sobrevila, C. (2008). *The role of indigenous peoples in biodiversity conservation* (84 pages). The Natural but Often Forgotten Partners.

Chapter 35
Multilevel Governance Roles in Land Use Change: Lessons for REDD+ from the Case Study in Nghe An Province, Vietnam

Nguyen Dinh Tien

Abstract Multilevel governance (MLG) is discussed in relation to influences over land use and forests, to identify who makes decisions, how those decisions are made, and the interaction between multiple levels and sectoral offices of government. Since 1994, the government system in Vietnam was deconcentrated, with established lower-level governments—provincial, district, and commune—largely carrying out national directives. Forests fall under the responsibility of the Ministry of Agriculture and Rural Development (MARD), whereas natural resources and land are the responsibility of the Ministry of Natural Resources and Environment (MONRE). There are conflicting interests and weak coordination between agencies and conflicting objectives between the agriculture and forestry sectors. Findings from the study identify the provincial government as a key decision-making body, relative to the commune and district. The province sets out overall land use planning, subject to central government approval, while commune and district governments are important for identifying and resolving accountability issues surrounding land use decisions. The commune government's role is as an intermediary for higher-level government decisions and being mandated to bring the voices of local people to the district level. Nevertheless, lower-level governments' ability to achieve these functions is constrained by weak capacity and technical skills, and limited human and financial resources as well as their ability to influence decisions of those in higher government both in implementation and for resolving local concerns. The research suggests that the outcomes of forest and land use policies are strongly driven by (i) will, interest, and attitudes of influential actors to promote forests over other land use goals; (ii) the effectiveness of coordination and coalitions among those actors and between these actors and local communities; and (iii) local people's understanding of pros and cons of these land uses, and their confidence and right to accept or reject these land use changes.

N. D. Tien (✉)
VNU University of Economics and Business, Vietnam National University, Hanoi, Vietnam

A. T. Nguyen, L. Hens (eds.), *Global Changes and Sustainable Development in Asian Emerging Market Economies Vol. 2*, https://doi.org/10.1007/978-3-030-81443-4_35

Keywords Multilevel governance · REDD+ · land use · Nghe An

1 Introduction

Managing forestry ecosystems requires a good understanding of the political, socio-economic, and ecological dynamics. Reduction in emissions from deforestation and degradation (REDD+) is a multilevel process in which different scales and multiple actors and institutions come together around decision-making processes on land use. Multilevel governance (MLG) is discussed in relation to influences over land use and forests, to identify who makes decisions, how those decisions are made, and the interaction between multiple levels and sectoral offices of government. Since 1994, the government system in Vietnam was deconcentrated, with established lower-level governments—provincial, district, and commune—largely carrying out national directives. Forests fall under the responsibility of the Ministry of Agriculture and Rural Development (MARD) , whereas natural resources and land are the responsibility of the Ministry of Natural Resources and Environment (MONRE) . However, there are conflicting interests and weak coordination between agencies and conflicting objectives between the agriculture and forestry sectors.

In Vietnam, the government structure and forestry sector have been evolving over the last 30 years. The Vietnamese governance system, following reforms in 1986, has been shifting from a central mono-unitary structure to a more decentralized structure with the emergence of lower government at the provincial, district, and commune levels (Trung, Phuong, Yang, & Hai, 2015). REDD+ presents an opportunity to encourage a "bottom-up" design and therefore involves more actors in forest governance. Larson and Petkova (2011) highlight, however, that REDD+ also risks driving power further into the hands of the elite often at the detriment of forest-dependent communities, such as through inequitable benefit-sharing mechanisms (BSMs) or land grabs. Equally, burdens can be both direct and indirect but are usually broken down into financial costs (e.g., lost opportunity costs), transaction costs, and reduced access to natural resources (Luttrell et al., 2013).

The report provides the description of the changes in land use and forest management in different levels including the analyses of key drivers and actors involving in land use change process and decision as well as the legal arrangement of power between different levels of local government and actual power practices and its effects to land use and forest management and in relation to benefit-sharing mechanisms (BSMs) of Nghe An Province. This report is based on an analysis of MLG land use decision making with a focus on the subnational level. A complete overview of all land uses was not investigated; instead, the focus was on land uses directly perceived as influencing forests. This case study approach enables a unique examination of decision making at various levels, in particular from the subnational to the local, to inform land use options and future land use decisions within Vietnam and globally.

2 Methodology

The methods include secondary data collection and a literature review, alongside primary data collection based on key informant interviews at multiple scales. The secondary data collection involved a literature review on legal and policy documents related to Vietnam's forestry sector and other relevant land use sectors. Data and information were also collected from reports and official documents at different government offices levels and NGOs, institutions in Nghe An Province including administrative offices, and department from province level to local government levels.

The researchers combined and adapted the interview guides as appropriate in order to conduct open-ended semi-structured interviews with multiple actors. Overall, interviews aimed at understanding the actors involved in land use decision making, the relationships among actors, the processes leading up to land use changes, agreements to distribute benefits, and results of land use decisions. The research team also interviewed key informants from provincial and district-level governments in all sites to capture their involvement in decision making on land use, coordination with other levels of government, and knowledge of REDD+ and other such initiatives. Primary data were collected at the provincial, district, and commune levels in 2014 and 2018. The key study sites included four communes in the two districts of Nghe An Province with interviews focused on Luc Da and Chi Khe communes in the Con Cuong District and Thach Giam and Yen Na communes in Tuong Duong District. A total of 52 official staffs and local people who involved in making and implementing policies on land and forest management were willing to interview and agree to share their information.

For the study site selection at commune level, the following selection criteria were applied:

- Significant land use or land use management changes within the last 20 years;
- At least two land use activities associated with deforestation and degradation drivers (perceived as "increasing carbon emissions" sites).
- At least two land use activities (BSMs) aimed at stopping/slowing deforestation and degradation (potentially decreasing carbon emissions).
- Accessibility and feasibility.

In the Nghe An Province, the two sites (Chi Khe and Yen Na communes) classified as "increasing emission" sites included hydropower plant (HPP) developments. The other sites considered "decreasing emissions," included afforestation BSMs, from private business enterprises in the Luc Da Commune, and government afforestation programs in Thach Giam Commune. There were no REDD+ initiatives in Nghe An at the time of the study although one had recently ended.

The interview data were stored, managed, and analyzed using qualitative data analysis software (e.g., NVIVO 10). Data were coded using a heretic node tree based on an initial literature review. The codes were also updated and adapted according to the data compiled. The coding process was iterative, so that the coding tree changes could be data-driven. Coding was specified within a coding guide and spot verified by a single coder. Queries were then conducted in order to assist in finding patterns for data analysis (Ravikumar, Kijazi, Larson, & Kowler, 2015).

3 Results

3.1 History of Land Use Change in Nghe an Province

Nghe An Province, formally known as Nghe Tinh Province, was established in 1991 when Nghe Tinh Province was divided into Ha Tinh and Nghe An provinces. Currently, Nghe An has total of 1649 km^2 natural land and bordering with Thanh Hoa to the north, Lao PDR to the west, and Ha Tinh Province to the south. For land use change, the purpose from agricultural land to industrial zone is observed in the districts in lowland nearby Vinh City such as Thanh Chuong, Nam Dan, Do Luong, and Quynh Luu districts (NA#12). At the upland areas, the province was allocated forestland to individual household based on the Decree 02/CP-1994[1] in the period from 1998 to 2000. The program affected villagers' land management and swiddening practice, and a sharp decrease in swidden area was observed after the forestland allocation (Nguyen, 2010).

The remarkable period of land use change was in the period of 1975 to 1987. The fact is that after Vietnam reunification in 1975, the quick impact of the development programs in the north area of Vietnam was encouraged particularly in agricultural production and forest logging. Ineffective cooperative model in this period caused low agricultural production, which leads to the serious food shortages, and the Communist Party leadership approved Resolution 100 in 1981, which sanctioned the household production contract system (McElwee, 2011). Primary forests were converted to upland rice, cassava, and maize cultivation. The forest area managed by SFEs caused serious deforestation during this time (Leisz, Ginburg, Nguyen, Tran, & Ramussen, 2011). The main cause of forest degradation is poor management system, which leads to unplanned logging and extensive shifting cultivation (Tran, Pham, & Rasmussen, 2006). The forest enterprises were established, and all rich forests were managed by them with the focus general only on exploitation of forest because of the economic development and poor management system. Consequently, forest declined fast in both area and quality (two third of forest area in Nghe An Province was logging and cut down by the State Forest Enterprises and forest) (Tran et al., 2006). For example, in 1973, the State Forest Enterprises of Con Cuong District was awarded the Uncle Ho's medal (the honor prize award by President Ho Chi Minh) for the highest production in logging forest. Mr. Nguyen Ngoc Lai, the director of Con Cuong State Forest Enterprise, was awarded a medal

[1] Decree No 02/CP dated on January 15, 1994, on allocated forestland to organization, household, and enterprises to use land in long term and stable use.

for the "labor hero" in 1985 because of his contribution to the State Forest Enterprise in logging timbers. The goverment was supporting forest production and logging activities during this time.

In addition, migration from lowland to upland areas lived in the mountainous districts of Nghe An was observed in this period. Some district towns were established at that time such as Con Cuong Town in Con Cuong District and Hoa Binh Town in Tuong Duong District in 1988.[2] As provided by the key informants, a larger area of rich primary forest had changed into poor forests in this time.

Land use in Nghe An undergone many changes during period of economic transformation since 1988 when the government has adopted a land use reform and new policies for agricultural development, including policies for land allocation and reforestation such as Decree 10 in 1981 allowed farmers to use land in turn for a fixed amount of crop produced, and the new national Land Law in 1988 (known as Resolution 10) on entrusting land to households and organizations for long-term use (Leisz et al., 2011). In the period 1988–1993, the land use change in Nghe An Province was influenced much by the Land Law in 1988. This was remarkable change in land use of Nghe An Province because the Land Law indicated that the agricultural land was allocated to household. Second, then, the 1993 Land Law, which has been amended three times in 1998, 2001, and 2003, granted five rights to the households (the right to transfer, exchange, inherit, rent, and mortgage) (Nguyen, 2010). The law also extended the lease term to 20 years for annual cropland and 50 years for perennial cropland (Land Law of Vietnam 1993, Article 20). The farmers have been allocated agricultural land for self-production and investment in their land (land use right certificate was granted to households) while it did not happen in the past (NA#21).

Another period of land use change in Nghe An was from beginning of 1991 to 2003. In this period, there was a split from Nghe Tinh Province to Nghe An and Ha Tinh provinces as recently. Vinh City was selected for a new provincial town of Nghe An Province, and a large area of paddy rice and agricultural land had to change into residential land and government offices. Even in the mountainous districts of Nghe An, under this process, primary forests in Nghe An were continuously undergone strong exploitation activities for the construction such as that road system. Especially, the Land Law in 1993 had to be revised and effective from 1998 to 2003, and forestland allocation policy had been implemented in the province. Forestland was allocated to individual household in long-term use of forestland (land title certificate). The individual household had been providing land title (red book) for using forestland. However, until 2010, the policy was fully effective, and local people really realized their rights in using forestland, and therefore, the forest protection management has been paying attention by householder.

Table 35.1 illustrates the land use changes in Nghe An Province. In 1997, a total natural area of land was 1.649 million ha. Of which, agricultural land was 899,283 ha; annual crop area was 126,384 ha. However, this figure increased by 9294 ha and

[2] http://btxvnt.org.vn/cms/?m=14&act=view&id=119

Table 35.1 Land use status in Nghe An Province in 1997, 2007, 2013, and projected 2020 (Unit: ha)

	1997*	2007	2010	2013	2020
1. Agricultural land	**899,283.0**	**1,034,338.1**	**1,239,677.3**	**1,249,779.6**	**1,438,701.1**
Land used for annual crops	126,384.0	135,768.0	105,151.2	107,237.8	96,083.1
Rice cultivation	87,540.1	87,540.1	87,540.1	88,887.5	84,000.0
Long-term plantation trees	15,586.4	26,566.8	66,300.2	81,141.4	90,697.0
Forestland area	663,703.5	777,482.2	972,425.5	963,691.0	1,146,377.0
Production forest	*60,782.0*	*88,282.1*	*501,163.0*	*492,948.4*	*581,841.0*
Protection forest	*602,799.0*	*689,077.6*	*302,055.3*	*301,263.3*	*392,036.0*
Special-use forest	*122.5*	*122.5*	*169,207.2*	*169,479.3*	*172,500.0*
Aquaculture production areas	5231.0	6143.0	7422.3	7984.1	9030.0
Salt production areas	837.98	837.98	837.98	837.78	837
2. Nonagricultural land	**14,859.0**	**21,347.0**	**125,251.6**	**128,568.1**	**174,207.0**
3. Unused land	**735,227.0**	**593,683.9**	**284,440.1**	**271,021.3**	**36,461.0**
Total natural area	**1649,369.0**	**1649,369.0**	**1649,369.0**	**1649,369.0**	**1649,369.0**

Source: * Department of Statistic of Nghe An; data of 2007 to 2020 are provided by DONRE of Nghe An

29,094 ha in 2007 and 2013, respectively (DONRE, 2014). This was an increase in both paddy rice area and upland rice, as the Nghe An authority encouraged local people in expanding paddy rice area for food security. Due to increase in orange tree plantation and other fruit trees, area of perennial trees plantation had increased by 2765 ha from 1997 to 2013. On the contrary, the unused areas of the province have been reduced.

Forestland area had increased significantly in the period of 1997 to 2013 because of increase and expansion by forest plantation development. Acacia, rubber, and bamboo plantation had been increasing quickly in this period from 60,782 ha to 92,347 ha (the 661 program).

The acacia plantation has been planted in most of the mountainous districts of the province such as Ky Son, Con Cuong, Tuong Duong, Anh Son, and Thanh Chuong. During the period from 2010 up to now, acacia plantation is expanding and it becomes an economic value tree plantation of upland areas. In addition, rubber plantation had been planting in the province since 2007. According to Nghe An DARD (Nghe An Provincial People Committee [NAPPC], 2014), the total rubber area in Nghe An was about 4000 ha, which was mostly planted in Thanh Chuong and Anh Son districts. In 2010, the provincial people committee of Nghe An retrieved 5354.9 ha of forest areas, which was managed by the Anh Son Forestry Company to the Nghe An Rubber Stock and Investment Company for rubber plantation. Because rubber plantation has been being promoted continuously, area of long-term plantation in the province continues to be increased. The land use planning of Nghe An indicated that the rubber plantation area will be increased to

25,000 ha in 2020 (Nghe An Provincial People Committee [NAPPC], 2014).[3] However, area of rubber plantation has increased slightly as local people are not interested in rubber plantation because of low price compensation to land and acacia trees on land (NA#13), and it is a cause of deforestation, loss of biodiversity, and reduced water recharge (NA#12, NA#14). Until 2013, the actual rubber area production in whole province was 3718 ha with the yield of about 1.2 tones/ha. Currently, Nghe An Rubber Stock and Investment Company has faced problem in compensation to landowners.

3.2 *Multilevel Governance and Land Use Decision Making*

While the Vietnamese central government defines institutional structures and provides regulations and guidelines to the lower government and land users, forestry policy outcomes vary at the local levels. We assess multilevel governance by outlining the key actors in forestry decentralization and then analyze the strengths and weaknesses of the decentralization process in Vietnam. This is followed by a discussion of actor influence on deforestation as identified by respondents in the case studies.

3.2.1 Decentralization of Forest and Land Use Powers to Lower Governments

Forests in Vietnam have undergone a major transition. A period of centralized control, from 1975 to 1986, coincided with a natural forest loss of 1.7 million ha (To, 2009). Government claims over forests dismissed the rights of an estimated 25 million local people, including mostly ethnic minorities who lived in or near forested areas (Sam & Trung, 2003: 15). In 1986, the legal system was gradually revised, with management responsibilities of land and forests transferred to "lower" levels of government (Trung et al., 2015). Since 1994, the government system in Vietnam was divided into two parts: (i) central government and (ii) local government (provincial, district, and commune levels). Each of these government units remains under the control of the central state following a form of administrative decentralization, known as "deconcentration," with established lower governments still largely following national directives.

The Ministry of Agriculture and Rural Development (MARD) controls forest sector policy in Vietnam. At the provincial level, forests are under the remit of the Department of Agriculture and Rural Development (DARD) within the provincial people's committee (PPC), and responsibilities are further delegated to the

[3] Resolution on Adjust and addition to general planning of social–economic development of Nghe An to 2020 in the meeting of Nghe An People Committee 2014.

district-level divisions. Below the district level, the commune people's committee (CPC) is accountable for forest management and protection on the ground. Forest protection monitoring is conducted by the communes' forest rangers mandated to enforce the Law on Forest Protection and Development within their territory. Forest rangers cooperate with local forest protection groups at the village level. Each village has an assigned "village head" selected by the community, who provides the communication link between the commune government and the community. In 1992, following the issuing of the forest protection and development law and its subsequent revisions, the central government began forestland allocation (FLA), providing increasing rights and responsibilities to communities, households, and individuals.

MONRE is another key land use ministry, responsible for natural resources, including land itself and national land use planning. For REDD+ coordination, MARD shares responsibilities with the Ministry of Natural Resources and the Environment (MONRE), with divided duties outlined by the prime minister in 2011[4] (Pham, Moeliono, Nguyen, Nguyen, & Vu, 2012). The prime minister also approves provincial land use plans developed by the provincial and district people committees (DPCs).[5] Central government provides the guidelines on how land use plans should be developed, and MONRE, on behalf of central government, has a mandate to prepare national land use plans. DONRE at the provincial level has mandated to prepare and implement provincial land use plans. Land use planning is critical to how and what decisions are made on land use, as how local governments decide to classify land impacts its protection and management status. Thus, decision-making power over land use plans at the provincial level arguably provides the provincial government a strong instrument through which to influence land use decisions.

Furthermore, as part of the planning process, the classification of forests, for example, as production or protection, is vital to determine landholders' rights of access and use, which means there are socioeconomic impacts for forest-dependent smallholders. Thus, any oversights or mistakes may lead to conflict between the local government and local people.[6] A respondent who works in land use planning and classification admitted risks when the land use plan is developed without local inputs (NA#13). He said, *"land use planning is not suitable for [local] realities. Top-down planning is not practical."*

---

[4] No. 282/VPCOQHQT (January 13, 2011)

[5] One of the differences in Land Law 2013 is that there is no requirement for planning and preparation of land use plans at the commune level (Trung et al., 2015).

[6] Forests were classified into three categories: special use, protection, and production, each with differing rights according to the type of user group as defined in the 1991 Forest Protection and Development Law (De Jong et al., 2006). In early 2013, it was estimated that of Vietnam's total 13.66 million ha of forest, two million hectares was classified as special-use forest, 4.68 million ha was classified as protection forest, and 6.96 million ha was classified as production forest (MARD, 2013).

3.2.2 Decentralization and legitimacy in Land Use Decision-Making Powers

The Vietnamese government issued Land Law in 1993, and it was revised and amended in 2003 and 2013, which stipulates the requirement of land use planning and land use plans. Land use is defined by 5 types of land use planning and plans, which are (a) national land use planning and plan; (b) provincial land use planning and plans; (c) district land use planning and plans; (d) land use planning and plan for national defense; and v) land use planning and plans for national security. The planning of land use is implemented in 10-year rotation, and land use plans for national and provincial scale and national defense are prepared for 5-year term, but district land use plan is required on the annual basis.[7] The responsibility of implementing land use planning and preparation of land use plans as well as power to review and approve the planning and plans are clearly stated in the Land Law. The government is responsible for national land use planning and plan, which was directly supported by MONRE. The provincial and district people committees take the responsibility for land use planning and plans at provincial and district levels. The National Assembly is the agency to make the decision on approving national land use planning and plans. The prime minister is responsible for approval of provincial land use planning and plans (including planning and plan for national defense and security).

The coordination among the organizations regarding the management of land and forests includes vertical and horizontal coordination. Vertical coordination is consistent guidance made by the government to responsible ministries and then to provincial and district people committees. Horizontal coordination is made among management agencies in the province and district levels. At the provincial level, DARD and DONRE are the key management organizations responsible for land and forest management. The DONRE at district level was separated from a land management unit of DARD since 2003 and is responsible for management of all natural resources such as mineral, water, environment, and land (NA#10). All land use planning has to follow the DONRE's planning, before DARD has more power in land use planning decision but now DONRE has more power (NA#47). DONRE is the leading organization to prepare land use planning and plan for the province, and DARD is the leading agency for the management of forestry and agriculture operation and development in the province.

The chairman of provincial people committee (PPC) is authorized to approve land use planning and plans for districts.[8] The district levels are approved for the land use of commune. Currently, there is a weakness in land use planning at commune level because of low capacity of staffs and limited funding used for collecting basic information and data (NA#3, NA#10). *The weakest capacity is at commune and village level. It needs to strengthen capacity for commune officers because they*

[7] Articles 36 and 37 of Land Law 2013.

[8] Article 45 of Land Law 2013.

are nondegree and low education, and lack technical support (#5). Besides, due to the weak management mechanism combined with the high level of dependence on forests of people leads to forest degradation and deforestation (NA#13). The commune level also has land use planning for 5 years and 10 years following the guideline from district and province levels (NA#43); however, the actual planning is different from document planning (map) because the land use change every year but the commune could not update this situation since they do not have enough human recourses (NA#43, NA#44). This issue also raised by Mr. Tuan Anh (NA#14) that currently the land use planning of forests (special-use forest, protection forest, and production forest) is not appropriate showing in the map and in the real is some time different (NA#14).

Land use planning process has involved different departments such as DARD, DONRE, Army, Department of Trade and Investment, and Department of Culture and Tourist (NA#47). Therefore, there is a significant conflict between departments because every department also has land use planning (NA#47). For example, DARD has land use planning itself, and DONRE also has land use planning, Department of Planning and Investment, Department of Education, and Department of Transportation. Hence, the general land use planning for whole province is to collect land use planning from different departments, and DONRE is the host organization to collect the land use planning from different departments and then submitted to the PPC who will make the final land use for whole province (NA#47).

The reaction of commune level and provincial level is very slow, and it takes long time to solve issues. *For example, the conflict between villagers with Forest Company and Pu Mat National Park has been happing since 2005 up to now, but it has no solution from higher level. Local people want to get the land from forest company because they said that these lands belong to their parents, but the company gets it from them (the FC has permission from province level)* (NA#44). The other issues are that the state and private forest companies occupy a large forest area, while the local people do not have land to work, which is the cause of conflict between local people and forest owners. The government needs to make land use planning according to local needs. The government can revoke the inefficient lands to allocate to households (NA#14).

3.2.3 Multiple Actors' Influence on Forests

Actors at different levels were identified as having varying influence on forests and forestland, with both complementary and conflicting interests within and among levels. In the case study locations, key land use change actors included state companies (HPP in Chi Khe and Yen Na communes) supported by national government and the afforestation programs (in Thach Giam Commune). The private sector (a paper mill company) and small business enterprises (focused on afforestation activities) were also important in the Luc Da Commune. All of these interventions and actors have impacted smallholder livelihoods and forests to varying degrees.

Land use change, to the detriment of forests, illustrates the influence of more powerful land use actors that can encourage or even force smallholders to convert forests as an alternative livelihood strategy. Respondents frequently cited HPP companies as highly influential and as drivers of deforestation and degradation (NA#3, NA#30).[9] HPP (in terms of the larger stations) must first have approval and support of the national government, under the Ministry of Industry and Trade, and the provincial government. Furthermore, each government level is responsible for identifying and planning the location of the HPP, as well as evaluating and regulating the HPP environmental impacts (Trung et al., 2015). While each government level has a responsibility for HPP development, the provincial government has substantial decision-making power. In the Yen Na Commune, the HHP recovered land from households for construction and flooded the area; complaints were made to the commune and district government, and a commune government employee admitted they *"did not understand why those people had not yet been compensated"* and that they had *"raised the issue several times"* (NA#42). This point implies that those at the commune level are more aware of the local impacts of these interventions than those in higher government levels, but still have little power when it comes to decisions; e.g., in this case, compensating affected people. Equally, the commune government has a role to bridge the voice of the local people to higher government levels, and findings indicate that this occurred in practice (Trung et al., 2015). Interviewed village heads commonly described the most frequent communication between the CPC and communal forest ranger (NA#8). Yet, communication to the commune government demonstrates that this may be to no avail if the district and provincial governments ignore their own responsibility to respond to local-level concerns.

The provincial government was recognized as one of the most influential actors for land use change. A respondent from the Nghe An PCP said that was because it has the ability to change *"approximately 20,000 hectares of forestland, converting protection forests to production forests"*[10] (NA#6). That is, the provincial government can change the classification from natural forest, with heavy restrictions on production, to increase rights to allow harvesting. Nevertheless, the central level still has a critical role, as all land use plans, such as these, must be approved by the prime minister (NA#6; Trung et al., 2015).

Environmental and development program fall under the responsibility of a multitude of ministries and agencies, which often have overlapping roles (Pham et al., 2012). A report by the World Bank (2011) stated that this was an issue at a national level: *"lack of inter-sector cooperation by MARD and MONRE in land use and development planning, and the lack of an agreed upon ... forest and land use classification threatens the attainment of overall forest sector policy goals"* (p.12). One key informant from DARD also observed the same difficulties between DONRE and DARD at the provincial level *"because each department has its own*

[9] Linked 22 times with deforestation in the interviews conducted.

[10] Under the Pursuant Decision No. 38 by prime minister.

management regulations" (NA#11). He also argued that DONRE was in fact more influential than DARD, as DONRE is responsible for land use plans, so it can make more powerful decisions. Analyzing these perspectives together, along with the legal context of land use governance in Vietnam, it appears that the agricultural sector has considerable freedom to facilitate the establishment of plantations, and the environment sector has limited scope for challenging such decisions, and rarely even attempts such contestations.

The role of government in land use decision making does not mean, however, that local people are without influence. In a number of examples, smallholders refused to bow to external interests. For example, a villager in the Chắn Nằn hamlet of Nghe An rebuked offers made by the HPP (NA#31). In a more unique example, the commune authority attempted to seize illegally harvested timber but local people refused by threatening violence (NA#24). Although these examples are more exceptions than the rule, they demonstrate that smallholders in some cases could influence the direction of change.

Pham et al. (2012) found that policies that withdraw productive assets from local people would often put those already marginalized, such as the poor, at a further disadvantage by failing to offer viable alternatives. Overall results from this study indicate that local people were often associated with deforestation and degradation as a result of (1) a lack of livelihood options or incentives; (2) agricultural lands being converted for HPP or rubber plantations (thus other indirect forces are evidently at play); (3) the government's failure to provide clear land rights and access to forests, as well as weak forest monitoring and inventory; and (4) a lack of legitimacy in procedural and distributive outcomes in BSMs and other government programs. Yet, smallholders were also widely recognized for their positive influence on forests. Smallholders were linked on 72 occasions with forest improvements. Four of the case study sites, e.g., PFES, government afforestation programs, and the REDD+ pilot project,[11] illustrate how BSMs focused on local people could be effective in promoting forest protection and development (Table 35.2).

3.2.4 The Role of Lower Governments in Practice

Forest protection and development are key features of Vietnam's land policy and structural reforms, leading to attempts to further include lower government and local people in the forestry sector. Respondents were asked to explain their perceptions of decentralization, including the relationships between government levels, their relative strengths and weaknesses, and whether decentralization had been effective.

The provincial government was recognized to have "*more power in making decisions on how to make use of resources and how to develop the province.*" Some respondents claimed these challenges were the same among the three levels of local

[11] Case study communes recognized as sites of decreasing carbon emissions.

Table 35.2 Actor influence in land use change in Luc Da Commune, Con Cuong District, Nghe An Province

Actor mentioned	Influence (1: lowest; 3: highest)	Interest (1: lowest; 3: highest)	Reason why
Villager/household	1	3	They are very interested in land use but they do not have enough power to force commune to follow their idea
Farmer associations, youth union	2	3	They are very interested in land use, but they do not have enough power to force commune to follow their idea
DARD at district level	3	3	They have power to do land use planning because they have right to approve
DONRE at district level	3	3	The same DARD because they get recommendation from DARD
Department of industry and trade at district level	2	2	Less care about land use and they only comment about road or building
Head of village/ village committee	2	3	He can talk to villager and explain their idea
Communal people committee	2	2	They received land use plan from village head and submit to the DARD at district level
Department of Finance	3	3	They do not care about land, but they have right to approve the budget in investing
District PC	3	3	They have highest power at district level, and they will approve all land use plan within district
Tan Hong paper mill company	1	3	The company invested seedling and fertilizer to villagers and made the contract to buy acacia products

government, while others highlighted variation among the levels. For instance, *"at provincial level [there are] enough power and resources to ..[act] effectively but at district levels and commune levels [there] are not enough. This needs to be changed" (NA#11).*

Respondents were specifically asked whether decentralization had been effective. In terms of the government objective for the forestry sector, "effectiveness" was interpreted as improvements in and development, as this has been the objective of devolving forestry management—to promote local government and local people as protectors and managers of forest resources (Trung et al., 2015). Hence, benefits from decentralization of the forestry sector were recognized, but almost 13 respondents in the study did not answer the question. This was both as a result of not knowing what to say, but also stemming from issues with the question itself.

Analyzing respondents' perceptions of decentralization reveals a broad consensus that the provincial governments, while theoretically implementing the central governments' mandates, in fact, have considerable influence and are consequently important centers of political negotiation around land use. The key benefits of decentralization in the forestry sector have been raising public awareness on forest protection. Overall, however, the obstacles to and weaknesses of decentralization were much more notable in the findings.

There is apparent variation in how responsive communes and districts are to local needs, and there is widespread agreement among respondents—and especially local people—that communes that are responsive can have positive impacts on procedural equity in land use. Despite many achievements in decentralizing the forestry sector, respondents commonly felt that, in addition to inherent constraints, efforts were hampered by a number of factors, namely capacity, human resources, unclear guidance, and budget. For example, government officials themselves frequently commented, "*local governments do not have good human resources, suitable budgets or good instructions for implementing their tasks effectively*" (NA#10). The ability of actors to make and enact decisions from the different levels seemed to be impeded by the distance of higher-level decisions from local reality and the capacity of lower levels (particularly at the district and commune levels).

The findings, however, were not unanimous. Opinions varied across the different levels and case study sites; for example, capabilities and resources had improved in one place, but were constrained in another (NA# 10, NA#12, NA#13, NA#14, NA#17). The obstacles identified reflect the common views of government respondents. The following subsections describe the summarized weaknesses and challenges to lower governments' efforts in the forestry sector as identified by respondents.

Poor cooperation and coordination between government levels and sectors: Poor cooperation, moreover "poor communication," between different levels of government and organizations, was also a barrier to decentralization (NA#47). Respondents explained that their own roles in land management and protection depended heavily on the cooperation with other actors, including the CPC, people's committee of the district (DPC), local people, DARD, forest ranger station, and DONRE. Nevertheless, poor cooperation was believed to have influenced the quality of land use plans prepared by different government levels, as well as FLA. One respondent said "*there is inefficient collaboration among the agencies in implementing allocated tasks. Land conflict is an increasing trend, particularly…in forest land areas*" (NA#47). These points suggest a need to build cooperation between government levels and sectors as well as within the sector to ensure effective information exchange.

Limited resources of lower government.

Weak technical skills.

It was fairly consistently stated in the interviews that weak capacity was a key issue in policy implementation and one of the biggest hurdles to "effective" decentralization (NA#5, NA#10, NA#12, NA#13, NA#40). This was not consistently associated with all local government levels but rather was mostly an issue for the district and commune levels. A member of the commune peoples' committee (CPC) admitted "officials of districts' and communal governments do not understand their mandates or responsibilities. They do not know how to carry out their assigned tasks. They also lack capacity to carry out the assigned tasks." This is a sweeping statement, and capacities undoubtedly vary. Nevertheless, numerous respondents from both provinces agreed that a lack of knowledge, education, and technical support was common challenge at the village, commune, and district levels (NA#5).

Lack of human resources.

The lack of human resources in lower government was further cited as an issue for decentralization. Insufficient human resources were linked with unrealistic policy expectations from upper government. For example, forest rangers are expected by central government to patrol large areas of forest with small budgets and perceived to be unrealistic by a number of forest rangers. A forest ranger in Nghe An Province explained as there were no enough staff to patrol assigned areas; hence, their abilities to monitor forests effectively were hindered (NA# 16).

Lack of financial resources

Oates (1998) argues that a centralized funding mechanism can be unfavorable to the establishment of integrated and responsive local policy-making. The issue of insufficient budgets was commonly noted as a weakness for lower government in the province (NA#15). This point suggests that the central government is aware of the magnitude of work requested, but still allocates inadequate resources. Land Laws and laws on forest protection and development, for instance, require that local governments allocate forestland to local households, communities, and so forth. However, there is no regulation on budget, human resources, and other resources in order to implement the allocation activities; this leads to poor or incomplete FLA, as evident in study communes. Budgets for forest protection programs and payments to local people were also argued to be insufficient. A member of the Forest Inventory and Planning Group (NA#12) noted the "*protection management fee ...of 200 thousand VND/ha/year is not [enough to] encourage local people to protect forest; they have to look for alternative income from forests.*" This response pointed out that policies for forest protection and development will be challenged if the incentives provided cannot provide for peoples' livelihoods.

Overall, respondents, but particularly provincial authorities, often cited lower-level government capacities as major concerns. Nevertheless, they were less likely to recognize contributing factors that limit the effectiveness of decentralization policies, such as conflicting interests or the political vision of central government, as will be seen below.

3.3 Risks Surrounding Land Use Decisions

Currently, land use planning is prepared by top-down from the provincial levels to district and commune levels. The chairman of provincial people committee has the highest power to approve land use planning of the whole province, which is submitted from departments. The existing process has several risks/weakness because of inadequate information and consideration during the preparation process, particularly the local condition and engagement of different stakeholders, and the land use plan is commonly not suitable to the real conditions of locality that causes low feasibility for the implementation of land use plan (NA#47). The lack of manpower at local levels is the main constraint of ineffective land use planning process (NA#44).

The goal of land use policies is to improve local livelihoods. The biggest challenge is to ensure food security of people who live near the forest because most of them are relying much on the forest (NA#12). Currently, food insecurity is often observed in the case studies of Nghe An; normally, they often lack food about 4–5 months a year (NA#27, NA#31, NA#35, NA#37, NA#40). The shortened fallow length and decrease in swidden fields have led to food insecurity to local people because the main source of income is from rice production (Nguyen, Tran, & Nguyen, 2011). Intensive agricultural production to ensure food security will reduce the pressure on forest (NA#45). The food insecurity of the households who lost their land by the HPP is clear evident (NA#23, NA#42).

Environmental pollutions such as water pollution are caused by the HPP and factories in the villages. Currently, many people in the Mac Village, Thach Giam Commune, are facing unclean water for their living. The water pollution caused by Tan Hong Company is reported on the local authorities. The HPP constructions and reduced forestland are the main cause of water pollution and lack of water. Ms. Anh (NA#25), a village leader of Mac Village, said that "*The people are very frustrated because they do not have domestic water, the river water is very polluted, but they have to take it. Water quantity loss is caused by the deforestation in the upper-stream.*"

4 Conclusions

Decentralization, in the form of deconcentration, has increased the responsibilities of lower levels of government in Vietnam. Lower-level governments largely implement higher-level government policies, but provincial governments are now seen as having important influence over decisions about land and land use, particularly through the land use planning and classification process. Although these plans must be approved by the central government, provincial governments are seen as playing an important role, at least, in determining *how* policies will be implemented locally. If policies from the central government are contradictory, this suggests that provincial governments may, thus, have very important influence on forests and local livelihoods. Other levels of local government are more limited, both in terms of their decision-making power and in terms of their technical, human, and financial capacities. Nevertheless, commune governments have a mandate to bring the voices of local people to the district level and are seen as intermediaries for higher-level government decisions (Trung et al., 2015). Commune and district governments are crucial for delivering and implementing policy but also in a bridging accountability of the government to the local people; however, their ability to do this will remain constrained if they afforded scarce power to influence such decisions, as will be seen in the next section.

With regard to coordination between and among levels of government, there are calls for clearer policy guidelines from higher to lower levels, while leaving room for local discretion to adapt policies appropriately—that is, for greater accountability. There are also calls for greater collaboration and information exchange and more effective land use planning. The outcomes of forest and land use policies appear to be driven by (i) will, interest, and attitudes of influential actors to promote forests over other land use goals; (ii) the effectiveness of coordination and coalitions among those actors and between these actors and local communities; and (iii) local people's understanding of pros and cons of these land uses, and their confidence and right to accept or reject these land use changes. Although decentralization in Vietnam has given more decision-making power to the provincial government, in the context of an important platform for land use negotiations, the real power still lies in the hands of the central government.

Acknowledgements This research is part of CIFOR's Global Comparative Study on REDD+ (www.cifor.org/gcs). The funding partners that have supported this research include the Norwegian Agency for Development Cooperation (NORAD).

References

Larson, A. M., & Petkova, E. (2011). An introduction to forest governance, people and REDD+ in Latin America: Obstacles and opportunities. *Forests, 2*(1), 86–111.

Leisz, S. J., Ginburg, R. F., Nguyen, T. L., Tran, D. V., & Ramussen, K. (2011). Geographical settings, Government Policies and Market Forces in Upland of Nghe An. In T. Sikor, N. P. Tuyen, J. Sowerwine, & J. Romm (Eds.), *At the upland transformations in Vietnam*. NUS press.

Luttrell, C., Loft, L., Gebara, M. F., Kweka, D., Brockhaus, M., Angelsen, A., & Sunderlin, W. D. (2013). Who should benefit from REDD+? Rationales and realities. *Ecology and Society, 18*(4).

McElwee, P. (2011). Who should manage the land? Common property and community response in Vietnam's shifting uplands. In T. Sikor, N. P. Tuyen, J. Sowerwine, & J. Romm (Eds.), *At the upland transformations in Vietnam*. NUS press.

Nghe An Provincial People Committee [NAPPC]. (2014). *Report on Social Economic in 2014 and the plan for 2015*. Date 05 Dec, 2014.

Nguyen, Q. T. (2010). *Land law reform in Vietnam- past & present*. ASLI working paper, no. 015, August 2010. Retrieved from www.law.nus.sg/asli/pub/wps.htm

Nguyen, D. T., Tran, D. V., & Nguyen, T. L. (2011). *Too much focus on forest conservation, too little on food*. Policy Brief, RECOFTC. Retrieved from http://forestindustries.eu/sites/default/files/userfiles/1file/RECOFTC_21July10_B2_web.pdf

Pham, T. T., Moeliono, M., Nguyen, T. H., Nguyen, H. T., & Vu, T. H. (2012). *The context of REDD+ in Vietnam: Drivers, agents and institutions (no. CIFOR occasional paper no. 75)*. Center for International Forestry Research (CIFOR), Bogor, Indonesia.

Ravikumar, A., Kijazi, M., Larson, A. M., & Kowler, L. (2015). *Project guide and methods training manual*. CIFOR.

Sam, D. D., & Trung, L. Q. (2003). Forest policy trends in Vietnam. In M. Inoue & H. Isozaki (Eds.), *People and Forest - policy and local reality in Southeast-Asia, the Russian Far East and Japan*. Kluwer Academic Publishers.

To, X. P. (2009). *Forestry in Vietnam: After 20 years accompanying with the national renovation*. MARD.

Tran, D. V., Pham, T. H., & Rasmussen, M. S. (2006). Social and environmental dimensions of land use changes in the ca river basin in Vietnam. *CARES, 2006*.

Trung, L.Q., Phuong, V.T., Yang, A.L., Hai, V.D. (2015) *The distribution of powers and responsibilities affecting forests, land use, and REDD+ across levels and sectors in Vietnam: A legal study*, CIFOR occasional paper no. 137, DOI: https://doi.org/10.17528/cifor/005743.

World Bank. (2011). Using strategic environmental assessment to inform the forestry master plan 2010-2020 of Vietnam. In *The International Bank for Reconstruction and Development*. Washington, DC.

Chapter 36
Technical Efficiency of Irrigation Water Use of Robusta Coffee Production in the Dong Nai River Basin (Vietnam): A Case Study of Lam Dong Province

Nguyen Dinh Tien, Tran Nhat Lam Duyen, Nguyen An Thinh, and Nguyen Truc Le

Abstract Irrigation water use of coffee farmers in the Central Highland of Vietnam has faced several challenges such as drought and insufficient irrigated water. Our paper analyzes the efficiency of irrigation water use for Robusta coffee (*Coffea canephora*) in Lam Dong Province using Cobb–Douglas production function. The Technical Efficiency (TE) and Irrigation Water Use Efficiency (IWUE) were analyzed using a Data Envelopment Analysis (DEA) model. The Cobb–Douglas results show that the volume of irrigation water and amount of working capital, labor, and farm size significantly influence coffee productivity. The study recommends that preventing and mitigating water shortages in coffee farms require not only national policy supports (access to credit and extension services, training, land management) but also improving household-level effort in farming practices that are needed.

Keywords Data envelopment analysis (DEA) · Efficiency · Irrigation water · Robusta coffee · Vietnam

1 Introduction

Coffee was exported for USD 3.74 billion in 2012, and it is the second largest export-earning crop in Vietnam (Amarasinghe, Hoanh, D'haeze, & Hung, 2015). The coffee production contributed significantly to poverty reduction in Vietnam's

N. D. Tien (✉) · N. A. Thinh · N. T. Le
VNU University of Economics and Business, Vietnam National University, Hanoi, Vietnam

T. N. L. Duyen
VNU School of Interdisciplinary Study, Vietnam National University, Hanoi, Vietnam

A. T. Nguyen, L. Hens (eds.), *Global Changes and Sustainable Development in Asian Emerging Market Economies Vol. 2*, https://doi.org/10.1007/978-3-030-81443-4_36

rural areas (Kotecha, Hieu, Kuit, Von Enden, & Swinkels, 2002; World Bank, 2004). Vietnam's coffee area is mainly concentrated in the Central Highland (Marsch, 2007) with 95% of Robusta coffee planting in Dak Lak, Gia Lai, Kon Tum, and Lam Dong (Giungato, Nardone, & Notarnicola, 2008; Haggar & Schepp, 2012).

Robusta coffee is profitable when it is grown intensively with large inputs of fertilizer, water, and labor (Marsch, 2007). However, the main production constraint of coffee producers in the Central Highlands is water which is insufficient to meet the increasingly competing demands of agriculture, industry, and households, as well as poor irrigation management (Amarasinghe et al., 2015; Dewbre, 2010). Farmers' practices of over-irrigation lead to depletion of groundwater reserves (Ahmad, 2001; Amarasinghe et al., 2015; D'haeze, Deckers, Raes, Phong, & Loi, 2005; Technoserve, 2013), while 56.6–95% of water for coffee irrigation is mainly generated from groundwater (Haggar & Schepp, 2012; Marsch, 2007). In 2016, about 470 hectares of coffee in the West Highland and southeast region had insufficient water during the dry season (Ministry of Natural Resource and Environment of Vietnam [MONRE], 2016). According to the West Highlands Agriculture and Forestry Science Institute (2016), only about 72% of the regions' households were able to secure sufficient amounts of water to irrigate their coffee farms.

Although Vietnam is one of the most competitive coffee producers worldwide, the sustainability of coffee production has been challenged by climate change, deforestation, and insufficient irrigation water (Luong & Tauer, 2006; Nguyen, Bokelmann, Ngo, Do, & Nguyen, 2019; Technoserve, 2013). National programs, such as the Vietnam–Netherlands Partnership on Water for Food & Ecosystems (WFE) and the German–Vietnam project—Integrated Water Resources Management Vietnam, Planning and Decision Support, have been implemented to promote sustainable use of water resources (Long, Gerardo, & Jake, 2009). These programs mainly determine the levels of water supply and demand and consequent water surpluses or deficits (Jolk, Greassidis, Jaschinski, Stolpe, & Zindler, 2010). Such studies have neither presented estimates for water use efficiency nor proposed key determinants that affect water use efficiency. It is necessary to undertake research on water use in coffee production, with the aim of understanding whether irrigation water use is efficient in hotpot areas or not.

This paper aims to calculate Technical Efficiency (TE) and proposes a measure for Irrigation Water Use Efficiency (IWUE) based on the concept of input-specific technical efficiency for Robusta coffee production in Lam Dong Province, Central Highlands Vietnam. The technical efficiency of coffee is fundamental to increase farm-level total factor productivity, returns to coffee farmers, and stabilizing the region's underling agroecology (Cheesman, Son, & Bennett, 2007).

2 Materials and Methods

2.1 Study Area

The selected Lam Dong Province is located in the southern part of Central Highlands Vietnam. The province has several fertile high plateaus, and a large percentage of the province is forested (Jolk et al., 2010). In 2017, the total cultivated area in the province was 279,000 ha, of which 160,000 ha (24.5% of total area) was planted with coffee. Since 2010, the total area of coffee plantations has increased by 12,000 ha. By 2020, the total area planted with coffee over 20 years of age will be 60,000 ha (Department of Agriculture and Rural Development of Lam Dong [DARD], 2017). Lam Dong Province with its warm tropical climate and distinct dry and rainy seasons influenced by the South Asian monsoon is suitable for coffee production. Coffee is a key cash crop of Vietnam with Lam Dong Province accounting for 23% of total national land area allocated to coffee production. However, coffee production is severely constrained by the lack of irrigation water, especially during the dry months when the level of groundwater drops significantly. Currently, the main water supply source is surface water (80%), while the remaining 20% is sourced from groundwater. However, surface water and groundwater have become polluted due to intensive agricultural production (Department of Agriculture and Rural Development of Lam Dong [DARD], 2017). During the dry season of November 2013 to April 2014, for example, the amount of coffee produced from about 3600 ha of planted area was reduced by 20%, while agricultural crop production from 36,200 ha dropped by 5% due to the shortage of irrigation water (Department of Agriculture and Rural Development of Lam Dong [DARD], 2017), which in Lam Dong Province is further aggravated by poor water management. Free or wild flooding which is inefficient is the usual method of distributing irrigation water in the province's coffee-growing areas. Only a few areas utilize more efficient systems of irrigation, such as furrow irrigation that would conserve water (Department of Agriculture and Rural Development of Lam Dong [DARD], 2017). The coffee beans are harvested during the dry season, from November to March. In Lam Dong Province, these are sun-dried in patios and hulled locally, while the green beans are sold to local traders and exporters.

Three districts (Di Linh, Lam Ha, and Bao Lam) were selected for this study because of the vast tracts of land devoted to Robusta coffee production, as well as total production. Together, they comprise approximately 71.51% of coffee area in Lam Dong Province (General Statistics Office (GSO) of Vietnam, 2013). The 6530 ha planted to coffee in this area is suffering from severe water shortages that may partly be attributed to weak management of water resources. Lam Dong Province has suffered from drought since 2010, with the majority of the coffee area (78%) using surface water (water from lakes and ponds) and the rest utilizing groundwater (Japan International Cooperation Agency [JICA], 2018). In 2015, the VnSAT project with funding from the World Bank in the amount of VND 197 billion (USD 8.5 million) was implemented in eight districts of Lam Dong. The objective was to save water for coffee production through the use of sensors and smart irrigation technology.[1] The project aimed to support 14,700 households in applying advanced technology to 16,000 ha of coffee within the province. However, until 2018, only 17 households applied this technology due to the high investment costs. The majority of coffee farmers in Lam Dong continue to use the overflow irrigation method, using pumps to harvest water from rivers and lakes.

2.2 *Sampling Selection*

A simple random sampling method was employed to selected coffee farmer respondents for the survey. The sample size for farmers was calculated based on (Farrell, 1957) the following:

$$n_0 = \frac{Z^2 pq}{e^2} \tag{36.1}$$

where n_0 is the sample size; Z^2 is the abscissa of the normal curve that cuts off an area α at the tails (1 - α) that equals the desired confidence level (e.g., 95%); e is the desired level of precision (sampling error); p is the estimated proportion of an attribute that is present in the population; and q is 1 - p. The value for Z refers to the area under the normal curve found in statistical tables.

Then, the finite population correction for proportions method to adjust n_0 achieved from eq. (36.1) is as follows:

$$n = \frac{n_0}{1 + \frac{(n_{0-}1)}{N}} \tag{36.2}$$

where n is the sample size after adjustment and N is the population size of Robusta coffee households in Lam Dong Province, Vietnam.

Total numbers of households engaged in Robusta coffee production in Di Linh, Lam Ha, and Bao Lam districts were 41,651, 40,215, and 25,098 households, respectively. As determined using eqs. 36.1 and 36.2, a total of 194 households were selected for this study. The respondents were willing to interview and agree to share their information.

[1] https://baogialai.com.vn/channel/1624/201903/lam-dong-tiet-kiem-50-nuoc-tuoi-ca-phe-bang-cong-nghe-thong-minh-5625343/index.htm

2.3 DEA Approach of IWUE

The level of efficiency may be determined by estimating the production function from the sample data, using either the parametric or nonparametric methods (Cochran, 1963). DEA deterministic analysis and Stochastic Frontier Analysis (SFA) are the two main approaches that can be used to calculate efficiency. The stochastic frontier production, in particular, the Cobb–Douglas stochastic frontier production, is a parametric method using stochastic estimations that impose an explicit functional form and distribution assumption on the data. However, the stochastic frontier production function is limited by its incorporation of a random error term (Aigner, Ameiya, & Poirier, 1976). The DEA is a linear programming-based technique for evaluating the relative efficiency of Decision-Making Units (DMUs) and is used to construct a piecewise frontier of the data. Terms like DMU are used to emphasize that the interest is centered on decision making by not-for-profit entities, rather than more customary firms and industries (Charnes, Cooper, & Rhodes, 1978). The best way to introduce DEA is via ratio form of all outputs to all inputs for each farm/DMUs. The optimal weight may be derived by specifying the mathematical programming problem.

In this study, the IWUE is defined as the ratio of effective water use to the water applied to the crop (McGockin, Gollehon, & Ghosh, 1992; Omezzine & Zaibet, 1998). The standard radial is not appropriate for measuring the individual efficiency of inputs used, as it measures the contribution of each input to productive efficiency equally (Reinhard, Lovell, & Thijssen, 1999). Therefore, it can be calculated via the subvector technical method for each individual input. Individual efficiency is a non-radial notion of input efficiency measure that allows for a differential reduction in the inputs applied (Ball, Lovell, Nehring, & Somwaru, 1994; Färe, Grosskopf, & Lovell, 1994; Lansink, Pietola, & Bäckman, 2002; Reinhard, Lovell, & Thijssen, 2000).

Mathematically, the input-oriented model for estimating IWUE can be written as shown in Eq. (36.3) (Färe et al., 1994; Frija, Chebi, Speelman, Buysse, & Van Huylenbroeck, 2009; Lilienfeld & Asmild, 2007) using the notion of the proposed subvector efficiency. The technical subvector efficiency for variable input k irrigation water is calculated for each firm's i by solving the following linear programming problem:

$$\theta^{t} = \min_{\theta,\lambda} \theta \qquad \textit{Subject to}: \tag{36.3}$$

$$\sum_{j=1}^{n} \lambda_j y_{m,j} \geq y_{m,i}$$

$$\sum_{j=1}^{n} \lambda_j x_{k-t,j} \leq x_{k-t,i.}$$

$$\sum_{j=1}^{n} \lambda_j x_{t,j} \le \theta^t x_{t,i}$$

$$\sum_{j=1}^{n} \lambda_j = 1$$

$$\lambda_j \ge 0$$

where θ^t is the input subvector technical efficiency score for input t for each DMU. The measure θ^t represents the maximum reduction in variable input t holding outputs and all remaining inputs (*n-t*) constant. The θ^t can have a value between 0 and 1, where a value of 1 indicates that the observation is a best performer located on the production frontier and has no reduction potential on irrigation water. Any value of θ smaller than 1, however, indicates water use inefficiency, i.e., that excessive irrigation water is being used. On the other hand, λj is a vector of n elements, representing the influence of each DMU in determining the efficiency of the DMU; x_t is the subvector of the inputs contracted for the production of outputs; and x_{k-t} is the vector of all other inputs. The term $\sum_{j=1}^{n} \lambda_j y_{m,j}$ is the weighted sum of outputs of all DMUs, which must be superior or equal to the output of DMUi ($x_{t,i}$, $y_{t,i}$).

The DEA was run on EXCEL-PC. The outputs of the DEA linear programming problem in model (36.3) were technical efficiencies and IWUEs.

3 Results

3.1 Characteristics of Coffee Farmers

The average land area of Kinh and indigenous coffee farmers in the selected study sites is 2.3 hectares per household. Of the 194 respondents interviewed, 139 were of Kinh origin and the rest were indigenous to the area. Most of the Kinh people migrated from other provinces, even from the north of Vietnam, following the government's migration policy in the 1970s and 1980s (Harvey et al., 2014).

A comparison of the socioeconomic characteristics of the Kinh and indigenous farmers in the study sites shows that most Kinh farmers are better educated with secondary education compared to indigenous farmers who only have elementary education (Table 36.1). The indigenous farmers have larger coffee areas and more experience in coffee cultivation than Kinh farmers. There was not much difference, however, in farm size and experience in coffee production between the two groups. Although the amount of irrigation water used by indigenous farmers (4766.8 m^3) was higher than that used by Kinh farmers (4719.6 m^3), there was no significant difference between the Robusta coffee productivity of indigenous farmers and those of Kinh farmers.

Table 36.1 Summary statistics of Robusta coffee farmers in Lam Dong Province, Vietnam, 2017

	Kinh group		Indigenous group		All	
Variables	Mean	S.D	Mean	S.D	Mean	S.D
Age (years)	44.0*	11.0	42.0*	12.0	44.0	12.0
Education level	3.0*	0.7	2.0*	0.8	2.6	0.8
Household size (people)	5.0*	1.4	6.0*	2.2	5.0	1.7
Experience (years)	18.9	6.1	20.7	6.9	19.4	6.4
Farm size (hectares)	2.2	1.3	2.3	1.3	2.3	1.3
Irrigation water (m^3)	4719.6	3321.2	4766.8	3167.7	4733.0	3270.3
Coffee output (kg)	5864.4	3487.6	5051.1	2857.3	5633.8	3334.2
No. of observations	139		55		194	

Note: * indicates difference between means of two groups is statistically significant at 95% confidence level in paired t-test; S.D.: standard deviation

3.2 *The Response of Robusta Coffee Productivity to the Level of Irrigation Water Used*

The Cobb–Douglas coffee production function was used to analyze the influence of the explanatory variables, namely quantity of irrigation water (m^3), farm size (hectares), labor (man-days), and capital (VND) on coffee productivity. These explanatory variables were selected based on a previous study of Atici and Podinovski (2014) and the estimated pairwise correlation coefficients ($r > 0.6$). Other factors such as fertilizers and pesticides were excluded because of the low value of their pairwise correlation coefficients and many outliers. Similarly, exogenous factors, such as weather conditions and government policy, were excluded in this model because farmers have no influence on these factors (Harvey et al., 2014). In addition, the variables were tested for multicollinearity and heteroscedasticity problems in the empirical model. Park's test was used to determine heteroscedasticity issues. Results indicated that homoscedastic errors were not rejected in all cases, indicating no serious heteroscedasticity issues.

The OLS regression results indicated that the four independent variables (quantity of water, labor, farm size, and working capital) were statistically significant and had a positive influence on the level of Robusta coffee output. The adjusted R-square value was 78.8%. This means that 78.8% of the change in the Robusta coffee output was explained by the changes in the quantity of irrigation water, labor, capital, and farm size (Table 36.2).

Our findings show that a 1% increase in irrigation water results in a 0.163% increase in coffee output (Shammout, Qtaishat, Rawabdeh, & Shatanawi, 2018; Tesfaye, Mohd, Kausar, Marziah, & Ramlan, 2013). Similarly, an equivalent increase in capital, labor, and farm size increases coffee output by 0.072%, 0.175%, and 0.536%, respectively. This means that coffee output is most responsive to the size of farm and least responsive to amount of capital. The low output response to water and capital might suggest that water use in the study area is below productive potential.

Table 36.2 OLS regression results for coffee production function, Lam Dong Province, Vietnam, 2017

Independent variable	Coefficient	Std. Err	T-value	P-value
Log constant	4.581*	0.690	6.63	0.059
Log capital	0.072***	0.038	1.90	0.005
Log labor	0.175***	0.062	2.84	0.001
Log water	0.163***	0.047	3.50	0.000
Log farm size	0.536***	0.063	8.56	0.000
R-square		0.792		
Adjusted R-square		0.788		
F (4, 189)		180.24		
Prob > F		0.0000		
Root MSE		0.2928		

Note: ***, **, and * are significant at 1%, 5%, and 10% probability level, respectively. NS is not significant at 10% probability level

The sum of all production elasticities of inputs (regression coefficients) in the Cobb–Douglas production model is 0.946. As this figure is less than 1, it indicates a situation of variable return to scale or decreasing return to scale. This implies a less than proportionate increase in output of coffee, given a certain level of input. This variable return to scale suggests that investments in new technologies would be a better alternative for increasing productivity, rather than increasing the quantity of inputs applied.

Results from the OLS econometric model could not prove whether resources were efficiently utilized or not. The results only reveal the functional relationship between the factors of production and output, with the assumption that all respondent farms were fully efficient (Coelli, 1998), which is not true in all cases. It is therefore necessary to complement these results with a technical efficiency analysis. Likewise, given the water shortage for coffee production in Lam Dong Province and the effect on coffee productivity, as shown by the regression analysis in this section, it is also necessary to analyze irrigation water use efficiency using DEA. The next section presents results for irrigation water use efficiency.

3.3 *The DEA Results: TE and IWUE Scores*

The overall TE and IWUE scores, given CRS and VRS in the sample and the two groups of Kinh and indigenous farmers, are summarized in Tables 36.4 and 5, respectively. The TE scores for all coffee farmer respondents ranged from 30% to 100% with an average of 72% for the VRS DEA model, while for the CRS DEA model, the TE scores ranged from 21% to 100% with an average of 66%. These results revealed that inputs for coffee production were not being efficiently utilized. The current level of coffee output could still be attained even if the amount of inputs used is reduced by 28% and 34% based on the VRS and CRS, respectively. The

Table 36.4 Frequency distribution of irrigated water use efficiency in Lam Dong Province, Vietnam, 2017

	IWUE					
	VRS			CRS		
Efficiency (%)	Kinh group	Indigenous group	All	Kinh group	Indigenous group	All
Summary statistics						
Mean	0.54	0.69	0.52	0.40	0.52	0.39
Minimum	0.14	0.11	0.13	0.10	0.10	0.12
Maximum	1.00	1.00	1.00	1.00	1.00	1.00
Std. dev	0.29	0.30	0.28	0.24	0.27	0.24
Efficiency interval						
100	15 (11)	14 (25)	17 (9)	5 (4)	6 (11)	9 (5)
90–100	14 (10)	7 (13)	15 (8)	7 (5)	3 (5)	6 (3)
80–90	8 (6)	5 (9)	9 (5)	1 (0)	0 (0)	2 (1)
70–80	3 (2)	1 (2)	11 (6)	4 (3)	3 (5)	5 (3)
60–70	8 (6)	4 (7)	10 (5)	6 (4)	8 (15)	8 (4)
50–60	18 (13)	6 (11)	24 (12)	10 (7)	6 (11)	13 (7)
40–50	14 (11)	4 (7)	24 (12)	16 (12)	7 (13)	24 (12)
30–40	21 (15)	9 (16)	32 (16)	28 (20)	8 (15)	37 (19)
<30	38 (27)	5 (9)	52 (27)	62 (45)	14 (25)	90 (46)
Total	139 (100)	55 (100)	194 (100)	139 (100)	55 (100)	194 (100)

Note: Figures in parenthesis are percentage of column totals

difference between the VRS and CRS measurements indicates that coffee farmers are not operating efficiently. The scale efficiency of 0.92 indicates that by operating at an optimal scale, the amount of inputs used by Robusta coffee farms in the study area could be reduced by as much as 8.0%.

Table 36.3 also provides a comparison of TE between Kinh and indigenous farmer groups. Results show that indigenous farmers produce more efficiently than Kinh farmers, under both VRS and CRS in the DEA model (81% vs. 76% in VRS and 75% vs. 67% in CRS). The results seem somewhat surprising given that Kinh farmers are more educated, better trained, and have better access to market information than indigenous farmers. This result, however, is consistent with the findings of Tran (2007) on the efficiency of coffee farming in Vietnam's Central Highlands. The reason is that farmers of Kinh origin, most of whom migrated from northern Vietnam in the 1980s, have less experience in coffee cultivation (18.9 vs. 20.7 years), smaller farm sizes (2.2 vs. 2.3 ha), and fewer family labor (66 vs. 78 man-days/ha) than indigenous farmers. Note that family labor is expected to be more efficient than hired labor due to the moral hazard problem.

Table 36.4 also shows that the average IWUE scores for the DEA frontiers for all respondents are much lower than TE scores at 52% for VRS and 39% for CRS. The results also show that variability for the estimated IWUE given the VRS assumption at 13% to 100% is less than that under the CRS assumption, at 12% to 100%. These

Table 36.3 Frequency distribution of technical efficiency of Robusta coffee production by group in Lam Dong, Vietnam, 2017

	TE					
	VRS			CRS		
Efficiency (%)	Kinh group	Indigenous group	All	Kinh group	Indigenous group	All
Summary statistics						
Mean	0.76	0.81	0.72	0.67	0.75	0.66
Minimum	0.31	0.30	0.24	0.21	0.29	0.21
Maximum	1.00	1.00	1.00	1.00	1.00	1.00
Std. dev	0.18	0.19	0.19	0.18	0.19	0.18
Efficiency interval						
100	23 (17)	17 (31)	21 (11)	9 (6)	9 (16)	16 (8)
90–100	14 (10)	9 (16)	24 (12)	9 (6)	6 (11.5)	10 (5)
80–90	27 (19)	4 (7)	24 (12)	15 (11)	11 (20)	17 (9)
70–80	16 (12)	8 (15)	31 (16)	20 (14)	6 (11.5)	31 (15)
60–70	28 (20)	9 (16)	32 (16)	34 (25)	9 (16)	40 (21)
50–60	23 (16)	6 (11)	44 (23)	35 (25)	9 (16)	48 (25)
40–50	4 (3)	1 (2)	11 (6)	11 (8)	4 (7)	21 (11)
30–40	4 (3)	0 (0)	6 (3)	6 (4)	0 (0)	8 (4)
<30	0 (0)	1 (2)	1 (1)	1 (1)	1 (2)	3 (2)
Total	139 (100)	55 (100)	194 (100)	139 (100)	55 (100)	194 (100)

Note: Figures in parenthesis are percentage of column totals

results imply that the level of productivity of Robusta coffee could still be maintained even if the amount of irrigation water used is reduced by 48% and 61% under VRS and CRS, respectively, ceteris paribus. Results also show that under VRS, the score for irrigation water use efficiency (IWUE) of 108 coffee farms is below 50%, between 50 and 80% for 45 farms, and just 41 farms score over 80%. These results indicate that the majority of farms could achieve significant savings in water use if their irrigation systems are improved. Table 36.4 also shows differences in IWUE scores between the Kinh and indigenous farmer groups. Similar to the results for TE, the average IWUE scores show that the indigenous farmer group produces more efficiently than the Kinh group, under both the VRS and CRS DEA models (69% vs. 54% under VRS and 52% vs. 40% under CRS, respectively).

The scale efficiency equal to 0.75 suggests that the amount of water used for the Robusta coffee farms in Lam Dong Province could be reduced by as much as 25% if utilized optimally. The estimated scale efficiency for Kinh and indigenous farmer groups was 0.74 and 0.75, respectively. This means that by operating at optimal scale, input use could be reduced by as much as 26% and 25% for the Kinh and indigenous farmer groups, respectively. In other words, farmers in both groups could be advised to increase their scale of operation to an optimal level. The efficiency level for the two groups differs under the VRS assumption. Results show that about 11% of the Kinh group and 25% of the indigenous farmer group were on the frontier (100% efficiency) (Fig. 36.1).

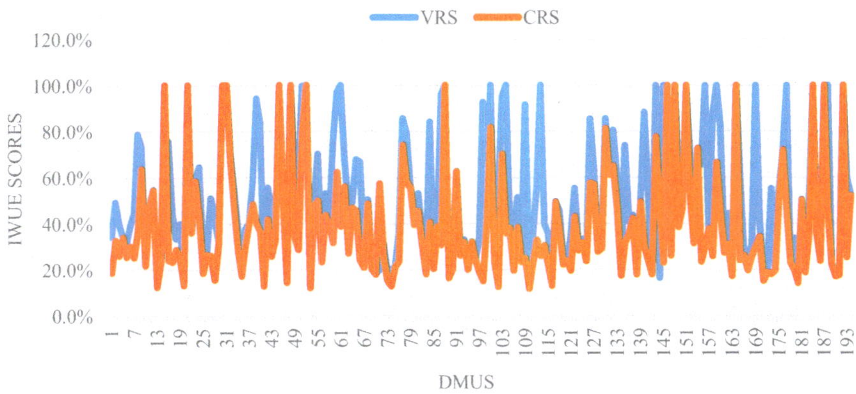

Fig. 36.1 IWUE under VRS and CRS

In sum, the DEA results for TE and IWUE scores indicated that for many Robusta coffee farmers in Lam Dong Province, the key inputs, especially irrigation water, could be reduced without affecting the levels of production.

4 Conclusions and Policy Recommendations

The analysis of the factors affecting technical efficiency of coffee farmers in the study areas shows that irrigation water has a very significant effect on coffee productivity findings which show, however, that local coffee farmers are very inefficient in utilizing irrigation water. Local farmers in fact can reduce the amount of irrigation water used by 25% without reducing the productivity of Robusta coffee. The results of the study also show that indigenous Robusta coffee farmers are more efficient in using irrigation water than Kinh farmers. Although farmers in Lam Dong Province felt the need to install water wells to meet their irrigation water requirements, they were constrained by the high investment costs. This means that financial and technical support from the government will be critical in improving IWUE.

The following are technical and institutional interventions to address the issues faced by farmers and improve their IWUE:

1. Enhance extension services (training on good agriculture practice (GAP) to teach farmers how to apply production inputs judiciously (pruning and irrigation techniques, water and soil management, fertilizer and pesticide usage, etc.).
2. Promote intensive collaboration among stakeholders (institutions, governmental extension departments, and farmer associations) to implement coffee-farming experiments and best management practices.
3. Increase access to credit, with favorable interest rates for coffee farmers. This may help farmers to overcome financial constraints, resulting in an increase in TE and IWUE.

4. Encourage farmers to apply water-saving irrigation technologies (sprinklers and drip irrigation) and farming practices through application tools (cellphones, computers, Internet, etc.).

Acknowledgments This study was supported by Vietnam National Project Code BĐKH.12/16-20 and SEAMEO Southeast Asian Regional Center for Graduate Study and Research in Agriculture (SEARCA).

References

Ahmad, A. (2001). *An institutional analysis of changes in land use pattern and water scarcity in Dak Lak Province, Vietnam*. NIAS Press. Retrieved July 25, 2019, from http://dlc.dlib.indiana.edu/dlc/bitstream/handle/10535/7903/ahmadviet.pdf?sequence=1

Aigner, D. J.; Ameiya, T.; Poirier, P. J. (1976). On the estimation of production frontiers: Maximum likelihood estimation of the parameters of a discontinuous density function, Int econ rev, 17: 2.

Amarasinghe, U. A.; Hoanh, C.; D'haeze, D.; Hung, T. (2015). Toward sustainable coffee production in Vietnam: More coffee with less water. *Agric Syst, 136*, 96–105.

Atici, K., & Podinovski, V. A. (2014). *Review of data envelopment analysis studies in agriculture*. Warwick Business School, University of Warwick.

Ball, V. E., Lovell, C. A. K., Nehring, R. F., & Somwaru, A. (1994). Incorporating undesirable outputs into models of production: An application to US agriculture. *Cahiers d'Economique et Sociologie Rurales, 31*, 59–73.

Charnes, A., Cooper, W., & Rhodes, E. (1978). Measuring the efficiency of decision-making units. North-Holland publishing company. *Eur J Op Res, 2*, 429–444.

Cheesman, J.; Son, T. V. H.; Bennett, J. (2007). *Valuing irrigation water for coffee production in Dak Lak, Vietnam: A marginal productivity analysis*. Crawford School of Economics and Government, The Australian National University, Canberra, Australia, No. 6.

Cochran, W. G. (1963). *Sampling technique* (2nd ed., p. 413). Wiley.

Coelli, T. (1998). *A guild to DEAP version 2.1. A data envelopment analysis computer program*. CEPA working paper, 96.08.

D'haeze, D., Deckers, J., Raes, D., Phong, T., & Loi, H. (2005). Environmental and socio-economic impacts of institutional reforms on the agricultural sector of Vietnam: Land suitability assessment for Robusta coffee in the Dak Gan region. *Agriculture Ecosystem Environment, 105*, 59–76.

Department of Agriculture and Rural Development of Lam Dong [DARD]. (2017). Retrieved December 21, 2017, from http://www.lamdong.gov.vn/vi-VN/Home/Pages/Default.aspx.

Dewbre, J. (2010). *Improving resource allocation and incomes in Vietnamese agriculture. A Case Study of Farming in the Dong Nai River Basin*. Environment and Production Technology Division.

Färe, R., Grosskopf, S., & Lovell, C. A. K. (1994). *Production Frontiers*. Cambridge University Press.

Farrell, M. (1957). The measurement of productive efficiency. *Journal of the Royal Statistical Society, 120*(3), 253–290.

Frija, A., Chebi, L. A., Speelman, S., Buysse, J., & Van Huylenbroeck, G. (2009). Water use and technical efficiencies in horticultural greenhouses in Tunisia, (in English). *Agricultural Water Management, 96*(11), 1509–1516.

General Statistics Office (GSO) of Vietnam. (2013). Area, population and population density in 2013 by province. Retrieved January 17, 2015, from http://www.gso.gov.vn/default_en.aspx?tabid=467&idmid=3&ItemID=15751.

Giungato, P., Nardone, E., & Notarnicola, L. (2008). Environmental and socio-economic effects of intensive agriculture. *The Vietnam Case, 47*, 135–151.

Haggar, J., & Schepp, K. (2012). *Coffee and climate change impacts and options for adaption in Brazil, Guatemala, Tanzania and Vietnam*. NRI working paper series. Climate change, agriculture and natural resources no. 4. Natural resources institute. Retrieved July 24, 2019, from https://www.nri.org/publications/working-paper-series/4-coffee-and-climate-change/file.

Harvey, C. A., Rakotobe, Z. L., Rao, N. S., Dave, R., Razafimahatratra, H., Rabarijijohn, R. H., Rajaofara, H., & Mackinnon, J. L. (2014). Extreme vulnerability of smallholder farmers to agricultural risks and climate change in Madagascar. *Philosophical Transactions of the Royal Society of London. Series B, Biological Sciences, 369*, 1639.

Japan International Cooperation Agency [JICA]. (2018). *Data collection survey on water resources management in central highlands*. Final Report, Nippon Koei, .

Jolk, C., Greassidis, S., Jaschinski, S., Stolpe, H., & Zindler, B. (2010). Planning and decision support tools for the integrated water resources Management in Vietnam. *Water, 2*, 711–725.

Kotecha, S.; Hieu, S. P.; Kuit, M.; Von Enden, J.; Swinkels, R. (2002). *Vietnamese coffee competitiveness and policy options to improve it*. Conference proceedings. Retrieved January 21, 2017, from http://ipsard.gov.vn/news/mispa/Nhom%20ca%20phe/Hop,%20Hoi%20thao/Hoi%20thao%20kha%20nang%20canh%20tranh%20cafe%20VN/Policy_brief__ver09__Hieu.pdf.

Lansink, O. A., Pietola, K., & Bäckman, S. (2002). Efficiency and productivity of conventional and organic farms in Finland 1994-1997. *European Review of Agricultural Economics, 29*(1), 51–65.

Lilienfeld, A., & Asmild, M. (2007). Estimation of excess water use in irrigated agriculture: A data envelopment analysis approach. *Agricultural Water Management, 94*, 73–82.

Long, T. K.; Gerardo, H.; Jake, B. (2009). Vietnam- Netherlands partnership. In: *Water for Food & Ecosystems*.

Luong, Q. V., & Tauer, L. W. (2006). A real options analysis of coffee planting in Vietnam. *Agricultural Economics, 35*, 49–57.

Marsch, A. (2007). *Diversification by smallholder farmers: Vietnam Robusta coffee*. FAO.

McGockin, J. T., Gollehon, N., & Ghosh, S. C. (1992). Water conservation in irrigated agriculture: A stochastic production frontier model. *Water Resources Research, 28*(2), 305–312.

Ministry of Natural Resource and Environment of Vietnam [MONRE]. (2016). *Climate change, sea level rise scenarios for Vietnam*. Vietnam Publishing House of Natural Resources.

Nguyen, H. A., Bokelmann, W., Ngo, T. T., Do, T. N., & Nguyen, V. M. (2019). Smallholders' preferences for different contract farming models: Empirical evidence from sustainable certified coffee production in Vietnam. *Sustainability, 11*, 3799.

Omezzine, A., & Zaibet, L. (1998). Management of modern irrigation systems in Oman: Allocative vs irrigation efficiency. *Agricultural Water Management, 37*(2), 99–107.

Reinhard, S., Lovell, C. A. K., & Thijssen, G. (1999). Econometric estimation of technical and environmental efficiency: An application to Dutch dairy farms. *Am J Agr Econ, 81*, 44–60.

Reinhard, S., Lovell, C. A. K., & Thijssen, G. J. (2000). Environmental efficiency with multiple environmentally detrimental variables; estimated with SFA and DEA. *Eur J Op Res, 121*, 287–303.

Shammout, M. W., Qtaishat, T., Rawabdeh, H., & Shatanawi, M. (2018). Improving water use efficiency under deficit irrigation in the Jordan Valley. *Sustainability, 10*, 4317.

Technoserve. (2013). *A Business Case for Sustainable Coffee Production*. Retrieved July 24, 2019, from www.sustainablecoffeeprogram.com/site/getfile.php?id=211.

Tesfaye, S. G., Mohd, R. I., Kausar, H., Marziah, M., & Ramlan, M. F. (2013). Plant water relations, crop yield and quality in coffee (Coffea arabica L.) as influenced by partial root zone drying and deficit irrigation. *AJCS, 7*(9), 1361–1368.

Tran, N. A. (2007). *Measuring the efficiency of coffee producers in Vietnam: Do outliers matter?* Purdue University West Lafayette.

World Bank. (2004). *The Socialist Republic of Vietnam*. Coffee sector report no. 29358-VN. Retrieved August 22, 2019, from https://openknowledge.worldbank.org/handle/10986/14405

Chapter 37
Climate Change Vulnerability of Agriculture in Coastal Communes of Quang Tri Province, Vietnam

Son Hoang Nguyen, Cham Dinh Dao, Hang Anh Phan, Quan Trong Nguyen, and Toai Anh Le

Abstract Due to its long coastline, Vietnam is regarded as the country that is seriously influenced by climate change. When the sea level increases by 1 meter, 40% area of the Mekong Delta and 10% area of the Red River Delta will be deluged, and this means that life of 20–30 million people will be promptly affected. Particularly, in Quang Tri Province, the coastal area is the place where people mainly live on agriculture and the climate change has made serious impacts on their life. Applying GIS and remote sensing allows the evaluation of the vulnerability of agricultural production activities in coastal communes in Quang Tri Province to be more effective. The research process has found the sensitivity index (S—sensitivity) (namely the index of traffic access, the effects of residential areas, the effects of industrial zones, the community dependence level), exposure index (E—exposure) (including sea-level rise until 2100; temperature change until 2100), and adaptive capacity index (AC—adaptive capacity) (including slope index; morphology), and afterward, summarizing the index of vulnerability due to the influences of climate change on agricultural production (V-vulnerability to climate change).

Keywords Climate change · Vulnerability level · Agricultural production · Coastal · Quang Tri

---

S. H. Nguyen (✉) · Q. T. Nguyen · T. A. Le
Institute of Open Education and Information Technology, Hue University, Thua Thien Hue, Vietnam
e-mail: nhsonsp@hueuni.edu.vn

C. D. Dao
VAST Institute of Geography, Hanoi, Vietnam

VAST Graduate University of Science and Technology, Hanoi, Vietnam

H. A. Phan
University of Science, Hue University, Thua Thien Hue, Vietnam

A. T. Nguyen, L. Hens (eds.), *Global Changes and Sustainable Development in Asian Emerging Market Economies Vol. 2*, https://doi.org/10.1007/978-3-030-81443-4_37

1 Introduction

Currently, the climate change and its effects have become a significant research area. Without proper adaptation strategies, climate change will lead to considerable environmental changes and has serious impacts on various countries all over the world. Besides, climate change also makes multidimensional impacts on humanity in several socioeconomic aspects such as agriculture, human health, tourism activities, labor shortage, and widespread epidemics. In particular, agricultural production is the most seriously affected area including changing the structure of crops; cultivation and husbandry; catching and aquaculture; and the risk of new epidemics affecting plants and animals. Thus, scientific researches on climate change and its effects on agricultural production should take into consideration the vulnerability in many areas, especially in coastal zones.

There have been several methods to evaluate the vulnerability and a host of reports that compare these methods over the past few decades, such as summary reports on vulnerable situation resulted from the climate change and impact assessment tools (Balangue, 2013); handbook about vulnerable situation of current and following generation and vulnerability assessment tools (Garg et al., 2007); assessment of the vulnerable situation: a summary of approaches (Morgan, 2011); and guideline for evaluating the vulnerable reality, effects, and capacity to alter to climate change (Provia, 2013) in USAID Mekong ARCC (2013), the report evaluates the vulnerability ranking in some provinces to find out the most vulnerable provinces in the lower Mekong Delta, a Mekong tributary (World Wide Fund For Nature [WWF], 2013), a simple ecosystem like Ramsar wetland (Meynell, Thongsavath, Xeuasing, Vannalath, & Glémet, 2014), or urban center (ICEM, 2015). These studies are almost complex, involving detailed assessment of vulnerability and implementation of large-scale interventions such as the whole region, the nation, and the area based on a combination of diversity of ecosystems, livelihoods, infrastructure, and economic assets. Meanwhile, many studies have shown the necessity to conduct vulnerability assessment and climate change adaptation based on social factors.

In Vietnam, between 2001 and 2005, the research on the vulnerability assessment of coastal areas in south-central Vietnam is considered a scientific foundation for decreasing disasters, sustainable use of land plan carried out in 2001–2002, or a project to examine and assess the vulnerability of Vietnam's natural resources, environment, and marine meteorological conditions. Since Vietnam became a member of the United Nations Framework Convention on Climate change (UNFCCC, 2008) in 1992 and the Kyoto Protocol in 1998, climate change has become a common topic for study. The Ministry of Natural resources and environment presides over a number of activities relevant to climate change. Up to now, the ministry has conducted three scenarios of the climate change and sea-level rise in Vietnam, respectively, published in 2009, 2012, and 2016. However, the research of assessing the vulnerability in Vietnam began in the end of the twentieth century. Studies approach different fields of natural systems, society, communities, and coastal resources on the scale of research from region/region to the coastal zone of Vietnam.

Natural disasters and the environmental pollution in these areas are forecasted and the vulnerability map for the total coastal area of Vietnam has been developed, namely north, central, southwest, south, and Truong Sa islands in 2015 and 2020 and the scenario of sea-level rise of 0.5 meter and 1 meter (Nhuan, 2011).

World Wide Fund For Nature (WWF) has carried out a research project named "Synthetic rapid assessment of vulnerability and adaptability to climate change in three coastal districts, Ben Tre Province" year 2012 and the project "Assessing the vulnerability to climate change of ecosystems in Vietnam" in 2013. WWF has the same approach view with author Nhuan, and vulnerability assessment method in this project is determined according to the determination of exposure, and sensitive and adaptive capacity. In recent years, some coastal communes in Quang Tri, Vietnam, are often directly affected by several strong types of natural disasters, including storms, floods, droughts, and saline intrusion, which lead to severe damage, especially on productivity and quality in agricultural production. It has direct impact on people's livelihoods: agricultural land area is abandoned or changing the purpose of use is increasing. This causes instability of local food security, poverty increase, and social evils in the area as well.

In order to not only improve the lives of residents but also find a new direction for the agricultural sector in coastal communes against the impacts of climate change, it is necessary to have territorial organizations of agricultural production and suitable livelihoods models to climate change to promote the development of socioeconomic aspect and improve the life quality for local people. Therefore, the evaluation of the vulnerability level of the production of agricultural activities because of climate change in coastal communes in Quang Tri makes a vital contribution to scientific and practical aspects.

2 Methodology

2.1 Data Collection and Analysis

Data and maps of natural conditions, climate change, and its effects on agricultural production; livelihood, socioeconomic information in the coastal communes in Quang Tri Province related to climate change; and documents of project or programs which are about socioeconomic development and agricultural development adapting to climate change in coastal area are collected. All informations related to research subjects and areas are approached and suitably applied in research process.

2.2 Mapping, Remote Sensing, and Geographic Information System (GIS)

Applying cartography, remote-sensing techniques on the basis of aerial photographic images and satellite images in different periods can assess the degree of changes in natural characteristics and agricultural production activities because of the effects of climate change. Geographic information system (GIS) is used to update meteorological and hydrological documents, information on natural environment changes on the surface, storage of database systems, parts of maps helping for research, proposing solutions to solve and adjust to climate change, and update documents conveniently and quickly.

2.3 Determining Vulnerability Due to Climate Change Impacts

(a) The components of vulnerability.

- Sensitivity: Ecological sensitivity is the ability of suffering a specific impact (such as environmental change) and recovering after suffering the impact. The smaller the resistance and the ability can keep the system balanced, the more sensitive it is, and vice versa.
- Exposure: The density of vulnerability is the density of vulnerable objects determined by the distribution and the role of the vulnerable objects. Another definition of exposure is the level of exposure of the study subject with the factors affecting it in different directions depending on the element.
- Adaptive capacity: According to IPCC, adaptive capacity is the capacity of a system to adapt to climate change (namely negative climate changes) and minimize damages, exploit beneficial elements (Intergovernmental Panel on Climate Change [IPCC], 2007) (Table 37.1).

(b) Standardized method of variables.

Standardize statistics: Using inherited statistics from relevant branches' data, and then quantifying and using a calculation formula to standardize and bring the index from 0 to 100.

Spatial analysis: Applying spatial analysis tools in GIS to develop variables for analysis and evaluation process and integrating information and multiplying weighted information layers: using variables which indexes have been standardized to 0–100 for integration through algorithms to synthesize and calculate key indexes and subindexes.

Value variables can be known as a quantity that consisted of the formula to calculate a value to be searched. The choice of variables to evaluate vulnerability relies on the theory and approach way, along with expert opinions. Choosing various variables will give different outcomes. For each variable, as it is tested

Table 37.1 Weight of indexes

TT	Index	Weight	Result
I	**Sensitivity index**	**3**	**0.3**
1	Traffic access index	3	0.15
2	Impact of residential areas	6	0.3
3	Impact of industrial parks	5	0.25
4	Dependence level of the community	6	0.3
II	**Exposure index**	**4**	**0.4**
5	Sea-level rise to 2050	4	0.5
6	Temperature change to 2050	4	0.5
III	**Adaptive capacity**	**3**	**0.3**
7	Slope	5	0.45
8	Shape	6	0.55

by various quantities (for instance, the temperature variable is tested by degrees Celsius, the effect degrees, or the AC index is tested by socioeconomic factors), we have to put quantities into one axis to be able to assess. The unit is the evaluation index. Therefore, we apply the below formula: (37.1and 37.2) (World Wide Fund For Nature [WWF], 2013).

$$\text{Index} = \frac{\text{X} - \text{Xmin}}{\text{Xmax} - \text{Xmin}} \times 100 \tag{37.1}$$

$$\text{Index} = 100 - \left(\frac{\text{X} - \text{Xmin}}{\text{Xmax} - \text{Xmin}} \right) \times 100 \tag{37.2}$$

For the variables shown as low as possible, the Formula (37.1) should be applied to standardize, whereas with the higher variables as possible, the Formula (37.2) should be applied to standardize.

(c) Building and standardizing variables.

Variables in evaluation of vulnerability are determined based on natural and socioeconomic impacts. The key indexes are identified based on the theoretical views of the IPCC and applied by many scientists. The variables used in the evaluation include:

- Sensitive index (S): Traffic access index; Impact of residential areas; Impact of industrial parks; Dependence level of the community.
- Exposure index (E): Sea level rise to 2100; Temperature change to 2100.
- Adaptive capacity index (AC): Slope; Shape.

(d) Determine the weight and calculate the vulnerability index.

* Determine the weight.

Evaluating the weight based on experts' opinions, the result is calculated by the Formula (37.3), for example, weight of traffic access: 3/(3 + 6 + 5 + 6) = 0.15.

$$\frac{X_i}{\sum_{i=1}^{n} X_i} \tag{37.3}$$

In which, X_i is the weight of index (i = 1; 2…n).

*Method of index in evaluating vulnerability.

V (vulnerability) can be seen as expressed as a function of the exposure level (exposure)—the degree to which climate change affects the system; S (sensitivity level)—the degree to which the system is affected; and the AC (adaptive capacity) is the ability of the system to be adjustable (37.4).

$$V = \frac{(E + S + AC)}{f} \tag{37.4}$$

$$V = \frac{E + S + (100 - AC)}{3} \tag{37.5}$$

In which,

V: vulnerability index.
E: exposure index.
S: sensitivity index.
AC: adaptive capacity index.

Based on the analysis of vulnerable indexes, the project conducts vulnerability evaluation through the synthetic formula of calculating vulnerability index proposed by IPCC (37.5) Intergovernmental Panel on Climate Change [IPCC], 2007).

3 Study Area

The coastal communes in Quang Tri have an area of about 14.193,93 ha, which account for 3.01% natural area of the province, including 12 communes of Hai Lang District (Hai An and Hai Khe communes), Trieu Phong District (Trieu Van, Trieu An, Trieu Lang, and Trieu Do communes), Gio Linh District (Trung Giang, Gio Hai, Gio Viet, and Gio Mai communes), and Vinh Linh District (Vinh Thai and Vinh Giang communes) (Fig. 37.1).

In Quang Tri Province, the coastal area is popular for plain, abrasives, accumulation, and sand dunes. The plain does not form a continuous band but sometimes splits because of the protruding branches or hills. Absolute height is about 20m or less, consisting of two types of terrain: accumulation plains and coastal sand dunes.

In 2018, the population of coastal communes of Quang Tri Province is 54,003 people, accounting for 8.3% of the population of Quang Tri Province; population density here is 236.3 people/km^2, 1.86 times higher than the province's population

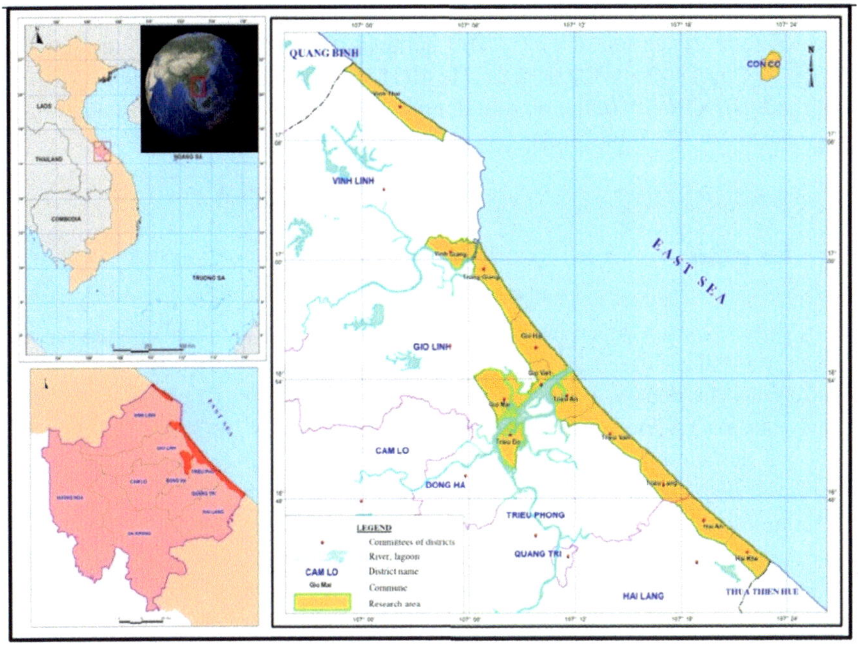

Fig. 37.1 Location of coastal communes in Quang Tri Province, Vietnam

Table 37.2 Average temperature in meteorological stations during 1993–2018 (°C)

TT	Station	1	2	3	4	5	6	7	8	9	10	11	12	Average temperature
1	Con co	20.6	21.1	22.1	24.8	27.7	29.7	29.5	29.4	28.1	26.5	24.8	22.1	25.5
2	Dong ha	19.6	20.8	22.7	26.0	28.3	29.9	29.5	28.8	27.0	25.3	23.1	20.4	25.1
3	Khe Sanh	18.1	19.5	21.6	24.5	25.7	26.0	25.4	25.0	24.3	23.1	21.2	18.6	22.8

density (126.7 people/km^2) (Quang Tri Statistical Office [QTGSO], 2019). Population structure by age: the study area has a young population structure, but for the agriculture–forestry–fishery production sector, according to the working age, the structure is aging.

3.1 Temperature

In the coastal communes in Quang Tri Province, the average temperature in the period of 1993–2018 is around 24.5 °C (Table 37.2).

Compared to the standard average temperature of 1975–2018, the average annual temperature in the period of 1993–2018 is mostly higher, from 0.1 °C in 1994 and 2013; 0.3 °C in 1997 and 2009; 0.2 °C in 2001 and 2007; 0.4 °C in 2003, 2006, and 2012; and 0.7 °C in 2010; the highest temperature was in 1998, higher than average temperature 0.9 °C (Table 37.3).

3.2 *Rainfall*

Due to differentiation depending on geographical location and local climate characteristics, rainfall at different stations in the coastal communes of Quang Tri Province is different. The data of rainfall collected from Vinh Linh, Gia Vong, Dong Ha, Thach Han, Cua Viet, Huong Hoa, and Ba Long stations in Quang Tri Province show that the annual rainfall is in the range of 2000–2800 mm. Rainfall in three months of rainy season accounts for 68–70% of annual rainfall (Table 37.4).

During the dry season from December to April, there are usually light rains from 7 to 8 days with rainfall from 20 to 30 mm. The rainy season starts from September to November; sometimes, the rainy season lasts until December. Due to the terrain characteristics, the rain in the rainy season is rarely equally spread all over the province.

The average annual rainfall during 1993–2018 in Quang Tri Province (the average rainfall across the province by weighting method) does not clearly show the increase or decrease trend. Compared to the standard period of 1975–2018 (2325 mm), the number of years with higher rainfall is 12 years on average, including 1995, 1996, 1998, 1999, 2001, 2002, 2005, 2007, 2008, 2009, 2011, 2013, and 2015. The year with the highest rainfall exceeded was 653 mm in 2011, followed by 594 mm in excess in 1999, and 469 mm in 2013. The year with the highest standard

Table 37.3 Characteristics of monthly and annually average temperature in the province in the period 1975–2018 (unit: °C, average temperature: AT)

Year	AT	Year	AT	Year	AT	Year	AT
1975	24.3	1986	24.3	1997	24.7	2008	23.8
1976	24	1987	25.1	1998	25.4	2009	24.4
1977	24.1	1988	24.4	1999	24.4	2010	24
1978	24.1	1989	24	2000	24.2	2011	24.2
1979	24.5	1990	24.5	2001	24.6	2012	24.4
1980	24.7	1991	24.9	2002	24.3	2013	24.5
1981	24.5	1992	24.2	2003	24.8	2014	24.0
1982	24.4	1993	24.4	2004	24.1	2015	25.4
1983	24.3	1994	24.5	2005	24	2016	24.4
1984	23.7	1995	24.2	2006	24.5	2017	24.6
1985	23.8	1996	24	2007	24.4	2018	24.3

Table 37.4 Average rainfall in some years (*unit: mm)*

Station	1	2	3	4	5	6	7	8	9	10	11	12	Year
Vinh Linh	129.9	83.3	48.6	51.9	100.5	97.8	94.3	125.3	420.2	766.0	462.3	227.0	2.614,1
Gia Vong	60.1	47.9	35.4	64.1	143.6	101.4	78.7	155.0	509.7	695.9	456.4	188.0	2.536,3
Dong ha	48.2	34.1	30.8	60.7	119.3	83.0	65.7	163.2	388.9	683.9	429.0	175.2	2.291,8
Thach Han	84.3	60.7	48.9	63.0	135.0	105.7	82.9	135.3	476.4	710.6	438.6	240.7	2.627,3
Cua Viet	57.6	48.6	33.1	50.8	102.6	63.4	68.1	150.3	398.6	574.3	415.7	219.6	2.187,8
Huong Hoa	83.6	61.7	47.8	97.8	191.5	171.7	148.9	219.1	585.8	778.0	227.7	95.7	2.779,9
Khe Sanh	16.7	19.2	29.7	89.8	158.9	210.8	187.8	295.9	376.7	455.0	175.8	64.7	2.118,6
Ba long	99.8	90.1	51.0	71.7	156.6	156.8	74.2	173.1	473.4	762.0	411.8	227.8	2.794,3

rainfall loss was 2004 with a decrease of 555 mm, followed by 1993 with a decrease of 513 mm, and in 1998, it decreased 498 mm.

3.3 *Sea-Level Rise*

As Quang Tri is a coastal province in the central region of Vietnam, it is generally affected by sea-level rise on a global scale in general and Vietnam in particular. By analyzing sea-level data at Hon Dau and Vung Tau from 1957 to the present, it is clear that in about 40 years, the increase trend of sea level is real with a rising water level of 2.3 mm/year on the big plains in Vietnam.

According to the scenario of Quang Tri climate change and sea-level rise by 2020 when the sea level rises from 8 to 9 cm, the national highways and provincial roads are not seriously influenced by sea-level rise. However, by 2100 when the sea-level rise is 51–63 cm, there will be about 2.67% of the national highway length and 8.23% of the provincial road length will be affected by frequent flood; the worst case is that railway will be affected 0.21%. Besides, it also affects the ability of flood in coastal roads.

3.4 *Extreme Weather Phenomena*

Climate change has some impacts on extreme weather events, and they can be divided into three groups as below:

- Extremes of weather and climate variables (temperature, rain, wind, etc.).
- Extreme weather and climate events (monsoon, El Nino, storm, etc.).
- The phenomena affect natural physical environment (drought, flood, extreme sea level, etc.).

In general, the identification and definition of weather and climate phenomena in terms of risk management are very complicated and depend on the specific purpose. This aspect focuses on the collection and synthesis of extreme weather data and is defined as the occurrence of values higher or lower than the threshold value of a weather or climate element, near the upper limits, or below the range of observed values of that element. The dataset for the study which is used based on actual monitoring data at meteorological and hydrological stations was updated to 2018. The mentioned phenomenal include:

– Absolute maximum temperature (Tx) and absolute minimum (Tm):

According to statistics in the period of 1993–2015, the annual average maximum temperature in Dong Ha station is about 29.5 °C, higher than the average period (1973–2015) of 0.2 °C, and higher than the average time in the period 1973–1992 of 0.4 °C (Ministry of Natural Resources and Environment [MONRE], 2016).

In particular, the average maximum temperature data in the period of 2003–2015 tend to increase slightly compared with the average period of 1993–2002. For annual average minimum temperature, at Dong Ha station, the value is about 22.6 °C, higher than the standard period 0.2 °C, and higher than in 1973–1992 of 0.5 °C. Similar to the peak temperature changes, the average minimum temperature in the period of 2003–2015 showed a slight increase compared with the average of 1993–2002 periods. Thus, both the annual average maximum and minimum temperatures in the period of 2003–2015 are higher than the average maximum and minimum temperatures in the rest of periods.

– Storms, tropical depressions:

The characteristic of storms and tropical depressions in Quang Tri varies greatly depending on the storm and the period when it lands. There are years without storms, but there are years with 2–3 storms (1964; 1996). On average, there are about 1.2 to 1.3 storms. Quang Tri coastal area has up to 78% of storms and tropical depressions in the East Sea, causing heavy rains and flood in rivers and flooding coastal plains of Quang Tri or valley areas or on some parts of the Thach Han River. Storm landing usually lasts from 8 to 10 h, but the accompanying rain usually lasts up to three days, causing floods and flash floods which lead to serious damage to people and property.

According to the statistical results, generally, the number of storms and tropical depression directly affecting Quang Tri Province tended to decrease slightly, but the level of decrease was not considerable. In some years, Quang Tri Province was not affected by any storms. Other years, this province was affected by 1–3 storms and tropical depressions. In the period of 1962–2009, in 1964 and 1984, there were two years that the province was directly affected by four storms and tropical depressions.

– Floods and flash floods:

Due to the high slope and short river system, floods occur quickly and fiercely; combining with heavy rains, vegetation cover and weak soil structure places can cause flash floods. Floods and flash floods greatly affected the province's economic development. For example, floods from September 29 to October 5, 2010, caused floods and flash floods for Ha Tinh, Nghe An, Quang Binh, Quang Tri, and Thua Thien Hue provinces. The heavy rain starting on September 29, 2010 caused floods and flash floods across the Ngan Sau–Ngan Pho River basin. Due to heavy rains, the upstream water level concentrates quickly. In Quang Tri, more than 2000 houses were flooded, many rice field areas in the Dong Ha City and Gio Linh District were flooded, one person died, and many infrastructure works were damaged.

– Thunderstorms, whirlwinds, rain and hail, and fog:

Due to the climate characteristics, the coastal communes in Quang Tri Province have relatively large number of thunderstorms. According to Dong Ha meteorological observation station from 1975 to 2013, on average, there are about 67.3 thunderstorms a year. Especially, there are years when the number of thunderstorms are more than 100 days (in 1980 and 1981, there were 104 thunderstorms).

Whirlwind is a phenomenon where the wind accelerates suddenly, the direction changes suddenly, the air temperature drops sharply, and the humidity increases rapidly with thunderstorms, showers, or hail. Tornado is vortices in which the wind in the circulation is small in the tens or hundreds of meters. Tornado is small swirling swirls, which often occurs when the atmosphere is turbulent and basically unpredictable.

According to statistics of the number of foggy days at Dong Ha station from 1975 to 2018, on average, there are about 17 days of fog a year. The foggiest time is usually in January, February, and March, while in June, July, and August there is not usually foggy. Incomplete statistics at Khe Sanh station averaged from 2007 to 2013 up to 126.6 days/year. This is also reasonable because Khe Sanh station represents the mountainous region with a mild climate.

4 Results

4.1 Sensitivity

– *Traffic access index:* Sensitive index of the production of agricultural activities and roads is conducted based on the division of the traffic road information of topographic maps. Calculating the distance with the maximum value of 10 km, the meaning of this index indicates that the closer to the road that the agricultural production activities are, the more sensitive and vulnerable they will be.
– *Impacts of residential areas:* Sensitive index of the production of agricultural activities and roads is conducted based on the division of the population and

urban information layers of the current state of land cover map, and then, calculating the distance with the maximum value of 15 km, the meaning of this index indicates that the closer to residential areas that agricultural activities are, the more sensitive and vulnerable they will be.

- *Impacts of industrial zones:* Sensitive index of the production of agricultural activities and roads is conducted based on the division of the information on the current status of industrial zones of the current land use map. Calculating the distance with the maximum value of 25 km, the meaning of this index indicates that the closer to industrial zone that agricultural production activity is, the more sensitive and vulnerable it will be.
- *Dependence level of the community*: The criteria of agriculture, forestry, and fishery labor shows the dependence on the affected field, the total laborers/population of localities show the economic dependence of the local agricultural sector. The ratio of agriculture, forestry, and fishery/total population is determined based on survey data of the General Statistics Office in each commune. The lower the rate is, the greater the level of community dependence on the production of agriculture, the more sensitive and vulnerable it will be (Table 37.5, Fig. 37.2).

Table 37.5 Sensitive index variables used in vulnerability evaluation

Main index	Subindex	Meaning	Source	Formula for standardization
Sensitivity index	Traffic access index	This index indicates that the nearer to the road the production of agricultural activities is, the more sensitive and vulnerable it will be.	Space analysis	(37.2)
	Impacts of residential areas	This index indicates that the closer to residential areas that the production of agricultural activities is, the more sensitive and vulnerable it will be.	Space analysis	(37.2)
	Impacts of industrial zones	This index indicates that the closer to industrial zone that the production of agricultural activities is, the more sensitive and vulnerable it will be.	Space analysis	(37.2)
	Dependence level of the community	This indicates the level of livelihood dependence on agricultural production (index is determined from the number of agricultural, forestry, and fishery/total population)	Statistics–general statistics office	(37.1)

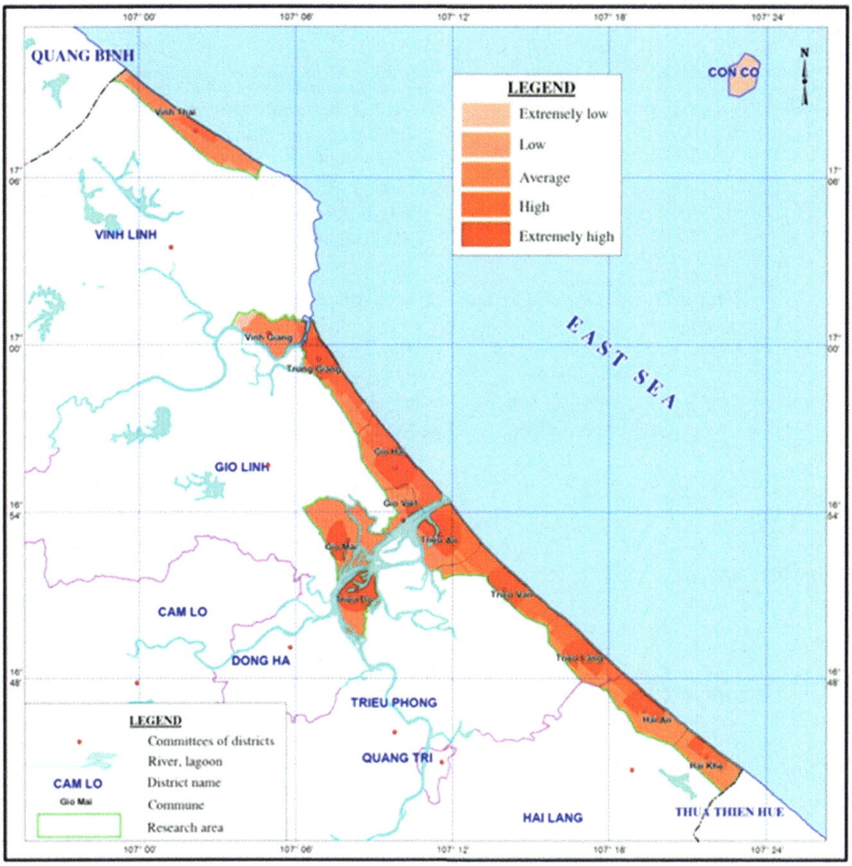

Fig. 37.2 Map of sensitivity (S) of agricultural production activities

4.2 Exposure

Sea-level rise and temperature change become two primary indexes of climate change. The IPCC organization has demonstrated that temperature increase leads to sea-level rise. Therefore, two main indexes are applied to evaluate vulnerability.

Sea-level rise to 2100: According to the Ministry of Natural Resources and Environment (2016), the climate change scenario is the foundation for the sea-level rise index. The data of average sea-level rise according to the scenario RCP 4.5 calculated by 2100 are 53 cm (32–75) for the Quang Tri area. Combined with the DEM elevation numerical model, it is possible to determine which areas will flood till 2100 (Ministry of Natural Resources and Environment [MONRE], 2016).

Temperature changes to 2100: In 2100, in Quang Tri Province according to RCP scenario 4.5, the temperature increase at different places in the region can range from 1.3 to 2.7 °C (period 2080–2099); the most common increase will be 1.9 °C

Table 37.6 Variables of the exposure index used in vulnerability assessment

Main index	Subindex	Meaning	Source	Formula for standardization
Exposure	Sea-level rise to 2100	Determine the level of impacts of sea-level rise on agricultural production activities	- sea-level rise scenario to 2050 (based on climate change scenario of Ministry of Natural Resources and Environment)—DEM elevation model	(37.1)
	Temperature changes to 2100	Determine the level of impact of temperature changes on agricultural production activities	Scenario of temperature change to 2050 (based on climate change scenarios of the Ministry of Natural Resources and Environment)	(37.1)

(Table 37.6) (Fig. 37.3) (Ministry of Natural Resources and Environment [MONRE], 2016).

4.3 Adaptive Capacity

- *Slope*: The slope index is built from a DEM height model; the meaning of this index is that the lower slope areas that agricultural production activities are, the higher the adaptive capacity will be.
- *Shape*: The shape index is evaluated by the following steps: calculating the area and circumference of agricultural production activities, separating the area by pi (3.14), and taking the square root to determine the radius of the circle. The standard corresponds to the area of the agricultural production activity and then calculates the circumference of the circle corresponding to the standard radius, and divides the circumference of agricultural activities by the circumference of the circle. The larger this index shows, the longer the shape of production of agriculture, and the more vulnerable it will be (Table 37.7, Fig. 37.4).

4.4 Vulnerability

Formula for calculating vulnerability index (37.4) (37.5) (World Wide Fund For Nature [WWF], 2013).

In Tables 37.5 and 37.8, it is shown that there are 2.498,66 ha, accounting for 17.60% of the area of production of agricultural activities in coastal communes in Quang Tri Province, which is highly vulnerable and concentrated in Trieu An, Gio

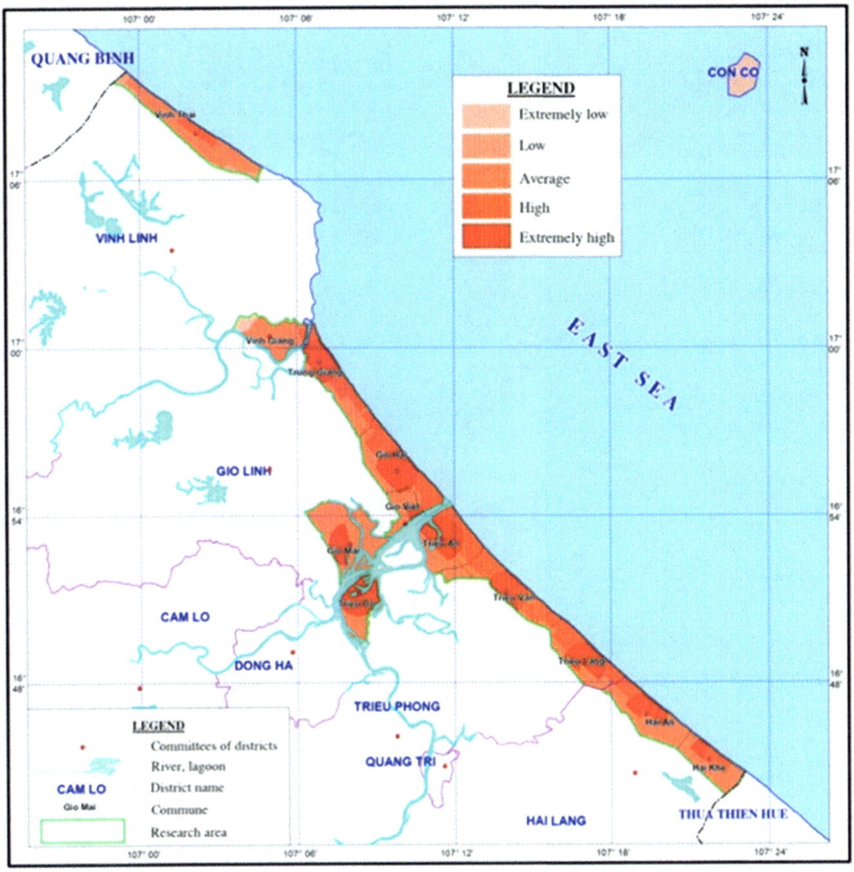

Fig. 37.3 Map of expose (E) of agricultural production activities

Table 37.7 Adaptive capacity index variables used in vulnerability evaluation

Main index	Subindex	Meaning	Source	Formula for standardization
Adaptive capacity (AC)	Slope	The higher slope in agricultural production of areas is, the more adaptable will be.	Analysis from DEM	(37.2)
	Shape	Agricultural production has a long and fragmental structure, which is less able to adapt to the core areas.	Analysis from the distribution map of ecosystems in the coastal area of Quang Tri Province	(37.1)

Fig. 37.4 Map of adaptive capacity (AC) of agricultural production activities

Table 37.8 Areas and vulnerability level rate of agricultural production activities in coastal communes in Quang Tri Province

Vulnerability	S (sensitivity)		E (exposure)		AC (adaptive capacity)		V (vulnerability to climate change)	
	(ha)	(%)	(ha)	(%)	(ha)	(%)	(ha)	(%)
Very low	141.01	0.99	6123.76	43.14	625.73	4.41	189.83	1.34
Low	165.63	1.17	1442.27	10.16	549.91	3.87	1462.14	10.30
Average	6925.2	48.79	507.45	3.58	1884.2	13.27	3943.86	27.79
High	3008.16	21.19	5971.17	42.07	6638.47	46.77	6099.44	42.97
Very high	3953.93	27.86	149.28	1.05	4495.62	31.67	2498.66	17.60
Total	**14,193.93**	100	**14,193.93**	100	**14,193.93**	100	**14,193.93**	100

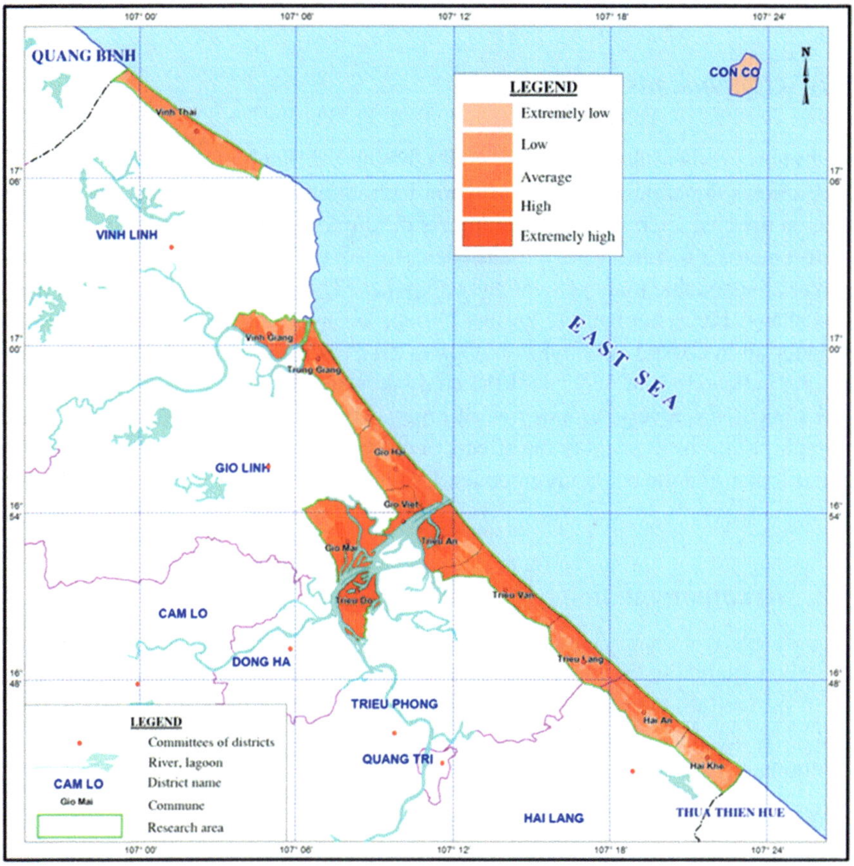

Fig. 37.5 Map of vulnerability (V) of agricultural production activities

Viet, Gio Mai, Trieu Do, and Trung Giang communes. The high level of vulnerability has an area of 6.099,44 ha, equivalent to 42.97% of Gio Mai, Gio Viet, Gio Hai, Vinh Giang, and Trieu Van communes. The average level of vulnerability accounts for 3.943,86 ha, equivalent to 27.79%. The very low level of damage, accounting for 189.83 ha, is equivalent to 1.34% of the communes of Hai An and Hai Khe. Thus, the area with high or higher vulnerability level makes up 60.57% of the total area of the research area (Table 37.8, Fig. 37.5).

5 Conclusions and Recommendation

5.1 Conclusions

Up to now, evaluating vulnerability still remains a difficult problem. Applying GIS tools allows the collection of information from multifield and standardized expression on the characteristics of space, meets the current and synchronous information requirements of vulnerability assessment to serve scientific research study, and makes a reasonable management and use policy. Through the evaluation process, it was shown because of the effects of climate change, the majority of agricultural production in coastal communes in Quang Tri Province has a high level of vulnerability, indicating that 42.97% of the study area rescue has a high degree of injury and 17.60% has a very high degree of injury. This raises the problem that requires suitable solutions to adapt to the effects of climate changes on agricultural activities in particular and the economy in general.

5.2 Recommendations

Based on analyzing the impacts of climate change, the adaptability and vulnerability of agriculture have many structural and nonstructural measures to adapt to climate change; however, it is essential to concentrate on the following primary solutions:

- Solutions to save surface water sources along with water source reserves:
  - Protecting and developing vegetation cover: Strict protection and development of coastal protection forests to prevent natural disasters, coastal erosion, riverbank erosion, antisand and sand phenomenon, and big flow restrictions in the rainy season.
  - Exploiting water sources from natural rivers and lakes in combination with the construction of suitable artificial lakes and dams to ensure enough supply for production activities.
  - Strictly and economically managing water sources used for agricultural production and completing irrigation water systems to limit water loss and leakage.
- Solutions to modernize agricultural production along the direction of adaptation:
  - Developing seed sources: Researching and developing a new seed group capable of adapting to the impacts of climate change.
  - Flexibility for the seasons and production objects: building a reasonable seasonal calendar based on the changing nature of the weather and climate and replacing objects of low productivity and poor adaptability to those of higher economic efficiency.

- Mechanizing production: Investment in mechanization in stages of the production process, especially for the cultivation industry in order to decrease dependence on natural conditions and to take initiative in seasonal activities and mitigate losses.
- Modernizing agricultural infrastructure and materials.
- Improving production techniques: Promoting research, piloting, and spreading mass deployment of techniques, forms, and models of production adapting to climate change in accordance with the targets of sustainable development.
- Reasonable production plans: Due to the trend of erratic movements of extreme weather events and climate change, it is crucial to have reasonable plans in proposing future agricultural production plans. In order to make long-term, sustainable, and reasonable between environment and agricultural production.
- Weather forecast and climate changes in a timely and accurate manner along with raising people's awareness and ability to adjust to and change climate.

References

Balangue, T. O. (2013). *Summary report on vulnerability caused by climate change and impact assessment tools* (168 pages). Climate Change Committee.

Garg, A., Rana, A., Shukla, P. R., Manmohan, K., Narayanan, K., Parthasarathy, D., & Patnaik, U. (2007). *Handbook of current and next generation vulnerabilities and vulnerability assessment tool* (98 pages). Routledge.

ICEM. (2015). *Assessment of vulnerability and adaptation planning of the construction industry in urban areas in the Mekong Basin* (75 pages). ADB and Nordic development Fund.

Intergovernmental Panel on Climate Change [IPCC] (2007). Fourth assessment report, working group II report. *Impacts: Adaptation and Vulnerability*, 987 pages.

Meynell, P. J., Thongsavath, O., Xeuasing, K., Vannalath, V., & Glémet, R. (2014). *Assessment of the vulnerability of Beung Kiat Ngong Ramsar site, Lao PDR* (127 pages). IUCN.

Ministry of Natural Resources and Environment [MONRE]. (2016). *Scenario of climate change and sea level rise for Vietnam* (190 pages). Environment - Resources Publisher and Vietnam Map.

Morgan, C. L. (2011). *Vulnerability assessments: A review of approaches* (31 pages). IUCN.

Nhuan, M. T. (2011). *Investigating and assessing the extent of vulnerability to Vietnam's marine resources and environment and meteorological conditions; forecasting natural disasters and environmental pollution in sea areas* (398 pages). Ministry of Natural Resources and Environment.

Provia. (2013). *Provia guidelines for assessing vulnerability, Adaptation and Adaptation Capacity. Consultation materials* (198 pages). United Nations Environment Program.

Quang Tri Statistical Office [QTGSO]. (2019). *Quang Tri Statistical Yearbook, 2018* (p. 416). Statistical Publishing House.

UNFCCC. (2008). *Climate change: Impacts, vulnerabilities and adaptation in developing countries* (68 pages). UNFCCC.

USAID. (2013). Mekong ARCC climate change impact and adaptation study. Retrieved March 25, 2019, from https://www.usaid.gov/asia-regional/documents/usaid-mekong-climate-change-study-main-report-2013. 294 pages.

World Wide Fund For Nature [WWF]. (2013). *Assessing the level of vulnerability to climate change of ecosystems in Vietnam* (70 pages). WWF.

Chapter 38
A GIS Application in Optimizing the Collection and Transportation Route of Domestic Solid Waste in Hue City, Vietnam

Chan Van Mai, Son Hoang Nguyen, Cham Dinh Dao, Tin Van Le, and Hang Anh Phan

Abstract The collection and transportation of domestic solid waste in Hue City in particular and other cities in Vietnam generally are still manually implemented, which require a lot of manpower and costly operation but get low economic efficiency and management difficulties in return. This article with the title of GIS application in optimizing the route of collecting and transporting domestic solid waste in Hue City aims to shorten the maximum distance of waste collection and transportation in order to reduce the number of waste trucks. From the research results, the number of daily trucks reduces by 7 from 20 trucks to 13 ones, which contributes to the dramatic reduction in collection and transportation operation cost, saves manpower, and minimizes potential environmental issues.

Keywords GIS · Location allocation · Vehicle routing problem · Routing optimization · Solid waste

---

C. Van Mai
Institute of Open Education and Information Technology, Hue University, Thua Thien Hue, Vietnam

The College of Environment and Planning of Henan University, Kaifeng city, Henan Province, China

S. H. Nguyen (✉) · T. V. Le
Institute of Open Education and Information Technology, Hue University, Thua Thien Hue, Vietnam
e-mail: nhsonsp@hueuni.edu.vn

C. D. Dao
VAST Institute of Geography, Hanoi, Vietnam

VAST Graduate University of Science and Technology, Hanoi, Vietnam

H. A. Phan
University of Science, Hue University, Thua Thien Hue, Vietnam

A. T. Nguyen, L. Hens (eds.), *Global Changes and Sustainable Development in Asian Emerging Market Economies Vol. 2*, https://doi.org/10.1007/978-3-030-81443-4_38

1 Introduction

Along with the fast-growing national economy, the management system of urban solid waste has been comprehensively improved. Besides, science and technologies have been gradually applied in many cities to quickly and efficiently sort, collect, transport, and dispose waste. However, outdated facilities and equipment, and limited techniques in solving issues of collecting, transporting, and disposing waste lead to manpower consumption, high cost of transportation, and under-expected economic and environmental results (Han & Ponce-Cueto, 2015; Hue People's Committee [HPC], 2018; Ministry of Natural Resources and Environment [MONRE], 2018).

In many cities in Vietnam in general and Hue City in particular, collecting and transporting solid waste from transfer stations to landfills are still being manually operated. Specifically, the amount of daily waste is calculated, waste collecting workers are assigned to collect at fixed gathering points along the roads into bigger waste containers or transport to waste transfer stations, and then waste trucks are sent to collect and transport to landfills. The schedule of waste trucks is completely done manually without applying software or algorithms to optimize the route. According to the statistics from Hue urban and environmental company, around 220 tons of domestic waste are daily discharged in Hue City and 20 trucks are needed from 6 pm to 2 am the next morning to collect and transport all of this quantity (Hue Urban Environment and Public Works State Company R [HEPCO], 2018), which can be inferred that the ability to collect and transport waste of the city is still limited and weak.

This paper suggests applying GIS into optimizing the route of collecting and transporting urban domestic waste to minimize the distance of collection and transportation, and then accordingly reducing the number of waste trucks and the operation cost.

2 Methodology

2.1 Data Collecting Method

Primary data include detailed data on road traffic of Hue City; location of garbage bins; location of transfer station; the average daily waste volume of the city; and the daily number of trucks used in collecting and transporting waste.

Secondary data include data from relevant research results in Hue City and data from Hue Urban Environment and Public Works State Company like annual assessment reports on the environmental impacts of waste collecting and transporting activities in Hue City, results of provincial scientific and technological studies on GIS application in management of drainage system infrastructure, and means of waste transportation in Hue City (Nguyen, 2014).

2.2 *Analyzing Method*

To optimize the route of collecting and transporting domestic waste in Hue City, the authors analyze and process the data in the following steps:

- Prepare a complete database and create Network Data in Catalog to analyze data system.
- Use the analysis function of location allocation in the Network Analyst toolbar of ArcGIS 10.3 software to establish parameters to analyze and adjust the location of waste transfer stations in the city.
- Use the function of analyzing Vehicle Routing Problem in Network Analyst toolbar of ArcGIS 10.3 software to set up parameters to run the model of waste collecting and transporting route.

3 Results

3.1 *Subsection*

3.1.1 Current Situation of Collecting and Transporting Domestic Waste in Hue City

Collection and transportation of domestic waste in Hue City can be summarized as follows (Nguyen, 2014):

Domestic waste of households is collected by handcarts, and the waste can be put into plastic bags or garbage bins which can be put in front of the house or delivered to the nearest public waste containers. Every day, the workers will push handcarts through residential areas and collect them to the nearest transfer station.

Wastes from institutions, schools, production, and business establishments are stored in public garbage bins, when the cans are full, and they will be transported to landfills.

Waste at markets and on streets is swept and collected by workers at where it is and then delivered to transfer station to get transported to landfills by trucks.

Waste in hospitals is only collected if it is domestic waste contained in 600-liter cans which are transferred by handcarts to transfer stations. Other types of waste will be handled by hospitals themselves.

Nonhazardous industrial waste is taken to transfer station and then transported to landfills by specialized trucks.

Means of transportation used for waste collection are as follows:

Handcarts are used to collect domestic waste from households in small lanes, streets, markets, and roads. After these handcarts are full, they will be moved to hubs to transfer waste to specialized trucks. The working shift is from 6 pm to 2 am the next morning.

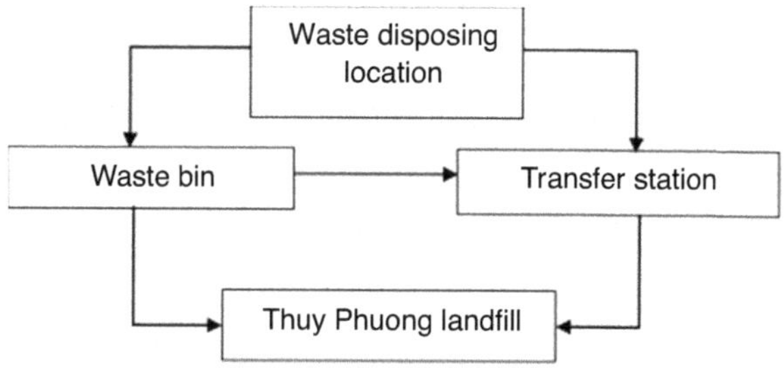

Fig. 38.1 Process of collecting and transporting domestic solid waste in Hue City (Le, 2012)

Specialized trucks are daily driven to fixed routes collecting waste on streets and in hubs and transferring it to landfills. The working shift is from 6 pm to 2 am the next morning.

Currently, Hue City system of solid waste collection and transportation does not use facilities like containers, container trucks, and waste transfer stations. The process of collection and transportation can be summarized as below (Fig. 38.1).

3.1.2 GIS Application to Analyze and Select the Location of Transfer Station

Derived from the fact that the trucks to collect and transport waste daily go through fixed roads and areas to collect waste at transfer stations and in roadside garbage bins, which is time-consuming and low efficient. The authors suggest that all wastes in garbage bins should be moved to the nearest transfer station by cleaning workers and handcart workers. Waste trucks will only go to transfer stations to collect waste and need not collect at different points on roadside, which can reduce quite a number of trucks and save a huge amount of time (Xiao, 2018; Zhu, 2018).

From the current data, there are 1755 locations of garbage bin and 92 transfer stations, and the authors use Location_Allocation tool to analyze the minimum distance from the location of each garbage bin to the nearest transfer station. The analysis is taken by the below steps:

- Network Analyst/New Location_Allocation.

Right click Facilities/Load Locations, uploading the data of current transfer stations (92 stations).

Right click Demand points/Load Locations, uploading the data of current garbage bin locations (1755 locations).

Layer Properties/Analysis Settings/Impedance/Length (Meters).

Layer Properties/Analysis Settings/U-Turns at Junctions/Allowed.

Layer Properties/Analysis Settings/Output Shape Type/Straight Line.

Layer Properties/Advanced Setting/Problem Type/Maximize Coverage.

Layer Properties/Advanced Setting/Facilities to Choose: Fill in the current number of transfer stations (92 in this case).

Layer Properties/Advanced Setting/Impedance Cutoff: Choose a suitable distance to ensure all garbage bins connected to the nearest transfer station (5000 m is chosen in this case).

Click Done to confirm the completion of data setup.

Finally, click Solve of Network Analyst toolbar to run the analysis (Fig. 38.2).

In Fig. 38.2, the green spots are the locations of 1755 garbage bins, the blue points are the locations of 92 transfer stations, and the purple lines are 1755 lines connecting garbage bins to the nearest transfer station.

Realizing that this result is still inadequate, the central area of the transfer stations is too dense, while it is too sparse in the suburb area, making the distance between garbage bins and transfer stations too far. When locations from garbage bins to transfer stations are too far, it will take more time and effort to collect waste from garbage bins to transfer stations, leading to low efficiency.

The authors have analyzed, adjusted, and added a number of transfer stations (this process always complies with the principle of selecting transfer station

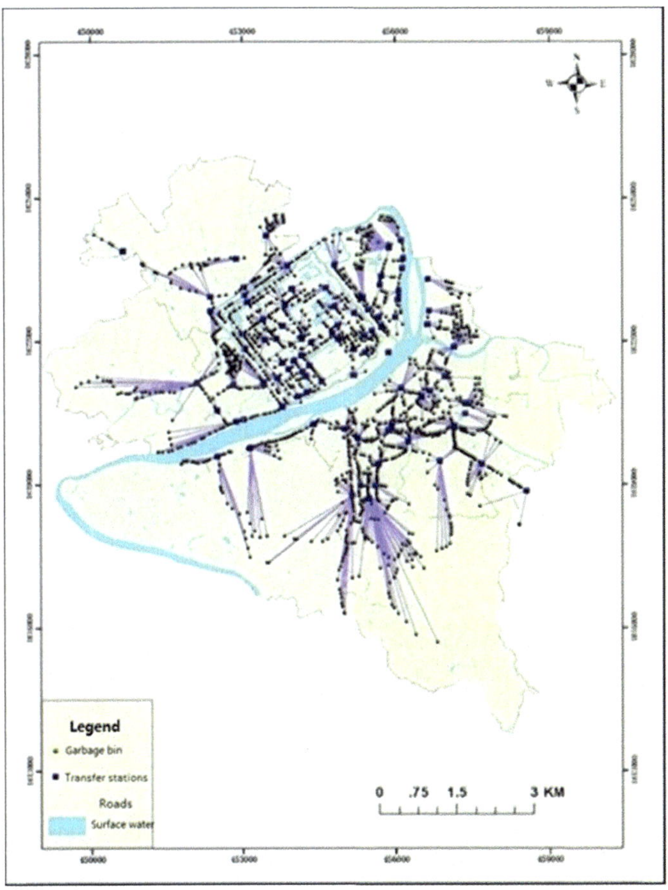

Fig. 38.2 Correlation between locations of current garbage bins and transfer stations

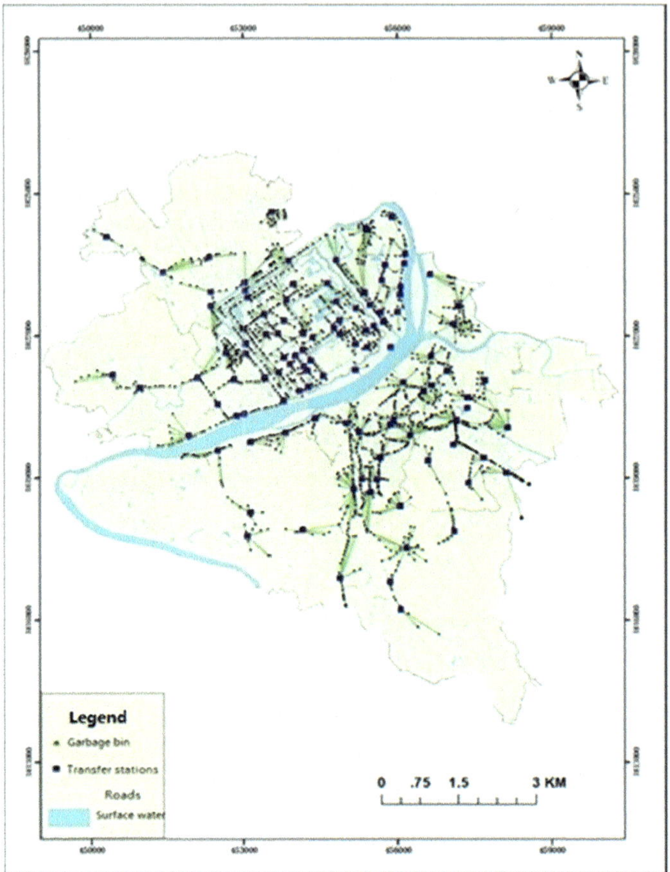

Fig. 38.3 Correlation between locations of garbage bins and transfer stations after adjustment

locations). The result is 16 stations adjusted and 18 stations added, increasing the total number of transfer stations from 92 to 110.

Continue using the Location_Allocation tool to perform the analysis steps as mentioned above and get the results (Fig. 38.3).

3.1.3 GIS Application in Optimizing the Route of Collecting and Transporting Domestic Solid Waste in Hue City

Scheduling and arranging for waste collecting trucks to go along fixed routes are mainly done manually, which cannot suggest the best optimal route, leading to far and even overlapping travel routes, longer collection, and transportation time, and required vehicles, as a result, increasing the total cost of the operating system.

In order to optimize the route of collecting and transporting domestic waste in Hue City, the authors have chosen to use the New Vehicle Routing Problem in Network Analyst toolbar of ArcGIS 10.3 software. The analysis process is done as follows:

Network Analyst/New Vehicle Routing problem.

On Vehicle Routing problem, right click on "Orders/Load Locations" to upload the data of transfer stations after added and supplemented.

Use Geometry/Search Tolerance to fill 5000 (this is the longest distance to the transfer station, meter unit, other parameters defaulted).

On "Vehicle Routing problem", right click to choose "Deports/Load Locations", add the starting point and ending point of waste trucks.

In this article, the starting point is the parking lot at Huong So ward, and the ending point is Thuy Phuong landfill. However, the distance from Hue City to Thuy Phuong landfill is quite far, while the map and transport layer to build network data are located in Hue City, so the ending point will be located in Hue City for easier analysis. The final result of each route will be added with the distance from this ending point to Thuy Phuong landfill.

- Vehicle Routing Problem/Router to set up parameters for routes includes starting time; ending time; starting point; ending point; maximum reaching transfer stations; maximum load of each truck; and the longest possibly moving distance.

Set up parameters in Layer Properties like below:

- Layer Properties/Analysis Setting/Time Attribute/Travel_time (Minutes).
- Layer Properties/Analysis Setting/Distance Attribute/Length (Meters).
- Layer Properties/Analysis Setting/Capacity Count/1.
- Layer Properties/Analysis Setting/Time Field Units/Minutes.

  - Layer Properties/Analysis Setting/Distance Field Units/Meters.
  - Layer Properties/Analysis Setting/U-Turns at Junctions/Allowed.
  - Layer Properties/Analysis Setting/Output Shape Type/True Shape with Measures.
  - Layer Properties/Analysis Setting/Distance Units/Meters.
  - Finally, click 1111(Inline fig tag)222 (Directions) in Network Analyst to analyze the result (Fig. 38.4).

A total of 110 transfer stations were collected with 39 trucks. Each truck can go for 03 turns in one shift, so the minimum number of needed trucks is 13 (reducing by 7 compared to actual operation).

From Table 38.1, the result is shortened like Table 38.2.

Therefore, from the results in Fig. 38.4, and Tables 38.1 and 38.2, it is totally possible to make schedule for 13 trucks to travel along the analyzed route and collect waste at transfer stations shown in Table 38.1, control time and volume that each truck can handle (Aremu, Mihelcic, & Sule, 2011; Faccio, Persona, & Zanin, 2011; Han & Ponce-Cueto, 2015; Katusiimeh, Mol, & Burger, 2012; Malandraki & Daskin, 1992; Mariagrazia & Nicola, 2017; Nayati, 2008; Vecchi, Surco, & Ademir, 2016).

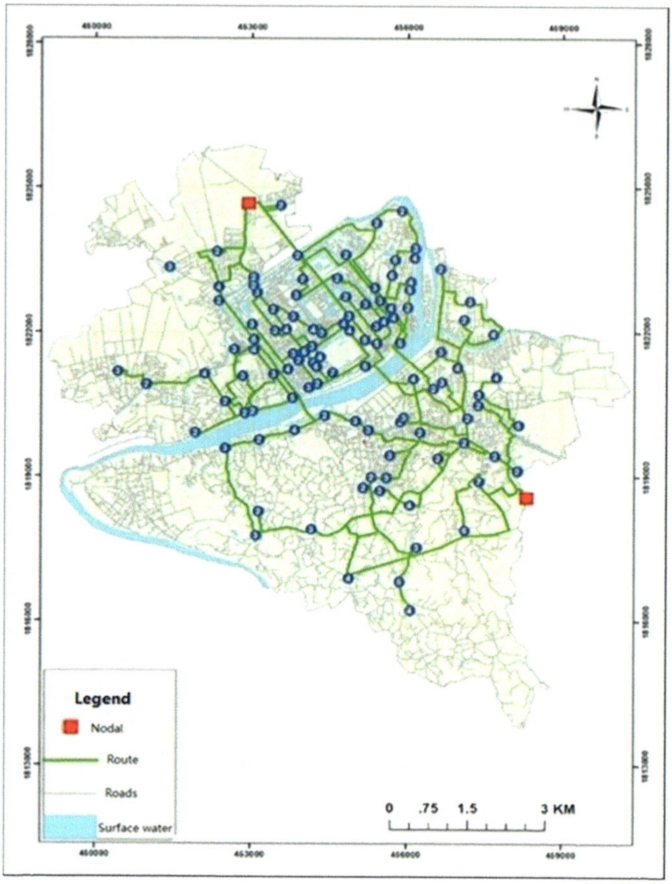

Fig. 38.4 Optimizing Hue City waste collection and transportation route

4 Conclusions and Discussion

Applying Network Analyst in ArcGIS 10.3 software to analyze and select the locations of waste transfer stations and optimize the route of solid waste collection and transportation in Hue City brings about favorable results shown by the fact that the number of trucks to collect waste are reduced from 20 to 13. A big reduction in waste trucks and time of collection and transportation will help to reduce the total operating cost of the city's waste collection and treatment system.

This model of optimizing the route of waste collection should be quickly applied and replicated in other cities throughout the country, contributing to saving manpower, operating costs, and minimizing environmental issues.

Table 38.1 Result of analyzing the route of waste collection and transportation in Hue City

No.	Truck turn	Time to travel (min)	Loading weight (kg)	Transfer station
1	Truck 1 turn 1	95	6000	11–30
2	Truck 1 turn 2	68	6000	81
3	Truck 1 turn 3	190	4750	95–101-107
4	Truck 2 turn 1	138	5875	62–15–33-31
5	Truck 2 turn 2	146	6000	45–42–40-41
6	Truck 2 turn 3	133	5875	80–75–93-94
7	Truck 3 turn 1	136	6000	37–35–20-60
8	Truck 3 turn 2	71	4500	86
9	Truck 3 turn 3	119	5750	48–76-68
10	Truck 4 turn 1	118	5250	8–34-32
11	Truck 4 turn 2	121	5250	99–97-74
12	Truck 4 turn 3	92	5000	82–69
13	Truck 5 turn 1	140	6000	39–16–36-59
14	Truck 5 turn 2	94	5875	72–78
15	Truck 5 turn 3	92	5125	71–83
16	Truck 6 turn 1	137	5875	29–57–47-58
17	Truck 6 turn 2	121	5375	65–67-110
18	Truck 6 turn 3	92	5625	92–85
19	Truck 7 turn 1	97	5750	63–38
20	Truck 7 turn 2	121	6000	18–19-70
21	Truck 7 turn 3	94	5750	73–84
22	Truck 8 turn 1	178	5875	5–26–25-3-4-23
23	Truck 8 turn 2	115	6000	108–77-105
24	Truck 8 turn 3	125	5125	44–61-109
25	Truck 9 turn 1	138	6000	14–13–50-49
26	Truck 9 turn 2	97	6000	90–87
27	Truck 9 turn 3	75	6000	66
28	Truck 10 turn 1	137	5750	10–9–1-2
28	Truck 10 turn 2	98	4000	79–98
30	Truck 10 turn 3	75	5750	91
31	Truck 11 turn 1	160	6000	53–52–51-55-56
32	Truck 11 turn 2	129	4000	89–106-100
33	Truck 11 turn 3	111	5625	96–88
34	Truck 12 turn 1	117	5625	6–7-12
35	Truck 12 turn 2	107	5875	43–24
36	Truck 12 turn 3	85	5875	54
37	Truck 13 turn 1	140	4875	64–102–27-28
38	Truck 13 turn 2	133	5750	104–103–46
39	Truck 13 turn 3	125	6000	17–21-22

Table 38.2 Result of analyzing waste collection and transportation route in Hue City

No.	Truck	Travel time (min)	Loading weight (kg)
1	Truck 1	353	16,750
2	Truck 2	417	17,750
3	Truck 3	326	16,250
4	Truck 4	331	15,500
5	Truck 5	326	17,000
6	Truck 6	350	16,875
7	Truck 7	312	17,500
8	Truck 8	418	17,000
9	Truck 9	310	18,000
10	Truck 10	310	15,500
11	Truck 11	400	15,625
12	Truck 12	309	17,375
13	Truck 13	398	16,625

References

Aremu, A. S., Mihelcic, J. R., & Sule, B. F. (2011). Trip time model for municipal solid waste collection applicable to developing countries. *Environmental Technology, 32*(15), 1749–1754.

Faccio, M., Persona, A., & Zanin, G. (2011). Waste collection multi-objective model with real time traceability data. *Waste Management, 31*(12), 2391–2405.

Han, H., & Ponce-Cueto, E. (2015). Waste collection vehicle routing problem: Literature review. *Promet - Traffic&Transportation, 27*(4), 345–358.

Hue People's Committee [HPC]. (2018). *Report on the current environmental situation of hue 2018* (132 pages). Hue province.

Hue Urban Environment and Public Works State Company R [HEPCO]. (2018). A summary report on planning to collect waste in hue city in 2018.. Thua Thien Hue province.

Katusiimeh, M. W., Mol, A. J., & Burger, K. (2012). The operations and effectiveness of public and private provision of solid waste collection services in Kampala. *Habitat International, 36*(2), 247–252.

Le, V.D. (2012). *Apply GIS (geography information system) into management of domestic solid waste in central districts of Thai Nguyen city, Thai Nguyen province*. Code: 60.44.03.01. Thai Nguyen University.

Malandraki, C., & Daskin, M. S. (1992). Time-dependent vehicle routing problems: Formulations, properties and heuristic algorithms. *Transportation Science, 26*(3), 185–200.

Mariagrazia, D. A., & Nicola, E. A. (2017). A vehicle routing technique for hazardous waste collection. *IFAC Papers Online, 50-1*(2017), 9694–9699.

Ministry of Natural Resources and Environment [MONRE]. (2018). *National Report on Environment 2018 - an overview of Vietnam's environment* (158 pages). Vietnam Natural Resources and Environment Publishing House and Map Company.

Nayati, M. A. K. D. (2008). *School bus routing and scheduling using GIS*. University of Gavle.

Nguyen, H.S. (2014). *Apply GIS into management of drainage system and means of waste transportation in Hue city*. Scientific report on provincial scientific and technological research results, code TTH.2012-KC.10. Thua Thien Thue province.

Vecchi, T. P. B., Surco, D. F., & Ademir, A. (2016). A sequential approach for the optimization of truck routes for solid waste collection. *Process Safety and Environmental Protection, 102*, 238–250.

Xiao, C. A. (2018). Site selection of rural household refuse reclamation transfer stations based on GIS. *China Academic Journal, 17*, 109–110.

Zhu, C. J. (2018). Evaluation and optimization of garbage transfer station in Anqing City based on GIS and AHP. *Anqing Normal University, 24*(1), 78–82.

Chapter 39
Water–Food–Energy Nexus in the Context of Climate Change: Developing a Water Security Index for Water Resource Management in Vietnam

Nguyen Mai Dang, Duong Tran Anh, and Thanh Duc Dang

Abstract Reasonable distribution of water for agricultural and electricity production is pivotal for economic development, environmental protection, and social equality, and this task is becoming more challenging. This is because the demand for water, food, and energy has increased significantly recently due to population growth, higher standards of living, and industrialization, especially in developing countries like Vietnam. Agricultural activities in Vietnam account for about 20% of GDP, and the total freshwater consumption is up to approximately 70% for the irrigation of agricultural cultivation, making it the largest water consumer. Food production and manufacturing processes also consume about 20% of the total energy production. Energy is necessary for food production, transportation, and distribution as well as extracting, pumping, collecting, transporting, and treating water. Consequently, Vietnam confronts with the problems of water, food, and energy shortages if these links are not thoroughly addressed. Additionally, climate change, causing the alteration of weather conditions, will exacerbate a range of risks to water, food, and energy. Study results to deliver a comprehensive review of water, food, and energy nexus in Vietnam and to propose using water security index and coupled water–energy–food modeling for better water resources management plans in the context of climate change.

Keywords Water–food–energy nexus · Climate change · Water resource management · Water security · Vietnam

N. M. Dang (✉) · T. D. Dang
Thuyloi University, Hanoi, Vietnam
e-mail: dang@tlu.edu.vn

D. T. Anh
Ho Chi Minh University of Technology, Ho Chi Minh City, Vietnam

A. T. Nguyen, L. Hens (eds.), *Global Changes and Sustainable Development in Asian Emerging Market Economies Vol. 2*, https://doi.org/10.1007/978-3-030-81443-4_39

1 Introduction

The water–energy–food nexus is a common concept used to umbrella the complex and interrelated nature of water usage for a variety of social, economic, and environmental targets (FAO, 2014). The three factors (water, energy, and food) are closely linked. For example, water is used to rotate turbines for hydropower production and water is also necessary for irrigation and aquaculture production (Ackerman & Fisher, 2013; Vogt, 2010) (as can be seen in Fig. 39.1). However, water supply is sometimes limited, which is caused by resource unavailability. Thus, there is water shortage for agricultural and energy production (Feng, Siu, Guan, & Hubacek, 2012; Pacetti, Lombardi, & Federici, 2015). This problem normally occurs in fast-growing regions with limited precipitation or inflow. The concept "water–energy–food nexus" helps scientists understand better the impact of human activities on the natural environment across different sectors and scales. Stakeholders can also identify and manage trade-offs and propose integrated and cost-effective plans (Haase, Schwarz, Strohbach, Kroll, & Seppelt, 2012; Lehmann, 2018).

Generally, water shortage is a global problem. In a report of Population Action International (PAI) , 2.8 billion people in 48 countries are predicted to face water scarcity by 2025. Most countries in the list are from the West and Southeast Asia, North Africa, and sub-Saharan Africa (Aaron, 2008). Among these countries, Vietnam is among the most vulnerable countries to water scarcity, especially on large river deltas and coastal lowlands such as the Red River Basin and Mekong River Basin (ADB, 2013).

Recently, the economy of Vietnam is growing rapidly, but fast development also contains economic, social, and environmental challenges. In the growth model of Vietnam, industrial manufacturing and FDI companies contribute a large percentage

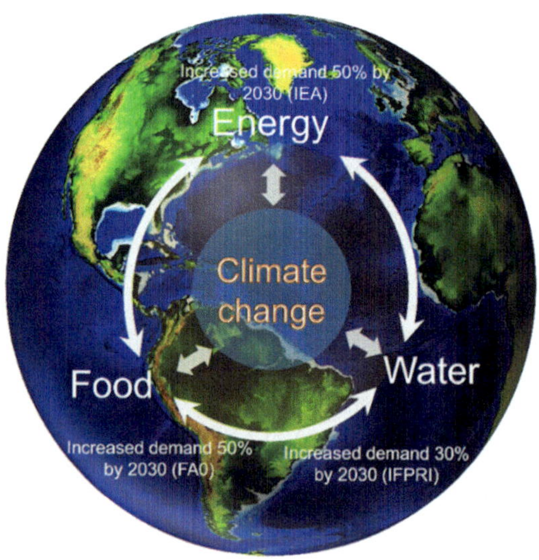

Fig. 39.1 Water–food–energy nexus in climate change context (developed based on Haase et al., 2012 and Lehmann, 2018)

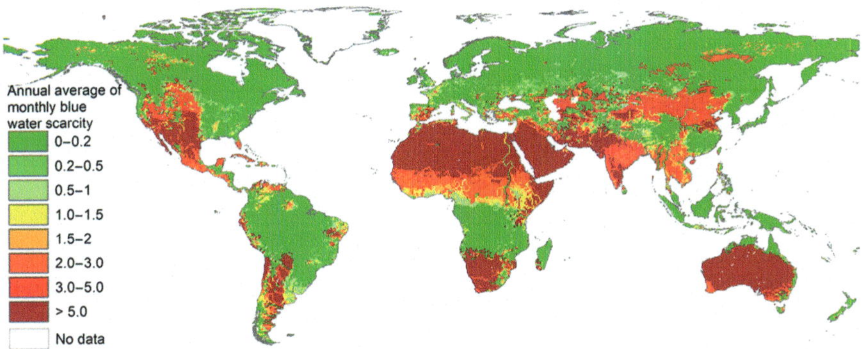

Fig. 39.2 Global annual average of monthly water scarcity, and the water scarcity in Vietnam is at level 3 (source: Mekonnen & Hoekstra, 2016)

of GDP (OECD, 2014). This model, however, requires a lot of natural resources, energy, and water. More seriously, in the last decade, the country has been facing unsustainable development because of burgeoning populations, increasing energy demand, and worsening water scarcity (Fig. 39.2). In the future, climate change is predicted to worsen the problem of water scarcity. In this context, water, food, and energy, which are the three most fundamental human needs and economic necessities, have been studied separately. However, academic researchers and policymakers have put more and more attention on the emergence of the water–energy–food (WEF) nexus since the interlink among these three factors becomes more apparent (Smajgl, Ward, & Pluschke, 2016). This paper attempts to provide a comprehensive overview of the water, energy, and food nexus in Vietnam.

2 Review of Water–Food–Energy Framework

In the nexus of water–food–energy, water plays a key role (Ripl, 2003), which can be expressed as (1) "water flowing through the veins of the economy" and (2) "water: the bloodstream of the biosphere." This exhibits that water acts as an important factor influencing and influenced by economic development and ecological sustainability (Fig. 39.3) (Hoff, 2011). In a water cycle, water is renewable and unlimited resource, but a burgeoning population and growing living standard are putting greater pressures on this resource. It is projected that water, food, and energy demand will rise by 30–50% in the next few decades (Swatuk & Cash, 2017). Water shortages, however, may result in geopolitical–social conflict and damage environmental integrity. The interrelated nature of water, food, and energy makes any strategies, which only focus on one part of component, will result in unintended consequences (World Economic Forum, 2012).

Since the WEF Bonn Conference in 2011, several nexus frameworks have been developed (European Report Development, 2012; ICIMOD, 2015; United Nations Economic Commission for Europe [UNECE], 2015). As described in McGrane

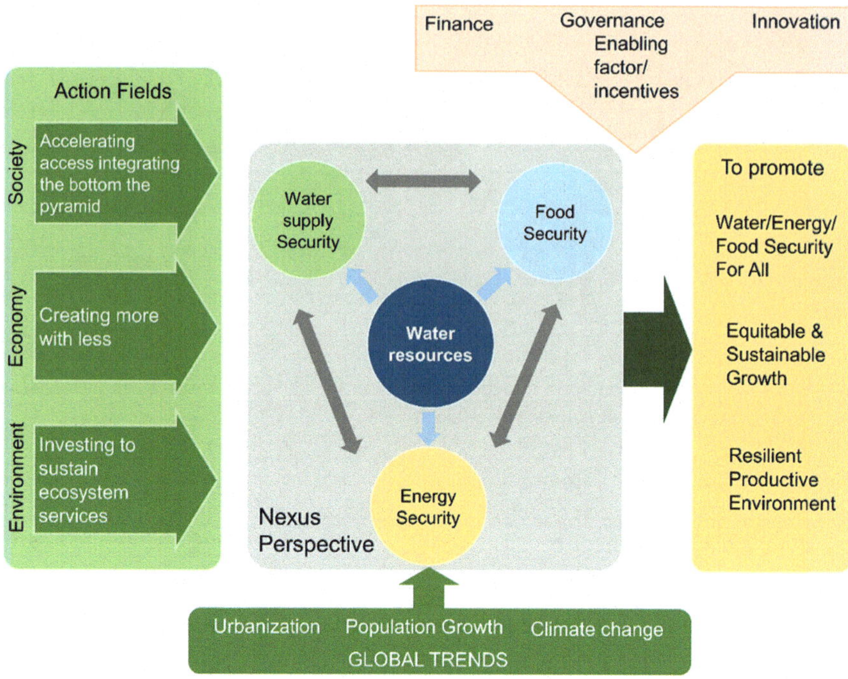

Fig. 39.3 Framework for water–food–energy security nexus (developed based on Hoff, 2011)

et al. (2018), different boundaries for frameworks have resulted from different purposes with constraints such as land and mineral resources. These frameworks have been built from regional scales to global scale (ICIMOD, 2015; World Economic Forum, 2012), of the complexity of these frameworks figures that there is a need for a hierarchical framework integrating across scales. However, a complex framework relies on many tools and corresponding data to represent the nexus from different perspectives and scales. This paper focuses more on the case study of Vietnam. The status of water consumption and production of food and energy in Vietnam is described and analyzed; the solutions to improve water use for food and energy are discussed and proposed.

3 Water

3.1 Water Availability

In terms of spatial distribution, Vietnam has dense river systems with an accumulative length of internal rivers up to 41,900 km (2030WRG, 2017; WEPA, 2002). There are nine major river basins with more than 10,000 km^2 of catchment areas.

Nevertheless, over half of the total water resources are from outside the country (ADB, 2009a). More specifically, around 323 km^3/year (38%) out of 848 km^3/year is generated in Vietnam's river basins. River basins such as Ca and Ma have about 40% of their basins outside the territory of Vietnam, when this rate for Dong Nai is 15%. Neighboring countries contribute around 524.71 km^3 into the total annual runoff, including 470 km^3 from the Mekong River, 44.1 km^3 from the Red River, 9.1 km^3 from the Ca and Ma rivers, and 1.41 km^3 from the Dong Nai River (Frenken, 2012). This dependence results in Vietnam's lack of effective solutions to manage water resources. Table 39.1 presents statistical data of the total area of basins and the percentage of basins in Vietnam and neighboring countries.

In terms of temporal distribution, the country has abundant surface water resources all around, but water availability varies largely over the year. When the total rainfall amount is considerably high, runoff in the dry season only contributes around 15–30% of the annual flow.

3.2 *Water Demand and Supply*

ADB reports that the total demand of the country is around 80.2 billion m^3/year in 2009 (ADB, 2009b), while IWRP (2015) estimates that water demand is at 80.6 billion m^3/year, and this number increases to 95 billion m^3/year in 2030. Water in Vietnam is mostly used for agricultural production (as shown in Fig. 39.4), especially rice production. In 2016, the total agricultural water use is at around 76 billion m^3/year and is projected to increase to 91 billion m^3/year by 2030. Water demand for rice changes over seasons, ranging between 10,000 and 12,000 m^3/ha in the winter–spring growing season and only 5000 m^3/ha in the summer–autumn growing season. Paddy rice fields in the Mekong Delta require almost 45% of Vietnam's irrigation

Table 39.1 Eight largest river basins in Vietnam (Frenken, 2012)

Name	Total basin area (km^2)	Basin area in Vietnam (km^2)	% of basin in Vietnam	% of basin area in Vietnam and total Vietnam area (331,052 km^2)
Mekong	795,000	63,600	8	19
Red-Thai Binh (includes Da River basin)	155,000	85,250	55	26
Dong Nai	44,100	37,485	85	12
Ma-Chu	28,400	17,608	62	5
Ca	27,200	17,680	65	5
Ba	13,900	13,900	100	4
Ky Cung-bang Giang	11,220	10,547	94	3
Thu bon	10,350	10,350	100	3
Total	1,086,170	256,420	–	77

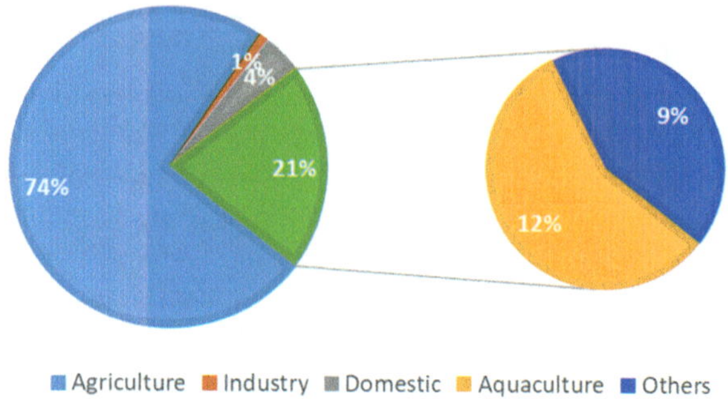

Fig. 39.4 Water demand distribution for different sectors in 2020 (developed based on data from Nhuan, Anh, Hoc, & Cu, 2014)

water (Fig. 39.5). Thus, to ensure water security, the rice production area is limited beyond 3.8 million ha (World Bank, 2016). Nevertheless, in order to ensure food security and export, rice production has to increase to 41 to 43 million tons/year in 2020 and 44 million tons/year in 2030.

Water demand in Vietnam not only increases due to the expansion of agricultural rice farms, but also increases due to urbanization, industrialization, tourism development, and population growth. Although in the future, a large fraction of water use will be for irrigation demand and aquaculture activities (86–93%) (ADB, 2009a, 2009b), economic development and population growth will result in further increase in domestic and industrial demand. Besides water for food production, hydropower generation is expanded drastically to fulfill the thirst of energy in Vietnam and its neighboring developing economies in Southeast Asia and China. This will definitely influence the availability of water for domestic and agricultural uses in downstream communities (Waibel, 2010). Water for food production and energy production will be discussed in the next sections.

4 Food

4.1 Food Production

Agricultural production and food industry are the two largest users of water globally, accounting for nearly 80–90% of consumptive water. When the world population is predicted to increase to 9.2 billion people in 2050, it is also projected that agricultural and domestic water consumption will increase by 70% and 40%, respectively. Consequently, the world may confront with a water shortage in the future. This situation may also happen in Vietnam although it is an agricultural exporter.

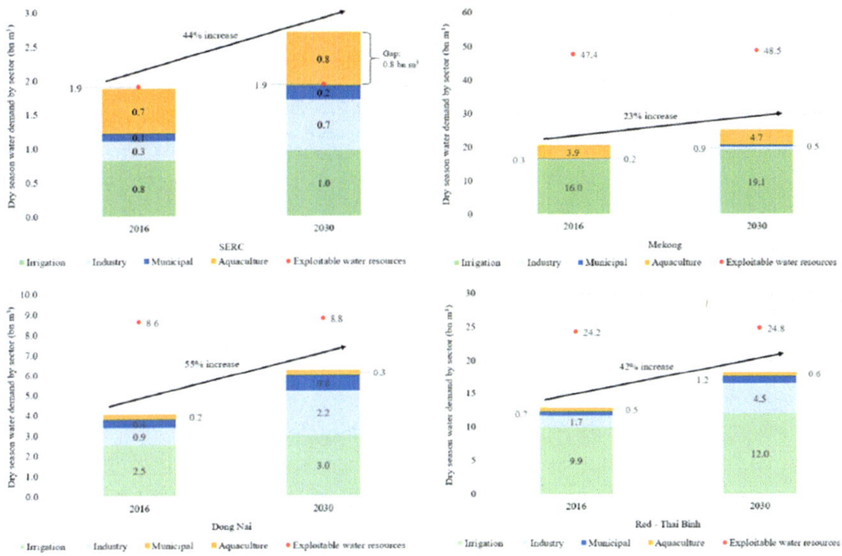

Fig. 39.5 Water demand projections (dry season only) for Mekong, Red-Thai Binh, Dong Nai, and SERC (Source: 2030WRG, 2017)

4.2 Food Demand and Security

The food crisis in 2007–2008 revealed a structural problem in the global food system. The crisis pushed around 100 million people into poverty and a half of this number in the latter half of 2010. This global problem seriously influenced Asia Pacific region. Especially, Vietnam was one of the biggest rice exporters worldwide and had to restrict its rice export, causing problems to many other countries that depend on its rice.

Although the industrial proportion in Vietnam's GDP is higher, the agricultural sector is still important in its economy. This sector accounts for 22% of the country's GDP and 52% of the country's labor force works on this sector (IFAD, 2012). Recently, agricultural growth has been accelerating by improving technology adoption and irrigation areas. The Mekong River Delta is the largest rice and fruit production area in Vietnam, accounting for over half of the country production. 90% of rice and 65% of aquaculture products are produced here for exports.

It is noted that major agricultural regions in Vietnam, for example, the Red River Delta (Sông Hồng) and the Mekong River Delta (Sông Cửu Long), are coping with impacts due to urbanization and climate change. Although the government keeps 3.8 million hectares of rice fields to fulfill domestic consumption and export demand, the total paddy area is predicted to decrease in the future compared to the current stage. Up to 32.2% of agricultural areas will be affected by climate change by the end of the twenty-first century (FAO, 2014; JICA, 2013).

Figure 39.6 shows that the concept of food security comprises four dimensions, including food availability, food accessibility, supply stability, and utilization. In the nexus of water–energy–food, FAO considers food security as an entry point. This concept and historical data reveal that the main problem of food security in Vietnam stays in utilization and stability.

5 Energy

5.1 Energy Production

The slope of most rivers in the country is quite steep, so there are high potentials for hydropower development (Anh, 2015). In the last few decades, hydropower plants of various sizes have been developed in Vietnam. For example, from 2002 to 2004, 17 medium- to large-sized plants (>100 MW) with a total installed capacity of approximately 3000 MW and about 20 small-sized plants with a total capacity of 500 MW are installed. In 2010, the total installed capacity of all power plants increased to over 20,600 MW, equivalent to 3.2 times in 2000 and 1.78 times in 2005. Meanwhile, the total energy production increased 3.2 times between 2000

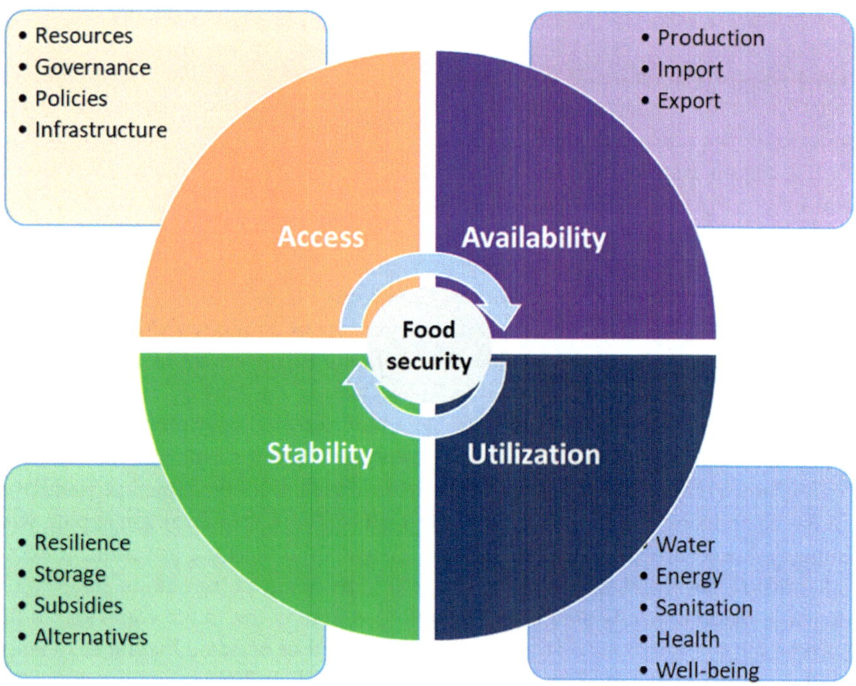

Fig. 39.6 A component of food security (developed from FAO, 2014)

and 2010 and 1.78 times between 2005 and 2010. According to the Electricity Planning Scheme VII, hydropower output will increase to 17,400 MW in 2020. In order to obtain this target, 1000 medium- and small-sized hydropower projects will be implemented with a total capacity of 7500 MW.

When hydropower contributes to 45.5% of electricity generation, Vietnam also has other forms of energy resources, including 615 million tons of crude oil, 600 billion m^3 of natural gas, and 5883 million tons of coal (Asia Pacific Energy Research Centre, 2012; Danish Energy Agency [DEA], 2017). In terms of petroleum, the country is the third largest producer in Southeast Asia. Natural gas is mostly extracted in the southern part of the country, and the discovery of 50 billion m^3 gas in the Red River Basin will lead to an increase in gas production in the coming year. In 2011, gas turbine power plants accounted for 33.6% and the number for coal power plants is 15.3%. Oil and diesel power plants contribute a limited power generation with a total installed capacity of 22 GW (5.6%). Table 39.2 and Fig. 39.7 show the installed capacity of each technology in 2011. These fossil resources are significant, but rapidly growing energy demand still puts more stresses on energy production. Subsequently, Vietnam has imported electricity from its neighboring countries by linking its energy network to Laos, Cambodia, and China.

5.2 Energy Consumption

In different sectors, energy consumption is often the opposite of water uses. For example, industrial areas use most of electricity (33%), while agricultural production only accounts for 1% of energy consumption in Vietnam. Residential areas and transportation activities account for 33% and 24% of energy consumption, respectively, when service sections only need 3%. Figure 39.8 shows the energy production in proportion by sector in 2012.

Table 39.2 Power generation by fuel type (December 2016) (source: Das, 2019)

Power source	Capacity (MW)	Shared (%)
Hydropower	15,857	37.6
Coal-fired power	14,448	34.3
Oil-fired power	1370	3.3
Gas-fired power	7502	17.8
Diesel, small hydropower, and renewables	2418	5.8
Import	540	1.2
Total	42,135	100

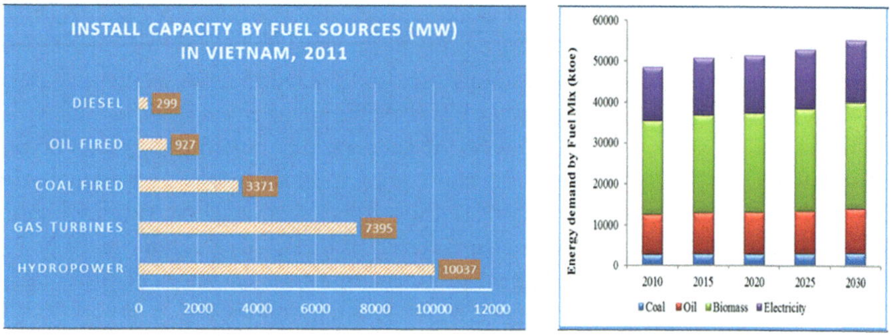

Fig. 39.7 (left) Energy generation capacity by different fuel sources in 2011 and (right) forecasted energy demand in 2030 by fuel sources in Vietnam (developed based on data from Thuy & Bundit, 2014)

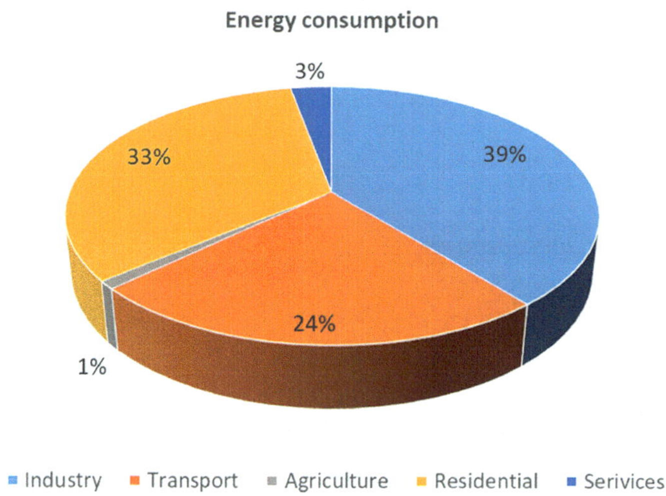

Fig. 39.8 Energy consumption by sectors

5.3 *Energy Development*

In 2015, hydropower plants (59.76%), gas turbine plants (22.45%), coal-fired power plants (33.45%), renewable energy (5.37%), and import (1.42%) with a total installed capacity of 45,600 MW are the main source of power generation in Vietnam. Among these resources, hydro- and renewable energy sources are constrained by water availability, limited construction sites, and high cost, when nuclear power raises public attention due to safety issues (Table 39.3).

Electricity demand is projected to increase at the rate of 10.5% per year during 2016–2020, and the rate may slightly reduce to 8.0% per year for the period 2021–2030. Correspondingly, electricity consumption is predicted to increase to

234.6 TWh in 2020 and up to 506.0 TWh in 2030, respectively. Electricity consumption in 2030 is about 4 times higher in comparison with that in 2014. The peak demand is expected to reach around 42.1 GW in 2020 and 90.7 GW in 2030, respectively. Table 39.4 presents important indicator of energy production and consumption in 2015, 2020, 2025, and 2030.

6 Water-Related Effect by Climate Change in Vietnam

Vietnam is among the top countries influenced by climate change, especially its coastal and river deltas. The main manifestations of climate change impacts on water resources include the modification of water availability and spatial distribution with more extreme events (Giang, Toshiki, Kunikane, & Sakata, 2012). This study projects that the amount of water availability will reduce by 96%, 91%, and 86%, respectively, compared with today's quantity. Currently, the Mekong Delta accounts for approximately 60% of surface water of the nation, and the remaining 40% is shared by other regions, having 80% of Vietnam's population and 90% of the national production. Among these regions, there are several mountainous river basins, such as the Central Coast and the Southeast, with very high flow gradients. Consequently, floods are one of the most dangerous disasters in the country. Besides flood risk, Vietnam also faces tropical typhoons due to 2000 km^2 of coastal zones' UNDP (2003). The country, which is located in the northwest of Pacific Ocean, is along the typhoon route and is ranked as one of the top ten countries most vulnerable to typhoons. Around seven typhoons hit the country every year, especially the northern and central Vietnam (MONRE, 2003). Reports from IPCC (2007) figure out that the number of typhoon events may fluctuate over year, but the intensity of typhoons has increased with higher wind speeds and rainfall intensification. Additionally, typhoons tend to move southwards in the recent decades (European Union, 2006). Consequently, this phenomenon changes local livelihoods and environmental protection activities.

Under the impact of climate change, floods occur more often than in the last century. With the increase in population and economic development, flood damage

Table 39.3 Forecasted energy generation by power sources in 2030 (source: Danish Energy Agency [DEA], 2017)

Power source	2015	2020	2025	2030
Renewable energy (%)	5.37	9.9	12.5	21
Coal (%)	33.45	42.7	49.3	42.6
Gas turbine (%)	22.45	14.9	15.6	14.7
Hydro (%)	37.31	31.1	21.1	16.9
Import (%)	1.42	2.4	1.5	1.2
Nuclear (%)	---	---	---	3.6
Total (MW)	45,600	60,000	96,500	129,500

Table 39.4 Electricity demand actual and projected in 2030 (source: Danish Energy Agency [DEA], 2017)

Item	2015	2020	2025	2030
Annual demand (TWh)	141.8	234.6	252.3	506.0
Annual generation (TWh)	161.3	265.4	400.3	571.8
Maximum demand (GW)	25.3	42.1	63.5	90.7
Per capita consumption (kWh)	1560	2545	3610	4950

Notes: GW: gigawatt; kWh: kilowatt hour; TWh: terawatt hour

is expected to excavate to 12–19% in 2070 (MONRE, 2003). In coastal areas, sea level rise will force the government to implement more costly hard solutions. Actually, sea level rise has been observed along the coastline of Vietnam with a rate of 3 mm/year during the period of 1993–2008. Sea level will rise by 30 cm in 2050 and 75 cm in 2100 compared with the 1980–1999 period (MONRE, 2009).

It is considerably difficult to develop long-term adaptation strategies due to the wide band of climatic projections for Vietnam. The 2012 report of MONRE shows that the average temperature will increase from 1.9 °C to 3.1 °C over the country with the highest increase in the central Vietnam. Days with high temperatures will be expected to increase by 10–20 days in large parts of the country. In addition, climate change also impacts the decrease of precipitation in the dry season, resulting in the higher risk of drought, while the increase of rainfall in the wet season will increase flood risk in the river deltas and coastal lowlands.

7 Water Security Index Method for River Basin

In order to consider the affordability of water and water demand, in 2000, the Global Water Partnership proposed the definition of water security (Cook & Bakker, 2012). Water security means reliable water source availability for humans, livelihoods, and production with a certain level of water shortage risks.

Water security assessment for basin scale can be carried out with the framework in Table 39.5. This framework includes five variables and eight indicators (Babel & Shinde, 2018). The variable that measures each indicator is then normalized to the range 1 to 5 using reference values from literature and expert opinion. These normalized values are aggregated together using equal weights. When there is one variable more significant than the others, weights are already considered in the normalization method. Using this procedure, indicators are calculated and then arranged into dimensions. Subsequently, dimensions are placed in the overall WSI:

$$WSI = (\text{Score for DIM1} + \text{Score for DIM2} + \text{Score for DIM2}$$
$$+\text{Score for DIM4} + \text{Score for DIM5}) / 5$$

where DIM1 = water availability, DIM2 = water productivity, DIM3 = water-related disasters, DIM4 = watershed health, and DIM5 = governance.

Table 39.5 Framework for basin-scale assessment of water security (Babel & Shinde, 2018)

Dimension	Indicator	Potential variables	Suggested ways to measure
Water availability	Sustainable basin exploitation	1. Per capita water availability	Surface runoff/population
		2. Water scarcity	Annual per capita water resource availability (Babel and Wahid, 2008)
		3. Water variation	The coefficient of variation of precipitation over the last 50 years (Babel and Wahid, 2008)
Water productivity	Economic value of water	1. Commercial/ industrial revenue per drop	Nonagricultural GPP/ nonagricultural water use in the basin (ADB, 2013)
		2. Agricultural, aquaculture, and livestock revenue per drop	Agricultural, aquaculture, and livestock GPP/ Agricultural, aquaculture, and livestock water use in the basin (ADB, 2013)
Water-related disasters	Drought factor	1. Drought damage	Economic damage caused by droughts
		2. Proportional area under drought	Drought area/total area
		3. Drought occurrence frequency	Number of drought occurrence per year
		4. Ratio of the area with water-saving irrigation to the total area of arable land	Area of irrigation/area of arable land
	Flood factor	1. Flood damage	Economic damage caused by floods
		2. Proportional area of flooding	Flooding area/total area
		3. Flood occurrence frequency	Number of flood occurrence per year
		4. Percentage of population living in hazard-prone areas	Population living in hazard-prone areas/total population
		5. Flood control capacity	Ratio of the water reserved in dams at the end of the year to the total water utilization

(continued)

Table 39.5 (continued)

Dimension	Indicator	Potential variables	Suggested ways to measure
Watershed health	Health of water bodies	1. Surface water quality factor	Dissolved oxygen concentration/ permissible limit
		2. Groundwater quality factor	Concentration of site-specific pollutants /permissible limits of these pollutants
		3. Average class water quality rivers	Country-specific conditions (ADB, 2013)
		4. Biochemical oxygen demand (BOD) in water bodies	BOD 5-day values of river water samples
	Vegetation cover	Natural vegetation factor	Natural vegetation area/basin area
Water governance	Overall, management of the water sector	Institution factor	Questionnaire
	Potential to adapt to future changes	Adaptability factor	Questionnaire

One of the most popular operational frameworks developed by Babel, Shinde, Sharma, and Dang (2015) evaluates the level of water security at city (administrative boundary) and basin (hydrological boundary) scales. The core component of this framework is the DPSIR technique, which is used to identify different dimensions and indicators at both scales. Both dimension and indicator have five levels, ranging from 1 to 5, which are determined based on either literature or expert opinion. Subsequently, indicators and dimensions are aggregated into an overall index for water security (note that each level (city and basin) has one index value). There is also an interpretation system to explain information behind the index.

The above research was also used to test the operational framework in three study areas, including India, Thailand, and Vietnam. These study areas included three basins and three corresponding cities, namely the Banas River Basin and Jaipur City in India, Chao Phraya River Basin and Bangkok City in Thailand, and Red River Basin and Hanoi City in Vietnam. Dang et al. (2016) presented an assessment of water security based on the developed framework for city scale, Hanoi. The governance and adaptation dimension are well developed, while the environment dimension is quite low level even when the improvement was done. Together with the effort of government and local people, the water security condition will be higher in the near future. Some results have not been normalized yet. However, the ratios of water supplied, water demand, and the treatment of wastewater and others help for better understanding of water security in Hanoi City. In 2017, the application for basin scale in the Red River of Vietnam was conducted. Five dimensions such as water availability, water productivity, water-related disaster, watershed health, and water governance are considered. It is indicated that the Red River has a moderate level of water security during 2010 to 2015. This requires the integrated water resource management in Red-Thai Binh River Basin to improve the control, monitoring, and management capacity (Dang et al., 2017).

Water security index will provide an overview of physical, socioeconomic, and governance dimensions of water management in the country. Coupling water–energy–food models should be considered to understand how climate change, population growth, and economic development affect water, food, and energy security in Vietnam. Recently, Chowdhury, Dang, and Galelli (2018) coupled a hydrological model named the variable infiltration capacity model (VIC) with a Network Constrained Unit Commitment (NCUC) model to study the impact of climate variability on energy shortfall in Laos.

In order to obtain safe solutions for the nexus of water, food, and energy, Integrated Water Resources Management (IWRM) should be implemented. IWRM is a concept that changes the water sector and encourages the coordinate management of natural resources to obtain economic and social benefits without compromising sustainable environment (Zeitoun, 2011). Figure 39.9 shows how social and natural processes are combined to establish or deny water security. This figure reveals that sustainable water security is related to the degree of equitability and balance among many socioeconomic and political factors.

8 Conclusions

The rapid growth of Vietnam's economy improves the living standard of people, but it also leads the country to confront with many problems such as water, food, and energy shortage. As the impact of climate change on every aspect of agricultural and industrial production activities becomes more apparent, it is predicted that the country will face more challenges in the future. Among the three factors, water is the key resource and plays a central role in producing food and energy. Thus, solutions for sustainable water resources management are needed. In order to improve water security, it requires a holistic understanding of the proportion of water uses for food and energy production and other activities. This can only be implemented by a comprehensive water–food–energy database for the whole country. A water security index assessment can be implemented to provide guidance for master plans. Then, coupled water–food–energy models can be considered in the implementation stages. It is expected that the outcome of this study will raise more public attention on the nexus of water–food–energy in the country.

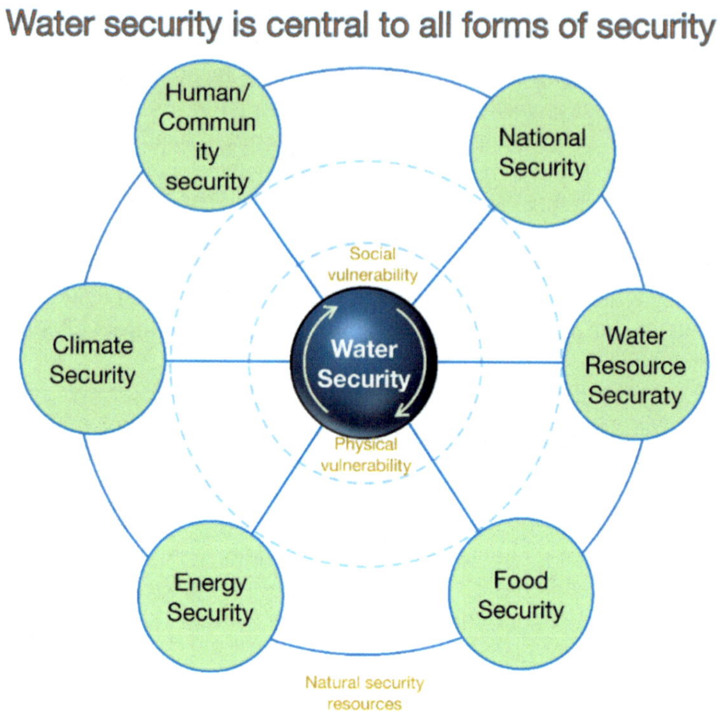

Fig. 39.9 Water security related to other key elements security (adapted from Zeitoun, 2011)

References

2030WRG. (2017). Vietnam hydro-economic framework for assessing water sector challenges. Retrieved May 26, 2019, from https://www.2030wrg.org/hydro-economic-framework-for-assessing-water-sector-challenges/.

Aaron, G. W. (2008). *Vital water graphics: An overview of the state of the World's fresh and marine waters* (2nd ed., pp. 1–88). United Nations Environment Program for Development.

Ackerman, F., & Fisher, J. (2013). Is there a water - energy nexus in electricity generation, long-term scenarios for the western United States. *Energy Policy, 59*, 235–241.

ADB (2009a). Water sector review. Asian Development Bank. . Retrieved June 26, 2019, from https://www.adb.org/sites/default/files/project-document/65077/40621-vie-tacr.pdf.

ADB. (2009b). *Water: Vital for Vietnam's future* (pp. 1–29). Asian Development Bank.

ADB. (2013). *Vietnam: Environment and climate change assessment* (pp. 1–69). Asian Development Bank.

Anh, L. T. (2015). *Vietnam's hydropower policy reform*. Technical Report, Lower Mekong public policy initiative, 1–26.

Asia Pacific Energy Research Centre (2012). *APEC peer review on energy efficiency: Vietnam follow-up evaluating & monitoring the implementation of national energy efficiency programs*. Retrieved July 1, 2019, from https://aperc.ieej.or.jp/file/2014/5/1/PREE_Follow-up_Viet_Nam_Workshop.pdf.

Babel, M. S., & Wahid, S. M. (2008). Freshwater under threat: South Asia. Vulnerability assessment of freshwater resources to environmental change. UNEP and AIT, 2008, ISBN: 978-92-807-2950-4.

Babel, M. S., & Shinde, V. R. (2018). A framework for water security assessment at basin scale. *APN Science Bulletin, 8*(1), 27–32.

Babel, M. S., Shinde, V. R., Sharma, D., & Dang, N. M. (2015). *Developing an operational water security assessment framework for application in diverse regions of Asia* (pp. 1–103). Asia-Pacific Network for Global Change Research (APN).

Chowdhury, A. K., Dang, T. D., Galelli, S. (2018). Coupling hydrologic and network constrained unit commitment models to understand the water-energy nexus in Laos. In *AGU fall meeting abstracts* (pp. 1-22). Washington DC.

Cook, C., & Bakker, K. (2012). Water security: Debating an emerging paradigm. *Global Environmental Change, 22*(1), 94–102.

Dang, N. M., Tu, V. T., Hai, N. M., Babel, M. S., Victor, R. S., & Devesh, S. (2016). Water security assessment framework for Hanoi Capital of Vietnam the data collection and analysis. *Proceedings of international conference: Water security and climate change - challenges and opportunities in Asia*, November 29th – December 1st, 2016, AIT, Bangkok, Thailand, 1–9.

Dang, N. M., Tu, V. T., Dao, N. V., Babel, M. S., Shinde, V. R., & Devesh, S. (2017). Water security assessment for red - Thai Binh River basin, Vietnam. In *Proceedings of international conference: Water security and climate change, September 18th - 21st 2017, Cologne, Germany* (pp. 1–18).

Danish Energy Agency [DEA]. (2017). *Vietnam Energy Outlook Report 2017*. Embassy of Denmark and Vietnam Ministry of Industry and Trade: Vietnam-Denmark Government Cooperation in the energy sector - Low Carbon Transition in the Energy Efficiency (LCEE). Retrieved June 26, 2019, from https://ens.dk/sites/ens.dk/files/Globalcooperation/Official_docs/Vietnam/vietnam-energy-outlook-report-2017-eng.pdf

Das, K. (2019). *Renewables in Vietnam: Current opportunities and future outlook*. Vietnam Briefing. Retrieved June 15, 2019, from https://www.vietnam-briefing.com/news/vietnams-push-for-renewable-energy.html/

European Report Development. (2012). *Confronting scarcity: Managing water, energy and land for inclusive and sustainable growth*. The 2011/2012 European Report on Development, 1–208.

European Union. (2006). *Linking climate change adaptation and disaster risk management for sustainable poverty reduction* (pp. 1–36). Vietnam Country Study. Government Statistical Office (GSO), Statistical Yearbook of Vietnam.

FAO. (2014). *The water-energy-food nexus at FAO: Concept note* (pp. 1–28). Food and Agriculture Organization of the United Nations.

Feng, K., Siu, Y. L., Guan, D., & Hubacek, K. (2012). Assessing regional virtual water flows and water footprints in the yellow river basin, China: A consumption based approach. *Applied Geography, 32*(2), 691–701.

Frenken, K. (2012). Irrigation in southern and eastern Asia in figures: AQUASTAT Survey-2011. Water reports 37. Food and agriculture organization of the United Nations, Rome, ISBN 978-92-5-107282-0, 1-512.

Giang, P. Q., Toshiki, K., Kunikane, S., & Sakata, M. (2012). Integrated water resources management in Vietnam under the challenges of climate change. *Environment and Natural Resources Journal, 10*(1), 28–41.

Haase, D., Schwarz, N., Strohbach, M., Kroll, F., & Seppelt, R. (2012). Synergies, trade-offs, and losses of ecosystem services in urban regions: An integrated multiscale framework applied to the Leipzig-Halle region, Germany. *Ecology and Society, 17*(3), 1–22.

Hoff, H. (2011). *Understanding the Nexus. Background Paper for the Bonn 2011 Conference: The water, energy and food security nexus* (pp. 1–42). Stockholm Environment Institute.

ICIMOD. (2015). Contribution of Himalayan ecosystems to water, energy, and food security in South Asia: A nexus approach. International Centre for Integrated Mountain Development (ICIMOD), Kathmandu, Nepal, 1-8.

IFAD. (2012). *Annual report 2012*. International Fund for Agricultural Development. Retrieved June 16, 2019, from https://www.ifad.org/documents/38714170/39625681/annual+report+2012+english+final+print.pdf/

IPCC. (2007). *Climate change 2007: Impacts, adaptation and vulnerability*. Contribution of Working Group II to the Fourth Assessment Report of the IPCC. Appendix I: Glossary, 869–883.

IWRP. (2015). *Report on conflict of water use in the downstream of reservoirs and adapting measures*. Institute of Water Resources Planning (IWRP).

JICA (2013). *Agricultural transformation & food security 2040 ASEAN*. Regional with a focus on Vietnam, Indonesia, and Philippines, Vietnam Country Report. Retrieved June 20, 2019, from http://open_jicareport.jica.go.jp/pdf.

Lehmann, S. (2018). Implementing the urban nexus approach for improved resource-efficiency of developing cities in Southeast-Asia. *City, Culture and Society, 13*, 46–56.

McGrane, S. J., Acuto, M., Artioli, F., Chen, P. Y., Comber, R., Cottee, J., Farr-Wharton, G., Green, N., Helfgott, A., Larcom, S., & McCann, J. A. (2018). Scaling the nexus: Towards integrated frameworks for analysing water, energy and food. *The Geographical Journal, 12*, 1–13.

Mekonnen, M. M., & Hoekstra, A. Y. (2016). Four billion people facing severe water scarcity. *Science Advances, 2*(2), 1–6.

MONRE. (2003). Vietnam initial National Communication under the United Nations framework convention on climate change. Vietnam Ministry of Natural Resources and Environment (MONRE), 1-135.

MONRE. (2009). *Climate change, sea level rise scenarios for Vietnam* (pp. 1–34). Hanoi.

Nhuan, M.N., Anh, T.N., Hoc, T.Q., Cu, P.V. (2014). *Water environment in Vietnam: Opportunities and challenges for sustainable management the context of climate change*. Water Environment Partnership in Asia, under the initiative of the Ministry of the Environment of Japan. Retrieved June 12, 2019, from http://www.wepa-db.net/pdf/1003forum/.

OECD. (2014). *Towards Green growth in Southeast Asia* (pp. 1–28). Organization for Economic Co-operation and Development (OECD).

Pacetti, T., Lombardi, L., & Federici, G. (2015). Water-energy nexus: A case of biogas production from energy crops evaluated by water footprint and life cycle assessment (LCA) methods. *Journal of Cleaner Production, 101*, 1–14.

Ripl, W. (2003). Water: The bloodstream of the biosphere. The Royal Society. *Phil Trans Roy Soc B, 358*(1440), 1921–1934.

Smajgl, A., Ward, J., & Pluschke, L. (2016). The water-food-energy nexus - Realising a new paradigm. *Journal of Hydrology, 533*, 533–540.

Swatuk, L. A., & Cash, C. (2017). *Water, energy, food and people across the global south: 'The nexus' in an era of climate change*. Springer.

Thuy V.T.H., Bundit L. (2014). Analysis of hydropower potential in Vietnam. *Proceedings of international conference: Management and innovation technology*, 17th - 19th December 2019, Pattaya, Thailand.

UNDP. (2003). *Reducing disaster risk, a challenge for development. A global report*. UNDP.

United Nations Economic Commission for Europe [UNECE]. (2015). *Reconciling different uses in transboundary river basins: Assessment of the water-food-energy-ecosystems nexus* (pp. 29–44). United Nations.

Vogt, K. (2010). *Sustainability unpacked: Food, energy and water for resilient environments and societies* (pp. 1–328). Routledge.

Waibel, G. (2010). *State management in transition: Understanding water resources management in Vietnam* (ZEF working paper series, no. 55). University of Bonn, Center for Development Research (ZEF).

WEPA (2002). State of water environmental issues. Water Environment Partnership in Asia (WEPA). Retrieved June 20, 2019, from http://www.wepa-db.net/policies/state/vietnam/overview.htm

World Bank. (2016). *Transforming Vietnamese agriculture: Gaining more from less* (Vietnam Development Report 2016) (pp. 1–148). Hong Duc Publishing House.

World Economic Forum. (2012). *Water security: The water-food-energy-climate nexus*. Island Press.

Zeitoun, M. (2011). The global web of National Water Security. Global policy. *London School of Economics and Political Science and John Wiley & Sons Ltd, 2*(3), 286–296.

Chapter 40
An Integrated Approach for Saltwater Intrusion Monitoring Based on Remote Sensing Combined with Multivariable Analysis: A Case Study of Coastal Zone in Southern Vietnam

Quoc Huy Nguyen, Tien Yin Chou, Mei Ling Yeh, Thanh Van Hoang, Xuan Linh Nguyen, Huyen Ai Tong, and Quang Thanh Bui

Abstract Saltwater intrusion is a basic concern in many parts of Vietnam relative to long-term dependable water supplies. It affects many sides of human life and the ecosystem. Remote sensing is a useful tool for saltwater intrusion monitoring. In this study, we proposed an integrated approach to estimate EC (electrical conductivity) value from multitemporal optical remote sensing data for monitoring saltwater intrusion of coastal zone in southern Vietnam. Multiple variable analysis was used to discover the relation between EC and different index groups, which were extracted from LANDSAT satellite images, including original band group, PCA (principal component analysis) group, brightness group, vegetation group, salinity group, ratio group, and combined group. All results were validated by field survey data. This research indicated that a group of combined indices from LANDSAT (EC6) had the highest correlation to EC index with $R^2 = 0.77$ and could be used for multitemporal saltwater intrusion monitoring. A set of maps from 2005 to 2019 were established for Ben Tre Province, which is one of the coastal zones in southern Vietnam to support policy managers to make decision for reducing damage from saltwater intrusion.

Q. H. Nguyen (✉) · X. L. Nguyen
PhD program in Civil and Hydraulic Engineering, Feng Chia University, Taichung, Taiwan
e-mail: st_huy@gis.tw

T. Y. Chou · M. L. Yeh · T. Van Hoang
GIS Research Center, Feng Chia University, Taichung, Taiwan

H. A. Tong
Space Technology Institute, Hanoi, Vietnam

Q. T. Bui
Faculty of Geography, Hanoi University of Science, Hanoi, Vietnam

A. T. Nguyen, L. Hens (eds.), *Global Changes and Sustainable Development in Asian Emerging Market Economies Vol. 2*, https://doi.org/10.1007/978-3-030-81443-4_40

Keywords Saltwater intrusion · Remote sensing · Multiple variable analysis · Mekong Delta · Vietnam

1 Introduction

Saltwater intrusion is an important problem concern for humans and the ecosystem. It is a factor that changes the properties of the soil, which adversely affects the development of crops, causing damage to agricultural production. Besides, it also causes human and animal health impacts and influences industrial production activities using natural water sources (Chhabra, 2017). There are many definitions of saltwater intrusion. In general, saltwater intrusion is a phenomenon of saltwater intruding deeply into the interior of areas when sea level rises at high tide. It is the process when saltwater seeps into coastal groundwater systems and mixes with freshwater. Salty land is an area that contains soluble salts at concentrations higher than usual and causes adverse effects on crops. Saltwater intrusion is a common soil degradation. It is a process of accumulation of salts and dissolved alkali metals on the soil surface and upper soil layers. This process usually starts from the lower soil layers and then slowly spreads to the surface.

In the past, saltwater intrusion monitoring was usually based on field measurement of salinity indicators, but it was expensive that need to find other methods. Remote sensing technology is one of the tools to solve this problem. There has been a lot of research on developing correlation models to observe saltwater intrusion through extracted indices from satellite images. An et al. (2016) used NIR and SWIR index data from LANDSAT7 and LANDSAT8 images to estimate soil salinity index (SSI) . Meanwhile, Scudiero, Skaggs, and Corwin (2015) found a correlation between canopy response salinity index (CRSI) from LANDSAT7-TM to monitor saltwater intrusion. And many other studies have demonstrated benefits in combining indicators extracted from remote sensing images to monitor multitemporal saltwater intrusion (El Harti et al., 2016; Elhag & Bahrawi, 2017; Fan, Liu, Tao, & Weng, 2015; Liu et al., 2016; Nawar, Buddenbaum, & Hill, 2015; Peng et al., 2019; Wu, Muhaimeed, Al-Shafie, & Al-Quraishi, 2020). However, the results of these researches are not applicable to all areas because of different local conditions in each region.

In this study, we proposed an integrated approach for saltwater intrusion monitoring in Mekong Delta Vietnam based on using multiple variable analysis to discover relation between EC and 6 index groups, which are extracted from LANDSAT satellite images. All of the results will be verified by collected sample data in our research area.

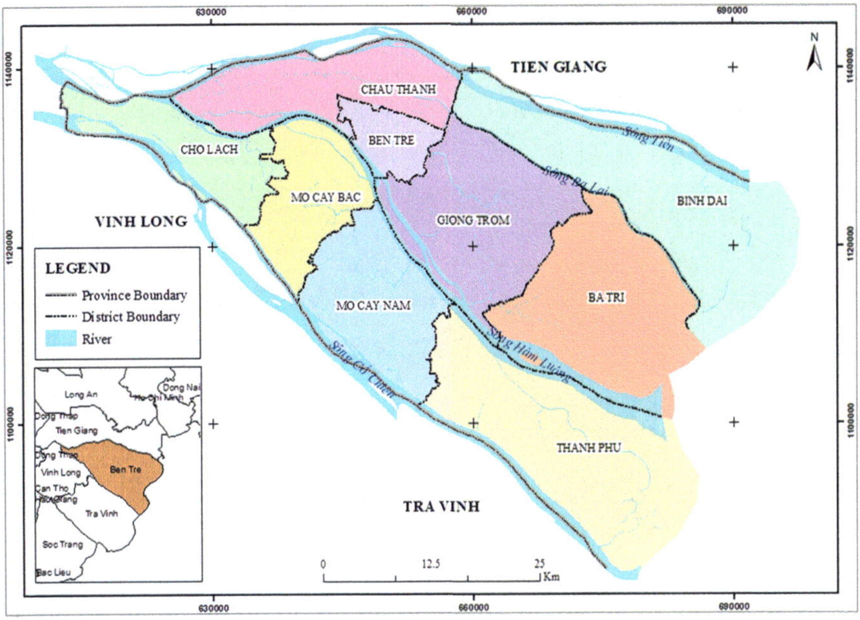

Fig. 40.1 Location of Ben Tre Province in the Mekong Delta, Vietnam

2 Methodology

2.1 Study Area

Ben Tre is selected for our study (Fig. 40.1). Its area is nearly 2360 km^2 and is formed by three islands (An Hoa, Bao, and Minh) that were deposited by alluvial from four branches of Mekong River (Tien, Ba Lai, Ham Luong, and Co Chien). The location of Ben Tre ranges from 9°48′ to 10°20′ latitude and from 105°57′ to 106°48′ longitude, and it is in the coastal zone in southern Vietnam. This region has natural factors, which are directly affected by the coastal zone. Especially, when water level in the river is low, saltwater can be pushed back into the river and canal system to create a salinization area with different concentrations because flows are not strong enough to prevent salt water from sea with high tide.

2.2 Materials

LANDSAT5-TM and LANDSAT8 satellite images used for this study because of advantages in time series analysis for regional with an average spatial and temporal resolution, the medium of spectral bands and free also (Table 40.1).

Table 40.1 Description of LANDSAT5-TM and LANDSAT8

Band		Wave length (μm)		Resolution (10 m)	
		LANDSAT8	LANDSAT5-TM	LANDSAT8	LANDSAT5-TM
1	Ultra blue (coastal/aerosol)	0.43–0.45		30	
2	Blue	0.45–0.51	0.45–0.52	30	30
3	Green	0.53–0.59	0.52–0.60	30	30
4	Red	0.63–0.67	0.63–0.69	30	30
5	Near infrared (NIR)	0.85–0.87	0.76–0.90	30	30
6	Shortwave infrared (SWIR) 1	1.56–1.65	1.55–1.75	30	30
7	Shortwave infrared (SWIR) 2	2.10–2.29	2.08–2.35	30	30
8	Panchromatic	0.50–0.67		15	
9	Cirrus	1.36–1.38		30	
10	Thermal infrared (TIRS) 1	10.6–11.2	10.40–12.50	100	120
11	Thermal infrared (TIRS) 2	11.5–12.5		100	

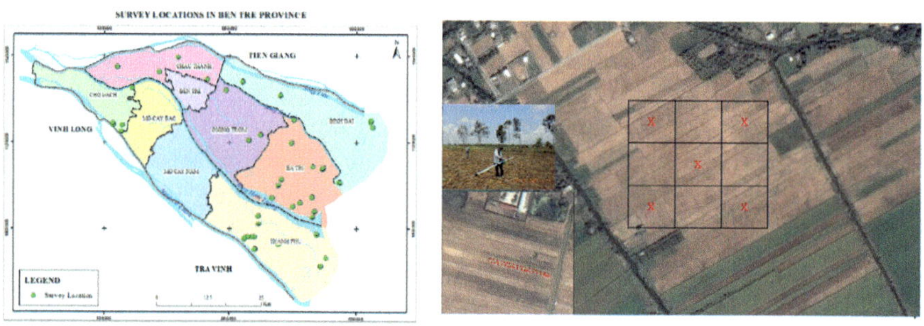

Fig. 40.2 Survey locations and design of sample locations

A field salinity survey was conducted by taking soil samples in the study site to verify results. One sample is designed in a 90 m x 90 m cell with five positions (Fig. 40.2).

EM31-MK2 device was used to measure EC data with 556 samples over different land uses. It will be standardized by reference to soil sample analysis (Table 40.2).

2.3 *Data Processing*

To discover the relation between index group of LANDSAT and EC, we divided them into six physical index groups including original index, principal component analysis index, brightness index, vegetation, salty index, and ratio index. See Table 40.3 for more details.

Table 40.2 Mean of EC over land uses

Land use	Number of samples	EC mean	Salty level
Salty land	18	19,3739	High
Mangrove	7	17,8200	High
Sand coastal	67	12,7965	Medium
Aquaculture	48	12,0464	Medium
Cropland (1c)	49	90,753	Medium
Cropland (2c)	103	81,309	Medium
Cropland (3c)	96	75,258	Low
Vegetation	77	38,513	Nonsalty
Plant	91	34,208	Nonsalty

After calculation, these indicators will be extracted based on measured EC locations to establish in regression analysis model, where EC is a dependent variable and others are independent variables. Criteria for regression assessment include Sig coefficient (P-value), Pearson correlation coefficient (R), root mean square error (R^2), and adjusted root mean square error (R^2_{adj}). Variables that have Sig > 0.05 and R ≈ 0 will be removed. If R^2 and R^2_{adj} are in (0, 1) range and values moved to near 1 that means model fits data set. Typically, model can be applied when $R^2 > 0.5$ (Abdullah et al., 2019).

Multiple variable regression model was applied for separate index groups to identify which group has the highest correlation with EC. One combined group of all indices will also be used in this progress assessment. About 70% of sample data will be used for model input and 30% for validation (Fig. 40.3).

3 Results

3.1 Relationship between EC and Index Group from LANDSAT

The results demonstrate that a combined group has the highest correlation with EC ($R = 0.783$ and $R^2 = 0.6126$) and will be used to estimate EC and establish salinization map for saltwater intrusion monitoring. Lowest correlation with EC is PCA group ($R = 0.675$ and $R^2 = 0.4839$). The brightness index group is removed because of having a reserve correlation with EC (Table 40.4, Fig. 40.4).

3.2 Validation

About 30% of sample data are used for validating EC estimation model from LANDSAT. The correlation between the estimated EC and validated EC is $R^2 = 0.77$. It is not really high but can be accepted. This model is applied in EC estimation

Table 40.3 Index group that are extracted from LANDSAT

Index group	Formula	References
Original index	B1, B2, B3, B4, B5, B6	
PCA index	PCA1, PCA2, PCA3, PCA4, PCA5, PCA6	
Brightness index	BI = sqrt (G^2 + NIR^2)	Yoo and Chung (2018)
	BI2 = sqrt (R^2 + NIR^2)	Gadal and Ouerghemmi (2019)
	INT = (G + R) / 2	Rahmati and Hamzehpour (2017)
Vegetation index	NDVI = (NIR – R) / (NIR + R) $\frac{(NIR-R)}{(NIR+R)}$	Silvia, Alexander, Anna, and Polina (2019)
	SAVI = (1 + L) * (NIR – R) / (NIR + R + L) $\frac{(NIR-R)}{(NIR+R+L)}$	Ren, Zhou, and Zhang (2018)
	EVI = G * (NIR – R) / (NIR + C_1R – C_2B + L) $\frac{(NIR-R)}{(NIR+C_1R-C_2B+L)}$	Li, El-Askary, Qurban, Allali, and Manikandan (2019)
	GDVI = (NIR^n – R^n) / (NIR^n + R^n) $\frac{(NIR^n-R^n)}{(NIR^n+R^n)}$	Avola et al. (2019)
Salty index	SI1 = sqrt (G * R)	Touhami, Bouraoui, and Berguig (2019)
	SI2 = sqrt (G^2 + R^2 + NIR^2)	Alexakis, Daliakopoulos, Panagea, and Tsanis (2018)
	SI3 = sqrt (G^2 + R^2)	Abdullah, Biswas, Chowdhury, and Billah (2019)
	SI4 = sqrt (((NIR * R) – (G * B)) / ((NIR * R) – (G * B)))	Samiee, Ghazavi, Pakparvar, and Vali (2018)
	SI5 = (R – NIR) / (R + NIR)	Samiee et al. (2018)
	SI6 = sqrt (R * NIR)	Samiee et al. (2018)
Ratio index	B/G	
	B/NIR	
	B/R	
	B/SWIR1	
	B/SWIR2	
	G/R	
	G/NIR	
	G/SWIR1	
	G/SWIR2	
	R/NIR	
	R/SWIR1	
	R/SWIR2	
	NIR/SWIR1	
	NIR/SWIR2	
	SWIR1/SWIR2	

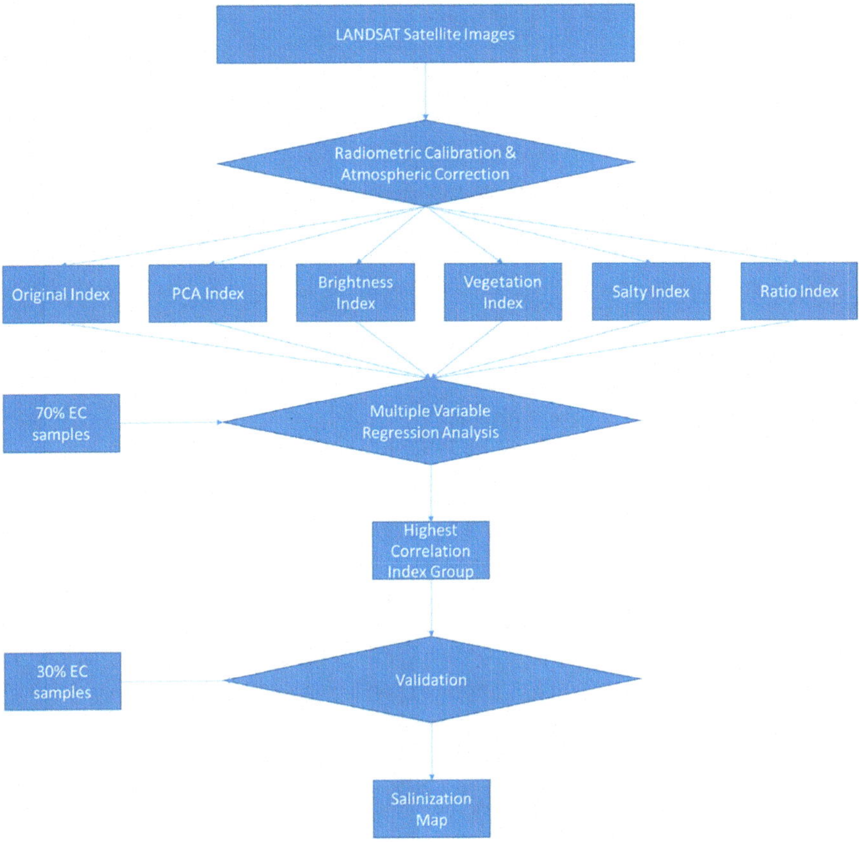

Fig. 40.3 Logical framework of study

from LANDSAT from 2005 to 2019 with four classification levels of EC values by the Ministry of Agriculture and Rural Development (Fig. 40.5, Table 40.5).

3.3 Saltwater Intrusion Mapping

After validation, saltwater intrusion maps were established by using LANDSAT and L_EC6 model to estimate EC value. In particular, salty regions are mainly distributed in three coastal districts over mangrove and aquaculture areas including Binh Dai, Ba Tri, and Thanh Phu (Figs. 40.6, 40.7, 40.8, and 40.9).

Table 40.4 Regression models from index group

Index group	Estimation model	R	R^2	R^2_{adj}	F-Static	DF
Original index	L_EC1 = 11,609 + 40,220B3–82,156B5 + 87,971B6	0.750	0.5545	0.559	137.955	3:388
PCA index	L_EC2 = 13,223–30,522 * PCA1 + 7836 * PCA2 + 50,068 * PCA3	0.675	0.4839	0.451	97.042	3:388
Vegetation index	L_EC3 = 18,822–13,088 * GDVI	0.696	0.4839	0.483	307.650	1:388
Salty index	L_EC4 = 16,816 + 8475 * SI5–65,088 * SI6 + 533,248 * SI1–314,669 * SI3 + 0,594 * SI4	0.741	0.5367	0.544	76.716	5:388
Ratio index	L_EC5 = 17,970 + 22,547 * (G/NIR) - 2638 * (G/R) - 10,087 * (SWIR1/SWIR2) + 2304 * (NIR/SWIR2) - 3879 * (B/G)	0.767	0.588	0.583	95.958	5:388
All combined index	L_EC6 = −6489 + 20,492 * (G/NIR) + 0,383 * T + 28,615 * B6 + 6247 * (NIR/SWIR1) - 20,053 * SI2–3505 * (SWIR1/SWIR2)	0,783	0,6126	0,607	75,305	6:388

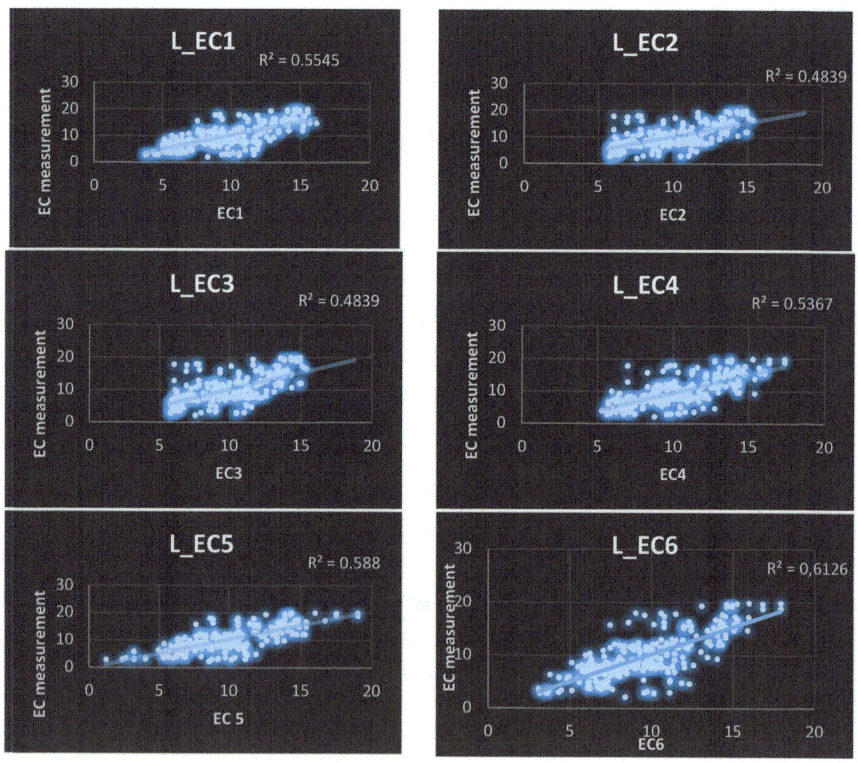

Fig. 40.4 Correlation of surveyed EC and index group

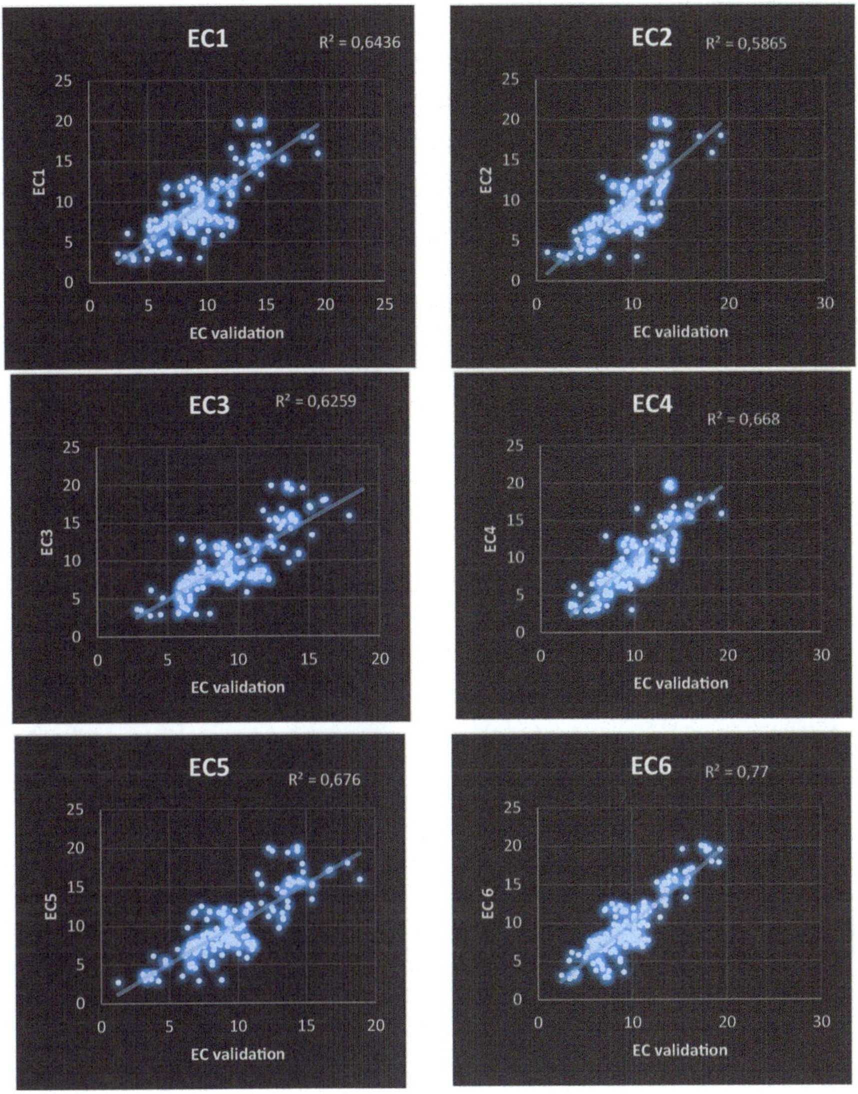

Fig. 40.5 Correlation of estimated EC and surveyed EC

Table 40.5 Salty level classification

Level	Description	EC (dS/m)
1	Nonsalty	< 4
2	Low	4–8
3	Medium	8–15
4	High	> 15

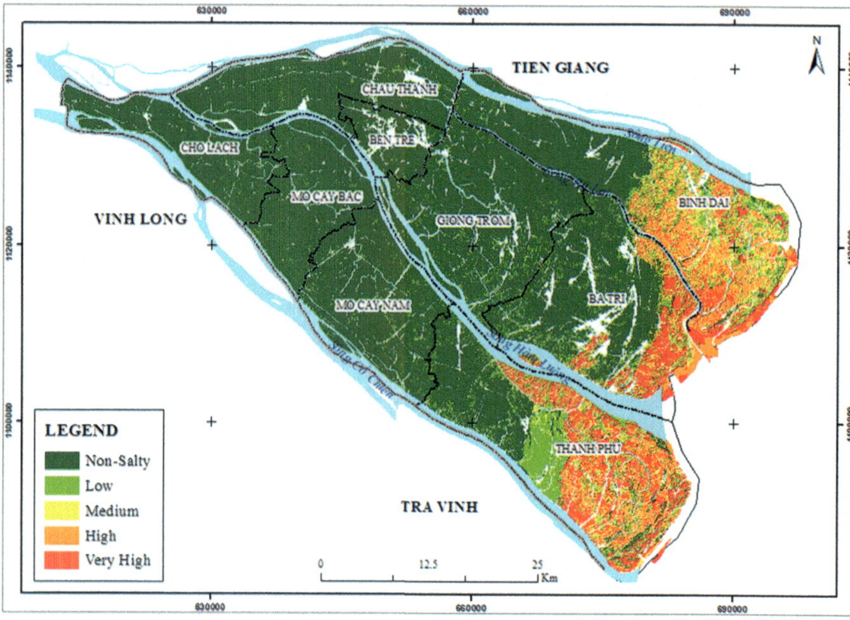

Fig. 40.6 Map of saltwater intrusion in Ben Tre in 2005

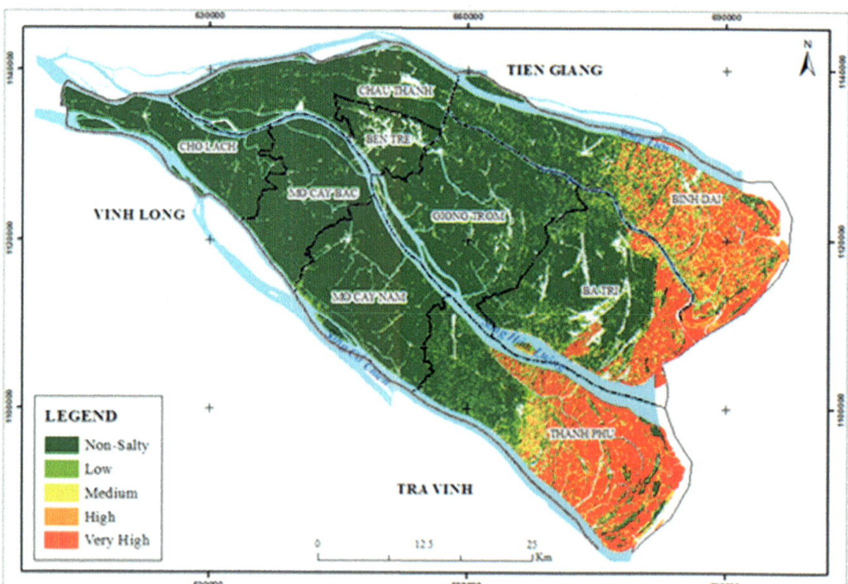

Fig. 40.7 Map of saltwater intrusion in Ben Tre in 2010

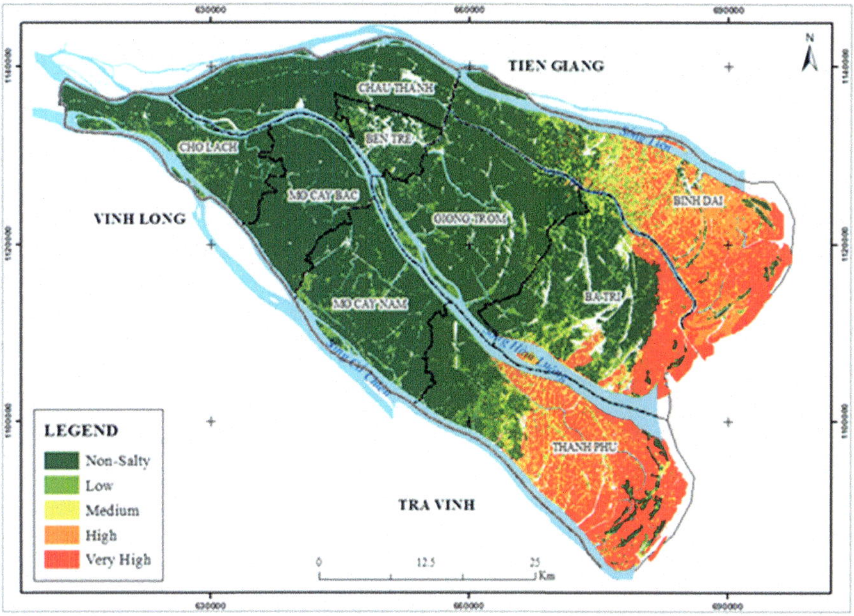

Fig. 40.8 Map of saltwater intrusion in Ben Tre in 2015

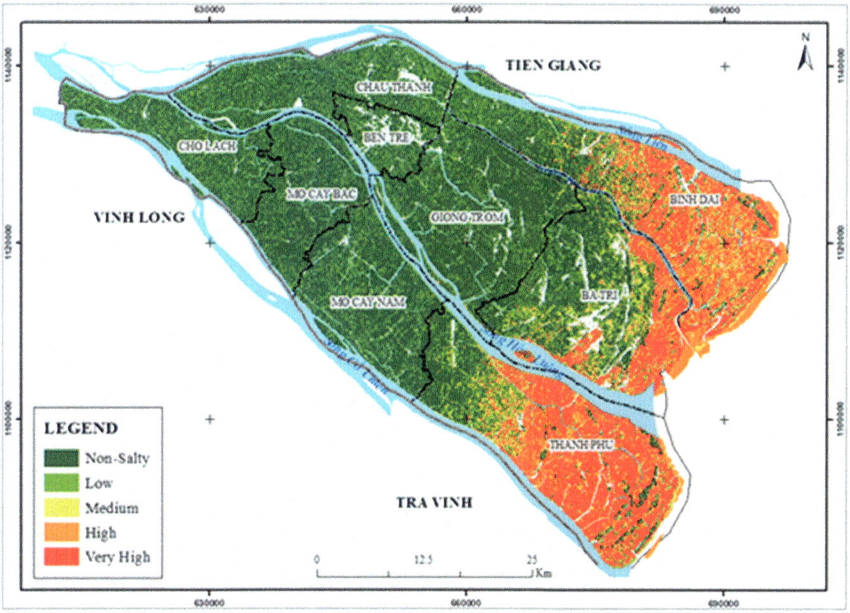

Fig. 40.9 Map of saltwater intrusion in Ben Tre in 2019

4 Conclusions and Discussion

The general trend of saltwater intrusion in Ben Tre Province showed an increase in salty area and a decrease in the nonsalty area, which is mainly concentrated in coastal districts. The conversion of land use from rice to aquaculture is one of the leading causes of saltwater intrusion in soils. Within 13 years from 2005 to 2019, a total of nonsalty areas were decreased by nearly 15824.84 ha, while salty area was increased where the lowest was 8169.98 ha and the highest was 10314.19 ha. During the 2010–2015 period, high salty area expanded fastest with 6861.46 (ha) due to the inefficient production of rice and aquaculture, which was converted to industrial shrimp farming. Another reason is that the alluvial land between large rivers such as Ham Luong, Co Chien, and Tien has also been changed from plant to aquaculture. The leading cause of saltwater intrusion in Ben Tre Province is an extension of aquaculture area to the land. Saltwater that is taken from canals will lead to saltwater intrusion deeper from the sea. Moreover, canal system is intertwined, while irrigation system is not closed and makes some places become saltwater holes in the dry season. Therefore, saltwater intrusion occurs every year along the river and agricultural land in Ben Tre Province.

Saltwater intrusion is a severe worldwide problem, and it directly affects the natural environment, agriculture, and food security. It is important to establish saltwater intrusion map, which provides useful information about salty area and can be useful for land use planning and management. Results demonstrated that index group is very important in estimating EC from optical satellite images. In this study, we reviewed remote sensing applications related to analysis and assessment of saltwater intrusion using index group from LANDSAT, which was integrated with field survey data, spatial analysis, and statistical methods. However, they are not really suitable for coastal zones such as Ben Tre because of its cloudy nature. In the future, radar satellite images may be considered to solve this problem.

Acknowledgments The data in this article were supported by project VT-UD.03/16-20, entitled "Studying, Assessing, and Zoning Soil Salinity by Using Multi-Temporal Satellite Imagery: A Case Study in Ben Tre Province," which belongs to the national program on Space Science and Technology (2016–2020).

References

Abdullah, A. Y. M., Biswas, R. K., Chowdhury, A. I., & Billah, S. M. (2019). Modeling soil salinity using direct and indirect measurement techniques: A comparative analysis. *Environmental Development, 29*, 67–80.

Alexakis, D. D., Daliakopoulos, I. N., Panagea, I. S., & Tsanis, I. K. (2018). Assessing soil salinity using WorldView-2 multispectral images in Timpaki, Crete, Greece. *Geocarto International, 33*(4), 321–338.

An, D., Zhao, G., Chang, C., Wang, Z., Li, P., Zhang, T., & Jia, J. (2016). Hyperspectral field estimation and remote-sensing inversion of salt content in coastal saline soils of the Yellow River Delta. *International Journal of Remote Sensing, 37*(2), 455–470.

Avola, G., Di Gennaro, S. F., Cantini, C., Riggi, E., Muratore, F., Tornambè, C., & Matese, A. (2019). Remotely sensed vegetation indices to discriminate field-grown olive cultivars. *Remote Sensing, 11*(10), 1242.

Chhabra, R. (2017). Chapter 1: Origin and distribution of salt-affected soils. In *Soil salinity and water quality* (2nd ed., pp. 1–25). Routledge.

El Harti, A., Lhissou, R., Chokmani, K., Ouzemou, J. E., Hassouna, M., Bachaoui, E. M., & El Ghmari, A. (2016). Spatiotemporal monitoring of soil salinization in irrigated Tadla plain (Morocco) using satellite spectral indices. *International Journal of Applied Earth Observation and Geoinformation, 50*, 64–73.

Elhag, M., & Bahrawi, J. A. (2017). Soil salinity mapping and hydrological drought indices assessment in arid environments based on remote sensing techniques. *Geoscientific Instrumentation, Methods and Data Systems, 6*(1), 149–158.

Fan, X., Liu, Y., Tao, J., & Weng, Y. (2015). Soil salinity retrieval from advanced multi-spectral sensor with partial least square regression. *Remote Sensing, 7*(1), 488–511.

Gadal, S., & Ouerghemmi, W. (2019). Multi-level morphometric characterization of built-up areas and change detection in Siberian sub-Arctic urban area: Yakutsk. *ISPRS International Journal of Geo-Information, 8*(3), 129.

Li, W., El-Askary, H. M., Qurban, M., Allali, M., & Manikandan, K. P. (2019). On the drying trends over the MENA countries using harmonic analysis of the enhanced vegetation index. In *Advances in remote sensing and geo informatics applications* (pp. 243–245). Springer.

Liu, G., Li, J., Zhang, X., Wang, X., Lv, Z., Yang, J., & Yu, S. (2016). GIS-mapping spatial distribution of soil salinity for eco-restoring the Yellow River Delta in combination with electromagnetic induction. *Ecological Engineering, 94*, 306–314.

Nawar, S., Buddenbaum, H., & Hill, J. (2015). Estimation of soil salinity using three quantitative methods based on visible and near-infrared reflectance spectroscopy: A case study from Egypt. *Arabian Journal of Geosciences, 8*(7), 5127–5140.

Peng, J., Biswas, A., Jiang, Q., Zhao, R., Hu, J., Hu, B., & Shi, Z. (2019). Estimating soil salinity from remote sensing and terrain data in southern Xinjiang Province, China. *Geoderma, 337*, 1309–1319.

Rahmati, M., & Hamzehpour, N. (2017). Quantitative remote sensing of soil electrical conductivity using ETM+ and ground measured data. *International Journal of Remote Sensing, 38*(1), 123–140.

Ren, H., Zhou, G., & Zhang, F. (2018). Using negative soil adjustment factor in soil-adjusted vegetation index (SAVI) for aboveground living biomass estimation in arid grasslands. *Remote Sensing of Environment, 209*, 439–445.

Samiee, M., Ghazavi, R., Pakparvar, M., & Vali, A. A. (2018). Mapping spatial variability of soil salinity in a coastal area located in an arid environment using geostatistical and correlation methods based on the satellite data. *Desert, 23*(2), 233–242.

Scudiero, E., Skaggs, T. H., & Corwin, D. L. (2015). Regional-scale soil salinity assessment using Landsat ETM+ canopy reflectance. *Remote Sensing of Environment, 169*, 335–343.

Silvia, L., Alexander, T., Anna, K., & Polina, K. (2019). Assessment of carbon dynamics in Ecuadorian forests through the mathematical spatial model of global carbon cycle and the normalized differential vegetation index (NDVI). In E3S web of conferences (Vol. 96, p. 02002). EDP sciences.

Touhami, M. K., Bouraoui, S., & Berguig, M. C. (2019). Contribution of satellite imagery to study salinization effect of agricultural areas at northern eastern oasis Algerian region. In *Advances in remote sensing and geo informatics applications* (pp. 235–238). Springer.

Wu, W., Muhaimeed, A. S., Al-Shafie, W. M., & Al-Quraishi, A. M. F. (2020). Using radar and optical data for soil salinity modeling and mapping in Central Iraq. In *Environmental remote sensing and GIS in Iraq* (pp. 19–40). Springer.

Yoo, H., & Chung, K. (2018). Heart rate variability based stress index service model using biosensor. *Cluster Computing, 21*(1), 1139–1149.

Chapter 41
Diversity of Non-Timber Forest Products (NTFPs) in Hoang Lien—Van Ban Nature Reserve (Lao Cai, Vietnam): Implications for Local Livelihood Improvement and Biodiversity Conservation

Nguyen Thi Phuong and Nguyen An Thinh

Abstract Non-timber forest products (NTFPs) are crucial for rural people's livelihoods in Hoang Lien-Van Ban Nature Reserve. Communities living in and around the nature reserve rely on a variety of non-timber forest products for their livelihood. This chapter aims to give an overview of the diversity of NTFPs in Hoang Lien-Van Ban Nature Reserve, focusing on medicinal plants and food. We collect data on NTFPs through sociological interviews. Interviews were conducted with households living in the Hoang Lien-Van Ban Nature Reserve. It is hypothesized that there is no significant difference between the amount of bamboo shoots harvested and their contribution to household income among studied villages. To identify the vulnerability and risks of threatened species, a rapid vulnerability assessment (RVA) was conducted. A total of 256 species of medicinal plants have been discovered in Hoang Lien-Van Ban Nature Reserve. Among them, 193 species have identified scientific names to species, 59 species have been identified to genera and 4 species have been identified to their families to be used as NTFPs of local people. The results show that there is a critical difference in NTFPs harvesting and sale affecting household income among villages. However, the contribution from NTFPs harvesting to per capita income is small, since they are mainly used for household demands. Vulnerability assessment shows that most medicinal species are moderately vulnerable, and most edible species have less vulnerable position. Management and conservation of NTFP species should be implemented to prevent overexploitation of these species.

N. T. Phuong (✉)
VNU School of Interdisciplinary Studies, Vietnam National University, Hanoi, Vietnam

N. A. Thinh
VNU University of Economics and Business, Vietnam National University, Hanoi, Vietnam

A. T. Nguyen, L. Hens (eds.), *Global Changes and Sustainable Development in Asian Emerging Market Economies Vol. 2*, https://doi.org/10.1007/978-3-030-81443-4_41

Keywords Biodiversity · Non-timber forest products (NTFPs) · medicinal plants · livelihood · Hoang Lien-Van Ban Nature Reserve · Lao Cai

1 Introduction

Non-timber forest products (NTFPs) are becoming a topic of international interests regarding poverty reduction, participatory development, natural conservation and food security (Ros-Tonen, 2000). It is argued that the exploitation of NTFPs for livelihoods of local people is less ecologically damaged than timber harvest or other land-use changes for agricultural purposes. NTFPs can increase forests' values and raise the incentives for protecting forests (Arnold & Perez, 2001; Hall & Bawa, 1993). Local people do not value local flora species equally; some particular species groups, genera and families are preferred than others (Bennett & Husby, 2008). Identifying herbal groups and their usages is fundamental in identifying and implementing priorities of conservation and sustainable management strategies (Vodouhê, Coulibaly, Greene, & Sinsin, 2009).

Lao Cai province has a total natural area of 636,403.3 hectares. The area of forests and forestry land is 429,536 ha, accounting for 64.49% of the total natural area, in which special use forest is 64,526 ha, protection forest 172,800 ha, production forest 192,210 ha. Lao Cai province is classified as a high biodiversity province with three nature reserves, namely, Hoang Lien National Park (National Park) (recognized by the Government); Hoang Lien-Van Ban Nature Reserve (NR); and Bat Xat Nature Reserve. These are the areas with the highest biodiversity of Vietnam, with many rare and precious species of plants and animals in the Red Book of Vietnam and the World. In Hoang Lien Son mountain range, Lao Cai has recorded 2847 species of vascular plants, with 7/17 rare plant species. Particularly, medicinal plants account for more than 700 species, especially valuable medicinal plants such as Ngoc Linh ginseng, Che Day, Giao Co Lam, Ba Kich, Cardamom, Amomum, Turmeric, Hoang Tinh, Hoang Lien, Amomum, Artichoke, etc. The number of endemic species in Lao Cai accounts for a quarter of Vietnam's endemic species. Along with the biodiversity, there are abundant indigenous knowledge sources, especially ethnic minorities in the harvest and use of medicinal plants to treat diseases and health care for the community.

Due to cultural characteristics and intellectual conditions, a fairly common phenomenon in many highland areas in Lao Cai is that people mainly exploit NTFPs in a natural way. Currently, Lao Cai province has about half of the population living in the uplands; there are many people earning their living from forestry and forest products. It is undeniable that there are thousands of farming households doing well and becoming wealthier thanks to exploiting forestry elements. But forest resources are increasingly scarce, natural NTFPs are still very limited in quantity, including some kinds of medicinal plants that have been exploited recently for making the Dao herbal bath for tourists in Sapa. Without active measures to protect and plant new trees, the number of NTFPs will be exhausted. A typical kind of high-value

NTFPs is cardamom, which is enrichment tree of highland people in some communes of Sa Pa, Bat Xat and Van Ban districts.

Particularly, local people living in Hoang Lien-Van Ban Nature Reserve (NR) use large quantities of forest products in their daily lives. These products can be classified as pharmaceuticals, foodstuffs, firewood and construction materials. The edible NTFPs can be divided into forest vegetables, mushrooms and bamboo shoots, cardamom, etc. This place has Sat bamboo and cardamom that have high economic value and some species of medicinal plants such as giao co lam, hoang dang, sa nhan, etc.

The hypothesis that the amount of collection and the contribution of bamboo shoots as a representative for commercially traded NTFP species is not significantly different between villages. Bamboo shoots and Cardamom were chosen as a representative NTFPs because it can gather reliable data on harvested and commercial quantities. As there is no assessment of sustainable NTFPs exploitation practices in the region, we also identified the most vulnerable NTFPs based on rapid vulnerability assessment (RVA) analysis to make recommendations for collecting, planting and sustainably preserving these species.

2 Methodology

2.1 Study Area

Hoang Lien-Van Ban Nature Reserve has a total natural area of 25,093 ha, located on the Hoang Lien Son mountain range, 40 km southeast of Fansipan summit, in the area of Nam Xe, Nam Xay and part of Liem Phu commune, Van Ban district, Lao Cai province. Before 2002, a number of domestic and international organizations conducted surveys of Hoang Lien mountainous area of Van Ban district (Vietnam Program FFI, BirdLife International, etc.). The results of these surveys have recorded a number of globally threatened animal and plant species in and around the Nature Reserve. Animals such as the black crested gibbon and Zebra civet, the black-backed climbing bird, and toadfish, which is a species currently only recorded in Northern Vietnam, are threatened with extinction. Regarding plants, *Fokienia hodginsii* and *Taiwania cryptomerioides* now have a population in Van Ban with more than 100 individuals. Besides, Hoang Lien-Van Ban Nature Reserve has been recognized as one of 63 important bird areas of Vietnam. With these values, in 2013, Hoang Lien-Van Ban Nature Reserve was approved by the People's Committee of Lao Cai Province to approve the Planning for Conservation and Sustainable Development of Hoang Lien-Van Ban Nature Reserve, in the period 2013–2020 according to Decision No. 972 / QD-UBND dated April 25, 2013. Since then, Hoang Lien-Van Ban Nature Reserve has had a positive impact on biodiversity conservation.

Rice cultivation and shifting cultivation are the main livelihood activities of Lao Cai ethnic minorities. In addition, raising cattle, poultry, aquaculture and handicrafts are considered as important sources of cash for households. However, production is self-sufficient, there are no typical examples of commodity production. The province has a large forest area, especially Hoang Lien-Van Ban Nature Reserve where four ethnic groups live in forest planting, and management is still an important livelihood activity of the people here. Currently, the project of reforestation for shifting cultivation in Lao Cai province was approved by the Provincial People's Committee in 2009, creating conditions for 11,356 households with shifting cultivation areas on forestry land with an area of nearly 21,000 ha. This is an opportunity for local ethnic minorities to participate in afforestation, change traditional and inefficient farming practices to new production methods for higher economic efficiency, creating favourable conditions for the local people to work in forest planting to stabilize and develop economy for ethnic minority people in remote areas. The livelihoods of afforestation and forest management have played an important role in the lives of the people in Lao Cai, such as providing firewood, bamboo shoots and forest products. The ethnic minorities here still practice forest hunting, harvesting, medicinal plants, slash and burn cultivation, affecting the forest biodiversity.

Due to the habit of settling and farming on sloping land, there are many potential risks due to floods, flash floods and landslides. The objects most affected are highland, poor areas and areas inhabited by ethnic minorities. When affected by natural disasters, some households can quickly recover their livelihoods and rebuild their assets, but many others will take longer to recover. Poor households in the ethnic minority areas in Lao Cai in general and Hoang Lien-Van Ban Nature Reserve, in particular, go to the forest to hunt, harvest timber, NTFPs, etc. to increase their income and ensure livelihoods.

2.2 Data Collection

Data on NTFPs was collected through the planning report of Hoang Lien-Van Ban NR. Semi-structured, free listing and household interviews were conducted from December 2018 to March 2019. Information was randomly selected for free listing and semi-structured interviews. Key information was determined by the snowball method (Berlin & Berlin, 2005; Quinlan, 2005; Stepp, 2005). All of them were interviewed similar questions. The household interviews were conducted in Nam Si Tan village (24 households) and Ta Nang village (50 households) in Nam Xé commune (Table 41.1) due to two direct impacts on the NR. Hoang Lien-Van Ban Nature Reserve is located in the administrative area of three communes of Van Ban district, namely, Nam Xay, Nam Xé and LiemPhu. This is the object that needs investment to support production development, encourage people to participate in the protection and sustainable development of forests.

Table 41.1 Details of rural selection of NTFPs collection interview

Villages	Ethnic minorities	Number of households	Number of NTFP harvesting interviews
Nam Si Tan	Tay, Kinh, H'Mông and Dao	24	15
Ta Nang	Tay, Kinh, H'Mông and Dao	50	20

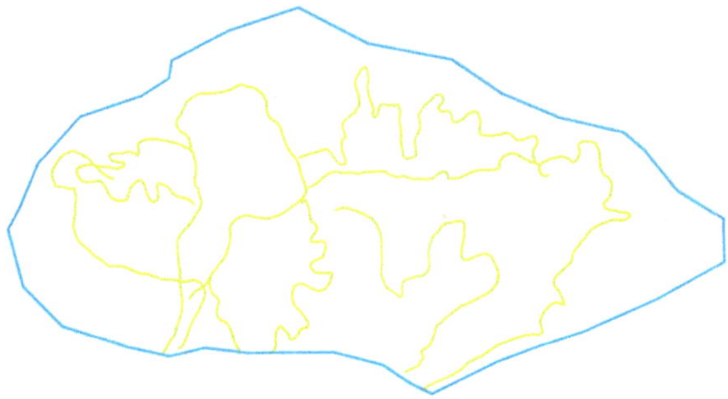

Fig. 41.1 Map of survey routes in the Hoang Lien-Van Ban Nature Reserve

Interviews were also conducted with NTFP traders. For the medicinal plant survey, the study area was conducted by linear survey method and collected information in the field, with key informant panels. Key informant panels (KIP) are knowledgeable about medicinal plants in Ta Nang village community; through field surveys they observe, interview and collect samples. The objective of the survey is to determine the species composition and usage of medicinal plants in the area. Steps to follow include:

1. Determining survey routes: Based on the actual status of vegetation, topography or distribution of medicinal plants in the area, the survey routes are determined. In order to ensure the objectivity in the investigation process, the survey line is designed according to different terrains and vegetation (high mountain route, primary forest, secondary forest, roadside, plantation forest, etc.). A total of 11 survey routes were implemented in the study area.
2. Collection of information in the field: Collect information on any tree encountered on the road or stop at each location where there is a change in vegetation and conduct interviews. Information to collect includes: names of local dialects, parts used, uses; collecting samples and taking photos of medicinal plants (Fig. 41.1).

Investigation by standard plots: Establishing standard plots with size of 100 m^2 (10 m ×10 m), determined by stratification method – randomly based on the reality of vegetation and topography, total of 100 standard plots have been located in the research area. Use a GPS to mark the coordinates of selected cells on the map. Investigative activities include:

1. Establishment of plots: Determine plot boundaries with tape measure, piling and stretching of coloured nylon rope.
2. Collecting information: Using the questionnaire set to collect information. Each study cell is a sample, consisting of two parts: (1) information about environmental conditions: coordinates, altitude, slope, exposure direction, vegetation type, cover exposed rocks, boulder cover, macadam cover, surface water regime, canopy cover, coverage of fresh carpets, main trees, the number of woody trunks with a diameter at breast height of 10 cm or more, height soaring; (2) information on medicinal plants: names of medicinal plants appear in the box. Collecting samples and taking photos of medicinal plants.
3. Information processing and analysis: List of species and species variables, including all medicinal plants and timber; science name; plant family; nameplate; common name; variables of the species, including: (i) life forms (trees, shrubs, vines, parasites, etc.), (ii) use. Survey plot data, including: ecological/environment/vegetation information (plot variables) including: (i) "direct" variables: field measurements, no treatment physics, calculation; (ii) "indirect" variables, which are determined through the calculation of direct variables: the number of tree species, number of medicinal plants, frequency of medicinal plants, basal area of woody plants, etc. Information regarding species of umbrella, including name of medicinal plants and timber (scientific name) (Fig. 41.2).

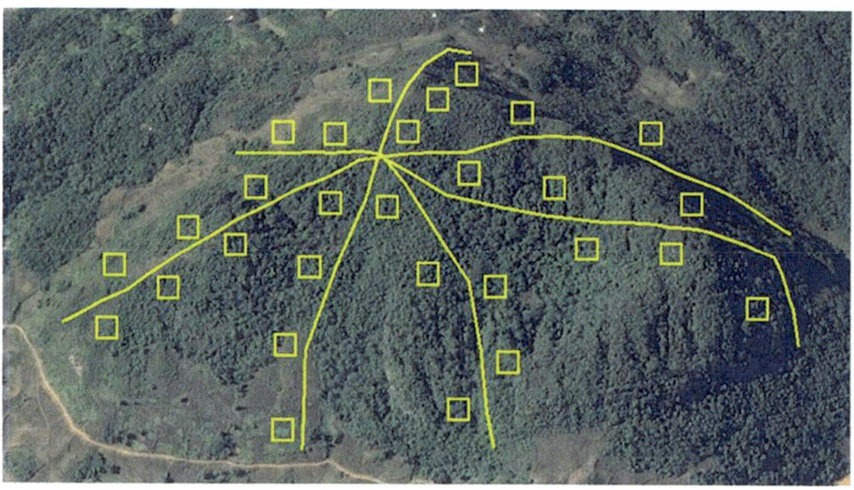

Fig. 41.2 Principle of line-based survey combined with standard plot survey

3 Results

3.1 Forest Products

Based on our ethnobotanical inventory in Hoang Lien-Van Ban Nature Reserve, a total of 256 medicinal plant species have been discovered at the reserve. Among them, 193 species have identified scientific names to species, 59 species have been identified to genera and 4 species have been identified to their families. The medicinal plants were identified in four plant branches, namely, Lycopodiophyta, Polypodiophyta, Gnetophyta and Magnoliophyta, 94 families and 189 different genera. Among plants, Magnoliophyta has the largest number of families, branches and species, respectively, 90 families, 183 genera and 250 species, accounting for 97.7% of the total number of species. The remaining species of plant branches account for only 2.3%. Of the 91 families in the study area, none of them have genera and species account for 10%. The family with the most genera and species (Orchidaceae) has only 7.25% and 7.81%. In the taxon, only 18 families (19.78%) have 3 (1.55%) or more, while up to 55 families (60.44%) have only one genus. In taxon species, there are 15 families (16.48%) with the number of species from 5 (1.95%) or more, but up to 44 families (48.35%) have only one medicinal plant species. The total number of species of 15 families (with 5 or more medicinal plants) is 121, accounting for 47.27% of the total species in the study area. Of the 89 genera of medicinal plants in the study area, 17 genera (19.1%) have species of three or more species. The total number of species of these 17 genera is 56, accounting for 21.88% of the total number of medicinal plants surveyed in this area. There are seven additional species to the Vietnamese flora. One species may be new to Vietnam's flora. There are 19 species in the Vietnam Red Book (7.4% of the medicinal plants in the area, accounting for 4.24% of the total number of species in the Vietnam Red Book). In the study area, 20 species of orchids were discovered. This is a valuable resource for making ornamental and ornamental plants, which is at high risk of being affected by medicinal collection and planting/selling activities. Among them, there are three species included in the Vietnam Red Book 2007: *Anoectochilus setaceus Blume*, *Dendrobium fimbriatum Hook*, and *Nervilia fordii (Hance) Schltr*. Medicinal plants in Hoang Lien-Van Ban Nature Reserve belong to eight different life forms: saprophytic, semi-parasitic, parasitic, epiphyte, wood, dust, vines, grass. The life forms with a lot of medicinal plants are vines (76 species, accounting for 29.69%), grass (86 species, accounting for 33.59%). In general, compared with other areas, there are a number of medicinal plants that are vines and epiphytes in the forests of Hoang Lien-Van Ban reserve. Although weeds and exotic species include a small percentage of the utilized species, some of them are regularly used by the villagers (sometimes daily) and contribute greatly to their diet (Table 41.2 and Fig. 41.3).

Table 41.2 Fast criteria for assessment of injury levels

Criteria	Degree of vulnerability	
	Low vulnerability = 2	High vulnerability = 1
Ecological	High abundance	Low abundance
	Fast growth	Slow growth
	Breed fast	Breed slowly
	Sexual and vegetative reproduction	Sexual reproduction
	Habitat is not specific	Unique living environment
	High distribution range	High distribution range low distribution range
	Response to harvest: Rapid regeneration	Response to harvest: Rapid regeneration
	Herbs, plants	Plants, shrubs and epiphytes
Use in life, harvesting and management methods	Leaves, flowers, bark and stems	Roots, rhizomes and tubers
	Choose according to specific size/age of harvested product	Do not choose by specific size / age of harvested product
	Harvest seasonally	Unlimited time
	Apply traditional conservation methods	Do not apply traditional conservation methods
	Pressure on habitat – No	Pressure on habitat – Yes
Economic	Mainly used in daily / less commercialized use	Main products for sale
	Demand (number of harvests + frequency) is low	Demand (number of harvest + frequency) is high
	Alternative species – Yes	Alternative species – No
	Accessing resources is not easy	Access to resources is easy
	Law enforcement capacity – Yes	Law enforcement capacity – No

3.2 Wood Products

NTFPs products are important income sources, providing essential needs and daily health care for poor households, contributing to poverty reduction in Lao Cai province in general and Van Ban district in particular, especially households of ethnic minority communities, living in Hoang Lien-Van Ban Nature Reserve.

The production and development of NTFP products plays a vital part in improving living condition of local people. These products are not only used on site to serve the lives of local people such as providing food, animal feed, medicine, they also contribute to ensuring food safety, health care, supply of raw materials, and fuels (Fig. 41.4).

NTFPs collected at Hoang Lien-Van Ban Nature Reserve can be grouped into bamboo shoots, mushrooms and medicinal plants. At higher elevations or economically poorer villages, the health system relies more on traditional medicine and medicinal plants.

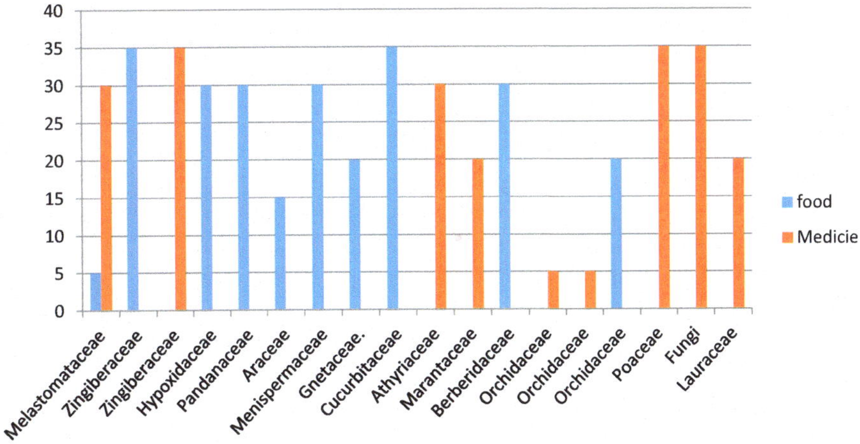

Fig. 41.3 Families representing a high number of medicinal and food plants in Hoang Lien-Van Ban Nature Reserve

Fig. 41.4 Photos of some medicinal plants

Food crops (wild vegetables, fruits, bamboo shoots and mushrooms; pickled bamboo shoots, bitter herbs, etc.) are harvested for daily needs and sold to markets (Minh Luong Commune Market, the central market of Van Ban district). Some important food crops are listed in Table 41.3.

All of these villages have easy access to the forest and are all located at an altitude of over 1400 m; the income of the villagers is mainly from raising cattle and other short-term agricultural crops. Collective forests have been under less pressure from land use changes, and villages have reforested some of their collective land over the last decade. The area around the nature reserve is practiced by free grazing of cattle, burning and deforestation for cultivation, leading to deforestation around the villages. Part of the collection comes from NTFPs in the NR's forests despite the prohibition. However, further survey shows that NTFPs contribute to a very low average household income mainly from livestock production and other livelihood activities.

Some medicinal plants are exploited and transported to the district market for sale, such as betel leaf 30,000–40,000 VND / kg, Hoang Dang 18,000–20,000 VND / kg, Thao Qua 140,000 VND / kg, Guam 100,000 VND / kg, Dao co lam, Hoa Tien,

Table 41.3 Important wild food and medicinal plants traded in Hoang Lien-Van Ban Nature Reserve (name of plants are ordered based on use frequency)

Number	Local name	Science name	Family	Harvested products	Harvest time	Frequency of use
1	Mua	Melastomataceae	Melastomataceae	Fruit	Yearly	Sometimes
2	Sa nhân	Amomum longiligulare	Zingiberaceae	Roots	Yearly	Sometimes
3	Thảo quả	Amomum tsaoko Crevost et Lem.	Zingiberaceae	Fruit	Sep-Nov	Sometimes
4	Sâm cau	Curculigo orchioides	Hypoxidaceae	Roots	Yearly	Sometimes
5	Dứa rừng	Pandanus tectorius sol.	Pandanaceae	Fruit	Yearly	Sometimes
6	Bã trầu	Epipremnum aureum	Araceae	Roots and leaves	Yearly	Regularly
7	Hoàng Đằng	Radix et caulis Fibraurea	Menispermaceae	Stems, roots	Yearly	Regularly
8	Gắm	Gnetum montanum Mgf	Gnetaceae	Stems, roots	Yearly	Regularly
9	Giảo cổ lam	Gynostemma Pentaphyllum Makino	Cucurbitaceae	Every part	Yearly	Regularly
10	Rau dớn	Diplazium esculentum	Athyriaceae	Leaves	Yearly	Regularly
11	Lá dong	Phrynium placentarium	Marantaceae	Leaves	Yearly	Regularly
12	Hoàng Liên gai	Berberis julianae C.K.Schneid.	Berberidaceae	Roots and leaves	Yearly	Regularly
13	Lan kim tuyến	Anoectochilus setaceus Blume	Orchidaceae	Roots	Yearly	Sometimes
14	Kim điệp	Dendrobium fimbriatum hook	Orchidaceae	Roots and leaves	Yearly	Regularly
15	Thanh thiên quỳ	Nervilia fordii (Hance) Schltr	Orchidaceae	Roots and leaves	Yearly	Regularly
16	Măng	Bambusa spp	Poaceae	Roots	Yearly	Regularly
17	Nấm	Fungi	Fungi	Roots	Yearly	Regularly
18	Mần tang	Litsea cubeba	Lauraceae	Leaves	Yearly	Regularly

Hoang Lien thorn, Dang ginseng; some kinds of bamboo shoots and medicinal plants of Red Dao bath such as Com Chay tree, Parachute Pagoda, Mang Tang, Ong Lao flowers, Lien Dang Hoa small, etc. for home use but not for sale.

The medicinal plants in Hoang Lien-Van Ban Nature Reserve are mostly discovered by indigenous people who still have access to the forest. An obvious interesting fact in Table 41.3 is that some of the medicinal plants are traded with very small frequency of use. In fact, some plants, such as Jiaozhi and Dang ginseng, are not commonly used in villagers' herbal medicines; Indeed, most villagers are unaware of the uses and medicinal properties of these species. This means that these species are collected primarily for commercial purposes; those species are on demand and through local people they are collected and brought to the central market for sale.

3.3 Forestry and Sustainable Development

Herbal ingredients are renewable resources through natural regeneration; however, regeneration does not meet the needs of human use and is also at risk of extinction. Plant resources, especially endangered and rare medicinal plant species, are seriously threatened and increasingly losing the rich diversity of species composition and number of individuals due to pressure from increase in the population of ethnic minorities living in forested areas; pressures from climate change; illegal medicinal plants exploitation; slash and burn cultivation; and forest fires, due to the difficult living habits of the majority of ethnic minorities. Due to the high reliance on forest resources, illegal exploitation, cultivation, cardamom cultivation in the forest still occur, resulting in degradation of forest resources, reduction of rare plant genetic resources and negative impacts on biodiversity. However, the development of hydropower projects deprives forest land, and the exploitation of medicinal plants is based only on demand and not on the ability to supply medicinal resources as well as the ability to recover. As a result, many forests have been exhausted and exploited, resulting in the loss of valuable and rare genetic resources and many species are at the risk of extinction unable to regenerate; this reduces the biodiversity of protection and special-use forests leading to the decrease in benefits of forests, and making people's lives near the forest more difficult.

The important regenerated medicinal plants have also been determined. The highest proportion of regenerated trees are Đìa bay, Puồng đìa bua, Đìa siêu, Puồng đìa diêm, Mà gầy khăng (Fig. 41.5).

Most of the interviewed households mentioned the diversity of medicinal plants in the past ten years due to habitat destruction and excessive harvesting. This can be attested to by the fact that for many medicinal plants traded in the area, underground parts such as rhizomes, tubers and roots can be exploited and used. This harvesting

Fig. 41.5 Photos of some regenerated medicinal plants

technique affects the ability of plants to grow and negatively affects the regeneration potential of the population. According to the 2017 Law on Forestry, the use of non-timber forest products must be permitted by the authorities, banning NTFP exploitation activities in biosphere reserves, nature reserves and national parks. However, these regulations were not enforced because villagers' compliance was low, especially regarding the exploitation of trees for medicinal purposes. This provision has not been implemented to date, and people living in mountainous areas do not have much access to the law, so almost never follow these rules.

4 Conclusions and Discussion

People living in Hoang Lien-Van Ban Nature Reserve use NTFPs for food and medicinal plants for their daily needs and generate additional income. Although only a few of the products directly contribute to household cash income, their livelihood is through other sources. Some NTFPs areas that directly contribute to very low per capita income are mostly shifting to agriculture and livestock production. However, in some villages, medicinal plants account for a high proportion of household income. In my research conducted in two communes in Hoang Lien-Van Ban Nature Reserve, a large number of economically important NTFP species will be harvested and cash income provided to households. Family. On the other hand, NTFPs are one of the ecosystem services that can contribute to encouraging forest conservation from other land use transitions.

Although NTFP collection is no longer important in some villages, it is still important in other areas in the buffer zone of Hoang Lien-Van Ban Nature Reserve. For example, in some central villages, vegetables and bamboo shoots are less important in people's daily lives because they earn extra income from livestock, other agricultural activities, buy vegetables from markets and working part-time jobs in enterprises in Van Ban district. Another reason to abandon NTFP collection may also be due to reduced resources due to conversion of forests for other purposes. Because households have begun to change their livelihoods to a more modern lifestyle, folk medicine, plant and wildlife food collection activities on the table seem to have disappeared. The development of modern medicine and other food needs are more attractive, especially for the younger generation.

Based on the RVA analysis, most important NTFPs are at risk of being overexploited. Strategies and plans for sustainable management and conservation of NTFPs should be implemented to prevent overexploitation. To well implement the conservation and sustainable development of medicinal plants, it is necessary to strengthen the leadership of local administrations at all levels with the management, conservation and protection of natural materia medica in the locality. Mechanisms and policies must be applied to support the cultivation of pharmaceutical materials suitable to the terrain of the fragmented land area of the province. Mechanisms must be developed to encourage socialization of conservation and sustainable use of medicinal genetic resources, establishment of botanical gardens, medicinal plant gardens, and conservation facilities invested by non-public organizations/individuals. Developing mechanisms and policies to attract community participation in conservation is associated with the development of short-term and long-term projects in conservation. Considering the increasing demand for medicinal plants used in traditional medicine and limited natural resources, planting medicinal plants in the region could reduce the pressure on plants and also diversify local livelihood strategies by generating additional income for households. To formulate a scheme on conservation and development of medicinal plant resources, with priority given to investigating medicinal plant resources, proceeding to make a list of rare and precious medicinal plants for export.

Acknowledgements This research is supported by a scholarship from McGill University, Canada. We would like to express our special thanks to the Management Board of Hoang Lien-Van Ban Nature Reserve for facilitating the research team. Also we are grateful to all the villagers, plant collectors and traders for sharing their knowledge with us.

References

Arnold, M. J. E., & Perez, M. R. (2001). Can non-timber forest products match tropical forest conservation and development objectives? *Ecological Economics, 39*, 437–447.

Bennett, B., & Husby, C. (2008). Patterns of medicinal plant use: An examination of the Ecuadorian Shuar medicinal flora using contingency table and binomial analyses. *Ethnopharmacology, 116*, 422–430.

Berlin, L. A., & Berlin, B. (2005). Some field methods in medical ethnobiology. *Field Methods, 17*(3), 235–268.

Hall, P., & Bawa, K. (1993). Methods to access the impact of extraction of non-timber tropical forest products on plant populations. *Economic Botany, 47*(3), 234–247.

Quinlan, M. (2005). Consideration for collecting freelists in the field: Examples from ethnobotany. *Field Methods, 17*(3), 219–234.

Ros-Tonen, M. (2000). The role of non-timber forest products in sustainable tropical forest management. *Holzals Rohund Werkstoff, 58*, 196–201.

Stepp, J. R. (2005). Advances in ethnobiological field methods. *Field Methods, 17*(3), 211–218.

Vodouhê, F. G., Coulibaly, O., Greene, C., & Sinsin, B. (2009). Estimating the local value of non- timber forest products to Pendjari biosphere reserve dwellers in Benin. *Economic Botany, 63*(4), 397–412.

Chapter 42
Quantify Forest Stand Volume Using SPOT 5 Satellite Image

Nguyen Thi Thanh Huong and Luong The Phuong

Abstract Stand volume is one of important attributes in forest management and quantification of forest resource value. This studies tested solutions to estimate forest stand volume for the large area using SPOT 5 data and field data. Firstly, the forests were stratified into four strata using maximum likelihood supervised classification methods based on SPOT 5 and field data. A set of 111 sample plots was distributed in these four forest strata, which represented four disturbed (by humans) forests under different levels in the tropical forest in Tuy Duc district of Dak Nong Province. Within the sample plots, DBH and tree height were measured to calculate stand volume using equation of stand volume from the previous study. The method of kNN (k-nearest neighbour) was applied to estimate the stand volume using SPOT 5 and field data. The estimates were tested on SPOT 5 bands, normalized difference vegetation index (NDVI), and combination of SPOT 5 bands and NDVI for the whole area and for each forest stratum. Quality of the predictions was assessed using leave-one-out cross-validation method. The results indicated that the accuracy of estimate was significantly improved when applying for each stratum compared to the combined SPOT 5 and NDVI data. The lower errors were found in the forest strata of less disturbances than the heavy degraded stratum. Among the image data, the estimates were based on the NDVI giving the lower accuracy compared to other.

Keywords Remote sensing · Forest strata · Stand volume · K-nearest neighbour · RMSE

N. T. T. Huong (✉)
Tay Nguyen University, Dak Lak, Vietnam

L. T. Phuong
Regional 4 Sub Forestry Protection Department, Dak Lak, Vietnam

A. T. Nguyen, L. Hens (eds.), *Global Changes and Sustainable Development in Asian Emerging Market Economies Vol. 2*, https://doi.org/10.1007/978-3-030-81443-4_42

1 Introduction

All decision-making demands information; in forestry, such information is collected by forest inventories that estimate forest characteristics over a defined area. The updated and quality of information will greatly contribute to the success of forest management (Nguyen, 2011). The mission of National Forest Inventory (NFI) is to produce and report timely and accurate estimates of forest resources (McRoberts & Tommp, 2007). The NFI monitoring in Vietnam has been conducted since the 1980s. As remote sensing data provide opportunity to estimate forest quality in vast areas with lower financial expenses compared to terrestrial inventory only, establishment of forest thematic maps using satellite image data is common application in forestry management (Nguyen, 2016). Accordingly, increasingly satellite imageries are considered as indispensable data in such national forest inventories in Vietnam. However, the main application is to map forest status under different level of disturbances (e.g. insignificant minor, medium, heavy impact). The forest attributes (e.g. volume) have been collected from the permanent system of sample plots distributed for the whole country. The inventory cycle is every 5 years; natural forests are always impacted by humans, and therefore, in addition to the inventory cycles a quick assessment should be done in timely and less costly manner. The advent of low-cost, widely available, remotely sensed data has been the basis for many of the important recent technological improvements. Remote sensing data have not only contributed to increasing speed, cost-efficiency, precision, and timeliness associated with inventories, but they have facilitated construction of maps of forest attributes with spatial resolutions and accuracies that were not feasible even a few years ago (McRoberts & Tommp, 2007); therefore, estimation of forest attributes using remotely sensed data is seen as a new potential for continuous management of natural resources (Mohammadia, Shataee, & Babanezhad, 2011). Since information about forest volumes is essential for forest management planning, large attempts have been given to estimate stand volume using different methods in order to improve estimation accuracy. Non-spatial modelling and spatial modelling are common methods to estimate the stand volume. A large quantity of literature has appeared to relate the estimation of stand volume using remote sensing data. Most of this literature has applied the regression method with linear or nonlinear regression (Awaya, Tsuyuki, Kodani, & Takao, 2004; Cohen, Maiersperger, Gower, & Turner, 2003; Fransson, Magnusson, & Holmgren, 2004; Rahman, 2004; Trotter, Dymond, & Goulding, 1997). Applying other regression types in forest attribute estimations and their spatial modelling using decision tree analysis such as regression tree has shown to be more useful compared to linear regression (Mohammadia et al., 2011). Geostatistics models have been given promising results (Meng, Cieszewski, & Madden, 2009; Nguyen, 2011; Tuominen, Holopainen, & Poso, 2006; Wallerman, Joyce, Vencatasawmy, & Olsson, 2002). Among the methods of forest attribute estimation, kNN is considered as a relatively simple and easy method to apply, becoming one of the most popular methods for conducting forest inventory using remote sensing (Meng et al., 2009). The kNN is known as a robust nonparametric method. It is used to estimate unknown values of data sets by means of similarity to reference data sets with known values (Scheuber, 2010) and to provide a very feasible tool for local to landscape level estimation (Gu et al., 2006); therefore, kNN has been applied widely in estimates of forest attributes since the first implementation of the multisource inventory based on the k-NN technique by the Finnish Forest Research Institute in 1990 (Tomppo, 2006).

The KNN method has two advantages in that it uses a nonparametric approach and allows for the use of robust to noisy training data. It is, therefore, becoming one of the most popular methods applied for forest inventory using large-area remote sensing data (Meng et al., 2009). kNN has been found as a useful tool for forest mapping over a large geographic area using a fine spatial resolution (Tokola, Pitkanen, Partinen, & Muinonen, 1996; Tomppo, Goulding, & Katila, 1999; Tomppo, Korhonen, Heikkinen, & Yli-Kojola, 2001; Holmström & Fransson, 2001; Tanaka et al., 2015; Lang, Gulbe, Traškovs, & Stepčenko, 2016). Although numerous studies of relationships between spectral responses and forest parameters have been conducted during the past several decades, conclusions about these relationships vary, depending on the characteristics of the study areas and the used data (Nguyen, 2011). Because of the complex forest stand structure and abundant vegetation species of tropical forests, the remote sensing spectral forest attributes relation is poorly understood (Lu, Mausel, Brondízio, & Moran, 2004). However, relatively less attention has been devoted to the moist tropical regions such as Asian forests; this might be due to difficulty in field data collection and complex biophysical environments. Meanwhile, a better understanding of forest stand parameters and spectral relationships is a prerequisite for effectively using appropriate image bands for developing spectral response-based estimation models (Lu et al., 2004).

2 Methodology

2.1 Study Area

The study was conducted in the district of Dak Nong Province. This forest site belongs to the Central Highlands of Vietnam. The study area is located between in 11⁰59′ to 12⁰16′ latitude North and 107°13′ to 107°28′ longitude East. The size of study area is about 500 square km (20 × 25 km). The forest is dominated by evergreen broad-leaved tropical natural forest but disturbed by humans over time at different levels. Many of valuable species trees have been selectively logged.

2.2 Data

Different data sources, including satellite image, digital data, and sample plots, were used under the study. The required satellite imagery product was SPOT 5 whose multi-spectral optical data was captured using the High Resolution Geometric

(HRG) instrument on board the satellite. The radiometric resolution is 256 digital levels, the spatial resolution is 10 m × 10 m and the images cover 60 km × 60 km. The SPOT 5 image was rectified using GCPs, and the elevation information was captured from a DEM created from an available 10 m contour line GIS shapefile. The SPOT image was projected to UTM 48 N, WGS84 to ensure compatibility between images and available digital data. A nearest neighbour resampling method was applied during this process with a pixel spacing of 10 m × 10 m in order to maintain the integrity of the pixel values. Because of the topographic effect of bright values of the images in some locations, some normalization algorithms were tested to remove this effect. The methods of Cosine, Minnaert method and C-correction were used to topographically correct the images (Blesius & Weirich, 2005; Jones, Settle, & Wyatt, 1988; Smith, Lin, & Ranson, 1980; Teillet, Guindon, & Goodenough, 1982). The C-correction was used since it presented the best one for topographic normalization for this area with the lowest determination of coefficient (R^2) (Nguyen, 2015).

A total of 111 sample plots with an approximate area of 0.1 ha with size similar to the SPOT 5 imagery pixel (10 x10 m) for each plot was sampled in the field. The stratified random sampling procedure was applied to assure that the sampling measurements captured all possible variability of forest conditions. Dense, moderate, open or/and very open forest structures were delineated during the field survey. Within the plots, the forest variables measured were breast height diameter (Dbh), tree height (H), tree density (N), and crown area (CA). Sample coordinates were recorded in the centroid position by GPSMap 60CSx. The standing volume equation was referred from previous research conducted by Nguyen (2011) for this area. This equation was then applied for all trees in all sample plots. For each plot, the mean forest parameters of sample plots and the 9-pixel means of SPOT 5 bands were calculated. The measured forest stand parameters were aggregated from 111 sample plots to represent forest stand conditions for forest classes. The standing volume equation was the following:

$$\begin{gathered} Ln(V) = -10.0094 + 1.066 \times Ln(Dbh) + 1.933 \times Ln(H) \\ With\, R^2 = 0.982, P < 0.05 \end{gathered} \tag{42.1}$$

where V is the stand volume; Dbh is the diameter at breast height; and H is the tree height.

2.3 Classifying SPOT 5 Image to Forest Strata

The stratification aims to divide forests into homogeneous units of one or a few specific indicators. Firstly, the forests were stratified using unsupervised classification algorithm of ISODATA (Iterative Self-Organization Data Analysis). Under forest-masked image, the forest was classified into maximum four classes as tested by Nguyen (2016) and considered as the first-phase sampling stage. Field sample plots were distributed on the classified image to measure forest attributes. In addition to the sample plots, field sample points were also taken based on prior knowledge to stratify the forest using supervised classification. The sample points and field data were distributed throughout the class to ensure the adequate representation of all the classes. The field data were chosen with respect to the size of the forest classes. One part of the field data was used to select training areas for the maximum likelihood supervised classification process and the another was employed to assess the classification accuracy. According to Congalton and Green (1999), the matrix is the most effective method to evaluate the accuracy. Matrix is the difference between the pixel has been classified and actual pixel matrix error of statistical results. Evaluation results are based on criteria of overall accuracy, producer's and user's accuracy. The values that participated in the accuracy assessment were computed through a method introduced by Congalton and Green (1999). Producer's accuracy is computed by looking at the reference data for a class and determining the percentage of correct prediction for these samples, whereas user's accuracy is computed by looking at the predictions produced for a class and determining the percentage of correct predictions.

2.4 Estimate Forest Volume Using the K-NN Algorithm

The k-nearest neighbours (kNN) algorithm, which is known to be the oldest and simplest approach, is regarded as non-parametric regression. The advantage of this method is due to the fact that no assumptions about the distribution of the variables involved are made (e.g. Efromovich, 1999; Linton & Härdle, 1998). The pixel-wise estimates were derived using the k-nearest neighbours (kNN) method, in which forest parameters (*v*) are calculated as weighted averages of the k-nearest field plots. The feature space distance (*d*) between a field plot and a pixel defines how close they are to each other. Feature space distances can be measured by arbitrary metrics. In this study, the Euclidean distance was used in the SPOT 5 spectral space.

For estimation with Euclidean distances, consider the spectral distance $d_{pi,p}$, which is computed in the feature space from the target pixel p (to be classified) to each reference pixel pi for which the ground data is known is as follows:

$$d_{p(pi)} = \left[\sum_{j=1}^{n} \left(x_{p,j} - x_{pi,j} \right)^2 \right]^{\frac{1}{2}} \tag{42.2}$$

where $x_{p,j}$ = digital number for the feature *j*, *n* = number of feature in the spectral space.

For each pixel *p*, take *k*-nearest field plot pixels (in the feature space) and denote the distances from the pixel *p* to the nearest field plot pixels by d_{pi}, $p_{,},\ldots,d_{pk},p$ ($d_{pi,p} \leq \ldots \leq d_{pk,p}$). The estimate of the variable value for the pixel *p* is then expressed as a function of the closest units; each such unit value is weighted according to a distance function in a particular feature space. A commonly used function for weighting distances is:

$$W_{(pi)p} = \frac{\frac{1}{d^{t}_{(p_i)p}}}{\sum_{i=1}^{k} \frac{1}{d^{t}_{(p_i)p}}} \tag{42.3}$$

where *k* describes the number of nearest neighbours and *t* is a distance decomposition factor, typically set to 0, 1, or 2. The sum of weighting $W_{(p_i)p}$ is always equal to 1.

With $t = 2$, the estimate of the variable *m* for pixel *p* is then:

$$m_p = \sum_{i=1}^{k} W_{(pi)p} m_{(pi)} \tag{42.4}$$

where $m_{(pi)}$ are the terrestrially recorded values of $i = 1, \ldots k$ pixels, which are located nearest to pixel *p* in the spectral space. The process is repeated for every pixel and results in intensive computations, depending on the resolution of the sensor and the size of the inventory area (Stümer, 2004).

The *kNN* software developed by Stümer (2004) was used in this study. For this application, two input files are necessary: an 'image file' and a 'field sample file' in ASCII format. The required image data, which are necessary for the *k*NN calculations, are converted from the corrected SPOT 5 bands and NDVI band within masked forest strata into ASCII files. For purpose of comparing between the different estimates, the predictions were run separately for each strum and for the whole forest area. The three input parameters were tested in this study including $t = 2$; and $r = 2$ and k (the number of nearest neighbours) = 5.

Leave-one-out cross-validation (LOOCV) was employed to evaluate the estimate results. This approach leaves one data point out of training data, that is, if there are n data points in the original sample then, n^{-1} samples are used to train the model and one point is used as the validation set. This is repeated for all combinations in which original sample can be separated this way, and then the error is averaged for all trials.

For every trial, accuracy of the predicted volume was evaluated using the root mean square error (RMSE):

$$\mathrm{RMSE} = \sqrt{\frac{(\hat{y}_i - y_i)^2}{n}} \tag{42.5}$$

where $\hat{y}_i$ was the estimated parameter on the i*th* observation and y_i was the field-measured parameter, respectively.

To facilitate a comparison with other forest strata, we also used relative RMSE (RMSE, %). The RMSE were calculated using the following:

$$RMSE\% = \frac{RMSE}{a} * 100 \tag{42.6}$$

The kNN was implemented in separate classes, which were obtained from forest stratification based on images classification. At the same time, the estimate was conducted for the whole area without stratification.

3 Results

3.1 Forest Stratification and Accuracy Assessment

The four forest classes were discriminated using MLC method in this study. These classes represented for very heavily degraded forest (Class 1), moderately disturbed forest (Class 2), insignificant disturbance (Class 3), and dense forest (Class 4). Based on field measurement, the forest strata were characterized in Tables 42.1 and the accuracy assessment of classification result was presented in Table 42.2.

The forest strata were distinguished with an accuracy of 85.69% and kappa of 0.79, indicating the substantial agreement between classification result and observations (Landis & Koch, 1977). The confusion matrix in Table 42.2 show most of individual classes had a relatively high accuracy with more than 75% excepting for Class 1 (UA = 68.85%). This indicated there is higher ability of misclassification in

Table 42.1 Confusion matrix of the maximum likelihood classification (for four classes)

	Class 1	Class 2	Class 3	Class 4	Row total	UA (%)
Class 1	42	19	0	0	61	68.85
Class 2	12	138	23	2	175	78.86
Class 3	0	2	67	12	81	82.72
Class 4	0	0	0	172	172	100
Column total	54	159	90	186	Overall accuracy = 85.69%	
PA (%)	77.78	86.79	74.44	92.47	Kappa = 0.79	

Table 42.2 Characteristics of stand volume of forest strata

Stand volume (m^3ha^{-1})	Class 1	Class 2	Class 3	Class 4
Mean	91.60	152.66	197.85	286.80
Standard deviation (SD)	22.53	27.71	33.51	56.96
Minimum (min)	45.76	110.63	142.77	212.97
Maximum (max)	135.34	204.76	279.02	412.70
Kurtosis standard	0.517	−0.969	0.847	1.236
Skewness standard	−0.984	−0.087	0.964	−0.398
Number of sample plot	25.00	28.00	39.00	19.00
Confidence level (95.0%)	9.30	10.74	10.86	27.46

the heavily disturbed forest compared to others. This may explain that the forest stand that was heavily impacted, forest structure are destroyed, leading to the high heterogeneity that was found in this stratum. Resulting the low accuracy presented in such stands (Šebeň & Bošeľa, 2010).

Based on field data and result of classification, summary statistics of standing volume for field sample data in the four forest classes was shown in Table 42.3. Skewness and kurtosis should be standardized with a constant, depending on the sample size. The standardized skewness and standardized kurtosis in this case were in the range from −2 to +2. This means the samples collected was able to represent for each stratum.

3.2 Estimates of Standing Volume for Forest Strata

The results shown in the Table 42.3 presented the estimate errors of the whole area and each stratum using spectral bands of SPOT 5 and combined SPOT 5 and NDVI data, while Table 42.4 gave those based on NDVI band. The estimate errors indicated the kNN method was quite promising, especially when the forest stands were stratified into the relatively homogeneous classes compared to the estimates

Table 42.3 RMSE (m^3ha^{-1}) and RMSE % for stand volume estimation using spectral SPOT bands and NDVI

	SPOT		SPOT + NDVI	
Forest stand	RMSE (m^3ha^{-1})	RMSE %	RMSE (m^3ha^{-1})	RMSE %
Estimate without stratification	47.06	26.73	46.81	26.10
Class1	32.72	37.27	25.83	26.82
Class2	18.09	11.52	20.84	13.67
Class3	26.87	13.35	27.55	14.54
Class4	30.87	10.68	28.66	11.03
General calculation of the four classes	27.70	18.64	26.70	16.71

Table 42.4 RMSE (m^3ha^{-1}) and RMSE % for stand volume estimation using NDVI

	RMSE (m^3)	RMSE %
Estimate without stratification	53.30	28.57
Class1	24.05	28.10
Class2	21.05	14.06
Class3	35.10	17.22
Class4	64.24	20.99
General calculation of the four classes	37.72	20.95

performed for the whole area without consideration on state of disturbed levels. This presented in both data set of SPOT and NDVI.

Among strata, the lowest accuracy of the estimation was found by Class1 in both SPOT and NDVI image. This suggests that the predictions from forest well managed were more accurate than the heavily degraded forest. There also was a small difference of RMSEs among the remaining classes. In almost all cases, the estimations performed using the SPOT or the SPOT + NDVI gave lower errors than the NDVI except for Class1. This may be the fact the forest stand that was strongly impacted the vegetation index variety becomes more optimal than using the SPOT image. The result also indicated that though the combination SPOT and NDVI improved the estimate results, the difference was not insignificant.

The difference of stand volume estimated and field volume ranged from 18.0 m^3ha^{-1} (estimated from SPOT) to 64.0 m^3ha^{-1} (estimated from NDVI); 12% and 21% of the mean measured value, respectively. The lowest error of 18.0 m^3ha^{-1} was given by Class2 which was classified as moderately disturbed forest and the highest was from Class4 which was discriminated as the dense forest. However the higher RMSE% was found in the estimate from Class2 with 11.52% comparing those of 10.68% in Class4. This was due to the mean stand volume of Class4 was of 286.8 m^3ha^{-1} while those of Class2 was of 152.66 m^3ha^{-1} (Table 42.2). Therefore, the comparisons should be considered among values of relative RMSE (%).

The worst predictions were found in all cases of estimate for the whole area without stratification with RSME of 47.06 m^3ha^{-1}, 46.81 m^3ha^{-1} and 53.30 m^3ha^{-1} corresponding to RMSE% of 26.72%, 26.10% and 28.57% for SPOT, SPOT+NDVI and NDVI, respectively. However, the results were significantly improved when the estimates were performed for each stratum. Among the strata, the lowest estimate was given by Class1 corresponding to heavily disturbed forest with RMSE of 32.72 m^3ha^{-1} (SPOT 5) and RMSE% of 37.27%. The difference between actual and estimated volume in other classes was impressively low with RMSE% <20%. Although the lowest value of RMSE% was obtained in Class4 both SPOT and SPOT+NDVI, the higher result was observed by NDVI. It may be the NDVI was saturated in the stable forest stands, for example, Class4. Conversely, the lower error was gained from Class1 using the NDVI compared to other (SPOT and SPOT+NDVI). In the forest stand which was heavily disturbed, NDVI data became a better choice in estimating the stand volume. In generally, the multi-spectral bands (e.g. SPOT or SPOT + NDVI) gave the better predictions compared to NDVI in this case. RMSE% of 18% and 16.71% were gained for the whole strata compared to the prediction of volume based only on NDVI with RMSE of 37.72 m^3ha^{-1} and RMSE% of 20.95%. Some authors compared the estimate errors of stand volume among sizes of area (e.g. Fazakas, Nilsson, & Olsson, 1999; Reese, Nilsson, Olsson, & Sandström, 2002). Reese et al. (2002) reported that when the accuracy of the estimates is assessed over larger areas, the errors are lower. The RMSE reduced to 10% RMSE over a 100 ha aggregation compared to 17% RMSE over an area of 19 ha aggregation. Meanwhile, Gu et al. (2006) showed the errors in the volume estimates by tree species were clearly higher than those of the total volume estimates. Specifically, the volume of Larix forest was estimated with a relative error of 51.7%,

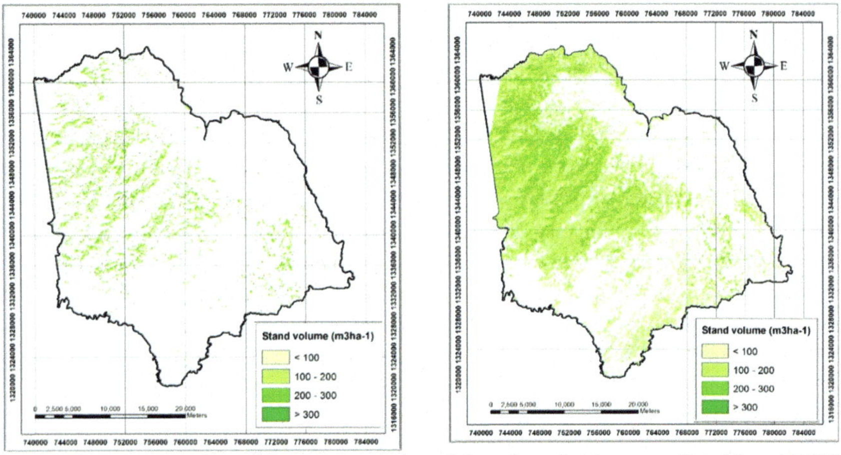

Stand volume map prediected from stratum of Class4 (insignificant disturbance)

Map of stand volume predicted from SPOT 5

Fig. 42.1 Forest stand volume estimated using SPOT 5

while the estimation errors for the Korean pine and broad-leaved tree species were over 71.7% and 88.19%, respectively. Franco-Lopeza, Ekb, and Bauerb (2001) reported RMSE forest volume was 48.68, 54.58 m^3ha^{-1} for coniferous forests, which are relatively few species. Reese et al. (2004) showed 33% RMSE (85 m^3ha^{-1}) for the estimates of total wood volume in the highly managed coniferous forest in Norway. In the current study, the output result provided another example of assessing the effect of volume estimate based on the homogenous forest stands rather than estimate for the whole forest area without stratification. Moreover, this result indicated that although the non-parametric method kNN is simple, the application presented as the promising method in estimating stand forest volume (Fig. 42.1).

4 Conclusions

Although there have been some studies on applying remote sensing data to the management of forest resources in Vietnam, most studies have focused on discrete variables such as the development of current status maps and land cover rather than estimating forest attributes, for example, producing forest volume map. Therefore, finding suitable solutions with updated information and low cost to quantify forest resources is essential especially in forest management. The non-parametric method applied in this study shows its potential in estimating forest continuous variables, for example, stand volume. The estimate was highly accurate, with an overall accuracy of around 80%. The better accuracy was found when the estimates were applied for separable stratum; the results showed a significant improvement compared to the estimate for the whole area. The overall errors for the whole area was around 25%,

26% and 28% for SPOT image, combined SPOT and NDVI image, and NDVI, respectively; while these ranged 16–20% for SPOT 5, SPOT + NDVI, and NDVI, respectively. Considering among forest stratum, except for the heavily impacted forest stand, the estimation errors were 37% (SPOT), 26% (SPOT+NDVI) and 28% (NDVI), the remaining classes (2,3,4) were predicted with quite high accuracies. The study provided an example to quantify forest resource using approach of forest strata. The use of non-parameter regression kNN to estimate the stand volume for strata forest in this study may be applicable potentially in complex structures of degraded forests in different levels such as Vietnamese natural forest stands.

References

Awaya, Y., Tsuyuki, S., Kodani, E., & Takao, G. (2004). Potential of woody carbon stock estimation using high spatial resolution imagery: A case study of spruce stands. In M. Shiyomi et al. (Eds.), *Global environmental change in the ocean and on land* (pp. 425–450). TERRAPUB.

Blesius, L., & Weirich, F. (2005). The use of the Minnaert correction for land-cover classification in mountainous terrain. *International Journal of Remote Sensing, 26*(17), 3831–3851.

Cohen, W. B., Maiersperger, T. K., Gower, S. T., & Turner, D. P. (2003). An improved strategy for regression of biophysical variables and Landsat ETM+ data. *Remote Sensing of Environment, 84*(4), 561–571.

Congalton, R. G., & Green, K. (1999). *Assessing the accuracy of remotely sensed data: Principles and practices*. Lewis Publishers.

Efromovich, S. (1999). *Nonparametric curve estimation - methods, theory, and applications* (411 pages). Springer.

Fazakas, Z., Nilsson, M., & Olsson, H. (1999). Regional forest biomass and wood volume estimation using satellite data and ancillary data. Agric. *Forest Meteorol, 98-99*, 417–425.

Franco-Lopeza, H., Ekb, A. R., & Bauerb, M. E. (2001). Estimation and mapping of forest stand density, volume, and cover type using the k-nearest neighbors method. *Remote Sensing of Environment, 77*, 251–274.

Fransson, J. E. S., Magnusson, M., & Holmgren, J. (2004). Estimation of Forest stem volume uing optical SPOT-5 satellite and laser data in combination. *IEEE Transactions on Geoscience and Remote Sensing: pp.*, 2318–2322.

Gu, H., Dai, L., Wu, G., Xu, D., Wang, S., & Wang, H. (2006). Estimation of forest volumes by integrating Landsat TM imagery and forest inventory data. *Science in China: Series E Technological Sciences, 49*, 54–62.

Holmström, H., & Fransson, J. E. S. (2001). Combining remotely sensed optical and radar data in kNN estimation of forest variables. *Forest Science, 49*(3), 409–418.

Jones, A. R., Settle, J. J., & Wyatt, B. K. (1988). Use of digital terrain data in the interpretation of SPOT-1 HRV multispectral imagery. *International Journal of Remote Sensing, 9*, 669–682.

Landis, R. J., & Koch, G. G. (1977). The measurement of observer agreement for categorical data. *Biometrics, 33*(1), 159–174.

Lang, M., Gulbe, L., Traškovs, A., & Stepčenko, A. (2016). Assessment of different estimation algorithms and remote sensing data sources for regional level wood volume mapping in Hemiboreal mixed forests. *Baltic Forestry, 22*(2), 283–296.

Linton, O., & Härdle, W. (1998). Nonparametric regression. In S. Kotz, C. B. Read, & D. L. Banks (Eds.), *Encyclopedia of statistical sciences* (Update vol) (Vol. 2, pp. 470–485). Wiley.

Lu, D., Mausel, P., Brondízio, E., & Moran, E. (2004). Relationships between forest stand parameters and Landsat TM spectral responses in the Brazilian Amazon Basin. *Forest Ecology and Management, 198*, 149–167.

McRoberts, R., & Tommp, E. (2007). Remote sensing support for national inventories. *Remote sensing and Environment, 110*, 412–419.

Meng, Q., Cieszewski, C., & Madden, M. (2009). Large area forest inventory using Landsat ETM+: A geostatistical approach. *ISPRS Journal of Photogrammetry and Remote Sensing, 64*, 27–36.

Mohammadia, J., Shataee, S., & Babanezhad, M. (2011). Estimation of forest stand volume, tree density and biodiversity using Landsat ETM+ data, comparison of linear and regression tree analyses. *Procedia Environmental Sciences, 7*(2011), 299–304.

Nguyen, T. T. H. (2011). *Forestry remote sensing: Using multidata sources for inventory of natural broad leaved ever-green forests in the central highlands of Vietnam*. Lambert Academic Publishing.

Nguyen, T. T. H. (2015). Topographical correction of SPOT 5 data. *Scientific Journal of Tay Nguyen University, 1859-4611*(14), 27–33.

Nguyen, T. T. H. (2016). Mapping tropical forest for sustainable management using SPOT 5 satellite image. *Int. arch. Photogramm. Remote Sens. Spatial Inf. Sci., 7*, 335–340.

Rahman, M.M. (2004). Estimating carbon pool and carbon release due to tropical deforestation using high resolution satellite data. Doctoral thesis. Faculty of Forest, Geo and Hydro Sciences, Dresden University of Technology, Germany. 191 pages.

Reese, H., Nilsson, M., Olsson, H., & Sandström, P. (2002). Applications using estimates of forest parameters derived from satellite and forest inventory data. *Computers and Electronics in Agriculture, 37*(1–3), 37–55.

Reese, H., Nilsson, M., Pahlén, T. G., Hagner, O., Joyce, S., Tingelöf, U., Egberth, M., & Olsson, H. (2004). Countrywide estimates of Forest variables using satellite data and field data from the National Forest Inventory. *AMBIO A Journal of the Human Environment, 32*(8), 542–548.

Scheuber, M. (2010). Potentials and limits of the k-nearest-neighbour method for regionalising sample-based data in forestry. *European Journal of Forest Research, 129*(5), 825–832.

Šebeň, V., & Bošeľa, M. (2010). Different approaches to the classification of vertical structure in homogeneous and heterogeneous forests. *Forest Science, 56*(4), 171–176.

Smith, J., Lin, T., & Ranson, K. (1980). The Lambertian assumption and Landsat data. *Photogrammetric Engineering and Remote Sensing, 46*, 1183–1189.

Stümer, W. (2004). Kombination von terrestrischen Aufnahmen und Fernerkundungsdaten mit Hilfe der kNN methode zur Klassifizierung und Kartierung von Wäldern. Dissertation, Fakultät für Forst- Geo- und Hydrowissenschaften, Technischen Universität Dresden, 153 pages.

Tanaka, S., Takahashi, T., Nishizono, T., Kitahara, F., Saito, H., Iehara, T., Kodani, E., & Awaya, Y. (2015). Stand volume estimation using the k-NN technique combined with Forest inventory data, satellite image data and additional feature variables. *Remote Sensing, 7*, 378–394.

Teillet, P. M., Guindon, B., & Goodenough, D. G. (1982). On the slope-aspect correction of multispectral scanner data. *Canadian Journal of Remote Sensing, 8*, 84–106.

Tokola, T., Pitkanen, J., Partinen, S., & Muinonen, E. (1996). Point accuracy of a non-parametric method in estimation of forest characteristics with different satellite materials. *International Journal of Remote Sensing, 17*(12), 2333–2351.

Tomppo, E. (2006). The Finnish multisource National Forest Inventory: Small-area estimation and map production. *Proceedings of the eighth annual Forest inventory and analysis symposium*, 341-349.

Tomppo, E., Goulding, C., & Katila, M. (1999). Adapting Finnish multi-source forest inventory techniques to the New Zealand preharvest inventory. *Scandinavian Journal of Forest Research, 14*, 182–192.

Tomppo, E., Korhonen, K. T., Heikkinen, J., & Yli-Kojola, H. (2001). Multisource inventory of the forests of the Hebei forestry bureau, Heilongjiang, China. *Silva Fennica, 35*, 309–328.

Trotter, C. M., Dymond, J. R., & Goulding, C. J. (1997). Estimation of timber volume in a coniferous plantation forest using Landsat TM. *International Journal of Remote Sensing, 18*(10), 2209–2223.

Tuominen, S., Holopainen, M., & Poso, S. (2006). Multiphase sampling. In A. Kangas & M. Maltamo (Eds.), *Forest inventory: Methodology and applications* (pp. 235–251). Springer.

Wallerman, J., Joyce, S., Vencatasawmy, C. P., & Olsson, H. (2002). Prediction of forest stem volume using kriging adapted to detected edges. *Canadian Journal of Forest Research, 32*, 509–518.

Chapter 43
Evaluating Impact of Climate Change to Fishing Productivity of Vietnam: An Application of Autoregressive Distributed Lag (ARDL) Regression Model

Nguyen Thi Vinh Ha

Abstract Fisheries are most affected by climate change. Yet, studies on impact of climate change to fisheries in Vietnam are still limited. This chapter uses production function with Autoregressive Distributed Lag (ARDL) regression and finds out that fishing productivity in Vietnam is negatively impacted by climate change. In the long run, if sea surface temperature increases by 1 °C, fishing productivity, as measured by catch per unit effort (CPUE), will decrease by 0.25 ton/CV. However, CPUE does not statistically significantly alter if there are changes in precipitation and number of storms.

Keywords Fishing productivity · Climate change · Impact assessment · Autoregressive Distributed Lag (ARDL) regression

1 Introduction

Climate change is having profound effects on environment, natural resources and economic, political and social life of economies around the world. Fisheries are most affected by climate change (Williams & Rota, 2010). The world's fish stock has been significantly affected by various impacts such as overfishing, pollution, loss of habitats, declining biodiversity, epidemics, etc. (Brander, 2010). Climate change exacerbates and has faster direct and indirect effects on aquatic species. The impacts of climate change to the oceans include increasing sea temperature, reducing dissolved oxygen, changing salinity, falling pH and shifting ocean currents, etc. (Brander, 2010; Sumaila, Cheung, Pauly, & Herrick, 2011). These changes positively

N. T. V. Ha (✉)
VNU University of Economics and Business, Vietnam National University, Hanoi, Vietnam

A. T. Nguyen, L. Hens (eds.), *Global Changes and Sustainable Development in Asian Emerging Market Economies Vol. 2*, https://doi.org/10.1007/978-3-030-81443-4_43

or negatively affect growth, reproductive productivity, behaviours and distribution of marine species (Brander, 2010; Pinnegar et al., 2013; Sumaila et al., 2011).

Aquatic invertebrates and fishes are thermoplastic; each species has different tolerance to temperature (Williams & Rota, 2010). Rising temperature reduces the ability of dissolving oxygen in water, restricting respiration and affecting the health of aquatic species. Therefore, when the water is warmer, fish will move to cooler areas to find their favourite temperature condition.

Aquatic species also prefer different salinity levels. The alteration of seawater salinity due to climate change is not significant so far. However, in the future, the salinity of the oceans is likely to upsurge due to the increasing amount of groundwater containing salt running into the sea. In the polar regions, salinity may fall due to augmented rainfall and low-salinity water flow from rivers (Roessig, Woodley, Cech Jr., & Hansen, 2004).

The indirect impacts of climate change via ecosystems include foods, competitors, predators and pathogens of aquatic species (Brander, 2010). The ocean acidification makes marine creatures such as molluscs, zooplankton difficult to create shells. This disturbs the food webs, thereby shifting species distribution, growth and structure, leading to influences on organisms in the oceans, estuaries, coral reefs, mangroves, and seagrass beds which are habitats of fishes. Under-optimal environmental conditions can lessen feed intake, foster competition for food and shelter. These changes deteriorate rate of growth, reproduction, behaviour, distribution, species structure, stocks, migration behaviour and survival of aquatic species. The likelihood of fish disease also worsens due to rise of temperature, sea level, storms and cyclones (Brander, 2010; Roessig et al., 2004).

According to Roessig et al. (2004) and Cheung et al. (2009), climate change negatively affect fisheries in low-latitude waters while it increases fishing benefits in high-latitude waters. However, fisheries in both sea areas can be negatively affected by the deterioration of water quality and the increasing likelihood of disease (Williams & Rota, 2010).

Vietnam locates in tropical region; its fishing industry is, therefore, more likely to be negatively affected by climate change. This chapter uses production function model with time series data to assess impact of climate change on fishing productivity in Vietnam.

The production function approach has been widely used in research on assessment of effects of changes in environmental quality.

The Economic Commission for Latin America and the Caribbean (ECLAC) (Kirton, 2011) assessed the relationship between fishery production (both capture and aquaculture) and seafood export price, sea surface temperature and annual average rainfall. The results showed that the sea surface temperature and the average rainfall were inversely proportional to the fishery production in Guyana. Damage to the fishery sector in the year 2050, under the A2 (high emission) scenario, was estimated to be at USD 15 million (4% social discount per annum). For the B2 (medium emission) scenario, the anticipated damage by 2050 was 12 million USD at 4% discount.

Caviedes and Fik (1992) showed that during the El Nino period 1997–1998, the pelagic fish catch yield in Peru and Chile decreased by 50% and 52%, respectively, leading to declines in export values and negative economic impacts (losses of job and income) in both countries. Catch per unit effort was modelled to depend on annual average sea surface temperature and El Nino events. MARMA regression was applied to correct time series problems such as autocorrelation and non-stationarity.

Asbjørn and Sygna (2000) used time series from 1980 to 1998 to examine the impact of El Nino and La Nina on tuna catches in Fiji and Kiribati. The study showed that the Southern Oscillation did not significantly affect tuna catch in Fiji while the catch increased with El Nino in Kiribati. However, Asbjørn and Sygna (2000) acknowledged that the regression model was so simple that it might not produce good estimation results.

In Vietnam, Pham et al. (2012) studied impacts of climate change on shrimp production in seven ecological regions of Vietnam. Research results showed that there was no correlation between shrimp productivity and temperature between 1990 and 2009 and seasonal rainfall from 1995 to 2009 in Phu Tho and Hoa Binh provinces. In the North Central Coastal region, temperature had an impact while rainfall had no influence on shrimp production in Nghe An and Thua Thien Hue provinces. Cao, Nguyen, and Nguyen (2013) quantified the variation in shrimp production due to changes in temperature and rainfalls in Thanh Hoa and Ha Tinh provinces. They found out that there was inverse correlation between shrimp production with temperature and precipitation, in addition to capital, labour and acreage of shrimp ponds. Nguyen, Nguyen, and Nguyen (2015) forecasted the impact of climate change on fisheries production in northern region of Vietnam. The study showed that the total damage of fishing industry in 2050 would be 584 billion VND at social discount rate of 3% in the medium (RCP4.5) climate change scenario.

2 Production Function Model and Data

2.1 The Production Function

Production function describes the relationship between inputs and outputs of a production process. In theory, there are two major inputs of production which are capital and labour. In sectors such as agriculture or fisheries, climate could be considered as an additional input. Impacts of climate change are measured as differences of outputs as results of variations in climate factors. Let Y denote production output, K capital, L labour and CC climate factor, the production function is expressed as formula (43.1):

$$Y = f\left(K,L,CC\right) \tag{43.1}$$

If climate change has impact on output, then δY/δCC is different to zero.

Adjusted Cobb-Douglas function (Zellner, Kmenta, & Dreze, 1966) is chosen for model specification, and the production function has the formula (43.2):

$$Y = A.K^{\alpha}.L^{\beta}.CC^{\gamma} \tag{43.2}$$

where α, β and γ are elasticities of output to capital, labour and climate factors, respectively. A is the impact of other factors. The logarithm of the two sides are applied to have the formula (43.3):

$$LnY = LnA + \alpha LnK + \beta LnL + \gamma LnCC \tag{43.3}$$

In this study, the output of fisheries sector is measured in terms of fishing productivity, represented by catch per unit effort (CPUE). Fishing productivity, as measured by catch per unit effort (CPUE), might be used as proxy to fish stock. Stable CPUE shows sustainable catch yield, while decreasing CPUE means that fish is over-exploited (Quirijns, Poos, & Rijnsdorp, 2008).

Since there is no available data on investment in fisheries sector, the study uses the variable of total fishing vessel capacity (measured in horsepower, or *cheval vapeur* – CV) as a proxy to capital.

In Vietnam, climate change manifests in increasing temperature and precipitation, which may have influence on fish growth and migration behaviour, affecting the fish stocks and then catches. Storms (with windspeed of level 8 or higher) cause damages to the fisheries activities and fishing vessels, losses of life and property, and deteriorating livelihoods of fishing communities. According to Ngo et al. (2013), wind speed from level 0 to level 6 is convenient for fishing activities at sea. El Nino and La Nina, which perform similarly to climate change in short run, also have some bearing on fishing. So, variables on temperature, precipitation, storms and El Nino are included in the production function.

In 1997, the Vietnamese Government encouraged fishermen to invest in offshore fishing via a preferential finance project (Decision No. 393-TTg dated 09 June 1997). In 2003, the National Assembly promulgated the Law of Fisheries. After these two milestones, there were major policy changes related to fishing activities in Vietnam, supposed to have positive effects on the catch yields. Therefore, we add two dummies to assess the impact of these policies in the production model.

The production function has the following form:

$$\begin{aligned} CPUE_t &= \beta_0 + \beta_1 LnCapacity_t + \beta_2 LnLabour_t + \beta_3 SST_t \\ &+ \beta_4 LnRainfall_t + \beta_5 Typhoon_t + \beta_6 SOI_t + \beta_7 D_1 + \beta_8 D_2 + \varepsilon_t \end{aligned} \tag{43.4}$$

2.2 Data

Data and sources of data for the regression models are described in Table 43.1.

Table 43.1 Data description

Variables	Description	Sources	Notes
T	Time in year	From 1976 to 2014	
$Catch_t$	Catch yield in year t	1976–2010: Ngo et al. (2013) 2011–2014: General Statistical Office (2016)	Including marine and inland catch
$Capacity_t$	Catch effort in year t (CV)	1976–2010: Ngo et al. (2013) 2011–2014: General Statistical Office (2016)	
$CPUE_t$	Catch per unit effort in year t (tons)	$CPUE_t = Catch_t/Capacity_t$	Fishing productivity
$Labour_t$	Total fishing labour in year t (persons)	1976–2010: Ngo et al. (2013) Missing values (in 1978, 2011–2014) are filled by interpolation.	
SST_t	Average Sea surface temperature in year t (°C)	The National Oceanic and Atmospheric Administration (NOAA), USA	Measured at Ha Long Bay
$Rainfall_t$	Total precipitation in year t (mm)	Climate change knowledge portal, the World Bank	
SOI	Southern oscillation index	NOAA	Difference in air pressure between Tahiti (southern Pacific) and Darwin (north to Australia).
$Typhoon_t$	Number of typhoons in year t	National Centre for hydro-meteorological forecasting, Vietnam Dinh (2010)	Number of storms in the Eastern Sea
D_1	Proxied to offshore fishing finance project in 1993	Value 0 for years before 1997, value 1 for years 1997 and later	
D_2	Proxied to the availability of law of fisheries	Value 0 for years before 2003, value 1 for years 2003 and later	
LnX	Logarithm of X		
β_i	Coefficients		
ε_t	White noise		

2.3 Model Specification and Testing

2.3.1 Autoregressive Distributed Lag Regression

In addition to the fishing inputs and other impacting factors of the same year, annual fishing productivity tends to rely on productivity and factors of previous years due to lagging impacts. Therefore, the Autoregressive Distributed Lag (ARDL) model (Pesaran & Shin, 1998) is chosen to demonstrate these dependencies. ARDL is ordinary least square (OLS) regression, which includes the lagged variables of dependent and independent variables. The ARDL model is appropriate when time series have different degrees of integration (e.g. I (0), I (1) or a combination of both) and especially when there is a single long-run relationship among variables. It is also suitable with small sample size ($n \leq 30$) (Nkoro & Uko, 2016).

ARDL model has the following form:

$$Y_t = c + \alpha_1 Y_{t-1} + \alpha_2 Y_{t-2} + \ldots + \alpha_p Y_{t-p} + \beta_0 X_t + \beta_1 X_{t-1} + \ldots + \beta_q X_{t-q} + u_t \quad (43.5)$$

where Y is dependent variable, X are explanatory variables, α and β are coefficients, p and q are number of lags of dependent and explanatory variables, respectively, c is intercept, t denotes time and u_t is white noise.

Several tests should be performed to confirm the appropriateness of the ARDL model, including selection of number of lags, tests for stationarity, long-run relationship among variables, model specification, autocorrelation, heteroscedasticity, multicollinearity, white noise (i.e. residual series are normal distribution and stationary), stability of the coefficients and convergence of long-run coefficients.

2.3.2 Test for Stationarity of Time Series

Normally, time series regressions require all series to be stationary, that is, mean, variance and covariance at different lags have constant values over time (Gujarati & Porter, 2009). Non-stationary series can lead to spurious regression. However, according to Nkoro and Uko (2016), ARDL regression is suitable with integrated time series of order zero or one. Augmented Dickey-Fuller (ADF) test and Schwarz information criteria (SIC) are applied to perform unit root tests of the time series. The test results (Table 43.2) show that *CPUE* and *LnLabour* are integrated of order one. They are difference stationary. *LnCapacity* is integrated of order zero, which is trend stationary, or its mean trends are deterministic. The other time series are integrated of order zero. Therefore, while traditional OLS regression is not applicable to this data set, ARDL regression can work.

Table 43.2 Unit root test results

				Critical value		
Variable	ADF statistics	*p*-value	*p*-value of difference	1%	5%	10%
CPUE	−1.933	0.618	0.072	−4.219	−3.533	−3.198
D(CPUE)	−6.277	0.000	–	−3.621	−2.943	−2.610
LnCapacity	−3.599	0.043	0.000	−4.219	−3.533	−3.198
LnLabour	−2.467	0.342	0.008	−4.260	−3.548	−3.209
D(LnLabour)	−4.276	0.002	–	−3.621	−2.943	−2.610
SST	−3.862	0.005	–	−3.616	−2.941	−2.609
LnRainfall	−5.447	0.000	–	−3.616	−2.941	−2.609
Typhoon	−5.039	0.000	–	−3.616	−2.941	−2.609
SOI	−4.358	0.001	–	−3.616	−2.941	−2.609

Where $D(x)$ denotes the first-order difference of x, that is, $D(CPUE_t) = CPUE_t - CPUE_{t-1}$.

2.3.3 Selecting Number of Lags for Regression Models

Vector autoregression (VAR) test and Akaike information criterion (AIC) are applied to select number of lags in ARDL model. The VAR results (Table 43.3) show that the ARDL should have three lags for all variables.

2.3.4 Test for Long-Run Relationship among Variables

To seek for the existence of long-run relationship among variables, the bound test is performed, using bound F-statistics and t-statistics to determine the cointegration among variables (Nkoro & Uko, 2016). The bound test has the following form:

$$\Delta Y_t = \delta_0 + \sum_{i=1}^{p} \alpha_{1i} \Delta X_{t-i} + \sum_{i=1}^{p} \alpha_{2i} \Delta Y_{t-i} + \delta_1 \Delta Y_{t-1} + \delta_2 \Delta X z_{t-1} + v_t \quad (43.6)$$

where Δ denotes the difference values, for example $\Delta Y_t = Y_t - Y_{t-1}$; p is the maximum lag chosen by author; $(\delta_1 - \delta_2)$ depicts the long-run relationship, while $(\alpha_1 - \alpha_2)$ depicts short-run one. Wald test is applied with null hypothesis is that all coefficients of lagged variables are zero.

F-statistics in Wald test does not follow normal distribution. It depends on: (1) integration orders of variables (I(0) or I(1)); (2) number of independent variables; (3) the existence of constant and trend variables in the model; and (4) sample size (Narayan, 2005). Narayan (2005) provided critical values for ARDL model with small sample size (from 30 to 80 observations). If the F-statistics is larger than the upper bound of the critical value, the null hypothesis is rejected.

Results of bound test are described in Table 43.4. F-statistics in model with dummies D_1 and D_2 is 7.4652, larger than the critical value 5.797 at significance level of 1%. F-statistics in model without dummies is 5.1777, larger than the critical value 4.324 at significance level of 5%. The t-statistics in model with dummies is −5.3166, less than the critical value −4.99 at significance level of 1%. The t-statistics in model without dummies is −4.5429, less than the critical value −4.38 at

Table 43.3 Selection of optimal number of lags using VAR

Lag	LogL	LR	FPE	AIC	SIC	HQ
0	0	−98.51	-	8.29e-07	5.86	6.17
1	1	68.02	259.04*	1.28e-09	−0.67	1.80*
2	2	116.54	56.61	1.81e-09	−0.64	3.98
3	3	191.46	58.27	1.19e-09*	−2.08*	4.69

aNumber of lags selected by criteria

Table 43.4 Bound tests

Test statistics	Value	Significance level	I(0)	I(1)
		Sample size $n = 40$		
F-statistics		5%	2.797	4.211
With dummies	7.4652	1%	3.800	5.643
Without dummies	5.1777	Sample size $n = 35$		
Number of degrees k	6	5%	2.864	4.324
Real sample size	36	1%	4.016	5.797
t-statistics				
With dummies	−5.3166	5%	−2.86	−4.38
Without dummies	−4.5420	1%	−3.43	−4.99

Table 43.5 Breusch-Godfrey test

	With dummies	Without dummies
n^*R^2	21.623	21.344
Chi^2	0.000	0.000

significance level of 5%. Therefore, we reject the null hypothesis and accept that there is long-run relationship among variables in the models. The application of ARDL regression is acceptable with the data set.

2.3.5 Test for Autocorrelation

Breusch-Godfrey test is applied to seek for autocorrelation. It gives *p*-values of Chi square at 0.000 for both models (Table 43.5). So, we reject the null hypothesis and accept that the models have autocorrelation. For OLS regression, when there is autocorrelation, the estimates are not biased but ineffective (non-smallest variance), leading to unreliable F and t tests. Newey-West estimates are used as a remedy. The correlograms after the Newey-West estimation show that the models are no longer autocorrelated (Fig. 43.1).

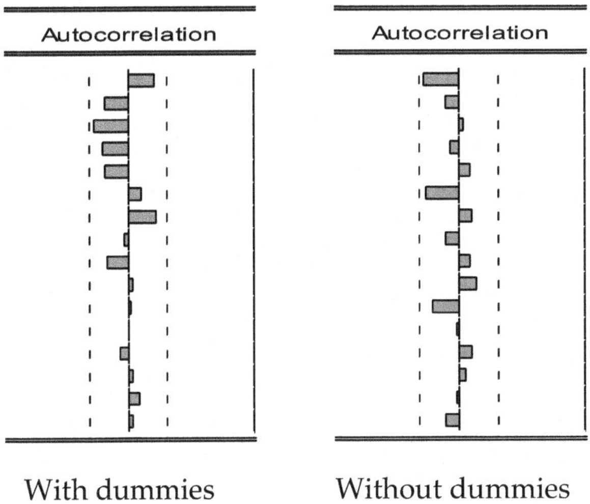

Fig. 43.1 Correlograms (Source: Author's estimation on Eviews 10)

2.3.6 Test for Suitability of Model Specification (Table 43.6)

Table 43.6 Ramsey reset test for the suitability of model specification

	With dummies	Without dummies
F-statistics	5.5179	4.3497
Degree of freedom *df*	(3.3)	(1.7)
p-value	0.0972	0.0775

p-value of the models are larger than $\alpha = 0.05$, so the model specification is suitable

2.3.7 Test for Heteroskedasticity (Table 43.7)

Table 43.7 Breusch-Pagan-Godfrey test for heteroskedasticity

	With dummies	Without dummies
n^*R^2	29.7270	26.3448
p-value of chi square	0.4277	0.4995

p-value of Chi square is larger than $\alpha = 0.05$; we accept the hypothesis that there is no heteroskedasticity in models. Harvey and Glejser tests provide the same results

2.3.8 Test for Multicollinearity

Most of explanatory variables of the models are statistically different from zero; Durbin-Watson d statistics is close to 2, so we can ignore the multicollinearity in the models.

2.3.9 Test for Stability of the Coefficients

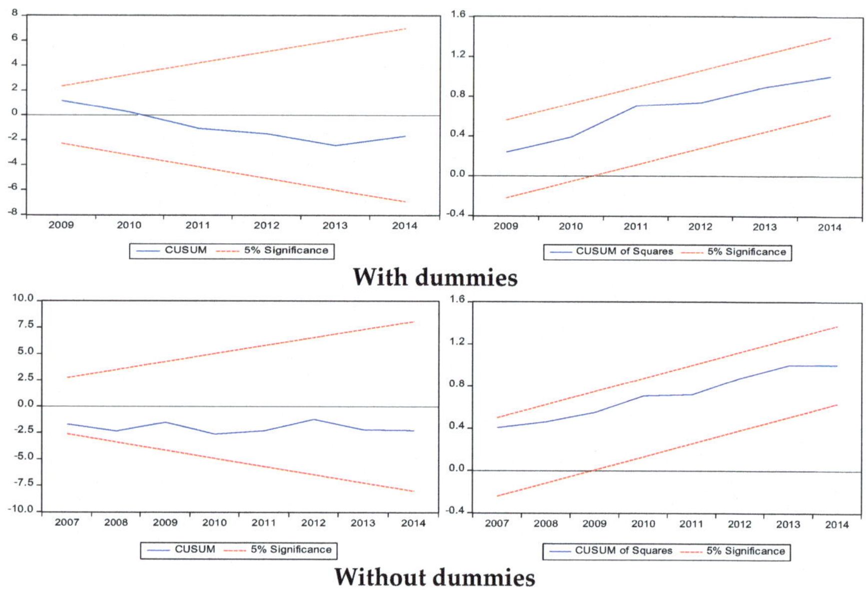

Fig. 43.2 CUSUM and CUSUMSQ control charts

Cumulative sum (CUSUM) and cumulative sum square (CUSUMSQ) control charts (Fig. 43.2) show that CUSUM and CUSUMSQ curves lie between the critical curves at significance level of 5%. This result confirms that there is long-run relationship among variables and the coefficients are stable.

Test for Normal Distribution of Residuals.

Using the Jarque-Bera test on the residual series of the models gives the p-values of all models greater than $\alpha = 0.05$. We accept the hypothesis that the residual series follows normal distribution (Fig. 43.3).

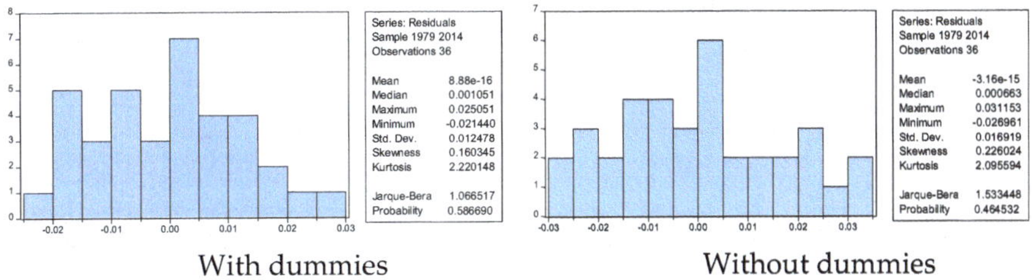

Fig. 43.3 Histograms and Jarque-Bera tests

Table 43.8 Unit root test for residual series

	With dummies	Without dummies	Critical value at significance level of 1%
ADF	−7.7960	−7.8574	−2.6327

2.3.10 Test for Stationarity of Residuals (Table 43.8)

ADF statistics are negative and less than the critical value at significance level of 1%, *p*-value is 0.000. Therefore, the residual series are stationary and normal distributed, that is, they are white noise.

The coefficient of the first lag of Error Correction EC(−1) of the Error Correction Model (ECM) is −1.0859 (i.e. negative and larger than −2), and statistically significant. It confirms that it is dynamically stable (Loayza & Ranciere, 2005).

The results show that there is long-run relationship among the variables; the model has no autocorrelation, no heteroskedasticity, no multicollinearity; the residuals are white noise, the coefficients of the models are stable and model specifications are suitable. Therefore, it can be said that regression models are appropriate and reliable.

3 Results

The ARDL model and its Conditional Error Correction model (ECM) can show the short-run and long-run relationship among variables. In case the number of observations of the model is small, the ECM model gives more reliable results (Nkoro & Uko, 2016). However, the coefficient of the ARDL model can be explained more easily and visually (Table 43.9).

Models with and without the dummies for policies give similar results, but the coefficients of the non-dummy model are less statistically significant, since the influence of policies are not controlled. Therefore, in the short run (within 3 years), we choose the dummy model as research results on the impact of climate factors on the fisheries production.

Table 43.9 shows that when the fishing productivity (CPUE) of the previous year is improved by 1 ton/CV, the productivity of the next year will rise by 0.68 ton/

Table 43.9 Results of ARDL models and ECM model

Variables	ARDL models						Variables	ECM models					
	With dummies			Without dummies				With dummies			Without dummies		
	Coef.		SE	Coef.		SE		Coef.		SE	Coef.		SE
Dependent variable			CPUE				Dependent variable			D(CPUE)			
CPUE(−1)	0.6838	***	0.2025	0.7368	***	0.1836	*D(CPUE(−1))*	0.7697	***	0.0909	0.8223	***	0.1142
CPUE(−2)	−0.4835	***	0.1643	−0.5697	**	0.1975	*D(CPUE(−2))*	0.2862	*	0.1097	0.2526	*	0.1272
CPUE(−3)	−0.2862		0.2259	−0.2526		0.2131							
LnCapacity	−0.5464	***	0.1769	−0.5985	**	0.2024	*D(LnCapacity)*	−0.5464	***	0.0744	−0.5985	***	0.0902
LnCapacity(−1)	0.6154	***	0.1984	0.5944	**	0.2478	*D(LnCapacity(−1))*	0.5763	***	0.0893	0.6147	***	0.1124
LnCapacity(−2)	−0.3742		0.2523	−0.3427		0.2555	*D(LnCapacity(−2))*	0.2021	*	0.0911	0.2720	**	0.1125
LnCapacity(−3)	−0.2021		0.2362	−0.2720		0.2619							
LnLabour	0.3009		0.1744	0.2198		0.1572	*D(LnLabour)*	0.3009	**	0.1112	0.2198	*	0.1163
LnLabour(−1)	−0.0785		0.2071	−0.0220		0.2023	*D(LnLabour(−1))*	0.0623		0.0937	−0.1866		0.1147
LnLabour(−2)	−0.2437		0.1935	−0.2674		0.2897	*D(LnLabour(−2))*	−0.1814	*	0.0743	−0.4539	***	0.0998
LnLabour(−3)	0.1814		0.1955	0.4539	*	0.2348							
SST	−0.0132		0.0384	−0.0111		0.0326	*D(SST)*	−0.0132		0.0170	−0.0111		0.0199
SST(−1)	−0.0936	*	0.0423	−0.1035	**	0.0383	*D(SST(−1))*	0.1758	***	0.0297	0.1600	***	0.0310
SST(−2)	−0.1068	**	0.0428	−0.1069	*	0.0522	*D(SST(−2))*	0.0690	**	0.0221	0.0530	*	0.0231
SST(−3)	−0.0690	**	0.0257	−0.0530		0.0327							
LnRainfall	−0.0715		0.1490	−0.1241		0.1160	*D(LnRainfall)*	−0.0715		0.0582	−0.1241		0.0704
LnRainfall(−1)	0.0134		0.1304	0.0011		0.1398	*D(LnRainfall(−1))*	−0.2588	***	0.0616	−0.2969	***	0.0807
LnRainfall(−2)	−0.0285		0.0916	0.0025		0.0997	*D(LnRainfall(−2))*	−0.2873	***	0.0504	−0.2944	***	0.0624
LnRainfall(−3)	0.2873	**	0.1106	0.2944	**	0.1064							
Typhoon	0.0010		0.0028	0.0028		0.0026	*D(typhoon)*	0.0010		0.0015	0.0028		0.0019
Typhoon(−1)	0.0017		0.0019	0.0021		0.0024	*D(typhoon(−1))*	0.0133	***	0.0023	0.0106	***	0.0027
Typhoon(−2)	−0.0062		0.0035	−0.0049		0.0041	*D(typhoon(−2))*	0.0071	***	0.0017	0.0057	*	0.0021
Typhoon(−3)	−0.0071	*	0.0031	−0.0057		0.0039							

Variables	ARDL models						Variables	ECM models					
	With dummies			Without dummies				With dummies			Without dummies		
	Coef.		SE	Coef.		SE		Coef.		SE	Coef.		SE
SOI	0.0171		0.0122	0.0085		0.0148	*D(SOI)*	0.0171	**	0.0050	0.0085		0.0060
SOI(−1)	0.0302		0.0160	0.0248		0.0134	*D(SOI(−1))*	−0.0598	***	0.0074	−0.0588	***	0.0093
SOI(−2)	0.0231		0.0120	0.0182		0.0111	*D(SOI(−2))*	−0.0367	***	0.0051	−0.0406	***	0.0067
SOI(−3)	0.0367	**	0.0114	0.0406	**	0.0145	*EC(−1)*	−1.0859	***	0.1062	−1.0855	***	0.1363
D1	−0.0073		0.0579				*D1*	−0.0073		0.0195			
D2	0.1726	**	0.0509				*D2*	0.1726	***	0.0257			
C	11.768	*	5.2219	10.443	**	4.118	*C*	11.7682	***	1.1504	10.4426	***	1.3094
Adjusted R^2			0.9903			0.9867	Adjusted R^2			0.9121			0.8615
Prob (F-statistics)			0.0000			0.0000	Prob (F-statistics)			0.0000			0.0000
Durbin-Watson stat.			2.5529			2.5626	Durbin-Watson stat.			2.5529			2.5626

n = 36 $^{*}p < 0.1$
$^{**}p < 0.05$
$^{***}p < 0.01$

CV. If there is enhancement in fish stock, more little fish will be borne, leading to further improvement in the following year's stock. The incremental investment of fishing capacity by 1% reduces CPUE by 0.55 ton/CV in the same year but increases CPUE by 0.62 ton/CV next year. This may be because fish stock is rather fixed in a year time; if we add more capacity, then catch per capacity will decrease. However, CPUE may improve later when fishing stocks have enough time to expand. Addition in labour does not statistically significantly alter CPUE.

D_1 does not affect CPUE significantly, but coefficient of D_2 is positive. This result shows that the offshore encouragement project in 1997 did not have impact on CPUE. However, the Law of Fisheries promulgated in 2003 and its subsequent policies have positive effect on fishing productivity.

The rise in sea surface temperature negatively effects CPUE, by 0.01, 0.09, 0.11 and 0.07 ton/CV in four consecutive years. There is a lag impact since migration mainly reduces fish stocks in the following years. Rainfall, typhoon and El Nino do not have statistically significant effect on CPUE in the short term.

ECM model explains the impact of production factors in terms of marginal productivity. When the marginal capacity goes up by 1%, marginal productivity decreases 0.55 tons/CV, reflecting the law of diminishing marginal productivity. Offshore fishing incentive policies implemented in 1997 have no effect on marginal CPUE while the Law of Fisheries 2003 increased the marginal CPUE.

If there are changes in marginal temperature, rainfall and typhoon, the marginal fishing productivity does not statistically significantly alter in the same year. However, marginal SST and typhoons have positive impact, while marginal rainfall has negative impact on marginal productivity next year.

Independent variable is CPUE, * $p < 0{,}1$ ** $p < 0{,}05$ ***$p < 0{,}01$

Table 43.10 shows that in the long run, if there is 1% more investment in capacity, CPUE will go down 0.57 ton/CV. If 1% more people join in the sector, CPUE will increase 0.35 ton/CV.

When sea surface temperature rises by 1 °C, CPUE will drop by 0.25 ton/CV. In global warming, marine species tend to migrate further off the coast, to the north, or to deeper water to find their favourite temperature condition (Brander, 2010). Water warming will also prevent the absorption of oxygen, affecting respiration and metabolism of species (Williams & Rota, 2010). In addition, researches also prove that water warming also reduces size of matured fish, increasing mortality rate

Table 43.10 The long-run form of production function

Variable	Coef.		SE	t-stat	p-value
LnCapacity	−0.5700	***	0.0472	−12.0781	0.0000
LnLabour	0.3540	***	0.0545	6.4925	0.0002
SST	−0.2529	***	0.0566	−4.4704	0.0021
LnRainfall	0.1601		0.2297	0.6973	0.5054
Typhoon	−0.0053		0.0037	−1.4344	0.1894
SOI	0.0848	***	0.0151	5.6238	0.0005

(Vietnam Institute of Fisheries Economics and Planning [VIFEP], 2012). These things lead to diminishing fish stock affecting fishing productivity.

Rainfall or number of typhoons does not statistically affect CPUE. When the southern oscillation index increases by 1 unit (i.e. El Nino decreases or La Nina increases), CPUE will go up 0.08 ton/CV. SOI acts similarly to climate change in short run. The rise of temperature negatively affects fish stock in Vietnam. These results are suitable with several studies in other countries (Caviedes & Fik, 1992; De Wit & Stankiewicz, 2006; ECLAC, 2011; Jeong & Lee, 2010; Lu, 2010; Parker Jr & Dixon, 1998; Roessig et al., 2004; Sun, Chiang, Tsoa, & Chen, 2006; Tseng & Chen, 2008) as well as in Vietnam (Cao et al., 2013; Nguyen et al., 2015; Pham et al., 2012). IPCC indicated that forecast number of tropical storms is low, likely to increase due to climate change (Intergovernmental Panel on Climate Change [IPCC], 2014).

4 Conclusion

In the long run, fishing productivity is affected by climate change, in which the increase in sea temperature will reduce fish stock and distribution, leading to decrease in fishing productivity. In addition, the fact CPUE decreases when capacity increases shows that there is overexploitation in Vietnam.

Vietnam should reduce fishing effort and catch yield to the maximum sustainable yield level, in order to preserve fish stock, which is diminishing dramatically due to foreseen increase in sea surface temperature in the context of climate change.

References

Asbjørn, A.H., Sygna, L. (2000). Economic impacts of climate change on tuna fisheries in Fiji Islands and Kiribati. Cicero Report, 21 pages.

Brander, K. (2010). Impacts of climate change on fisheries. *Journal of Marine Systems, 79*(3-4), 389–402.

Cao, L. Q., Nguyen, T. H., & Nguyen, N. H. (2013). *Evaluation of climate impacts on infrastructure, area, productivity and yield of aquaculture in coastal areas to develop intergrated response and piloting model for aqualculture* (142 pages). Institute of Fisheries Economics and Planning.

Caviedes, C.N., Fik, T.J. (1992). The Peru-Chile eastern Pacific fisheries and climatic oscillation. In Glantz (ed.), climate variability, climate change and fisheries. Cambridge university press, 355–376.

Cheung, W. W. L., Lam, V. W. Y., Sarmiento, J. L., Kearney, K., Watson, R., & Pauly, D. (2009). Projecting global marine biodiversity impacts under climate change scenarios. *Fish and Fisheries, 10*, 235–251.

De Wit, M., & Stankiewicz, J. (2006). Changes in surface water supply across Africa with predicted climate change. *Science, 311*, 1917–1921.

Dinh, V. U. (2010). Variation and landing of Tropial storms into Vietnamse coast. *Natural Science and Technology, Vietnam National University-Hanoi, 26*, 479–485.

ECLAC (2011). An assessment of the economic impact of climate change on the agriculture sector in Guyana. United Nation, Economic Commission for Latin America and the Caribbean.
General Statistical Office. (2016). *Statistical yearbook 2015*. Statistical Publishing House.
Gujarati, D. N., & Porter, D. C. (2009). *Basic econometrics* (5th ed., 909 pages). Tata McGraw-Hill Education.
Intergovernmental Panel on Climate Change [IPCC] (2014). Climate change 2014: Synthesis report. Contribution of working groups I, II and III to the Fifth Assessment Report of the Intergovernmental Panel on Climate Change, 151 pages.
Jeong, I.J., Lee, S.H. (2010). *Korea's effective approach to adapting climate change in the fisheries sector*. In the economics of adapting fisheries to climate change. OCED publishing, 367–376.
Kirton, C. (2011). *An assessment of the economic impact of climate change on the agriculture sector in Guyana*. Economic Commission for Latin America and the Caribbean (ECLAC).
Loayza, N., Ranciere, R. (2005). *Financial development, financial fragility, and growth*. IMF working paper, WP/05/170.
Lu, H. J. (2010). *Chinese Taipei: The impact of climate change on coastal fisheries* (pp. 377–387). OECD Publishing.
Narayan, P. K. (2005). The saving and investment nexus for China: Evidence from cointegration tests. *Applied Economics, 37*, 1979–1990.
Ngo, A. T., et al. (2013). *50 years of fisheries in Vietnam* (244 pages). Agricultural Publishing House.
Nguyen, N.T., Nguyen, V.T, Nguyen, T.V.H., Nguyen, Q.V. (2015). Impact of climate change on fisheries in the north of Vietnam. Vietnam National University - Hanoi Publishing House, 201 pages.
Nkoro, E., & Uko, A. K. (2016). Autoregressive distributed lag (ARDL) cointegration technique: Application and interpretation. *Journal of Statistical and Econometric Methods, 5*, 63–91.
Parker, R. O., Jr., & Dixon, R. L. (1998). Changes in a North Carolina reef fish community after 15 years of intense fishing - global warming implications. *Transactions of the American Fisheries Society, 127*, 908–920.
Pesaran, M. H., & Shin, Y. (1998). An autoregressive distributed-lag modelling approach to cointegration analysis. *Econometric Society Monographs, 31*, 371–413.
Pham, Q. H., et al. (2012). *Impact assessment and identication of response in agriculture and fisheries*. Institute of Agricultural Environment.
Pinnegar, J. K., Cheung, W. W. L., Jones, M., Merino, G., Turrell, B., & Reid, D. (2013). Impacts of climate change on fisheries. *MCCIP Science Review, 2013*, 302–317.
Quirijns, F. J., Poos, J. J., & Rijnsdorp, A. D. (2008). Standardizing commercial CPUE data in monitoring stock dynamics: Accounting for targeting behaviour in mixed fisheries. *Fisheries Research, 89*, 1–8.
Roessig, J. M., Woodley, C. M., Cech, J. J., Jr., & Hansen, L. J. (2004). Effects of global climate change on marine and estuarine fishes and fisheries. *Reviews in Fish Biology and Fisheries, 14*, 251–275.
Sumaila, U.R., Cheung, W.W.L., Lam, V.W.Y, Pauly, D., Herrick, S. (2011). Climate change impacts on the biophysics and economics of world fisheries. Nature Climate Change, 1: 449–456.
Sun, C. H., Chiang, F. S., Tsoa, E., & Chen, M. H. (2006). The effects of El Niño on the mackerel purse-seine fishery harvests in Taiwan: An analysis integrating the barometric readings and sea surface temperature. *Ecological Economics, 56*, 268–279.
Tseng, W. C., & Chen, C. C. (2008). Valuing the potential economic impact of climate change on the Taiwan trout. *Ecological Economics, 65*, 282–291.
Vietnam Institute of Fisheries Economics and Planning [VIFEP] (2012). Summary report of overall fisheries plan of Vietnam up to 2020, a vision to 2030. General Department of Fisheries, 110 pages.
Williams, L., Rota, A. (2010). Impact of climate change on fisheries and aquaculture in the developing world and opportunities for adaptation. Fisheries Thematic Paper: Tool for project design.
Zellner, A., Kmenta, J., & Dreze, J. (1966). Specification and estimation of cobb-Douglas production function models. *Econometrica: Journal of the Econometric Society, 12*, 784–795.

Chapter 44
An Experimental Study on Using Biogas Slurry to Improve the Water Quality of Aquaculture Systems in Acid Sulfate Soil Areas

Nhat Long Duong, Hoang Thanh Nguyen, and Vo Chau Ngan Nguyen

Abstract This study is aimed at testing a new method to improve fishpond water quality in the acid sulfate soil areas by using biogas slurry (BS). Two experiments were implemented on-site at the fishpond to test for quantity of biogas slurry applied and fish growing status. In the first experiment, treatment of 75% BS had water pH increased from 3.3 to 6.5 after 1.5 months; DO values and Chlorophyll-a concentrations were reasonably high (5.2 ppm and 143.05 μg/L). In the second experiment, the daily weight gain of Snakeskin gourami (0.002 g/day) in treatment 1 was significantly lower than that in treatment 2 (0.051 g/day) and treatment 3 (0.049 g/day); however, the survival rate and the fish yield of fingerling nursing in treatment 2 (22.19% and 270 kg/1000 m^2) were similar to those in treatment 3 (22.44% and 264 kg/1000 m^2). For Climbing perch, the daily weight gain of the fingerlings in treatment I (0.004 g/day) was significantly lower than those in treatment II (0.045 g/day) and in treatment III (0.048 g/day). In conclusion, treatment 75% BS (153 m^3/1000 m^2) is considered the best rate to improve water pH and produce good yield of fish for farmer's income.

Keywords Acid sulfate soil area · Biogas slurry · Fish nursing · Water pH treatment

N. L. Duong · H. T. Nguyen · V. C. N. Nguyen (✉)
Can Tho University, Can Tho, Vietnam
e-mail: nvcngan@ctu.edu.vn

A. T. Nguyen, L. Hens (eds.), *Global Changes and Sustainable Development in Asian Emerging Market Economies Vol. 2*, https://doi.org/10.1007/978-3-030-81443-4_44

1 Introduction

Within the field of agriculture, many of national and international scientists have continually researched and implemented the aspects of design, construction and soil improvement in reality, especially on the acid sulfate soil areas. The research results are of different levels and time depending on the variety of soil compositions. Accordingly, there have been previous studies in Vietnam such as the use and management of acid sulfate soil in aquaculture (Singh, 1985), the construction of fishery–forestry–agriculture model on the acid sulfate soil due to the ecological function of cajuput forest (Duong, Ton, & Edward, 2001), etc. Such findings have made a great contribution to the enlargement of culture areas, the productivity and profit increase for manufacturers on the Mekong delta acid sulfate soil.

It is obvious that to decrease the amount of acid sulfate farmers have always used the conventional methods like powdered lime or the mixture of lime and organic fertilizer resulting in a considerable increase of water pH. The use of 15 kg/100 m^2 lime increased water pH from 4.1 to 7.3 (Edwards, Kaewpaitoon, McCoy, & Chantachaeng, 1986). Besides, the organic fertilizer, especially animal manure is also one of effective materials for pond treatment. The organic fertilizer, a nutritious source, consumed by plankton becomes the natural food source available in water. This food source helps fish grow up quickly to create high productivity. Moreover, the decrease of disease transmission has also contributed to the development of fish yield. However, due to the growth of intensive agriculture, the environment gets more and more polluted because of untreated waste in an appropriate manner. There are varied methods of treating and reusing the waste like composting so as to directly and indirectly feed aquatic species. In particular, the use of biogas slurry is the most efficient treatment in the aquaculture development.

Biogas slurry is a product of the decomposition of organic matter. It is also considered an alternative energy source used in each household for cooking, lighting and heating as well. Biogas slurry includes solid and liquid waste involving a high degree of organic substances and inorganic nutritious salinity such as nitrogen and phosphorus. Biogas slurry was used in aquaculture and algae ponds and rice paddy. The organic substance usage can treat many kinds of waste, reuse nutrients, create valuable products and reduce pollution (Le & Nguyen, 2015). Nevertheless, there is little research on the use of nutritious source included in biogas slurry for pond treatment on the acid sulfate soil areas. Therefore, the study aims to find out effective methods of improving water quality in the acid sulfate soil areas by using biogas slurry. In addition, the experimental study on nursing some appropriate species is expected to enhance the effectiveness of soil treatment for production and the development of aquaculture models in each household, increasing the income for local communities.

2 Methodology

2.1 Design of the Experiment

The experiments on pond treatment and fish nursing were conducted on Hoa An Center, Can Tho University, located at the most acid sulfate soil area of the Mekong delta. The research was carried out in dry season. Each pond covers an area of 4 m^2 (2.0 m × 2.0 m) and 1.2 m depth.

Experiment 1. The treatment of water pH infected ponds by biogas slurry.

The experiment consists of six treatments with three repeated and randomly arranged.

- Treatment 1 (NT I): 100% organic fertilizer (pig manure) use (30 kg/100 m^2).
- Treatment 2 (NT II): biogas slurry use with 75% TN in organic fertilizer.
- Treatment 3 (NT III): biogas slurry use with 100% TN in organic fertilizer.
- Treatment 4 (NT IV): biogas slurry use with 125% TN in organic fertilizer.
- Treatment 5 (NT V): biogas slurry use 150% TN in organic fertilizer.
- Treatment 6 (NT VI): pond treatment without using biogas slurry.

All the above treatments were applied using CaO powdered lime (150 g/m^2). The time for all treatments showed the stability of water pH and Chlorophyll-a concentrations in the ponds.

Experiment 2. Nursing the fingerlings of fish in treated ponds.

After experiment 1, the treated fish pond was applied in nursing the fingerlings of Snakeskin gourami (*Trichopodus pectoralis*) and Climbing perch (*Anabas testudineus*). Experiment on each species includes three treatments with three repeated in 60 days.

- Experiment on nursing the fingerlings of Snakeskin gourami.

  Treatment 1: using biogas slurry without food supply.
  Treatment 2: using biogas slurry and 50% of food supply.
  Treatment 3: using biogas slurry and 100% of food supply.

- Experiment on nursing the fingerlings of Climbing perch.

  Treatment I: using biogas slurry without food supply.
  Treatment II: using biogas slurry and 50% of food supply.
  Treatment III: using biogas slurry and 100% of food supply.

Three day-old fingerlings of Snakeskin gourami and Climbing perch were nursed in the practical field of the College of Aquaculture and Fisheries. The nursing density is 400 fish/m^2. The food supply followed the procedure of fingerlings nursing of Department of Freshwater Aquaculture, College of Aquaculture and Fisheries, Can Tho University, Vietnam. All experiments were carried out in accordance with national guidelines on the protection of animals and experimental animal welfare in Vietnam (Vietnam National Assembly, 2015).

2.2 Sample Collection and Analysis

The cycle of collecting sample: water temperature, pH, dissolved oxygen (DO), total nitrogen (TN), total phosphorus (TP) and Chlorophyll-a in experiments 1 and 2 were collected and analyzed every 4 days. During the process of nursing, the fingerlings were examined every 15 days on their growth. The survival rate and productivity were also collected.

Collecting and analyzing organic fertilizer (pig manure): the pig manure of 30 days was put into PVC bags and kept at 4 °C for analysis. The analysis criteria include TN and TP concentrations.

- TN was determined by Kjeldahl method.
- TP was determined by Kjeldahl and by ascorbic acid photometric method.

Collecting and analyzing the water sample of biogas slurry: the water temperature, pH and DO were on-site measured by HANNA meter. TN and TP were analyzed at the laboratory of College of Aquaculture and Fisheries. The sample was kept at 4 °C.

- TN was determined by Kjeldahl method.
- TP was determined by Kjeldahl and by ascorbic acid photometric method.
- Chlorophyll-a concentrations were determined by acetone extraction method.

Collecting and analyzing the fingerlings sample: the sample was collected by rackets and analyzed every 15 days. The average of 30 fish/pond were kept in plastic buckets. The sample was weighted by the electronic scale (± 0.001) and then put into the ponds.

The applied formulas are as follows:

$$\text{The daily weight gain DWG}\left(\text{g / day}\right) = \frac{W_t - W_0}{t} \tag{44.1}$$

$$\text{Survival rate}\left(\%\right) = \frac{\text{Total of last fingerlings}}{\text{Total of beginning fingerlings}} \times 100 \tag{44.2}$$

$$\text{Fish yield}\left(\text{kg / m}^2\right) = \frac{\text{weight}\left(\text{kg}\right)}{\text{pond area}\left(\text{m}^2\right)} \tag{44.3}$$

$$\text{Total income}\left(\text{VND}\right) = \text{fish yield}\left(\text{kg / area}\right) \times \text{price}\left(\text{VND / kg}\right) \tag{44.4}$$

2.3 Data Analysis

All the data was collected and statistically analyzed by SPSS 13.0 software.

3 Results

3.1 *Water Quality of Biogas Slurry and Organic Fertilizer (Pig Manure)*

The experiment results are the basis for the evaluation and measurement of nutritious content. As can be seen from the tables, the amount of TN and TP in biogas slurry is so high and unstable. Samples from three analyzing turns at different times of a day showed that TN content in biogas slurry fluctuates between 69.12 and 87.32 mg/L, whereas the fluctuation in TP is between 146.60 and 194.73 mg/L. Concerning organic fertilizer, TN through the analysis accounts for 6.500–7.000 mg/g and at 6.787 mg/g on average while TP fluctuates between 0.602 to 1.306 mg/g, on average 0.958 mg/g. Therefore, this is considered valuable material resulting in lower production cost as well as higher income for farmers in the region of the Mekong delta (Tables 44.1 and 44.2).

3.2 *Pond Treatment with Biogas Slurry*

3.2.1 Water Temperature

Fig. 44.1 illustrates no wild fluctuation of water temperature among the treatments using biogas slurry in the acid sulfate-infected ponds ($p > 0.05$). The water temperature fluctuates from 27.3 to 31.2 °C, on average 29.3 °C. The figures demonstrate stable temperature of treated ponds. In comparison with previous research in the Mekong delta, there were no disadvantages for the living and developmental process of aquatic system (Pekar, Be, Dung, & Cong, 1998).

3.2.2 Water pH

The research on water in acid sulfate soil in Long Xuyen, An Giang province (Duong, Son, Oanh, Lien, & Ngoc, 1999) points out that water pH fluctuates from 4.57 to 6.35 in natural environment. After 2 years of pond treatment by using poultry manure through the integrated chicken-fish farming system, farmers have gradually applied the treated ponds in developing forms of aquaculture. According to

Table 44.1 TN and TP concentrations in biogas slurry for acid sulfate treatment

	TN (mg/L)				TP (mg/L)			
Testing time	7 h00	11 h00	15 h00	Average	7 h00	11 h00	15 h00	Average
1	67.27	84.67	55.42	69.12	119.7	194.9	125.2	146.60
2	63.52	90.47	100.17	84.72	238	176.4	169.8	194.73
3	97.67	81.37	82.92	87.32	169.1	161.2	138.1	156.13

Table 44.2 TN and TP concentrations in organic fertilizer (pig manure) for acid sulfate treatments

	TN		TP	
Sample	(mg/g)	(g/kg)	(mg/g)	(g/kg)
Pig manure 1	6.860	6.860	0.966	0.966
Pig manure 2	6.500	6.500	1.306	1.306
Pig manure 3	7.000	7.000	0.602	0.602
Average	6.787	6.787	0.958	0.958

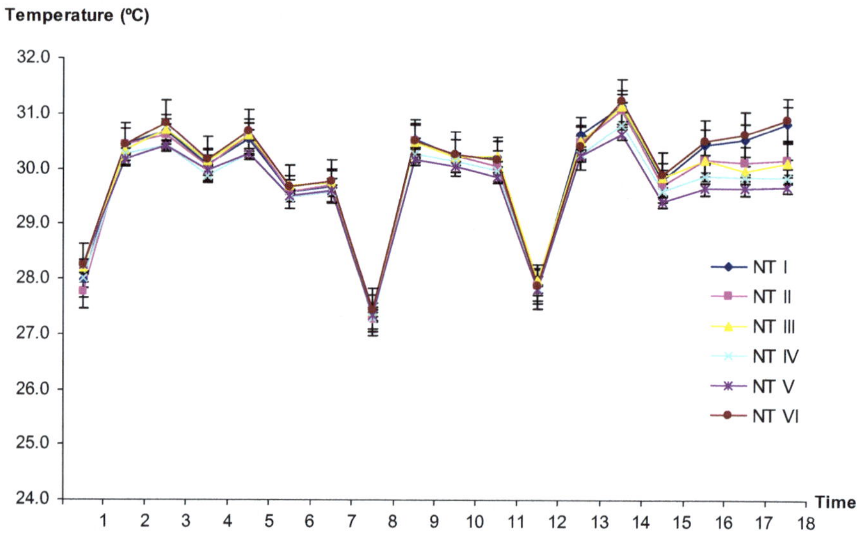

Fig. 44.1 Water temperature among the treatments

Boyd (1998), the fluctuation of 7.5–8.5 is the value of water pH suitable for the growth of plankton and fish (Fig. 44.2).

The research on biogas slurry usage in acid sulfate-infected ponds shows that water pH in experiment ponds in different treatments causes a high fluctuation. Before treatment, water pH is around 2.64–3.25. However, data from the experiments indicate a steady increase in water pH from 3.04 to 7.61 ($p < 0.05$). Therefore, it takes 45 days from the beginning of using biogas slurry in treated ponds until water pH reaches the stable value. The water pH in all treatments has been dramatically improved: treatment I of 6.35, treatment II of 6.50, treatment III of 6.29, treatment IV of 6.48, treatment V of 6.53, and control treatment VI of 4.5 significantly lower ($p > 0.05$). It can be explained that rain helps clear out acid sulfate in the infected ponds. Besides, the high level of waste from biogas slurry including decomposed organic matter causes an upward trend in TN, TP, and Chlorophyll-a concentrations in ponds. This condition creates a buffer system increasing the absorption of ion H^+ but limiting the acid sulfate formation at the water body of aquatic areas.

To sum up, water pH recorded from the treatments in fluctuation of 6.05–7.61 supports the previous studies of Boyd (1990) and Pekar et al. (1998) on the

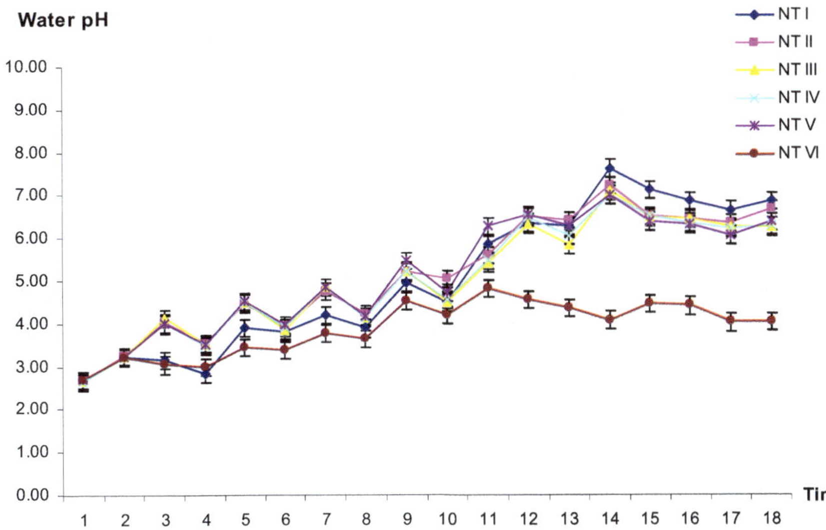

Fig. 44.2 Water pH among the treatments

development of aquaculture in naturally aquatic environment as well as in acid sulfate-infected areas at the Mekong delta.

3.2.3 Dissolved Oxygen

The analysis of dissolved oxygen in water during the treatment of acid sulfate-infected ponds is described in Fig. 44.3. The survey on the oxygen dissolved in nursing ponds shows that there is a dramatical fluctuation ($p < 0.05$) of oxygen content between untreated and treated ponds. Before the experiments, oxygen content in the treatments fluctuates between 1.47 and 2.6 ppm. At the starting time of all treatment, dissolved oxygen fluctuates between 3.07 and 6.63 ppm, the lowest value is demonstrated in treatment VI (3.07 ppm). Until the end of treatment and nursing process (the fingerlings of Snakeskin gourami and Climbing perch), dissolved oxygen in treated ponds reaches a stable value fluctuating between 4.03 and 5.20 ppm. This value totally meets the demand of DO for respiratory process and metabolism of aquatic system. However, as running the system of nursing ponds, there is a considerable fluctuation ($p < 0.05$) of DO among the treatments in the experiment. DO content varies from 1.83 to 5.57, and sharply declines ($p < 0.05$) at treatments IV (1.83–4.8 ppm) and V (1.83–4.9 ppm).

Compared with DO demand of aquatic species, the quality of water in the experiments is not very good. According to Wedemeyer (1996) and Post (1987), the demand of DO suitable for warm water fish in intensive culture is at 4 ppm. Meanwhile, Boyd (1998), Egna and Boyd (1997), Blakely and Christopher (1989), and Barnabe (1994) hold the view that suitable DO content fluctuates between 3.5

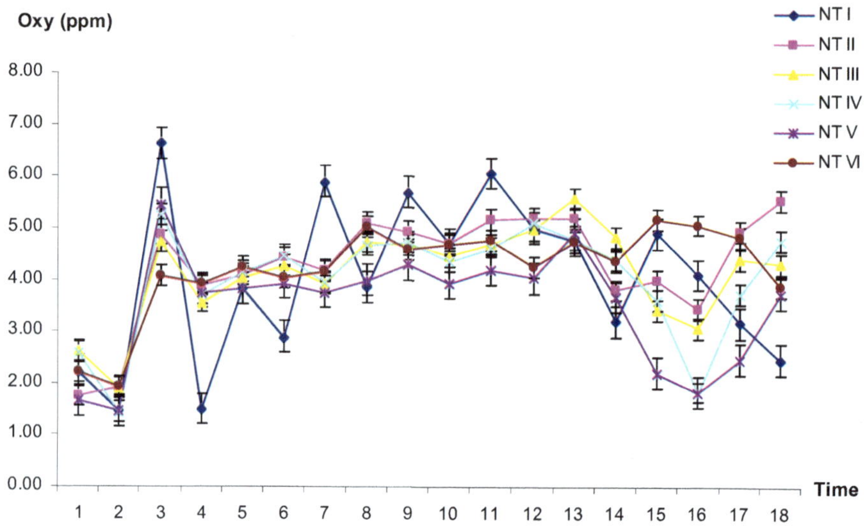

Fig. 44.3 Dissolved oxygen fluctuation in treated ponds

and 6.5 ppm. The sharp increase of TN in treatments IV and V (16.18 ppm and 15.69 ppm) is the main reason of the drop in DO in treatments IV of 125% and V of 150%. Therefore, water changing (30–50%) is an important solution to adjusting TN and maintaining the good quality of DO content in water. This method greatly improves the respiratory process and metabolism of fish nursing.

3.2.4 TN

Figure 44.4 indicates a dramatical fluctuation of TN in the treatments, especially treatments I (0.28–1.02 ppm) and VI (0.19–1.05 ppm) compared with treatments II (0.26–8.82 ppm), III (0.31–10.85 ppm), IV (0.10–16.18 ppm) and V (0.20–15.69 ppm) ($p < 0.05$). The experiment results point out that besides the advantages of inorganic and organic nutrition supply, the increase in TN causes a sharp drop in DO in treated ponds. This condition badly affects the respiratory process, metabolism as well as consumption of nursing fish. Duong and Lam (2004) found out that suitable TN content in ponds' nursing freshwater prawn is below 2 ppm. In reality, therefore, manufacturers have to adjust the waste from biogas slurry based on the water quality of nursing ponds so as to maintain the stability of water quality and improve the yield of nursing species.

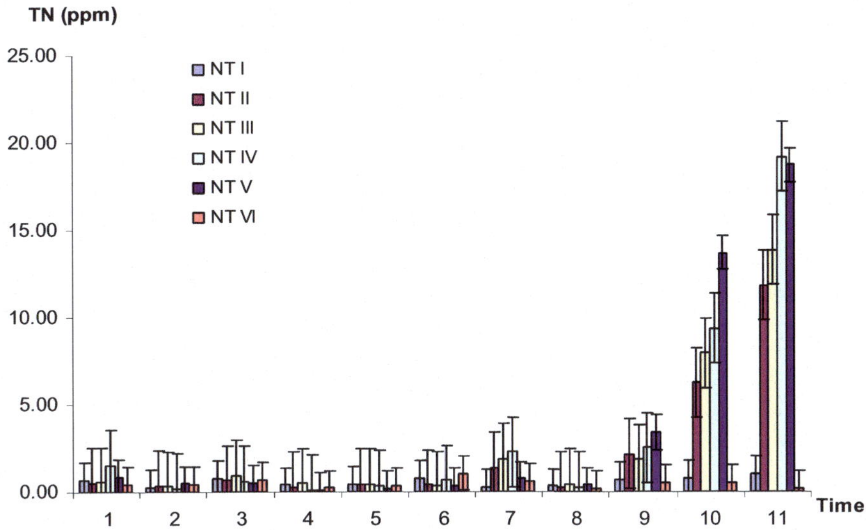

Fig. 44.4 TN fluctuation in the treatments

3.2.5 TP

Figure 44.5 shows the fluctuation of TP in six treatments using biogas slurry in acid sulfate-infected ponds for nursing. Appropriate TP concentrations are nutritious for the development of plankton building a buffer system for the stable growth of nursing species. However, the rise of TP content leads to the pollution of water (Boyd, 1990). The experiments indicate TP content in all the treatments accounts for a steadily high value of 0.01–0.98 ppm. Especially, after the treatment ended, the average content of TP varies from 0.20 to 0.95 ppm whereas it reaches a peak at 0.95 ppm ($p < 0.05$) in treatment I. In practice, 36–45 days after pond treatment, manufacturers can start nursing without using biogas slurry in the ponds unless they intend to feed some kinds of fish consuming compost.

3.2.6 Chlorophyll-a Concentration

There is a great change of Chlorophyll-a concentration ($p < 0.05$) among six treatments (Fig. 44.6). Chlorophyll-a concentration fluctuates between 1.37 μg/L and 6.90 μg/L before experiment. However, owing to the influence of organic fertilizer together with biogas slurry, Chlorophyll-a makes a difference after conduct the treatments. After 45 days, Chlorophyll-a concentration approaches a low level of 2.48–4.85 μg/L in treatments I and VI. Conversely, in treatments II, III, IV and V, Chlorophyll-a is higher from 11.07 to 11.90 μg/L. At the end of the experiment, there is a considerably high concentration of Chlorophyll-a in treatments II

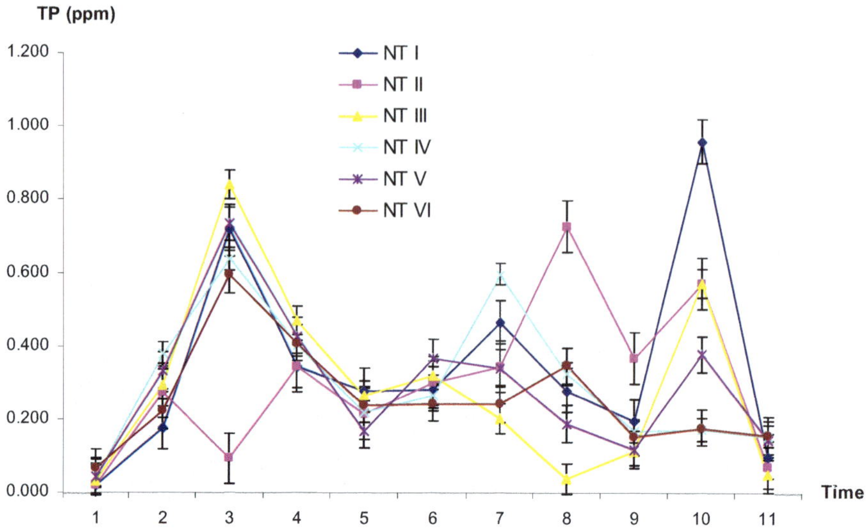

Fig. 44.5 TP fluctuation in the treatments

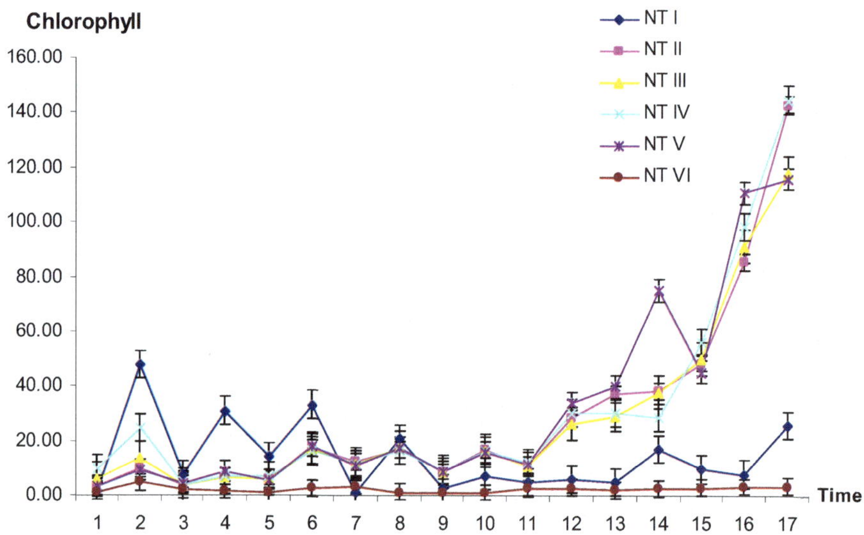

Fig. 44.6 Chlorophyll-a fluctuation in the treatments

(143.05 μg/L) and IV (144.80 μg/L) compared with a low Chlorophyll-a concentration in treatments VI (3.50 μg/L) and I (26.01 μg/L) ($p < 0.05$).

These figures indicate that biogas slurry plays a key role in providing organically and inorganically nutritious salinity, an essential part of increasing natural food supply in aquatic areas. Besides, the positive influence of biogas slurry on water quality increase in acid sulfate-infected ponds results in the improvement of water pH (2.65–6.67), oxygen (1.77–5.53 ppm), TN (0.55–8.82 ppm) and Chlorophyll-a (3.20–143.05 μg/L). This condition is caused by the organic materials covering the pond bed to limit acid sulfate release into water. The result of treatment II (75% biogas slurry use) improves water quality in the infected ponds. Therefore, the result of experiment 1 is continually applied in experiment 2.

3.3 Experiments on Fingerlings Nursing in Treated Ponds

3.3.1 The Growth of Snakeskin Gourami and Climbing Perch Fingerlings in the Nursing Process

The values from Tables 44.3 and 44.4 indicate the fluctuation of water pH (6.33–7.21) and DO (3.43–5.53 ppm) in the nursing ponds. This is such a beneficial environment that the fingerlings of Snakeskin gourami and Climbing perch can develop in treated ponds. For Snakeskin gourami (60 nursing days), the weight and weight gain in treatment 2 (3.05 g/fish and 0.051 g/day) and treatment 3 (2.94 g/fish and 0.049 g/day) are higher than those in treatment 1 (0.14 g/fish and 0.002 g/day) ($p < 0.05$). Compared with the weight gain of the fingerlings in uninfected ponds, this value is relatively lower, yet still acceptable. Moreover, the appropriate concentration of DO (2.47–5.20 ppm) along with Chlorophyll-a, a natural food supply for the fingerlings of Snakeskin gourami are the reason for the difference in their weight and growth in the two treatments (Pillay, 1990). For Climbing perch, the results showed the weight and weight gain in treatment II (2.69 g/fish and 0.045 g/day) and treatment III (2.88 g/fish and 0.048 g/day) are higher than those in treatment I (0.24 g/fish and 0.004 g/day) ($p < 0.05$). Accordingly, 100% industrial food supply with a high

Table 44.3 The growth of the fingerling of Snakeskin gourami in treated ponds

Treatments	Time (day)	Weight (g)	Daily weight gain (g/day)
		0.001	–
1 *(biogas without food supply)*	1–15	0.04 ± 0.01	0.003
	16–30	0.1 ± 0.05	0.003
	31–45	0.12 ± 0.07	0.003
	46–60	0.14 ± 0.08 [b]	0.002 [b]
2 *(biogas and 50% food supply)*	1–15	0.11 ± 0.05	0.007
	16–30	0.78 ± 0.33	0.026
	31–45	2.08 ± 0.95	0.045
	46–60	3.05 ± 0.87 [a]	0.051 [a]
3 *(biogas and 100% food supply)*	1–15	0.09 ± 0.04	0.006
	16–30	0.78 ± 0.44	0.026
	31–45	1.79 ± 1.07	0.040
	46–60	2.94 ± 0.94 [a]	0.049 [a]

Table 44.4 The growth of the fingerling of Climbing perch in treated ponds

Treatments	Time (day)	Weight (g)	Daily weight gain (g/day)
		0.001	–
I *(biogas without food supply)*	1–15	0.04 ± 0.02	0.003
	16–30	0.13 ± 0.12	0.004
	31–45	0.21 ± 0.14	0.005
	46–60	0.24 ± 0.19 [b]	0.004 [b]
II *(biogas and 50% food supply)*	1–15	0.2 ± 0.16	0.013
	16–30	0.9 ± 0.82	0.030
	31–45	2.07 ± 1.32	0.045
	46–60	2.69 ± 1.02 [a]	0.045 [a]
III *(biogas and 100% food supply)*	1–15	0.1 ± 0.04	0.007
	16–30	0.6 ± 0.5	0.020
	31–45	1.83 ± 1.37	0.041
	46–60	2.88 ± 1.27 [a]	0.048 [a]

Notes: *The characters of the same column followed by different letters are of statistical value ($p < 0.05$)*

Table 44.5 The survival rate and the fish yield of fingerlings nursing (Snakeskin gourami and Climbing perch)

Samples	Treatments (T)	Survival rate (%)	Yield (kg/1000 m^2)
Snakeskin gourami	T 1 (biogas without food supply)	15.83 ± 3.2 [b]	8.9 [b]
	T 2 (biogas and 50% food supply)	22.19 ± 2.6 [a]	270 [a]
	T 3 (biogas and 100% food supply)	22.44 ± 2.9 [a]	264 [a]
Climbing perch	T I (biogas without food supply)	10.81 ± 1.8 [c]	10.4 [c]
	T II (biogas and 50% food supply)	12.58 ± 3.2 [b]	135 [b]
	T III (biogas and 100% food supply)	14.06 ± 2.7 [a]	161 [a]

Note: *The characters of the same column followed by different letters are of statistical value ($p < 0.05$)*

protein content influences more on the growth of fingerlings nursing in treatment III (0.048 g/day) than that in treatment II (0.045 g/day).

3.3.2 The Survival Rate and Yield of Snakeskin Gourami and Climbing Perch Fingerlings

As can be seen from Table 44.5 and Fig. 44.7, there is a significant difference ($p < 0.05$) in the survival rate and the yield of Snakeskin gourami in treated ponds. In the experiment on nursing Snakeskin gourami, the survival rate (15.83%) and yield (8.9 kg/1000 m^2) in treatment 1 (biogas without food supply) are at lowest position ($p < 0.05$) compared with those in treatment 2 (22.19%, 270 kg/1000 m^2) and treatment 3 (22.44%, 264 kg/1000 m^2). However, there is no significant

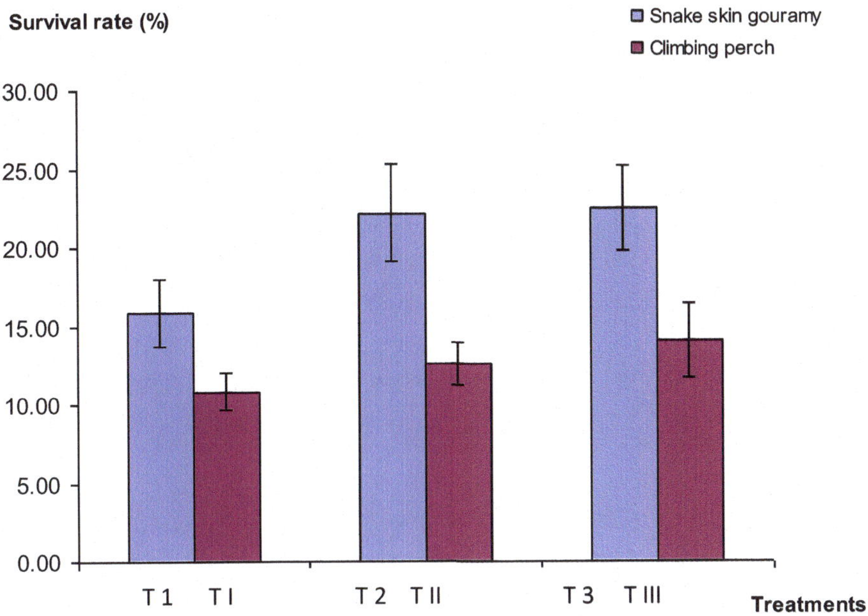

Fig. 44.7 The survival rate of the fingerlings of experimental fish in treatments

difference ($p > 0.05$) between treatment 2 (biogas and 50% food supply) and treatment 3 (biogas and 100% food supply) in terms of the survival rate and yield.

For Climbing perch, from Table 44.5 and Fig. 44.7, there is also a significant difference ($p < 0.05$) in treatments I, II, and III in terms of fish survival rate and fish yield. The highest point appears in treatment III (14.06%, 161 kg/1000 m^2), followed by the values in treatment II (12.58%, 135 kg/1000 m^2), and the lowest values is in treatment I (10.81%, 10.4 kg/1000 m^2). The difference in eating style (Climbing perch is one of omnivores) and the percentage of industrial food with high nutritious content (32–35% protein) are the main reasons for the different survival rate and fish yield as well (Pillay, 1990).

In short, after the treatment of acid sulfate-infected ponds by biogas slurry usage, manufacturers can entirely use the treated ponds for fingerlings nursing. The weight gain of fingerlings initially in treated ponds is not as high as that in uninfected ponds (pH = 7.5–8.0). However, the survival rate recorded from treatment 2 for Snakeskin gourami and treatment III for Climbing perch is quite high without much difference in comparison with the fingerlings nursing in uninfected ponds. Consequently, the treatment of 75% biogas slurry use (153 m^3/1000 m^2) will make a great condition to the fingerling nursing activities in the acid sulfate soil areas in the Mekong delta.

3.4 Analysis and Comparison of the Treatment Effectiveness of Acid Sulfate-Infected Ponds

As figured out in Table 44.6, using 75% biogas slurry is the most effective method of improving the water quality, followed by the use of organic fertilizer (pig manure). For treatment using lime, more time of water treatment is expected to maintain the stability of technical and environmental parameters that leads to the success in running the nursing ponds in the field of aquaculture.

4 Conclusions and Recommendations

4.1 Conclusions

Using biogas slurry with nutrients content of TN (69.12–87.32 mg/L) and TP (146.60–194.73 mg/L) considerably improves the water quality in acid sulfate-infected ponds for fish nursing.

Among the six tested treatments, treatment II (biogas slurry with 75% TN in organic fertilizer) after 45 days leads to good results of water pH (6.5), dissolved oxygen (5.2 ppm) and Chlorophyll-a (143,05 μg/L).

The fingerlings of Snakeskin gourami and Climbing perch grow well in treated ponds after 45 days by using biogas slurry with water pH (6.33–7.21) and DO (3.43–5.53 ppm). For Snakeskin gourami of 60 nursing days, the weight and weight gain in treatment 1 (0.14 g/fish, 0.002 g/day) are lower than those in treatment 2 (3.05 g/fish, 0.051 g/day) and treatment 3 (2.94 g/fish, 0.049 g/day). In treatment 2, the survival rate and yield (22.19%, 270 kg/1000 m^2) are equivalent to those in treatment 3 (22.44%, 264 kg/1000 m^2). For Climbing perch, there is a difference ($p < 0.05$) in the weight as well as weight gain in treatment I (0.24 g/fish, 0.004 g/day) compared with those in treatments II (2.69 g/fish, 0.045 g/day) and III (2.88 g/

Table 44.6 The effectiveness of the treatments (1000 m^2)

No.	Used materials	Treatments Lime (CaO) (kg/1000 m^2)	Organic fertilizer (kg/1000 m^2)	75% biogas slurry (m^3/1000 m^2)
1	Lime (CaO)	900	900	900
2	Organic fertilizer	–	1.800	–
3	Biogas slurry	–	–	115 m^3
4	Total	VND 1,800,000	VND 3,600,000	VND 1,890,000

Note: *1 kg organic fertilizer (pig manure) = 6.79 g; TN = 85 L biogas slurry; 1 L biogas slurry = 0.08 g TN lime = VND 2000/kg, pig manure = VND 1000/kg, cost of water pumping = VND 15,000/20 m^3.*

fish, 0.048 g/day). The highest values of the survival rate and fish yield appear in treatment III (14.06%, 161 kg/1000 m^2).

The treatment of using 75% biogas slurry (153 m^3/1000 m^2) has brought great benefits to the maintenance of good water quality and nursing techniques, increasing the income to local communities as well.

4.2 Recommendations

Owing to the continual change of nutritious concentrations in biogas slurry, the treatment on acid sulfate-infected ponds needs repeated experiments in order to lay out the foundations of the treatment process with biogas slurry use.

References

Barnabe, G. (1994). *Aquaculture biology and ecology of cultured species* (403 p). Ellis Horwood.

Blakely, D. R., & Christopher, T. H. (1989). *Inland aquaculture development handbook* (183 p). St Edmundsbury Press Ltd.

Boyd, C. E. (1990). *Water quality in ponds for aquaculture* (482 p). Birmingham Publishing.

Boyd, C. E. (1998). *Water quality for pond aquaculture* (37 p). International Center for Aquaculture and Aquatic Environments Alabama Agriculture. Experiment Station Auburn University.

Duong, N. L., & Lam, M. L. (2004). *The integrated aquaculture system* (p. 87). Can Tho University. (in Vietnamese).

Duong, N. L., Son, V. N., Oanh, D. T. H., Lien, N. T. K., & Ngoc, L. B. (1999). *Assessment of the development of aquatic fauna and flora in problem soil areas in relation with aquaculture aspects. Scientific report* (p. 28). Can Tho University.

Duong, V.N., Ton, P.T., and Edward M. (2001). *Integrating Melaleuca cajuputi with rice in an agro-forestry trial on acid sulphate soil in the Mekong Delta, Vietnam*. Retrieved July 21, 2018, from https://www.researchgate.net/publication/308118285.

Edwards, P., Kaewpaitoon, K., McCoy, E. W., & Chantachaeng, C. (1986). *Pilot small-scale crop/livestock/ fish integrated farm. Asian Institute Technology Research Report No.*, 184, 131 p.

Egna, H. S., & Boyd, C. E. (1997). *Dynamics of pond aquaculture* (411 p). Lewis Publishers in an imprint of CRC Press.

Le, H. V., & Nguyen, V. C. N. (2015). *Organic waste management and re-use*. Can Tho University Publishing House. (in Vietnamese).

Pekar, F., Be, N. V., Dung, D. T., & Cong, N. V. (1998). *The eco-technological analysis of fish farming households in the Mekong delta, Vietnam. WES scientific report* (p. 16). Can Tho University.

Pillay, T. V. R. (1990). *Aquaculture: Principles and practices* (575 p). Fishing News Books, A Division of Blackwell Publishing Ltd, Oxford.

Post, G. B. (1987). *Textbook of fish diseases* (288 p). TFH Publications.

Singh, V. P. (1985). *Management and utilization of acid sulfate soils for aquaculture* (105 p). University of the Philippines in the Visayas.

Vietnam National Assembly (2015). *Law on Animal Health No. 79/2015/QH13*.

Wedemeyer, G. A. (1996). *Physiology of fish in intensive culture systems* (225 p). ITP International Thompson Publishing.

Chapter 45
Assessment of Saltwater Intrusion Vulnerability in the Coastal Aquifers in Ninh Thuan, Vietnam

Quy Nhan Pham, Thi Thoang Ta, Thanh Le Tran, Thi Thu Pham, and The Chuyen Nguyen

Abstract As surface water has declined under climate change conditions and human activities, groundwater has become the most important source of water supply for Ninh Thuan province, which is one of the most arid regions of Vietnam. The increase of groundwater extraction and sea-level rise has brought about increase in seawater intrusion, which has become the biggest concern for the groundwater resources of this coastal area; thus, it is necessary to evaluate the saltwater intrusion vulnerability in aquifers for planning and sustainable exploitation of groundwater in the province. In this study, an overview of the previous fresh–saltwater interface studies and updated geophysical investigation (Vertical Electrical Sounding – VES) was carried out to understand the situation or status of the salinity of groundwater. Besides, chemical compounds of water sources were also analysed to clarify the recharge sources and origin of saltwater in the aquifers. Based on collected and field survey data, GALDIT method was used to assess the saltwater intrusion vulnerability in the coastal aquifers of the study area with six factors incorporated: groundwater occurrence, aquifer hydraulic conductivity, depth to groundwater level, distance from shore, impact of existing seawater intrusion, and aquifer thickness. The weight of each GALDIT parameter was calculated by analytic hierarchy process (AHP) based on the results of surveys of over 50 hydrogeological experts in Vietnam to calculate the GALDIT index for the final saltwater intrusion vulnerability map of Ninh Thuan coastal aquifers.

Keywords Seawater intrusion · GALDIT method · climate change · groundwater extraction · Ninh Thuan province

Q. N. Pham (✉) · T. T. Ta · T. Le Tran
Hanoi University of Natural Resources and Environment, Hanoi, Vietnam
e-mail: pquynhan@hunre.edu.vn

T. T. Pham · T. C. Nguyen
Center for Water Resources Quality and Protection (WARAPO), Hanoi, Vietnam

A. T. Nguyen, L. Hens (eds.), *Global Changes and Sustainable Development in Asian Emerging Market Economies Vol. 2*, https://doi.org/10.1007/978-3-030-81443-4_45

1 Introduction

The central coastal plains of Vietnam are narrow and small area plains that were filled by sediments of short and steep rivers cross the region, so central coastal aquifers of Vietnam in general and Ninh Thuan coastal aquifer in particular characterized by small and scattered in distributed areas and thickness. Saltwater intrusion in aquifers has become the biggest problem to human life in this area where limited surface water and groundwater are often the main sources of drinking water and agriculture activities. Besides the investigation to assess the status or occurrence of saltwater intrusion in the aquifer system, the assessment of saltwater vulnerability may also helpful for management of groundwater source in the area.

There are a number of methods to assess vulnerability caused by the continuous extent of seawater intrusion in coastal aquifers in which the GALDIT index model developed by Chachadi and Ferreira in 2001 was widely used over the world based on rating scores and weights of six distinct influencing hydrogeological parameters as follows: (1) the groundwater occurrence G; (2) the aquifer hydraulic conductivity A; (3) level of the groundwater level above sea level L; (4) the distance from shoreline D; (5) the impact of existing status of seawater intrusion in the area I; and (6) thickness of the mapped aquifer T. The weights of GALDIT parameters are assigned value from 1 (the least significant parameter) to 4 (the most significant parameter) from the statistics of survey results of scientists and experts (Chachadi and Ferreira, 2001).

In this study, an overview on occurrence of saltwater intrusion was carried out based on previous investigation results, geophysical surveys (VES measurements), and water sample analysis. Then, the GALDIT parameters weights were modified by analytic hierarchy process (AHP) method that is based on questionnaire survey results of over 50 hydrogeological experts in Vietnam, and the range of each factor rating is slightly changed to meet the condition of coastal unconsolidated Quaternary aquifers of Ninh Thuan province. Accordingly, GALDIT index mapping obtained with the factors values gathered from previous studies and this study, and the modified weights and parameter ratings.

2 Study Area

Ninh Thuan is located in the south central coast of Vietnam lying between latitude of 11°18'28" and 11°50'52" North, and longitudes of 108°45'19" and 109°13'54" East with total area of 3358 km^2. It is bounded by Khanh Hoa, Lam Dong, and Binh Thuan provinces to the North, West and East respectively, and faces the sea in the East. The province has three types of topography as mountains, hills, semi-mountains, and coastal plain in which the hill and mountain occupy 77.6% of the total land with the elevation ranges from 200 to 1000 meters, and the coastal plain accounts only for 22.4% of the total province area. The province has a tropical monsoon climate characterized by hot dry and windy weather with temperature ranges between 24.7 and 29.1 °C. There are two distinct seasons, that is, the rainy season lasts for 3 months from September to November which contributes to total of 40% to 80% annual rainfall, and the dry season lasts from December to August. The average annual potential evaporation is 1844.1 mm (around 153.66/month) and the highest evaporation rate is in January (about 166.2 mm) while the average annual rainfall is about 1131.4 mm (i.e. about 50% evaporation); this makes Ninh Thuan one of the most arid regions of Vietnam making groundwater resources a cause of concern in this province.

The aquifer system of Ninh Thuan coastal plain consists of Quaternary unconsolidated aquifers, fracture igneous rock and sedimentary rock aquifers formed in between Middle Jurassic and Pleistocene era. The fracture aquifers have very low productivity and are distributed at great depth; limited study has been carried out to investigate the quality and quantity of those aquifers and they are not the subject of this study as well. The Quaternary unconsolidated aquifers consists of Pleistocene, Holocene, and undivided Quaternary aquifer in which the undivided Quaternary aquifer is distributed in transition zone between the mountainous area and the plain with distributed area of around 65 km^2; the Holocene aquifer is widely distributed in Phan-Rang plain and along Cai river valley with total area of 315 km^2, and the Pleistocene aquifer is found at Tan Hai ward, the southern area of Phan Rang plain, and from Phuoc Hoa to Quang Sơn ward with a total area of 364 km^2 (Nguyen Truong Giang et al., 1998).

*According to Ngo Tuan Tu and Nguyen Ton (*2015*), seawater was intruded to aquifer system of Ninh Thuan coastal plain due to current sea water throughout a long coastline of around 105 km and ancient seawater in depth aquifers through tectonic faults and groundwater exploitation activities. The distribution of saline groundwater depends on the aquifer system characteristics and the natural conditions: (i) the saline groundwater can be found over the whole aquifer system; (ii) saline groundwater can be found on top of aquifer system (Holocene aquifer); (iii) the saline groundwater can be found in the lower part of aquifer system (Pleistocene aquifer). Total complete saline groundwater area is around 280* km^2*, and partly saline groundwater area is around 420* km^2.

In this study, the current saltwater intrusion status and the origin was investigated by 16 survey lines of vertical electrical sounding (VES) measurements in which 15 survey lines were arranged perpendicular to the coastlines in northwest-southeast direction; only survey line No.16 was arranged parallel to the coastlines in north-south direction. Forty -ive groundwater samples were collected during sampling trip for chemical analysis (both conventional and stable isotopic analysis). The ES investigation was arranged in the area of current saltwater intrusion; the results showed that the fresh–salt water interface moved toward the land at the distance up to 13 km from the shoreline and the saltwater covered almost length of all investigated lines of both Holocene and Pleistocene aquifers. The arrangement of VES survey lines, the fresh–salt (F-S) water interface and groundwater samples are presented in Fig. 45.1.

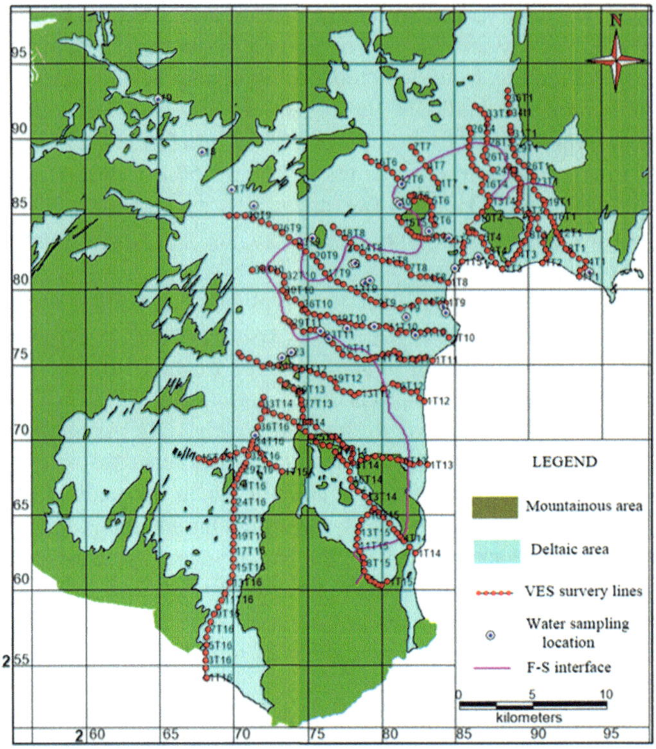

Fig. 45.1 ES surveys arrangement, F-S interface, and water sampling locations in Ninh Thuan coastal plain

One example of salt–fresh water interface analysis of the survey line No.1 can be seen in Fig. 45.2. It is clear that the saline water covered whole survey line No.1 except some part of the bedrock.

The chemical analysis showed that water samples in 40 out of 45 locations to be saltwater, and the origin of saltwater differs from place to place depending on the hydrogeological structures, groundwater flow regime; the origin can be from current saltwater intrusion found at the coastal area or from ancient seawater found in the inland area.

3 Saltwater Intrusion Vulnerability

According to Chachadi and Ferreira (2001), the final decision criteria of vulnerability to saltwater intrusion are based on GALDIT index value; the higher the vulnerability index value, the more vulnerable the aquifer is to seawater intrusion, and it

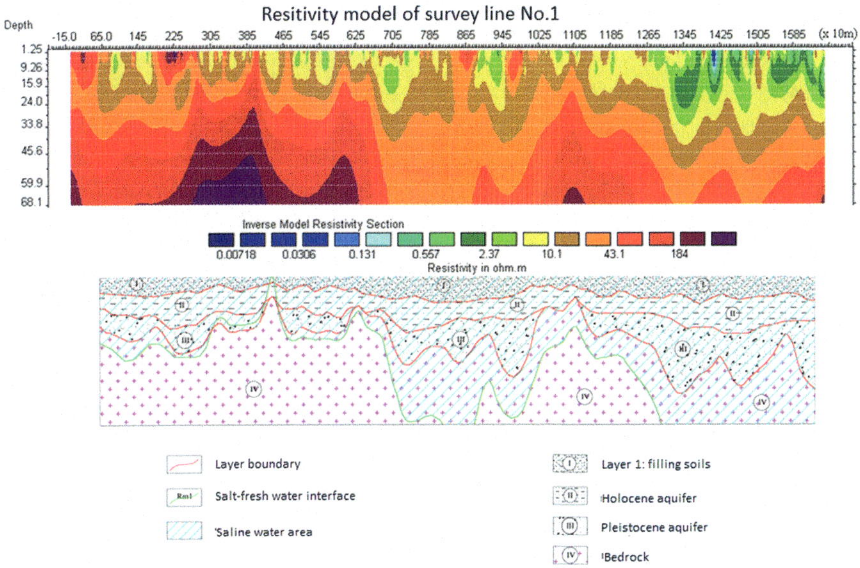

Fig. 45.2 Salt–fresh water interface analysis for VES survey line No.1

was obtained by evaluating each factor score and summing them in the following formula:

$$\text{GALDIT index} = \frac{\sum_{i=1}^{6} \text{WiRi}}{\sum_{i=1}^{6} \text{Wi}} \tag{45.1a}$$

or

$$GALDIT_{Index} = \frac{(W_1 \times G) + (W_2 \times A) + (W_3 \times L) + (W_4 \times D) + (W_5 \times I) + (W_6 \times T)}{\sum_{i=1}^{6} W_i} \tag{45.1b}$$

where:

W_i is the weight and R_i is the importance rating of the ith factor.
G is the groundwater occurrence.
A is the aquifer hydraulic conductivity.
L is the level of the groundwater level above sea level.
D is the distance from shoreline.
I is the impact of existing status of seawater intrusion in the area.
T is thickness of the mapped aquifer.

3.1 Calculating the Weight of GALDIT Parameters

The weight of each GALDIT parameter was calculated using AHP method that was proposed by Thomas Saaty (1980) on the basis of pairwise relative evaluation of both the criteria and the decision-maker's experience. The process of AHP method and its application to calculate GALDIT parameter for Ninh Thuan coastal aquifers are described as follows.

The criteria weight vector is calculated based on the pairwise comparison matrix A, the m x m real matrix with a_{ij} entry of the matrix A represents the importance with respect to the goal of the ith parameter relative to the jth parameter. In this study, pairwise comparisons were conducted to six GALDIT parameters with effect to seawater intrusion based on survey results of 56 hydrogeological scientists and experts in Vietnam (the pairwise comparison results was presented in Table 45.1), and accordingly the A matrix for six GALDIT parameters with effect to seawater intrusion was built, as presented in Table 45.2.

A normalized pairwise comparison matrix, A*norm,* was constructed with each entry b_{jk} a result of dividing each element in every column by the sum of that column as presented in eq. (45.2), and based on A*norm* matrix, the weight of each parameter was calculated as Eq. (45.3):

$$b_{jk} = \frac{a_{ij}}{\sum_{j=1}^{m} a_{ij}} \tag{45.2}$$

Table 45.1 Pairwise comparison results for GALDIT parameters of Ninh Thuan coastal aquifer

Pairwise	Compared result	Interpretation
G and A	1/3	G is slightly less important than A
G and L	1	G and L are equally important
G and D	1/3	G is slightly less important than A
G and I	1/7	G is strongly less important than I
G and T	1	G and T are equally important
A and L	3	A is slightly more important than L
A and D	5	A is more important than D
A and I	1	A and I are equally important
A and T	5	A is more important than T
L and D	1/3	L is slightly less important than D
L and I	1/3	L is slightly less important than I
L and T	3	L is slightly more important than T
D and I	1	G and I are equally important
D and T	3	D is slightly more important than T
I and T	5	I is more important than T

Table 45.2 Pairwise comparison matrix for GALDIT parameters of Ninh Thuan coastal aquifer

Parameters	G	A	L	D	I	T
G	1.000	0.333	1.000	0.333	0.200	1.000
A	3.000	1.000	3.000	5.000	1.000	5.000
L	1.000	0.333	1.000	0.333	0.333	3.000
D	3.000	0.200	3.000	1.000	1.000	3.000
I	5.000	1.000	3.000	1.000	1.000	5.000
T	1.000	0.200	0.333	0.333	0.200	1.000
Sum	**14.000**	**3.067**	**11.333**	**8.000**	**3.733**	**18.000**

Table 45.3 The normalized pairwise comparison (A*norm*) matrix and the calculated weight of GALDIT parameters for Ninh Thuan coastal aquifer

Parameter	G	A	L	D	I	T	**Calculated weight**
G	0.071	0.109	0.088	0.042	0.054	0.056	**0.42**
A	0.214	0.326	0.265	0.625	0.268	0.278	**1.98**
L	0.071	0.109	0.088	0.042	0.089	0.167	**0.57**
D	0.214	0.065	0.265	0.125	0.268	0.167	**1.10**
I	0.357	0.326	0.265	0.125	0.268	0.278	**1.62**
T	0.071	0.065	0.029	0.042	0.054	0.056	**0.32**

$$w_i = \sum_{j=1}^{m} b_{jk} \tag{45.3}$$

where: a_{ij} is the entry of matrix A; b_{jk} is the entry of matrix A*norm*. w_i is weight of ith parameter; m is number of comparison parameter

The A*norm* matrix and the weight of GALDIT parameters for Ninh Thuan coastal aquifer were calculated and presented in Table 45.3.

To check the consistency of the calculated weight, the consistent index (CI) and RI (random index) ratio was calculated; If $CI/RI < 0.1$, the result is reliable. The CI is calculated by eq. (45.4) and RI values for small problems ($m \leq 10$) are shown in Table 45.4.

$$CI = \frac{x - m}{m - 1} \tag{45.4}$$

Where: x equal to average value of the weighted sum value divided to the calculated weight.

m is number of compared parameter.

The weighted sum value is sum of each value in the row of a multiplied weight matrix that is constructed by taking multiple of each calculated weight to the A matrix column.

The calculation of consistency for GALDIT parameter weight of Ninh Thuan aquifer is presented in Table 45.5.

Table 45.4 Values of the random index (RI) for small problems (Thomas Saaty, 1980)

m	2	3	4	5	6	7	8	9	10
RI	0	0.58	0.90	1.12	1.24	1.32	1.41	1.45	1.51

Table 45.5 Multiplied weight matrix and consistency for the weights of GALDIT parameters

Parameter	G	A	L	D	I	T	Weighted sum value	Calculated weight		CI
G	0.419	0.659	0.566	0.368	0.324	0.317	2.652	**0.42**	6.327	
A	1.257	1.976	1.698	5.519	1.619	1.584	13.653	**1.98**	6.910	
L	0.419	0.659	0.566	0.368	0.540	0.951	3.502	**0.57**	6.187	
D	1.257	0.395	1.698	1.104	1.619	0.951	7.023	**1.10**	6.363	
I	2.096	1.976	1.698	1.104	1.619	1.584	10.076	**1.62**	6.225	
T	0.419	0.395	0.189	0.368	0.324	0.317	2.011	**0.32**	6.348	
x value									6.394	0.079

As the CI = 0.079 < 0.1, the calculation of GALDIT parameter weight is reasonably consistent so that the result can be used to process calculation of GALDIT index for this study.

3.2 *Vulnerability Map of Ninh Thuan Coastal Area*

The indicator variables of ratings for GALDIT parameters were slightly modified based on the questionnaire surveys and summary with calculated weights as presented in Table 45.6.

Vulnerability map was constructed based on GALDIT index that was calculated using new GALDIT factor weight, and the data gathered from previous study (type of aquifer, hydraulic conductivity, height of groundwater level above mean sea level, distance from shore, thickness of the aquifer) and this study data (status of seawater intrusion) are the basis to assign the ratings of each GALDIT factor. The final result showed that the vulnerability of Ninh Thuan unconsolidated aquifers can be classified into three levels: the high vulnerability, medium vulnerability, and low vulnerability and the distribution is around 40%, 50 and 10% to the high, medium and low vulnerability, respectively. The high vulnerability was found in the central coastline and extends landward along with Cai river where the most concentrated population and economic activities are seen as shown in Fig. 45.3.

Table 45.6 Calculated weights and ratings for GALDIT parameters

Factor	Weight	Indicator variables		Rating
Groundwater occurrence / aquifer type, (G)	0.42	Confined		10
		Unconfined		7.5
		Leaky confined		5
		Bounded		2.5
Aquifer hydraulic conductivity, (A) (m/day)	1.98	High	>40	10
		Medium	10–40	7.5
		Low	5–10	5
		Very low	<5	2.5
Height of groundwater level above mean sea level, (L) (m)	0.57	High	<1	10
		Medium	1.0–1.5	7.5
		Low	1.5–2.0	5
		Very low	>2.0	2.5
Distance to the shore, (D) (m)	1.1	Very small	<2500	10
		Small	2500-5000	7.5
		Medium	5000-10,000	5
		Far	>10,000	2.5
Impact of the existing status of seawater intrusion, (I)	1.62	High	>2.0	10
		Medium	1.5–2.0	7.5
		Low	1.0–1.5	5
		Very low	<1.0	2.5
Thickness of aquifer (saturated), (T) (m)	0.32	High	>10	10
		Medium	7.5–10	7.5
		Low	5–7.5	5
		Very low	<5	2.5

4 Conclusions

Saltwater intrusion has always been a serious concern for coastal areas. Therefore, the need of mapping groundwater vulnerability to saltwater intrusion is important in order to plan and manage this precious resource. Six factors in GALDIT method incorporated allow us to assess, numerically rank and map areas from low to high vulnerability. For more precise vulnerability map, more data is required. For specific water use such as irrigation, industry etc. important rating is required to modify to suit this purpose.

Acknowledgement This work was supported by Science and Technology Program in Climate change adaptation and Resources management and environment (BDKH/16-20), Ministry of Natural Resources and Environment under project "Investigation of saltwater intrusion mitigation strategies for the Central coastal aquifers of Vietnam in the context of climate change: an application to Ninh Thuan province-BDKH16/16-20".

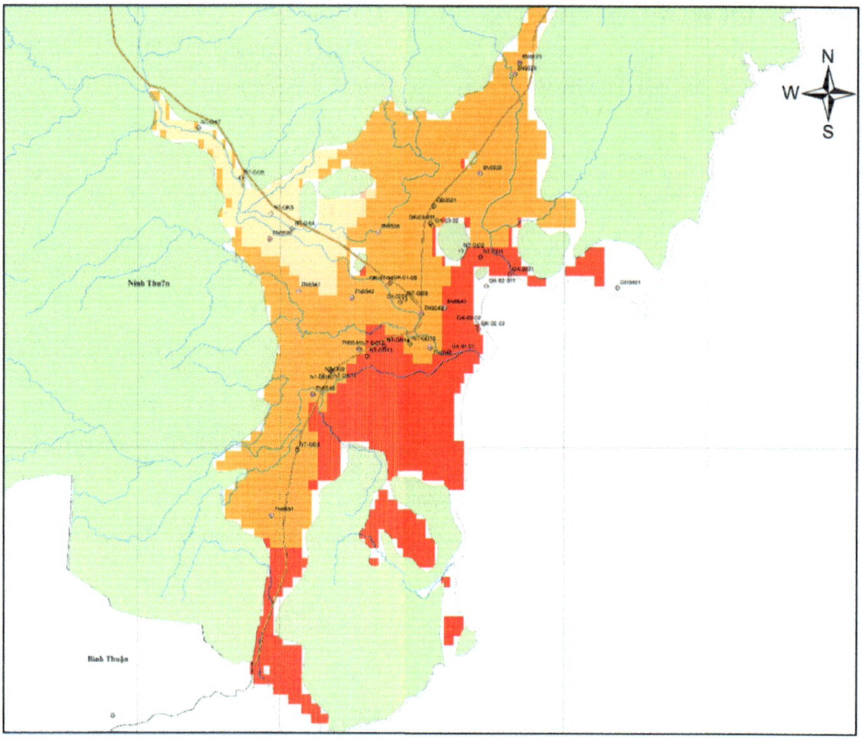

Fig. 45.3 Vulnerability map of Ninh Thuan coastal aquifers

References

Chachadi, A.G., Ferreira, L.J.P. (2001). Assessing aquifer vulnerability to sea-water intrusion using GALDIT method: Part 2 - GALDIT indicators description.

Giang, N. T., Nghiep, V. C., On, D. H., & Tran, V. N. (1998). *Groundwater of the south central Coastal Plains*. Department of Geology and Mineral of Vietnam, Ministry of Industry. (in Vietnamese).

Saaty, T. L. (1980). *The analytic hierarchy process*. McGraw-Hill.

Tu, N. T., & Ton, N. (2015). *Report on mapping of hydrogeological scale 1:50.000 in NinhThuan-BinhThuan area*. Division for water resources planning and investigation for Central region of Vietnam. (in Vietnamese).

Chapter 46
Integrating Climate Change Adaptation into Urban Planning of Vietnamese Coastal Towns toward Sustainable Development

Pham Thanh Huy

Abstract This chapter deals with the content of urban planning and sustainable development for Vietnamese coastal towns adapting to climate change. We use some methodologies of the research such as specialist method, method of investigation and assessment, etc. in order to summarize, investigate, and analyze several existing urban structures of these coastal towns in the context of climate change, then classify these towns into some groups of urban characteristic in the research location like seafront towns, littoral towns and coastal mangrove towns with some different criteria. The chapter also analyzes the impacts of climate change on urban structure and vice versa, integrating into sustainable urban development. As a result, the methodology of systematization models is used in order to propose three models of urban spatial structure and solutions of urban land use adapting to climate change for the coastal towns toward sustainable development.

Keywords Urban planning · Climate change · Adaptation · Sustainable development · Vietnamese coastal towns

1 Introduction

Climate change, which is the most challenging factor for the living environment of human, has recently impacted heavily on global temperatures. The climate change, with sea level rise, has affected seriously the process of urban development of Vietnam's coastal areas. The process of urbanization and the expansion activities of residences into the areas of natural disasters have many risks while infrastructure has still not been developed for urban development. The process of development of

P. T. Huy (✉)
Institute of Tropical Architecture, Hanoi Architectural University, Hanoi, Vietnam

A. T. Nguyen, L. Hens (eds.), *Global Changes and Sustainable Development in Asian Emerging Market Economies Vol. 2*, https://doi.org/10.1007/978-3-030-81443-4_46

Vietnamese coastal cities has still lacked in mainstreaming climate change into urban planning.

According to the Orientation for development of the Vietnamese urban system to 2025 and the vision to 2050 under Decision No. 445/QD-TTg, it is predicted that population is approximately 52 million of urban people, accounted for 50% of Vietnam population and reach at 1000 of cities to 2005 year (Vietnam Government [VG], 2009a, 2009b, p. 2). There are six urban zones that are distributed based on six national economic-social regions, including Northern Midland and Mountainous, Red River Delta, North Central and Central Coastal region, Central Highlands, South East and Mekong Delta. The Mekong Delta that consists of one city (Can Tho city) and 11 provinces: Dong Thap, Vinh Long, Ben Tre, Tra Vinh, An Giang, Tien Giang, Hau Giang, Soc Trang, Kien Giang, Bac Liêu, and Ca Mau under Decision No. 1581/QD-TTg on approving the Regional Planning for the Mekong Delta to 2020 year, the vision to 2050 (Vietnam Government [VG], 2009a, 2009b). The Mekong Delta is identified to become the one sixths of Vietnam's main urbanization zones. These urban systems would be developed as concentrated-multipolar model that integrates with the economic corridors of cities along the systems of principal rivers and head roads.

The research location and scope include 15 coastal cities and towns in the Western zone of the Mekong Delta (belongs to the Western coastal zone from Ca Mau Cape to Ha Tien town of Kien Giang province and Ca Mau province) (Fig. 46.1). Kien Giang province includes Rach Gia city, and towns: Ha Tien, Kien Luong, Hon Dat, Minh Luong, An Bien and An Minh. Ca Mau province includes Ca Mau city, and towns: U Minh, Song Doc, Tran Van Thoi, Cai Doi Vam, Cai Nuoc, Nam Can and Rach Goc. This research location is also the zone No.7 of the Climate Change, Sea Level Rise Scenarios for Vietnam published in 2012 (MONRE, 2012), which is the heaviest impacted by sea level rise in Vietnam. These 15 coastal cities can be easily vulnerable, harmful economy and affects in urban sustainable development if they are not adjusted the process of urban planning adapting to climate change and integrated with the Climate Change, Sea Level Rise Scenarios for Vietnam following to each stage of the next years (2030, 2050, etc.). Therefore, it is necessary to have an effective method for urban planning adapting to climate change for the master plan of the cities. The contents of urban master plan propose the solutions to the towns adapting to climate change, such as urban structure, spatial planning, land use planning and infrastructure planning.

The object of the research is urban planning, mainly the type of master plan, particularly concentrating to create models of spatial structure for the case study.

These cities and towns, which have three identities, are classified into three groups of seafront towns, littoral towns and coastal mangrove towns by each category of criteria (Table 46.1).

The process of development of Vietnam coastal cities has still lacked in mainstreaming climate change into urban planning, and the expanding activities of residences into the areas having risks of natural disasters contain many risks while infrastructure has still not developed for urban development (MONRE, 2012).

Fig. 46.1 The system of coastal cities and towns in the western zone of the Mekong Delta

Table 46.1 Criteria of classification for the seafront towns, littoral towns, and coastal mangrove towns

Urban locations	Named	Significant identities	Distances from the coastline
Seafront towns	Ha Tien, Kien Luong, Rach Gia, and song Đoc	With or without the river mouth	Approach to intertidal area
Littoral towns	Ca Mau, hon Đat, minh Luong, an Bien, an minh, U minh, and Tran Van Thoi	Higher ground than seafront towns, crowded population	Higher 10 km far from the coastline
Coastal mangrove towns	Cai Đoi Vam, Cai Nuoc, Nam can, and Ngoc Hien	Into Ca Mau coastal mangrove forest	From coastline to inside Ca Mau coastal mangrove forest

Source: Pham Thanh Huy (2016)

The impacts of climate change are likely to affect urban spaces as flooding, erosion, land decreasing, infrastructure and ecosystem destroying (VIUP, 2015) (Table 46.2). Climate change can also impact urban spatial structure, such as urban center and system of public services, residences, green space, industrial zones, transportation and infrastructure, in some aspects. Climate change affects individual town and each group of towns in the coastal zone of the Vietnam's South West.

Table 46.2 The summarization of climate change risks in Kien Giang and Ca Mau

Zones	Climate change phenomenon						
	Storms	Flooding	Coastal erosion	Highland erosion	Sea level rise & tides	Drought	Salinization
Kien Giang	++	+++	+++	o	+++	++	+++
Ca Mau	++	o	+++	o	+++	++	+++

Notes: +++ max impacts; ++ medium impacts; + minimum impacts; o without impacts
Source: VIUP (2015)

Table 46.3 Sea Level Rise Scenarios for Vietnam in 2012 at medium level (measuring unit: centimetres)

Zones	Each stage in twenty-first century								
	2020	2030	2040	2050	2060	2070	2080	2090	2100
From Ca Mau province to Kien Giang province	9–10	13–15	19–22	25–30	32–39	39–49	47–59	55–70	62–82

Source: MONRE (2012)

The research location is in the West of the Mekong Delta zone, which is predicted to be heavily impacted by sea level rise in Vietnam. This also is in the zone No.7 area of the Climate Change, Sea Level Rise Scenarios for Vietnam (MONRE, 2012), following to each stage of the next years (2030, 2050, etc.) (Table 46.3).

In this context, however, the cities and towns have significantly developed economic aspects, and therefore, these towns have not still considered toward sustainable development and climate change adaptation. The existence of these cities and towns is assessed following to each group.

1.1 The Group of Seafront Towns

The seafront towns include Rach Gia, Ha Tien, and Kien Luong (Kien Giang province); Song Doc (Ca Mau province). These towns symbolize some urban structures' characteristic as urban land runs along coastline, urban center is at river mouth, the main roads run along coastline and riverside (Fig. 46.2).

1.2 The Group of Littoral Towns

The littoral towns consist of Ca Mau, U Minh, and Tran Van Thoi (Ca Mau province); Hon Dat, Minh Lương, An Bien, and An Minh (Kien Giang province). These towns symbolize some urban structures' characteristic as locating approximately 10 km from coastline, the higher ground than seafront towns, urban center gather at

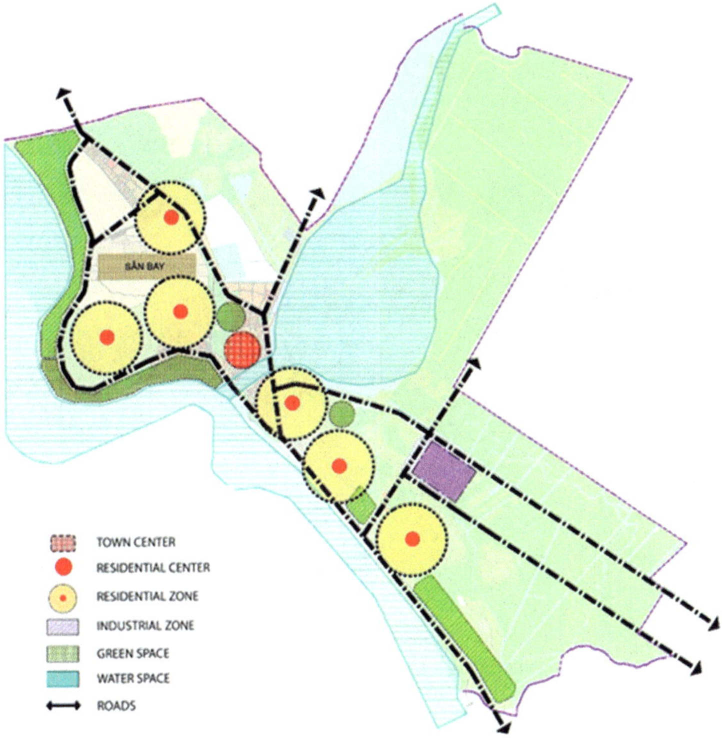

Fig. 46.2 Ha Tien's urban structure of the master plan in 2008 (KGG, 2008)

the interaction of rivers and main roads with surrounding urban functional areas (Fig. 46.3).

1.3 The Group of Coastal Mangrove Towns

The coastal mangrove towns consist of Cai Doi Vam, Cai Nuoc, Nam Can, and Ngoc Hien towns located in the Ca Mau mangrove system. These towns symbolize some urban structures' characteristic as locating into Ca Mau coastal mangrove forest, high density of canals, urban centers expand in each interaction of main canals with main roads, urban functional areas surround these interactions (Fig. 46.4).

Most of the urban planning in the research location, especially master plans, implemented from 2000 to present have not integrated with climate change in urban structure, land use, green space, and infrastructure. Therefore, it is difficult to adapt to climate change if local governments still implement these master plans. There are some problems in research and application of urban planning for the western zone of the Mekong Delta:

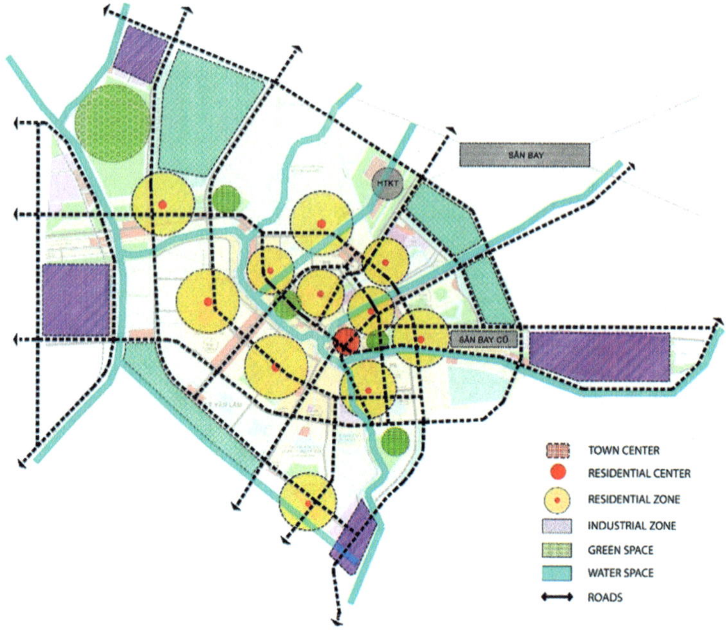

Fig. 46.3 Ca Mau's urban structure of the master plan in 2008 (CMG, 2008a)

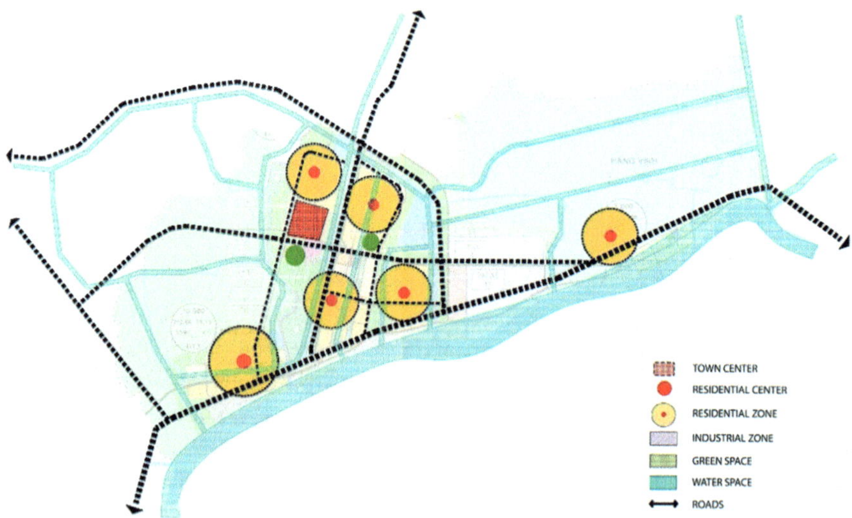

Fig. 46.4 Nam Can's urban structure of the master plan in 2008 (CMG, 2008b)

- Urban planning methodology: these urban planning significantly concentrated in the aspects of technology and artistry to organize urban spatial structure without protecting environment and achieving the benefit of urban economy that support to adapt to climate change.
- Urban planning contents: There are still lack of assessing, analyzing the impacts of climate change and sea level rise in urban planning. Also, urban spatial structure, land use planning, and infrastructure have not applicated for climate change adaptation. The projects of urban development focused significantly on increasing land budget and urban resources without concerning sustainable development and conserving eco environment.

If the urban planning is not adjusted to integrate with climate change, especially mainstreaming with the Climate Change, Sea Level Rise Scenarios for the western zone of the Mekong Delta for each stage, the coastal cities and towns would be vulnerable, increase urban economy, and affect sustainable urban development.

2 Vision for the Urban Planning of the Coastal Cities and Towns Adapting to Climate Change

The IPCC (2007) identified one of the ways to increase climate resilience by implementing sustainable development planning in land use planning and infrastructure design and disaster risk reduction measures. Coastal regions have the most complex and vulnerable environmental systems in the face of climate change and sea level rise. Therefore, coastal climate change adaptation must be integrated into the sustainable development plan and maximize future economic benefits from the coastal areas (Isaac, 2010).

In Vietnam, due to the important transformation dynamics of growth and change, sustainable development has become a new national standard and vision of the Agenda 21 in Vietnam (Vietnam Government [VG], 2004). Facing the challenges of climate change, the key criteria of sustainable urban development need to ensure adaptation to climate change in terms of economic development, environmental responsibility, social progress, urban management and urban infrastructure. Sustainable urban development must work towards climate change mitigation and adaptation. In the current context, climate change has a direct impact on a wide range of areas, ecosystems and resources. The current ecosystem is under pressure from many climate change series (Yohe, 2007). Therefore, research on ecological urban and sustainable development in the context of climate change is now important in the orientation of developing urban climate change adaptation, especially for coastal cities. According to Richard (2013), sustainable ecological cities are low-spread, low-density urban areas, which are transformed into a network of medium- or high-density urban residential areas of limited scale separated by green spaces; most people live and work within walking and biking distance.

Urban elements such as urban structure, urban spatial organization, land use, building density, green space, transportation, and energy use are impacted by climate change, and vice versa; these elements also affect an increase in climate change (Ralf, 2012). The principal elements of city like urban form, land use, and developing frame can be impacted by climate change. Kahn (2006) pointed out that the progression of urban spatial structure in the context of global climate change illustrated that urban structure ensures climate change adaptation successfully. Therefore, the urban planning in the research location need to be adjusted in order to adapt to climate change, integrating with each stage of the Climate Change, Sea Level Rise Scenarios for Vietnam to mitigate urban risks, economic decrease, and impact to urban sustainable development. Lujia (2009) also supports that the spatial structure of coastal town affects both the increase and mitigation of climate change, thus urban planning identifies urban structure and CO_2 emission.

A general vision of urban planning for the western zone of the Mekong Delta is to create an urban spatial structure toward sustainable development and eco-city to adapt climate change. Models of urban spatial structure for climate change adaptation and mitigation are proposed by distributing, linking and mixing urban functional zones: urban center and system of public services, residences, green space, industrial zones, transportation and infrastructure, etc. and integrating with the control of land use density with three levels as high density, medium density, and low density to adapt climate change effectively.

3 Solutions

3.1 Models of Urban Spatial Structure for Climate Change Adaptation in the Western Zone of the Mekong Delta

The mainstreaming of urban functions and infrastructure with the natural characteristics of the research location will lead to the distinctive development of urban spatial structures for the western zone of the Mekong Delta. Proposing models of these structures adapting to climate change is to distribute reasonably urban functions among main urban elements; it also satisfies several solutions such as protecting, adapting and avoiding impacts of climate change. These elements include:

- Urban center, resident center, and the system of public services: Adding more news commercial and tourism service center is to reinforce new multipolar centers which will support present urban center and prevent the risk from climate change.
- Districts and new neighborhood units: Identifying inner city and suburban area and interleaving green space among urban functions is to create buffer zones in order to adapt effectively to climate change.

- The system of green space: Enhancing the areas of green space into urban structures, which can result in room for water and corridors for flooding drainage.
- Transportation system and infrastructure.
- Industrial zones.

These elements can be integrated to control land use at high density, average density, and low density for climate change adaptation. Models of urban spatial structure are proposed based on the classification of urban character as model of spatial structure for the seafront towns, model of spatial structure for the littoral towns, and model of spatial structure for the coastal mangrove towns in order to adapt to optimal climate change.

3.2 *Proposing the Model of the Seafront Towns' Spatial Structure*

The identity of the group of seafront towns is that most of urban functions runs along the coastline and river mouth. The elements of urban spatial structure are located below (Fig. 46.5). Linear urban structure spreads gradually from inner city to suburban area.

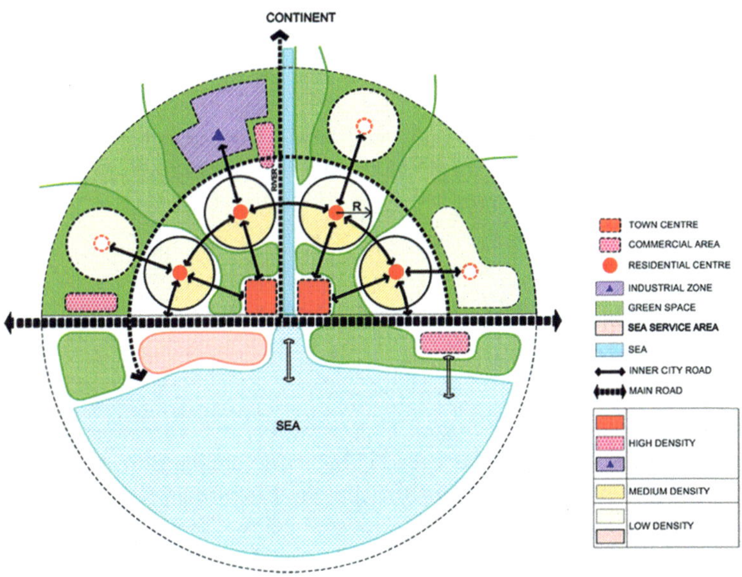

Fig. 46.5 The model of the seafront towns' spatial structure adapting to climate change

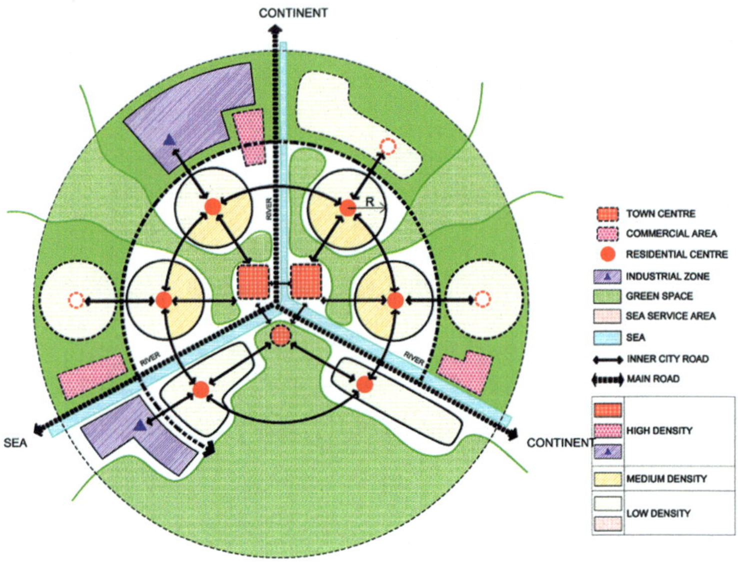

Fig. 46.6 The model of the littoral towns' spatial structure adapting to climate change

3.3 Proposing the Model of the Littoral Towns' Spatial Structure

The identity of the group of littoral towns is that the towns locate 10 kms far from the coastline, higher ground than the seafront towns, crowded population at the interaction between main rivers and main roads. The elements of urban spatial structure are located below (Fig. 46.6). Linear urban structure spreads gradually from inner city to suburban area.

3.4 Proposing the Model of the Coastal Mangrove Towns' Spatial Structure

The identity of the group of coastal mangrove towns is that most towns locate inside Ca Mau coastal mangrove forest. Urban structure tends to spread along rivers and canals. The elements of urban spatial structure are located below. The elements of urban spatial structure are located below (Table 46.4 and Fig. 46.7). This structure represents distinctive form of coastal ecosystem and green space.

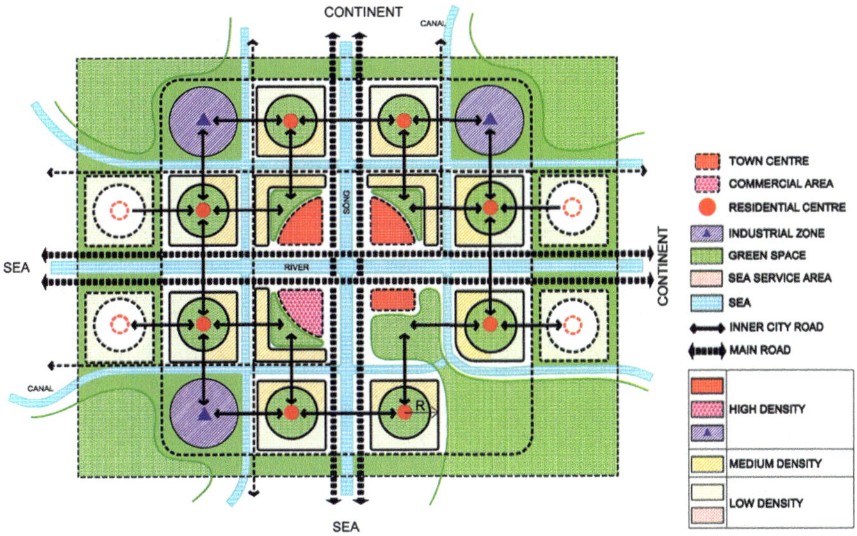

Fig. 46.7 The model of the coastal mangrove towns' spatial structure adapting to climate change

4 Land Use Planning and Controlling Urban Development for Climate Change Adaptation in the Western Zone of the Mekong Delta

4.1 Land Use Planning

The measure of urban structural reform integrating with the land use management is to reinforce the adaptive capacity to climate change by some orientations:

- Planning the value lands for investment, and developing infrastructure is to make more competitive ability despite this urban land are impacted by climate change.
- Planning the safety areas and unsafe areas is to manage better the using of urban citizens in public services, such as health care, education, entertainment, etc.
- Identifying the land areas that can be affected by climate change is to construct city. Also having the strategy planning is to prevent and mitigate climate change risks.

Land use planning for the western zone of the Mekong Delta relates to the process that controls the density of land use, reduces the land use density where there are risks of erosion or high tides. This is to be replaced by green buffer zones along coastline and riverside or canal bank without building expansion in location that can have the risk of natural disasters. The methods of the land use planning highly integrates with the aspect of urban economy also each type of coastal towns (seafront

Table 46.4 Comparing and contrasting the distinctive aspects of three models of urban spatial structures

Order	Urban functional elements	The models of urban spatial structures		
		The seafront towns	The littoral towns	The coastal mangrove towns
	The model of the spatial structures adapting to climate change	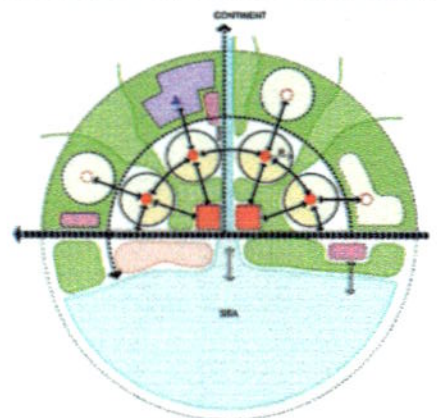 The model of the seafront towns' spatial structure adapting to climate change	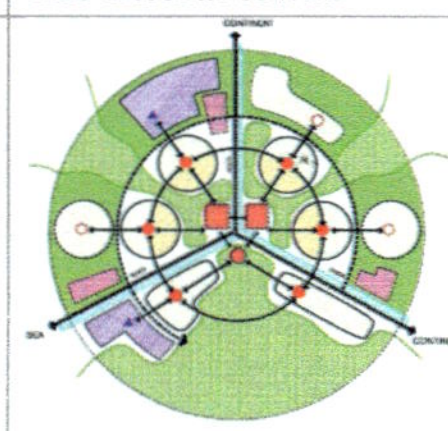 The model of the littoral towns' spatial structure adapting to climate change	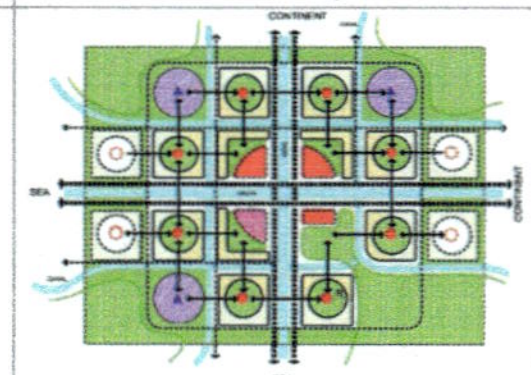 The model of the coastal mangrove towns' spatial structure adapting to climate change
1.	Urban center, resident center, and the system of public services	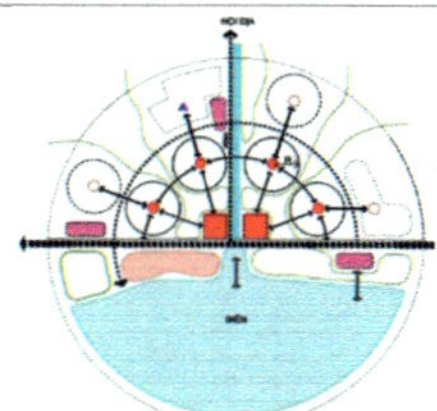 Locating at seafront land and river mouth is to catch sea's strength. Implementing the solution that protects completely urban center to face the negative impacts of climate change. Buildings of public service in resident center have ≤500 meters radius (for average towns) and ≤ 1000 meters radius (for small towns).	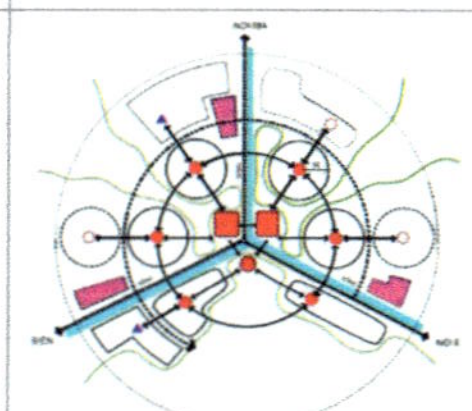 Locating at the interaction between main roads and rivers. Implementing the solution that protects completely urban center to face the negative impacts of climate change. Buildings of public service in resident center have ≤500 meters radius (for average towns) and ≤ 1000 meters radius (for small towns).	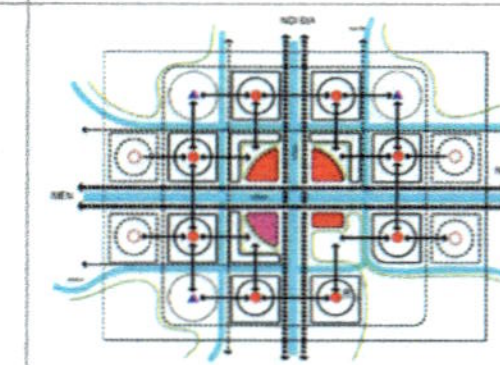 Locating at the interaction between main roads and canal system. Implementing the solution that protects completely urban center to face the negative impacts of climate change. Buildings of public service in resident center have ≤1000 meters radius (for average towns) and ≤ 2000 meters radius (for small towns).

2.	Districts and new neighborhood units	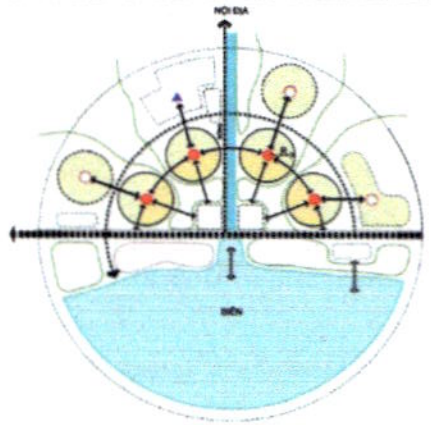 The medium level of building density will apply for districts that are closer to urban center. The low level of building density will apply for districts that are further from urban center.	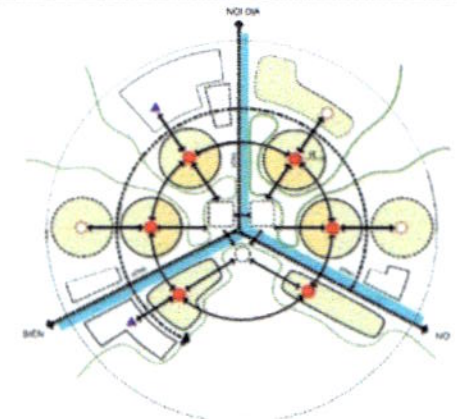 The medium level of building density will apply for districts that are closer to urban center. The low level of building density will apply for districts that are further from urban center.	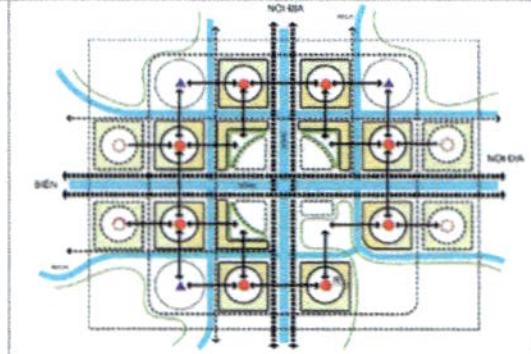 The medium level of building density will apply for districts that are closer to urban center. The low level of building density will apply for districts that are further from urban center.
3.	Green space	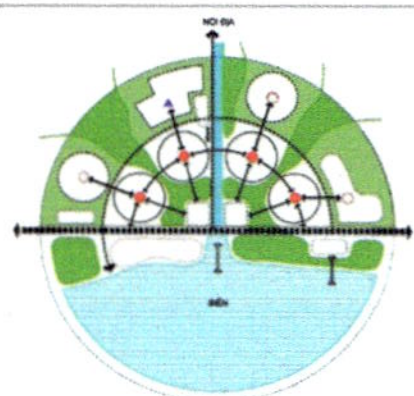 The area of green space accounts for small proportion of urban land. Creating green buffer zones between towns and sea. Creating room for water and flooding.	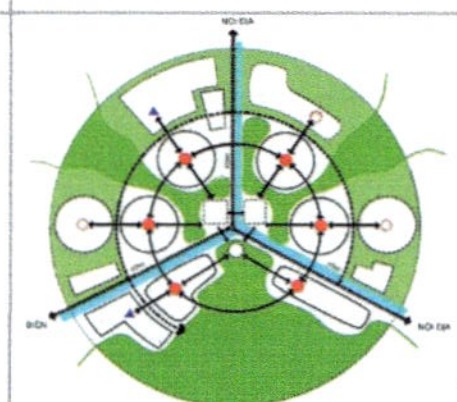 The area of green space accounts for remarkable proportion of urban land. Creating green buffer zones between towns and rivers. Creating room for water and flooding.	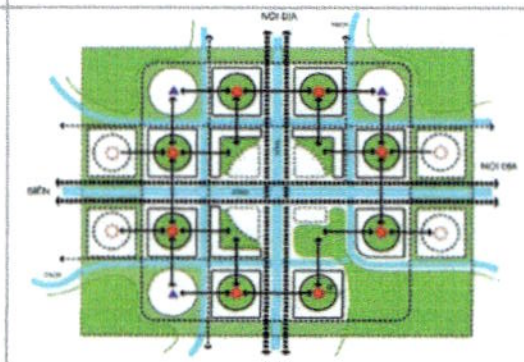 The area of green space accounts for significant proportion of urban land. Creating green buffer zones between towns and canals. Creating room for water and flooding.

(continued)

Table 46.4 (continued)

Order	Urban functional elements	The models of urban spatial structures		
		The seafront towns	The littoral towns	The coastal mangrove towns
4.	Transportation system and infrastructure	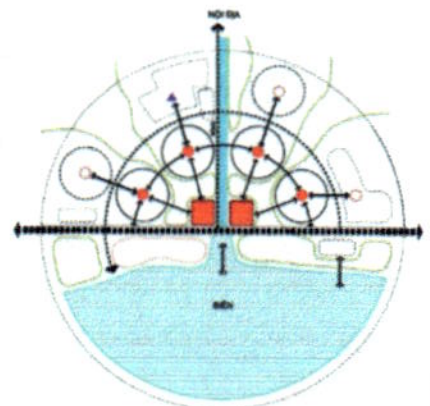 Beside the main road runs along coastline, adding several new main roads that parallel to the coastline are backward compares to the coastline. Adding new radial roads is to connect urban functions to sea. These result in shorter movement that can reduce CO_2 emission.	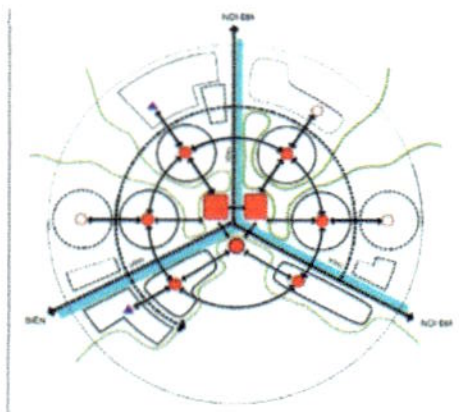 The main road runs along principal riverside. Adding new radial roads and ring roads is to connect urban functions to river. These result in shorter movement that can reduce CO_2 emission.	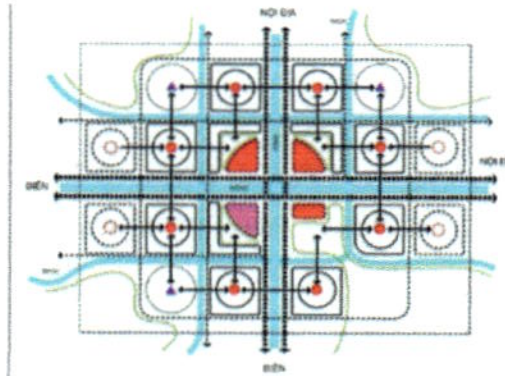 The main roads are parallel and perpendicular to canal system. Limiting diagonal roads that tend to run radially. Waterway is used like the main transport.
5.	Industrial zones	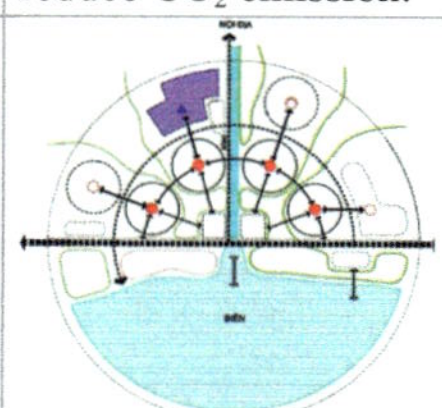 Including heavy industry, light industry and supplementary industry. Implementing the level that protects completely for heavy industry, light industry to resolve the negative impacts of climate change	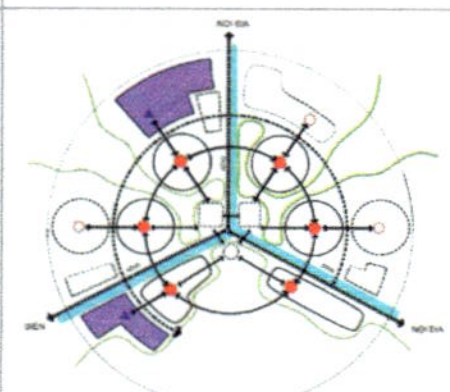 Including heavy industry, light industry and supplementary industry. Implementing the level that protects completely for heavy industry, light industry to resolve the negative impacts of climate change	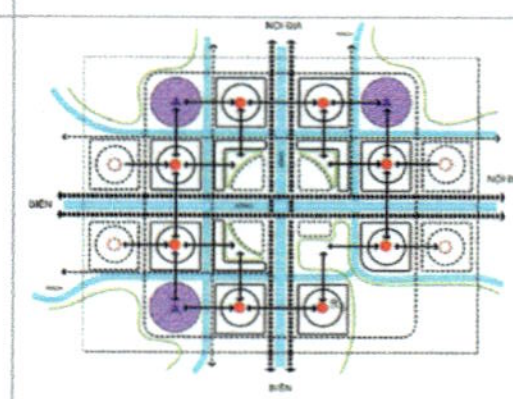 Including supplementary industry. Implementing the level that adapts to climate change for supplementary industry

towns, littoral towns, and coastal mangrove towns) which be impacted by climate change in order to evaluate the cost-benefit in each concept of urban planning. For instance, the way to choose the seafront towns' land use is likely to affect to the high level of ground, urban infrastructure, buildings, sea dikes, seaports, and so on. An assessment frame of land use can make an essential role to adjust land use in coastal areas that are possible to be vulnerable. Existing land use and new exploitation land fund must be consistent with this frame as well as the level of coastal risk, environment issues and the Climate Change (CC), Sea level rise (SLR) Scenarios (Table 46.5).

Table 46.5 Land use control for the western zone of the Mekong Delta

Stage of the term in the CC, SLR Scenarios 2012	The levels of sea rise in the CC, SLR Scenarios 2012	Distances from the coastline	Planning of coastal land use	Type of building level
Now to 2020	9–10	≥ 50–100 m	Zone I: Construction for coastal protection, ecological buffer zones, marine tourist	Avoid building because of less effect.
To 2030	13–15	≥ 100–200 m	Zone II: Unsettled buildings, auxiliary buildings, tourism service buildings in short term; unsettled houses and coastal public services; buildings for coastal protection	The type of level-IV buildings should be used in the short time, nearly 20 years
To 2050	25–30	≥ 200–300 m	Zone III: Tourism buildings, residential buildings, and coastal public services	The type of level-III buildings should be used in a period of 20 years to nearly 50 years
To 2070	39–49	≥ 300–500 m	Zone IV: New districts, administrative buildings, cultural buildings, hospitals, schools, kindergartens, elder health care center	The type of level-II buildings should be used in a period of 50 years to nearly 100 years
To 2100	62–82	≥ 500 m	Zone V: New districts, administrative buildings, cultural buildings, hospitals, schools, kindergartens, elder health care center.	The type of level-I buildings should be used over 100 years

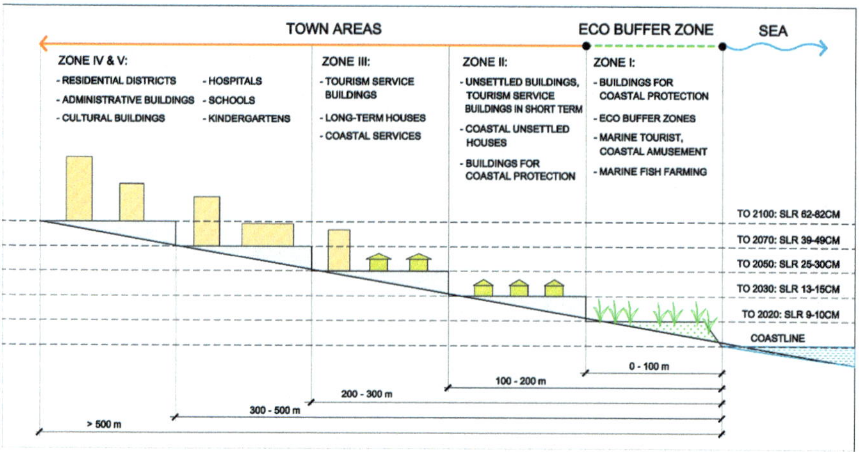

Fig. 46.8 The selection for land use of the coastal spatial zone

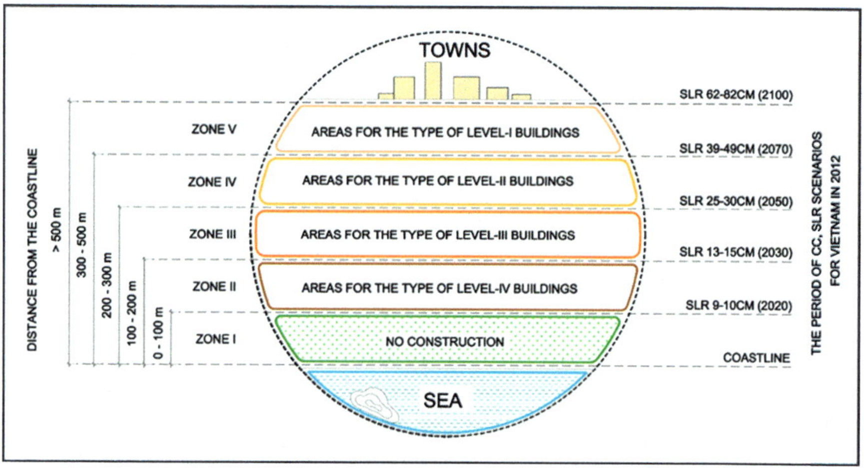

Fig. 46.9 Distributing the land use planning of coastal area

4.2 *Controlling Land Use*

Controlling the land use is applied to for majority of coastlines which face many risks. It is proposed that the constructive boundary of coastal buildings is similar the constructive boundary inside city, must to move backward from coastline. These coastal constructive boundaries are identified in short term until 2020 to long term between 2020 and 2100 (Figs. 46.8 and 46.9).

- Zone I: In the scope of the distance from 0.0 m to 100 m (from the coastline), construction in this area needs to be banned because of its negative effects. According to the CC, SLR Scenario for Vietnam 2012 to 2020, the western zone of the Mekong Delta will be directly affected by sea level rise, high tides, and erosion. It can be built for coastal protection, ecological buffer zones, marine tourist, marine fish farming, etc. within the scope of ≥50–100 m (to 2020).
- Zone II: In the scope of the distance from 100 m to 200 m (from the coastline), this land area should be used in the short time until 2030 because of sea level rise, high tides and erosion; only the type of level-IV buildings is permitted. Unsettled buildings, auxiliary buildings, tourism service buildings in short term; unsettled houses and coastal public services; buildings for coastal protection within the scope of ≥100–200 m (to 2030) are permitted.
- Zone III: In the scope of the distance from 200 m to 300 m (from the coastline), this land area should be used in the mid-term until 2050 exception to apply higher constructive technology adapting to CC because of sea level rise, high tides, and erosion; only the type of level-III buildings or less are permitted. Tourism buildings, residential buildings, and coastal public services within the scope of ≥200–300 m (to 2050) are permitted.
- Zone IV: In the scope of the distance from 300 m to 500 m (from the coastline), this land area can be used in the long term until 2070 for buildings in each next stage because of sea level rise, high tides and erosion; only the type of level-II buildings or less are permitted.
- Zone V: In the scope of the distance ≥500 m (from the coastline), this land area can be used in the long term until 2100 for most buildings because of sea level rise, high tides, and erosion; only the type of level-I buildings or less are permitted. New districts, especially hospitals, healthcare center and schools, should be located outside of coastal risk areas within the scope of ≥300 m (to 2070) and ≥ 500 m (to 2100) in order to avoid the increase of facing some potential risks in coastal areas.

4.3 Controlling Urban Development

Some solutions of controlling land use are to control the urban development for the coastal towns in the western zone of the Mekong Delta for climate change adaptation:

- There are measures for new districts: concentrating to build the apartments having medium height; controlling density for residential buildings, offices, and so on; reducing the split of residential areas into each small unit which can lead to waste of coastal land.
- Proposing technical criteria integrating with the criteria of land use value is to pressure the investors that must adapt to financial effects in order to make the good solutions of land use which suit to most used demands.

4.4 *Mainstreaming the Risks of Natural Disasters into Land Use Planning*

The hazards of natural disasters and the risks of climate change in the research location need to be integrated into urban planning, particularly for land use planning in each stage and field:

- Mainstreaming the hazards when planning the land use for economic-social development and other fields in Kien Giang province and Ca Mau province.
- Mainstreaming the risks when assessing the environment issues for land use planning which can be affected by climate change and vice versa.
- Integrating the risks to some solutions that can prevent and adapt the natural disasters such as sea dikes, supporting infrastructure, lakes for waterfall and flooding and coastal protective forest.
- Integrating the risks to adjust the present urban planning to update the new Climate Change, Sea Level Rise Scenarios. It is also necessary to mainstream and supervise climate change adaptation in the process of implementing urban planning.

5 Conclusions and Discussion

Proposing to renew the methodology of urban planning which integrates with climate change. Adding new research is to improve the content of mainstreaming for climate change adaptation in urban planning via methodology and the process of mainstreaming. Making the research methodology of climate change impact on cities and urban planning. Mainstreaming climate change adaptation for cities via the planning measures for spatial structure, orientation of spatial development, and land use planning.

Proposing the models of spatial structure and the solutions of urban planning adapting to climate change for the coastal cities and towns of the Vietnam's South West (belongs to the West coastal zone from Ca Mau Cape to Ha Tien town of Kien Giang province and Ca Mau province). These are the models of seafront urban spatial structure, the model of littoral urban spatial structure and the model of spatial structure of coastal mangrove towns which are applied into the groups of seafront towns, the groups of littoral towns, the groups of coastal mangrove towns, respectively. Proposing the planning measure of spatial development and land use for urban center, inhabited areas, green space, transport and infrastructure, industrial zones, etc. adapting to climate change.

Proposing three models of urban spatial structure for climate change adaptation in the Western zone of the Mekong Delta, which links with three groups of urban characteristics, is a fundamental measure to implement the type of master plan for the cities and towns adapting to climate change. The method of systematization models is used in order to propose three models of urban spatial structure and solutions of urban land use adapting to climate change for coastal towns toward sustainable development. These models would be referenced for Vietnam coastal cities that have similar natural conditions and identities.

References

Ca Mau government [CMG] (2008a). *Master plan of Ca Mau city, Ca Mau province.*

Ca Mau government [CMG] (2008b). *Master plan of Nam Can town, Ca Mau province.*

IPCC. (2007). Climate change 2007: Impacts, adaptation and vulnerability. In M. L. Parry, O. F. Canziani, J. P. Palutikof, P. J. van der Linden, & C. E. Hanson (Eds.), *Contribution of working group II to the fourth assessment report of the intergovernmental panel on climate change* (976pp). Cambridge University Press.

Isaac, B. (2010). *Spatial planning in coastal regions: Facing the impact of climate change* (pp. 50–54). FIG.

Kahn, M. E. (2006). *Green cities - urban growth and the environment* (160pp). Brookings Institution Press.

Kien Giang Government [KGG]. (2008). *The project of master plan for ha Tien town. Giang province.*

Lujia (2009). *Spatial planning in shenzhen to build a low carbon city.* 45th ISOCARP Congress, 1–2.

MONRE. (2012). *The climate change, sea level rise scenarios for Vietnam in 2012* (pp. 7–77). NARENCA.

Ralf, K. (2012). Cities and the potential for climate change adaptation. *BTU Co'bus - Department for Urban Planning and Spatial Design, 12*, 22–33.

Richard, R. (2013). *Eco cities: Rebuilding cities in balance with nature* (Vol. 2013, p. 85). New Society Publishers.

Vietnam Government [VG]. (2004). Decision No. 153/QD-TTg dated August 17, 2004 by Prime Minister on approving the Vietnam Sustainable Development Strategy (Agenda 21 in Vietnam):5–14.

Vietnam Government [VG]. (2009a). Decision No. 1581/QD-TTg dated October 9, 2009 by Prime Minister on approving the Regional Planning for the Mekong Delta to 2020 year, the vision to 2050:2–3.

Vietnam Government [VG]. (2009b). Decision No. 445/QD-TTg dated April 7, 2009 by Prime Minister on approving the Orientation for development of the Vietnamese urban system to 2025 year, the vision to 2050:1–2.

VIUP. (2015). *The outline of preparing the process of implement in planning concentrated economic zones in Vietnam integrating climate change* (pp. 104–106). VIUP.

Yohe, G. W. (2007). Perspectives on climate change and sustainability. Climate Change 2007: Impacts, adaptation and Vulnerability. In M. L. Parry, O. F. Canziani, J. P. Palutikof, P. J. van der Linden, & C. E. Hanson (Eds.), *Contribution of Working Group II to the Fourth Assessment Report of the Intergovernmental Panel on Climate Change* (pp. 811–841). Cambridge University Press.

Chapter 47
MIKE FLOOD Application for Solving Inundation Issues for Ho Chi Minh City in the Context of Climate Change: A Case Study in District 1

Pham Thanh Long, Tran Tuan Hoang, and Nguyen Phuong Dong

Abstract In recent years, urban flooding has caused significant impacts on people's activities, socio-economic development, environmental pollution and traffic congestion in Ho Chi Minh City. In the context of climate change, there have been extreme rainfall events in, along with the impacts of sea level rise on the tides that increase the water level of canals and rivers; heavy flooding has occurred, on roads without any rains. With the MIKE FLOOD model, intensity-duration-frequency (IDF) curve is used for calculating the flooding level and time, taking the terrain, drainage system, sewer and the size of a culvert into account for Ho Chi Minh City, a pilot for District 1. The results can be used for technical infrastructure planning, construction of flood warning and forecasting models, especially for flooding management in connection with the smart city for Ho Chi Minh.

Keywords Inundation · Climate change · Mike flood · Ho Chi Minh City

1 Introduction

Over the years, the economic growth rate of Ho Chi Minh City (HCMC) has increased sharply, but not synchronously. Many problems such as urban flooding, saline intrusion, environmental pollution, etc. are mentioned in the annual reports of departments. The whole city had an area of 2061.4 km^2 in 2016 but nearly 60% of the total area is located below 1.5 m above sea level. So, the city faces frequent flooding during the rainy season (from June to November each year). In the last decade, the urbanization rate of Ho Chi Minh City happened very fast with large

P. T. Long (✉) · T. T. Hoang · N. P. Dong
Sub-Institute of Hydrometeorology and Climate change, Ho Chi Minh City, Vietnam

A. T. Nguyen, L. Hens (eds.), *Global Changes and Sustainable Development in Asian Emerging Market Economies Vol. 2*, https://doi.org/10.1007/978-3-030-81443-4_47

works constantly appearing in the downtown area, especially in District 1. This has put more pressure on infrastructure (Larsen et al., 2008) as well as the drainage system (Mai, Tran, Ho, & Nguyen, 2019; Nie, Lindholm, Lindholm, & Syversen, 2009). Consequently, the central roads have often been flooded when heavy rains occurred continuously.

Currently, Vietnam is also one of the countries vulnerable to the impacts of climate change with the manifestation of extreme weather events increasing in frequency and unpredictable. The frequency of heavy rains causes (Tran, 2015) more inundation and directly affects traffic and people's lives. Urbanization of an area will, however, increase the magnitude of impacts on runoff and hydraulic performance (Tait, Ashley, Cashman, & Blanksby, 2008). There have been many studies on related issues. Grum, Jørgensen, Johansen, and Linde (2006) simulated and analyzed the effects of rainfall on urban drainage systems using the ECM model. The impact of climate change on rainfall in an urban basin (Willems, Arnbjerg-Nielsen, Olsson, & Nguyen, 2012)) and establishing standards for design and operating urban drainage systems (Arnbjerg-Nielsen et al., 2013) have been investigated. Olsson, Berggren, Olofsson, and Viklander (2009) also analyzed the characteristics of the urban drainage system when considering climate change scenarios B2 and A2. Based on this analysis, the previous studies assessed the impact of climate change (Berggren, Olofsson, Viklander, Svensson, & Gustafsson, 2011; Berggren, Packman, Ashley, & Viklander, 2014) and provided a feasible approach for urban drainage systems to adapt to climate change (Neumann et al., 2015). The influence of urban development and global climate change on increasing rainfall intensity in HCMC area was also studied and discussed in an international conference on the impact of climate change on inundation (Luong, 2008). Constructing the IDF curve based on extreme rain events according to each rain period was developed in the study of Nguyen, Pham, and Dao (2017).

This chapter aims to review, evaluate and determine the intensity of rain following the frequency of rain, serving the calculation of designed rain and floods for water drainage and urban planning works. Therefore, we chose an approach to build intensity-duration-frequency (IDF) of rainfall at present and future climatic conditions under the impact of climate change. Subsequently, we are developing future IDF scenarios to assess the degree of change in rain events and the impact of existing drainage systems in District 1, HCMC.

2 Methodology

In this study, we apply the IDF curve to assess the intensity and frequency of rain for serving water drainage system in HCMC, the pilot area is District 1. Collected data include water level, meteorology-hydrology data, drain size data, etc. and comprehensive inheritance of the database in the research area. The models used for simulation take account of different scenarios with flooded periods such as MIKE FLOOD, MIKE 11 HD, MIKE 21 FM and MIKE URBAN (DHI, 2014a, 2014b). The ArcGIS 10.1 software suite is used to present District 1 flood calculation results on the base map.

2.1 Establishing Models and Data

2.1.1 MIKE 11 HD Model

- Boundary conditions (DHI, 2014c, 2014d):
  - Upstream boundary: Hourly water level boundary at stations Tan An (Vam Co Tay river), Go Dau (Vam Co Dong river) from 0:00 on 01/01/2016 to 23:00 on 30/11/2016 and daily flow margin from 01/01/2016 to 30/11/2016 of the lakes Dau Tieng (Saigon River) and Tri An (Dong Nai River).
  - Downstream boundary: Data of water level to be monitored at Vung Tau station from 0:00 on 01/01/2016 to 23:00 on 30/11/2016.

2.1.2 MIKE URBAN Model

From the collected data, we filtered and removed some data such as small sewer lines, insufficient data. Besides, the data needs to be homogenized with the coordinate system into WGS 84 UTM Zone 48 N (EPSG: 32648). After aggregating the complete data set, the digitization into MIKE URBAN is done by importing from the GIS data set via the Import tool (DHI, 2014e, 2014f).

To ensure the system works well with the model, it is necessary to check digitized data. The cases leading to the model generating errors may include missing ground elevation data, lack of diameter data, the bottom elevation being higher than surface elevation, the slope in the drain being unreasonable, the system being broken, etc. "Project Check Tool" is a tool to find out where the errors are. It is easy to fix errors and perfect the system after digitizing, including 1902 tunnel, 37 outlets, 1343 round drain and 717 square culverts.

2.1.3 MIKE 21 FM Model

Digital elevation data (LiDAR) is generated using the latest LiDAR technology of the Department of Science and Technology, HCMC, with a resolution of 5x5 m. We need to convert LiDAR digital elevation data to * .xyz file format in

MIKE ZERO into the Mesh Generator tool which will be terrain in the model MIKE 21 FM.

2.1.4 MIKE FLOOD Model

Using MIKE FLOOD model we connect with MIKE 11 HD model, MIKE 21 FM and MIKE URBAN to calculate floods for District 1 area. The paper made a connection at 1902 underground station and 37 outlets of the current status of District 1 drainage into the model follow these steps:

- Network connection in MIKE 11 HD model and terrain in the MIKE 21 FM model with Link river branch to Mike 21.
- Connect river network in MIKE 11 HD model to 37 outlets of MIKE URBAN with Link river branch to MIKE URBAN tool.
- Connect 1902 manhole from MIKE URBAN model to MIKE 21 FM model terrain using Link urban node to MIKE 21 (Figs. 47.1, 47.2, 47.3 and 47.4).

Fig. 47.1 Diagram of the river network in hydraulic model MIKE 11 HD

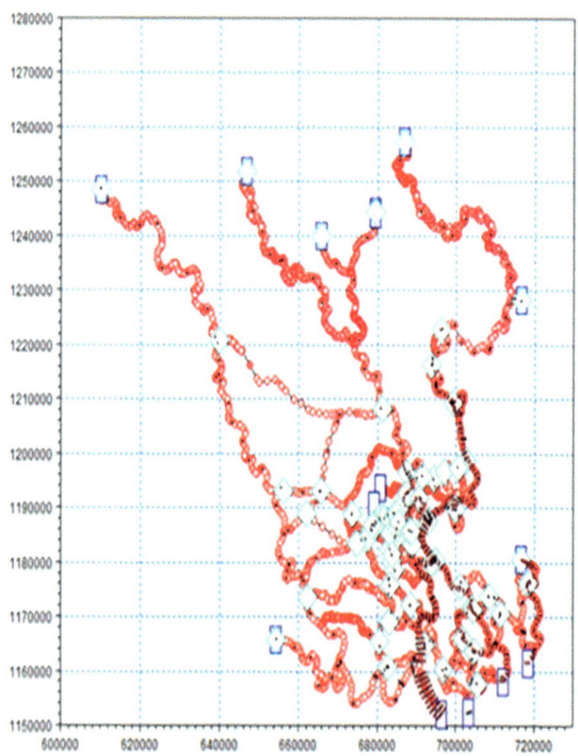

Fig. 47.2 Current situation of drainage in District 1 in the MIKE URBAN model

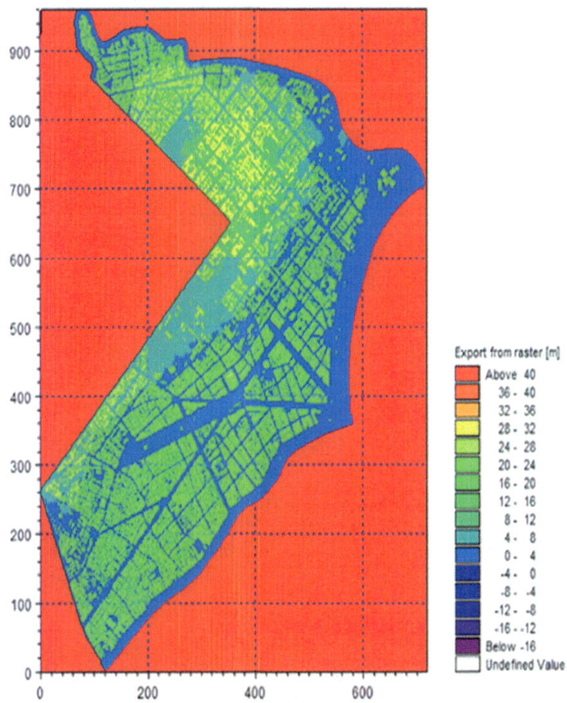

Fig. 47.3 The topography of District 1 in MIKE 21 FM model

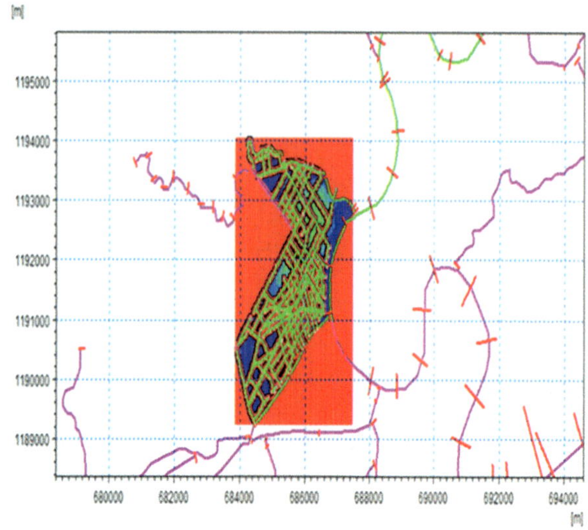

Fig. 47.4 Diagram of the model connection between modules (MIKE 11 HD, MIKE 21 FM and MIKE URBAN) in MIKE FLOOD model interface

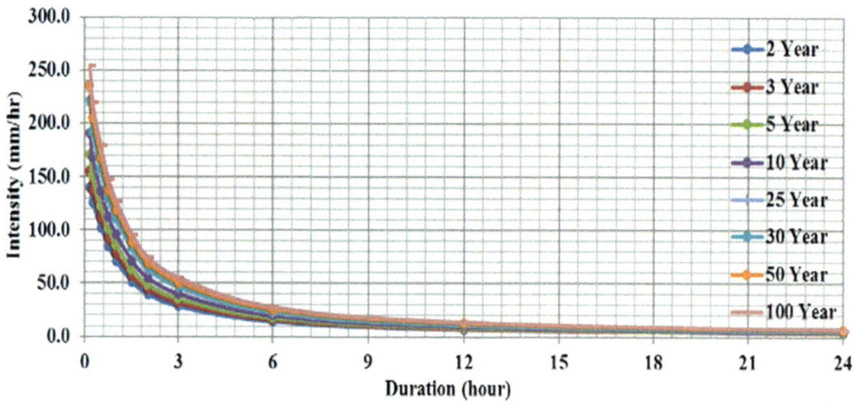

Fig. 47.5 IDF line for rainfall based on short-term monitoring data (1971–2016) at Tan Son Hoa station

2.2 Design Rain Chart

The design rain chart is built for a period of 180 minutes, impulse rain 5 minutes/1 data. The rain chart was built by the method of instant rain intensity based on the delegate rain (on September 26, 2016, the amount of rain was 132 mm, 3-hour period) causing heavy flooding in District 1. Combining with IDF curve of rain is built for Ho Chi Minh City according to the short-term rain data at Tan Son Hoa station (1971–2016) and the IDF curve is constructed according to a future scenario under climate change condition. Design rain charts are used as the input rain margin for the MIKE URBAN model (Figs. 47.5, 47.6, 47.7, 47.8, 47.9, and 47.10).

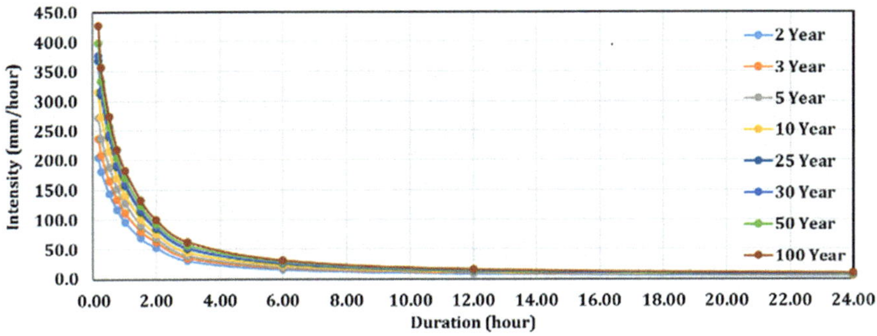

Fig. 47.6 IDF road rains under average scenario at Tan Son Hoa station

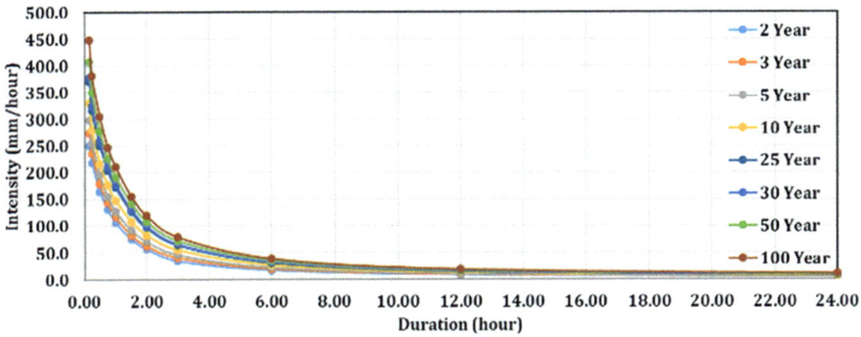

Fig. 47.7 IDF road under high scenario rain at Tan Son Hoa station

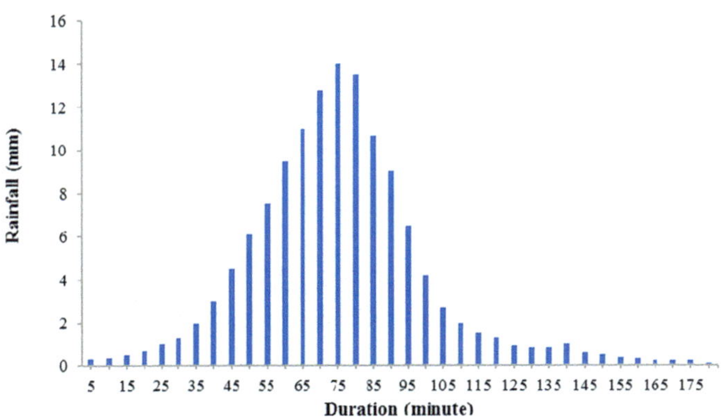

Fig. 47.8 Representative rainfall chart of September 26, 2016 (3-hour period)

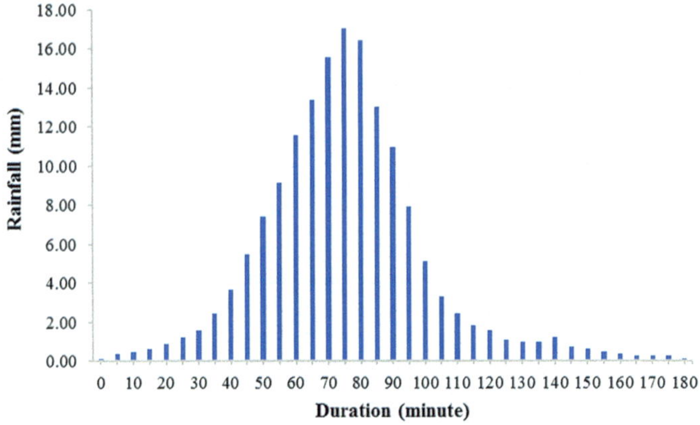

Fig. 47.9 Design rain chart of the 10-year repeated rain cycle according to the average scenario (3-hour period)

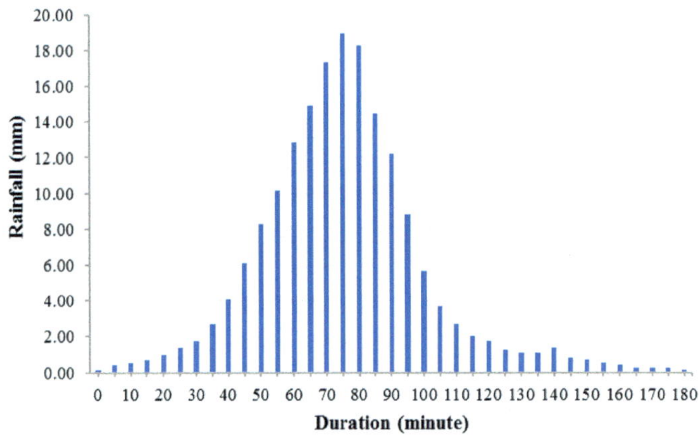

Fig. 47.10 Design rain chart of the 10-year repeated rain cycle according to the high scenario (3-hour period)

2.3 Develop Flood Calculation Scenarios

During the rainy season, to ensure proper inter-reservoir operating procedure for the safety of the lake, regulatory authorities release regulated water through the spillway. Under the impact of climate change, this overflow will change. So, in this study, the upstream boundaries are assumed to be the running volume plus the overflow discharge at Dau Tieng reservoir, and Tri An has an increase in rainfall flow.

The selected scenarios to calculate for the 2016 status quo and the proposed scenarios include: medium and high scenario for 2030 (early century)

Table 47.1 Model calculation scenarios

No.	Scenarios	Period	Water level downstream	Precipitation on-site in Ho Chi Minh City area	Upstream flow
1	Status quo	2016	Vung Tau in 2016	Rain chart designed for the delegate rain on September 26, 2016 (rainfall 132 mm)	Data of Ho Dau Tieng reservoir Tri An in 2016
2	Average RCP4.5	2030 early stages of the century	Increase 12 cm	Rain chart designed according to the early stages of the century + repeated cycle 10 years (160.5 mm rainfall)	Dau Tieng increase: 33 m^3/s Ho Tri An increase: 90 m^3/s
3	High RCP8.5	2030 early stages of the century	Increase 12 cm	Rain chart designed according to the early stages of the century + repeated cycle 10 years (precipitation 178,5 mm).	Dau Tieng increase: 30 m^3/s Tri An increase: 80 m^3/s

corresponding to the 10-year repeated rain period (according to Scenario of Climate Change and Sea Level Rise for Vietnam developed by the Ministry of Natural Resources and Environment with the latest version in 2016). The applied flood calculation for District 1 area is presented in Table 47.1.

3 Results

3.1 Calibration Results and Test the Models

3.1.1 MIKE 11 HD Model

Based on the input data set and input conditions, boundary conditions have been established, the set of parameters of the 1-D flow model is calibrated with the actual water level data measured at Phu An station, Nha Be from 9:00 September 28, 2015 to 09:00 hours September 30, 2015, with the Nash index all over 0.90 (Figs. 47.11 and 47.12).

To test a hydraulic model, we use a modified set of parameters and change the calculation time. Testing time: 3 days from 9:00 September 17, 2016 to 9:00 September 19, 2016. Model test results are also evaluated by Nash index above 0.90 (Figs. 47.13 and 47.14).

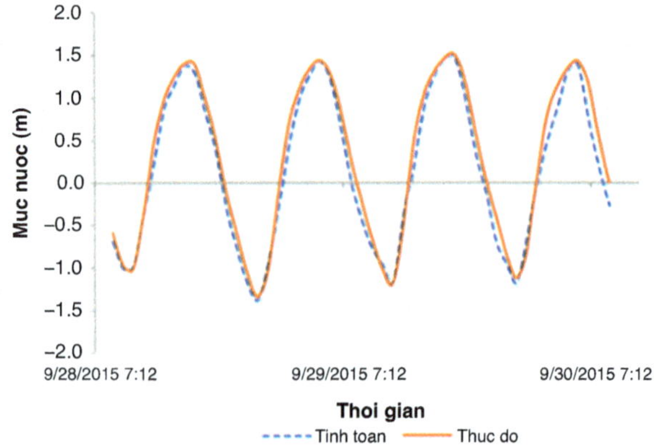

Fig. 47.11 The water level of Phu An station in September post-calibration (Nash = 0.97)

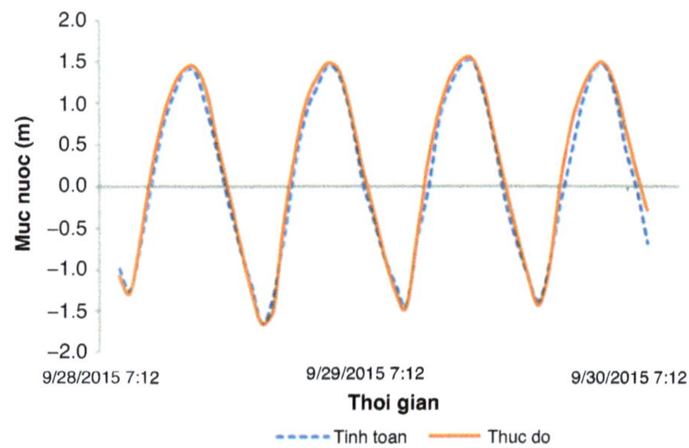

Fig. 47.12 The water level of Nha Be station in September after calibration (Nash = 0.98)

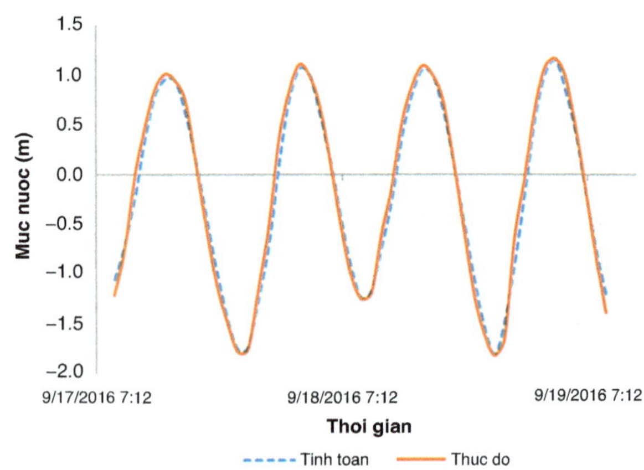

Fig. 47.13 The water level of Phu An station in September after a test (Nash = 0.98)

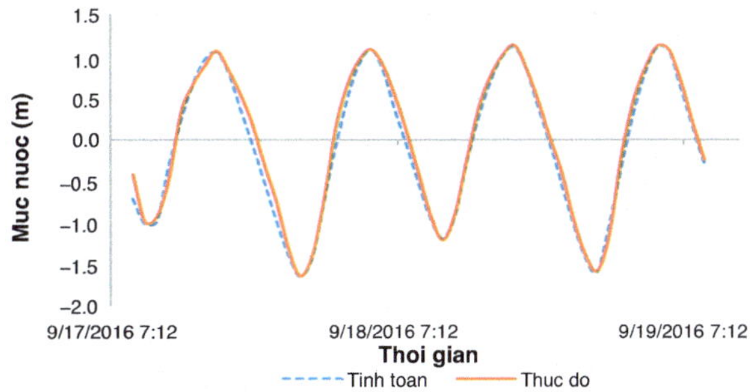

Fig. 47.14 The water level of Nha Be station in September after a test (Nash = 0.98)

Table 47.2 Comparison of the flooded depth for model and observation in District 1

Numerical order	Street names	Flooded location		Actual flood depth (m)	Calculated flooded depth (m)
		From	To		
1	Cong Quynh	Thai Binh Market	Nguyen Cu Trinh	0.20	0.25
2	Nguyen Cu Trinh	Tran Hung Dao	Tran Dinh Xu	0.30	0.30
3	Tran Hung Dao	Ho Chi Minh City Police Headquarters	Nguyen Van Cu	0.30	0.33
4	Mai Thi Luu	Dien Bien Phu	SN 99	0.30	0.30
5	Nguyen Van Cu	Tran Hung Dao	SN 160	0.25	0.30

3.1.2 MIKE URBAN Model

In the MIKE URBAN model, the parameters that are used to adjust in the calibration process are width of the manhole, sewer diameter, as well as the size of the drain (length and area of the section), Manning's roughness coefficient, loss coefficient, impermeability coefficient of sub-basin collecting water. Considering the rain on September 26, 2016, the model was adjusted to the above parameters to get results consistent with the reality. For residential areas, the waterproof coefficient is 65% and roads are 80%.

3.2 Results of Submergence Using MIKE FLOOD Model

3.2.1 Corrected Flood Calculation Results

According to 2016 statistics in District 1, flooded areas were caused by extreme rain, i.e. extreme rain occured on September 26, 2016 (132 mm arms, 3-hour period exceeding the design standards of HCMC drainage system according to Decision 752 / QD-TTg with 95.91 mm, 3-hour period). The depth of flood results calculated from the model is compared with the statistics of flood spots (from the Executive Center of the Ho Chi Minh City Flood Control Program). The model results are quite consistent with the rain on September 26, 2016 (presented in Table 47.2).

3.3 *Results of Flood Calculation in District 1 According to the Scenarios*

According to 2016 flood statistics, District 1 has about five flooding points. Floodplains are concentrated on Nguyen Thi Minh Khai roads, Mai Thi Luu, Nguyen Cu Trinh, Tran Hung Dao and Nguyen Van Cu. Compared with the results of flooding, the current situation is relatively accurate with the depth of flooding fluctuating 0.2 ~ 0.4 m. Flooding time is 60–90 minutes (Fig. 47.15).

Fig. 47.15 Flood map of District 1 area in 2016

Fig. 47.16 The highest flood map for District 1 in 2030 under the medium scenario

Flood simulation in the study area of District 1 under future scenarios for 2030 with a 10-year repetition cycle under two medium scenarios (design rainfall 160.5 mm) and high scenario (design rainfall is 178.5 mm) has also been studied. The results of flood simulation according to the scenarios are detailed as shown in Figs. 47.16 and 47.17:

According to the calculation results in 2030, in both RCP4.5 and RCP8.5 scenarios, the level of flooding increased significantly compared to the simulation

Fig. 47.17 The highest flood map for District 1 in 2030 under the high scenario

results of the 2016 rainfall. The depth of flood ranges from 0.2 to 0.53 m, flooding time is 70–100 minutes (according to the medium scenario). The depth of immersion ranges from 0.25 to 0.6 m, flooding time is 70–120 minutes (according to the high scenario).

Analysis of flood results in RCP4.5 and RCP8.5 scenarios at the beginning of the century, corresponding to the design rain according to the 10-year repetition cycles, shows:

Table 47.3 Results of District 1 flooded area statistics according to the calculated scenarios

Repetitive scenario/period of rain	Flooded area according to flood level (ha)						Area (ha)	Flood rate (%)
	0,1–0,15 (m)	0,15–0,2 (m)	0,2–0,25 (m)	0,25–0,3 (m)	0,3–0,5 (m)	0,5–0,8 (m)		
Status quo	3,68	2,99	2,58	1,41	2,17	0,22	12,83	**1,66**
Medium scenario	5,39	3,77	3,46	2,06	3,06	0,35	17,74	**2,30**
High scenario	6,06	4,48	3,86	2,61	3,75	0,49	20,77	**2,69**

- The flooding of District 1 roads has increased. Specifically, the flooded area increases with each scenario. The area flooded with rain on September 26, 2016 was 12.83 ha; according to the 10-year cycle of repeated rains in the average scenario is 17.74 ha and according to the 10-year cycle of repeated rain in the high scenario, it is 20.77 ha (see Table 47.3).
- Extreme rain in the future will reduce the responsiveness of many tertiary sewers, some secondary sewers; as a result of the statistics, the flooded area of the county leads to the effect of drainage.
- Extreme rain has increased floods; high flooding has affected traffic and drainage capacity of the whole District 1 drainage system.

4 Proposed Solutions

In fact, due to rain in urban areas, the buffer element plays a very important role. The park or vacant land is the ideal place to drain rainwater which will reduce the possibility of flooding due to rain in the city in general and of District 1 in particular. The rest are residential areas, road surface, pavement mainly made of concrete, brick, rock; due to poor water permeability, encountering heavy rainfall for short time (extreme rain) would cause flooding. In the future, District 1 will invest in developing waterproof bricks for use on the pavement; this will reduce flooding and increase drainage capacity for District 1.

After adjusting the permeability coefficient suitable for works such as pavement, which has been linked with bricks, it is good waterproof, the heavy flooding points have reduced as in the Nguyen Thi Minh Khai section, the flooded area has decreased from 211 m^2 to 89 m^2. The sections of Mai Thi Luu and Cong Quynh were no longer flooded, the average depth of inundation decreased from 0.5 m to 0.3 m (Figs. 47.18 and 47.19).

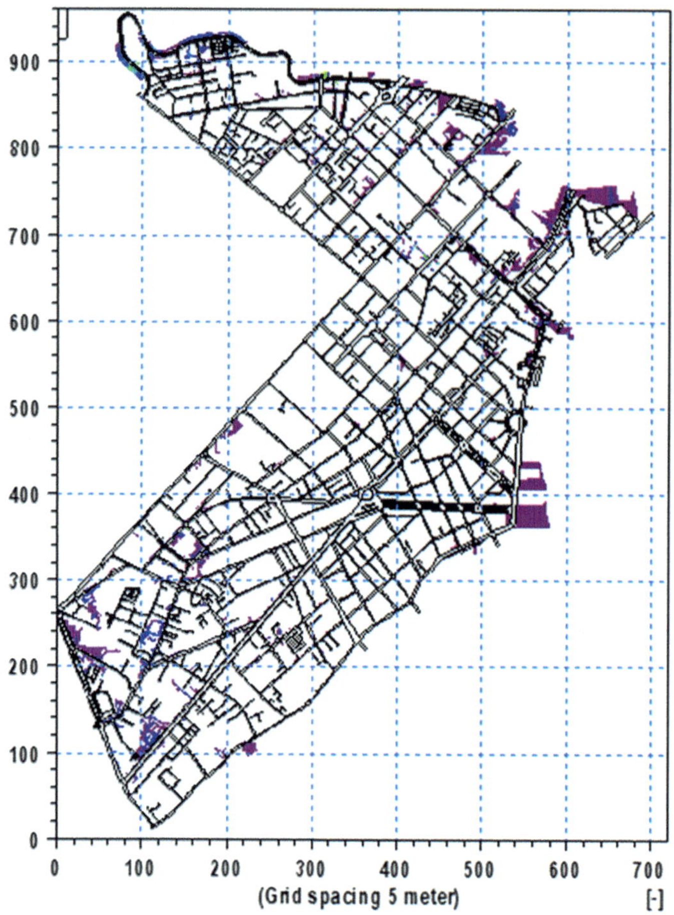

Fig. 47.18 Flooding after the rain on September 26, 2016

5 Conclusions

Results from the model for rain event on September 26, 2016 with the highest depth of 0.41 m, the inundated area is 12.83 ha. For future design rains with a 10-year repeat period of medium and high emission scenarios, at the beginning of the century, the situation of flooding in District 1 is more complicated. As the inundated time increases, the highest depth of flood increases by 13 cm (according to RCP4.5 scenario) and 20 cm (according to RCP8.5 scenario).

The compared results between the simulation and observed results of floods are suitable in the pilot area. Therefore, it is suggested to apply the calculation to the drainage basins and calculate the flood level and duration for heavy inundated areas in Ho Chi Minh City.

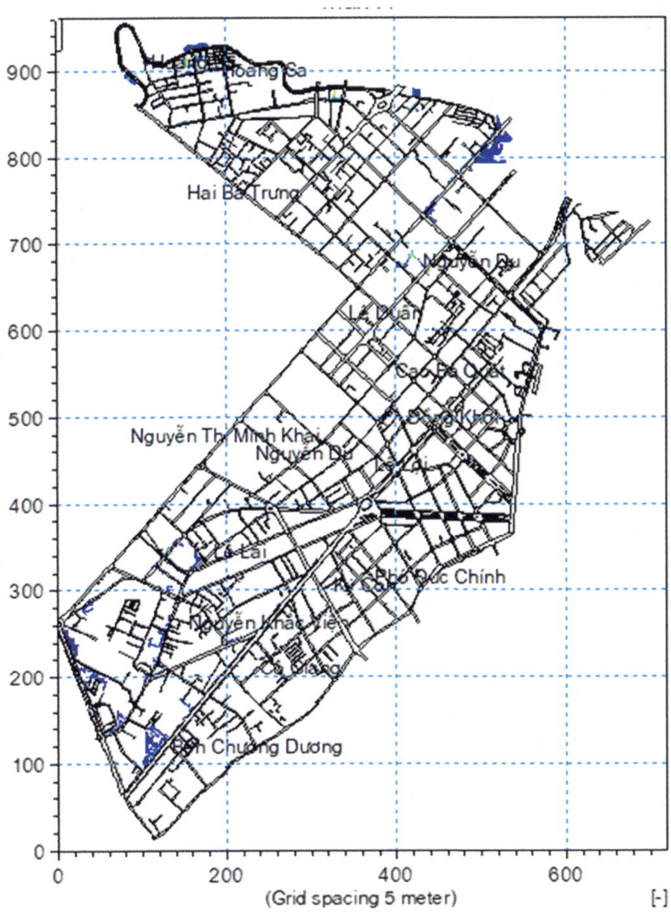

Fig. 47.19 Flood District 1 after using absorbent brick lining solution for sidewalks

Acknowledgement This study was done and completed with the support of the project "Studying the responsiveness of the drainage system in Ho Chi Minh City in the context of climate change".

References

Arnbjerg-Nielsen, K., Willems, P., Olsson, J., Beecham, S., Pathirana, A., Bülow Gregersen, I., Madsen, H., & Nguyen, V. T. V. (2013). Impacts of climate change on rainfall extremes and urban drainage systems: A review. *Water Science & Technology, 68*(1), 16–28. https://doi.org/10.2166/wst.2013.251

Berggren, K., Olofsson, M., Viklander, M., Svensson, G., & Gustafsson, A. (2011). Hydraulic impacts on urban drainage systems due to changes in rainfall caused by climatic change. *Journal of Hydrologic Engineering, 17*, 92–98.

Berggren, K., Packman, J., Ashley, R., & Viklander, M. (2014). Climate changed rainfalls for urban drainage capacity assessment. *Urban Water Journal, 11*, 543–556. https://doi.org/10.1080/1573062X.2013.851709

DHI. (2014a). *MIKE FLOOD user manual*. Denmark.

DHI. (2014b). *MIKE OPERATIONS User Guide*. Denmark.

DHI. (2014c), *MIKE URBAN COLLECTION SYSTEM*. Denmark.

DHI. (2014d). *MIKE URBAN TUTORIALS*. Denmark.

DHI. (2014e). *MOUSE Pipe Flow Reference Manual*.

DHI. (2014f). *MOUSE Runoff Reference Manual*.

Grum, M., Jørgensen, A. T., Johansen, R. M., & Linde, J. J. (2006). The effect of climate change on urban drainage: An evaluation based on regional climate model simulations. *Water Science and Technology, 54*, 9–15.

Larsen, P. H., Goldsmith, S., Smith, O., Wilson, M. L., Strzepek, K., Chinowsky, P., & Saylor, B. (2008). Estimating future costs for Alaska public infrastructure at risk from climate change. *Global Environmental Change, 18*(3), 442–457.

Luong, V. V. (2008). *Building a database of meteorological and hydrological characteristics for serving of flood control in the Ho Chi Minh City area*, Sub-Institute of Hydrometeorology and Climate change. (Vietnamese version).

Mai, V.K., Tran, T.H., Ho, C.T., Nguyen, P.D. (2019). Study the responsiveness of the drainage system in Ho Chi Minh City in the context of climate change. A research project at Ho Chi Minh City level.

Neumann, J. E., Price, J., Chinowsky, P., Wright, L., Ludwig, L., Streeter, R., Jones, R., Smith, J. B., Perkins, W., Jantarasami, L., & Martinich, J. (2015). Climate change risks to US infrastructure: Impacts on roads, bridges, coastal development, and urban drainage. *Climatic Change, 131*, 97–109. https://doi.org/10.1007/s10584-013-1037-4

Nguyen, T. Q., Pham, T. T. N., & Dao, N. K. (2017). Developing IDF curve of extreme rainfall at Tan Son Hoa station for the period of 1980–2015. *Science & Technology Development, 20*(M2), 17.

Nie, L., Lindholm, O., Lindholm, G., & Syversen, E. (2009). Impacts of climate change on urban drainage systems - a case study in Fredrikstad, Norway. *Urban Water Journal, 6*, 323–332.

Olsson, J., Berggren, K., Olofsson, M., & Viklander, M. (2009). Applying climate model precipitation scenarios for urban hydrological assessment: A case study in Kalmar City, Sweden. *Atmospheric Research, 92*, 364–375.

Tait, S.J., Ashley, R.M., Cashman, A., Blanksby, J., and Saul, A. J (2008). Sewer system operation into the 21st century, study of selected responses from a UK perspective. Urban Water Journal, 5 (1): 77–86.

Tran, T.H., Ngo, N.T (2015). Study on the calculation of flood inundation in district 12 - Ho Chi Minh City using the MIKE FLOOD model. A research project at Ho Chi Minh City level.

Willems, P., Arnbjerg-Nielsen, K., Olsson, J., & Nguyen, V. T. V. (2012). Climate change impact assessment on urban rainfall extremes and urban drainage: Methods and shortcomings. *Atmospheric Research, 103*, 106–118.

Chapter 48
Enhancing the Efficiency of Land Dispute Mediation for Ethnic Minorities in Van Canh District (Binh Dinh, Vietnam)

Hang Thi Pham, Doi Trong Nguyen, Hien Dieu Thi Bui, Thuy Le Thi Phan, and Anh Vu Pham

Abstract This paper deals with the status of land dispute mediation in grassroots organizations for ethnic minorities living in Van Canh district. Land disputes in the mountainous district of Van Canh are quite complicated, and the proportion of disputes related to ethnic minorities is quite high, mainly because of disputes over forest land and rights boundary disputes. It originates from economic benefits and conflicts between the ethnic minority households in the production and cultivation areas and the influence of religious and customary law on people's lives. Analyzing the facts of land dispute reconciliation and studying customary laws and aspirations of local ethnic minorities are integral for local officials proposing solutions to improve the efficiency of land dispute settlement, which relates to ethnic minorities in the direction of effectively combining the law and the customary law in the study area.

Keywords Land dispute · Mediation · Ethnic minority · Van Canh District · Vietnam

1 Introduction

Van Canh is a mountainous district, located in the southwest of Binh Dinh province. The total land area is 77,310.3 ha, and the population of Van Canh district is 28,168. The population density of Van Canh is over 36 per square kilometer. Ethnic

H. T. Pham (✉) · D. T. Nguyen · H. D. T. Bui · T. L. T. Phan · A. V. Pham
Quynhon University, Binh Dinh, Vietnam
e-mail: phamthihang@qnu.edu.vn

A. T. Nguyen, L. Hens (eds.), *Global Changes and Sustainable Development in Asian Emerging Market Economies Vol. 2*, https://doi.org/10.1007/978-3-030-81443-4_48

minorities make up 40% of the Van Canh population, concentrated in communes such as Canh Lien, Canh Hoa, Canh Thuan, Canh Hiep, and Van Canh town. The impact of low levels of education of ethnic minorities, the lack of legal awareness, adding the traditional religious and cultural beliefs, and the customs and traditions of each ethnic minority have important influences on their lives. This is leading to many bottlenecks of land use administration.

The study shows that the land use process of ethnic minorities arises from land use conflicts, which cause political and social instability. Unresolved conflicts have led to land disputes between ethnic minorities and state-owned agriculture, forestry farms, between ethnic minorities and Kinh and within ethnic groups. If good land dispute conciliation at the grassroots contributes to decreasing the number of letters to courts and district-level People's Committees at least this will ensure political stability and people's lives. At the same time, this will help to convert and persuade the people, to guarantee the strictness of the law and to conform with the customs of the ethnic, to maintain the national solidarity and social, political stability.

The survey results show that the effective land dispute mediation in Van Canh district, Binh Dinh province, is not high owing to the influence of customary laws of the Cham and Ba Na ethnic groups. In order to contribute to improving the efficiency of land dispute conciliation in Van Canh district in particular and provinces throughout the country in general, where there are a large number of ethnic minorities living, the authors have implemented the project "Enhancing the efficiency of land dispute mediation for ethnic minorities in Van Canh district, Binh Dinh province."

2 Methodology: Data Collection

2.1 Secondary Data Collection

Secondary information was collected and processed using document research methods and sources of information including: legal documents and law policies on land, customary laws of the Cham and Ba Na ethnic minorities, statistical data, papers, and research documents. These data sources are collected from state agencies, libraries, and the internet. Secondary data are exploited for generalizing research-related issues such as the influence of customary law, the role of village elders and village heads in land use management of the Cham and Ba Na ethnic minorities, and the grassroots land conciliation process. Secondary data help to support design research and analysis during the research process.

2.2 *Primary Data Collection*

The study was conducted in Van Canh town and communes (Canh Hoa, Canh Hiep, Canh Thuan, Canh Lien). The authors designed the questionnaire for two groups of officials, including cadres and residents, as follows.

2.2.1 Investigation Officers

This study conducted a survey of 50 interviews, include the following components:

- The Head of the Department of Natural Resources and Environment of Van Canh district, the expert in charge of resolving land disputes.
- Chairperson, Vice Chairperson, Judicial Officer, Cadastral Officer, President of the Women's Union, the Farmer's Union, the People's Committee of Van Canh Town, the People's Committee of Canh Thuan, Canh Hiep, Canh Hoa, and Canh Lien.
- Specialist of the People's Court of Van Canh district.
- Van Canh Protection Forest Management Board.
- Ha Thanh Forestry Limited Company.
- Van Canh District Ethnic Minority Office, etc.

The quality of data collection is high and reliable because the people surveyed play an important role and take part directly in mediation and land dispute resolution. Moreover, most of them have professional qualifications.

The questionnaire is designed as a semi-structured form, mainly focusing on qualitative information. The collected information plays a key role, helping the analysis process to be comprehensive and close to reality in the study area.

Contents of investigation: focusing on the issues of customary law, the situation of disputes, the conciliation of land disputes, and the cooperation of ethnic minorities when mediating and handling land disputes.

2.2.2 Household Survey

Basis for selecting sample size: according to Hatcher (1994), the sample should be five times larger than the number of observed variables. As a social research topic on land dispute mediation, the Hatcher sampling method was used with 200 samples were surveyed in five study areas including Van Canh town and communes (Canh Hoa, Canh Thuan, Canh Hiep. and Canh Lien). Based on properties and relevances of land users to land dispute conciliation, it is possible to divide the household survey into four main groups, specifically as follows:

Group 1: the village elder, village head, and deputy head of the village, the head of family.

Group 2: Cham and Ba Na people among whom land disputes have arisen in the study area.
Group 3: Cham and Ba Na people is living in disputed parts of the study area.

3 Results

3.1 Customary Law in Land Management and Use

"Ethnic minorities" is a scientific concept commonly used in the world. Western scholars say that this is an ethnic ethnography that refers to ethnic minorities. In some cases, people equate the meaning of "ethnic minorities" with "underdeveloped people," or "backward people." There are many cases, including those governed by the political views of the ruling class in each nation (Vietnamese Committee for Ethnic Minorities and Mountainous Areas [CEMMA], 2001). Customary law (Thai people called Ho khong, Ede people called Khkkad, M'nong people called Khkkau, Ma people called N'Ri, Gia Rai called Oiphiphian, Ba Na people called kroni nid string, etc.), basically, this is a treasure trove of indigenous knowledge and local knowledge (Indigenous Knowledge, Local Knowledge) of ethnic groups (Vietnamese Committee for Ethnic Minorities and Mountainous Areas [CEMMA], 2017).

Customary law has both legal content and sanctions (stipulating criminal acts, types and severity of crimes, adjudication and criminals), and the nature of customs and traditions (conventions, commandments, moral advice, behavioral guidance for each individual, social opinion to regulate human behaviors). In Customary Law, the Cham consists of eight chapters covering the contents of customs, community relations, leadership relations, inheritance, etc. The things in the chapters and roles of the village elders are considered to be the leader:) "The largest in the flat land has sea, the largest in the sky has the sun, the largest in the village has a leader (village elder)" is quoted in the Cham customary law to affirm the great role of the "village eldest" in the life of the Cham people.

The customary laws of ethnic minorities contribute to the protection of traditions and customs, natural resources (perennial forests, watershed forests, sacred forests, and water sources, etc.). Van Canh ethnic minorities consider the forest to be an invaluable resource of the village; the forest has a close relationship with the community so Customary laws clearly specify the importance of forest protection. They respect the rules of the community about establishing sovereignty over forests and forest land of each family and clan.

Van Canh district has ten ethnic groups living in: Kinh, Bana, Cham, Thai, Muong, Nung, K'ho, E De, Nguoi, and Tay with 28 ethnic minority villages in which there are two main ethnic minorities, accounting for nearly 20% of the district's population. That is, the Cham ethnic group living mainly in Canh Hoa commune, Van Canh town, and the Bana ethnic group mainly in Canh Lien, Canh Hiep, and Canh Thuan. Specific data on the number and distribution of ethnic minority habitats in Van Canh are shown in detail in Table 48.1.

The Bana in Van Canh are called Bơhna-Chamroi, or Bana Bang huon. In the study area, Cham people live together with with Bana and Kinh people. Cham people in the Van Canh district, Binh Dinh province, have many different names: Cham Hroi, Hroi, A roi, Cham Dac Ray, Cham Hoang, Cham Deo, etc. According to the Cham in Van Canh, Cham Hroi (or Hroi, A roi) there are Cham in the highlands and Cham people in the mountains. Cham ĐacRay or Cham Ho is also called Cham towards the sunrise. Currently, the Cham in Van Canh are very proud to associate the meaning of Cham with the meaning "Cham sunrise." In Van Canh, The Cham is named Cham Deo after the Cham in Phu yen.

Table 48.1 shows that the largest number of people were concentrated in Canh Thuan commune (27.8% of the total percentage of ethnic minority households in Van Canh district), Van Canh town, Canh Lien commune, and are evenly distributed in the remaining communes.

Table 48.1 Total number of ethnic minorities in Van Canh district

No	Town, commune	Ethnic	Total number of ethnic minorities in Van Canh					
			2014		2015		2016	
			Number of households	Number of people	Number of households	Number of people	Number of households	Number of people
1	Van Canh town	Cham	375	1,275	380	1,322	387	1,331
		BaNa	315	1,071	318	1,106	323	1,111
		Different	25	90	26	98	28	97
		Total	715	2,436	724	2,526	738	2,539
2	Canh Hiep	Cham	315	1,171	335	122	317	125
		BaNa	150	558	161	588	162	601
		Different	5	21	5	21	6	23
		Total	470	175	501	1,829	505	1,874
3	Canh Thuan	Cham	395	1,444	417	1,501	425	1,525
		BaNa	382	1,401	404	1,454	410	1,471
		Different	7	30	8	35	8	37
		Total	784	288	829	299	843	3,033
4	Canh Hoa	Cham	277	1,016	297	1,081	307	1,105
		BaNa	197	723	210	765	218	785
		Different	3	11	4	16	4	18
		Total	477	175	511	1,862	529	1,908
5	Canh Lien	Cham	30	111	34	120	36	127
		BaNa	580	2,146	640	2,284	648	2,284
		Different	3	13	3	13	3	15
		Total	613	227	677	2,417	687	2,426

Source: Ethnic Minority Office of Van Canh district (2019)

3.2 *Types of Land Disputes and the Causes of Disputes Related to Cham and Ba Na People*

In the study area, common types of land disputes include: disputes over land use boundaries; disputes related to the transfer of land use right; disputes related to land acquisition, compensation, assistance, and resettlement; and disputes over subjects being granted land use right certificates (LURCs). The main type of dispute are the land disputes of "Ca Than"[1] (In the terminology of the Cham people, Ba Na, the land of Ca Than is forest land originating from grandparents of the people who have declared it for a long time, leaving it to their family descendants).

The above land disputes arise between the following entities.

First, land disputes arise between land users: disputes between ethnic minorities and disputes between ethnic minorities and the Kinh ethnic group. Second, disputes between ethnic minorities and state-owned agriculture, such as forestry farms (mostly in Canh Lien commune). The specific causes of land disputes related to ethnic minorities are based on the survey data of 50 officials and 200 households. The specific figures are shown in Table 48.2.

Survey results show that the main causes of land disputes are customary law, poverty, and a lack of productive land. Despite having many preferential policies on productive land for the Cham, the BaNa follows the State's land and forest allocation policies, but many ethnic minorities have not been able to access them yet. A number of cases stemming from land management violations, both in terms of people and management systems, such as the lack of digital cadastral maps, lack of

Table 48.2 Causes of disputes related to Cham and Ba Na people

Numerical order	Causes	Opinion (%)
1	Lack of productive land	80
2	The belief about the land of "ca than" of the grandparents	96
3	Poverty, intentional violations	10
4	Failure to understand the laws on land in land use	68
5	Lack of exemplary and legal violations when conducting land management of cadres and civil servants	4
6	Lack of accuracy of cadastral map measurement	20
7	The unreasonability of process and procedure in land acquisition	16
8	Low efficiency of propagation and dissemination of land law education	80

Source: Survey results (2019)

[1]This is the type of forest land that all interviewees (including officials of Van Canh district People's Court, officers of Van Canh Town People's Committee, CPCs of Canh Hoa, Canh Hiep, Canh Lien, Canh Thuan; village elders; the object of the dispute) refer to when talking about the forest land dispute.

up-to-date information, and the improperness of the land change monitoring book lead to many disputes and irregularities in land use.

In many cases, many ethnic minority people have been granted LURCs with the same number of parcels and maps, leading to many people being issued with an LURC at the same land location, causing disputes. Equally important, many ethnic minority households fail to grasp the state's policies on land legislation, leading to land disputes. According to Point h, Clause 1, Article 64 *"Annual crop land must not be used for 12 consecutive months; land for perennial crops must not be used for 18 consecutive months; Afforestation land is not used for 24 consecutive months"* and shall be recovered without compensation for land by the State.

According to the survey results, 100% of ethnic minority households do not know the above regulations; thus, the ethnic minorities often do not use the land continuously. After reclamation, they often go away in the form of shifting cultivation and a nomadic life; therefore, the area will be recovered by the State. The local land management officials[2] say that the awareness of land laws of ethnic minorities is increasing. However, according to the survey by the authors: customary law on the origin of exploitation and ownership of land is deeply imprinted in the life of the Cham and Ba Na ethnic people. They[3] said that the their ancestors had reclaimed the forest land, marking the territory by using machetes, knives, and slitting of large trunks of trees belonging to their family or clan. Even after the reclamation of land, they left for 20–30 years and later returned, believing that the "Ca Than forest" could still be used by the family. It is the above thought that caused the dispute between the children and grandchildren who have reclaimed the land and the people who are currently cultivating that land (the State recovered it, with no compensation for households to exploit and allocate land for other households). For ethnic minorities who do not understand the State's legal policies, the authors think that officials need to explain gently to ethnic minorities so that they understand the State's legal policies: each land user only has the right to use the land for a period of time prescribed by law; otherwise, it will be revoked continuously by the State.

The period of assignment of land use rights of ethnic minorities by the State is 10 years from the date of issuance of the land allocation decision, as prescribed in Clause 3, Article 192 of the 2013 Land Law. In fact, there are two different opinions: 55% of the interviewed households do not transfer, because they do not understand the law; they partly implement the wills of their grandparents: keeping the land for subsistence production and not for selling. The other group, accounting for 45%, believes that they have the right to sell their land for which the State granted an LURC, and decide whenever they want. Thus, according to the provisions of the law, if ethnic minorities have transferred land for less than 10 years, this will lead to petitions and complaints.

---

[2] They are the Chairman of Commune People's Committee, the cadastral officials and the judicial officials who were interviewed in the study area.

[3] Including the Cham, Bana were surveyed and interviewed.

Regarding land and forest ownership: 80% of ethnic households do not distinguish between "forest ownership"[4] and "forest use rights",[5] which also causes land disputes with state officials. This number of households stated that the forest belongs to them, including the right to dispose of the forest land being cultivated. This awareness is not legal because according to the current law, users only have the right to use, not to own the land and forest resources, as prescribed in Clause 10, 11, Article 2, Forest Law 2017. This means that ethnic minority are still not fully aware of their rights to use land and forests. The serious consequence of unawareness of their rights to forest land is the land disputes between the people and the official agencies of the State. Therefore, the propagation and dissemination of the law for the Cham and Ba Na people needs to be done more often, in order to raise awareness.

The survey results on land disputes of the Cham and Ba Na ethnic minorities with the Forestry Corporation show that although the Cham and Ba Na ethnic minorities in the locality lack land to use, the land of the Van Canh District Protection Forest Management Board and Ha Thanh forestry corporation has a high rate of unused land. Ha Thanh Forestry Co., Ltd. is managing and using 18,778 ha of forests, including: 13,074.7 ha of natural forest land, 2196.4 ha of planted forest land and 3471.3 ha of non-forested land. In total, 3471.3 ha of forest (nearly 46%) has been occupied and claimed by local people. In total, 232.4 ha (including 84.6 ha of nonforest land and 147.8 ha of shifting land) managed by Van Canh District Protection Forest Management in 11 lots in plot 2, sub-zone 341 and plot 1, plot 3, sub-area 349 Canh Lien commune have not yet been put into use, whereas the Cham and Ba Na people lack land for cultivation. According to the survey results, as the State has allocated the land so far, the Van Canh District Protection Forest Management Board and Ha Thanh Forestry Company Limited have not planted forests in large areas. Besides, Ha Thanh Forestry Limited Company land acquisition of people but the competent state agency has implemented the decision allocation. This led to the situation in which ethnic minorities did not agree to hand over 260 ha of land in Canh Lien commune to Ha Thanh Forestry Company Limited. The cause of this problem is the local people thinking about the land of their grandparents who have been cultivating it for a long time, "Ca Than", as well as a difficult life, a lack of productive land, and no stable job. Moreover, 80% of people in Canh Lien commune are ethnic minorities of Cham and Ba Na. Thus, these are the causes leading to the illegal logging, encroachment, and common dispute over forest land in Canh Lien commune. Up to now, more than 17 ha of forest for productive function planning and protected forest have been destroyed; over 13 ha of forestry land planned for production and protection functions have been illegally encroached.

---

[4] Planted productive forest ownership includes the right to own, the right to use, and discretion of the forest owner toward plants, animals, and other property in the forest invested in by the forest owner during the allocation/lease term of afforestation (Clause 10, Article 2, Law On Forestry 2017).

[5] Forest use rights means the rights of the forest owner to utilize the forest and enjoy benefits arising therefrom. (Clause 11, Article 2, Law on Forestry 2017).

Overall, the issue of land dispute conciliation is affected by many groups of natural factors, land production means, economic issues, people's livelihoods, social factors, and legal factors (policies on land and forest allocation). More importantly, the customary law should be taken seriously when resolving disputes for ethnic minorities.

In Van Canh district, there are 1663 poor households. There are 334 households/6.68 ha lacking residential land; 312 households/40.56 ha lacking land for production; 277 households were without land for production and 112 households have neither residential land nor productive land, accounting for 6.73% of the total ethnic minority households in Van Canh. In order to improve people's lives, to reduce encroachment and land disputes, the State has issued many preferential policies to help ethnic minorities to obtain productive materials (land), at the same time supporting the people to settle down, supporting loans for production and employment. However, Cham and Ba Na in Van Canh town, Canh Hoa commune, Canh Hiep commune, Canh Thuan commune, and Canh Lien commune have not been able to fully access these State policies (specific results in Table 48.3).

There are three reasons why policies are not implemented for poor ethnic minority households who are in difficulties, lacking residential land and productive land: first, the difficulties with the local budget; second, the bottleneck regarding the land fund; third, there are still many illegal issues in supporting productive land for poor

Table 48.3 The level of access to preferential policies on residential land and productive land for ethnic minorities who lack productive land

No	Policy	Access (%)	No access (%)
1	Specific policies to support socio-economic development in ethnic minority and mountainous areas on resolving productive land and changing jobs under decision no. 2085/QD-TTg	45.0	55.0
2	Circular no. 02/2017/TT-UBDT solving production land, changing occupations, and preferential credits for poor ethnic minorities and poor households in areas with exceptional difficulties and lacking productive land	60.0	40.0
3	The project of concentrated sedentary farming according to the approved plan to continue arranging the population in order to stabilize the life and develop production for nomadic ethnic minority households in accordance with Decision No.1342/QD-TTg dated 15 August 2009	10.0	90.0
4	Policies on the support of residential land, productive land, daily-life water, and preferential credit incentives, as well as policies to arrange and stabilize the population for ethnic minority households who still practice shifting cultivation	5.0	95.0
5	Supporting policy on forest protection and development according to decree no. 75/2015/ND-CP (example, planting production forests: The support level ranges from 5,000,000 to 10,000,000 VND/ha for buying seedlings and fertilizer; support for forest protection is VND 400,000/ha/year, etc.)	40.0	60.0

Source: Survey results (2019)

ethnic minority households. The process of implementing land policy for ethnic minorities, the land is not allocated to the right beneficiaries in some cases, causing loss of confidence for the local people with the guidelines and policies of the State.

According to the authors' group, in order for the land fund to support ethnic minorities, Van Canh District People's Committee needs to organize meetings and direct the specialized Natural Resources and Environment Office to directly guide the communes in detail. The town shall review, draw up statistics, make a general list of ethnic minority households lacking residential and productive land, and send them to the provincial People's Committee for approval. The Ethnic Minorities Committee should coordinate closely with departments, branches, and People's Committees of communes and districts with regard to settling residential and productive land for people, regularly organizing cross-sectoral inspection teams, reviewing and allocating land funds appropriately and with the right subjects. It is necessary to handle those who intentionally violate the law in a strict manner, causing loss of confidence in the government, the Party, and the State.

3.3 *Conciliation of Land Disputes*

Conciliation in land dispute resolution is significant. The characteristics of land disputes are sensitive and complex; if land disputes are not handled promptly, they will increase the risk of socio-political instability. Thus, land dispute conciliation is the best way to support and help to reduce the pressure on competent state agencies in land dispute resolution. Land dispute conciliation at the grassroots level is only conducted for land disputes over rights and obligations between land users.

For land disputes arising between land users and public authorities in land management (land disputes of an administrative nature), under current law, conciliation is not required, under the authority of competent state agencies. In the study area, land disputes between households in Canh Lien commune and the Protection Forest Management Board, Ha Thanh Forestry Co., Ltd. are an example of these types of disputes.

3.3.1 The Process of Land Dispute Conciliation for Cham and Ba Na

The process of land dispute conciliation for the Cham and Ba Na people in the study area is in accordance with the Land Law 2013, Decree 43/2014/ND-CP and the Law on grassroots conciliation in 2013, including three main steps as follows (Fig. 48.1):

Under the current land law, the state encourages parties to resolve disputes through self-conciliation or grassroots conciliation. If the conciliation is unsuccessful, the dispute will continue to be mediated at the communal level under the coordination of social organizations at the commune. However, the procedure for conducting a grassroots conciliation is not specified by law.

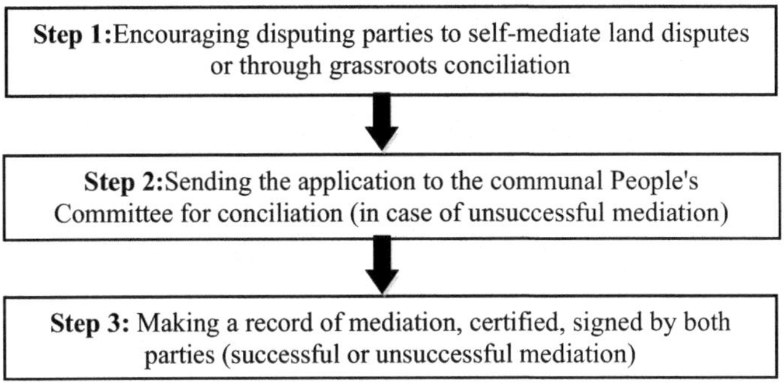

Fig. 48.1 Procedures for land dispute conciliation

Fig. 48.2 The processing of grassroots land dispute conciliation

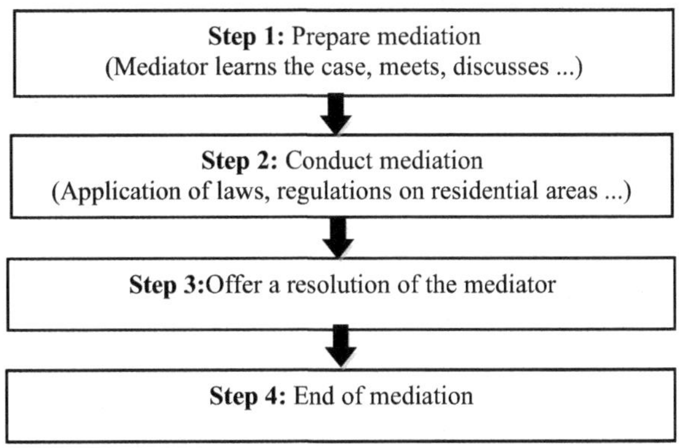

According to research results, the process of conducting a mediation process includes four main steps (Fig. 48.2):

When preparing for conciliation, the mediator must find out the contents of the case, the causes of conflicts, disputes, and interests that each party is seeking. At the same time, it is necessary for the mediator to understand the land law provisions that help to analyze the case and make the parties understand the issue. At the study sites, the conciliators in the villages are the elders, the heads of the villages, the deputy heads, and the ones who receive the trust of the villagers. According to Clause 2, Article 20 of the Law on grassroots conciliation, the time for conciliators to reach agreement with the parties to the conflict is 3 days from the date of being assigned to mediate. The time prescribed by law is very difficult because the study area is mountainous and rugged, and the study subjects are mainly co-workers. Thus, communication and meeting, removing as well as imparting information to people face many obstacles. The Cham and BaNa ethnic people in the locality often mediate at communal houses and cultural houses.

The mediation process is based on experience with people. After stating the purpose and meaning of the mediation, the conciliators agree with the parties on a number of conventions, ways of working, and try to create an atmosphere of intimacy, openness, and sincerity, fairness between the parties, helping them to easily exchange and negotiate. After the mediator has invited each party to present the incident, the parties have the right to supplement their opinions, and present their arguments and views. Persons with related rights and obligations (chiefs, prestigious people in the village, in their clan) express their views on the issue in dispute or conflict. The mediator sums up the dispute issues in a case analysis: application of legal provisions on land and civil to each disputing party; analysis of customs, factors, customary law, and social and ethical traditions; application of regulations on residential areas with contents related to land use management (each village has a residential village convention). The head of the unions (women's union, youth union, farmers' union, veterans' union) of the village coordinate with the village chief and deputy chief in organizing the mobilization of the people to implement the contents of the convention.

After analyzing the dispute for the parties concerned, the mediator suggests options for resolving it and invites the parties to present their dispute settlement plan. The mediator continues to analyze the benefits of successful mediation, and the legal consequences that the parties may face if they continue to dispute and commit misconduct.

Concluding the mediation, on the basis of plans to resolve disputes by mediator or parties, the parties discuss and agree on a dispute settlement plan, the specific responsibilities of each party, and the time limit for the performance of the responsibility. Conciliators finalize the contents of an agreement, responsibilities, and obligations of each party. If the parties agree to prepare a written record of successful mediation, the conciliator prepares the document and the parties and the mediator sign it. In cases in which the parties fail to reach an agreement, which means that mediation fails, the conciliator guides the parties with the right to request conciliation or request competent agencies or organizations to settle according to the provisions of law. If the parties request that a written mediation be unsuccessful, the mediator prepares a written unsuccessful mediation. In particular, in many cases the parties only agree on a part of the disputed issue, and the mediator continues to persuade the parties to discuss and continue the agreement to reach a final conclusion.

Thus, in order for grassroots conciliation to achieve a high success rate, conciliators play a key role. The conciliator must have a firm grasp of the land law and civil law system, understanding of customary laws and conventions, at the same time as being flexible, compliant, supple when resolving conflicts between the two parties to the dispute. Results of interviews with 40 grassroots mediators (Table 48.4) in the study area showed that the capacity and qualifications were not high. College and university graduation rates were very low. The level of updating of legal policies on land is not taken seriously (0.0% of conciliators update their knowledge of the law on a regular basis). However, notably, 95% to 100% of the mediators master the conventions and customary laws of ethnic minorities. This is also an important basis in land dispute conciliation, which also shows that customary law plays a more important role in the life of the people than the legal policies of the State.

An intermediary plays a key role in helping the mediator and the conciliation. The rate of successful mediation is very high, but the payment level for mediators according the law is very low; the maximum is VND 200,000/case (Decree

Table 48.4 Competence and qualifications of grassroots conciliators

Competence and qualifications of grassroots conciliators	Level	Result (%)
Graduation	Graduated high school	12.5
	Graduated junior high school	55.0
	Graduated primary school	25
Qualification	Graduated intermediate, college, university	7.5
Understanding land law, conciliation law, civil law	Have a grasp and have the capability to apply	0.0
	Understanding basic knowledge	37.5
	Do not exactly understand the policy	62.5
Updating knowledge of current law	Regularly	0.0
	Low frequency	5.0
	Seldom	5.0
	Never	90.0
Understanding the convention of residential areas		95.0
Understanding local laws, customs, and practices		100.0

Source: Survey results (2019)

No.15/2014/ND-CP). The above level of expenditure is not commensurate with the efforts of the mediator. It is noticeable that the 40 conciliators[6] said that they did not receive any money after the conciliation was conducted; some of the respondents said that they only received petrol and transportation assistance.

In the research team's opinion, although grassroots conciliation is a self-governed and voluntary activity, to encourage mediators to keep up to date on land law policies and have enthusiasm for the work, the fund for motivating mediators is very important. In fact, reconciliation takes time and effort.

As stated in the study, conventions of residential areas play an important role in grassroots land dispute conciliation. Currently, regulations and conventions have become increasingly important in supplementing the law when it is necessary to handle specific issues arising from the life of the community. Van Canh town has 11/11 villages that have built conventions; the contents of the convention are consistent with the traditions and customs and the rules of building a cultural lifestyle; Canh Lien has 8/8 conventions. The convention was created from four strict steps: establishing a drafting group and a drafting organization; organizing the collection of opinions of agencies and people in the draft convention; discussing conventions; and approving conventions. However, survey results show that many conventions are still general, do not yet have the characteristics of villages and ethnic groups,

[6] Including elders, heads of villages, heads of quarters, deputy heads of villages in the study area (oldest village of Canh Thanh village, Canh Hoa commune; heads and deputy heads of Dak Dam quarter, Van Canh town; Mr. Doan Van Bac, chief of Hon Meo village Canh Thuan commune; Head of Hiep Tien village, Canh Hiep commune; etc.)

and the feasibility is not high. The process of checking the practicing convention of officials has not been frequent.

The authors believe that we need to raise awareness about the role of conventions in the villages, from officials to local people. If the conventions are made to ensure the promotion of the village and hamlet's relationship and encourage the community's self-discipline in observing the law, such as the issue of deforestation, forest burning, land encroachment, significantly contributing to the settlement of land dispute conciliation in the villages, promoting the effectiveness of the reconciliation teams in 11/11 villages in Van Canh town. It will come as no surprise that over 90% of conflicting land and housing cases will be settled, ensuring stable political issues at the local level, minimizing complaints in court and in the district People's Committee. Currently, funding for organizing the review, amendment, and supplementation of conventions and regulations is not available. This is also one of the main causes of difficulties in implementation.

3.3.2 Land Dispute Conciliation from 2013 to 2019

Figure 48.3 shows that in the 7 years from 2013 to 2019, in Van Canh district, the number of land disputes is low, on average ranging from 10 to 33 cases. Van Canh Town, Canh Hoa Commune, Canh Thuan, Canh Hiep Commune, and Canh Lien Commune are the places where land disputes involving ethnic minorities account for a large number. Cham and Ba Na people in the study area account for over 70% of the local disputes. However, the average rate of successful conciliation at the

Fig. 48.3 Land dispute conciliation from 2013 to 2019 in Van Canh district

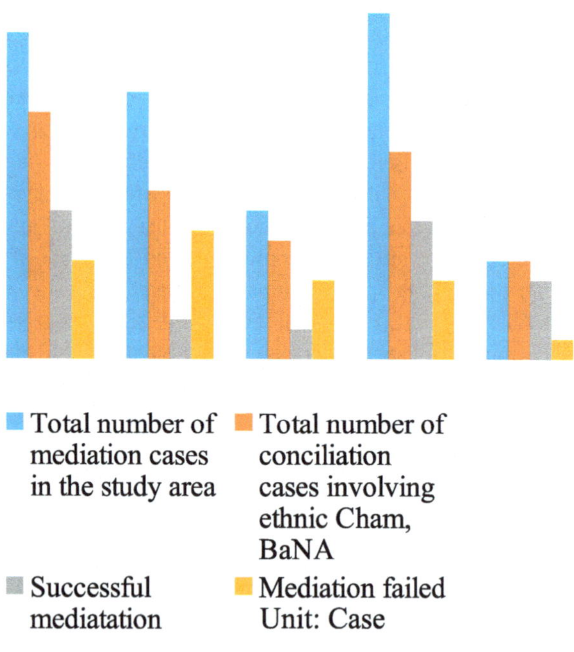

commune level is less than 50%. Canh Hiep is the locality with the highest number of land disputes, but the locality with the highest number of successful conciliation cases is Van Canh town.

From 2013 to the first 9 months of 2019, in Van Canh town, there were 33 land dispute conciliation cases at the commune level. Although the number of local land disputes was not high, up to 75.6% of land disputes related to the Cham and Ba Na ethnic people. Results of local reconciliation show that the successful conciliation rate was 60%. However, the number of unsuccessful conciliation cases still accounted for a high proportion and tended to increase. In 2019, there were 3/5 cases, accounting for 60.0% of unsuccessful land disputes in urban areas. Specific figures are shown in Table 48.5.

In the locality, 37.54% are ethnic minorities including Cham, BaNa, living in 11 villages. Poor households account for 39.32%. Households lacking productive land account for 25% of the population. The deforestation, encroachment, and forest disputes in the town still take place; notably, the "village elder" also participates in land encroachment (results of the interview with households in Van Canh town).[7]

The number of disputes in the town is low, and the results of successful conciliation are not high either, only 15/25 cases, accounting for 60% of the number of disputes. The reason is that up to now, the 2003 cadastral map of the town has not been re-measured because there is no budget; paper maps are not updated or revised; the data of the land parcel are not updated and adjusted in the inventory list and the cadastral book.

In Dak Dam village, Van Canh town, after reconciling, the village elders asked each disputing party to pay VND 200,000 as athreat, to avoid disputes taking place. In Van Canh town, generally there are disputes between land users, but not between the household and the Forest Management Board.

Table 48.5 Results of land dispute conciliation in Van Canh town from 2013 to 2019

No.	Year	Total number of local land disputes (cases)	Land dispute between the Cham and Ba Na people; between the Cham and Ba Na people and the Kinh ethnic group (cases)		
			Total	Successful mediation (cases)	Unsuccessful mediation (cases)
1	2013	4	2	0	2
2	2014	1	1	1	0
3	2015	5	4	3	1
4	2016	5	5	4	1
5	2017	4	3	2	1
6	2018	7	5	3	2
7	2019	7	5	2	3
Total		33	25	15	10

Source: People's Committee of Van Canh town (2019)

[7] Including Mr. Gio Lui Pham, Mr. Tai Van Doan, Mr. Dam Van Doan in Van Canh town.

From the results of the research, the authors found that the forest allocation and protection in the locality were carried out strictly linking from the Forest Management Board to the people. As follows, according to the Dak Dam village elder,[8] in the town, The natural forest area in the town had 996 ha in 2005. Then it decreased by 9.6 ha in 2009 due to landslides. The current natural forest area had 4 ha zoned off for regeneration, 982.4 ha the State assigned to 47 contracted households, 30 ha for each. Once every 6 months a check is carried out, when people are assigned to manage the forests. Six months after clearing, the person assigned to protect the forest will report back to the head of the quarter and the head of the street reports to the forest management. If a dispute arises, the mediation team consists of 7 to 11 people including Vietnam Fatherland Front Committee of a commune, the party secretary, the head of the village, the village patriarch (old quarter), the women's union, the veterans' union, the youth union, the agricultural association (Table 48.6).

Disputes related to compatriots accounted for 63.0% of disputes in Canh Hoa commune, of which up to 76.5% of conciliation cases failed. The survey results show that disputes mainly occurred in Canh Thanh village, Canh Hoa commune, in the 375B area, in natural forests. In addition, inadequacies arose when households without productive land were officially provided with land, making LURCs, causing discontent in people's lives, leading to the main cause of the land disputes and unsuccessful local mediation.

For the compatriots in Canh Thuan, the forest is an integral part of the lives of the Cham, BaNa. The village elders often represent the villagers to worship the god of the forest on the second of the first lunar month. The offerings include a chicken and a bottle of wine in the forest. They read the names of mountain spirits, village names, praying for the god to protect the villages. In Canh Thuan, there are very few

Table 48.6 Results of land dispute conciliation in Canh Hoa Commune from 2013 to 2019

No.	Year	Total number of local land disputes (cases)	Land dispute between the Cham and Ba Na people; between the Cham and Ba Na people and the Kinh ethnic group (cases)		
			Total (cases)	Successful mediation (cases)	Unsuccessful mediation (cases)
1	2013	3	2	0	2
2	2014	4	3	0	3
3	2015	9	7	2	5
4	2016	1	1	0	1
5	2017	0	0	0	0
6	2018	2	1	0	1
7	2019	8	3	2	1
	Total	27	17	4	13

Source: People's Committee of Canh Hoa commune (2019)

[8] Interviewees would not like their names to be made public.

conflicts taking place with the ethnic households, only 1.2 cases a year. However, when a dispute arises, it is often difficult to settle successfully. The rate of unsuccessful conciliation related to disputes of ethnic minorities accounts for 8/12 cases (66.7%). Specific data are shown in Table 48.7.

In Canh Thuan commune, although the propagation of land law in the locality was concerning, the village elders often came to every house to call for meetings, propagandize the land policy in the communal house, with legal aid once a month. When there are new official instructions, the village management board gathers the households by playing games between 5/6 o'clock and 7 o'clock meetings with the people. The issue of hand-writing papers in the locality is still popularl, causing disputes in Canh Thuan commune.

The village patriarch, the village head, and deputy head of the commune[9] say that the local people in the village lacked productive land, but that the State did not care or support them. According to the research results, although the government has policies to support ethnic minorities who lack land to use, the policy does not reach the people owing to the deliberate wrongdoing in land management by competent officials. Specifically, to solve the shortage of productive land for people, the government has created favorable conditions for the People's Committee of Canh Thuan Commune to receive VND 300 million, in order to recover 29.2 ha of land, solving production problems for 30 ethnic minority households. Instead of allocating land to households of ethnic minorities, the People's Committee has incorrectly allocated land to three Kinh households, causing land disputes in the locality.

In Canh Hiep commune, disputes related to ethnic minorities made up 66.7%; 38.1% of mediations were unsuccessful. Specific figures are shown in Table 48.8.

Table 48.7 Results of land dispute conciliation in Canh Thuan Commune from 2013 to 2019

			Land dispute between the Cham and Ba Na people; between the Cham and Ba Na people and the Kinh ethnic group (cases)		
No.	Year	Total number of local land disputes (cases)	Total	Successful mediation (cases)	Unsuccessful mediation (cases)
1	2013	2	2	1	1
2	2014	3	1	0	1
3	2015	4	3	2	1
4	2016	1	1	0	1
5	2017	3	2	0	2
6	2018	1	1	0	1
7	2019	1	1	0	1
	Total	15	12	3	8

Source: People's Committee of Canh Thuan commune (2019)

[9] Interviewees did not want to be publicly identified because this is a sensitive answer; the answer refers to the negative side of local officials in land management. If their name were revealed, it would actively affect their lives.

Table 48.8 Results of land dispute conciliation in Canh Hiep Commune from 2013 to 2019

No.	Year	Total number of local land disputes (cases)	Land dispute between the Cham and Ba Na people and between the Cham and Ba Na people and the Kinh ethnic group (cases)		
			Total (cases)	Successful mediation (cases)	Unsuccessful mediation (cases)
1	2013	3	2	1	1
2	2014	4	2	2	0
3	2015	8	5	3	2
4	2016	12	8	6	2
5	2017	4	3	2	1
6	2018	3	2	1	1
7	2019	1	1	0	1
	Total	35	21	14	8

Source: People's Committee of Canh Hiep commune (2019)

Table 48.8 indicates that the number of disputes taking place in Canh Hiep is quite high compared with the communes and town in Van Canh district. Although the number of disputes is large, the rate of successful conciliation in the commune is quite high compared with the communes and towns in Van Canh district.

According to the results of the survey, the authors found that despite the violations in land management, the locality has quickly adopted policies to amend and grant residential land and make "Certificates of land use rights and ownership over houses and other assets attached to land" for people, contributing to increasing the rate of successful conciliation in land disputes in the locality. Specifically: 37.5 ha of land in Canh Hiep Commune, Van Canh District, which had been previously recognized incorrectly, the authorities quickly recovered and handed the land back to the rightful beneficiary households. Owing to participation of the consulting unit, the commune, ranger, and households are allocated land directly for use. In 2012, in the Hiep Giao village residential area, 36 households were considered for housing land. The local government has thoroughly mastered the allocation of residential land, which it is strictly forbidden to transfer in any form within 10 years from the date of land allocation. This creates conditions for people to settle down, abandon the nomadic way of living, and encroach on the land of other people, causing disputes. In 2013, the town council approved and allocated residential land in the detailed planning area of Hiep Giao village. Of the total number of households: 37/76 poor households will be granted 37 lots with an area of 4351 m^2. Of the 34 households that have been issued Certificates of land use rights and ownership over houses and other assets attached to land, 3 households have not yet fulfilled their financial obligations. At the same time, the town has implemented the project of giving legal rights to citizens. A total of 100 Certificates of land use rights and ownership over houses and other assets attached to land were granted to 71 households, 114 land plots with a total area of 43.65 ha of the Law Society in Binh Dinh Province.

Canh Lien is a highland commune of the Van Canh mountainous district, with a very sparse population. The total area of the commune is 38,416.58 ha, but only 635 households live there consisting of 2427 people. In the structure of land use of the commune, non-agricultural land accounts for only 1.44%; forestry land accounts for 95.22% of the natural area of the whole commune. With low qualifications and poverty, ethnic minorities make up 80% of the population in the region. In Canh Lien commune, disputes among ethnic households are very rare: from 2013 to 2019 there were only 10 disputes; the rate of successful conciliation reached 80% (specific figures are shown in Table 48.9).

The rate of successful mediation in Canh Lien commune is very high, because 100% of the disputes that occurred in communes related to compatriots. The mediators mainly live in the locality and have a good understanding of the life, customs, and practices of the people, persuading the parties to mediation. Notably, in Canh Lien commune, there is a land dispute between ethnic minorities and the Protection Forest Management Board, Ha Thanh Forestry Company Limited, stemming from conflicts about land. Although there is a shortage of productive land for local people, the area of land allocated to the Forest Management Board of the State has not been used effectively. Although the recovery of forest land "Ca Than" has not been completed, the State still issued the decision to allocate such forest land to Ha Thanh Forestry One Member Limited Liability Company, resulting in a long-standing dispute between the Cham, Bana, and the company. Although the recovery of forest land "Ca Than" has not been completed, the State still issued the decision to allocate such forest land to Ha Thanh Forestry One Member Limited Liability Company, resulting in a long-standing dispute between the Cham, Bana, and the company.

Table 48.9 Land dispute conciliation results from 2013 to 2019 in Canh Lien commune

No.	Year	Total number of local land disputes (cases)	Land dispute between the Cham and Ba Na people; between the Cham and Ba Na people and the Kinh ethnic group (cases)		
			Total (cases)	Successful mediation (cases)	Unsuccessful mediation (cases)
1	2013	1	1	1	0
2	2014	2	2	1	1
3	2015	2	2	2	0
4	2016	1	1	1	0
5	2017	2	2	1	1
6	2018	1	1	1	0
7	2019	1	1	1	0
	Total	10	10	8	2

Source: People's Committee of Canh Lien commune (2019)

Table 48.10 Difficulties in conciliation of land disputes at the communal People's Committee for local Cham and BaNa people

No.	Difficulties faced by cadres who go to conciliation of land disputes	Rate (%)
1	Officials have to explain many times to help the Cham, Ba Na ethnic minorities understand the problem	96.0
2	In many cases, the Cham, Ba Na ethnic minorities understood but still tried to break the law	60.0
3	Difficulties in contacting and sending invitations to the ethnic minorities present for mediation	40.0
4	Officials must visit homes to get information about the land dispute	30.0
5	Do not cooperate to resolve mediation relating to "ca than"	80.0
6	Lack of human resources, database, cadastral file system in servicing of land dispute resolution	98.0
7	Legal propagation is not effective	84.0

Source: Result survey (2019)

3.3.3 Assessing the Difficulties in Conciliation of Land Disputes

The difficulties of officials in land dispute conciliation, after grassroots conciliation failed, were assessed. Based on the survey results of 50 officials, the authors discovered their difficulties in solving land disputes related to ethnic minorities in Table 48.10, as follows:

Difficulties stem from ethnic minorities of Cham, BaNa. In many cases, the ethnic minorities of Cham, Ba Na, understood but still intentionally offended the law: for example, seeing that the cultivated land was not regularly managed by the Kinh people, their compatriots (Cham, Ba Na) still intentionally planted and exploited acacia. They still know that the Kinh people will sue them for using their land, but they still do it. They suppose that if the Kinh people want to get back their productive land, they must compensate for assets on the land, i.e., acacia planted on Kinh land.

In many cases, according to the results of interviews with officials at the Van Canh District People's Court, the Kinh people won lawsuits and the ethnic minorities, unsatisfied, intentionally destroyed the Kinh's trees by cutting half of the trunk of the trees, then allow the wind make the trees break. Sometimes, they let the cows into the gardens of the Kinh people to eat small trees, damaging seedlings. If the Kinh do not offer some particular benefits and have a good relationship with the Cham, Bana, they are difficult to use land even if they won the settlement of a land dispute. Because the "cham still shall encroach on land, or cut down crops on the land of the king".

The authors conducted interviews with 50 officials, and 100% said that the land dispute conciliation work complies with the law, not related to the elements of customary law of ethnic minorities. The authors believe that conciliation activities are important in resolving land disputes. However, whether or not the conciliation is effective, not only depends on the state legal system but is also reliant on elements of customary law, customs, and traditions in places in which a large number of ethnic minorities live. Besides, the Kinh conciliation activities will succeed with a high rate when the mediator tries to act in analytical ways, at the same time, having a grasp of the language, customs, and traditions of ethnic minorities.

Difficulties in dealing with disputes stemming from incomplete land database and inaccurate information. The paper cadastral map was surveyed in 1998; thus, the information is inaccurate, out of date, and is not revised regularly. Currently, it is not used in the management of complaints and land disputes. In measurement and cadastral mapping, in many cases when the land user is not identified correctly, the surveying officer considered it to be land of the People's Committee. At the commune level, this resulted in a lot of deviations in the areas of land, boundaries, and land users.

Currently, in the study area, a digital cadastral map has not been established, causing difficulties in management. In 2006, the project of setting up cadastral maps, granting LURCs to eight northern mountainous provinces, in the North Central coastal region, and ninein the Central and Southeast coastal provinces, was approved by the Prime Minister in Decision No. 672 / QD-TTg dated 26 April 2006 (hereinafter referred to as Project 672). Project 672 sets a roadmap for localities to basically complete the land allocation, land lease, and issuance of LURCs in 2008, but so far Van Canh District has not been completed. Stemming from many poor households no funds for measurement and certification, many ethnic minority households migrate spontaneously, do not have ID cards, household registration books, etc. When people cannot make LURCs, State agencies are also unable to update the information on land users and land plots to serve the management and settlement of land disputes. In addition, the lack of human resources and having to handle too many jobs also constitute one of the obstacles to resolving land disputes.

Legal propagandation is not effective. When asked about the propaganda and dissemination of the law on land and houses, 95% of households said that communes and towns regularly held meetings for people and propagation. However, when interviewing officials directly about the response, participation as well as the people's understanding of land law, 42/50 officials (84%) said that the propagation was not effective. Meetings to propagate land policy for the people were mainly held at night. The village chief had to go to people's gathering; loudspeakers of the communes and towns were almost damaged. When the people gathered at cultural houses, instead of listening to officials conveying legal information, the people talk, and do not focus on listening to common issues; thus, the incident occurs, and the people go back to the committee to ask for an explanation.

Little cooperation when resolving land disputes "Ca Than Forest". 80% of officials said that most of the compatriots' disputes arise from Ca Than land. The government has invited them many times but the people who encroach on the land still do not come, even though it takes nearly 1 month for officials to communicate and meet the disputing parties to resolve the case. According to the authors, in addition to the factors of rugged terrain, the main reason still comes from the ideology of the ethnic minorities themselves, *"Ca Than Forest"* is the land of their grandparents who discovered and struck an ancient territorial seal.

4 Conclusions and Discussion

Customary laws on land administration and use: Van Canh district has a large number of Cham and BaNa ethnic minorities living there, each of which has its own customary laws. In order to achieve a high efficiency in land use administration, including conciliation and land dispute resolution, it is necessary to link customary law and law.

Causes of disputes: there are two main types of disputes in the study area: disputes between land users (Cham and Ba Na ethnic minorities; disputes between ethnic minorities and Kinh) and disputes between land users and state agencies (disputes between the people and the Protection Forest Management Board, Ha Thanh Forestry Company). The main cause is the intentional violations in the management by officials, the lack of updating land information in the cadastral records, unmeasured digital cadastral maps, non-adjusted or out of date paper cadastral maps. The process of allocating land to local people has not been implemented, and there are still many violations; att the same time, there is a lack of productive land, and people's awareness of the law is not high. That is also one of the causes of many disputes in "Ca Than" land.

Regarding the grassroots and communal conciliation work: traditional customs, practices, ethics, village rules, customary laws, cultural village conventions, etc., play key roles in the mediation of land disputes at grassroots level (especially in areas where a large number of ethnic minorities live). However, the current Land Law does not specify the position, the role of customs, practices, ethics, and how to establish village conventions. Village regulations and conventions play important roles in grassroots land conciliation, but funding for revision, amendment. and supplementation is not available.

Although the number of local land disputes is not high, the successful conciliation rate is only just over 60%. The capacity and legal knowledge of village elders and village chiefs and the grassroots conciliators are not high. The issue of allowances for the conciliators has not been given due attention. In order to improve the quality of land dispute conciliation, the State needs to regularly carry out the work of fostering and raising the level of law knowledge for the conciliators, and the law needs to specify on financial regimes and policies for conciliators.

There are many difficulties in the conciliation of land disputes. However, according to the research team, the greatest is still the difficulty in propagating awareness for people to understand the law. The research results show that mainly disputes arise from "Ca Than" land. When the propagation is really effective, the conciliation rate is high, and this will reduce complaints to the People's Committee or the People's Court, contributing to political stability and stability of people's lives.

Acknowledgments To complete the research, the authors received the cooperation and support of agencies and departments in Van Canh district. The authors would like to thank the organizations as well as the individuals who were interviewed and assisted the group in completing the study.

References

Vietnamese Committee for Ethnic Minorities and Mountainous Areas [CEMMA]. (2001). *On the issue of ethnicity and ethnic affairs in our country*. National Political Publishing House. 53 pages (in Vietnamese).

Vietnamese Committee for Ethnic Minorities and Mountainous Areas [CEMMA]. (2017). Circular No. 02/2017 / TT-UBDT guiding the implementation of policies on support of residential land, productive land, daily-life water, preferential credit and arrangement policies population stabilization for ethnic minority households living in shifting cultivation areas specified in the Prime Minister's Decision No. 2085/QD-TTg of 31 October 2016 approving specific policies to support socio-economic development Ethnic Minority and Mountainous Area Associations for the period of 2017–2020 (in Vietnamese).

Chapter 49
A Stakeholder Delphi Study on the Adaptive Capacity of Local Communities to Climate Change in the Coastal Area: Case Study in An Duong District (Hai Phong, Vietnam)

Ha T. T. Pham

Abstract Hai Phong, a coastal city in the Vietnam Red River Delta, is considered to be one of the ten cities most threatened by climate change in the world. This paper presents the impacts of climate change - related to hazards and adaptive capacity of local communities of the An Duong district, Hai Phong city. The study deals with applying the Delphi technique combined with the PSR (Pressure-State-Response) framework. The research established a catalog of 18 questions and 18 statements, indicating the PSR constituents. Delphi questionnaires allow the agreement to be identified among a stakeholder group's respondents. 40 panel members engaged in a two-round Delphi process. The results indicate that the establishment of advanced agricultural production models, the intensification of training courses on farming techniques and response to climate change, and the economical use of energy should be the main responses in An Duong. The value of Kendall's W in the second round is 0.738, gaining "very strong agreement" and "very high confidence" from the panel members. The contribution of Delphi results achieves a significant impact on local socio-economic development, namely in ecosystem-based management, sustainability, and climate change-resilient goals.

Keywords Climate change · Adaptative capacity · Local communities · Delphi technique · PSR framework · An Duong district · Hai Phong city

H. T. T. Pham (✉)
University of Science, Vietnam National University, Hanoi, Vietnam

A. T. Nguyen, L. Hens (eds.), *Global Changes and Sustainable Development in Asian Emerging Market Economies Vol. 2*, https://doi.org/10.1007/978-3-030-81443-4_49

1 Introduction

Vietnam is among the countries most seriously affected by climate change and sea level rise, especially in coastal regions (Cruz et al., 2007; MONRE, 2012; Thayer, 2007; United Nations [UN], 2009). Hai Phong is a coastal city on the Red River Delta that has undergone the consequences of climate change, especially sea level rise, extreme weather events (tropical storm, flood, drought, etc.), and annual average temperature increases (Hai Phong Government [HPG], 2015; MONRE, 2010). According to the Organization for Economic Co-operation and Development (OECD), Hai Phong is one of the ten cities in the world that has been threatened the most by climate change. The climate change scenario given by the Vietnamese Ministry of Natural Resources and Environment in 2016 shows that the temperature of Hai Phong can increase by 0.6 to 1.4 °C; its heatwaves have increased as well over a period of 20 years (from 2016 to 2035) (MONRE, 2016). According to the meteorological observation of the North East Hydro-meteorological Station, the end of January in 2016, which is the coldest month, has the lowest temperature among months of 4.5 °C at Phu Lien C station. This has been the lowest temperature in Hai Phong for nearly 50 years (since 1968), which caused considerable impacts on the socio-economic situation in the region, especially on the agricultural ecosystem of Hai Phong city, such as damage to plants, animals, and humans (Hai Phong Statistical Office [HSO], 2018).

The Delphi technique contributes ideas aiming for early prediction of the effects of climate change, giving warning situations about the consequences of climate change (Biloslavo & Grebenc, 2012), eliciting adaptive solutions (Biloslavo & Dolinšek, 2010). In addition, this technique supports analyzing the trend and vulnerability of affected populations (Yang & Kim, 2013; Yoon et al., 2013). The study integrated stakeholder Delphi techniques with the PSR framework (Pressure-State-Response) to evaluate how climate change impacts agriculture in An Duong district, Hai Phong city, and adaptive solutions of local communities to climate change. These provide information to managers, policymakers, and local communities to help them to improve their adaptability to climate change in a sustainable way.

2 Methodology

2.1 Study Area

The Red River Delta (Vietnam) has ten provinces including four Gulf of Tonkin-bordered provinces: Hai Phong, Thai Binh, Nam Dinh, and Ninh Binh. An Duong district is located in the northwest of Hai Phong city, with 15 communes and 1 town. The terrain of the district is not smooth, sloping from North to South, with average height compared with sea level ranging from +0.3 to +0.7 m. The total area of natural land in An Duong district is 97.6 km^2, in which agricultural land occupies about

10.67% (104,125.5 ha) (Hai Phong Statistical Office [HSO], 2014). An Duong district, which is located in the economic triangle of Hanoi–Hai Duong–Hai Phong, especially with Highway 5, has favorable conditions for agricultural development. Therefore, An Duong district is one of the major agricultural districts of Hai Phong city today. According to the plan of the city to 2020, An Duong district was planned to be one of the agricultural production belts in the direction of specialized production to provide food and foodstuff for the city and nearby areas (Vietnam Government [VG], 2001). The main crops of the district are rice (39,870.58 tons) and vegetables (41,247.08 tons); common livestock is pigs (37,069) and poultry (572,400) (An Duong People's Committee [ADC], 2018). In agricultural activities, the district has been experimenting an agricultural model that applies new farming methods and advances in science and technology to adjust farming techniques, and the model is replicated throughout the whole city. According to the Department of Agriculture and Rural Development of Hai Phong city, An Duong is one of the three districts that are most affected by climate change (An Duong Department of Agriculture and Rural Development [ADDA], 2018).

2.2 *PSR-Based Stakeholder Delphi*

The Delphi technique was first developed by the Rand Corporation in the United States in the early 1950s. This technology has been applied flexibly in many areas of research such as medicine (Sinha, Smyth, & Williamson, 2011), social policy (Adler & Ziglio, 1996), tourism (Donohoe & Needham, 2009), sustainable development (Hugé, Le, Pham, Kuilman, & Hens, 2010), and has been comprehensively assessed in many places (Linstone, 1975; Lock, 1987; Parenté & Anderson-Parenté, 1987; Stewart, 1987). Membership of Delphi techniques focused on solving a problem through organized surveys (Hasson, Keeney, & McKenna, 2000). Delphi techniques have at least two rounds of inquiry corresponding to at least two structured questionnaires (Rowe & Wright, 1999). Members respond to questions in the form of anonymous feedback. This technique can be used to predict future problems (Dalkey & Helmer, 1963; Paliwoda, 1983) and to solve the problems (Martin et al., 2012; McBride et al., 2012). The Delphi Group is more efficient than the statistical groups and the standard interactive groups (Rowe & Wright, 1999).

This study integrates the Stakeholder Delphi technique with a PSR model. In this study, the Delphi process established 18 questions for the first round and 18 statements for the second round of the investigation. The list of these questions is presented in the PSR model as follows: five questions on pressure (P; the main cause of local environmental damage, pressure leading to environmental damage and change in agriculture), eight questions on state (S; time, observation and magnitude of climate change occurring in the local area, effects of climate change on local agriculture), and five questions on response (R; agricultural solutions that local communities use to cope with climate change). There have been 40 panel members in total who were randomly picked from four stakeholder groups in this study. They represent expert groups: local authorities (12), farmers (18), agricultural engineers (5), and agricultural product traders (5). The 12 panel members of the authorities work at the People's Committee of An Duong district (1), the Department of Natural Resources and Environment of An Duong district (6), the Department of Agriculture and Rural Development of An Duong district (3), the Department of Culture and Information of An Duong District (1), the Economic and Infrastructure Department of An Duong District (1). Five agricultural engineers are officials of An Duong District Agricultural and Fisheries Center. The 23 panel members who are agricultural product traders and farmers live in communes severely affected by natural risks, namely Dai Ban (3), Le Thien (3), Tan Tien (3), An Hung (2), Hong Phong (2), An Hoa (2), Quoc Tuan (2), An Dong (2), Dong Thai (1), Hong Thai (1), Nam Son (1), and Dang Cuong (1). Expert groups here are selected as they are knowledgeable and prestigious in the field of research. Approval was obtained from the local ethics committee.

In this study, the Delphi process is conducted through three main steps:

Step 1. Preliminary. Defining objectives and developing a comprehensive set of questions based on the content of climate change impacts on agriculture, feedback ability of the agricultural system, and the responses to climate change by local communities. Establishing sample size and selecting panel members (experts). The preliminary phase took place between 4 February and 2 March 2018.

Step 2. Round 1. The steps for this round include: preparation, pre-test and revising the open questions in accordance with the locality; introduction of the initial questionnaire (including open questions) to the panelists; submitting the questionnaire consist of 18 questions to 40 experts; getting feedback from panel members, and analyzing the collected data. The data set up the foundation for a new closed questionnaire, which would be used for the second round. This round took place from 2 to 20 May 2018.

Step 3. Round 2. This step constitutes the second round of the Delphi survey, which was conducted during the field trip between 5 and 27 August 2018. Based on the results of Round 1, the study selects highly valued answers to transfer into statements. Eighteen statements were established. Experts used a Likert scale (10 points, from 1 (total disagreement) to 10 (total agreement) to assess the degree of agreement for 18 statements. For each statement, the average score, standard deviation, and quartile were computed. The reliability of the responses was assessed using Kendall's coefficient of concordance (Schmidt, 1997) (Table 49.1). In this study,

Table 49.1 Interpretation of the agreement and confidence of Kendall's W (Schmidt, 1997)

Kendall's W	Agreement	Confidence in ranks
>0.7–1.0	Very strong	Very high
>0.5–0.7	Strong	High
>0.3–0.5	Average	Average
> 0.1–0.3	Weak	Low
0.0–0.1	Very weak	None

round two has a value of 0.738 ($p < 0.001$), which refers to "very strong" agreement and a "very high" degree of confidence (Schmidt, 1997). Hence, the Delphi process was terminated after round 2. Finally, the recorded results were reported back in favor of providing information for all panel members.

3 Results

3.1 *Round 1*

3.1.1 Pressure

Table 49.2 presents the main causes of environmental damage in An Duong district. There are two main factors being mentioned in this section: agricultural activity and non-agricultural activities. The results show that the over-use and abuse of plant protection chemicals and fertilizers (chosen by 37/40 respondents) and agricultural waste, which have not been rationalized (36/40 choices), are the two main causes of environmental damage. These elements were integrated and cited in the statement "S_1.1" in round 2. In off-farm activities, the rapid development of local industry (34/40 choices) has damaged the environment. This cause is transferred to round 2 with content code "S_1.2".

Industrial energy consumption and waste generation are the principal sources of stress from economic activities that change in agriculture. Thirty-eight out of 40 respondents assert that industrial waste was the main contributor to these problems; 32/40 panel members indicated that the changes in agriculture were due to energy consumption by industry. These choices are shown in the statement "S_1.3" in round 2. The reasons for agriculture variation stemmed from climate change (38/40), indicated in the statement "S_1.4". Land use change (35/40), and market demand and price for agricultural products (37/40) also led to changes in agricultural production at the local level. These two elements are combined and presented in the statement "S_1.5".

3.1.2 State

This section contains eight questions relating to the occurrence of extreme weather events, their manifestations, the impacts of climate change risks on agricultural production, and affected groups (Table 49.3). About 5–10 years ago, extreme weather events were more regular and more destructive (33/40 choices). Climate change occurs locally through the following manifestations: extreme weather events (storms, prolonged heat) occur more frequently and more severely (37/40 choices) and the increase in annual average temperature (31/40 choices). Twenty-eight out of 40 agreed with the opinion of the agricultural sector that it was heavily influenced

Table 49.2 Questions about Pressure (P) and response by the respondents in round 1

Code	Questions	Answers	Number of responses/ total responses	Most frequently chosen alternative	Symbols presented in round 2 statement (S code)
Q_1.1	What are the principal dynamics of agriculture adversely affecting the environment?	Abuse of plant protection chemicals and fertilizers	37/40	- Abuse of plant protection chemicals and fertilizers - Agricultural waste not properly treated	S_1.1
		Over-farming	17/40		
		Agricultural waste not properly treated	36/40		
Q_1.2	What are the principal dynamics of non-agricultural activities adversely affecting the environment?	Industry	34/40	Industry	S_1.2
		Transport	20/40		
		Tourism	13/40		
		Activities of dwellers (cooking, littering, etc.)	13/40		
Q_1.3	What are the principal pressures from economic activities that lead to changes in agricultural production?	Energy consumption by industry	32/40	- Energy consumption by industry - Waste generation (industry, agriculture, etc.)	S_1.3
		Energy consumption by transport	19/40		
		Waste generation (industry, agriculture, etc.)	38/40		
		Other sources (economic development of the surrounding areas, etc.)	5/40		
Q_1.4	What are the main biophysical pressures that lead to changes in agricultural production?	Climate change	38/40	Climate change	S_1.4
		Land degradation	22/40		
		Natural disease outbreaks (insect epidemics, animal epidemics, etc.)	20/40		

(continued)

Table 49.2 (continued)

Code	Questions	Answers	Number of responses/ total responses	Most frequently chosen alternative	Symbols presented in round 2 statement (S code)
Q_1.5	What are the main social pressures that lead to changes in agricultural production?	Market demand and price for agricultural products	37/40	- Market demand and price for agricultural products - Land use change	S_1.5
		Change in state agricultural policy	21/40		
		Land use change	35/40		

by climate change. The above three answers are transferred to round 2 corresponding to the statements "S_2.1," "S_2.2," and "S_2.3."

The impact of climate change on the productivity of crops and livestock (33/40 choices), increasing epidemics, and emerging new diseases (31/40 choices). These two elements are transferred into the content of the statement "S_2.4." Climate change has altered the cultivated area (narrow area; 34/40 choices), changed the crop's seasons (34/40 choices), changed the structure of crops and livestock (32/40 choices), and changed agricultural techniques (31/40 choices). These are the contents of the statement "S_2.5." Food crops (33/40 choices) and poultries (29/40 choices) were affected by the increase in extreme weather and natural disasters, appearing in "S_2.6" and "S_2.7." Climate change also affects the community. Most panel members agree that the group of people affected by climate change are farmers (32/40 choices) and local people (29/40 choices) who are present in the statement "S_2.8" of the second round.

3.1.3 Response

The second round's statement "S_3.1" contains the contents of establishing advanced agricultural production models (36/40 choices), intensifying training courses on farming techniques and responding to climate change (35/40 choices), and mainstreaming climate change responses into agricultural development policy (30/40 choices). These factors are selected from question "Q_3.1" of round 1. Thirty-five out of 40 members presented that adaptation measures locally were inadequate ("S_3.2"). In an effort to adapt to the detrimental effects of climate change, panel members indicated that governments locally provide financial support to farmers to overcome the consequences of climate change (35/40 choices), support of seed sources and livestock (32/40 choices), and seasonal changes in agricultural production and farming practices (31/40 choices) are the appropriate solutions. The three solutions are presented in the "S_3.3" statement of round 2. In other areas, the use of energy saving in socio-economic activities (38/40 choices) and

Table 49.3 Questions about States (S) and responses by the respondents in round 1

Code	Questions	Answers	Number of responses/ total responses	Most frequently chosen alternative	Symbols presented in round 2 statement (S code)
Q_2.1	Since when have the extreme weather events been occurring more regularly and more destructively until now?	5–10 years	33/40	5–10 years	S_2.1
		10–20 years	4/40		
		20–30 years	3/40		
		Unknown	1/40		
Q_2.2	How has climate change manifested during past years?	Extreme weather events (storms, floods, droughts, prolonged heat, etc.) occur more frequently and more severely	37/40	- Extreme weather events (storms, prolonged heat) occur more frequently and more severely - Annual average temperatures follow upward trend	S_2.2
		Annual average temperatures follow upward trend	31/40		
		Seasons of the year change the time period of appearance	19/40		
		The change in rainfall	19/40		
Q_2.3	How is the magnitude of change in agricultural production due to climate change impacts?	Relatively considerable change	28/40	Relatively considerable change	S_2.3
		Negligible change	10/40		
		Stay unchanged	2/40		
Q_2.4	What are the major impacts of climate change on local plants and animals?	Impact on growth and development	23/40	- Impact on productivity - Rise of diseases and breeding of new types of diseases	S_2.4
		Impact on productivity	33/40		
		Rise of diseases and breeding of new types of diseases	31/40		
		Impact on the quality of agricultural products	22/40		

(continued)

Table 49.3 (continued)

Code	Questions	Answers	Number of responses/ total responses	Most frequently chosen alternative	Symbols presented in round 2 statement (S code)
Q_2.5	What is the change in agricultural production owing to the climate change impacts in the locality?	Change in the area (narrowing or expanding) or shift of agricultural production location	34/40	- Change in the area (narrowing) of agricultural production location - Change in a crop's seasons - Change in the structure of crops and livestock - Change in agricultural techniques	S_2.5
		Change in a crop's seasons	34/40		
		Change in the organization of crops and livestock	32/40		
		Change in agricultural techniques	31/40		
		Improving and building an irrigation system	12/40		
Q_2.6	How does the increase in extreme weather events and natural disasters impact mostly on crops?	Food crops	32/40	- Food crops	S_2.6
		Vegetables	4/40		
		Fruits	3/40		
		Others	1/40		
Q_2.7	How does the increase in extreme weather events and natural disasters impact mostly on livestock?	Cattles	8/40	- Poultries	S_2.7
		Poultry	29/40		
		Seafood	3/40		
Q_2.8	Which groups of people in society were most susceptible to climate change catastrophes?	Farmers	32/40	- Farmers - Local people	S_2.8
		Local people	29/40		
		Agricultural merchants	14/40		

Table 49.4 Questions about Responses (R) and responses by the respondents in round 1

Code	Questions	Answers	Number of responses/ total responses	Most frequently chosen alternative	Symbols presented in round 2 statement (S code)
Q_3.1	What are local government solutions that support farmers and minimize the impact of extreme weather events and natural disasters on agriculture?	Establishing advanced agricultural production models	36/40	- Establishment of advanced agricultural production models - Intensifying training courses on farming techniques and responding to climate change - Mainstreaming climate change response policies into agricultural development policy	S_3.1
		Intensifying training courses on farming techniques and responding to climate change	35/40		
		Mainstreaming climate change response policies into agricultural development policy	30/40		
		Strengthening the construction of the reservoir system, ensuring water supply during the dry season and water retention during the rainy season	19/40		
Q_3.2	What about the assessment of the current adaptation solutions to handle climate change issues?	Inadequate	35/40	Inadequate	S_3.2
		Sufficient	5/40		
Q_3.3	What are solutions to help farmers acclimatize to climate change?	Financial support	35/40	- Financial support - Support the source of seeds and livestock - Changes in agricultural production and farming practices	S_3.3
		Support the source of seeds and livestock	32/40		
		Changes in agricultural production and farming practices	31/40		
		Use of plants and animals that can adapt to harsh conditions (prolonged heat, salty soil, etc.)	25/40		

(continued)

Table 49.4 (continued)

Code	Questions	Answers	Number of responses/ total responses	Most frequently chosen alternative	Symbols presented in round 2 statement (S code)
Q_3.4	How to reduce greenhouse gas emissions into the atmosphere?	Using energy saving in socio-economic activities	38/40	- Using energy saving in socio-economic activities - Planting trees, growing forests	S_3.4
		Planting trees, growing forests	32/40		
		Redirecting using other forms of energy (solar, wind, etc.)	25/40		
Q_3.5	What kind of renewable energy should be developed in the future in Hai Phong?	Solar energy	34/40	Solar energy	S_3.5
		Wind energy	3/40		
		Tidal energy	3/40		

planting trees and growing forest (32/40 choices) are the two most commonly chosen options. Thus, they were moved to round 2 with the content in the statement "S_3.4". Solar energy development (34/40 choices) is the most recommended solution in the future in Hai Phong ("S_3.5"). The results of this section are shown in Table 49.4.

3.2 *Round 2*

Table 49.5 presents the results of the second round. Forty experts assessed 18 statements using a Likert scale. Overall, the average values of 18 statements was very high (8.98 to 10). The standard deviation is relatively low (0.00 to 2.43). This proves that the consensus of experts on these 18 statements is very strong.

Figure 49.1 depicts the average scores of 18 statements. Expression S_2.6 has an absolute mean score (mean = 10.0), which is the consensus of 40 members on the Delphi board. There were three statements with the lowest mean scores of S_1.1, S_1.5, and S_2.3 (mean = 8.98).

3.2.1 Pressure

The abuse of pesticides and chemical fertilizers and the agricultural waste that has not been properly treated are agricultural activities that cause environmental damage. Statement S_1.1 was rated high, with an average score of 8.98, standard

Table 49.5 Statistical analysis data for perceived statements on climate change in round 2

Statements	Mean	Minimum	Maximum	Standard deviation	Quartiles		
					25%	50%	75%
S_1.1	8.98	7	10	0.85	8.00	9.00	10.00
S_1.2	9.08	7	10	0.74	9.00	9.00	10.00
S_1.3	9.00	7	10	0.87	8.00	9.00	10.00
S_1.4	9.45	5	10	1.89	10.00	10.00	10.00
S_1.5	8.98	6	10	2.23	7.25	10.00	10.00
S_2.1	9.20	7	10	0.78	9.00	9.00	10.00
S_2.2	9.45	7	10	0.51	9.00	10.00	10.00
S_2.3	8.98	7	10	1.15	8.00	9.00	10.00
S_2.4	9.28	7	10	0.77	9.00	10.00	10.00
S_2.5	9.25	6	10	1.06	9.00	10.00	10.00
S_2.6	10.00	10	10	0.00	10.00	10.00	10.00
S_2.7	9.33	8	10	0.48	9.00	9.00	10.00
S_2.8	9.33	4	10	2.43	10.00	10.00	10.00
S_3.1	9.35	7	10	0.90	9.00	10.00	10.00
S_3.2	9.68	8	10	0.48	10.00	10.00	10.00
S_3.3	9.40	7	10	0.76	9.00	10.00	10.00
S_3.4	9.60	8	10	1.53	10.00	10.00	10.00
S_3.5	9.78	9	10	0.28	10.00	10.00	10.00

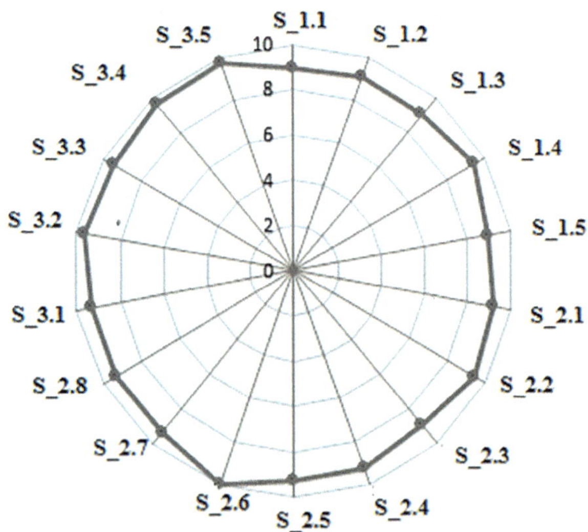

Fig. 49.1 Mean results of the Delphi survey statements in the second round

deviation 0.85. Industrial activity is thought to be the cause of the environmental damage. This is in the statement S_1.2, with an average score of 9.08 and a standard deviation of 0.74.

In the opinion of experts, the energy consumption in industry and the generation of waste in industry and agriculture have changed the agricultural activity (S_1.3;

mean = 9.00). Climate change is the natural pressure to change agriculture (S_1.4) with an average of 9.45. Social activity also exerts pressure on agriculture through changing market demands for agricultural products and changes in land use. This is S_1.5 with an average score of 8.98.

3.2.2 State

Climate change has been occurring locally. According to experts, extreme weather events are more frequent and more devastating than 5–10 years ago (S_2.1; mean = 9.20). Its local manifestations include extreme weather events (storms, floods, droughts, prolonged heat, etc.) occurring more often and more severely, annual temperatures increasing considerably (S_2.2; mean = 9.45). These phenomena have a great influence on agricultural production. This content is in S_2.3 with an average score of 8.98.

Experts show that climate change affects agriculture. They argue that climate change affects plant productivity, growth, and development, and increases disease (S_2.4, mean = 9.28). Climate change also changes the area of agricultural land, production seasons, crop structure, and farming practices (mean = 9.25). In statements S_2.6 and S_2.7, experts agree that food crops and poultry are most severely affected by climate change, with mean scores of 10.00 and 9.33 respectively. Statement S_2.8 shows that farmers and people living and working in the locality are also affected by climate change, with an average score of 9.33.

3.2.3 Response

The mean values of response measures have been rated very high (9.35–9.78). Almost all of panel members agreed with the five statements of this section because they realized the feasibility of the measures. Options to mitigate the impacts of climate change on agriculture can be applied locally, including the establishment of advanced production models, the intensification of training courses on farming techniques, adaptation to climate change, and insertion policies to cope with climate change in agricultural development policy. This statement has an average score of 9.35. With an average score of 9.68, the S_3.2 statement shows that most of the members have found that the government taking measures to adapt to climate change is not enough. Experts have come up with a high degree of consensus on the solutions to help farmers acclimatize to climate change (S_3.3; mean = 9.4). Efficient and energy-saving use in socio-economic activities, tree planting, the forest growth and the development of solar energy are the most valued solutions to reduce greenhouse gas emissions that are the cause of climate change (S_3.4; mean = 9.6; S_3.5; mean = 9.78).

Table 49.6 Kendall's W and the level of consensus assigned in round 2

n	Kendall's W	*P*	Agreement	Confidence in ranks
40	0.738	< 0.001	Very strong	Very high

3.3 *Kendall's W Test*

The feedback of the panel members collected in the second Delphi round allows a Kendall's W of 0.738 to be calculated. It could be seen from Table 49.6 that the panel members reached consensus with a "very strong agreement" and obtained "very high confidence" in ranks. Therefore, the Delphi process stops round 2.

4 Conclusions and Discussion

Climate change has had a significant impact on the coastal communities of Vietnam (Cruz et al., 2007; Thayer, 2007; United Nations [UN], 2009). Adapting and mitigating these impacts is becoming a matter of concern to the public. Adaptation to climate change, however, depends on many other social factors, such as injustice, environmental pollution, or famine (Colagiuri, Boylan, & Morrice, 2015; Eriksen et al., 2011).

Hai Phong is the largest port city in the North of Vietnam. In addition to industrial and transportation development, the city's agriculture is also important. An Duong is an example of agricultural production in Hai Phong. However, climate change and a number of other factors are threatening the district's agriculture. In order to identify the main factors affecting agriculture and to propose appropriate solutions, an integrated approach would allow the complexities surrounding this relationship to be removed (Doria, Boyd, Tompkins, & Adger, 2009; EEA, 1995; Newton & Weichselgartner, 2014). The combination of the Delphi technique between the stakeholder and the PSR model in this study demonstrates the effectiveness of the integrated approach in determining the causes, effects of climate change, and appropriate solutions to respond to climate change in An Duong district, Hai Phong. The results of the study are as follows:

- Pressure: industrial and agricultural waste, industrial energy consumption, climate change, land-use alteration, and the variation of market demand and price for agricultural products are the key dynamics that lead to changes in agriculture.
- State: extreme weather events are more regular and more detrimental compared with 5–10 years ago. The manifestations of climate change in the locality include extreme weather events (storms, prolonged heat) that appear at an increasingly serious level, the annual average temperature has increased considerably, and affects agriculture. Climate change impacts on the productivity of crops and livestock with increased disease and changes in the area of agricultural land. The two agricultural factors that are most heavily affected by climate change are food

crops and poultry. Farmers and people living and working in the area are also affected by climate change.

- Response: in agriculture, people should adopt advanced agricultural models and improve their knowledge of climate change issues and acclimatize to climate change. Agricultural development policies should integrate climate change and local governments should support agricultural materials (seeds, livestock, fertilizers, etc) and financial aid for farmers to overcome the consequences of climate change. Residents should carry out environmental protection and energy conservation activities in socio-economic activities.

Eighteen statements of the three PSR components are subject to evaluation by 40 experts on the review panel. Planning is determined by a number of factors, in which the consensus of stakeholders on the research issue is an important factor (Linstone, 1975). Most experts believe that the statements content of this study is consistent with the real situation in An Duong district, Hai Phong. Therefore, they rated a very strong degree of agreement with these statements (mean = 8.98–10.00). According to the plan of Hai Phong city up to 2020, An Duong district's agriculture was to develop in the direction of specialization (Vietnam Government [VG], 2001). However, at present, agricultural production has not effectively applied the advanced production models and technology. Agriculture in the district still faces many difficulties in investing capital for production, implementation of advanced farming techniques, and especially in responding to climate change. The misuse of both pesticides and chemical fertilizers still continues in the district. This is a major threat to the environment of An Duong district. During the implementation of the Delphi survey rounds, the study looked at farmers' perceptions of the use of chemicals in agriculture. About 75% of farmers use more than just drugs and fertilizers. In the period 2010–2018, the weather in An Duong district as well as Hai Phong has complex occurrences. Average annual temperatures increased significantly (Hai Phong People's Committee [HPC], 2018; Haiphong's Department of Natural Resources and Environment [HPE], 2012). The number of typhoons affecting Hai Phong is also increasing and the average annual rainfall is decreasing (Haiphong's Department of Natural Resources and Environment [HPE], 2012). In addition, diseases in plants and animals also occur abnormally and are quite serious. In 2005 and 2013, bird-flu epidemics occurred and spread throughout Hai Phong city in general and An Duong in particular. In the period of 2006–2012, the outbreaks were controlled to only small ones, but they also severely affected poultry farming in An Duong district, which resulted in the destruction of 100,000 poultry per year (An Duong Department of Agriculture and Rural Development [ADDA], 2018). The cause was determined to be abnormal weather changes, which facilitated the outbreaks.

In addition to the effects of climate change, the agricultural sector of An Duong district is also affected by many factors such as economic restructuring, land-use change, urbanization, industrialization, etc. According to the statistics of An Duong district, in the period 2010–2015, the area of agricultural land decreased significantly (from 10,246.1 ha to 8374.14 ha) (An Duong People's Committee [ADC],

2015). This decline in agricultural land is mainly due to the conversion of agricultural land to land for industry and livelihoods. The orientation of economic development of Hai Phong city to the year 2025 is to become a large and competitive industrial service center and the key to the development of the national economy (Vietnam Government [VG], 2018). Following the general trend of the city, An Duong district gradually shifted its economic structure to the development of industry and services to replace the agricultural sector, for example, industrial parks such as An Duong Industrial Park (812.62 ha), Nomura Industrial Park (153 ha), Trang Due Industrial Park (600 ha) (Hai Phong People's Committee [HPC], 2018). Demand for land use in the area of An Duong and Hai Phong city increased owing to the population increase. Along with that, the speed of urbanization in Hai Phong also increased sharply. According to the Government's plan, in 2020, Haiphong's population should have reached about 2.1 million people with an urbanization rate of 50–55%; by 2025, this would be about 2.25 million people with an urbanization rate of 60–65%; by 2030, about 2.4 million people with an urbanization rate of 65–70% (Vietnam Government [VG], 2018). Increased industrialization and urbanization combined with climate change have reduced the area of agricultural land in the district. This has greatly affected the local agricultural output.

Local authorities have implemented a number of measures to mitigate the consequences of climate change such as: building modern agricultural models and programmes, training people with the knowledge on how to respond to climate change, etc. (An Duong Department of Agriculture and Rural Development [ADDA], 2016). However, the actions taken were not synchronous and persistent; thus, the efficiency was not high. In addition, experts have expressed that the solutions proposed are not enough; more solutions are needed such as planting trees, adjusting crop seasons according to weather conditions, encouraging the development of clean energy sources, as shown in the results. A study on the effects of climate change on the coastal areas of Ha Tinh, Vietnam also pointed out similarities in local adaptation capacity (Nguyen, Vu, Dang, Hoang, and Hens, 2017).

The contents of this study may become a reference for policymakers in setting up measures for responding to the impacts of climate change on agriculture and contributing to solving the problems that the authorities and people in An Duong district are encountering.

Acknowledgments The author would like to thank the households, and the district and commune authorities, as well as the local people, for their assistance and cooperation in completing the questionnaires and providing a discussion of the survey findings.

References

Adler, M., & Ziglio, E. (1996). *Gazing into the Oracle: The Delphi method and its application to social policy and public health*. Jessica Kingsley Publishers.

An Duong Department of Agriculture and Rural Development [ADDA]. (2016). *Announcement on enhancing the direction of agricultural production and aquaculture in the spring in 2016*. Hai Phong.

An Duong Department of Agriculture and Rural Development [ADDA]. (2018). *Situation and results of agricultural production in 2011–2015: Directions, tasks for 2012–2018*. Hai Phong.

An Duong People's Committee [ADC]. (2015). *Report of the business years 2010–2015*. Hai Phong.

An Duong People's Committee [ADC]. (2018). *Report on the implementation of socio-economic, defense-security tasks in 2018, targets, tasks and solutions in 2019 in an Duong province*. Hai Phong.

Biloslavo, R., & Dolinšek, S. (2010). Scenario planning for climate strategies development by integrating group Delphi, AHP and dynamic fuzzy cognitive maps. *Foresight, 12*(2), 38–48.

Biloslavo, R., & Grebenc, A. (2012). Integrating group Delphi, analytic hierarchy process and dynamic fuzzy cognitive maps for a climate warning scenario. *Kybernetes, 41*(3/4), 414–428.

Colagiuri, R., Boylan, S., & Morrice, E. (2015). Research priorities for NCD prevention and climate change: An international Delphi survey. *International Journal of Environmental Research and Public Health, 12*(10), 12941–12957.

Cruz, R.V., Harasawa, H., Lal, M., Wu, S., Anokhin, Y., Punsalmaa, B., et al. (2007). Asia: Climate Change 2007: Impacts, Adaptation and Vulnerability. In: Contribution of Working Group II to the Fourth Assessment Report of the Intergovernmental Panel on Climate Change. Parry M.L. et al., eds. : Cambridge University Press; 469–506.

Dalkey, N., & Helmer, O. (1963). An experimental application of the Delphi method to the use of experts. *Management Science, 9*(3), 458–467.

Donohoe, H. M., & Needham, R. D. (2009). Moving best practice forward: Delphi characteristics, advantages, potential problems, and solutions. *International Journal of Tourism Research, 11*(5), 415–437.

Doria, M. F., Boyd, E., Tompkins, E. L., & Adger, W. N. (2009). Using expert elicitation to define successful adaptation to climate change. *Environmental Science and Policy, 12*(7), 810–819.

EEA. (1995). *Europe's environment: The Dobris assessment* (8 p). European Environmental Agency.

Eriksen, S., Aldunce, P., Bahinipati, C. S., Martins, R., Molefe, J. I., Nhemachena, C., O'Brien, K., Olorunfemi, F., Park, J., Sygna, L., & Ulsrud, K. (2011). When not every response to climate change is a good one: Identifying principles for sustainable adaptation. *Climate and Development, 3*(1), 7–20.

Hai Phong Government [HPG]. (2015). *Green growth promotion plans of Hai Phong city*. Hai Phong.

Hai Phong People's Committee [HPC]. (2018). *Report on the implementation of socio-economic, defense-security tasks in 2018: Targets, tasks and solutions in 2019 in Hai Phong City*. Hai Phong.

Hai Phong Statistical Office [HSO]. (2014). Hai Phong statistical yearbook 2014. *Hai Phong*, 7–169.

Hai Phong Statistical Office [HSO]. (2018). *Economic and social situation report in 2016*. Hai Phong.

Haiphong's Department of Natural Resources and Environment [HPE]. (2012). *Report on the action plan to respond to climate change and sea level rise of Hai Phong city to 2015*. Hai Phong.

Hasson, F., Keeney, S., & McKenna, H. (2000). Research guidelines for the Delphi survey technique. *Journal of Advanced Nursing, 32*(4), 1008–1015.

Hugé, J., Le, T. H., Pham, H. H., Kuilman, J., & Hens, L. (2010). Sustainability indicators for clean development mechanism projects in Vietnam. *Environment, Development and Sustainability, 12*(4), 561–571.

Linstone, H. A. (1975). Eight basic pitfalls: A checklist. In H. Linstone & M. Turoff (Eds.), *The Delphi method: Techniques and applications*. Addison-Wesley.

Lock, A. (1987). Integrating group judgments in subjective forecasts. In G. Wright & P. Ayton (Eds.), *Judgmental forecasting*. Wiley.

Martin, T. G., Burgman, M. A., Fidler, F., Kuhnert, P. M., Low-Choy, S., McBride, M., & Mengersen, K. (2012). Eliciting expert knowledge in conservation science. *Conservation Biology, 26*(1), 29–38.

McBride, M. F., Garnett, S. T., Szabo, J. K., Burbidge, A. H., Butchart, S. H. M., Christidis, L., et al. (2012). Structured elicitation of expert judgments for threatened species assessment: A case study on a continental scale using email. *Methods in Ecology and Evolution, 3*(5), 906–920.

MONRE (2010). Water environment. In Vietnam environmental report 2010: Overview of Vietnam environment.

MONRE. (2012). *Climate change, sea level rise scenarios for Vietnam* (96 p). NARENCA Publishing.

MONRE. (2016). *Climate change, sea level rise scenarios for Vietnam* (188 p). NARENCA Publishing.

Newton, A., & Weichselgartner, J. (2014). Hotspots of coastal vulnerability: A DPSIR analysis to find societal pathways and responses. *Estuarine, Coastal and Shelf Science, 140*, 123–133.

Nguyen, A. T., Vu, A. D., Dang, T. H. G., Hoang, A. H., & Hens, L. (2017). How do local communities adapt to climate changes along heavily damaged coasts? A stakeholder Delphi study in Ky Anh (Central Vietnam). *Environment, Development and Sustainability, 20*(2), 749–767.

Paliwoda, S. J. (1983). Predicting the future using Delphi. *Management Decision, 21*(1), 31–38.

Parenté, F. J., & Anderson-Parenté, J. K. (1987). Delphi inquiry systems. In G. Wright & P. Ayton (Eds.), *Judgmental forecasting*. Wiley.

Rowe, G., & Wright, G. (1999). The Delphi technique as a forecasting tool: Issues and analysis. *Forecast, 15*(4), 353–375.

Schmidt, R. C. (1997). Managing Delphi surveys using nonparametric statistical techniques. *Decision Sciences, 28*(3), 763–774.

Sinha, I. P., Smyth, R. L., & Williamson, P. R. (2011). Using the Delphi technique to determine which outcomes to measure in clinical trials: Recommendations for the future based on a systematic review of existing studies. *PLoS Medicine, 8*(1), 1000–1393.

Stewart, T. R. (1987). The Delphi technique and judgmental forecasting. *Climatic Change, 11*(1–2), 97–113.

Thayer, C. A. (2007). *Climate change and regional security: Vietnam in 2030, workshop on climate change and regional security*. U.S. Naval Postgraduate School.

United Nations [UN] (2009). Vietnam and climate change: A discussion paper on policies for sustainable human development. Ha Noi, 47 p.

Vietnam Government [VG]. (2001). *Decision approving the adjustment of general planning of construction of Hai Phong city to 2020*. Hai Phong.

Vietnam Government [VG] (2018). *Decision approving the adjustment and supplement of master plan for socio-economic development of Hai Phong city up to 2025, orientation to 2030*. Vietnam. No 821/QĐ-TTg.

Yang, J. S., & Kim, I. H. (2013). Development of drought vulnerability index using Delphi method considering climate change and trend analysis in Nakdong River basin. *The Korean Society of Civil Engineers, 33*(6), 2245–2254.

Yoon, S. T., Lee, Y. H., Hong, S. H., Kim, M. H., Kang, K. K., Na, Y. E., & Oh, Y. J. (2013). Vulnerability assessment of cultivation facility by abnormal weather of climate change. *Korean Journal of Agricultural and Forest Meteorology, 15*(4), 264–272.

Chapter 50
Assessing Flash Flood Risks Based on Analytic Hierarchy Process (AHP) and Geographic Information System (GIS): A Case Study of Hieu Catchment (Nghe An, Vietnam)

Tran Thi Tuyen and Hoang Thi Thuy

Abstract Flash flood hazards from heavy rainstorms are common in Hieu catchment (Nghe An province). This study presents a flash flood hazard assessment for Hieu catchment using the combination of Analytic Hierarchy Process (AHP) and Geographic Information System (GIS). A total of seven parameters to control flash flood occurrence were calculated using AHP and overlaid in GIS. Flash flood risk hazard map of Hieu catchment is divided into five risk levels: very high (2.01%), high (20.32%), medium (43.78%), low (30.59%), very low (3.3%). A hazard map was produced to help local authorities implement land use planning that reflects the hazards, and prepare to increase disaster response capacity to human and environmental resilience to flash floods.

Keywords Flash food · Hazard analysis · Analytic Hierarchy Process (AHP) · Geographic Information System (GIS) · Hieu catchment · Nghe An · Vietnam

1 Introduction

Flash flood is a dangerous phenomenon occurring in many river basins in the world, typically associated with hot summers and heavy convective rainfall (Maysa et al. 2017). Flash flood incidence is perceived by residents of the subject province to be increased. An accurate and credible assessment is needed to minimize flood impacts and inform land-use planning (Youssef et al., 2011).

T. T. Tuyen (✉) · H. T. Thuy
School of Agriculture and Resources, Vinh University, Nghe An, Vietnam
e-mail: tuyentt@vinhuni.edu.vn

A. T. Nguyen, L. Hens (eds.), *Global Changes and Sustainable Development in Asian Emerging Market Economies Vol. 2*, https://doi.org/10.1007/978-3-030-81443-4_50

The factors that form flash floods include topography, rainfall, vegetation cover and human changes to watershed function. Saleh (1989) described a set of influencing factors, including rainfall characteristics, evaporation infiltration, drainage networks shape, drainage orders, hillslope drainage characteristics, and environmental and human changes. Minea (2013) identified factors that lead to flash floods in terms of physiographic characteristics from the catchment, including terrain, slope, profile curvature, land use, and soil texture.

Smith (2003), in order to identify areas that promote flooding, suggests an indicator called Flash Flood Potential Index (FFPI). Its estimation is realized using GIS techniques and is based on four grids overlapping the physiographic features of the hydrologic catchment (terrain slope, land use, forest density, and soil texture); these four are considered to have a major hydrologic influence on surface runoff processes and flood occurrence.

Remote sensing tools combined with hydrological, geomorphologic data can be combined in GIS to provide a flash flooding hazard based on our best current understanding of the influencing factors.

Seven parameters were selected as being the primary determinants of the hazard. The selected parameters were: soil properties, geology, drainage density, flow direction, land use, and density of vegetation. The study identifies high flash flood hazard areas using GIS and an analytic hierarchy process (AHP) method. The AHP is used for multi-criteria decision, based on expert weighting of the importance of influencing factors. The AHP gives comparison of design criteria and elements in a pairwise technique of comparison of various parameters, decreasing the complexity of decision-making process and increasing its transparency to users. Hieu is a headwater catchment of the Ca River basin in Nghe An Province of Vietnam, and was selected as study area where flash floods are known to occur frequently, causing large losses of life and property.

2 Methodology

2.1 Study Area

Hieu is a headwater catchment of the Ca basin. Hieu catchment covers an area of approximately 4.935km^2, forming a basin with a rounded shape. It can be divided into three sub-basins: Upper Hieu basin, Song Con basin and Ban Mong-Cay Chanh. Local conditions (e.g., lithology, relief, and climate) along with the anthropic influence (land use) constitute conditional and control factors for runoff. Through northwest to southeast, it comprises a diversity of terrains, including Bu Khang formation (Proterozoi age), and Dong Trau formation rocks and deposits of Quaternary of Song Ca formation. The geology is primarily composed of granitic rocks, migmatites, gneisses, synorogenic granitoids, and gabbrodiorite intrusions that are further intruded by post-orogenic granites. Hieu River has a subtropical monsoon climate characterized by hot summers and warm winters, with average rainfall of 1630 mm, and a mild climate, with an annual average temperature of 16.3 °C. The highest is 2421 m at Pu Hoat mountain. Heavy rainfall can produce flash flooding in Hieu River and its tributaries, primarily during the summer months, from June to August. For the period 1990–2015, the average rainfall was 1593 mm in Upper Hieu basin, 1619 mm in Song Con basin, and 1587 mm Ban Mong-Cay Chanh, in which Upper Hieu catchment accounts for the largest proportion of the area (North Central Vietnam Hydro-meteorological Center, Vietnam).

2.2 Data Processing

The hierarchical analysis method supports expert group determination and weight of factors considered to be most important to the generation of flash flooding. After calculating the weight of the factors, integration in GIS will give us a flash flood index. Using the overlap tool in ArcGIS for re-edited maps, new maps were created and weighted to form a flash flood potential map. After being divided according to the appropriate influence levels, a flash flood potential map is created. The whole process of developing a flash flood risk map for the river basin is shown in Fig. 50.1.

AHP method was used to determine the important coefficient of parameters which were evaluated by designing a hierarchy of main criteria and sub-criteria. Multi-criteria decision problems were solved by establishing the pairwise comparison matrix, which reflects the relationship between the components of a level with

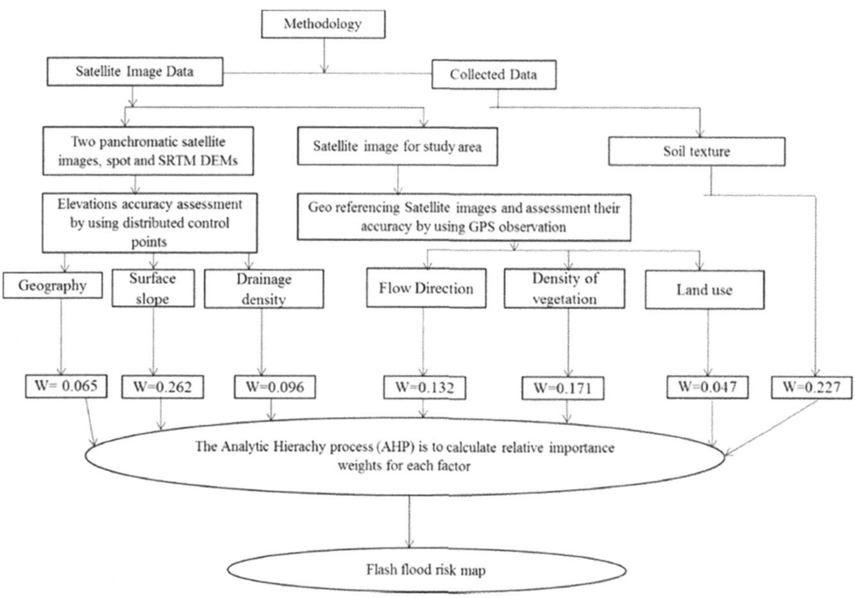

Fig. 50.1 Methodology for flood hazard mapping (w = the weight of each flood causative factor)

properties of a higher level. This technique was implemented in a comparison matrix with two criteria at a time. The comparison of the selected parameters was determined for calculating weight of multicriteria in Hieu catchment. The matrix classification is based on 1–9 scale relative importance of pairwise criteria, where level 1 represents an equal importannce and level 9 shows extreme importance. Then Saaty method was applied to calculate weights and eigenvalues. The efficiency criteria of AHP was evaluated by consistency relationship (CR), which is measured by eq. CR = CI/RI where CI represents consistency index and RI represents random index.

An Analytical Hierarchy Process (AHP) is used to set weight for factors causing flash floods: geology, surface slope, drainage density, flow direction, land use, density of vegetation, soil influences (texture). Saaty (1980) developed AHP to standardize as a support method decision making when there are many factors affecting flash flood. AHP provides a structure hierarchy by reducing the choice between many factors into pairwise comparison and priorities in each pair based on users' opinions. In AHP, factors are compared with other factors to determine the importance of each element in the general purpose. The value calculated for each pair of principle uses is published in Satty's standard measure. Details of the AHP method include a sequence of steps in order.

3 Results

3.1 Drivers of Flash Floods Risks

Soil influences: The process of hillslope runoff and flooding is dominated by soil surface conditions. Surface conditions affect the amount of slow runoff that do not contribute to flash flooding. Slow runoff influences include infiltration, surface depressions, vegetation cover, and evaporation. Soil type and structures are important factors in determining water retention and permeability characteristics of an area affecting the susceptibility to flash floods. Runoff from intense rainfall is likely to be more rapid and greater with clay soils than with sandy soils. In the study area, soil structure is mainly fine loamy to coarse loamy with relatively good permeability.

Geology: Geological factors determine the nature and characteristics of the catchments, including the soils, the risk of landslides, and the morphology of the river network.

Surface slope: The slope of a basin is considered to be of hydrological significance. Steep slopes have high surface runoff during rainfall that exceeds infiltration capacity, accelerating runoff delivery to channels, and soil erosion. Sediment loads tend to be highest in dryer watersheds, where slopes are overgrazed and barren.

Drainage density: Drainage density (Dd) is the ratio of total length of streams of all orders to the basin area. Dd value refers to the proximity of the channel spacing; therefore, it is a quantitative measure for relief analysis, runoff potential, and thus, in turn the drainage efficiency of the river basin (Yahya and Atef, 2017). A low drainage density indicates poor drainage basin with a slow hydrological response while a high drainage density shows a highly separated basin with a relatively rapid hydrologic reaction to rainfall (Melton 1957). Drainage density is the one of the factors controlling the surface flow and consequently affecting sediment and water production from the basin (Chorley, 1969). High Dd values denote high flow and low penetration rates due to the presence of waterproof base materials, spare vegetation, and hilly relief. Conversely, low drainage density implies low runoff, high infiltration, and groundwater recharge (Yahya and Atef, 2017). Dd value for Hieu watershed ranges from 0.035 to 0.35.

Flow direction: The higher the slope of the flow, the faster the ability to concentrate water, causing a high risk of flood formation.

Land use: Surface is a factor that is also quite important for flash floods where soil is the main component of the surface. Rain is a necessary condition, while the surface is sufficient. Surface conditions dominate the process of flood formation. Surface affects the amount of flood runoff loss. Land use is a continuous, wide and varied process. It not only changes the physical properties of the soil, but also changes the cover, even the surface terrain. For instance, the smooth surface makes the runoff concentration time and runoff rate increase after construction. Consequently, land use will affect the process of forming flash floods. For urbanized construction and traffic areas, the surface is usually poured with concrete, the soil is tightened to prevent water permeability and increase the flow concentration. It is easy to generate flash floods. However, natural and plantation forest will not only prevent water but also absorb water well because of the humus upper layer, reducing the risk of flood formation. Natural forest land that occupies the largest area (55.81% of the area) can limit the flow, reduce the potential of flash floods.

Density of vegetation: Vegetation plays a role in stabilizing the surface thanks to the mechanical effects of the roots associated on soil components and regulating the sudden change in soil moisture. The coverage ratio protects the soil from erosion, helps regulate the flow, and transforms part of surface water into groundwater. When it rains, not all of the rain water falls to the forest ground and it is partially retained. The amount of water retained in the forest canopy depends on the factors: forest type, age, species composition, meteorological conditions, precipitation and rainfall intensity, vegetation cover, moisture, air temperature, weather and season. Generally, the amount of water retained in the canopy is about 30–50% of the total rainfall. The surface runoff depends on the length the slope, the intensity and duration of the rain, soil texture, and terrain. In the forest that creates favorable conditions to convert surface flow into seepage, the surface flow accounts for about 2% of the total rainfall. While in the area of compacted soil where humus, litter is destroyed, the surface flow is very large.

Low forest density will increase flood peaks, shorten the time of flood concentration. It helps to affirm that heavily exploited forests make bare surfaces one of the causes of flash flood formation. Deforested forests will enhance erosion and landslides, causing mud and flash floods. The protection of forest rehabilitation, afforestation and rational exploitation of forests on the surface of the basin in general and the watershed in particular is one of the measures to prevent and limit the destruction of flash floods. Most of the area has an average cover of 30–40%, which is relatively sensitive to flash floods.

3.2 Weight of Parameters in the Model

Based on the AHP principle, the priority order of the elements will be compared one by one. The comparison results are shown in Table 50.1. The weight of the factors is determined by the average value in Table 50.2. Moreover, in order for the matrix to be reliable, the AHP has also calculated consistency ratio (CR), that is, the ratio between consistency index (CI) and random index (RI) (Table 50.3).

3.3 Generation of Component Maps

All parameters were rated on a scale of 1–10, divided into 10 different categories for mapping. The sub-criteria are divided into different scales in which, score 1 is for the lowest value and score 10 is the highest value (Table 50.4).

3.4 Flash Flood Risk Map

Study results indicate that very high risk of flash flood is 4796.23 ha (2.01%), high risk of flash flood is 48487.31 ha (20.32%), 104467.24 ha medium risk flash flood (43.78%), low risk (30.59%), very low risk (3.3%).

Table 50.1 Comparing the priority of the elements

No	Criteria	Surface slope	Soil Influences	Density of vegetation	Direct drainage	Drainage density	Geography	Land use
1	Surface slope	1	1.3	1.4	2.4	3.1	3.9	4.3
2	Soil influences		1	1.5	1.9	2.9	3.4	3.8
3	Density of vegetation			1	1.8	1.8	2.4	3.0
4	Flow direction				1	1.6	2.5	3.6
5	Drainage density					1	2.0	2.5
6	Geography						1	1.9
7	Land use							1

Table 50.2 Weight values of the elements

No	Criteria	Surface slope	Soil Influences	Density of vegetation	Direct drainage	Drainage density	Geography	Land use	Weight
1	Surface slope	1							0.262
2	Soil influences		1						0.227
3	Density of vegetation			1					0.171
4	Flow direction				1				0.132
5	Drainage density					1			0.096
6	Geography						1		0.065
7	Land use							1	0.047

Table 50.3 RI index with $n = 10$

N	1	2	3	4	5	6	7	8	9	10
RI	0	0	0.52	0.94	1.14	1.19	1.27	1.46	1.63	1.86

Very high risk area: About 4796.23 ha (2.01%) land falls in this category. This area has high slope (20–87^0), dry soil, rocky terrain, high drainage density, residential land, poor vegetation layer. Thus, it is necessary to take measures to warn people and not allow construction.

High risk area: About 48487.31 ha (20.32%) land falls in this category. Characteristics of this area are high slope, dry soil, barren land, rocky soil. These areas have a high risk of flash flooding. If they are residential areas, immigration is required.

Medium risk area: About 104467.24 ha land (43.78%) falls in this category. Characteristics of this area are gentle to moderate slopes (15°–20^0), forest restoration land, ferralitic soil and moderate coverage density. This area can be planned for construction works and residential areas, but it is necessary to have solutions to cope with the risk of flash floods in the rainy season.

Low risk area: Land in this category falls on gentle to stiff slopes (8°-15°), medium soil, moderate coverage density, at relatively low elevation. The area has low flash flood potential, reliable for human socio-economic activities.

Very low risk area: About 7874.42 ha land (3.30%) falls on this category. Characteristics of this area are gentle slopes (0°-8°), loamy sand, low drainage density, natural forest land, high coverage density. This area has a very low flash flood potential (Fig. 50.2).

Table 50.4 Statistics results area of decentralized criteria and sub-criteria

Criteria	Sub-criteria	Acreage (ha)	Portion (%)	Score
Surface slope (°)	0-3	20702.54	8.68	1
	3-8	90108.5	19.74	3
	8-15	28509.8	3.46	5
	15-20	18509.12	1.14	7
	20-25	73503.3	2.29	9
	25-87	7285.39	0.11	10
Geography	– Quaterary formation (Aq)	1143.079	0.47	10
	– Limestone rock of La Khe formation (C1lk)	3049.531	1.26	2
	– Limestone rock of Muong Long formation (Cpml)	13127.38	5.42	4
	– Sedimentary rock of Nam Tan formation (d12nt1)	551.765	0.23	3
	– Sedimentary rock of Nam Can formation (d1frnc)	315.894	0.13	4
	– Sedimentry and metamorphic rock of Huoi Loi formation (d2hl)	76612.86	31.66	1
	– Granit rock of Dai Loc complex (gad1dt1)	28523.47	11.79	7
	– Granit rock of Dai Loc complex (gad1dt2)	567.6453	0.23	7
	– Eruptions rock of Song Ma complex (G/t2-3sm1)	728.625	0.30	1
	– Eruptions rock of Muong Hinh complex (jmh)	14119.7	5.83	6
	– Sedimentary and metamorphic rock of Song Ca formation (o3s1sc2)	33956.48	14.03	8
	– Sedimentary and metamorphic rock of Song Ca formation (o3s1sc3)	3300.699	1.36	8
	– Metamorphic and sedimentary rock of Bu Khang formation (pr3e1bk)	10299.82	4.26	5
	– Metamorphic and sedimentary rock of Bu Khang formation (pr3e1bk2)	6440.4	2.66	5
	– Quaterary formation (Q)	153.162	0.06	10
	– Eruptions rock of Huoi Nhi complex (s2d1hn)	862.664	0.36	3
	– Sedimentary rock of Dong Trau formation (T2adt1)	39734.66	16.42	7
	– Sedimentary rock of Dong Trau formation (t2adt2)	1945.665	0.80	7
	– Sedimentary rock of Dong Trau formation (t3nrdd1)	6575.73	2.72	9

(continued)

Table 50.4 (continued)

Criteria	Sub-criteria	Acreage (ha)	Portion (%)	Score
Soil Influences	– Rock mountain	3712.409	1.56	1
	– River	708.5092	0.30	10
	– Fine loamy to Coarse loamy	163324.9	68.65	7
	– Gravelly loamy	48038.73	20.19	3
	– Coarse loamy	22114.3	9.30	5
Drainage density	0–0.035	8048.214	3.37	10
	0.035–0.07	18418.29	7.72	9
	0.07–0.105	28133.7	11.79	8
	0.105–0.14	38372.14	16.09	7
	0.14–0.175	34270.52	14.37	6
	0.175–0.21	32041.98	13.43	5
	0.21–0.245	31806.66	13.33	4
	0.245–0.28	23153.04	9.71	3
	0.28–0.315	13619.73	5.71	2
	0.315–0.35	10682.1	4.48	1
Flow Direction	1	18593.99	7.79	2
	2	41143.95	17.24	3
	4	22534.11	9.44	4
	8	30449.83	12.76	5
	16	28997.43	12.15	6
	32	18340.61	7.69	7
	64	40655.74	17.04	8
	128	37899.15	15.88	9
Land use	– Natural forest land	131716.6	55.81	1
	– Planted forest land	36339.71	15.40	3
	– Forest restoration land	26549.49	11.25	7
	– Agriculture land	15902.39	6.74	9
	– Settlement land	25514.13	10.81	10
Density of vegetation	0–10%	9855.491	4.13	1
	10–20%	13570.4	5.69	2
	20–30%	15131.92	6.34	3
	30–40%	165323.1	69.26	4
	40–50%	10271.68	4.30	5
	50–60%	11456.98	4.80	6
	60–70%	10639.12	4.46	7
	70–80%	1856.934	0.78	8
	80–90%	464.7491	0.19	9
	90–100%	120.1987	0.05	10

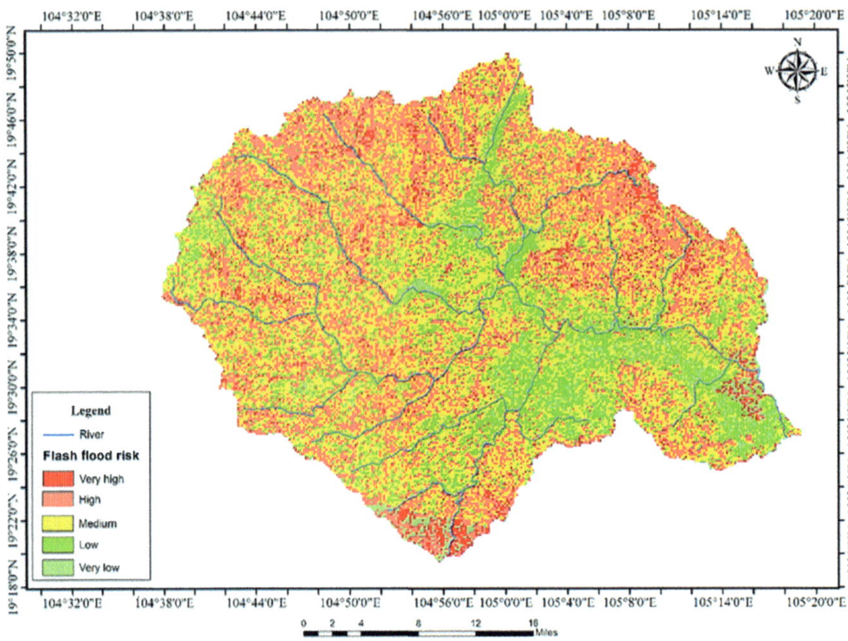

Fig. 50.2 Map of flash flood risk in Hieu catchment

4 Conclusions and Discussion

That integrating hierarchical analysis results into determinative factors in GIS method to develop yielded flash flood potential hazard maps that can be used in land-use and disaster response planning to reduce loss of life and property in Nghe An Province is an effective approach in researching natural hazards. The calculation process for mapping, risk partitioning and risk level of flash flood in Hieu catchment is carried out according to a logical and scientific assessment system based on GIS technology. Weighting for each element with quantitative values has quantified our subjective in the assessment of risk of flash flood hazard.

The weights of factors causing flash floods were determined by the AHP method. The results show that sevens factors affecting on flash flood in Hieu catchment river consist of geology, surface slope, drainage density, flow direction, land use, density of vegetation, soil influences (texture) with corresponding weight: 0.065; 0.262; 0.096; 0.132; 0.047; 0.171; 0.227. Flash flood risk hazard map of Hieu river basin is divided into five risk levels: very high (2.01%), high (20.32%), medium (43.78%), low (30.59%), very low risk (3.3%).

The present approach of AHP-GIS applied in this study area can help in the environmental protection management of other areas. However, this study has some limitations of suitability of selection of criteria for flash flood risk mapping depending on the local condition and variation in climatic condition.

References

Chorley, R. J. (1969). *Introduction to physical hydrology* (211p). Methuen and Ltd..

Maysa, M. N., Taha, Samah, M. E., Doaa, M. N., & Ibrahim, E. (2017). Flash flood hazard zonation based on basin morphometry using remote sensing and GIS techniques: A case study of Wadi Qena basin, Eastern Desert, Egypt. *Remote Sensing Applications: Society and Environment, 8*, 157–167.

Melton, M. A. (1957). *An analysis of the relations among the elements of climate, surface properties, and geomorphology. Technical report 11*. Department of Geology, Columbia Univ.

Minea, G. (2013). Assessment of the flash flood potential of Bâsca River catchment (Romania) based on physiographic factors. *Central European Journal of Geosciences, 5*(3), 344–353.

Saaty, T. L. (1980). *The analytic hierarchy process*. RWS Publications.

Saleh, A. S. (1989). Flash floods in deserts. A geomorphic study of desert Wadis. *Special Studies Series, 51*, 1–93.

Smith, G. (2003). *Flash flood potential: Determining the hydrologic response of FFMP basins to heavy rain by analyzing their physiographic characteristics*. Retrieved from http://www.cbrfc.noaa.gov/papers/ffp˙wpap.pdf.

Yahya, F., & Atef, A. (2017). Assessment of flash-flood Hazard in arid watersheds of Jordan. *Geographic Information System, 9*, 717–751.

Youssef, A. M., Pradhan, B., & Hassan, A. M. (2011). Flash flood risk estimation along the St. Katherine road, southern Sinai, Egypt using GIS based morphometry and satellite imagery. *Environmental Earth Sciences, 62*(3), 611–623.

Chapter 51
Analysis and Prediction of Noise Pollution from Wind Turbines: A Case Study of Loi Hai Wind Power Plant (Ninh Thuan, Vietnam)

Cuong Tran Thien, Huan Nguyen Quoc, An Thinh Nguyen, and Minh Tran

Abstract The demand for electric power in Vietnam has increased over the past 5 years at annual growth rates of 10–12%, and the challenge is to promote renewable energy sector. One of these sustainable energy sources is to harness energy from the wind through wind turbines. However, a significant hindrance preventing the widespread use of wind turbines in Vietnam is the noise they produce, because noise significantly contributes to the annoyance experienced by residents living near wind farms. The prediction of noise impacts for new wind farms is one of the many aspects of the environmental impact assessment process in Vietnam. This chapter deals with a predictive software approach for the calculation and simulation of the acoustical noise produced by 11 wind turbines of Loi Hai wind power plant in Ninh Thuan Province, Vietnam. In the software framework, several noise resonance simulations of 11 wind turbines and National Highway No.1A have been performed, giving a first description of the results that can be achieved in terms of noise mapping in more complex configurations of wind farm.

Keywords Wind farms · Noise pollution of wind turbine · Noise impacts · Prediction model · Modeling

C. T. Thien (✉) · H. N. Quoc · M. Tran
Faculty of Environmental Sciences, VNU University of Science, Vietnam National University, Hanoi, Vietnam
e-mail: tranthiencuong@hus.edu.vn

A. T. Nguyen
VNU University of Economics and Business, Vietnam National University, Hanoi, Vietnam

A. T. Nguyen, L. Hens (eds.), *Global Changes and Sustainable Development in Asian Emerging Market Economies Vol. 2*, https://doi.org/10.1007/978-3-030-81443-4_51

1 Introduction

With the potential and advantages of wind resource, Vietnam is placing a center role on this vital one of the alternative energy sources in the energy development strategy. The country is currently focusing on mobilizing resources, creating favorable conditions to attract investors who have experience, financial capacity and modern technology in developing wind power plants. This process helps contribute to ensuring national energy security, creating a breakthrough to promote socioeconomic development as well as realizing Vietnam's green development and environmental protection strategy (The Prime Minister of Vietnam Government [VG], 2015).

In response to the Vietnamese Government's commitment in the 21st Conference of the Parties to the United Nations Framework Convention on Climate Change (COP21) November 2015, Decision No. 2068/QD-TTg dated 25/11/2015 of the Prime Minister approving Vietnam's renewable energy development strategy to 2030, and a vision to 2050 (The Prime Minister of Vietnam Government [VG], 2015), stressed that "Development priority sources of electricity from renewable energy, creating a breakthrough in energy security assurance, contributing to the conservation of energy resources and minimizing environmental pollution."

According to the Revised Vietnam Power Plan for the period of 2011–2020, with a vision to 2030 approved by the Prime Minister in Decision No. 428/QD-TTg dated March 18, 2016 (The Prime Minister of Vietnam Government [VG], 2016), the development of wind power is set to total wind power capacity of 800 MW in 2020, 2000 MW in 2025 and 6000 MW in 2030. Wind power will account for 0.8% of the total in 2020, about 1% in 2025 and about 2.1% in 2030 (The Prime Minister of Vietnam Government [VG], 2016).

In short, the development of wind power brings great benefits in terms of energy and socio-economy, becoming the main driving force for the development of this industry in many countries around the world, as well as in Vietnam (The Prime Minister of Vietnam Government [VG], 2015). However, one of the significant obstacles preventing wind power development is the noise generated by wind turbines and its impacts on local communities where the wind power projects are located (Ofelia Jianu, Rosen, & Naterer, 2011), besides other obstacles on the policy mechanism and electricity tariff in Vietnam (MoIT of Vietnam, 2018). In fact, the development speed of wind power in Vietnam is still slow; only seven wind power projects with a total capacity of about 190 MW have been put into operation (The Ministry of Industry and Trade of Vietnam, 2019a, 2019b, 2019c).

Noise generated by the operation of wind turbines comprises two main sources:

1. Mechanical noise generated during the working of mechanical components such as gearboxes, generators, drives, cooling fans, and ancillary equipment (Rogers, Manwell, & Wright, 2006).
2. Aerodynamic noise generated during the interaction of turbine blades and wind turbine towers with blowing air. The level of noise generated by wind turbines also increases with the wind speed, usually from 4 to 12 m/s (Tickell, Ellis, & Bastasch, 2004).

Residents at a distance of 3 km or more from the nearest wind turbine can hear low-frequency noise emitted from wind farms (Hansen, Doolan, & Hansen, 2017; Hansen, Hansen, & Zajamsek, 2017; Hansen, Zajamsek, & Hansen, 2013; Katinas, Marčiukaitis, & Tamašauskienė, 2016; Rogers et al., 2006).

It has been proven that people exposed to excessive noise will suffer from health problems such as hearing impairment, headaches and fatigue (due to sleep disorders). (Alberts, 2006; McCunney et al., 2014). Extremely high noise exposure may even cause constricted arteries and a weakened immune system (Alberts, 2006).

According to a research by the Institute of the Counties & Municipalities in Denmark (AKF-Institute of Local Government Studies, Denmark [AKF], 1996), based on interviews with people living near wind power plants, the medical costs incurred by noise are estimated at nearly 0.0012 Euro/kWh.

Overall, in recent years, with the progress of science and technology, while wind turbines have been designed and manufactured with much lower noise level, the noise from wind turbines is still a serious problem. It is necessary to quantify and evaluate the extent of its impact on the local community as well as the workers operating the wind power plants.

2 Methodology

2.1 *Wind Farm Location (Figs. 51.1 and 51.2)*

Loi Hai Wind Power Plant is located in Loi Hai Commune, Thuan Bac District, Ninh Thuan Province, Vietnam; Plant location is located in an area of 523 ha, near the North-South railway and National Highway 1A.

The plant has a capacity of 30 MW, including 11 turbines (capacity of 2625 MW) manufactured by Gamesa firm (Spain) with a noise level at rotor of 102.9 dB (according to the manufacturer's catalog) (Tables 51.1, 51.2 and 51.3; Figs. 51.3

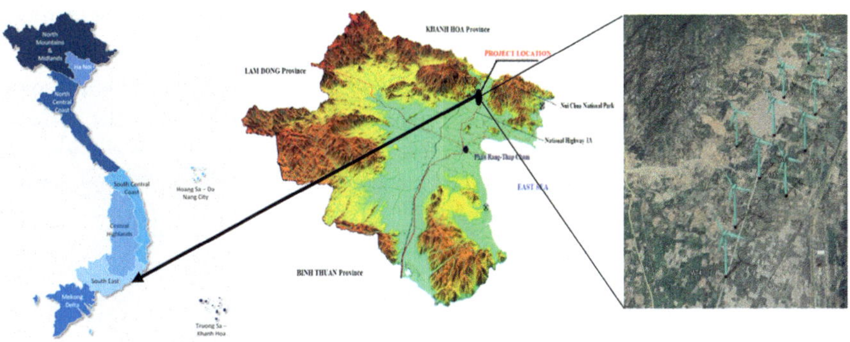

Fig. 51.1 Location of the project in Vietnam

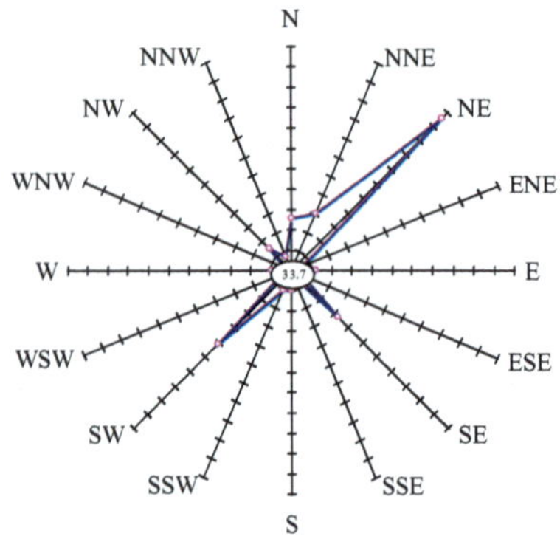

Fig. 51.2 Wind roses summarizing 16 wind directions (Power Engineering Consulting Joint Stock Company 3 [PECC3], 2017)

Table 51.1 Frequency of wind appearance toward 16 directions in the project area

Direction	Windless	N	NNE	NE	ENE	E	ESE	SE	SSE
Frequency (%)	33.7	5.13	6.00	20.9	1.28	2.38	1.11	6.44	0.609
Direction		S	SSW	SW	WSW	W	WNW	NW	NNW
Frequency (%)		1.96	2.06	10.1	1.19	1.92	0.562	3.06	1.51

Remark: N: North, NNE: North Northeast, NE: Northeast, ENE: East Northeast, E: East, ESE: East Southeast, SE: Southeast, SSE: South Southeast, S: South, SSW: South Southwest, SW: Southwest, WSW: West Southwest, W: West, WNW: West Northwest, NW: Northwest (Power Engineering Consulting Joint Stock Company 3 [PECC3], 2017).

and 51.4). The height of the turbine tower is 102 m (Power Engineering Consulting Joint Stock Company 3 [PECC3], 2017).

The average wind speed at the height of 50 m in the project area is 6.85 m/s (NASA), with the prevailing wind direction being the Northeast. Noise levels measured in the project center area and National Road 1A are 52 dBA and 68.4 dBA, respectively (Power Engineering Consulting Joint Stock Company 3 [PECC3], 2017).

2.2 Methods

This study was conducted on the basis of applying a modeling method to calculate and simulate the noise level generated from 11 wind turbines of Loi Hai wind power plant and National Highway 1A (the road section running through the project), simultaneously simulating sound resonance between sources using a combination of German Cadna/A software (Computer Aided Noise Abatement) (https://www.datakustik.com) and Dutch Inoise software (https://dgmrsoftware.com).

The calculation and prediction methods in both Cadna/A and Inoise software applied to wind turbines are based on the international standards of acoustic transmission: International Standard ISO 9613 (1996a, 1996b), and International Standard ISO 17534-3 (2015).

Table 51.2 A summary of wind farm information

Type of turbine	No. of turbines	Power capacity (MW)	The height of the turbine tower (m)	Noise at rotor (dBA)
Gamesa G126–2.625 MW	11	30	102	102.9

Source: Feasibility study report of Loi Hai wind power plant project (Power Engineering Consulting Joint Stock Company 3 [PECC3], 2017)

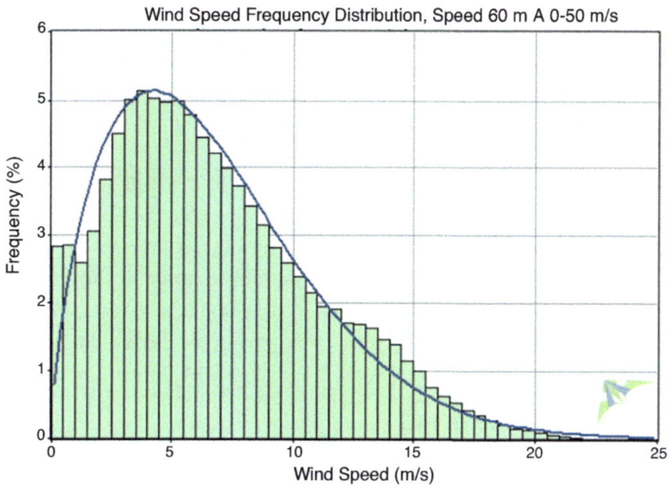

Fig. 51.3 Wind speed frequency distribution (Power Engineering Consulting Joint Stock Company 3 [PECC3], 2017

Table 51.3 Noise level of wind turbines by one octave frequency

Frequency (Hz)	31.5	63	125	250	500	1000	2000	4000	8000	$L_{WA(eq)}$ (dBA)/C
Noise level (dBA) (turbine Gamesa G126 2.625 MW)	95	85	104	102	105	85	81	76	67	102,9

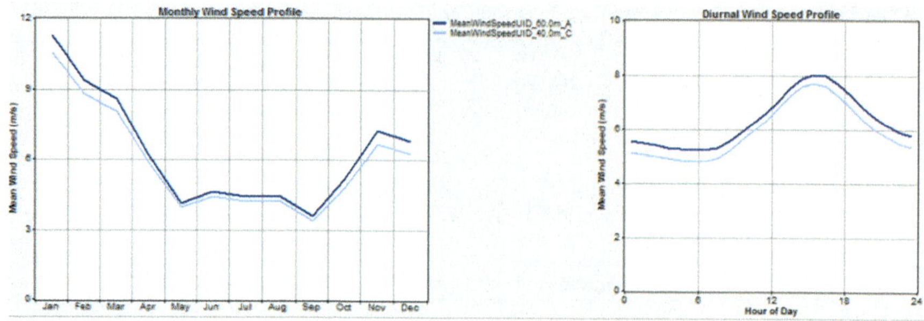

Fig. 51.4 Monthly and diurnal wind speed profile (Power Engineering Consulting Joint Stock Company 3 [PECC3], 2017)

Accordingly, each wind turbine is a noise source and is considered a point source. The sound pressure level at a position is determined by subtracting the sound pressure drop from the external elements from the noise source (wind turbine) in the octave frequency range. The noise level in the octave range is shown by the following equation:

$$SPL_{Turbine} = L_{W(eq)} + D - A_{geo} - A_{atm} - A_{gr} - A_{bar} - A_{misc} - 2\ dB$$

SPL: Sound pressure.

$L_{W(eq)}$: Noise level at source (rotor).

A_{geo}: Decreased noise level due to the spread of sound waves in spherical form in the free field from the sound source $\{A_{geo} = 20 \times \text{Log}d + 11$ (d: Distance from turbine to the recipient point)$\}$.

D: For wind turbines, the noise level is measured and calculated following the wind direction, so there is no need to adjust the amount of noise. Therefore, D is usually equal to 0.

A_{atm}: Noise level reduction by atmospheric absorption = d (Distance from wind turbine) × a (atmospheric absorption coefficient 0,003 dB/m).

A_{gr}: Noise level decrease due to absorption by earth surface = 0,5 dB (absorbing mixed earth).

A_{bar}: Noise level decrease due to absorption by obstructing objects (depends on the size of the object).

A_{misc}: Additional noise reduction level (spreading effect through foliage, plants, houses).

2 dB is the corrected noise level, used to convert the Lweq levels according to the background intensity parameter L_{A90}.

The resonant noise level of the whole wind power plant is calculated as follows:

$SPL_{Wind\ farm} = \sum(SPL_{Turbine}$ within a radius of 3 km from the wind turbine).

3 Results

We have selected three noise sensitive points (NSP) to quantitatively calculate the level of resonant noise level from wind turbines of Loi Hai wind power plant.

- NSP1: Ba Rau Primary School, Loi Hai Commune, 2 km from the wind power plant to the South.
- NSP2: Boundary of Nui Chua National Park, 1 km to the East from the plant.
- NSP3: Residential area of Suoi Da village, Loi Hai commune, 1.3 km northeast of the plant.

Calculation results are presented in the following Tables 51.4, 51.5 and Figures 51.5, 51.6 and 51.7:

Research results demonstrate that:

- Each wind turbine is considered a source of continuous noise, in point form, and NH1A is considered as a discontinuous noise source, and is in line form.
- The noise intensity generated by wind turbines depends mainly on aerodynamic phenomena and the processes that generate mechanical noise. Wind turbine noise increases due to the negative reflection processes and the ground surface of surrounding buildings, while its absorption is affected by air density, humidity and by airflow kinetics of ambient elements.
- Noise level at rotor of wind turbines and along National Road 1A exceeds the permissible standards (QCVN 26: 2010/BTNMT) regulation of noise limits in normal areas from 6 am–9 pm is 70 dBA and from 21 h–6 h is 55 dBA (National technical regulation QCVN 26:2010/BTNMT, 2010). The noise level at the base of the wind turbine tower meets QCVN 26: 2010/BTNMT during the daytime while it exceeds the permissible limit in QCVN 26: 2010/BTNMT at night. Outside the range of 130–150 m from the base of the tower, the noise level generated from the turbines is considered within QCVN 26: 2010/BTNMT (National technical regulation QCVN 26:2010/BTNMT, 2010).
- When there is a resonance between 11 turbines and NH1A road, the noise level increases from 6.3–7.3 dBA.
- The diffusion of noise in general depends on many factors, especially the speed and wind direction.
- The resonant noise level at night is usually 0.3–0.5 dBA higher than that in daytime.

Table 51.4 Noise levels at three noise-sensitive spots during the day

Noise calculation point	VN-2000 coordinate system (Projection 3^0)		Wind velocity (m/s)	Noise level due to wind turbines (dBA)	Resonance noise level (dBA)
NSP 1	1296274.37	586986.91	6.85	45.1	52.4
NSP 2	1298454.85	588818.08	6.85	47.4	53.7
NSP 3	1300100.32	589760.69	6.85	44.8	52.0

Table 51.5 Noise level at three noise-sensitive spots at night

Noise calculation point	VN-2000 coordinate system (Projection 3^0)		Wind velocity (m/s)	Noise level due to wind turbines (dBA)	Resonance noise level (dBA)
NSP 1	1296274.37	586986.91	6.85	45.5	52.8
NSP 2	1298454.85	588818.08	6.85	47.9	54.2
NSP 3	1300100.32	589760.69	6.85	45.1	52.3

- In case the project applies World Bank standards (limited to 45dBA), people and livestock are not allowed to live permanently within the contour of 45dBA. Households living within the yellow area must be relocated and resettled.
- Indeed, for the Loi Hai wind power plant project, due to the large number of affected households (197) due to the scope of the isoline noise of 45dBA, the project was not entitled to preferential loans from international financial institutions.
- Wind power projects are generally opposed by the local community for the reason that people are at risk of adverse physiological and psychological symptoms related to wind turbine noise. And it is because of the current acoustic nuisance that has caused a backlash from residents in social and health impact surveys on people living near wind turbines.
- To minimize the negative impacts due to the noise from wind turbines to the environment and human health, it is necessary to apply a number of synchronous solutions such as technological solutions (using new modern turbines with generating low noise and regularly servicing turbines, etc.), managerial solution (recommended for local people not to live in areas with a noise level of 45dBA or higher, etc.) and to implement a well-prepared plan on compensation and resettlement for 197 local households to help them feel secure to live in their new resettlement sites.

4 Conclusions

In this chapter, the authors presented a summary of scientific basis, modeling methods and results of calculation, prediction and simulation of the noise intensity distribution due to wind turbines of Loi Hai wind power plant in Vietnam, also considered the case of acoustical resonance with other sources in the region.

The application of the modeling method to quantify the noise level from wind turbines in particular and the combination of noise sources in general is highly feasible, presenting visual results. This correspondingly serves as an important basis for the environmental authorities and project investors in decision-making right from the project screening stage.

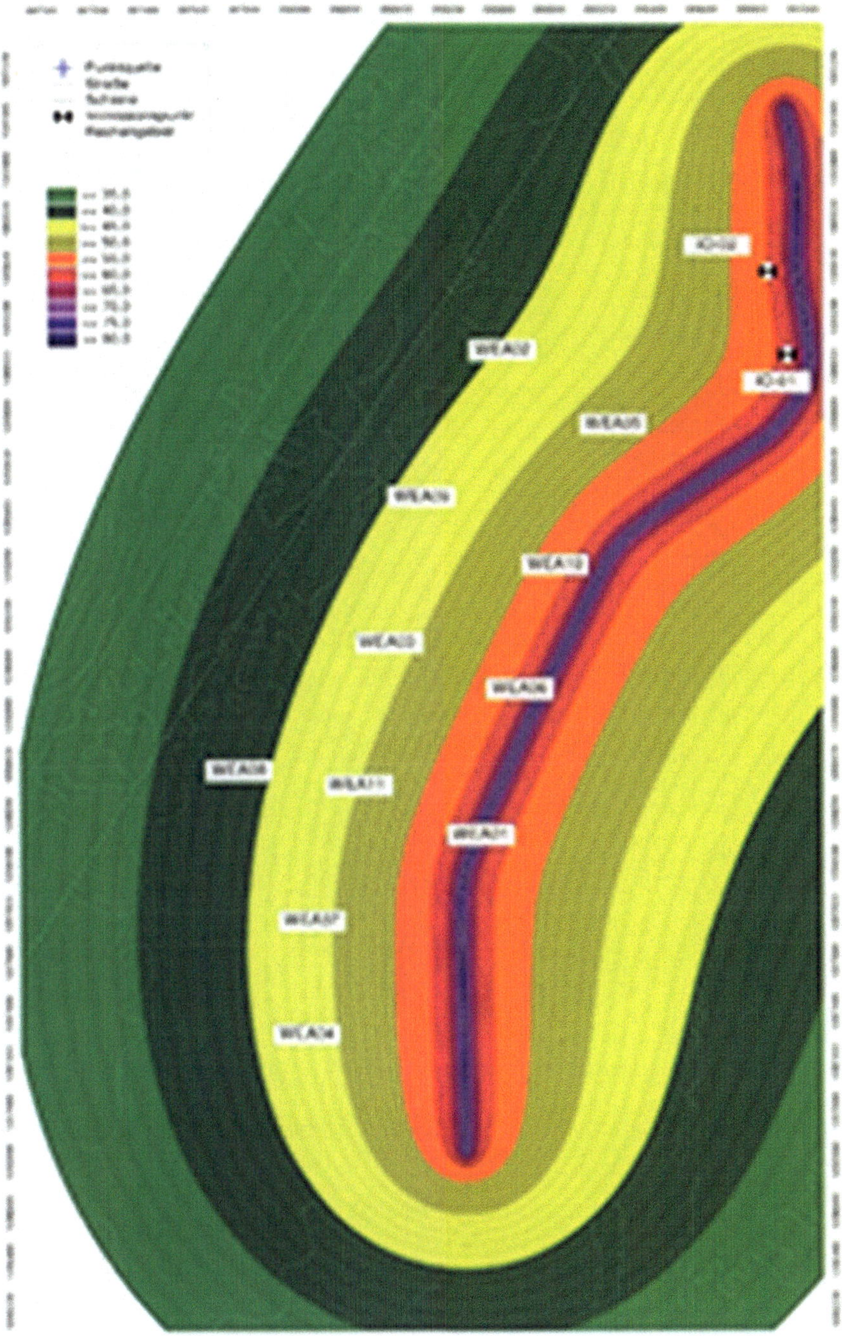

Fig. 51.5 Noise level due to traffic activities on NH1A

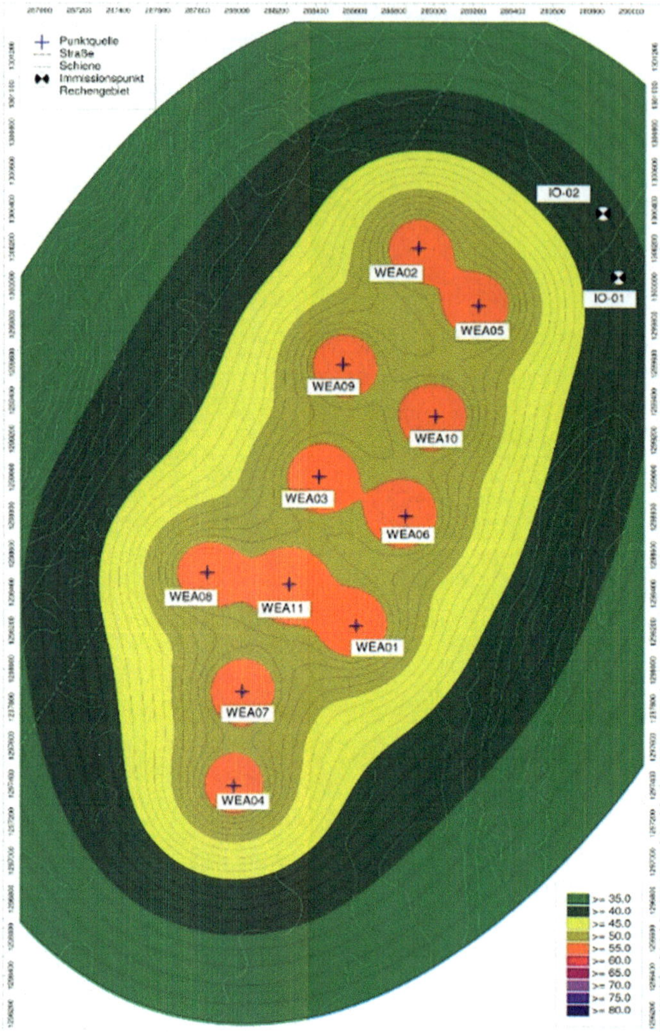

Fig. 51.6 Noise level generated by 11 wind turbines

Acknowledgments The authors extend their thanks to the Faculty of Environmental Sciences, VNU University of Science, Vietnam National University, Hanoi for their invaluable support for this research.

All data and information related to Loi Hai Wind Power Plant is retrieved and used with permissions from the leader of the Plant Management Company for scientific purposes. We also wish to extend our gratitude to the Company for this generosity.

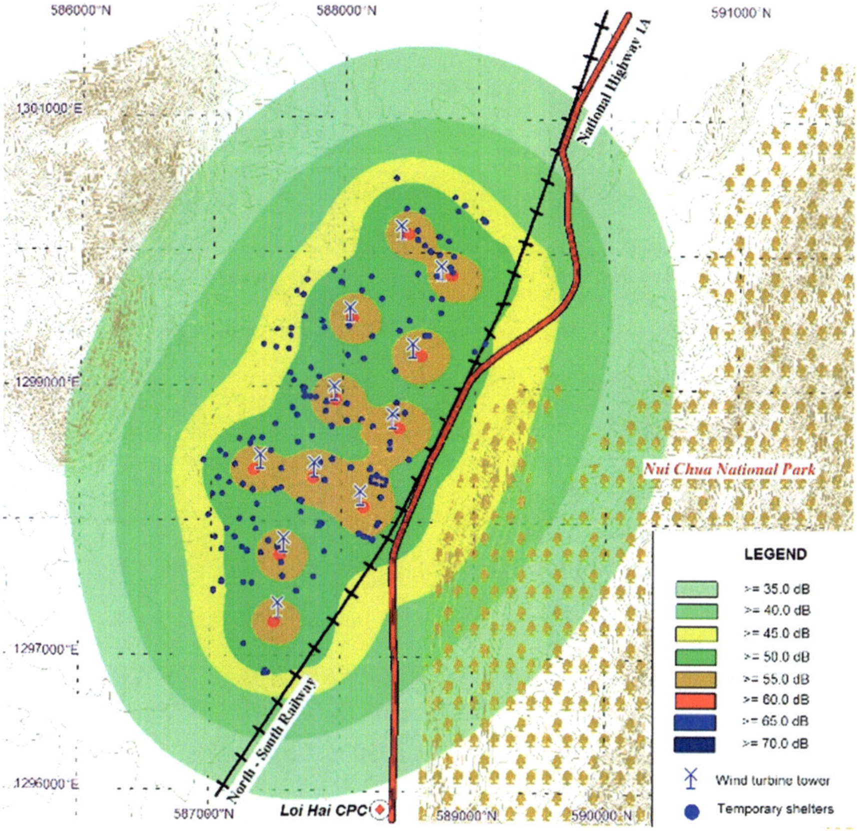

Fig. 51.7 Resonance noise level between 11 wind turbines of the plant and NH1A

References

AKF-Institute of Local Government Studies, Denmark [AKF]. (1996). Research Institute of the Counties & municipalities in Denmark., April 1996.

Alberts, D. J. (2006). *Primer for addressing wind turbine noise* (19 pages). Report from Lawrence Technological University.

Hansen, C. H., Doolan, C. J., & Hansen, K. L. (2017). *Wind farm noise: Measurement, assessment and control* (624 pages). Wiley.

Colin H. Hansen, Kristy L. Hansen, B. Zajamsek. (2017). *The occurrence of nocturnal wind farm rumbling noise*. 7th International Conference on Wind Turbine Noise. Rotterdam, The Netherlands, 2nd - 5th May 2017. pp. 123–130.

Hansen, Kristy, Branko Zajamsek, and Colin Hansen. (2013). Analysis of unweighted low frequency noise and infrasound measured at a residence in the vicinity of a wind farm. Australian Acoustical Society, Proceedings of Acoustics, 17–20 November 2013. 9 pages.

International Standard ISO 17534-3. (2015). Acoustics - Software for the calculation of sound outdoors - Part 3: Recommendations for quality assured implementation of ISO 9613-2 in software according to ISO 17534-1.

International Standard ISO 9613. (1996a). Acoustics - Attenuation of sound during propagation outdoors - Part 2: General method of calculation.

International Standard ISO 9613. (1996b). Acoustics - attenuation of sound during propagation outdoors - part 1: Calculation of the absorption of sound by the atmosphere. .

Jianu, O., Rosen, M. A., & Naterer, G. (2011). *Noise pollution prevention in wind turbines: Status and recent advances* (14 pages).

Katinas, V., Marčiukaitis, M., & Tamašauskienė, M. (2016). Analysis of the wind turbine noise emissions and impact on the environment. *Renewable and Sustainable Energy Reviews., 58*, 825–831.

McCunney, R. J., Mundt, K. A., David Colby, W., Dobie, R., Kaliski, K., & Blais, M. (2014). Wind turbines and health - a critical review of the scientific literature. *American College of Occupational and Environmental Medicine, 56*, 112–125.

National technical regulation QCVN 26:2010/BTNMT. (2010). December 16th, 2010 of the Minister of Natural Resources and Environment on noise.

Power Engineering Consulting Joint Stock Company 3 [PECC3]. (2017). *Feasibility study report of Loi Hai wind power plant project*. 111 pages.

Rogers A.L, Manwell JF, and Wright S. (2006). Wind turbine acoustic noise. In *Renewable energy research laboratory*. Department of Mechanical and Industrial Engineering University of Massachusetts at Amherst, Amended.

The Ministry of Industry and Trade of Vietnam. (2019a). Retrieved September 28, 2019, from http://www.moit.gov.vn/CmsView-EcoIT-portlet/html/print_cms.jsp?articleId=12101.

The Ministry of Industry and Trade of Vietnam. (2019b). Retrieved September 10, 2019, from https://www.datakustik.com

The Ministry of Industry and Trade of Vietnam. (2019c). Retrieved September 12, 2019, from https://dgmrsoftware.com.

The Prime Minister of Vietnam Government [VG]. (2015). *Decision No. 2068/QD-TTg dated 25 November 2015 on the approval of renewable energy development strategy of Vietnam up to 2030 with vision to 2050*. 15 pages.

The Prime Minister of Vietnam Government [VG]. (2016). Decision No. 428/QD-TTg dated 18 March 2016 of the Prime Minister, the approval of revisions to the national power development plan from 2011 to 2020 with visions extended to 2030. 14 pages.

Tickell, C. E., Ellis, J. T., & Bastasch, M. (2004). *Wind turbine generator noise prediction—Comparison of computer models* (Proceedings of Acoustics 2004) (pp. 45–50). Wiley.

Chapter 52
An Environmental Zoning for Sustainable Development in Thua Thien Hue Province, Vietnam

Hang Anh Phan, Thang Van Le, Tuan Anh Tran, and Son Hoang Nguyen

Abstract Based on the natural, socio-economic and environmental characteristics, Thua Thien Hue province is divided into six environmental regions. They include the environmental region of O Lau river basin, environmental areas of A Sap river basin and upstream of Dakrong river, the environmental region of Bo river basin, the environmental region of Huong river basin, and the environmental region of Truoi river basin and the environmental region of Bu Lu river basin. Each region was divided into three sub-regions corresponding to upstream, middle, and downstream river basins. The study results are considered scientific bases to propose solutions for socio-economic sustainable development for each sub-region in Thua Thien Hue province.

Keywords Environmental zoning · Sustainable development · Environmental functions · GIS · Thua Thien Hue

1 Introduction

Environmental zoning was the division of territory into regions and sub-regions based on spatial division of natural, socio-economic characteristics and environmental status according to the objectives of protection and preservation, development, and adaptation to climate change. Environmental partitioning is one of the bases of spatial development plans such as environmental protection planning, land-use planning, urban planning, etc. (Lindsey, Man, Payton, & Dickson, 2004; Taylor, 1998) and the ultimate goal is to ensure sustainable development socio-economic

H. A. Phan · T. Van Le · T. A. Tran
University of Science, Hue University, Hue, Vietnam

S. H. Nguyen (✉)
Institute of Open Education and Information Technology, Hue University, Hue, Vietnam
e-mail: nhsonsp@hueuni.edu.vn

A. T. Nguyen, L. Hens (eds.), *Global Changes and Sustainable Development in Asian Emerging Market Economies Vol. 2*, https://doi.org/10.1007/978-3-030-81443-4_52

sustainability of the territory (Elizabeth, 1992; McGregor, Miller, & Zirnhelt, 1999; Montgomery, 2004).

In the world, functional zoning has been used in urban planning and land-use adjustment. In Europe, since the late nineteenth century there has been a division of areas in cities to control development. In the USA, in 1916, New York City implemented the first partition (Buckingham, 1996; Edington & Edington, 1977). Environmental zoning has been conducted and according to various methods such as functional zoning in Europe, USA, eco-approach zoning in China, Australia, Brazil, Peru (Minaya, Gallardo, & Baud, 2012), Ecuador, Venezuela, and others (Presidency of the Republic House Civil Cabinet Subcommittee for Legal Affairs [PRHCCSLA], 2002), or environmentally sensitive partitions in Malaysia, India. In 2016, Leman et al. conducted an environmental-sensitive area assessment for land-use planning in Langkawi, Malaysia. The study to assess the environmental sensitivity of Langkawi also uses a multi-criteria evaluation model. The set of indicators used in the study include indicators on disaster risks (slope, vegetation, rainfall, seismic, etc.), heritage values and life support indicators. The study classified environmental sensitivity into four levels: high, medium, low and non-sensitive (ADB, 1991; Fang, Zhang, Hong, Zhang, & Bristow, 2008; Hall, 2002).

In Vietnam, there have been many studies on the theoretical basis and methodology of environmental planning or environmental protection planning, confirming the role of environmental zoning. At the provincial scale, our country has about 20 localities studying and developing environmental planning or environmental protection planning. At the regional scale, we have environmental planning for the basin of Cau, Nhue-Day and Dong Nai river systems by 2015 and orientations to 2020, the Hong River Delta environmental planning. Most of these plans have adopted an integrated, systematic approach and are aimed at sustainable development.

Thua Thien Hue, located in the North Central of Vietnam, on the North-South road transport axis and the East-West corridor axis linking Thailand–Laos–Vietnam along the ninth road. The province's economy is growing well, the average annual economic growth rate in the period 2000–2018 reached 8.8%. The economic structure has been shifted towards increasing the proportion of the service sector, reducing the proportion of agriculture. GDP in the province per capita in 2018 reached US$ 1793 / person. Total state budget revenue is 7236 billion Dong (Thua Thien Hue General Statistical Office [TTHGSO], 2019). In addition to economic achievements, resource degradation and environmental pollution are taking place particularly in production areas and urban areas. One of the causes of the above problem is that Thua Thien Hue has not implemented the planning of economic development spaces associated with environmental protection. To elaborate on an environmental protection plan which contains contents of environmental partitioning necessary to delimit territorial space into areas for natural and environmental conservation and restoration, give priority to development and ensure the sustainable development of the economy and society.

2 Methodology

2.1 Principles and Criteria of Zoning

- *Principles of environmental zoning:* Environmental zoning in Thua Thien Hue follows the principles: (1) respect the objectivity of the region, (2) accept the relative homogeneity of the region, (3) in accordance with the nature of the region, (4) in accordance with the management method (Dang, Le, et al., 2017).
- *Criteria of environmental zoning:* Environmental zoning is based on natural factors, people's livelihood and environmental status. Natural factors include geographical location, geology, topography, land cover, hydrology, and creatures. Group of livelihood factors including living, production activities (agriculture, industry, services), urbanization process, etc.

2.2 Data Collection and Processing

- ***Method of collecting, analysing, synthesizing documents:*** Collecting, systematizing, analysing and comparing documents on environmental zoning, including theoretical documents on environmental zoning, figures and data on natural, socio-economic characteristics, environmental status, climate change, development plans of Thua Thien Hue Province provided by some offices and departments.
- ***Field word study:*** The work of field survey aims to verify information related to environmental zoning; in addition, it also aims to collect and supplement new documents and data from province and districts departments. Field routes include Batch 1: Along the coast from Phong Dien to Lang Co; Batch 2: From downstream to upstream of Huong River; Batch 3: Follow the Ho Chi Minh road; Batch 4: From Hue city along National Highway 1A to Bach Ma. The subjects of survey include natural characteristics, natural resources, current status of natural resource exploitation and current protection of natural resources and environment. Socio-economic characteristics that are studied include the status of development of economic sectors (agriculture, industry and services), population distribution, population, settlement, local socio-economic development plan. Besides, it also studies problems of resource degradation, environmental pollution, environmental protection, climate change adaptation, and consults experts, management agencies and local people about problems related to environmental zoning.
- ***Map and GWAS methods:*** The map method was used to analyse the component maps (maps of topography, soil, vegetation) and edit the environmental zoning map of this province . The map method gives visual results about objects according to territorial space (Hall, 2002).

- ***Survey of experts:*** Environmental zoning needs to have comprehensive information on various fields; it was necessary to consult experts, professional staff of departments' opinions in order to get the result of zoning and orient reasonable use of resources, protect the environment, and develop economic suitably for each environmental zoning unit.

3 Selected Study Area

3.1 Natural, Socio-Economic and Environmental Characteristics

3.1.1 Natural Characteristics

(a) *Geology, minerals:* The geological structure of the territory was very diverse, including 16 stratigraphic units and 7 intrusive magma complexes. There are various kinds of magmatic, metamorphic and sedimentary rocks, accounting for three-fourths of the natural area, distributing in the western, southwest and southern mountains of the province. Unconsolidated sediments making up one-fourth of the territory areas distributed in the coastal plain in the East. There are 120 mines, mineral spots with 25 different types of minerals discovered in the territory (People's Committee of Thua Thien Hue Province [PCTTH], 2005).

(b) *Topography:* Thua Thien Hue has all kinds of terrain from mountains, hills, plains and coastal areas. The mountainous terrain ranges from 250 to 1800 m, accounting for one-fourth of the province's area, and distributing in the west of the territory. Hill topography ranges from 10 to 250 m, accounting for one-half of the territory. The plain was an abrasive, agglomerated, sandy and lagoon plain with a total area of about 1400 km^2 (People's Committee of Thua Thien Hue Province [PCTTH], 2005).

(c) *Climate:* The climate was tropical monsoon climate. The annual average temperature was 23.7 °C. The temperature decreases from the plain to the mountains, the annual average temperature was 24–25 °C in the plains and mounds lower than 100 m; the temperature drops to 20–22 °C when it rises to 500–800 m and to 180 °C in mountainous areas over 1000 m. Average rainfall over 2600 mm, some places up to 4000 mm (Bach Ma, Thua Luu). Average annual rainfall increases gradually from East to West and from North to South. The coastal plain has an average annual rainfall of 2700–2900 mm; the hilly areas in the west and the plain from Phu Bai to Truoi have an average annual rainfall of 2800–3200 mm; in the West of A Luoi-Ngai cave, the average annual rainfall was over 3400 mm; in Nam Dong - Bach Ma - Phu Loc, the average annual rainfall was 3400–4000 mm. The rainy season was from September to February of the following year, the rainiest months last from September to December having the highest rainfall (accounting for 30% of the annual rainfall). The average humidity was from 85 to 86%. This province was affected by two types of monsoon: the southwest monsoon and the northeast monsoon (Le, 2018).

(d) *The river system:* The total area of the river basin of the whole province was about 4200 km^2. The density of rivers and streams was from 0.3–1 km / km^2, in some places 1.5–2.5 km / km^2. In the territory of this province from the North to the South the main rivers include O Lau, A Sap, Bo, Huong, Nong, Truoi, Cau Hai and Bu Lu rivers.

(e) *Tam Giang–Cau Hai lagoon:* This lagoon system has length of 68 km, with the water surface area about 22,000 ha, belonging to one town and four districts including Huong Tra town, Quang Dien, Phu Vang, Phu Loc and Phong Dien. The lagoon system consists of four lagoons connecting with each other, including Sam, Tam Giang, Cau Hai, and Ha Trung–Thuy Tu Tam Giang lagoon has a water area of 5200 ha, 27 km long from O Lau estuary to Huong river estuary, which was open to the sea via Thuan An estuary. Sam lagoon, which was a closed lagoon, has an area of 1520 ha. Ha Trung–Thuy Tu lagoon, which was also a closed lagoon, has a water surface of 3600 ha. The Cau Hai lagoon was the largest one with a water surface of about 11,200 ha, connecting to the sea through Tu Hien estuary. Lap An lagoon (also known as An Cu lagoon or Lang Co lagoon) has an area of 1500 ha, connecting to the sea through Lang Co estuary (People's Committee of Thua Thien Hue Province [PCTTH], 2005).

(f) *Soil:* There are many types of soils in the plain, including sea sandy soil (accounting for 10% of the natural land area of the whole province), saline soil (accounting for 2.7%), and alluvial soil (accounting for 8.9%). Types of soils in the midlands and mountainous areas include yellow-red feralite (accounting for 66.3%), yellow-brown feralite on ancient alluvium, alluvial soils in the valleys, infertile soils, and inert eroded feralites. Rocky gravel, and humus feralite soil on mountains (People's Committee of Thua Thien Hue Province [PCTTH], 2005).

(g) *Creature:* The vegetation of the rainforest was various with native plants such as ironwood tree, sindoer, white meranti, etc., northern legume; migratory trees such as chestnut, cypress, pines, and southern oil cropsthe forest cover was about 57.3% (Center for Natural Resources and Environment Monitoring of Thua Thien Hue Province, 2018). Due to war and indiscriminate exploitation, forest area is decreasing. Rich forests are mainly distributed in the high mountains of the West, Southwest, and Southern territories; the rest are medium and poor forests. Besides, newly planted forests are being promoted. The natural animals of Thua Thien Hue are plentiful and have high economic value. In addition to the common animals in the forest, such as monkeys, deers, peacocks, wild chickens, many rare animals have been discovered in Nam Dong, Phu Loc, A Luoi such as elephants, tigers, pheasants, flying ferrets, bear dogs.

3.1.2 Socio-Economic Characteristics

(a) *Residential:* According to the Thua Thien Hue's Statistical Yearbook in 2018, the province's population was about 1,163,608 thousand people; the urban population was 568,552 thousand people, accounting for 48.86% of the population; the rural population was 595,056 thousand people. Natural population growth rate was 10.8‰. The average population density was 230.23 people/km^2. The population was not equally distributed, mainly concentrated in Hue city, towns and townships, in plains and sparsely in rural and mountainous areas. In Hue city, the population density was 5040 people/km^2, whereas in A Luoi district, the population density was 41 people/km^2. In terms of ethnic composition, the Kinh accounts for the majority of the population. Ethnic minorities in Thua Thien Hue, namely, Co Tu, Pa Koh, Ta Oi, Bru – Van Kieu living in the western mountainous region of the province.

(b) *Economy*: Regarding the size of the economy, the gross domestic product per capita in 2018 reached $1793. In terms of economic structure, the agriculture, forestry and fishery accounted for 11.6%, the service industry accounted for 31.2%, the service accounted for 50.2%, the product taxes minus product subsidies accounted for 7.0%.

Regarding agriculture, the area of annual crops was 66,268.4 ha, of which rice area was 54,731.7 ha, accounting for 82.6% of the annual crop area; 4680.3 ha of vegetable area; 3269.4 ha of peanut area; 1943.5 ha of sweet potatoes; 1643.5 ha of maize. The area of perennial crops was 11,523.3 ha, including 8875.1 ha of perennial industrial trees; fruit trees area was 2648.2 ha. For animal husbandry, in 2018, there were 178,811 pigs; 24,339 cows; buffalo 20,847 children; 9442 goats; 2,969,000 poultry; 2459 beehives; 3,72,000 quails; 16,000 doves (Thua Thien Hue General Statistical Office [TTHGSO], 2019).

In terms of forestry, the forest cover in 2018 was 57.3%, the newly concentrated planted forest area was 6.14,000 ha; production forest area was 5.84000 ha, protective forest area was 0.22000 ha, special-use forest area was 0.09000 ha. The exploited wood output was 585.6000 m^3; firewood production was 316.65000 ster.

Regarding industry, the main production sectors include food, beverages and apparel. Industrial production index in 2018 reached 108%, ranked 44th out of 63 cities and provinces in the country (the highest position was Ha Tinh at 188.5%, ranked fifth out of six regional provinces). The main industrial products are construction materials, beverages and consumer goods, in particular limestone and other calcium. Regarding the distribution of points, industrial zones (IPs), the North has Phong Dien, Tu Ha industrial zones, Dong Lam cement; in the South, there is Phu Bai Industrial Zone, Chan May-Lang Co urban economic zone; the West has built Ta Trach, Huong Dien, Binh Dien, A Luoi hydroelectric factories and Nam Dong cement factory.

In terms of service activities, total retail sales of goods and consumer services at actual prices in 2018 reached 38,230.7 billion VND, ranked 36th out of 63 provinces and cities in all over the country (ranked first was Hanoi with 469,502 billion

VND); ranked fourth among six provinces in the North Central region (Thanh Hoa ranked first with VND 93,734.8 billion). In 2018, revenue from travel services reached 175 billion VND, from accommodation services 1606.4 billion VND, from dining 4765.8 billion VND, from other consumer services 2000.9 billion VND. The tourism sector has outstanding strengths and brings large revenue to the province.

3.1.3 Environmental Quality

(a) *The current state-of-the-water environment.*

- The monitoring points of power of hydrogen parameter of rivers in the whole province are valid to meet the permitted standard according to QCVN 08-MT: 2015/ BTNMT column A1 (6–8.5). The pH of the monitoring points on the Huong River system ranges from 6.6 to 6.9; on O Lau, Bo, Truoi and Bu Lu rivers 6.6–7.8; in Tam Giang-Cau Hai lagoon from 6.5 to 8.1.
- The concentration of Biochemical Oxygen Demand 5 (BOD_5) is mostly in the allowed standard of column A1 (4 mg/l). BOD_5 concentration of the monitoring points on the Huong River system is from 1.1 to 2.5 mg/l; on O Lau, Bo, Truoi and Bu Lu rivers 1.1–3.7, the slot area receives the water source of the frozen factory CP-Phong Dien Industrial Park (symbol N_{SOL6}) with BOD_5 content up to 11 mg/l. In the downstream area of Phu Bai river, the Biochemical Oxygen Demand content exceeds the standard (values from 4.8 to 6.6 mg/l).
- COD (Chemical Oxygen Demand) content is mostly in the allowed standard in column A1 (10 mg/l). On Huong River, O Lau River, Bo River, Truoi River, Bu Lu River COD from 90 to 91, some points are beyond the standard such as the location is 1 km Tuan three-way crossroads in the upstream (95 mg/l), NSOL6 (30 mg/l), 1 km downstream of Truoi market (14.4 mg/l), area at the end of Truoi river (at Truoi sluice) (10.4 mg/l), lower area Phu Bai river basin (15.4 mg/l), Phu Tu bridge of Dai Giang river (19.2 mg/l). In Tam Giang–Cau Hai lagoon area, COD ranges from 7.6 to 9.3 mg/l, within the permitted standard limit.
- Turbidity in rivers in Thua Thien Hue has many points that exceed the permitted level in column A1, the average turbidity in rivers in Thua Thien Hue is 5.8 mg/l.
- The content of TSS (turbidity and suspended solids) at the monitoring points is within the permitted limit according to the water quality standard in column A1 (20 mg/l). Turbidity and suspended solids on Huong River is from 4.1 to 11.4 mg/l; on O Lau, Bo, Truoi and Bu Lu rivers, from 2.7 to 11.9 mg/l. At the point N_{SOL6}, the TSS content is up to 118.2 mg/l.
- Coliform rivers within permissible limit at column A1 (2500 MPN / 100 ml) such as Huong River (63–830 MPN / 100 ml), O Lau River, Bo River, Truoi and Bu Lu Rivers (10–673 MPN)/100 ml), Tam Giang–Cau Hai lagoon (15–1100 MPN/ 100 ml). Some spots with higher levels of Coliforms are NSOL6 (10,333 MPN/100 ml), Phu Bai River, 500 upstream from the receiving point of Phu Bai Industrial Area (11,000 MPN/100 ml).

(b) *The current state-of-the-air environment.*

Total of hanging dust (TSP) in urban areas and industrial parks and production facilities exceeds QCVN 05 published on November 11, 2013 by Ministry of Natural Resources and Environment in terms of the national technical regulation on ambient air quality. The level of dust pollution at the points in Hue City includes the Northern Bus Station Area – symbol K1 (482 μg/m^3), the South Bus Station – K2 (350 μg/m^3), the Pham Van T-junction area Dong and Tung Thien Vuong – K3 (899 μg/m^3), Dan Nam Giao – K4 area (192 μg/m^3), around Tinh Tam lake – K5 (119 μg/ m^3). Phong Dien District: At the junction of National Highway 1A and Provincial Highway 6 of Phong Dien Town – K7 (170.6 μg/m^3), around Dong Lam Cement Plant – K23 (154.5 μg/m^3), residential area residing on provincial road No. 6 north of Phong Dien industrial area – K24 (163.6 μg/m^3). Phu Vang District: Residential area of Thuan An town – K9 (132.6 μg/m^3). Huong Thuy district: Cultural area of Huong Thuy town – K10 (138.88 μg/m^3), the intersection of Da Le street and Hue bypass – K21 (1689 μg/ m^3), area near Phu Bai industrial area and Phu Ba Airport – K26 (103 μg/m^3). Nam Dong District: Residential area of Khe Tre Town – K14 (972 μg/m^3). Quang Dien district: Residential area of Sia town – K15 (136 μg/m^3). Huong Tra District: Tu Ha Town Residential Area – K8 (166.7 μg / m^3), Tu Ha Industrial area – K19 (471.8 μg / m^3), Oxi – Nito Tu Ha Factory, conveyor area Luks – K20 cement limestone (483 μg / m^3), Phu Loc District: Chan May Lang Co area – K11 (116 μg / m^3), Lang Co town residential area – K12 (221 μg / m^3), area around La Son industrial area – K18 (225.6 μg / m^3) (Center for Natural Resources and Environment Monitoring of Thua Thien Hue Province, 2018).

Content of nitrogen dioxide and Sulphur dioxide at monitoring locations in Thua Thien Hue is within the limit of QCVN 05: 2013 / BTNMT.

(c) The environment in industrial area and industrial clusters.

The environmental components around industrial area and clusters tend to be polluted. When the wastewater of industrial area and clusters have not yet been treated or processing not yet achieved, it is discharged directly into receiving sources, thus polluting the surface water. The amount of solid waste is increasing, but the problem of collection and treatment of solid waste is still inadequate. The management, transportation and registration of hazardous waste generators in facilities have not been focused. Boilers and incinerators have almost no treatment system to discharge smoke and dust and pollute the air.

4 Results

In this province, the hydrological system is considered the dominant factor, divides the territory into six environmental regions corresponding to six major river basins of the province. The regions are divided into sub-regions based on the dominant element of typographic hierarchy (mountainous areas with elevations from 250 to 1800 m, hilly areas with elevations between 10 and 250 m, and low plains of under

10 m). The system of environmental zones and sub-regions is divided (Figs. 52.1 and 52.2).

4.1 Environmental Region of O Lau River Basin

4.1.1 Environmental Sub-Region in the upstream of O Lau River

- *Characteristics of the sub-region:* Environmental sub-region covers an area of 120.54 km^2. The low mountainous terrain has height of 250–750 m; there are some places in the West above 750 m high, with a slope of more than 25. The main soil types of the sub-region including pale yellow soil on sandy rock with 101.73 km^2, accounting for 84.4% of the sub-region's land, red-yellow loam on metamorphic rocks (10.21 km^2 area, accounting for 8.47% of the sub-region's land), yellow-red soil on shale (7.17 km^2, accounting for 5.95% of the area). There is also yellow-red soil on metamorphic clay (accounting for 1.14% of the area), alluvial soil in streams (0.03%), alluvial soil deposited annually (0.01%). Mechanical components from light meat to heavy meat. This area has a small population (population density of 60 people / km^2). This area is watershed protection forest, mainly special-use forest land (area of 120.17 km^2, accounting for 99.69% of the total area); protective forest land (0.21 km^2, accounting for 0.17% of the area); production forest land (0.16 km^2, accounting for 0.13%).
- Environmental functions of the sub-region.
- The main environmental functions of the sub-region include biodiversity conservation in Phong Dien Nature Preservation zone; watershed protection function; water regulation in Hoa My irrigation lake.
- *Environmental problems of the sub-region* can be listed as the decrease of forest area, risk of soil erosion and degradation.
- Socio-economic sustainable development solutions for the sub-region.

- Strictly protect Phong Dien natural preservation zone; planting and protecting special-use and protection forests; development of production forests; erosion control.

4.1.2 Environmental Sub-Region in the Middle hill of O Lau River

- *Characteristics of the sub-region:* Environmental sub-region covers an area of 288.44 km^2. The hill has the shape of a bowl face down, flat top, gentle slopes (5–15^0). Soils occupying the largest area of the sub-region include light yellow soil on sandy rocks (area of 181.94 km^2, accounting for 63.08% of the total area of the sub-region), yellow-red soil on clay shale (82.47 km^2, accounting for 28.59% of the area). In addition, there also have alluvial soil without accretion (accounting for 2.16% of the area), alluvial soil deposited annually (1.78%), white sand dune (1.57%), water surface (1.07%), alluvial soil with red-yellow variegated soil (0.94%), alluvial soil in streams (0.39%), sea sandy soil (0.25%), alluvial soil covered with sea sand (0.18%). Soil composition is from light to medium meat, sand, mixed sand. The population density of this area is 90 people/km^2. The main economic activity is agricultural and forestry production with 128.72 km^2 special-use forest land, accounting for 44.63% of the total land-use area of the whole sub-region. Production forest land is 75.13 km^2, accounting for 26.05% of the area; protective forest land is 23.7 km^2, accounting for 8.22% of the area; land for planting perennial trees is 20.8 km^2, accounting for 7.21% of the area. Other land-use types include water surface (occupying 4.36% of the area), unused land (3.24%), residential land (2.25%), another annual cropland (1.45%), rice cultivation land (1.39%), special purpose land (1.06%), aquaculture land (0.11%), another agricultural land (0.03%).

- *The environmental functions* of the sub-region are to conserve the biodiversity of Phong Dien nature conservation zone; provide space for developing forestry-agricultural production; rural accommodation; provide resources and space for industrial development (providing minerals for construction materials, mainly sand and gravel); solid waste container.
- *The environmental problems of the sub-region* are the decrease of forest resources due to overexploitation. Besides, due to loss of vegetation cover and unreasonable cultivation, the land is eroded and degraded.
- *Socio-economic sustainable development solutions for the sub-region*: Protecting biodiversity in Phong Dien Nature Preservation zone; planting and protecting protection forests; planting production forests; preventing soil erosion; solving the problem of centralized collection and treatment of domestic wastewater; building solid waste treatment facilities; developing tourism; developing agriculture, mainly focusing on crops with the advantage of perennial crops and fruit trees; construction of industrial parks and industrial clusters.

4.1.3 Environmental Sub-Region in the Lower delta of O Lau River

- Characteristics of the sub-region: Environmental sub-region covers an area of 186.55 km^2. The delta has the height less than 15 m. Types of land occupying the largest area of the sub-region includes white sand dune 105.83 km^2, accounting for 56.73% of the land area of the whole sub-region; alluvial soil covered with sea sand 17.94 km^2, accounting for 9.62%, sea sandy soil 15.36 km^2, accounting for 8.23%. There are also other types of soil such as gleyic fluvisols (accounting for 7.82%), water surface (6.48%), alluvial soil without accretion (5.06%), acid sulphate soil operating deep, medium salinity (4.04%), alluvial soils with red and yellow variegated layers (1.15%), average saline soils (0.84%), high saline soils (0.03%). Mechanical components include sand, sand mixed, and light to medium meat. Population density is 99 people/km^2. Economic activities are mainly agricultural and aquatic production. Regarding the current land-use status of the sub-region, production forest land is 41.89 km^2, accounting for 22.46% of the total land-use area; rice cultivation land is 33.51 km^2, accounting for 17.96%; special purpose land is 25.73 km^2, accounting for 13.79%; water surface is 20.55 km^2, accounting for 11.02%; unused land is 17.2 km^2, accounting for 9.22%. Other land-use types include residential land (occupying 8.16% of the area), protective forest land (7.89%), another annual cropland (5.23%), aquaculture land (2.32%), land for planting perennial trees (1.73%), special-use forest land (0.21%), another agricultural land (0.02%).

- Environmental function of the sub-region: Rural residence; providing space for agricultural development (wet rice, vegetables, aquaculture).
- Environmental issues of the sub-region: Flood risk; pollution risk from living and production activities.
- Socio-economic sustainable development solutions for the sub-region: Priority in aquaculture development; developing small handicraft production places.

4.2 Environmental Areas of a Sap River basin and Upstream of Dakrong River

4.2.1 The Mountainous Environmental Sub-Region of the upstream a Sap River

- Characteristics of the sub-region: Environmental sub-region covers an area of 389.35 km^2. Average mountain terrain has height between 500 and 1800 m. Types of soil occupying the largest area of the sub-region include reddish yellow soil on metamorphic clay is 288.18 km^2, accounting for 74.02% of the sub-region's land area, red-yellow humus on acid magma rocks is 33.6 km^2, accounting for 8.63%. There is also red-yellow soil on acid magma rock (accounting for 6.12% of the sub-region's area), red-yellow loam on metamorphic rocks (accounting for 5.76%), yellow-red soil on shale (3.18%), yellow-brown soil on ancient alluvial (2.04%), water surface (0.26%). The population is sparsely distributed with a population density of 30 people/km^2. The main economic activity is forestry. Regarding the current land-use status of the sub-region, production forest land is 172.22 km^2, accounting for 44.23% of the total land-use area of the sub-region, protective forest land 116.25 km^2, accounting for 29.86%; special-use forest land 46.52 km^2, accounting for 11.95% of the area. Other land-use types include unused land (accounting for 7.25% of the area), land for planting perennial trees (2.06%), special purpose land (1.74%), rocky mountains without trees (1.15%), rice cultivation land (0.55%), residential land (0.49%), another annual cropland (0.45%), water surface (0.25%), aquaculture land (0.02%).
- The environmental function of the sub-region is to conserve biodiversity; regulate the flow; rural mountainous areas; providing mineral resources, timber, firewood, and forestry products.
- The environmental problems of the sub-region are the risk of declining watershed areas due to unreasonable exploitation of human beings.

- Socio-economic sustainable development solutions for the sub-region: Planting, protecting and restoring special-use forests and protection forests; developing production forests; improving the material and spiritual life of rural people; developing ecotourism.

4.2.2 Environmental Sub-Region of a Sap River Valley

- *Characteristics of the sub-region:* Environmental sub-region covers an area of 97.26 km^2. The topography has height from 250–750 m. Types of soil occupying the largest area of the sub-region include yellow-brown soil on ancient alluvial soil is 52.93 km^2, accounting for 54.42% of the sub-region land area; yellow-red soil on metamorphic clay is 38.98 km^2, accounting for 40.08%. There is also red-yellow soil on acid magma rock (accounting for 2.69% of the sub-region land area), yellow-red soil on shale (1.95%), water surface (0.82%), red-yellow humus on acid magma rocks (0.03%). The population density is 50 people/km^2, and A Luoi town is the urban area of the sub-region. Main economic activities are agriculture and forestry production. Regarding the current land-use status of the sub-region, production forest land is 39.22 km^2, accounting for 40.32% of the total land-use area of the sub-region; residential land is 16.43 km^2, accounting for 16.89%; special-purpose land is 12.05 km^2, accounting for 12.39%; rice cultivation land is 10.49 km^2, accounting for 10.79%. Other land-use types include another annual cropland (accounting for 6.63% of the whole sub-region), land for planting perennial trees (6.32%), unused land (2.49%), water surface (2.27%), protective forest land (1.58%), aquaculture land (0.29%), special-use forest land (0.01%), another agricultural land (0.01%).
- *Environmental functions of the sub-region include*: Conservation biodiversity; rural accommodation; providing resources and space for agricultural production and handicraft; providing mineral resources, timber, firewood, and forestry products; regulating water for reservoir.
- *Environmental problems of the sub-region* can be mentioned are landslides, especially along the Ho Chi Minh road corridor passing through A Luoi district, flash floods along rivers and streams threaten the safety of mountainous residential areas. The problem of dioxin toxins left behind after the war in A Luoi airport area so far has not completely solved all the dangers.
- *Socio-economic sustainable development solutions for the sub-region*: Planting and protecting special-use forests and protection forests; conservation biodiversity; developing production forests; preventing soil erosion; protecting and renovating population roads and spots at risk of landslides and flash floods; developing A Luoi town to become an economic, political and cultural centre of the region; implementing new rural construction; developing eco-tourism; developing agriculture with models of long-term industrial crops and fruit trees along the direction of commodity production and rice cultivation to meet local needs.

4.2.3 Environmental Sub-Region in the upstream of Dakrong River

- Characteristics of the sub-region: Environmental sub-region covers an area of 115.85 km^2. The mountainous terrain has height from 250 to 750 m. Types of soil occupying the largest area of the sub-region include yellow-red soil on metamorphic clay is 68.05 km^2, accounting for 58.74% of the sub-region land area; yellow-red soil on shale is 39.39 km^2, accounting for 34%. There is also red-yellow loam on metamorphic rocks (accounting for 4.57% of the sub-region land area), red-yellow humus on acid magma rocks (2.01%) red-yellow soil on acid magma rock (0.65%), light yellow soil on sandstone (0.03%). The population density is 38 people/km^2. Main economic activities are forestry production and agriculture. Regarding the current land-use status of the sub-region, special-use forest land is 38.72 km^2, accounting for 33.42% of the total land-use area of the sub-region; production forest land is 38.14 km^2, accounting for 32.92%; unused land is 29.61 km^2, accounting for 25.56%. Other land-use types include protective forest land (accounting for 3.96%), another annual cropland (1.79%), land for planting perennial trees (1.11%), residential land (0.69%), rice cultivation land (0.35%), water surface (0.17%), special-purpose land (0.02%).
- Environmental function of the sub-region: Preserving biodiversity.
- Environmental issues of the sub-region: The risk of deforestation due to overexploitation activities of human beings.
- Socio-economic sustainable development solutions for the sub-region: Watershed protection.

4.3 The Environmental Region of Bo river's Basin

4.3.1 Mountain Environment Sub-Region of Bo river's Upstream

- Characteristics of the sub-region: Environmental sub-region covers an area of 400.35 km^2. The topography of the sub-region is from 250 to 900 m in height. The soils occupying the largest area of the sub-region include red-yellow soil on metamorphic clay 211.18 km^2, accounting for 52.75% of the total area of the sub-region; red-yellow soil on acid magma rock is 69.94 km^2, accounting for 17.47%; yellow-red soil on shale 56.83 km^2, accounting for 14.2%; light yellow soil on sandstone is 37.16 km^2, accounting for 9.28%. Besides, there are also other soils including red-yellow humus on acid magma rocks (accounting for 4.02% of the sub-region land area), yellow-brown soil on ancient alluvial (1.3%), red-yellow humus on metamorphic rocks (0.9%), water surface (0.08%). Population density is 90–95 people/km^2. Economic activities of the resident are mainly agriculture and forestry. About status of using the current land of the sub-region, special-use forest land is 188.38 km^2, accounting for 47.05% of the total

land-use area of the sub-region, protective forest land is 96.02 km^2, accounting for 23.98%; production forest land is 77.58 km^2, accounting for 19.38%. In addition, the sub-region also has other annual cropland (accounting for 5.08% of the area), unused land (1.55%), special purpose land (0.83%), rice cultivation land (0.77%), land for planting perennial trees (0.7%), water surface (0.33), residential land (0.32%).
- Environmental functions of the sub-region: Biodiversity conservation; supplying wood, firewood and other forest products; rural accommodation.
- Environmental issues of the sub-region: Forest area degradation due to unreasonable exploitation.
- Socio-economic sustainable development solutions for the sub-region: Protection and restoration of special-use forests and protection forests; planting production forests; biodiversity conservation; erosion control; building a new countryside; eco-tourism development.

4.3.2 The Environmental Sub-Region of Bo river's Middle Hill

- Characteristics of the sub-region: Environmental sub-region covers an area of 427.98 km^2. The terrain has an altitude of 15 to 250 m. The soils occupying the largest area of the sub-region include yellow-red soil on shale 216.81 km^2, accounting for 50.66% of the total area of the sub-region; red-yellow soil on acid magma rock is 63.03 km^2, accounting for 14.73%; light yellow soil on sandy rock of 58 km^2, accounting for 13.55%; yellow-red soil on metamorphic clay 37.46 km^2, accounting for 8.75%. Besides, there are other soils such as unalluvial soil (accounting for 2.99% of the entire sub-area), sea sandy soil (2.32%), gleyic fluvisols (2.19%), valley soils due to steep slope products (1.21%), water surface (1.04%), alluvial soils with red and yellow variegated layers (0.76%), white sand dunes (0.66%), alluvial soil deposited annually (0.48%), gray soil on acid magma (0.46%), yellow-brown soil on ancient alluvial soil (0.15%), alluvial soil covered with sea sand (0.05%). The average population density is 200 people/km^2. Economic activities of the residents are mainly agricultural and industrial production. Regarding the current land-use status of the sub-region, production forest land is 204.13 km^2, accounting for 47.7% of the total land-use area of the whole sub-region; protective forest land is 50.28 km^2, accounting for 11.75%; another annual cropland is 45.97 km^2, accounting for 10.74%. Other land-use types include special purpose land (accounting for 7.05% of the area), water surface (5.83%), special-use forest land (5.31%), rice cultivation land (4.68%), residential land (3.79%), land for planting perennial trees (2.24%), unused land (0.85%), aquaculture land (0.07%).
- Environmental functions of the sub-region: Providing space for agriculture, producing development; living space; providing mineral resources (sand and gravel); solid waste.

- Environmental issues of the sub-region: Environmental pollution from residential areas, from production activities (agriculture, handicraft, industry).
- Socio-economic sustainable development solutions for the sub-region: Planting and protecting protective forests; planting production forests; preventing soil erosion; collecting and treating concentrated domestic wastewater; building solid waste treatment facilities; developing models of agricultural production such as planting of perennial industrial trees, fruit trees, short-term industrial plants, developing husbandry; building industrial areas and industrial clusters; to develop Tu Ha town into an economic, political and cultural centre of the region.

4.3.3 Environmental Sub-Region of Bo River downstream's Delta

- Characteristics of the sub-region: Environmental sub-region covers an area of 230.13 km^2. The terrain has an altitude of less than 10 m. Types of land occupying the largest area of the sub-region include white sand dunes 58.41 km^2, accounting for 25.38% of the land area of the whole sub-region; water surface is 43.96 km^2, accounting for 19.10% of the whole sub-region area; alluvial soil without accretion 39.27 km^2, accounting for 17.06%; gleyic fluvisols 28.44 km^2, accounting for 12.36%; sea sandy soil 21.68 km^2, accounting for 9.42%; acid sulphate soil operating deep, medium salinity 19.42 km^2, accounting for 8.44%. Other soils occupying a small area include alluvial soil covered with sea sand (accounting for 2.84%), yellow-brown soil on ancient alluvial soil (2.15%), alluvial soils with red and yellow variegated layers (1.49%), average saline soils (1.33%), high saline soils (0.18%), valley soils due to steep slope products (0.13%), yellow-red soil on shale (0.12%). Population density 95–110 people/km^2. Main economic activities are agriculture, fisheries, industrial production. About status of using land of the sub-region, rice cultivation land is 63.37 km^2, accounting for 27.54% of the total land using area of the whole sub-region; water surface is 47.48 km^2, accounting for 20.63%; residential land is 31.4 km^2, accounting for 13.64%; special purpose land is 27.61 km^2, accounting for 12%; another annual cropland is 20.1km^2, accounting for 8.73%. Types of land use with small areas include production forest land (accounting for 6.25% of the sub-area), unused land (4.22%), protective forest land (4.12%), aquaculture land (2.55%), land for planting perennial trees (0.28%), grassland used for husbandry (0.02%), another non-agricultural land (0.02%).
- The environmental function of the sub-region: Rural residence, providing resources and space for agricultural development (wet rice, vegetables, aquaculture).
- Environmental issues of the sub-region: Flood risk; environmental pollution from living and production activities.
- Socio-economic sustainable development solutions for the sub-region: Completing the waste treatment system; developing agricultural models such as aquaculture, wet rice, vegetables, cattle and poultry breeding; construction of

small industrial and production facilities, causing less pollution; developing small handicraft production establishments; tourism development.

4.4 The Environmental Region of Huong river's Basin

4.4.1 Environment Sub-Region of Huong river's Basin

- Characteristics of the sub-region: Environmental sub-region covers an area of 576.03km^2. The terrain has an altitude of 250 to 1800 m. The soils occupying the largest area of the sub-region include red-yellow soil on acid magma rock 405.07 km^2, accounting for 70.32% of the sub-regions land area; yellow-red soil on metamorphic clay is 138.67 km^2, accounting for 24.07%. In addition, there are red-yellow humus on acid magma rocks (accounting for 4.7%), yellow-red soil on clay shale (0.62%), pale yellow soil on sandstone (0.26%), water surface (0.02%), stream alluvium soil (0.01%). The population density is from 45 to 55 people / km^2. Main economic activities are agriculture and forestry. Regarding the current land-use status of the sub-region, special-use forest land is 285.62 km^2, accounting for 49.58% of the total land-use area of the sub-region; protective forest land is 139.88 km^2, accounting for 24.28%; production forest land is 95.24 km^2, accounting for 16.53%; another annual cropland is 50.19 km^2, accounting for 8.71%. Types of land use with smaller areas include land for planting perennial trees (accounting for 0.32%), unused land (0.21%), rocky mountains without trees (0.18%), water surface (0.14%), special purpose land (0.02%), residential land (0.1%).
- Environmental functions of the sub-region: Conservation of biodiversity in Bach Ma National Park, Nature Reserve of Sao La; regulating water for Binh Dien and Ta Trach lakes; provision of mineral resources (pyrite, iron), timber, firewood, forest products; rural accommodation.
- Environmental issues of the sub-region: The decrease of forest area due to over-exploitation; risk of erosion, landslides due to deforestation, construction of roads.
- Socio-economic sustainable development solutions for the sub-region: Planting and protecting special-use forests, protection forests and developing production forests; erosion control; developing agriculture and forestry with models of forest gardens; building mineral mining and processing industry points; ecotourism development.

4.4.2 The Middle of Huong river's Environment Sub-Region

- Characteristics of the sub-region: Environmental sub-region covers an area of 931.91 km^2. The hilly terrain is from 15 to 250 m high. Soils occupying the largest area of the sub-region include red-yellow soil on acid magma rock 498.13 km^2, accounting for 53.45% of the sub-regions land area; yellow and red soil on shale 290.19 km^2, accounting for 31.14%; yellow-red soil on metamorphic clay is 60.62 km^2, accounting for 6.5% of the total area of the region. Other soils comprising a small percentage include pale yellow soil on sandy rock (accounting for 2.76% of the sub-regions land area), golden brown soil on ancient alluvial soil (1.35%), stream alluvial soil (1.22%), water surface (1.11%), alluvial soil without accretion (0.97%), sea sandy soil (0.6%), alluvial soil with yellowish red patch (0.38%), eroded soil and gravelly soil (0.33%), gleyic fluvisols (0.19%) and alluvial soil are deposited annually (0.01%). The population density is 225 people/km^2. Activities of the resident are mainly agricultural, forestry and industrial production. About the status of using the current land of the sub-region, productive forest land is 325.66 km^2, accounting for 34.95% of the total land-use area of the sub-region; protective forest land is 229.97 km^2, making up 24.68%. Other types of land use include special-use forest land (7.19% of the sub-area), other annual crops (6.93%), land for planting perennial trees (6.35%), water surface (5.66%), special purpose land (5.2%), residential land (4.09%), grassland for animal husbandry (2%), rice cultivation land (1.63%), unused land (1.18%), aquaculture land (0.11%), rocky mountains without trees (0.05%).

- The environmental function of the sub-region: Living space; providing mineral resources, most construction materials (sand, gravel); providing resources and space for the development of production of forest gardens and perennial trees; space for industrial production development; contains CTR; centralized waste storage and handling area.
- Environmental issues of the sub-region: Risks associated with the operation of reservoirs and dams (Binh Dien, Ta Trach); environmental pollution from living and production activities; the land is eroded, degraded due to deforestation, irrational cultivation.
- Socio-economic sustainable development solutions for the sub-region: Planting and protecting protective forests; development of production forests; keep hydropower reservoirs safe; building a new countryside; modernize urban areas (Huong Thuy and Phu Bai); collection and treatment of domestic wastewater; tourism development; develop agriculture with models of planting industrial crops, fruit trees, and raising cattle; building and modernizing industrial parks; building concentrated solid waste treatment facilities, industrial wastewater treatment.

4.4.3 The Environment Sub-Region of Huong river's Downstream Delta

- Characteristics of the sub-region: Environmental sub-region covers an area of 336.11 km^2. Plain terrain has altitudes below 10 m. Soils occupying the largest area of the sub-region include unalluvial soil is 88.74 km^2, accounting for 26.4% of the total area of the sub-region; water surface 68.76 km^2, accounting for 20.46% of the entire sub-region area; average saline soils 44.6 km^2, accounting for 13.27%; sea sandy soil 42.68 km^2, accounting for 12.7%. Other soils include alluvial soils with red and yellow variegated layers (accounting for 6.18% of the entire sub-region), yellow-red soil on clay shale (6.09%), white sand dunes (5.91%), acid sulphate soil operating deep, medium salinity (4.56%), alluvial soil covered with sea sand (2.6%), gleyic fluvisols (0.86%), highly saline soil (0.49%), alluvial soil is accreted annually (0.49%). The resident is the most densely concentrated in the province, with the urban city of Hue with a population density of 5040 people/km^2; Other areas with a population density of 650–700 people/km^2. The main economic sectors are agriculture, services and industry. Regarding the current land-use status of the sub-region, rice cultivation land is 94.5 km^2, accounting for 28.12% of the total land-use area of the sub-region; water surface is 77.31 km^2, accounting for 23%; residential land 68.7 km^2, accounting for 20.44%; special purpose land 40.6 km^2, accounting for 12.08%. Other types of using land include aquaculture land (accounting for 5.72% of the sub-region area), other annual cropland (3.78%), protective forest land (2.12%), unused land (1.95%), production forest land (1.55%), special-use forest land (0.63%), land for planting perennial trees (0.45%), grassland used for animal husbandry (0.1%), other agricultural land (0.04%), other non-agricultural land (0.03%), rocky mountains without forests (0.01%).

- The environmental function of the sub-region: Living space; providing resources and space for agricultural development (wet rice production, vegetables and aquaculture).
- Environmental issues of the sub-region: Flood risk; pollution from living and manufacturing activities from urban areas.
- Socio-economic sustainable development solutions for the sub-region: To develop Hue city into an economic, political and cultural centre of the province and raise its status in the Central region and the whole country; preserving the heritage of architectural works belonging to Hue ancient vestige complex; tourism development; wastewater collection and treatment; complete the wastewater treatment system, solid waste; to develop models of agricultural production (aquaculture, rice, vegetables, cattle and poultry breeding) toward green, clean and modern production.

4.5 *The Environmental Region of Truoi River Basin*

4.5.1 Sub-Region Environment Sub-Region upstream of Truoi River

- Characteristics of the sub-region: Environmental sub-region covers an area of 88.17 km^2. Mountain topography is from 250–1000 m. Soils occupying the largest area of the sub-region include red-yellow soil on acid magma rock of 79.28 km^2, accounting for 89.92% of the sub-regions land area; red-yellow humus on acid magma rock of 7.71 km^2, accounting for 8.74% of the entire sub-region. There are also other soil types including stream alluvial soil (accounting for 0.58% of the sub-regions land area), yellow-red soil on shale (0.23%), soil erosion and inert gravel (0.17%), sandy soil (0.16%), yellow-red soil changed due to wet rice cultivation (0.16%), water surface (0.05%). The population is sparsely distributed with a density of 60 people/km^2. Main economic activities are agriculture and forestry. Regarding the current land-use status of the sub-region, special-use forest land 68.79 km^2, accounting for 78.02% of the total land-use area of the whole sub-region; protective forest land is 7.68 km^2, accounting for 8.71% of the land area used throughout the sub-region, production forest land is 6.99 km^2, accounting for 7.93%. Other types of land use include unused land (accounting for 3.35% of the sub-region area), water surface (1.2%), special-use land (0.52%), land for planting perennial trees (0.19%), rice cultivation land (0.03%), residential land 0.03%), another annual cropland (0.01%).

- The environmental function of the sub-region: Protection; regulate the flow; biodiversity conservation; providing resources including timber, firewood, and other forest products.
- Environmental issues of the sub-region: Forest loss due to excessive exploitation.
- Socio-economic sustainable development solutions for the sub-region: Planting and protecting special-use forests and protection forests; development of production forests; tourism development.

4.5.2 The Environment Sub-region of Truoi river's Basin

- Characteristics of the sub-region: Environmental sub-region covers an area of 306.14 km^2. The hilly terrain is from 15 to 250 m high. Types of soil occupying the largest area of the sub-region include red-yellow soil on acid magma rock is 84.74 km^2, accounting for 27.68% of the land area of the sub-region; red-yellow soil on clay shale 81.64 km^2, accounting for 26.67%; eroded soil and gravelly soil is 46.89 km^2, accounting for 15.32%; sea sandy soil is 29.44 km^2, accounting for 9.62%. There are also other types of soil including yellow-brown soil on ancient alluvial (accounting for 6.71%), alluvial soil deposited annually (3.81%), gleyic fluvisols (1.84%), alluvial soil without accretion (1.72%), water surface (1.71%), average saline soils (1.28%), yellow and red soil changed due to wet rice cultivation (1.02%), alluvial soil with yellow-red variegated soil (0.99%), alluvial soil in streams (0.95%), light yellow soil on sandstone (0.32%), alluvial soil covered with sea sand (0.22%), gleysols (0.15%). Population density is 100 people/km^2. Main economic activities are agriculture and forestry. Regarding the current land using status of the sub-region, production forest land is 160.53 km^2, accounting for 52.44% of the total land-use area of the sub-region; residential land is 30.24 km^2, accounting for 9.88%; special-use forest land is 29.76 km^2, accounting for 9.72%; special purpose land is 23.86 km^2, accounting for 7.79%;

rice land is 19.82 km^2, accounting for 6.47%. Other types of land use include water surface (accounting for 4.25% of the natural area), protective forest land (4.11%), land for perennial crops (2.06%), other annual crops (1.95%), unused land (0.81%), aquaculture land (0.51%).

- The environmental function of the sub-region: Living space; providing resources and space for product development of forest gardens, perennial trees, fruit trees; providing mineral resources, including iron, minerals for construction materials (sand, gravel); supply resources and space for industrial production development; centralized waste storage and treatment area;
- Environmental issues of the sub-region: Environmental pollution from residential areas, from agricultural and industrial production activities. Depletion of exploited resources for economic activities.
- Socio-economic sustainable development solutions for the sub-region: Planting and protecting protection forests; development of production forests; building a new countryside; urban development; collection and treatment of domestic wastewater; building concentrated solid waste treatment facilities, industrial wastewater treatment; developing models of agricultural production such as planting industrial crops, fruit trees, and raising cattle; building industrial zones; tourism development.

4.5.3 V.C. the Environment Sub-region Downstream of Truoi River Delta

- Characteristics of the sub-region: Environmental sub-region covers an area of 269.57 km^2. Plain terrain with altitudes below 10 m. Types of land occupying the largest area of the sub-region includes the water surface of 119.91 km^2, accounting for 44.48% of the total area of the sub-region; sea sand and soil 76.25 km^2, accounting for 28.29%; saline soil averaged 24.76 km^2, accounting for 9.18%. There are also other types of soil including white sand dunes (accounting for 4.47% of the entire sub-area), acid sulphate soils of deep activity, average salinity (4.32%), alluvial soil without deposition (2.6%), alluvial soils with red and yellow variegated layers (2.42%), red-yellow humus on acid magma rocks (1.43%), gleyic fluvisols (1.19%), high saline soils (0.73%), alluvial soil covered with sea sand (0.57%), alluvial soil deposited annually (0.32%). The population is crowded with a density of 250 people/km^2. Main economic activities are agriculture, fisheries and tourism. Regarding the current land-use status of the sub-region water surface is 105.71 km^2, accounting for 39.21% of the total land-use area of the whole sub-region; rice land is 67.13 km^2, accounting for 24.9%; residential land is 36.9 km^2, accounting for 13.69%; aquaculture land is 19.95 km^2, accounting for 7.4%; special purpose land is 19.15 km^2, accounting for 7.1%. Other land-use types include protective forest land (accounting for 2.47% of the sub-region area), another annual cropland (1.57%), productive forest land (1.28%), land for planting perennial trees (1.25%), unused land (1.12%).

- Environmental functions of the sub-region: Biodiversity conservation in Cau Hai lagoon area; rural accommodation; providing resources and space for agricultural development and services.
- Environmental issues of the sub-region: Flooding; Environmental pollution from domestic and production wastewater.
- Socio-economic sustainable development solutions for the sub-region: Planting coastal protection forests; biodiversity conservation of Cau Hai lagoon; development of aquaculture; develop sea and lagoon tourism.

4.6 Environmental Region of Bu Lu river's Basin

4.6.1 Upland Environmental Sub-Region upstream of Bu Lu River

- Characteristics of the sub-region: Environmental sub-region covers an area of 79.88 km^{2}. Mountain terrain has altitude from 250 to over 1000 m. Soils occupying the largest area of the sub-region include red-yellow soil on acid magma rock of 72.9 km^2, accounting for 91.26% of the land area of the sub-region; red-yellow humus on acid magma rocks of 6.75 km^2, accounting for 8.45% of the area. There is also sandy soil (accounting for 0.25% of the sub-regions land area), alluvial soil is deposited annually (0.04%). The population is sparsely distributed with a density of 70 people/km^2. The main economic activities are forestry and agriculture. Regarding the current land-use status of the sub-region, protective forest land is 54.08 km^2, accounting for 67.7% of the total land-use area of the whole sub-region; production forest land is 21.44 km^2, accounting for 26.84%. Other land-use types include unused land (accounting for 4.98% of the sub-area), special purpose land (0.24%), water surface (0.13%), residential land (0.05%), rice cultivation land (0.04%), land for perennial crops (0.03%).
- The environmental function of the sub-region: Preserving biodiversity.
- Environmental issues of the sub-region: The risk of forest area decline.
- Socio-economic sustainable development solutions for the sub-region: Biodiversity conservation; ecotourism development.

4.6.2 The Environment Sub-Region of the middle of Bu Lu River Hill

- Characteristics of the sub-region: Environmental sub-region covers an area of 80.72 km^2. The hilly terrain is from 10 to 250 m high. Soils occupying the largest area of the sub-region include red-yellow soil on acid magma rock of 46.37 km^2, accounting for 57.45% of the area of the sub-region; sea sandy soil of 22.38 km^2, accounting for 27.73% of the area. Besides, there are other types of soil such as gleyic fluvisols (accounting for 6.64% of the area), alluvial soil deposited annually (4.91%), water surface (2.39%), average saline soil (0.51%), yellow-red soil changes due to wet rice cultivation (0.38%). Population density is 100 people/ km^2. The main economic activities are agriculture and tourism. Regarding the current land-use status of the sub-region, production forest land is 36.22 km^2, accounting for 44.87% of the total land-use area of the sub-region; rice cultivation land is 11.66 km^2, accounting for 14.44%; protective forest land is 10.6 km^2, accounting for 13.13%; residential land is 9.14 km^2, accounting for 11.32%. Other land-use types include unused land (accounting for 6.69% of the sub-region area), specialized land (3.26%), river surface (2.8%), land for planting perennial trees (2.28%), land for other annual crops (1.03%), aquaculture land (0.17%).

- Environmental functions of the sub-region: Protection, green belt for Chan May industrial park; providing mineral resources (pyrite, placer gold).
- Environmental issues of the sub-region: Reduced protective forest area.
- Socio-economic sustainable development solutions for the sub-region: Tourism development; develop agriculture with models of growing perennial industrial trees and fruit trees; constructing an industrial zone.

4.6.3 The Environment Sub-Region of downstream of Bu Lu river's Delta

- Characteristics of the sub-region: Environmental sub-region covers an area of 79.2 km^2. Plain terrain with altitudes below 10 m. Soils occupying the largest area of the sub-region include sea sandy soil of 38.26 km^2, accounting for 48.31% of the land area of the whole sub-region; water surface is 17.66 km^2, accounting for 22.30% of the area; red-yellow soil on acid magma 12.02 km^2, accounting for 15.18%. Other types of soil include white sand dunes (accounting for 7.41% of the area), average saline soils (4.1%), gleyic fluvisols (2.1%), marsh soil (0.61%). The population density is 200 people/km^2. Economic activities are mainly fisheries, tourism, and services. Regarding the current land-use status of the sub-region, water surface is 20.06 km^2, accounting for 25.33% of the total land-use area of the whole sub-region; residential land 13.22 km^2, accounting for 16.69%; rice land is 11.19 km^2, accounting for 14.13%; special purpose land is 7.8 km^2, accounting for 9.85%, production forest land is 7.49 km^2, accounting for 9.46%; land for planting perennial trees is 6.59 km^2, accounting for 8.32%; protective forest land is 6.11 km^2, accounting for 7.71%. Other types of land use include another annual crop land (accounting for 5.82% of the entire sub-region), unused land (2.36%), aquaculture land (0.33%).
- The environmental function of the sub-region: Living space; provide space for agricultural production development; providing resources and space for Chan May industrial development; contain waste.
- Environmental issues of the sub-region: Biodiversity loss; environmental pollution from production and daily life activities.

– Socio-economic sustainable development solutions for the sub-region: Planting coastal protective forests; developing aquaculture and catching aquatic products; tourism development; developing seaport (Chan May).

4.7 *Figures, Tables and Schemes*

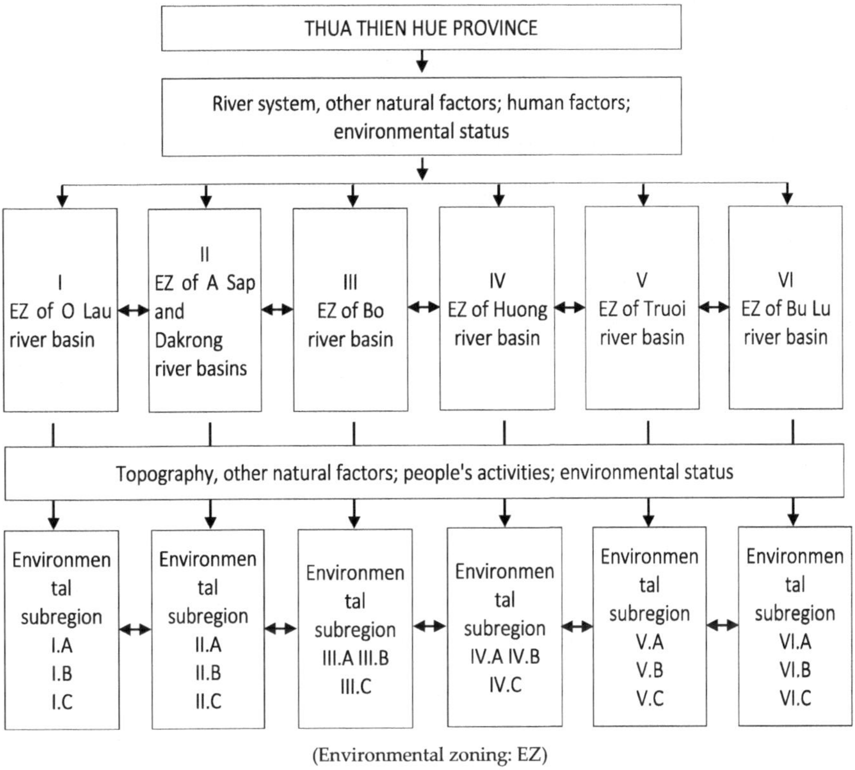

Fig. 52.1 Process diagram of environmental zoning in Thua Thien Hue

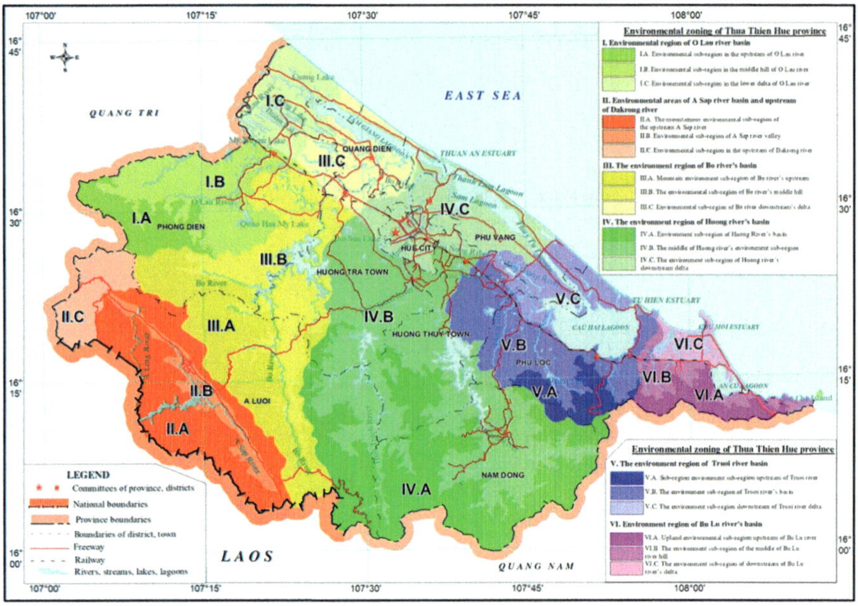

Fig. 52.2 Environmental zoning map of Thua Thien Hue province

5 Conclusions and Discussion

The environmental partitioning results have divided the province's space into.

6 regions and 18 sub-regions. Each region has its characteristics of nature, economy, society, and environment, but it has a close relationship with the unification of the whole territory of Thua Thien Hue province. The characteristics of each sub-region, environmental functions, environmental issues, and proposed solutions have been described as the basis for all levels, departments, sectors, and localities to consider the appropriate exploitation of territories, environmental protection for each sub-region in particular and the whole territory in general, ensuring the sustainable socio-economic development.

References

ADB. (1991). *Guidelines for Integrated Regional Economic-com-Environmental Development Planning - A Review of Regional Environmental Development Planning Studies in Asia*. Environment paper No 3, 125 pages.

Buckingham, S. (1996). *Environmental planning and sustainability* (208 pages). John Wiley & Sons.

Center for Natural Resources and Environment Monitoring of Thua Thien Hue Province. (2018). *Report on results of environmental monitoring in Thua Thien Hue Province in 2018* (288 pages).

Dang, T. T., Le, T., et al. (2017). *Environmental protection plan based on theory and practical experience* (361 pages). Science and Technology Publishing House.
Edington, J. M., & Edington, M. A. (1977). *Ecology and environmental planning* (246 pages). Chapman & Hall.
Elizabeth, B. (1992). *Planning for development and management* (p. 14).
Fang, Q., Zhang, L., Hong, H., Zhang, L., & Bristow, F. (2008). Ecological. *Functioning Zoning for Environmental Planning and Sustainability., 10*(1), 41–49.
Hall, P. (2002). *Urban and regional planning* (4th ed., p. 248). Routledge.
Le V. T. (2018). Proposed land use planning in Thua Thien hue and Phu Yen provinces in the climate change context. Ministerial-level scientific research topic, topic code: B2016-DHH-22, 133 pages.
Lindsey, G., Man, J., Payton, S., & Dickson, K. (2004). Property values, recreation values, and urban greenways. *Journal of Park and Recreation Administration, 22*(3), 69–90.
McGregor, C., Miller, D., & Zirnhelt, D. (1999). *Prince George land and resource management plan* (77 pages). The University of British Columbia.
Minaya, L. C. M., Gallardo, W. M., & Baud, I. (2012). *The ecological-economic zoning process in the Constitutional Province of Callao*. A case of synergy between public sector, private sector and organized civil society. 5 pages.
Montgomery, C. W. (2004). *Environmental geology* (576 pages). Northern Illinois University.
People's Committee of Thua Thien Hue Province [PCTTH]. (2005). *Thua Thien hue chorography - part nature* (307 pages). Social Sciences Publishing House.
Presidency of the Republic House Civil Cabinet Subcommittee for Legal Affairs [PRHCCSLA]. (2002). Decree No. 4297, Regulates article. 9, item II, of Law No. 6,938, of August 31, 1981, establishing criteria for ecological zoning in Brazil - EEZ, and other measure, 679 pages.
Taylor, N. (1998). *Urban planning theory since 1945* (192 pages). Sage.
Thua Thien Hue General Statistical Office [TTHGSO]. (2019). *Statistical yearbook 2018* (513 pages). Thuan Hoa Publishing House.

Chapter 53
Climate Change Vulnerability of Urban Development in Phanrang-Thapcham (Ninh Thuan, Vietnam)

Tuy Bao Van, Tho Tran Quang, and Kien Nguyen The

Abstract The impacts of climate changes and extreme weather events in recent years have had destructive impacts on urban development. Climate change increases the frequency and severity of floods, and as we are well aware, the impacts of flooding include loss of human life, damage to property, destruction of roads, schools, hospitals, markets, irrigation channels, dams, crops, loss of livestock, and deterioration of health conditions owing to waterborne diseases. This chapter recognizes key aspects of urban development that could be adversely affected by climate change, and develops measures and standards for assessing the urban vulnerability, develops an urban database of the sensitive variables consistent with vulnerability assessment measures, and applies the criteria in a study of vulnerability of urban Phanrang-Thapcham. Key findings highlight that approximately 50% of the area has a medium level of vulnerability and 10% of the area is highly vulnerable. Areas along the Dinh river and the central urban area have low vulnerability.

Keywords Climate change · Vulnerability assessment · Impacts · Urban · Phanrang-Thapcham · Vietnam

---

T. B. Van (✉)
HCM City University of Natural Resources and Environment, HCM City, Vietnam
e-mail: bvtuy@hcmunre.edu.vn

T. T. Quang
Southern Institute of Water Resources Research, HCM City, Vietnam

K. N. The
VNU University of Economics and Business, Vietnam National University, Hanoi, Vietnam

A. T. Nguyen, L. Hens (eds.), *Global Changes and Sustainable Development in Asian Emerging Market Economies Vol. 2*, https://doi.org/10.1007/978-3-030-81443-4_53

1 Introduction

All human activities such as land use change and fossil fuel burning increase the release of greenhouse gases into the atmosphere, which results in climate change. The main characteristics of climate change are changes in precipitation, increases in average temperature on a global scale (global warming); melting of ice and glaciers inducing rises in the sea level (UNFCCC, 2007). According to The Fourth Assessment Report of the IPCC, over the last century, the average global temperature rose by 0.74 °C and is predicted by 2100 to range from 1.8 °C to as much as 4 °C. During the twentieth century, sea levels rose by 0.17 m and by 2100, they are projected to rise between 0.18 and 0.59 m (IPCC, 2007). There are many negative impacts of climate change, such as water resources for human use will be exhausted because of the ice and glaciers melting in places such as Greenland in recent years (UNEP, 2007). Moreover, the type, frequency, and intensity of extreme events, such as hurricanes, typhoons, heavy precipitation events, floods, and droughts are also expected to rise (Greenough et al., 2001). If the average global temperature were to rise around 2 °C, approximately 59% of population in the world would be exposed to water shortage (Rockstrom et al., 2009).

Vietnam is one of the countries predicted to be most severely affected by climate change owing to its long coastlines, the high concentration of population, and economic activity in coastal areas, as well as a heavy reliance on agriculture, natural resources, and forestry (Adger, 1999). The impacts of climate change such as typhoons, floods, prolonged droughts, and sea level rises would increase risks to properties, livelihoods, and urban infrastructure assets (MONRE, 2012).

Phanrang-Thapcham is the urban coast of the Ninhthuan province, which is often affected by natural disasters and is predicted to be severely affected by climate change in the coming decades. Recently, droughts and depletion of water resources have frequently occurred, which have severely affected agriculture, aquaculture, urban water supply, and the environment. Moreover, the frequent heavy rains and floods have greatly damaged the urban area and its surroundings. Salinity intrusion, bank erosion, and coastal erosion are also severe impacts in these areas. All consequences have a negative effect on urban development (NinhThuan Provincial People Committee [NinhThuan PPC], 2012).

Vulnerability is a concept to describe a weakness in a system; its susceptibility to physical harm or damage. This concept is used across multiple disciplines, which are often place- or sector-specific. Many literature reviews on climate change have assessed these definitions of vulnerability (Ribot, 1995). The most common use is that the exposure and sensitivity to a hazard and the capacity to adjust to the hazard impacts are essential parts of the vulnerability (Brooks, Adger, & Kelly, 2005). From a social perspective, vulnerability can be considered as the exposure of people to livelihood stress caused by impacts of climate extremes or environmental change (Kelly & Adger, 2004). As such, the vulnerability can be a combination of social factors and environmental risk (Adger, 2006). The vulnerability of climate change is known as a function of biophysical and socio-economic factors (O'Brien, Eriksen,

Nygaard, & Schjolden, 2007). Vulnerability is also an equation of the character, magnitude, rate of climate variation to which a system is exposed, sensitivity, and capacity to adjust to the impacts of climate change (McCarthy, Canziani, Leary, Dokken, & White, 2001). Vulnerability to climate change is also defined as the degree to which a system is susceptible to and unable to cope with the negative impacts of climate change, including climate variability and extremes (Adger et al., 2007). The definition of the IPCC is criticized for being too vague and the resulting difficulty in making it operational (Hinkel, 2011). But the definition could be considered as an integrative concept that can link the social and biophysical dimensions of environmental change (Turner et al., 2003).

The general objective of this research study is to calculate the vulnerability index of urban development to climate change, to discover which urban assets are vulnerable, and to classify the major risks concerning resource stress, growth pressure, and management ability. The results of the vulnerability index computed from these basic components should provide the administrators with an estimation of the prevailing situation, modify current policies, and implement mitigation and adaptation measures for sustainable management of water resources in the study region (Figs. 53.1 and 53.2).

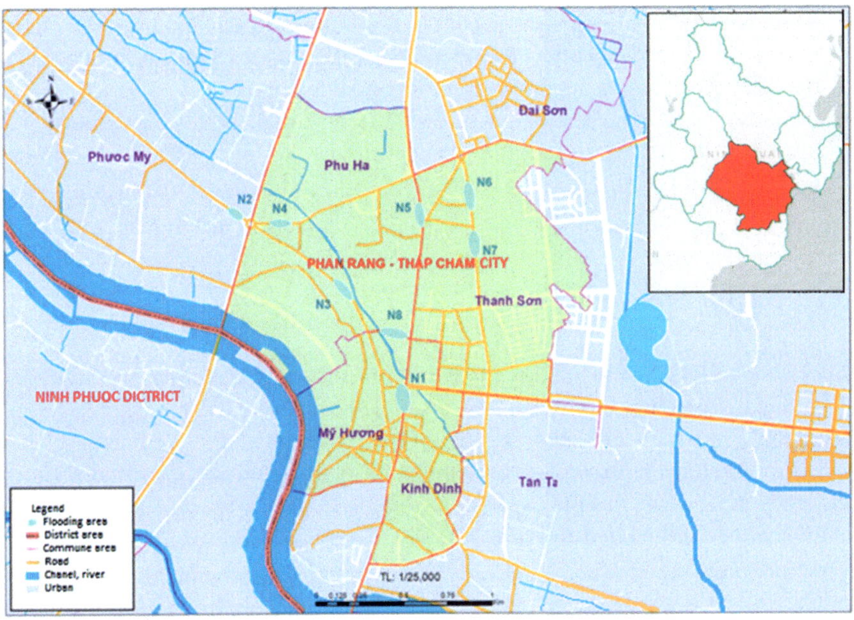

Fig. 53.1 Map of Phan Rang-Thap Cham city and surroundings

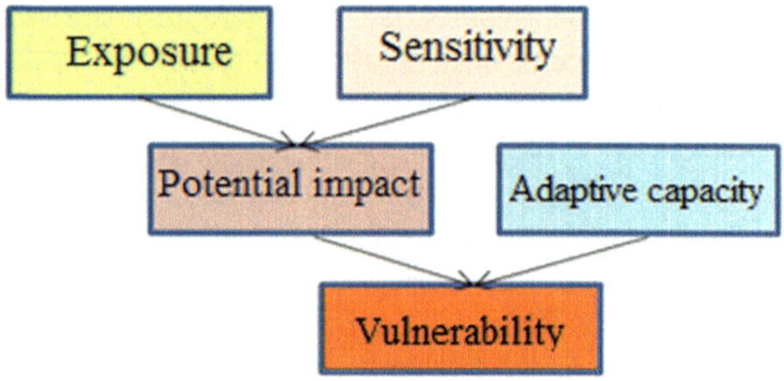

Fig. 53.2 Framework for vulnerability assessment (IPCC, 2007)

2 Methodology

Vulnerability is the degree to which a system is susceptible to the adverse effects of climate change, including climate variability and extreme events. Vulnerability is a function of the character, magnitude, and rate of climate change and variation to which a system is exposed, its sensitivity, and its adaptive capacity.

$$V = f(E,S,AC)$$

Of which:

- Exposure (E): the extent to which a system is exposed to the climate change. E depends on the threats (threat intensity, frequency, duration) and location of the system with respect to the threat (how far between the system and the threat).
- Sensitivity (S): the degree to which exposure to a threat arising from climate change will negatively affect the operation of the system. S may be influenced by the integrity of assets under threats and related factors.
- Adaptive capacity (AC): a measure of the potential, ability, or opportunities available to decrease exposure or sensitivity of a system to a climate-induced stress.

2.1 Set up Indicator Set

According to IPCC, vulnerability index is a function of E, S, and AC (IPCC, 2007). To calculate these primary indicators, a set of secondary indicators should be constructed. In this study, the set of secondary indicators are made based on the characteristics of natural resources—environmental, socio-economic, data availability of the study area, as well as the expert consultation results. The set of indicators E, S, and AC are described in detail in Table 53.1.

Table 53.1 The indicator set of exposure (E), sensitivity (S), and adaptive capacity (AC)

Indicator	No.	Sub-indicator	Explanation
Expose	1	Flood depth	Flood depth is measured in meters. It is classified into five categories from very low to very high
	2	Frequency	Frequency reflects the number of flood events in a year
	3	Percentage of affected areas	Percentage of affected areas in a commune/ward. It is classified into five categories: Very high (>30%), high (10–30%), medium (5–10%), low (2–5%), and very low (<2%).
Sensitive	1	Population density	Population density is measured in people/km^2)
	1	Percentage of resident area	Percentage of area per capita, allowed to build houses (%)
	3	Percentage of park/tree areas	Percentage of park/planting tree areas per capita (%)
	4	Percentage of surface water areas	Percentage of surface water areas per capita (%).
	5	Percentage of ethnic minority groups	Percentage of ethnic minority groups among the groups (%)
	6	Percentage of permanent house	Percentage of households having semi-permanent and permanent houses (%)
	7	Percentage of houses before 2000	Percentage of houses built before 2000 (%)
	8	Percentage of threatened areas	Percentage of households in natural disaster-prone areas (%)
Adaptive capacity	1	Life expectancy	High life expectancy is high AC to extreme climate events
	2	Percentage of females	Percentage of the population that are female (%)
	3	Percentage of working population	Percentage of the population of working age (%).
	4	Percentage who are poor	Percentage of the total population that are poor (%)
	5	Percentage with health insurance	Percentage of people who have health insurance in the total population (%)
	6	Percentage of medical staff	Number of medical staff percentage per 10,000 inhabitants
	7	Percentage of literate people	Percentage of literate people over 15 years old (%)
	8	Percentage with the internet	Percentage of households that have the internet (%)
	9	Percentage with a TV	Percentage of households that have TV (%)
	10	Percentage with radio broadcasting	Percentage of wards/commune that have a radio broadcasting system (%)
	11	Income	Income per capital (mil VND/person/year)
	12	Percentage of trained employees	Percentage of trained employees in the total population (%)
	13	Financial support	Percentage of the household supported by a credit loan (%)

Table 53.2 Determine indicators of (E), (S), and (AC)

Indicator	Equation	Explanation
Exposure (E)	$E = \frac{\sum_{i=1}^{n} E_i}{n}$	E_i is the standardized value of the exposure of i n is the number of (E) sub-indicators
Sensitivity (S)	$S = \frac{\sum_{i=1}^{n} S_i}{n}$	S_i is the standardized value of the sensitivity of i n is the number of (S) sub-indicators
Adaptive capacity (AC)	$AC = \frac{\sum_{i=1}^{n} AC_i}{n}$	AC_i is the standardized value of the adaptive capacity of i n is the number of (AC) sub-indicators

2.2 *Determine Vulnerability Index*

2.2.1 Determine Indicators (E), (S), and (AC)

Indicators (E), (S), and (AC) were three main elements to determine vulnerability index. These indicators were calculated in both RCP 4.5 and RCP 8.5 in 2030 and estimated according to the methodology and equations in Table 53.2.

The value of indicators is classified into five categories as below:

Value	0.0–0.2	0.2–0.4	0.4–0.6	0.6–0.8	0.8–1.0
Categories	Very low	Low	Average	High	Very high

2.2.2 Determine Vulnerability Indices

As shown in Fig. 53.1, the potential impact is calculated from (E) and (S), based on the assessment matrices in Tables 53.3 and 53.4. As Fig. 53.2, after (E) and (S) are calculated from Tables 53.1 and 53.2, the potential impact is combined between (E) and (S), based on the follow assessment matrix.

Vulnerability (V) is calculated based on the Potential Impact (I) and Adaptive Capacity (AC) ($V = I \times AC$).

2.3 *Climate Change Scenarios*

Vulnerable indices are an important element to develop the adaptive plan to climate change. Therefore, the vulnerable indices calculated should cover both average and severe climate change impacts in the short future. It is the reason the RCP 4.5 and RCP 8.5 scenarios in 2030 are chosen.

Table 53.3 The assessment matrix of (I)

	Exposure (S)					
Sensitivity (S)		Very low	Low	Average	High	Very high
	Very high	Average	Average	High	Very high	Very high
	High	Low	Average	Average	High	Very high
	Average	Low	Average	Average	High	Very high
	Low	Low	Low	Average	Average	High
	Very low	Very low	Low	Low	Average	High

Table 53.4 The assessment matrix of (AC)

	Impact (I)					
Adaptive capacity (AC)		Very low	Low	Average	High	Very high
	Very low	Average	Average	High	Very high	Very high
	Low	Low	Average	Average	High	Very high
	Average	Low	Average	Average	High	Very high
	High	Low	Low	Average	Average	High
	Very high	Very low	Low	Low	Average	High

3 Results

3.1 Vulnerability Indices under the Baseline Scenario

The results of the vulnerability assessment points out that the vulnerability of several communes is average (15/32 ward/commune). Only three communes, including Phuoc Dan, An Hai, and Phuoc Hai, have high vulnerability because these communes have high exposure and low adaptation. Low vulnerability is identified in communes along the Dinh River owing to low exposure and high adaptation. These communes are located in the urban center (Table 53.5 and Fig. 53.3).

3.2 Vulnerability Indices under the RCP 4.5 Scenario in 2030

In 2030, the vulnerable index of most communes shows an upward trend. There are five communes and one town with a high vulnerability index, Nhon Son, Phuoc Thuan, Phuoc Huu, Phuoc Hai, Phuoc Dan, and An Hai communes, whereas most in the urban areas suffer only a low to average index. This can be explained by the fact that the urban area has the lowest to average impact level and the highest adaptation compared with other areas (Table 53.6 and Fig. 53.4).

Table 53.5 The vulnerable index of each ward/commune under the baseline scenario

Ward/commune	Vulnerability	Ward/commune	Vulnerability
Do Vinh Ward	Medium	My son commune	Medium
Phuoc my Ward	Low	Nhon son commune	Medium
Bao an Ward	Medium	Khanh Hai town	Medium
Phu ha Ward	Low	Xuan Hai commune	Medium
Thanh son Ward	Low	Ho Hai commune	Medium
My Huong Ward	Low	Phuoc dan town	High
Tan tai Ward	Low	Phuoc son commune	Medium
Kinh Dinh Ward	Low	Phuoc Thai commune	Medium
Dao long Ward	Medium	Phuoc Hau commune	Medium
Dai son Ward	Low	Phuoc Thuan commune	Medium
Dong Hai Ward	Medium	An Hai commune	High
My dong Ward	Medium	Phuoc Huu commune	Medium
Thanh Hai commune	Medium	Phuoc Hai commune	High
Van Hai Ward	Medium	Phuoc Vinh commune	Medium
My Binh Ward	Medium	Phuoc Nam commune	Medium
My Hai Ward	Low	Phuoc Ninh commune	Medium

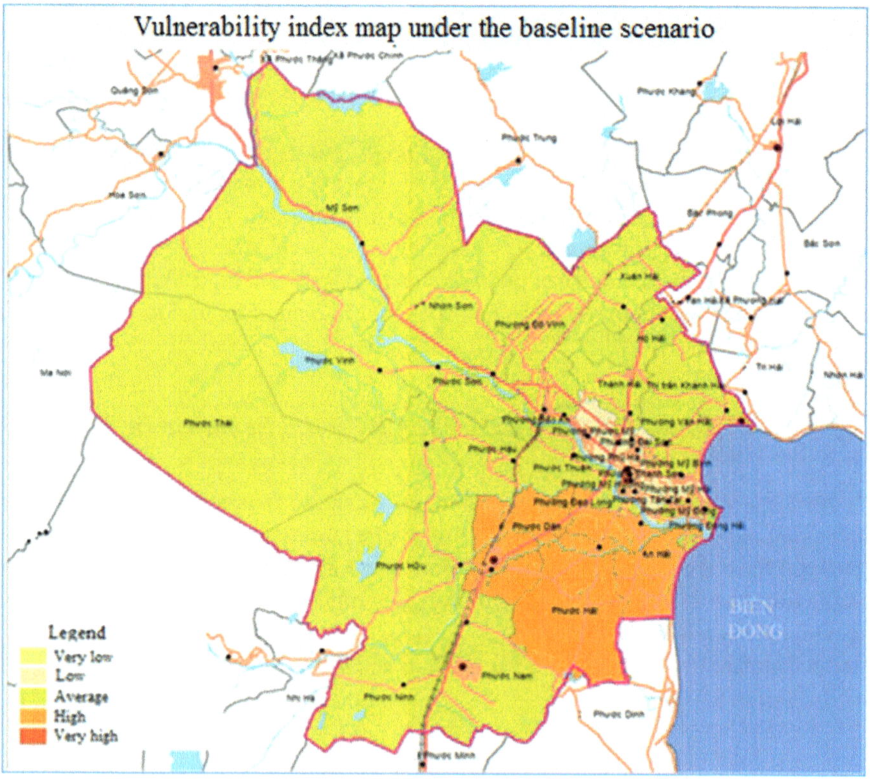

Fig. 53.3 The vulnerability index map under the baseline

Table 53.6 The vulnerability index of each ward/commune under the RCP 4.5 scenario

Ward/commune	Vulnerability	Ward/commune	Vulnerability
Do Vinh Ward	Medium	My son commune	Medium
Phuoc my Ward	Low	Nhon son commune	High
Bao an Ward	Medium	Khanh Hai town	Medium
Phu ha Ward	Low	Xuan Hai commune	Medium
Thanh son Ward	Low	Ho Hai commune	Medium
My Huong Ward	Low	Phuoc dan town	Medium
Tan tai Ward	Low	Phuoc son commune	Medium
Kinh Dinh Ward	Low	Phuoc Thai commune	Medium
Dao long Ward	Medium	Phuoc Hau commune	Medium
Dai son Ward	Low	Phuoc Thuan commune	High
Dong Hai Ward	Medium	An Hai commune	High
My dong Ward	Medium	Phuoc Huu commune	High
Thanh Hai commune	Medium	Phuoc Hai commune	High
Van Hai Ward	Medium	Phuoc Vinh commune	Medium
My Binh Ward	Medium	Phuoc Nam commune	Medium
My Hai Ward	Low	Phuoc Ninh commune	Medium

3.3 *Vulnerability Indices under the RCP 8.5 Scenario in 2030*

Vulnerability in the RCP 8.5 scenario tends to be higher than in the RCP 4.5. Two communes (Phuoc Dan and An Hai) have a very high level of vulnerability and five communes reached high vulnerability (Nhon Son, Phuoc Thuan, Phuoc Huu, Phuoc Hai, and Phuoc Nam). The vulnerability of the urban wards is ranges from a low to an average level. Unlike the RCP 4.5 scenario, the vulnerability in some areas is higher than the level in the RCP 8.5 scenario. In detail, the vulnerability index of Phuoc Dan town and An Hai communes rise from a high to a very high level and Phuoc Nam commune will go from a medium to a high level (Table 53.7 and Fig. 53.5).

4 Discussion and Conclusions

4.1 *Discussion*

Owing to the steep topography upstream, flooding downstream happens quickly (about 4–5 h). The flood risk map (Fig. 53.6) shows that the flooding occurs almost in the urban area, mainly along the banks of the Dinh river. The districts of Ninh Hai, Ninh Son, Thuan Nam, and PR-TC (Phan Rang-Thap Cham), especially Ninh Phuoc district, are the areas that are often severely flooded (accounting for nearly 50% of the district area). The areas with a high flooding risk are mainly located in Nhon Son (Ninh Son District); Phuoc Vinh; the neighboring areas of Phuoc Hau,

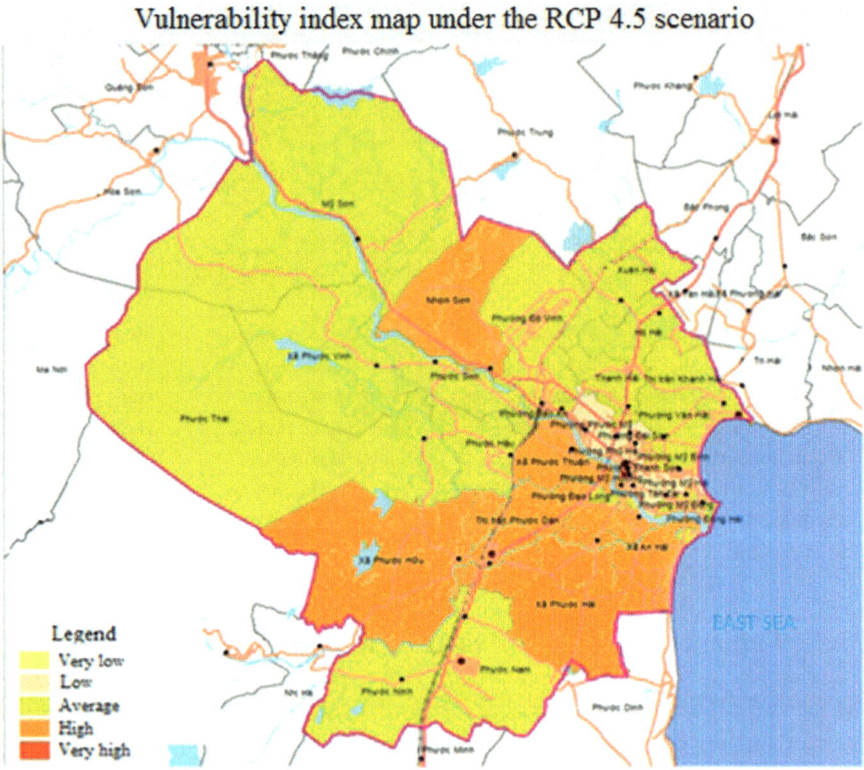

Fig. 53.4 The vulnerability index map under the RCP 4.5

Table 53.7 The vulnerability index of each ward/commune under the RCP 8.5 scenario

Ward/commune	Vulnerability	Ward/commune	Vulnerability
Do Vinh Ward	Medium	My son commune	Medium
Phuoc my Ward	Low	Nhon son commune	High
Bao an Ward	Medium	Khanh Hai town	Medium
Phu ha Ward	Low	Xuan Hai commune	Medium
Thanh son Ward	Low	Ho Hai commune	Medium
My Huong Ward	Low	Phuoc dan town	Very high
Tan tai Ward	Low	Phuoc son commune	Medium
Kinh Dinh Ward	Low	Phuoc Thai commune	Medium
Dao long Ward	Medium	Phuoc Hau commune	Medium
Dai son Ward	Low	Phuoc Thuan commune	High
Dong Hai Ward	Medium	An Hai commune	Very high
My dong Ward	Medium	Phuoc Huu commune	High
Thanh Hai commune	Medium	Phuoc Hai commune	High
Van Hai Ward	Medium	Phuoc Vinh commune	Medium
My Binh Ward	Medium	Phuoc Nam commune	High
My Hai Ward	Low	Phuoc Ninh commune	Medium

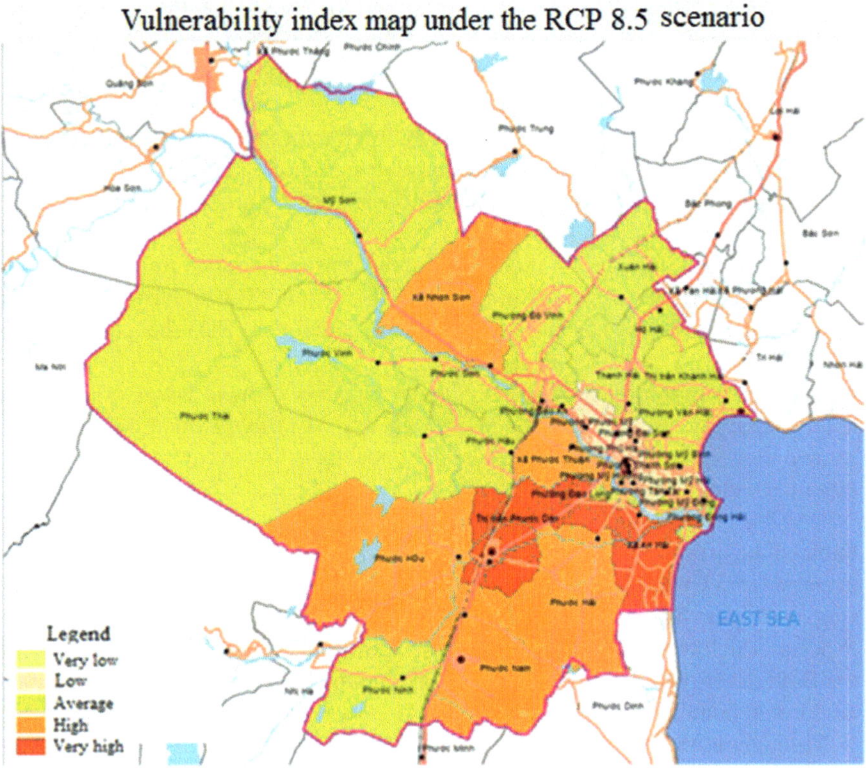

Fig. 53.5 The vulnerability index map under the RCP 8.5

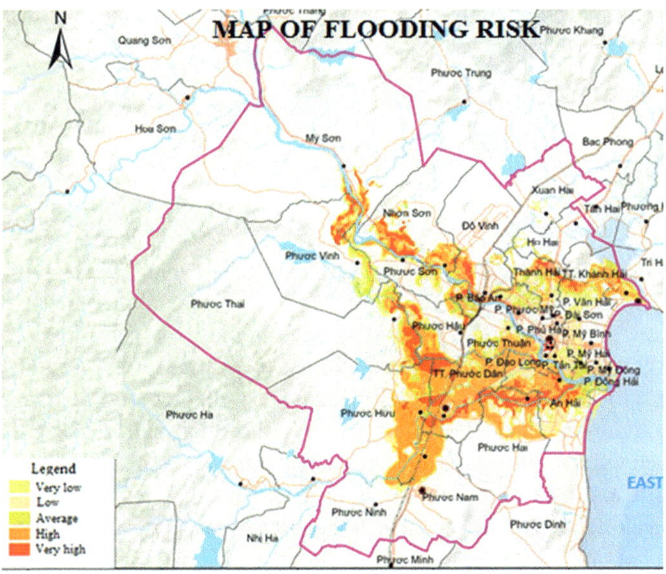

Fig. 53.6 The map of flooding risk

Phuoc Huu, and Phuoc Thai; and Phuoc Dan and An Hai of Ninh Phuoc district. The urban area is not severely affected because of the surrounding banks, but if extreme flooding occurs upstream, the urban area is affected more severely than other surrounding areas.

The exposure assessment shows that the high exposure occurs in Phuoc Dan, Phuoc Hai, and An Hai communes (Ninh Phuoc district), Bao An and Thanh Hai ward (the urban area). Other communes and wards of the urban area have different exposure levels, from very low to low.

The sensitivity assessment indicates that the sensitivity varies from low to high. In particular, My Son (Ninh Son) and Phuoc Ninh (Thuan Nam) are highly sensitive. The urban are has low sensitivity to the flood risk. The other communes in the flood-affected area have an average level.

In contrast to the sensitivity, the adaptive capacity of most wards of the urban area is high, except for Dong Hai, My Dong, and My Binh wards (average level), located near the coastal areas. The adaptive capacity is low in communes located in the surrounding areas.

Based on the indicators of exposure, sensitivity, and adaptation of ward/commune in the urban and surrounding areas, the vulnerability of several communes is average (15/32 ward/commune). Only three communes, namely, Phuoc Dan, An Hai, and Phuoc Hai, have a high vulnerability because these communes have high exposure and low adaptation. The low vulnerability is identified in communes along the Dinh River owing to their low exposure and high adaptation. These communes are located in the urban center.

Throughout the time to 2030, the climate change scenarios in Ninhthuan show that extreme weather events happen frequently and intensely. Therefore, the exposure and sensitivity indicators will be higher, leading to high vulnerability. The highly vulnerable areas are almost located in the surrounding areas, especially along the southern bank of the Dinh River, in the areas near the coast, etc., owing to high flooding and low adaptive capacity.

4.2 Conclusions

Recently, extreme weather events such as drought, water depletion, and floods have negatively impacted the Phan Rang-Thap Cham urban area. Therefore, the vulnerability of this area is very high. Throughout time, the vulnerability tends to be higher from the baseline to the RCP 8.5 scenarios. If in the baseline scenarios, there are only three communes with high vulnerability, then in the RCP 8.5, there are two communes with very high vulnerability and five communes with high vulnerability. Generally, approximately 10% of the area has very high vulnerability, 15% of the area has high vulnerability, mostly in the surrounding areas. Thirty percent have low vulnerability, and located mainly in the urban center. The remaining have average vulnerability, and are located near the urban area or are part of the urban area. One of the reasons why the urban area has low vulnerability is because it has high adaptive capacity, although the exposure is almost the same as that of other areas.

Acknowledgments The research described in this article is the result of the Cooperation Program between the Kingdom of Belgium and the Government of Vietnam. The intended target of the Program is to contribute to sustainable development through integrated water resources and urban development in relation to climate change in three provinces: Ha Tinh, Ninh Thuan, and Binh Thuan. We appreciate the support and assistance with this research from the Vietnamese and Belgian experts. We also thank the experts from Ninhthuan province, who contributed to this research.

References

Adger, N. W. (1999). Social vulnerability to climate change and extremes in coastal Vietnam. *World Development, 27*, 249–269.

Adger, W. N. (2006). Vulnerability. *Global Environmental Change, 16*, 268–281.

Adger, W. N., Agrawala, S., Mirza, M. M. Q., Conde, C., O'Brien, K., Pulhin, J., Pulwarty, R., Smit, B., & Takahashi, K. (2007). Assessment of adaptation practices, options, constraints and capacity. In M. L. Parry, O. F. Canziani, J. P. Palutikof, P. J. van der Linden, & C. E. Hanson (Eds.), *Climate change 2007: Impacts, adaptation and vulnerability; contribution of working group II to the fourth assessment report of the intergovernmental panel on climate change* (pp. 717–743). Cambridge University Press.

Brooks, N., Adger, W. N., & Kelly, P. M. (2005). The determinants of vulnerability and adaptive capacity at the national level and the implications for adaptation. *Global Environmental Change, 15*, 151–163.

Greenough, G., McGeehin, M., Bernard, S. M., Trtanj, J., Riad, J., & Engelberg, D. (2001). The potential impacts of climate variability and change on health impacts of extreme weather events in the United States. *Environmental Health Perspectives, 109*, 191–198.

Hinkel, J. (2011). Indicators of vulnerability and adaptive capacity — Towards a clarification of the science-policy interface. *Global Environmental Change, 21*, 198–208.

IPCC (2007). AR4—Impacts, Adaptation Working Group II Report and Vulnerability.. Retrieved October 14, 2019, from https://www.ipcc.ch/report/ar4/wg2/.

Kelly, P. M., & Adger, W. N. (2004). Theory and practice in assessing vulnerability to climate change and facilitating adaptation. *Climatic Change, 47*, 325–352.

McCarthy, J. J., Canziani, O. F., Leary, N. A., Dokken, D. J., & White, K. S. (2001). *Climate change: Impacts, adaptation, and vulnerability*. Cambridge University Press.

MONRE. (2012). *Vietnam climate change scenario and sea level rise*. The publishing house of Natural resources and map.

NinhThuan Provincial People Committee [NinhThuan PPC]. (2012). National program of climate change adaptation in Ninhthuan province. Project report.

O'Brien, K., Eriksen, S., Nygaard, L., & Schjolden, A. (2007). Why different interpretations of vulnerability matter in climate change discourses. *Climate Policy, 7*, 73–88.

Ribot, J. (1995). The causal structure of vulnerability: Its application to climate impact analysis. *GeoJournal, 35*, 119–122.

Rockstrom, J., Falkenmark, M., Karlberg, L., Hoff, H., Rost, S., & Gerten, D. (2009). Future water availability for global food production: The potential of green water for increasing resilience to global change. *Water Resources Research, 45*, 1–16.

Turner, B. L., Kasperson, R. E., Matsone, P. A., McCarthy, J. J., Corell, R. W., Christensen, L., Eckley, L., Kasperson, J. X., Luerse, A., & Martello, M. L. (2003). A framework for vulnerability

analysis in sustainability science. *Proceedings of the National Academy of Sciences of the United States of America, 100*, 8074–8079.
UNEP. (2007). *Global outlook for ice and snow*. United Nations Environment Programme. Retrieved October 14, 2019, from http://www.unep.org/geo/geo_ice/
UNFCCC (2007). Climate change: impacts, vulnerabilities and adaptation in developing countries.. Retrieved October 14, 2019, from https://unfccc.int/resource/docs/publications/impacts.pdf.

Index

A. T. Nguyen, L. Hens (eds.), *Global Changes and Sustainable Development in Asian Emerging Market Economies Vol. 2*, https://doi.org/10.1007/978-3-030-81443-4

S

Printed by Printforce, the Netherlands